High-Efficient Exploration and Development of Oil & Gas from Ocean

High-Efficient Exploration and Development of Oil & Gas from Ocean

Guest Editors

Mianmo Meng
Wenming Ji
Guodong Cui

MDPI

Basel • Beijing • Wuhan • Barcelona • Belgrade • Novi Sad • Cluj • Manchester

Guest Editors
Mianmo Meng
Hubei Key Laboratory of Marine Geological Resources
China University of Geosciences
Wuhan
China

Wenming Ji
School of Geosciences
China University of Petroleum (East China)
Qingdao
China

Guodong Cui
Faculty of Engineering
China University of Geosciences
Wuhan
China

Editorial Office
MDPI AG
Grosspeteranlage 5
4052 Basel, Switzerland

This is a reprint of the Special Issue, published open access by the journal *Journal of Marine Science and Engineering* (ISSN 2077-1312), freely accessible at: https://www.mdpi.com/journal/jmse/special_issues/2ITQJY2BY5.

For citation purposes, cite each article independently as indicated on the article page online and as indicated below:

Lastname, A.A.; Lastname, B.B. Article Title. *Journal Name* **Year**, *Volume Number*, Page Range.

ISBN 978-3-7258-3391-7 (Hbk)
ISBN 978-3-7258-3392-4 (PDF)
https://doi.org/10.3390/books978-3-7258-3392-4

Contents

Journal of
Marine Science and Engineering

Editorial

High-Efficient Exploration and Development of Oil & Gas from Ocean

Mianmo Meng [1,2,*], Wenming Ji [3,*] and Guodong Cui [4,*]

1 Hubei Key Laboratory of Marine Geological Resources, China University of Geosciences, Wuhan 430074, China
2 College of Marine Science and Technology, China University of Geosciences, Wuhan 430074, China
3 School of Geosciences, China University of Petroleum (East China), Qingdao 266580, China
4 Faculty of Engineering, China University of Geosciences, Wuhan 430074, China
* Correspondence: mengmianmo@cug.edu.cn (M.M.); jiwenming@upc.edu.cn (W.J.); cuiguodong@cug.edu.cn (G.C.)

Citation: Meng, M.; Ji, W.; Cui, G. High-Efficient Exploration and Development of Oil & Gas from Ocean. *J. Mar. Sci. Eng.* **2024**, *12*, 1930. https://doi.org/10.3390/jmse12111930

Received: 17 October 2024
Accepted: 20 October 2024
Published: 28 October 2024

The ocean takes up 71% of the Earth's surface, and hosts abundant resources, such as oil/gas. At present, oil/gas exploitation from the strata under the ocean is still a challenge, especially in unconventional reservoirs. Unconventional types of oil/gas mainly include shale oil/gas, tight sandstone oil/gas, coalbed gas, and gas hydrate.

This Special Issue focuses on the highly efficient exploration and development of oil and gas, especially in relation to unconventional oil/gas resources. At present, there are still some critical problems in the exploration and development of unconventional oil and gas. Firstly, the selection of sweet points from geological perspectives is still challenging [1–3]. The use of geophysical data (wireline logs and seismic data) to identify the most favorable layers remains controversial. Secondly, prospecting well can allow for the direct recovery of the sample from the target layers, as well as determining detailed information about fluid and reservoir benefits to select the best layer for development [4]. Evaluating the hydrocarbon occurrence is also problematic, especially when investigating the movability in in situ conditions [5]. Lastly, hydraulic fracturing is a basic and widely used method to develop most types of unconventional oil/gas. However, there remain unsolved problems with generating the ideal fracture network due to the complex natural cracks and artificial fractures [6]. As for unconventional offshore oil/gas, such as gas hydrate, their highly efficient development is still challenging and attracts extensive attention. Our Special Issue concentrates on the aforementioned problems, obtaining some interesting conclusions. The main contributions of each article are presented below.

Contribution 1 focused on the organic matter (OM) accumulation mechanism of the Taiyuan and Shanxi Formations, which formed in a marine—continental transitional environment. This research pointed out that the main controlling factors for OM accumulation are redox conditions, paleoproductivity, and sedimentary rate.

Contribution 2 investigated the natural fracture properties of the Lower Jurassic shale oil reservoir in the Sichuan Basin. Natural fracture plays a significant role in shale oil migration, which is a high-throughput flow channel. It was found that the permeability sensitivity of natural fracture is obviously different among fractures with different widths.

Contribution 3 studied the influence of gravel size on the mechanical properties of conglomerate reservoirs. The samples were selected from the Triassic Baikouquan Formation in Mahu sag, Junggar Basin. With a low gravel content, the cracks are mainly controlled by stress, while a high gravel content makes the fractures more complex and discrete.

Furthermore, Contribution 4 focused on the reservoir characterization, depositional evolution, and sequence stratigraphy of the Cretaceous Mishrif Formation in South Iraq, which is located in the Mesopotamian Basin. In this formation, several facies were identified, such as deep marine, shallow marine, rudist bioherm, shoal, and lagoon. The development of the Mishrif Formation was controlled by intrashelf basinal deposits, clinoform-like shoals, and tidal channels.

Contribution 5 investigated the influence of quartz on the preservation and abundance of the organic pores of marine shale reservoirs in the Cambrian Shuijingtuo Formation, South China. It was found that early euhedral quartz benefits the preservation of organic matter (OM)-hosted pores, and late microquartz is detrimental to the organic pores.

Contribution 6 monitored oil recovery during the process of heavy water displacing the sample with saturated oil. It indicated that a high-pore-volume displacement can increase the oil recovery comprehensively. The porosity at the front end increases, while that one at the other end decreases, which is caused by particle migration.

Moreover, Contribution 7 focused on the effect of shale composition on the pore structure of marine shale. It indicated that the pore structure has a good relationship with clay minerals and organic matter. Quartz is a rigid skeleton, which is prone to retaining pores, but its influence is complicated because quartz has various origins.

Contribution 8 investigated the sequence stratigraphy of marine shale in the Sichuan Basin, which aims to analyze the sedimentary fill history. It showed that deglaciation and tectonic events lead to the special characteristic of the sequence stratigraphy of the deep Longmaxi Formation. It also indicated that the regional subsidence of the Weiyuan and Luzhou areas may lead to shale gas accumulation.

Contribution 9 appraised the properties of shale oil reservoirs in Bohaiwan Basin, which is located in the Zhanhua and Dongying Sags. Multiple methods were combined to conduct this research, which indicated that these types of reservoirs have the potential to hold vast volumes of shale oil.

Contribution 10 focused on a de-noising method for hydrocarbon exploration. Using symplectic geometry mode decomposition (SGMD) can greatly improve the signal, obtaining high-quality data. Organic matter has a positive relationship with pore structure complexity.

Contribution 11 examined the pore structure of marine shale reservoirs, mainly focusing on their heterogeneity and complexity. Deep siliceous marine shale has a complex pore structure, which is more complex than that of shallow marine shale.

Contribution 12 proposed a method to evaluate the permeability in porous samples. Usually, the permeability can be obtained using the three-dimensional reconstruction of multiple pictures. However, the two-dimensional picture can also be used to obtain the sample permeability, with good results.

Contribution 13 focused on the hydrocarbon generation characteristics and potential of marine carbonates in the Odovician Majiagou Formation, Ordos Basin. Its tested samples were selected from Nordic carbonates, which were similar to the target formation. The main experiment was a gold tube thermal simulation, which showed that when the temperature is 600 °C, it still produces methane. Hydrocarbon production is mainly related to the type of organic matter.

Contribution 14 clarified the pore structure of shale in the Upper Permian Longtan Formation, western Guizhou province, South China. Both scanning electron microscopy (SEM) and nitrogen adsorption (NA) were used as testing methods. This study indicated that OM-hosted pores are irregular polygonal variants, bubble-like, and elliptical, while the number of organic pores is lower.

Contribution 15 investigated the organic matter (OM) accumulation of the Cambrian Qiongzhusi Formation, which is located in Yangtze Platform in South China. It was suggested that a redox environment of high-quality source sediment and a lack of upwelling are good conditions for OM accumulation in the Qiongzhusi Formation.

Contribution 16 clarified the controlling factors for the poor shale gas in the Longmaxi and Niutitang Formations in northwestern Hunan. The factor accounting for the Lower Silurian Longmaxi Formation is that the evolution time of hydrocarbon generation is too short. On the other hand, the factor accounting for the Lower Cambrian Niutitang

Formation is that organic matter endures excessive evolution at a deep paleo-burial depth, which makes the organic matter approach graphite.

Contribution 17 researched the pore structure of the Jurassic Ziliujing Formation in the lower Sichuan Basin. The shales can be divided into argillaceous, mixed, and carbonate shales. The average pore size of carbonate shale is higher than that of mixed and argillaceous shale. Pores between clay mineral platelets have a great contribution to nanopores that are less than 30 nm in size, indicating that the organic matter has less influence on the pore complexity.

Contribution 18 investigated the organic matter enrichment and sedimentary environment of the Cambrian Shuijingtou Formation, South China, which was divided into three intervals. Interval I has a high TOC content, Interval II has a moderate TOC content, and Interval III has a low TOC content. Comparing all three intervals, Interval III is characterized by a lower paleoproductivity and poor preservation conditions.

Contribution 19 points out the disadvantages of existing tri-linear analytical solutions in predicting production because the uncertainty can be caused by numerical inversion and the Laplace transform. A new general analytical solution can solve this problem by combining the average pressure substitution and integral transform.

For further research about the highly efficient development of unconventional oil/gas, we suggest that more attention should be paid to the pore structure characterization, oil occurrence and mobility, artificial and natural fracture complexity, and so on.

Acknowledgments: As Guest Editors of the Special Issue "Highly Efficient Exploration and Development of Oil and Gas from the Ocean", we wish to extend our sincere gratitude to all the authors whose valuable contributions made the publication of this Special Issue possible.

Conflicts of Interest: The authors declare no conflicts of interest.

List of Contributions:

1. Wang, Y.; Cheng, X.; Fan, K.; Huo, Z.; Wei, L. The Paleoenvironment and Mechanisms of Organic Matter Enrichment of Shale in the Permian Taiyuan and Shanxi Formations in the Southern North China Basin. *J. Mar. Sci. Eng.* **2023**, *11*, 992. https://doi.org/10.3390/jmse11050992.
2. Bai, X.; Huang, S.; Wang, X.; Wang, Z.; Wang, Y.; Ma, W.; Zhu, Y.; Sun, M.; Liu, B.; Fu, X.; et al. Microscopic Analysis of Natural Fracture Properties in Organic-Rich Continental Shale Oil Reservoirs: A Case Study from the Lower Jurassic in the Sichuan Basin, China. *J. Mar. Sci. Eng.* **2023**, *11*, 1036. https://doi.org/10.3390/jmse11051036.
3. Liu, P.; Shen, Y.; Meng, M.; Luo, S.; Zhong, Y.; Cen, Q. Experimental Study of Mechanical Properties and Fracture Characteristics of Conglomerates Based on Mohr–Coulomb Criteria. *J. Mar. Sci. Eng.* **2023**, *11*, 1219. https://doi.org/10.3390/jmse11061219.
4. Zhang, L.; Zhang, W.; Li, Y.; Song, B.; Liu, D.; Deng, Y.; Xu, J.; Wang, Y. Sequence Stratigraphy, Sedimentology, and Reservoir Characteristics of the Middle Cretaceous Mishrif Formation, South Iraq. *J. Mar. Sci. Eng.* **2023**, *11*, 1255. https://doi.org/10.3390/jmse11061255.
5. Wei, S.; Hu, M.; He, S.; Shu, Y.; Dong, T.; He, Q.; Yang, W.; Cai, Q. Effects of Quartz Precipitation on the Abundance and Preservation of Organic Matter Pores in Cambrian Marine Shale in South China. *J. Mar. Sci. Eng.* **2023**, *11*, 1267. https://doi.org/10.3390/jmse11071267.
6. Liu, J.; Li, H.; Xu, J.; Liu, S.; Liu, R.; Hou, L.; Tan, Q. Exploring the Unique Characteristics of High-Pore-Volume Waterflooding and Enhanced Oil Recovery Mechanisms in Offshore Sandstone Reservoirs Using Nuclear Magnetic Resonance Technology. *J. Mar. Sci. Eng.* **2023**, *11*, 1296. https://doi.org/10.3390/jmse11071296.
7. Xu, L.; Pan, R.; Hu, H.; Meng, J. Contribution of Various Shale Components to Pore System: Insights from Attributes Analysis. *J. Mar. Sci. Eng.* **2023**, *11*, 1327. https://doi.org/10.3390/jmse11071327.
8. Xu, L.; Meng, J.; Pan, R.; Yang, X.; Sun, Q.; Zhu, B. Sequence Stratigraphy and Implications for Shale Gas Exploration in the Southern Sichuan Basin, South China. *J. Mar. Sci. Eng.* **2023**, *11*, 1393. https://doi.org/10.3390/jmse11071393.
9. Ma, C.; Fang, X.; Ma, X.; Liu, X.; Xu, B.; Ge, X. Properties and Model of a Lacustrine Shale Oil Reservoir: A Case Study of the Upper Fourth Member and Lower Third Member of the Shahejie Formation in Dongying Sag and Zhanhua Sag, Jiyang Depression. *J. Mar. Sci. Eng.* **2023**, *11*, 1441. https://doi.org/10.3390/jmse11071441.

10. Chen, Y.; Guo, Z.; Gao, D. Marine Controlled-Source Electromagnetic Data Denoising Method Using Symplectic Geometry Mode Decomposition. *J. Mar. Sci. Eng.* **2023**, *11*, 1578. https://doi.org/10.3390/jmse11081578.
11. Zhu, B.; Meng, J.; Song, C.; Pan, R.; Zhu, Z.; Jin, J. Complexity and Heterogeneity Evaluation of Pore Structures in the Deep Marine Shale Reservoirs of the Longmaxi Formation, China. *J. Mar. Sci. Eng.* **2023**, *11*, 1613. https://doi.org/10.3390/jmse11081613.
12. Lei, G.; Liu, T.; Liao, Q.; He, X. Estimating Permeability of Porous Media from 2D Digital Images. *J. Mar. Sci. Eng.* **2023**, *11*, 1614. https://doi.org/10.3390/jmse11081614.
13. Wang, Y.; Xu, Y.; Huang, J.; Shi, J.; Zhao, H.; Wang, Q.; Meng, Q. Experimental Characteristics of Hydrocarbon Generation from Scandinavian Alum Shale Carbonate Nodules: Implications for Hydrocarbon Generation from Majiagou Formation Marine Carbonates in China's Ordos Basin. *J. Mar. Sci. Eng.* **2023**, *11*, 1616. https://doi.org/10.3390/jmse11081616.
14. Zhang, M.; Hu, M.; Wei, S.; Cai, Q.; Fu, W.; Shi, F.; Zhang, L.; Ding, H. Factors Controlling the Pore Development of Low-Mature Marine–Continental Transitional Shale: A Case Study of the Upper Permian Longtan Shale, Western Guizhou, South China. *J. Mar. Sci. Eng.* **2023**, *11*, 1862. https://doi.org/10.3390/jmse11101862.
15. Zhang, Q.; Liu, E.; Pan, S.; Wang, H.; Jing, Z.; Zhao, Z.; Zhu, R. Multiple Controls on Organic Matter Accumulation in the Intraplatform Basin of the Early Cambrian Yangtze Platform, South China. *J. Mar. Sci. Eng.* **2023**, *11*, 1907. https://doi.org/10.3390/jmse11101907.
16. Miao, Y.; Wang, P.; Li, X.; Huang, H.; Jin, C.; Gao, W. Shale Gas Exploration and Development Potential Analysis of Lower Cambrian Niutitang Formation and Lower Silurian Longmaxi Formation in Northwestern Hunan, South China, Based on Organic Matter Pore Evolution Characteristics. *J. Mar. Sci. Eng.* **2023**, *11*, 1910. https://doi.org/10.3390/jmse11101910.
17. Wang, Y.; Li, W.; Wang, X.; Wang, Z.; Ma, W.; Zhu, Y.; Sun, M.; Liu, B.; Cheng, L.; Fu, X. Nanopore Structure and Multifractal Characteristics of Continental Shale Oil Reservoir: A Case Study from Ziliujing Shales in the Sichuan Basin. *J. Mar. Sci. Eng.* **2023**, *11*, 1989. https://doi.org/10.3390/jmse11101989.
18. Wei, S.; Hu, M.; He, S.; Yang, W.; He, Q.; Cai, Q.; Li, P. Sedimentary Environment Interpretation and Organic Matter Enrichment of the Lower Cambrian Shuijingtuo Shale in the Yichang Slope, South China: Insight from Sedimentary Geochemical Proxies with Major/Trace Elements. *J. Mar. Sci. Eng.* **2023**, *11*, 2018. https://doi.org/10.3390/jmse11102018.
19. Qiu, K.; Fan, K.; Chen, X.; Lei, G.; Wei, S.; Navik, R.; Li, J. A New Approach for Production Prediction in Onshore and Offshore Tight Oil Reservoir. *J. Mar. Sci. Eng.* **2023**, *11*, 2079. https://doi.org/10.3390/jmse11112079.

References

1. Wang, Q.; Li, Y.; Utley, J.E.P.; Gardner, J.; Liu, B.; Hu, J.; Shao, L.; Wang, X.; Gao, F.; Liu, D.; et al. Terrestrial dominance of organic carbon in an Early Cretaceous syn-rift lake and its correlation with depositional sequences and paleoclimate. *Sediment. Geol.* **2023**, *455*, 106472. [CrossRef]
2. Luo, Q.; Zhang, L.; Zhong, N.; Wu, J.; Goodarzi, F.; Sanei, H.; Skovsted, C.B.; Suchý, V.; Li, M.; Ye, X.; et al. Thermal evolution behavior of the organic matter and a ray of light on the origin of vitrinite-like maceral in the Mesoproterozoic and Lower Cambrian black shales: Insights from artificial maturation. *Int. J. Coal Geol.* **2021**, *244*, 103813. [CrossRef]
3. Huang, H.; Chen, L.; Dang, W.; Luo, T.; Sun, W.; Jiang, Z.; Tang, X.; Zhang, S.; Ji, W.; Shao, S.; et al. Discussion on the rising segment of the mercury extrusion curve in the high pressure mercury intrusion experiment on shales. *Mar. Pet. Geol.* **2019**, *102*, 615–624. [CrossRef]
4. Wang, Q.; Yang, W.; Li, Y.; Jiang, Z.; Wen, M.; Zuo, R.; Wang, X.; Xue, Z.; Wang, Y. In-situ fluid phase variation along the thermal maturation gradient in shale petroleum systems and its impact on well production performance. *J. Earth Sci.* **2023**, *34*, 985–1001. [CrossRef]
5. Meng, M.; Zhang, Y.; Yuan, B.; Li, Z.; Zhang, Y. Imbibition behavior of oil-saturated rock: Implications for enhanced oil recovery in unconventional reservoirs. *Energy Fuels* **2019**, *37*, 13759–13768. [CrossRef]
6. Wang, J.; Zhou, F.; Bai, H.; Li, Y.; Yang, H. A Comprehensive method to evaluate the viscous slickwater as fracturing fluids for hydraulic fracturing applications. *J. Pet. Sci. Eng.* **2020**, *193*, 107359. [CrossRef]

Journal of
Marine Science and Engineering

MDPI

Article

The Paleoenvironment and Mechanisms of Organic Matter Enrichment of Shale in the Permian Taiyuan and Shanxi Formations in the Southern North China Basin

Yanan Wang [1,2,3], Xiulei Cheng [4], Kai Fan [5], Zhipeng Huo [3,6,*] and Lin Wei [1,2,*]

1 School of Energy Resources, China University of Geosciences, Beijing 100083, China
2 Key Laboratory of Strategy Evaluation for Shale Gas of Ministry of Land and Resources, China University of Geosciences, Beijing 100083, China
3 School of Resources and Materials, Northeastern University-Qinhuangdao Campus, Qinhuangdao 066004, China
4 No.1 Data Acquisition Branch Company, No.1 Geo-Logging Company, Daqing Drilling & Exploration Engineering Corporation, Daqing 163000, China
5 CNOOC EnerTech-Drilling & Production Co., Ltd., Tianjin 300452, China
6 National Engineering Research Center of Offshore Oil and Gas Exploration, Beijing 100028, China
* Correspondence: huozhipeng521@163.com (Z.H.); linwei@cugb.edu.cn (L.W.); Tel.: +86-134-8873-0863 (Z.H.); +86-155-1030-9590 (L.W.)

Abstract: The organic-rich shale of the Permian Taiyuan Formation (TYF) and Shanxi Formation (SXF) in the Southern North China Basin (SNCB) is considered a potential shale gas source. The shale was formed in a marine-continental transitional sedimentary environment, which has rarely been studied, with the enrichment mechanisms of organic matter (OM) remaining unclear. This study investigated the controlling factors and enrichment mechanisms of OM by analyzing the total organic carbon (TOC) content, paleoclimate, paleoproductivity, sedimentation rate, redox, and paleosalinity. The TOC of the TYF ranged from 0.92 to 7.43 wt.%, with an average of 2.48 wt.%, which was higher than that of the SXF (TOC = 0.36–5.1 wt.%, average of 1.68 wt.%). These geochemical indices suggest that both the TYF and SXF were deposited in warm and humid paleoclimates, with relatively high biological productivity and sedimentation rates. During the deposition process, the TYF experienced frequent transgression and regression events, leading to an enhancement of water reducibility, a relatively high sedimentation rate, reduced OM oxidation, and rapid deposition of OM, which were conducive to the preservation of OM. Moreover, a high biological productivity increased respiratory oxygen consumption in the water column, which could lead to OM accumulation. However, the regression event experienced by the SXF reduced the paleoproductivity and sedimentation rate and increased water oxidation, leading to a decrease in OM. The main controlling factors for the enrichment of OM in the TYF and SXF were the sedimentation rate, paleoproductivity, and redox conditions, thus establishing the enrichment models for OM in the TYF and SXF. This study is conducive to understanding shale enrichment mechanisms and guiding shale gas exploration.

Keywords: paleoenvironment; organic matter enrichment; enrichment model; shale; marine-continental facies; southern north China basin

Citation: Wang, Y.; Cheng, X.; Fan, K.; Huo, Z.; Wei, L. The Paleoenvironment and Mechanisms of Organic Matter Enrichment of Shale in the Permian Taiyuan and Shanxi Formations in the Southern North China Basin. *J. Mar. Sci. Eng.* **2023**, *11*, 992. https://doi.org/10.3390/jmse11050992

Academic Editors: Wenming Ji, Guodong Cui and Mianmo Meng

Received: 4 April 2023
Revised: 27 April 2023
Accepted: 2 May 2023
Published: 7 May 2023

1. Introduction

Organic-rich shale is not only a source rock of high-quality conventional oil and gas reservoirs but also a focus of shale oil and gas exploration. It can record important paleoclimatic and paleoenvironmental information (such as paleotemperature, paleoredox, and paleoproductivity) relating to when it is deposited, thus possessing significant value from a scientific research perspective [1–3]. Furthermore, as noted above, study of organic-rich shale also contributes critical information to oil and gas exploration activities, including both understandings of source rock in conventional petroleum systems [4] and of reservoir rock in unconventional shale gas and shale oil systems [5], and is critical to the evolving understanding of potentially exploitable gas hydrate resources [6,7]. Currently, significant progress has been made in marine and continental shale gas research; however, there is little research on marine-continental transitional shale gas. Marine-continental transitional facies are widely distributed in China, often with multiple lithologies coexisting and rapid changes in lithofacies, and the source and composition of organic matter are complex [8]. Therefore, compared with marine or terrestrial strata, they have distinct characteristics and are a potentially important area of unconventional oil and gas exploration in China.

The accumulation of organic matter (OM) is a complex physicochemical process [9,10]. Previous studies have shown that owing to the complex geological, oceanographic, and climatic histories, it is impossible for a single factor to explain OM accumulation in a sedimentary environment [11]. The accumulation and preservation of OM are controlled and affected by paleoclimate, paleoproductivity, sedimentation rate, redox, and paleosalinity [12–14]. Generally, elemental geochemistry [2] and organic geochemistry [15] are used to qualitatively reconstruct paleoenvironment [16–18]. Several redox-sensitive elements (such as V, U, and Ni) and their ratios (such as U/Th, V/Cr, Ni/Co, and V/(V + Ni)) have been widely used to explain ancient redox conditions [2,19]. Important geochemical indicators that qualitatively characterize paleoclimate include the chemical index of alteration (CIA) and the C-value [20–22].

Factors controlling OM enrichment in marine environments have been widely studied. From different influence mechanisms, they can be divided into the "productivity model" and the "conservation model" [23,24]. In the productivity model, organic matter input is the main reason for its enrichment. In the preservation model, it is believed that the factors of high salinity and anoxic sedimentary environments are the main conditions for the enrichment of organic matter. Unlike marine shale, marine-continental transitional shale is deposited in a sea-and-land interaction environment, with rapid facies changes on the plane, and complex enrichment mechanism of OM, leading to large differences in OM enrichment factors between marine shale and marine-continental transitional shale.

The SNCB is the largest sedimentary basin in China. The Permian is widely distributed and well developed, and its sedimentary types are diverse and rich in biological fossils [25]. The main strata of shale gas in the SNCB are the TYF and SXF, and the shale kerogen is generally type III [26,27]. There are few studies on Permian shale in the SNCB; the geological conditions are complex, and the sedimentary and tectonic evolution are changeable [25]. Presently, the amount of research on shale gas in this area is low [25], and an OM enrichment model has not been established, restricting the objective evaluation of shale gas resources in the basin [28].

This study measured and analyzed TOC, major elements, trace elements, and rare-earth elements to reconstruct the paleoclimate, paleoproductivity, sedimentation rate, paleoredox, and paleosalinity of the Permian TYF and SXF in the SNCB to reveal the enrichment mechanisms of OM in the SNCB, which provide an important basis for shale exploration in this area.

2. Geological Background

Located south of the North China Block and its southern margin (Figure 1), the SNCB is mainly controlled by the structure of the Qinling–Dabian orogenic belt [29]. From the distribution characteristics of the Mesozoic and Cenozoic in the basin, the SNCB is divided from north to south into five structural units: the Kaifeng Depression, the Taikang Uplift, the Zhoukou Depression, the Bengbu Uplift, and the Xinyang Hefei Depression (Figure 1) [29–31].

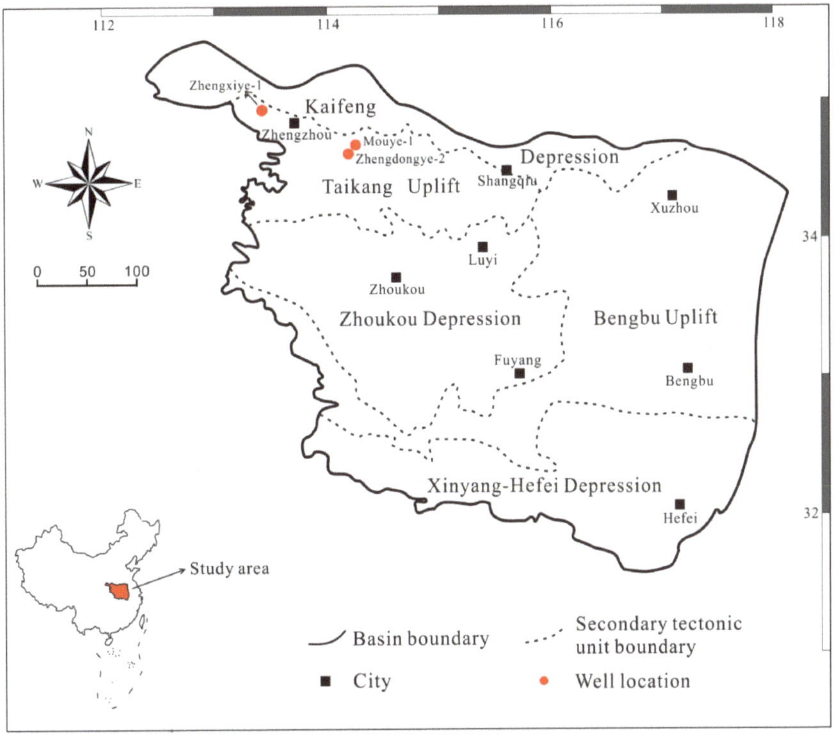

Figure 1. Simplified structural block map of the SNCB with investigated well location.

From the Late Ordovician to the Middle Carboniferous, the SNCB was uplifted due to the Caledonian movement. Owing to long-term weathering and denudation, most parts of the region lack Upper Ordovician, Silurian, Devonian, and lower Carboniferous strata [31,32]. Subsequently, Lower Permian marine-continental transitional strata were immediately deposited on the Upper Ordovician limestone [32]. Controlled by regional tectonic movements, the sedimentation process can be subdivided into three stages. First, Late Carboniferous seawater invaded from the northeast to southwest, forming a unified North China epicontinental sea basin. In this environment, the thickness of the Early Permian sediments is 30–175 m. Second, the collision between the North China and Siberian plates caused large-scale tectonic movement and regression from north to south. The main sedimentary environment in the Permian was a shallow water delta, which transitioned from the epicontinental sea through regression. Finally, the North China plate was uplifted again by compression during the Late Permian, resulting in the complete withdrawal of seawater from the North China platform and the formation of a continental sedimentary environment [26,31–33]. In general, the sedimentary environment of the North China platform from the Early to the Late Permian experienced both marine-continental transitional and lacustrine environments.

The Mouye-1 (MY1) well, drilled in 2014, was the first shale gas exploration well in the SNCB, and shale gas was found in the Upper Paleozoic strata. The Weican-1 well, drilled in the same year, encountered OM-rich shale with a cumulative thickness of 465 m in the Shihezi, TYF, SXF, and Benxi Formations. The Zhengxiye-1 (ZXY1) well, drilled in 2015, also saw good shale gas in the Permian SXF and TYF [34,35]. As of July 2020, the MY1, Zhengdongye-2 (ZDY2), and ZXY1 wells had been drilled northwest of the SNCB to explore for shale gas (Figure 1) [35]. The shale thicknesses in the TYF of the MY1, ZDY2, and ZXY1 wells are 54 m, 31 m, and 12 m, respectively, and the thicknesses of the SXF are 44 m, 14 m, and 16 m, respectively. From the cores, it is apparent that the lithology of the TYF is mainly grayish black mudstone, dark gray limestone, grayish black sandstone, dark gray sandstone, and dark gray shale, with thin layers of siltstone, coal lines, and barrier coastal facies. The lithology of the SXF is dominated by grayish black mudstone, carbonaceous mudstone, silty mudstone, fine sandstone, and sandstone, which are delta facies (Figure 2).

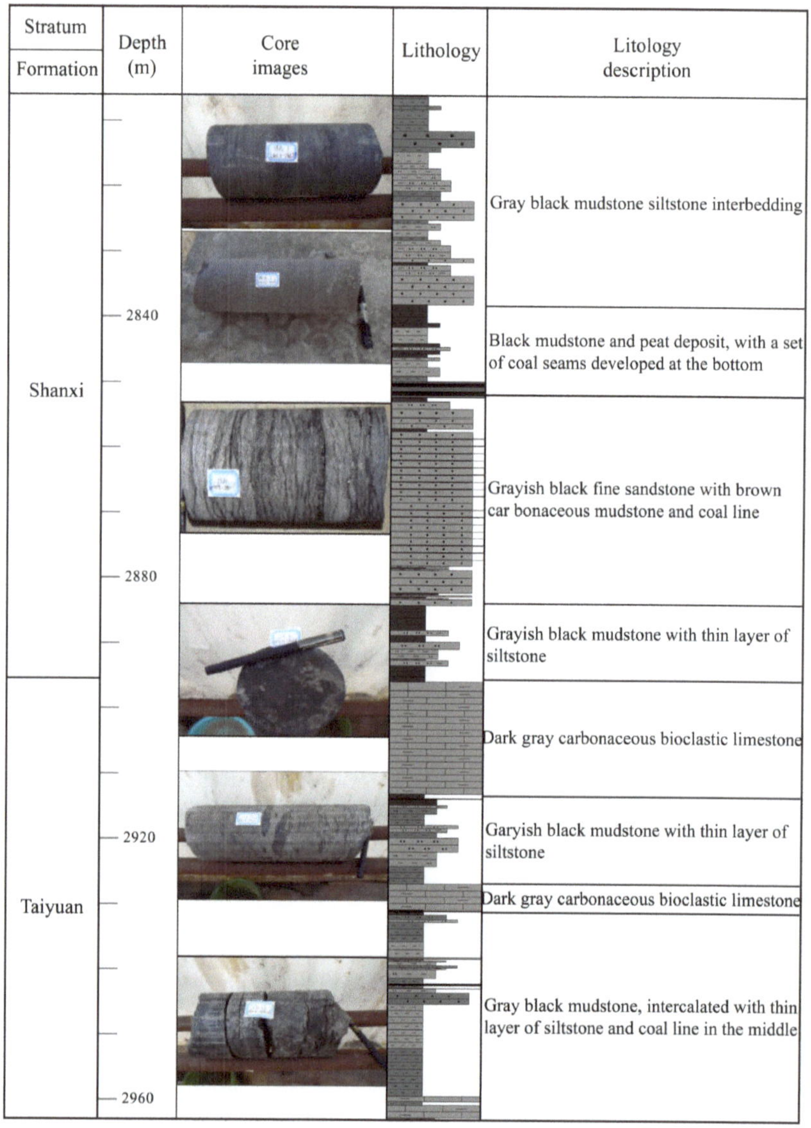

Figure 2. The stratigraphic column of the Lower Permian Shanxi and Taiyuan Formations (MY1 well) in the Southern North China Basin. (The stratigraphic column data mainly come from core logging and later artificial correction based on the core).

3. Sampling and Methods

Overall, 62 core samples (39 from MY1, 16 from ZDY2, and 7 from ZXY1) were selected from the study area, including 33 samples from the TYF and 29 samples from the SXF. The samples from Well MY1 included both TYF and SXF. The samples from Well ZDY2 were mainly from the TYF, while those from Well ZXY1 were mainly from the SXF. The lithology is mainly shale with a small amount of coal. The TOC content, major elements (ME), trace elements (TE), and rare-earth elements (REE) of all samples were measured and analyzed.

3.1. TOC Analysis

First, powder with the size of 200 meshes was treated using chemical methods in accordance with the Chinese Standard GB/T19145-2003, and then the sample (100 mg) was treated with 5% HCl for 24 h to remove the inorganic carbon component. Finally, a Leco® CS744 analyzer (Equipment source: LECO, St. Joseph, MI, USA) was used to measure the organic carbon content with an accuracy of ±0.5%.

3.2. Major Element Analysis

The primary elements (Si, Al, Fe, K, Na, Ca, Mg, Mn, Ti, and P) were tested using the XRF (X-ray fluorescence) method. The XRF tester was a Rigaku ZSX-100e. To mix the organic matter-removed sample (powder) and $Li_2B_4O_7$ evenly, they were poured into a platinum crucible, placed in a high-frequency heating-dissolving machine, and heated to 1150 °C. The melted mixed sample was poured onto a glass sheet and finally placed into an XPF instrument for testing.

3.3. Trace and Rare-Earth Element Analysis

The TE and REE contents were measured using an Agilent 7500a inductively coupled plasma mass spectrometer (ICP-MS), with an analysis error lower than 5%. The sample (powder) with the removed OM was subjected to acid dissolution ($HF/HNO_3/HClO_4$ = 2:2:1). The treated samples were then heated in a pressure-tight Teflon bomb at 200 °C for 48 h, and the resulting liquid was analyzed using inductively coupled plasma mass spectrometry (ICP-MS). The analytical procedures followed the Chinese National Standards GB/T 14506.1~14-2010 (2010) and GB/T 14506.30-2010 (2010).

4. Results

4.1. TOC

The TOC content of the TYF shale was 0.92–7.43 wt.%, with an average of 2.48 wt.% and a median of 2.24 wt.%. The TOC of coal was 54.45–66.55 wt.%, with an average of 60.5 wt.%. Approximately 51.52 wt.% of the samples were >2 wt.%, which indicates a good source rock [4] (Table 1; Figure 3). The TOC of the SXF shale was 0.36–5.1 wt.%, with an average of 1.68 wt.% and a median of 1.73 wt.%, and the TOC of coal was 58.95–61.29 wt.% (average of 60.5 wt.%), of which approximately 35.48 wt.% of the samples were greater than 2 wt.% (Table 1; Figure 3). Overall, the TOC content of the Lower Permian shale in the SNCB is relatively high, with the TOC content of the shale in the TYF being higher than that in the SXF.

Table 1. ME contents (wt%) of the TYF and SXF in the SNCB.

Well	Sample	Depth (m)	TOC (wt.%)	MEs (wt.%)									
				SiO_2	Al_2O_3	TFe_2O_3	MgO	CaO	Na_2O	K_2O	TiO_2	P_2O_5	MnO
MY1	JX2	2804.81	4.19	41.52	33.71	0.73	0.22	0.10	1.27	2.43	1.15	0.09	0.22
MY1	JX3	2807.74	1	55.00	30.00	1.44	0.36	0.17	1.56	4.55	1.02	0.14	0.36
MY1	JX4	2810.32	0.76	49.12	25.90	6.66	0.73	0.27	1.81	3.85	0.80	0.23	0.74
MY1	JX6	2815.3	0.55	46.92	10.30	22.52	1.46	0.56	1.24	0.72	0.28	0.37	1.46
MY1	JX7	2819.27	0.53	56.31	27.11	2.43	0.50	0.18	1.99	5.40	1.08	0.09	0.50
MY1	JX8	2821.74	1.23	51.32	25.41	3.00	0.90	0.27	3.37	4.99	1.03	0.18	0.89
MY1	JX10	2827.13	1.09	49.31	31.40	2.20	0.40	0.25	2.67	3.93	0.93	0.18	0.40
MY1	JX11	2830.72	0.52	50.81	26.72	2.76	0.76	0.22	4.58	5.66	0.95	0.09	0.76
MY1	JX12	2831.66	0.44	48.82	24.51	3.62	1.03	0.46	3.59	5.35	1.03	0.23	1.05
MY1	JX14	2838.32	0.89	49.12	30.00	0.00	1.04	0.28	3.07	6.53	0.87	0.14	1.05
MY1	JX15	2840.23	0.48	52.61	29.31	2.65	0.53	0.24	3.32	5.61	0.97	0.09	0.53
MY1	JX16	2841.88	2.37	48.63	24.30	3.47	0.85	0.29	2.59	5.08	0.90	0.27	0.85
MY1	JX17	2844.32	2.39	50.92	21.11	4.59	1.08	1.76	3.59	2.99	0.85	0.23	1.08

Table 1. *Cont.*

Well	Sample	Depth (m)	TOC (wt.%)	MEs (wt.%)									
				SiO_2	Al_2O_3	TFe_2O_3	MgO	CaO	Na_2O	K_2O	TiO_2	P_2O_5	MnO
MY1	JX18	2847.16	0.62	56.80	26.30	2.60	0.56	0.22	3.53	5.81	0.75	0.09	0.57
MY1	JX19	2848.82	1.98	48.63	29.91	2.32	0.65	0.32	5.12	4.77	0.70	0.18	0.65
MY1	JX20	2849.51	54.45	2.18	1.79	0.67	0.15	0.21	0.19	0.22	0.03	0.00	0.15
MY1	JX21	2855.01	66.55	0.58	0.55	0.14	0.02	0.07	0.08	0.02	0.02	0.00	0.03
MY1	JX22	2852.36	2.21	42.72	25.72	0.44	0.32	0.06	1.46	10.65	0.92	0.09	0.32
MY1	JX24	2857.02	4.24	41.03	24.81	4.16	0.76	0.17	1.08	11.18	0.82	0.18	0.76
MY1	JX31	2884.3	5.1	44.73	22.71	5.52	1.16	0.43	1.29	9.08	0.88	0.55	1.16
MY1	JX32	2886.27	1.97	44.73	22.71	5.52	1.16	0.43	1.29	9.08	0.88	0.55	1.16
MY1	JX33	2889.34	2.56	45.03	23.11	4.27	1.01	0.13	1.70	8.50	0.73	0.14	1.01
MY1	JX34	2894.31	1.78	47.13	21.90	0.01	1.23	0.85	1.35	8.60	0.72	0.37	1.24
MY1	JX36	2913.2	3.74	43.54	21.41	4.73	1.16	0.34	1.62	5.11	0.77	0.32	1.16
MY1	JX37	2917.42	0.92	47.43	24.01	3.60	1.01	0.22	1.54	6.02	0.83	0.27	1.02
MY1	JX39	2923	1.87	48.52	21.31	0.01	1.08	0.70	1.99	6.05	0.75	0.23	1.08
MY1	JX40	2925.92	1.93	48.52	21.31	3.79	1.08	0.70	1.99	6.05	0.75	0.23	1.08
MY1	JX42	2931.5	1.75	43.54	20.61	1.56	0.65	0.15	1.78	4.10	0.55	0.09	0.65
MY1	JX43	2933.86	4.14	42.42	22.60	5.45	0.76	0.50	1.48	3.93	0.77	0.37	0.76
MY1	JX44	2935.4	61.29	54.21	28.61	0.97	0.60	0.29	1.51	4.89	1.12	0.09	0.61
MY1	JX45	2939.27	1.69	44.52	21.71	4.62	1.08	0.46	1.81	4.58	0.88	0.32	1.08
MY1	JX46	2941.28	2.25	48.63	24.30	4.39	1.08	0.39	1.78	5.13	0.92	0.27	1.08
MY1	JX47	2943.35	0.66	0.34	0.19	0.97	0.00	0.10	0.00	0.02	0.02	0.05	0.00
MY1	JX49	2948.32	58.95	38.53	19.50	4.60	0.70	0.57	1.21	4.41	0.50	0.73	0.70
MY1	JX50	2950.09	5.06	42.94	18.80	5.56	1.08	0.99	1.48	4.00	0.67	0.23	1.08
MY1	JX51	2953.09	1.67	44.73	21.01	5.20	1.13	1.02	2.21	4.07	0.72	0.41	1.14
MY1	JX52	2955.71	1.88	48.22	22.01	4.72	1.48	4.49	2.21	4.99	0.70	0.27	1.48
MY1	JX53	2958.16	2.55	53.31	20.20	0.10	1.06	2.27	2.05	5.40	0.67	0.37	1.06
MY1	JX54	2960.13	1.89	37.74	16.61	2.96	1.13	15.68	1.59	3.93	0.58	0.23	1.14
ZDY2	ZY2-SX-2	2841.67	2.58	56.82	18.48	6.51	1.53	1.02	1.75	4.65	0.93	0.50	0.14
ZDY2	ZY2-TY-8	2849.59	2.27	51.69	23.33	5.30	1.33	0.84	2.16	5.64	0.87	0.41	0.15
ZDY2	ZY2-TY-11	2850.78	1.67	44.18	18.93	12.44	2.02	2.20	1.73	4.82	0.67	1.60	0.44
ZDY2	ZY2-TY-24	2861.54	2.5	55.37	19.50	7.01	1.11	0.29	1.32	4.65	0.92	0.27	0.09
ZDY2	ZY2-TY-33	2868.8	2.27	55.90	19.29	5.93	1.24	0.74	1.73	5.85	0.80	0.27	0.06
ZDY2	ZY2-TY-37	2870.02	1.68	56.71	20.92	4.40	1.19	0.87	2.21	6.34	0.77	0.23	0.05
ZDY2	ZY2-TY-46	2878.45	2.47	51.58	18.54	10.52	0.65	0.57	1.35	3.40	0.78	0.46	0.06
ZDY2	ZY2-TY-48	2880.01	3.52	54.55	20.78	5.13	1.09	0.85	1.27	3.81	0.78	0.37	0.14
ZDY2	ZY2-TY-52	2882.25	7.43	37.97	15.29	5.78	1.63	11.64	1.21	3.20	0.55	0.18	0.15
ZDY2	ZY2-TY-61	2886.59	1.79	55.41	20.29	6.05	1.26	0.99	1.21	4.10	0.87	0.23	0.14
ZDY2	ZY2-TY-69	2894.67	3.48	61.33	20.48	1.93	0.61	0.21	1.29	3.32	0.78	0.09	0.01
ZDY2	ZY2-TY-75	2897.14	1.32	52.65	15.78	6.36	2.37	4.73	1.64	2.89	0.60	0.32	0.21
ZDY2	ZY2-TY-80	2900.52	1.31	66.51	14.34	3.60	0.73	1.06	1.35	4.14	0.55	0.55	0.04
ZDY2	ZY2-TY-87	2902.32	1.42	61.98	17.12	3.07	1.13	1.74	1.67	4.82	0.58	0.27	0.05
ZDY2	ZY2-TY-92	2905.69	1.25	55.92	17.08	8.61	0.66	1.32	2.18	4.65	0.57	0.23	0.05
ZDY2	ZY2-TY-98	2910.56	2.24	33.80	4.16	2.62	0.68	28.96	0.43	1.33	0.15	0.23	0.01
ZXY1	JX9	3267.87	0.36	60.61	23.01	4.02	0.91	0.22	2.67	6.77	1.03	0.18	0.92
ZXY1	JX13	3274.1	0.8	58.38	27.81	0.80	0.38	0.22	4.07	6.17	1.32	0.09	0.37
ZXY1	JX19	3288.17	0.9	57.61	21.84	5.59	1.18	0.46	0.94	9.56	0.97	0.37	1.19
ZXY1	JX39	3318.5	3.27	58.34	23.77	3.09	0.66	0.15	2.51	8.65	0.83	0.09	0.67
ZXY1	JX41	3321.08	1.57	53.83	16.36	13.20	1.46	0.90	1.16	5.93	0.53	0.46	1.47
ZXY1	JX44	3324.8	2.44	50.83	23.35	9.22	0.93	0.21	2.94	6.84	0.83	0.14	0.94
ZXY1	JX51	3340.75	3.88	53.91	22.50	6.62	0.53	0.83	0.92	4.60	0.97	0.37	0.54

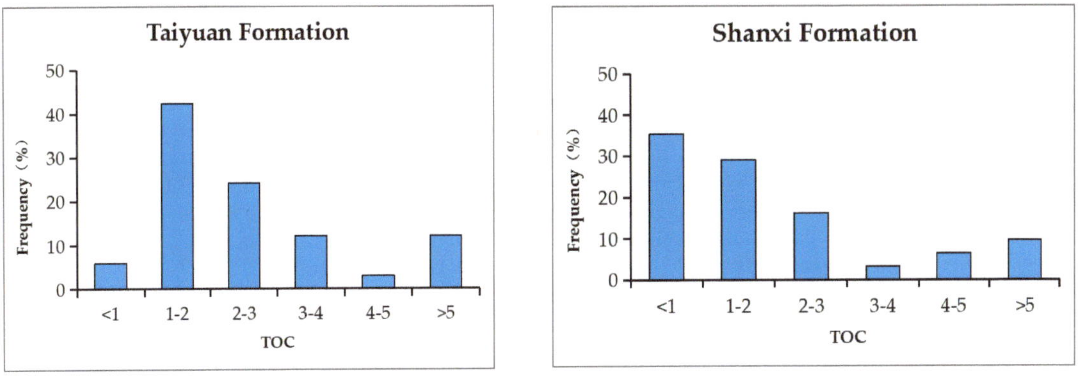

Figure 3. TOC—frequency distribution histogram of the TYF and SXF.

4.2. Major Elements

The primary ME oxides of the TYF and SXF shale samples were SiO_2, Al_2O_3, TFe_2O_3, K_2O, CaO, MgO, MnO, Na_2O, TiO_2, and P_2O_5. The main ME oxides of the TYF were SiO_2, Al_2O_3, TFe_2O_3, K_2O, and CaO, and their contents ranged from 0.34 to 66.51 wt.% (average of 48.1 wt.%), from 0.19 to 28.61 wt.% (average of 19.27 wt.%), from 0.01 to 12.44 wt.% (average of 4.78 wt.%), from 0.02 to 6.84 wt.% (average of 4.45 wt.%), from 0.1 to 28.96 wt.% (average of 2.63 wt.%), respectively. The main ME oxides of the SXF were SiO_2, Al_2O_3, K_2O, and TFe_2O_3, and their contents ranged from 0.58 to 60.61 wt.% (average of 47.23 wt.%), from 0.55 to 33.71 wt.% (average of 23.12 wt.%), from 0.02 to 11.18 wt.% (average of 5.74 wt.%), and from 0.01 to 22.52 wt.% (average of 3.96 wt.%), respectively (Table 1). In general, the TYF and SXF are rich in SiO_2 and comprise siliceous mudstone and argillaceous and siliceous shales (Figure 4).

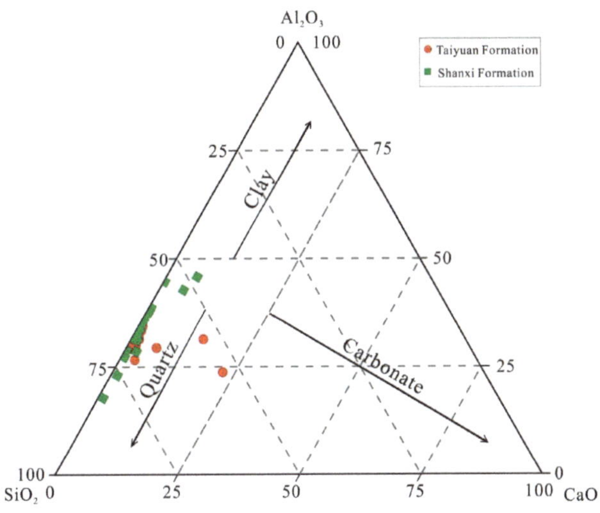

Figure 4. $SiO_2-Al_2O_3-CaO$ ternary diagram for the TYF and SXF shale samples from the MY1 well.

4.3. Trace Elements

The enrichment factor (EF) values can be used to establish the degree of TE enrichment [16]. The equation for EF is as follows:

$$EF = X/Y \quad (1)$$

where X is the content of TEs in minerals or rocks, and Y is the content of TEs in the corresponding basic rocks.

For the TYF, TFs with obvious enrichment (EF > 1) included Sr (EF = 0.47–14.99, average of 1.58), Cr (EF = 0.58–3.63, average of 1.14), and TEs with deficit (EF < 1) were V, Co, Ni, Ba, Cu, and Zn (Table 2). For the SXF, Ni, Co and Cr showed obvious enrichment (EF > 1), which included EF (Ni) = 0.23–18.35, with an average of 2.21, EF (Co) = 0.35–3.04, with an average of 1.86, and EF (Cr) = 0.08–3.04, with an average of 1.07 (Table 2). The TES showing deficits (EF < 1) were V, Sr, Cu, and Zn. As shown in Figure 5, compared with the SXF, the shale samples from the TYF show higher EF values in most TEs, except for Co, Ni, and Ba, and the TYF is obviously enriched with Cr and Sr.

Table 2. TE contents of the TYF and SXF in the SNCB.

Well	Sample	Depth (m)	TOC (wt.%)	TEs (ppm)								
				V	Cr	Co	Ni	Sr	Ti	Ba	Cu	Zn
MY1	JX2	2804.81	4.19		15			113	69			38
MY1	JX3	2807.74	1					159	61		70	55
MY1	JX4	2810.32	0.76		137			157	48			154
MY1	JX6	2815.3	0.55		169			517	17			175
MY1	JX7	2819.27	0.53		143			152	65	447		
MY1	JX8	2821.74	1.23		214			222	62	986		46
MY1	JX10	2827.13	1.09		155			274	56		55	
MY1	JX11	2830.72	0.52		156			213	57	482		
MY1	JX12	2831.66	0.44					234	62	694		63
MY1	JX14	2838.32	0.89			114		258	52	813		129
MY1	JX15	2840.23	0.48	246				219	58	652	52	
MY1	JX16	2841.88	2.37					219	54			107
MY1	JX17	2844.32	2.39		175		77	184	51			71
MY1	JX18	2847.16	0.62					217	45	577		
MY1	JX19	2848.82	1.98		138		58	349	42			56
MY1	JX20	2849.51	54.45				41	123	2			27
MY1	JX21	2855.01	66.55			27	41	59	1			
MY1	JX22	2852.36	2.21	129	193		55	134	55			
MY1	JX24	2857.02	4.24				81	203	49	1090		105
MY1	JX31	2884.3	5.1		169		81	212	53	708		111
MY1	JX32	2886.27	1.97	187			77	256	53	791		125
MY1	JX33	2889.34	2.56				138	251	44	1220		94
MY1	JX34	2894.31	1.78				70	279	43	948		113
MY1	JX36	2913.2	3.74		133		69	218	46	506		61
MY1	JX37	2917.42	0.92	327			57	251	50	510		106
MY1	JX39	2923	1.87		417		94	240	45	434	54	83
MY1	JX40	2925.92	1.93				94	240	45	434	54	83
MY1	JX42	2931.5	1.75				60	331	33	473		56
MY1	JX43	2933.86	4.14				85	295	46			189
MY1	JX44	2935.4	61.29		166		58	280	67			
MY1	JX45	2939.27	1.69				58	230	53			68
MY1	JX46	2941.28	2.25		135			264	55			69
MY1	JX47	2943.35	0.66					53	1			
MY1	JX49	2948.32	58.95				67	274	30			82
MY1	JX50	2950.09	5.06					256	40			84
MY1	JX51	2953.09	1.67		147		78	323	43			84
MY1	JX52	2955.71	1.88					459	42	495		53
MY1	JX53	2958.16	2.55	194	156			429	40			90
MY1	JX54	2960.13	1.89		191		111	984	35			137
ZDY2	ZY2-SX-2	2841.67	2.58	125	77.86	17.5	30.2	197.9	56	675	18	88.4
ZDY2	ZY2-TY-8	2849.59	2.27	212	90.22	19.4	60.8	198.9	52	712	31.2	
ZDY2	ZY2-TY-11	2850.78	1.67	219	85.56	19	80.9	301.9	40	773	27.4	162.2
ZDY2	ZY2-TY-24	2861.54	2.5	117	79.58	15.6	28.8	163.9	55	381	23.1	91
ZDY2	ZY2-TY-33	2868.8	2.27	106	72.8	21.3	37.2	190.9	48	514	25.8	86.6

Table 2. *Cont.*

Well	Sample	Depth (m)	TOC (wt.%)	TEs (ppm)								
				V	Cr	Co	Ni	Sr	Ti	Ba	Cu	Zn
ZDY2	ZY2-TY-37	2870.02	1.68	123	78.54	10.7	24.8	193.9	46	582	23.9	90.4
ZDY2	ZY2-TY-46	2878.45	2.47	137	66.24	28	47.3	251.9	47	318	30.3	50
ZDY2	ZY2-TY-48	2880.01	3.52	155	85.78	12.6	25.2	264.9	47	326	22.3	97.5
ZDY2	ZY2-TY-52	2882.25	7.43	117	60.84	16.2	42.6	439.9	33	680	24.2	123.4
ZDY2	ZY2-TY-61	2886.59	1.79	108	87.4	11.2	23.4	210.9	52	336	24	125.5
ZDY2	ZY2-TY-69	2894.67	3.48	92	64.06	3.6	8.6	227.9	47	308	16.1	96.1
ZDY2	ZY2-TY-75	2897.14	1.32	71	52.1	15.3	16.5	340.9	36	352	17.7	60.6
ZDY2	ZY2-TY-80	2900.52	1.31	94	74.18	16.8	22.8	360.9	33	415	13.3	99.8
ZDY2	ZY2-TY-87	2902.32	1.42	114	80.73	9.2	23.2	379.9	35	516	12	44
ZDY2	ZY2-TY-92	2905.69	1.25	125	79.81	18.2	43.7	367.9	34	566	27	90.3
ZDY2	ZY2-TY-98	2910.56	2.24	51	80.61	4.4	61.8	727.9	9	137	18.4	93.9
ZXY1	JX9	3267.87	0.36	145.6	83.58	19.07	41.8	246	62	769.8	33.1	97.98
ZXY1	JX13	3274.1	0.8	162.3	95.7	11.12	26.16	309	79	890.2	9.37	11.54
ZXY1	JX19	3288.17	0.9	142.4	80.71	13.44	24.77	185.69	58	908	29.1	102.72
ZXY1	JX39	3318.5	3.27	198.1	98.38	36.7	69.22	250.6	50	1355.8	15.02	68.96
ZXY1	JX41	3321.08	1.57	159.8	68.58	109.62	180.66	211	32	905.8	20.6	103.92
ZXY1	JX44	3324.8	2.44	192.6	89.34	31.92	100.52	292	50	1376.2	21.48	44.78
ZXY1	JX51	3340.75	3.88	194.1	88.68	15.33	48.08	268.6	58	534.8	25.3	36.28

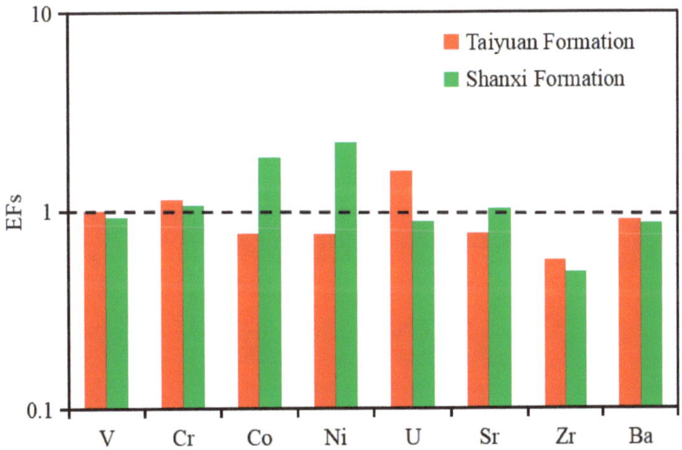

Figure 5. EF diagram of selected TEs for the TYF and SXF shales. The horizontal line (EF = 1) separates the enrichment (above the line) or depletion (below the line) of an element.

4.4. Rare-Earth Elements

REEs and their related parameters for the TYF and SXF in the study area are listed in Table 3. Total rare-earth concentration of the shale in the TYF (ΣREE) ranged from 90.31 to 381.436 ppm, with an average of 224.36 ppm; the range of ΣREE in the SXF was 206.12–376.104 ppm, with an average of 304.21 ppm (Table 3). The average REE concentrations of the shale samples from the TYF and SXF were higher than those of the North American shale composite (NASC, 173.21 ppm) [36] and the post-Archean average Australian shale (PAAS, 184.77 ppm) [37] (Table 3).

Table 3. REE contents (ppm) of the TYF and SXF in the SNCB.

Well	Sample	La	Ce	Pr	Nd	Sm	Eu	Gd	Tb	Dy	Ho	Er	Tm	Yb	Lu	L/H	∑REE	$(La/Yb)_N$
ZDY2	ZY2-SX-2	55.52	98.32	12.19	43.34	8.13	1.48	6.90	1.07	5.99	1.16	3.51	0.56	3.60	0.55	13.74	242.32	1.14
ZDY2	ZY2-TY-8	45.96	82.09	10.91	38.61	7.43	1.21	5.41	0.78	5.00	1.04	3.24	0.52	3.40	0.52	13.22	206.12	1.00
ZDY2	ZY2-TY-11	63.48	136.85	16.22	60.50	13.00	2.67	13.50	2.02	10.51	1.79	4.75	0.69	4.39	0.66	12.34	331.03	1.07
ZDY2	ZY2-TY-24	57.20	100.16	13.80	47.72	8.50	1.52	7.71	1.12	6.47	1.23	3.67	0.53	3.52	0.55	13.84	253.70	1.20
ZDY2	ZY2-TY-33	54.44	95.00	12.68	46.84	8.95	1.36	7.37	1.11	6.44	1.32	3.81	0.60	3.87	0.60	12.77	244.39	1.04
ZDY2	ZY2-TY-37	51.72	86.11	12.24	42.46	7.46	1.33	6.37	0.84	5.01	1.01	3.13	0.49	3.11	0.52	14.72	221.80	1.23
ZDY2	ZY2-TY-46	47.16	86.14	10.72	38.72	7.15	1.81	6.42	0.92	4.92	0.95	2.77	0.40	2.72	0.43	15.11	211.23	1.28
ZDY2	ZY2-TY-48	47.00	86.71	10.78	37.40	7.57	1.69	6.26	0.89	5.12	1.02	3.05	0.48	3.21	0.49	13.84	211.67	1.08
ZDY2	ZY2-TY-52	52.68	104.65	11.45	38.39	6.80	1.32	5.83	0.79	4.53	0.87	2.65	0.40	2.68	0.41	17.93	233.45	1.45
ZDY2	ZY2-TY-61	49.92	88.44	10.65	36.52	6.93	1.42	5.90	0.81	4.88	0.97	2.84	0.43	2.81	0.44	15.16	212.96	1.31
ZDY2	ZY2-TY-69	48.72	93.61	9.66	33.55	5.66	1.40	4.95	0.62	3.17	0.58	1.61	0.22	1.38	0.21	25.36	205.34	2.61
ZDY2	ZY2-TY-75	49.32	89.70	10.11	34.10	5.98	1.29	5.25	0.72	4.06	0.76	2.28	0.35	2.33	0.37	18.01	206.62	1.56
ZDY2	ZY2-TY-80	64.56	131.10	14.14	52.14	9.86	1.90	7.42	0.89	4.94	0.93	2.84	0.42	2.77	0.44	21.25	294.35	1.72
ZDY2	ZY2-TY-87	60.24	120.42	13.34	47.30	8.20	1.44	6.30	0.84	4.85	0.99	2.91	0.43	2.85	0.43	19.34	270.54	1.56
ZDY2	ZY2-TY-92	66.00	140.30	14.84	52.25	8.52	1.31	6.64	0.87	5.17	1.01	3.06	0.46	2.96	0.45	20.73	303.84	1.65
ZDY2	ZY2-TY-98	26.16	31.86	5.34	20.68	5.31	0.96	4.65	0.73	3.88	0.72	1.96	0.26	1.65	0.26	10.04	104.42	1.17
ZXY1	JX9	75.76	151.34	15.44	60.22	12.65	3.26	10.47	1.29	7.60	1.48	4.42	0.62	4.17	0.60	16.31	349.33	1.34
ZXY1	JX13	64.50	120.72	11.40	39.36	5.65	1.24	4.42	0.60	4.01	0.87	2.87	0.44	3.12	0.46	19.99	259.65	1.53
ZXY1	JX19	82.50	165.12	16.82	64.28	11.07	1.95	8.10	1.05	6.51	1.32	4.11	0.59	4.03	0.58	19.22	368.03	1.51
ZXY1	JX39	88.84	181.50	17.13	56.54	7.34	0.90	5.62	0.89	6.27	1.30	4.18	0.62	4.35	0.62	19.63	376.10	1.51
ZXY1	JX41	59.34	133.20	13.67	53.46	10.10	1.82	9.26	1.25	7.47	1.46	4.42	0.63	4.33	0.63	13.90	301.05	1.01
ZXY1	JX44	90.62	192.42	18.60	67.62	10.95	1.23	10.72	1.77	11.73	2.33	6.91	0.97	6.40	0.90	12.64	423.17	1.04
ZXY1	JX51	51.80	118.68	11.04	40.98	7.04	1.37	5.79	0.84	5.40	1.12	3.57	0.54	3.76	0.54	15.02	252.46	1.02

As shown in the diagram of the PASS-normalized REE distribution patterns (Figure 6), the two curves show nearly the same trend, which may indicate that the REEs of the shale samples from the TYF and SXF come from similar sources. The REE distribution patterns have the characteristics of a relatively flat LREE (from La to Gd) and HREE (from Tb to Lu). The range of L/H (LREE/HREE) for the TYF shale was 10.04–25.36, with an average of 15.96. The L/H (LREE/HREE) of the SXF shale ranged from 12.34–19.99, with an average of 16.04 (Table 2). Ce anomaly (δ Ce) and Eu anomaly (δ Eu) was defined as $\delta Ce = Ce_N/(La_N \times Pr_N)^{1/2}$ and $\delta Eu = Eu_N/(Sm_N \times Gd_N)^{1/2}$, respectively [37]. The range of δ Ce and δ Eu for the TYF shale was 0.62–1.15 (average of 0.92) and 0.54–1.26 (average of 0.97), respectively; the δ Ce and δ Eu of the SXF shale ranged from 0.85 to 1.08 (average of 0.99) and from 0.66 to 1.34 (average of 0.98), respectively (Table 3), indicating that the Ce and Eu anomalies for the TYF and SXF shale were weakly negative.

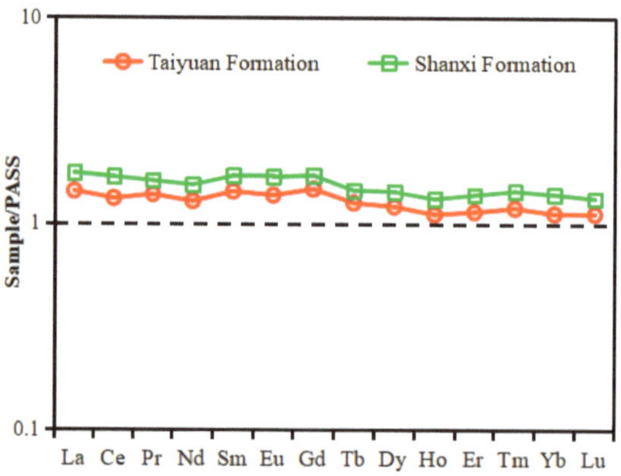

Figure 6. PASS-normalized REE distribution patterns of the TYF and SXF shale samples [37].

5. Discussion

5.1. Paleoclimate

Paleoclimate can affect parent rock weathering, erosion, and sediment transport, and ultimately control the input of terrestrial nutrients and sediment into an ocean [38,39]. The CIA is a typical parameter for reconstructing paleoclimatic conditions [20,21] and for judging the degree of paleoweathering [40,41].

The formula is as follows:

$$CIA = 100 \times Al_2O_3/(Al_2O_3 + CaO^* + Na_2O + K_2O) \quad (2)$$

The MEs used in the calculations are converted into mole fractions. CaO^* is CaO in silicate, that is, the mole fraction of CaO in the whole rock minus the chemically deposited CaO. For CaO^* correction, we used

$$CaO^* = CaO - (10/3 \times P_2O_5) \quad (3)$$

After correction, the mole number of CaO^* is the minimum value of Na_2O and CaO [42].

Previous studies have shown that the greater the CIA value, the greater the chemical alteration and the climate changes to warmer and wetter conditions. When the CIA value ranges from 50 to 65, the degree of petrochemical weathering is low, indicating a cold and arid climate; when the CIA value ranges from 65 to 85, the degree of petrochemical weathering is medium, indicating a semi-arid and semi-humid climate; and when the CIA value ranges from 85 to 100, the degree of petrochemical weathering is high, indicating a warm and humid climate [20,40]. The CIA range of the TYF shale samples was 74.12–90.02, with an average of 85.58; The CIA range of the SXF shale samples was 80.49–95.03, with an average of 84.88 (Table 4). The paleoclimate and weathering degree of a source rock can be clarified by using the $Al_2O_3-(CaO^* + Na_2O)-K_2O$ ternary diagram (Figure 7). Both the TYF and SXF indicate a warm and humid paleoclimate with moderate to strong weathering (Figures 7 and 8).

Table 4. Paleoenvironmental condition analysis.

Environment	Ratios	Geological Units		Index Limit			Reference
		Taiyuan Formation	Shanxi Formation				
Paleoclimate	CIA	74.12–90.02 (85.58)	80.49–95.03 (84.88)	Cold and arid (50 < CIA < 65)	Semi-arid and semi-humid (65 < CIA < 85)	Warm and humid (85 < CIA < 100)	[20,40,41]
	C-value	0.91–1.08 (1.01)	0.99–1.09 (1.01)	Cold and arid (C-value < 0.4)	Semi-arid and semi-humid (0.4 < C-value < 0.6)	Warm and humid (C-value > 0.6)	[21,22]
Paleo-productivity	P/Ti	0.02–1.83 (0.24)	0.22–0.45 (0.11)				
	Ba_{bio}	136.98–1376.11 ppm (507.71 ppm)	446.89–1355.71 ppm (828.38 ppm)				
Sedimentary Rate	$(La/Yb)_N$	0.99–2.6 (1.35)	1.01–1.53 (1.34)				
Paleoredox	Cu/Zn	0.13–0.7 (0.34)	0.2–1.27 (0.48)	Anoxic: Cu/Zn < 0.21	Dysoxic: 0.21 < Cu/Zn < 0.63	Oxic: Cu/Zn > 0.63	[2]
	V/(V + Ni)	0.45–0.91 (0.77)	0.47–1 (0.77)	Anoxic: V/(V + Ni) > 0.6	Dysoxic: 0.46 < V/(V + Ni) < 0.6	Oxic: V/(V + Ni) < 0.46	
Paleosalinity	Sr/Ba	0.21–5.31 (0.78)	0.18–0.44 (0.29)	Fresh water: Sr/Ba < 0.5	Brackish water: 0.5 < Sr/Ba < 1	Saline water: Sr/Ba > 1	[42,43]
	Ca/(Ca + Fe)	0.02–0.98 (0.26)	0.02–0.35 (0.11)	Fresh water: Ca/(Ca + Fe) < 0.4	Brackish water: 0.4 < Ca/(Ca + Fe) < 0.8	Saline water: Ca/(Ca + Fe) > 0.8	

Note: The format of the data is minimum–maximum, and the data in brackets represent the average values.

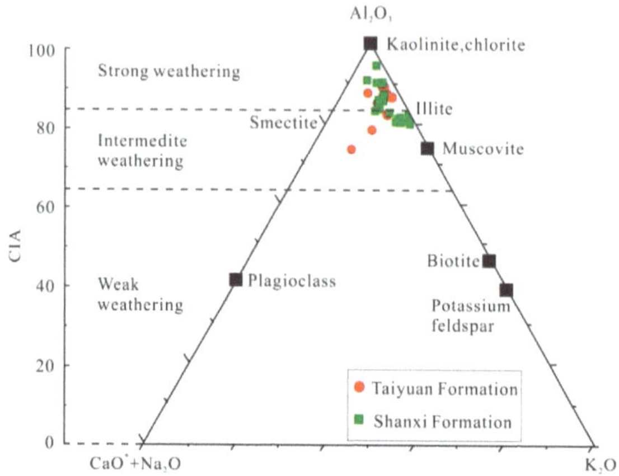

Figure 7. Ternary diagram of molecular proportions $Al_2O_3 - (CaO^* + Na_2O) - K_2O$. The left side shows the CIA scale. The TYF and SXF shale samples display intermediate to strong weathering [20,40].

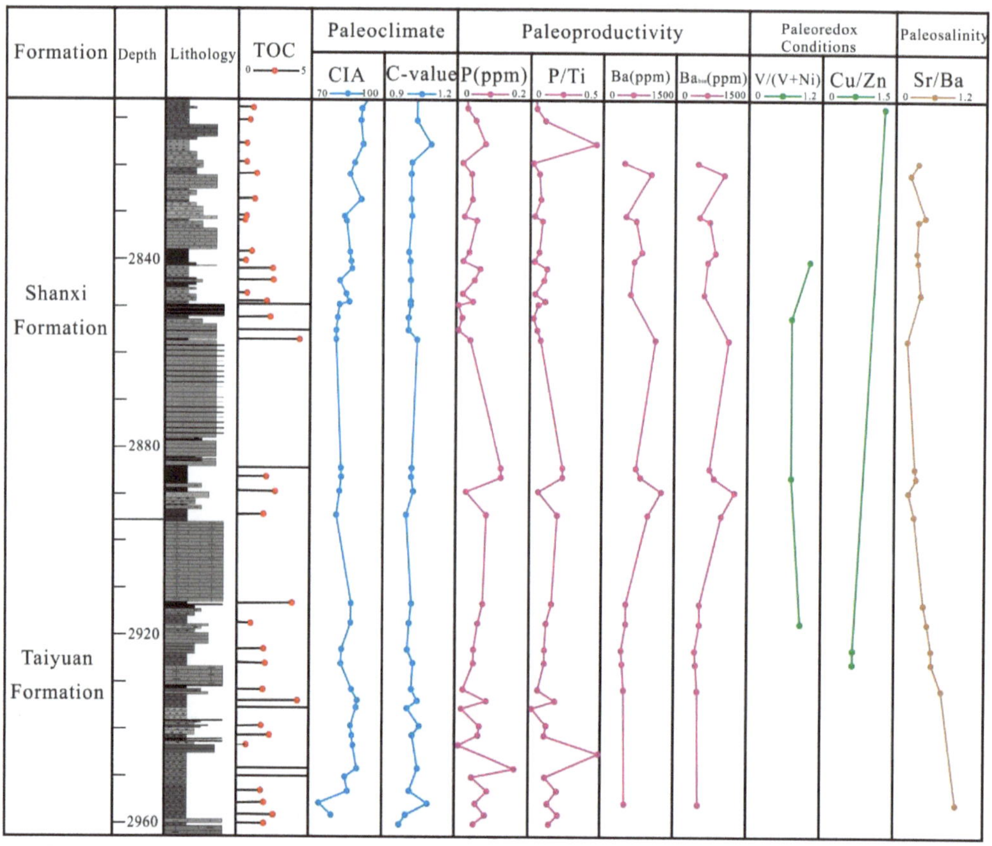

Figure 8. Stratigraphic distributions of TOC, paleoclimate indicators (CIA and C-value), paleoproductivity-related proxies (P, P/Ti, Ba, and Ba_{bio}), paleoredox indexes (V/(V + Ni) and Cu/Zn), and paleosalinity (Sr/Ba and Ca/(Ca + Fe)) in the core of the MY1 well.

The C-value, which is a ratio of transition metals to alkali elements, is also a common proxy for paleoclimate and is defined by the following formula [22]:

$$\text{C-value} = \Sigma(V + Ni + Mn + Fe + Cr + Co)/\Sigma(Ca + Mg + Ba + Sr + Na + K), \quad (4)$$

When the C-value < 0.4, it indicates a cold and arid climate; when the C-value range is 0.4–0.6, it indicates a semi-humid semi-arid climate; and when the C-value > 0.6, it indicates a warm and humid climate [21,22]. In this study, the C-value of the TYF shale samples ranged from 0.91 to 1.08, with an average of 1.01; the C-value of the SXF shale samples ranged from 0.99 to 1.09, with an average of 1.01 (Table 4). Based on the C-value results, the TYF and SXF were deposited in a warm and humid paleoclimate, which was conducive to the accumulation of OM (Figures 8–10).

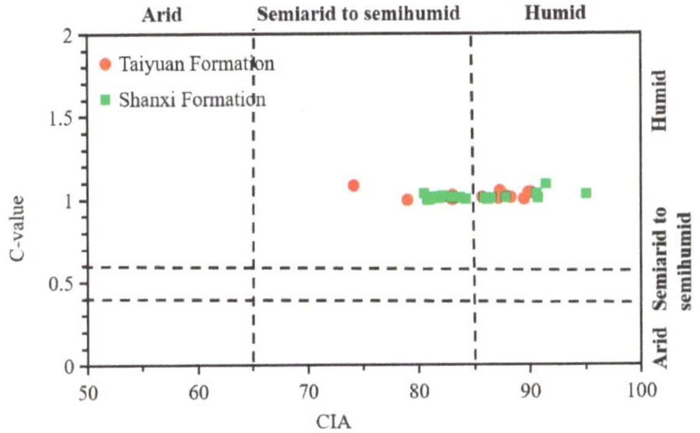

Figure 9. CIA versus C-value of shale from the TYF and SXF.

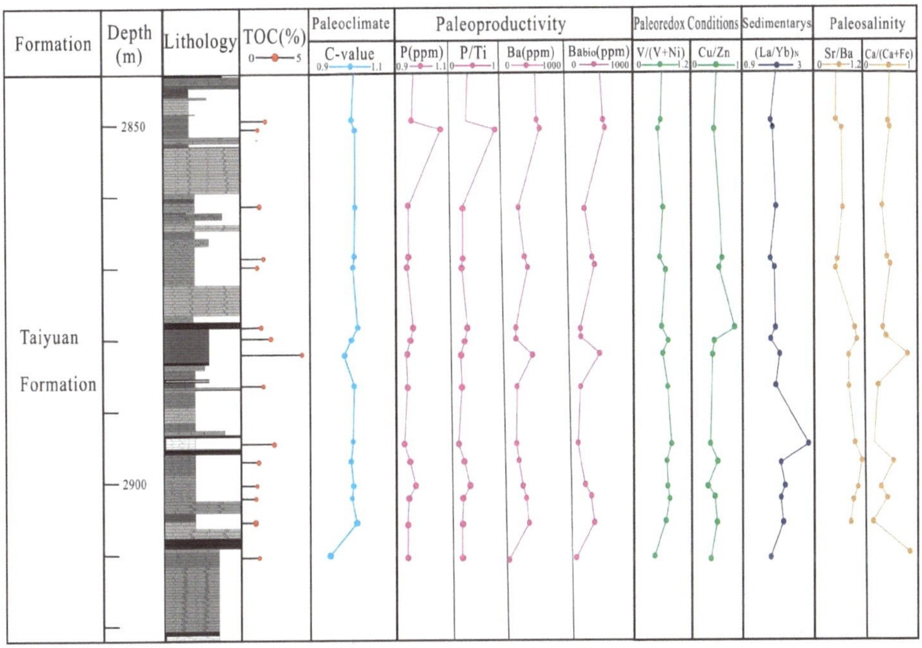

Figure 10. Stratigraphic distributions of TOC, paleoclimate indicators (C-value), paleoproductivity-related proxies (P, P/Ti, Ba, and Ba_{bio}), paleoredox indicators (V/(V + Ni and Cu/Zn), sedimentation rate ($(La/Yb)_N$), and paleosalinity indicators (Sr/Ba and Cu/(Cu + Zn)) in the core of the ZDY2 well.

The above paleoclimate indicators (CIA and C-value) collectively indicate that the SNCB experienced a warm and humid paleoclimate during the TYF and SXF deposition period (Figures 7–9). Moreover, there are no obvious differences in the CIAs and C-values for the TYF and SXF shale samples (Figure 8).

5.2. Paleoproductivity

Paleoproductivity, which is affected by water nutrients and biological activities, is a major factor in the accumulation of OM in sediments. Al, Fe, P, Ba, Cu, Zn, Ca, and Mn can be used to estimate the primary productivity [43,44]. However, firm restrictions exist for the selection of elements. P and Ba are commonly used to qualitatively assess biological productivity [2,16,17]. The elements Ba and P provide an effective assessment of the degree of primary productivity [16,17,43]. Ba is preserved as barite, while P is related to algal development [2,17].

Ba is generally combined with SO_4^{2-} and exists in sediments or water bodies in the form of barite ($BaSO_4$). Barium in sediments can be divided into terrestrial, hydrothermal, and biological sources [45]. However, only the source of Ba (Ba_{bio}) accurately reflects the level of primary productivity. It is necessary to extract abiotic elements from the total elements before evaluating the primary productivity [46,47]. Schroeder et al. [47] proposed Ba_{bio}'s applicable formula as follows:

$$Ba_{bio} = Ba_{total} - Ba_{alusilicate} = Ba_{sample} - Al_{sample} \times (Ba/Al)_{alusilicate} \quad (5)$$

where Al_{sample} and Ba_{sample} are the Al and Ba contents of the measured samples, respectively, and $(Ba/Al)_{alusilicate} = 0.0075$ is a correction factor, which is used to exclude the influence of barium in terrigenous aluminosilicates. Some scholars have also used Ti instead of Al as a correction factor [48].

As shown in Figure 8, the Ba_{bio} content of the TYF shale samples is 136.98–1376.11 ppm, with an average of 507.71 ppm (Table 4), while the Ba_{bio} content of the SXF shale samples is 446.89–1355.71 ppm, with an average of 828.38 ppm (Table 4), indicating that the primary productivity of the SXF is higher than that of the TYF. The value of La/Ce in ancient seawater is 2.8, which is much lower than that in the hydrothermal sediments. The value of La/Ce in hydrothermal sediments is only 0.25; in comparison, the value in normal sea water is usually greater than one. The La/Ce ratio of the shale samples in the TYF is less than 1, ranging from 0.44 to 0.82, with an average value of 0.54; the La/Ce ratio of shale samples in the SXF ranges from 0.45 to 0.56, with an average value of 0.51. This indicates that the shale was affected by strong hot-water sedimentation during the deposition process and is the product of the mixture of hot-water sedimentation and normal seawater sedimentation. Because of the influence of hydrothermal deposition in the study area, it was not appropriate to use Babio to judge the paleoproductivity.

The element ratios, P/Ti, Cu/Ti, P/Al, and Cu/Al, are not affected by the supply of terrigenous clastic sediments and can indicate biological productivity [14,49,50]. A higher ratio suggests a higher level of paleoproductivity [51]. The P/Ti of the TYF shale samples ranges from 0.02 to 1.83 (average of 0.24) (Table 4), which is slightly higher than the P/Ti value of PAAS (0.12 [37]), indicating that biological productivity is relatively high. The range of P/Ti of the SXF shale samples is 0.02–0.45 (average of 0.11) (Table 4), which is slightly lower than PAAS, indicating that the paleoproductivity level is medium. According to the P/Ti value, the primary productivity of the TYF is higher than that of the SXF (Figures 8 and 11).

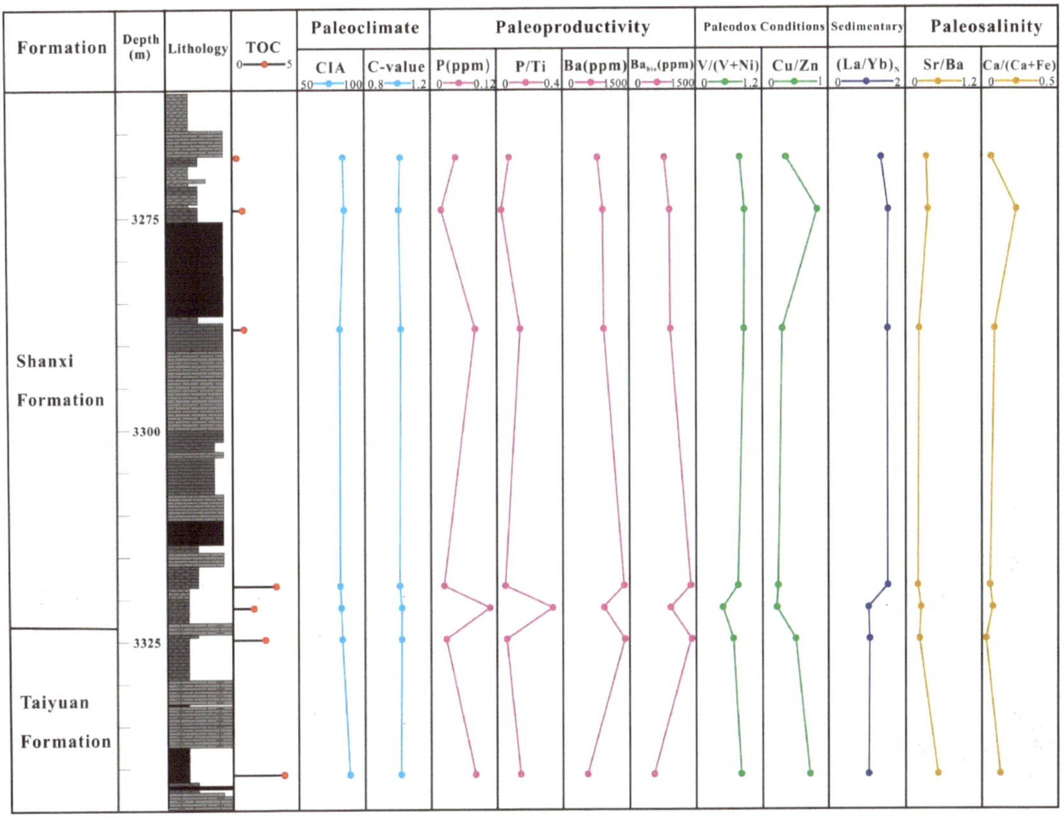

Figure 11. Stratigraphic distributions of TOC, paleoclimate indicators (CIA and C-value), paleoproductivity-related proxies (P, P/Ti, Ba, and Ba_{bio}), paleoredox indicators (V/(V + Ni) and Cu/Zn), sedimentation rate ($(La/Yb)_N$), and paleosalinity indicators (Sr/Ba) in the core of the ZXY1 well.

5.3. Sedimentation Rate

OM enrichment is affected by sedimentation rate to some extent [48,52]. A low sedimentation rate and sustained residence time of sediments in a water column would lead to significant differences between LREE and HREE, which can be characterized using the $(La/Yb)_N$ ratio and normalized REE distribution curve [52–55]. The degree of REE fractionation and distribution patterns can be used to establish the sedimentation rate [52,53]. REEs exist in water in combination with debris and suspended solids. Owing to the different residence times in water, the degree of REE fractionation is different [54,55]. When the sedimentation rate is low, REEs have sufficient time to be adsorbed by clay minerals, resulting in a high degree of fractionation, and when the sedimentation rate is high, the situation is the opposite, meaning the sedimentation rate can be deduced from the degree of REE fractionation. Since $(La/Yb)_N$ is a reliable indicator of the degree of REE fractionation, when the value of $(La/Yb)_N$ is close to one, it indicates a high sedimentation rate and a low degree of REE fractionation, and when the $(La/Yb)_N$ value is greater than or less than one, it indicates a low sedimentation rate and a high degree of REE fractionation [56,57]. Besides that, a weak differentiation would lead to a flat REE distribution curve in sediments [54]. As shown in Figure 6, the REE distribution curve trends of the TYF and SXF are relatively gentle, indicating that the shale sedimentation rate was relatively stable during the deposition of the TYF and SXF. The $(La/Yb)_N$ value range of the TYF is 0.99–2.6, with an average of 1.27, while the $(La/Yb)_N$ value of the SXF ranges from 1.01 to 1.53, with an average of 1.33, showing a relatively high sedimentation rate (Table 4). Because the sedimentation rate of the TYF is closer to one, its sedimentation rate is higher than that of the SXF (Figures 10 and 11). The high sedimentation rate of the TYF shortens the contact time between OM and bacteria, which is conducive to the preservation of OM in oxic water.

5.4. Paleoredox Conditions

TEs, such as the metallic elements U, Th, V, Cr, Ni, Cu, and Zn, are sensitive to the redox environment of water bodies because their solubility is controlled by the redox state of water bodies [19]. U, V, Cr, Ni, Cu, Zn, and other elements are easily soluble in water when the water body provides an oxidizing environment, but they are not soluble in water under reducing conditions [2]. When the water body is an oxygen-poor environment, it is enriched in sediments and almost does not migrate. This provides an indication of ancient water environment and can be used to judge the oxidation–reduction environment of its water body [2,19]. In this study, V/(V + Ni) and Cu/Zn were used to determine the redox conditions of the sedimentary environment.

When the value of V/(V + Ni) is greater than 0.6, it indicates an anoxic environment; when the value of V/(V + Ni) is between 0.46 and 0.6, it indicates a dysoxic sedimentary environment; and when the V/(V + Ni) value is less than 0.46, it indicates an oxygen deposition environment [58,59]. When the Cu/Zn value is less than 0.21, it indicates an anoxic environment; when Cu/Zn ranges from 0.21 to 0.63, it indicates a dysoxic environment; and when the value is greater than 0.63, it indicates an oxic sedimentary environment [2,60]. The V/(V + Ni) value of the TYF samples was between 0.45 and 0.91, with an average of 0.77, while the V/(V + Ni) value of the SXF samples was between 0.47 and 1, with an average of 0.76 (Table 4), indicating that the paleosedimentary environment of both the TYF and SXF was dysoxic–anoxic (Figures 10–12). The Cu/Zn value of the TYF samples ranged from 0.13 to 0.7, with an average of 0.34, while the Cu/Zn value of SXF samples was between 0.2 and 1.27, with an average of 0.48 (Table 4), indicating a dysoxic–anoxic environment (Figures 10–12). The values of V/(V + Ni) and Cu/Zn show that the SNCB was in a dysoxic–anoxic environment during the TYF and SXF periods (Figure 12), while the reducibility of the sedimentary environment of the TYF was slightly greater than that of the SXF.

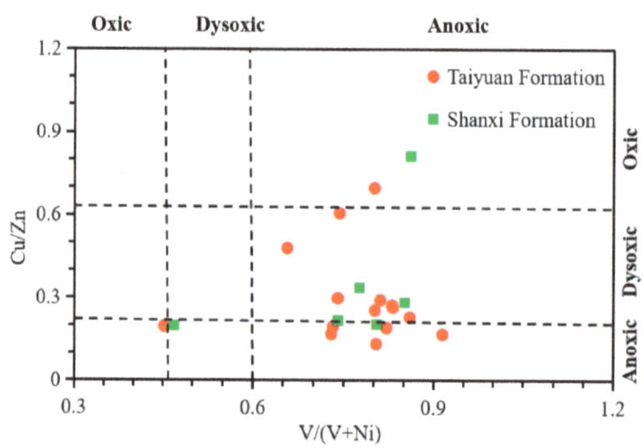

Figure 12. A diagram of V/(V + Ni) versus Cu/Zn of the shale samples from the TYF and SXF.

5.5. Paleosalinity

The Sr/Ba ratio is an important index for judging the salinity of ancient water. Sr is more soluble in water than Ba, meaning Sr migrates further into water, and the Sr/Ba ratio can indirectly reflect continental and marine sediments [60–64]. A Sr/Ba value less than 0.5 indicates a freshwater environment; a value of 0.5–1.0 indicates a brackish water environment; and a value greater than 1.0 suggests a saline water environment [43]. The Sr/Ba value of the TYF samples ranged from 0.21 to 5.31, with an average of 0.78 (Table 4), indicating a freshwater–brackish water environment; the Sr/Ba value of the SXF samples ranged from 0.18 to 0.44, with an average of 0.29 (Table 4), indicating a freshwater environment, implying that the TYF belonged to marine-continental transitional facies while the SXF was continental (Figures 8, 10 and 13).

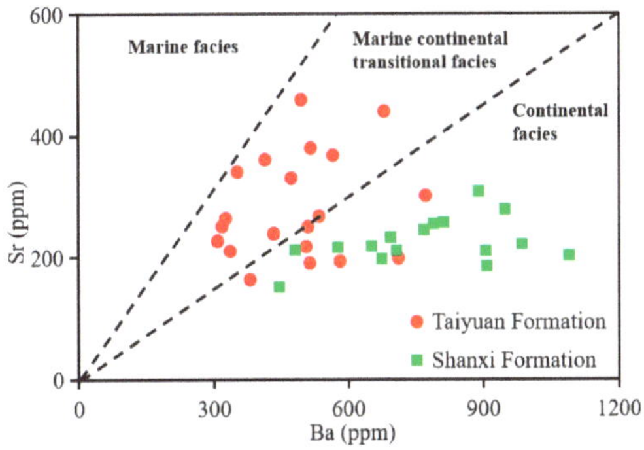

Figure 13. A diagram of Sr/Ba of the shale samples from the TYF and SXF [43].

Moreover, Ca/(Ca + Fe) can be used to reflect the salinity level of a water body [11,65]. Generally, the Ca/(Ca + Fe) ratio in a seawater environment is greater than 0.8, the ratio in a marine-continental transitional environment is 0.4–0.8, and the ratio in a continental freshwater environment is less than 0.4 [66]. The Ca/(Ca + Fe) value of the TYF ranged from 0.02 to 0.98, with an average of 0.26, indicating that the sedimentary environment was freshwater–brackish water, while the Ca/(Ca + Fe) value of the SXF ranged from 0.02 to 0.35, with an average of 0.11, indicating that the sedimentary environment was fresh water (Table 4, Figures 8 and 10). The change in the paleosalinity indices (Sr/Ba and Ca/(Ca + Fe)) both show that the paleosalinity of the TYF is higher than that of the SXF, and that OM is more easily preserved in saline water environments with higher salinity.

5.6. Controlling Factors and Formation Mechanisms of OM Accumulation

OM enrichment is usually affected by many factors, such as primary productivity, redox, paleosalinity, and sedimentation rate, which are complex physical and chemical processes [9,16,18,65]. According to the paleosedimentary environmental indicators, the paleoclimate conditions of the TYF and SXF in the SNCB were similar, both of which were deposited in a warm and humid environment, with relatively high biological productivity and sedimentation rates. Furthermore, the paleosalinity of the TYF and SXF was low during the deposition period, with the environment dominated by a dysoxic–anoxic set of circumstances belonging to marine-continental transitional and continental facies.

From the cross diagram of TOC and the above geochemical indexes (Figure 14), it can be observed that for the TYF and SXF, TOC has a clear correlation with paleoproductivity (P/Ti) (Figure 14a), sedimentation rate ($(La/Yb)_N$) (Figure 14b), and redox index (Cu/Zn) (Figure 14c), and a weak correlation with paleosalinity (Sr/Ba) (Figure 14d) and paleoclimate (C-value and CIA) (Figure 14e,f), indicating that the main controlling factors of the enrichment of OM in the TYF and SXF are paleoproductivity, sedimentation rate, and redox conditions, while paleoclimate and paleosalinity have relatively little impact on OM accumulation. The correlation between the TOC content of the TYF and these three indicators was better than that of the SXF, indicating that the OM concentration of the TYF was higher (Figure 14).

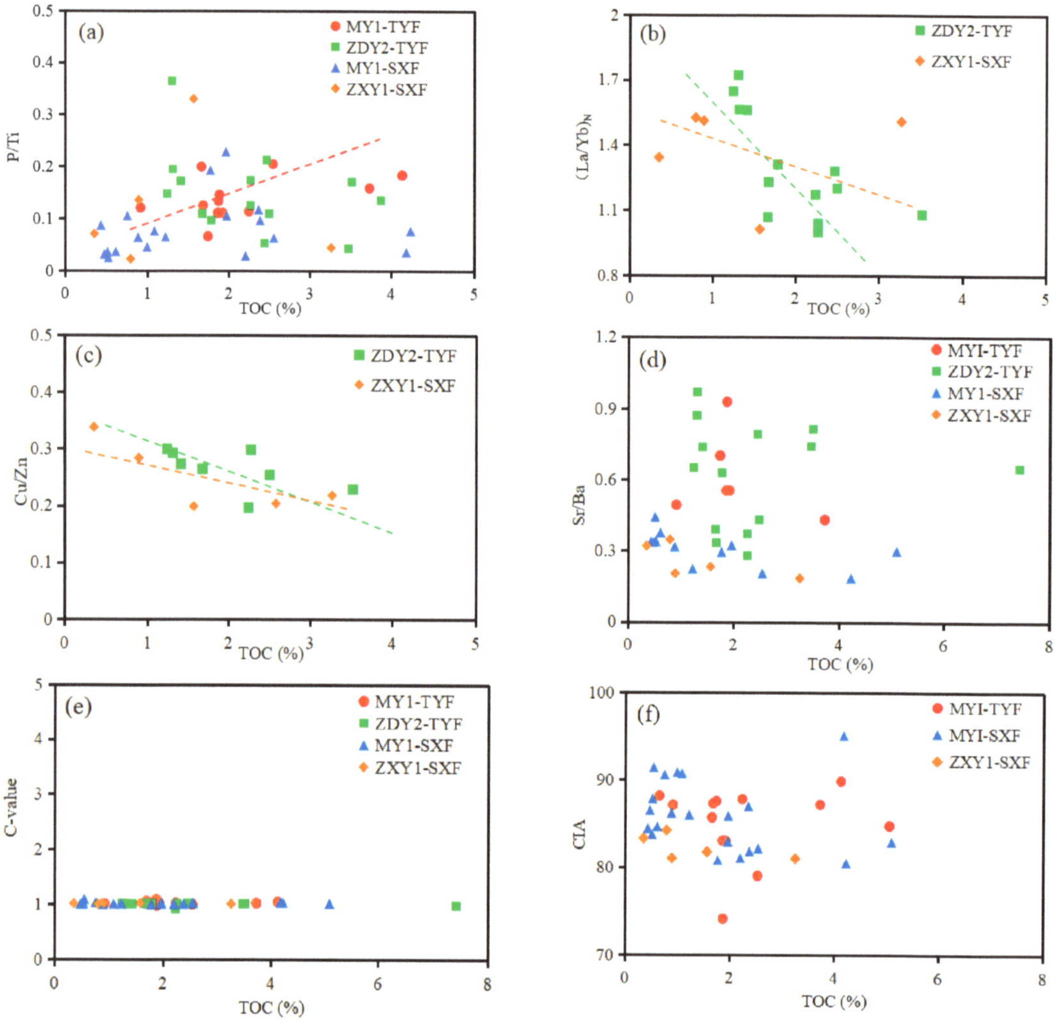

Figure 14. Correlations between TOC contents and productivity index (P/Ti), sedimentation rate index $(La/Yb)_N$, redox index (Cu/Zn), paleosalinity (Sr/Ba), and paleoclimate (C-value, CIA): (**a**) TOC vs. P/Ti; (**b**) TOC vs. $(La/Yb)_N$; (**c**) TOC vs. Cu/Zn; (**d**) TOC vs. Sr/Ba; (**e**) TOC vs. C-value; and (**f**) TOC vs. CIA.

Using the study of the paleoclimate, paleoproductivity, sedimentation rate, redox, and paleosalinity of the TYF and SXF, an OM enrichment model of the Permian TYF and SXF in the SNCB was developed (Figure 15). Two large-scale transgressions occurred in the early and late stages of TYF sedimentation, while the middle stage was the regression stage [67]. Overall, the TYF sea level decreased. The frequent transgression and regression events led to low-energy hydrodynamic conditions, enhanced the reduction of the water body, increased the salinity, reduced the decomposition of OM, and saved more OM. At the same time, the relatively high deposition rate shortened the residence time of aerobic bacteria in OM, reducing the decomposition of OM (Figure 15). Moreover, a high biological productivity increased respiratory oxygen consumption in the water column, leading to OM accumulation. However, during the deposition of the SXF, the seawater retreated from the SNCB and the lithofacies transitioned from marine to continental. The increase in terrestrial input led to a gradual decrease in salinity; the sedimentary environment changed from an anoxic environment to a poor-oxygen environment, and there was a decrease in paleoproductivity and sedimentation rate, which were not conducive to the deposition and preservation of OM and also reduced TOC. Both the TYF and SXF were deposited in a warm and humid paleoclimate, which was conducive to the growth and proliferation of plants and provided favorable conditions for the later formation of coal seams (Figure 15). In summary, the OM-rich shale of the TYF and SXF in the SNCB is not determined by a single factor but is due to the interaction of many factors, such as paleoproductivity, sedimentation rate, and redox, all of which directly or indirectly affect OM enrichment.

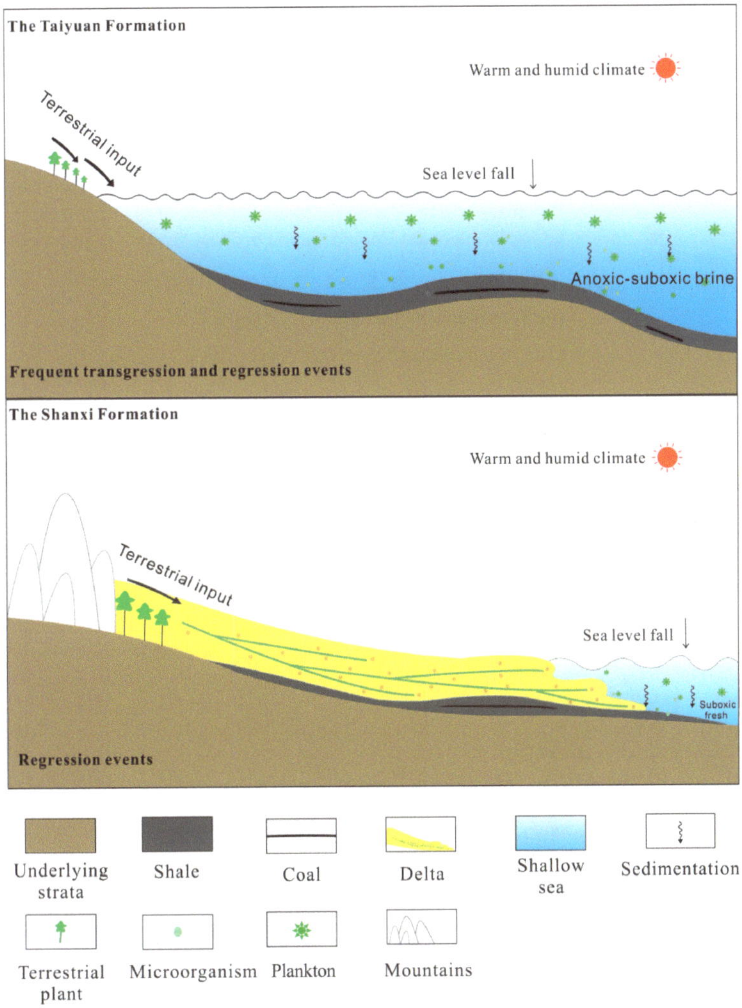

Figure 15. Organic matter enrichment model of the TYF and SXF in the SNCB.

6. Conclusions

1. The TYF and SXF were both deposited in warm and humid environments, with relatively high productivity and sedimentation rates. Moreover, the TYF and SXF were mainly characterized by dysoxic–anoxic environments, belonging to marine-continental transitional and continental facies.

2. The TYF shale mainly developed in a dysoxic–anoxic water environment, with a relatively high sedimentation rate, which reduced the oxidation of OM, caused OM to settle rapidly, and was conducive to OM accumulation. In addition, the high biological productivity increased respiratory oxygen consumption in the water column, leading to OM accumulation. However, the regression event of the SXF reduced the paleoproductivity and sedimentation rate, resulting in a decrease in OM.

3. An OM enrichment model of the TYF and SXF is established. This study is conducive to understanding shale enrichment mechanisms and guiding shale gas exploration in the region. The study could also have broader, global implications for its contributions to the general understanding of shales formed in marine-continental transitional environments.

Author Contributions: Literature search: Y.W. and X.C.; Figure: Y.W. and K.F.; Study design: Y.W., Z.H. and L.W.; Data collection: X.C., K.F. and Z.H.; Data analysis: Y.W.; Data interpretation: Y.W.; Writing: Y.W.; Supervision: X.C. and K.F.; Writing—Review & Editing: Z.H. and L.W. All authors have read and agreed to the published version of the manuscript.

Funding: The National Natural Science Foundation of China (41702152), Special Grant of China Postdoctoral Science Foundation (2018T110124), Foundation of State Key Laboratory of Petroleum Resources and Prospecting, China University of Petroleum, Beijing (PRP/open-1808), and Research on Exploration and Demonstration of Shale Gas in Henan Province (151100311000).

Institutional Review Board Statement: Not applicable.

Informed Consent Statement: Not applicable.

Data Availability Statement: Data supporting the results can be found in the manuscript text, tables, and figures.

Acknowledgments: This work was co-supported by the National Natural Science Foundation of China (41702152), the Special Grant of China Postdoctoral Science Foundation (2018T110124), the Foundation of State Key Laboratory of Petroleum Resources and Prospecting, the China University of Petroleum, Beijing (PRP/open-1808), and the Research on Exploration and Demonstration of Shale Gas in Henan Province (151100311000).

Conflicts of Interest: The authors declare that they have no conflict of interest.

References

1. Zhang, S.; Wang, X.; Wang, H.; Bjerrum, C.J.; Hammarlund, E.U.; Costa, M.M.; Connelly, J.N.; Zhang, B.; Su, J.; Canfield, D.E. Sufficient oxygen for animal respiration 1,400 million years ago. *Proc. Natl. Acad. Sci. USA* **2016**, *113*, 1731–1736. [CrossRef] [PubMed]
2. Tribovillard, N.; Algeo, T.J.; Lyons, T.; Riboulleau, A. Trace metals as paleoredox and paleoproductivity proxies: An update. *Chem. Geol.* **2006**, *232*, 12–32. [CrossRef]
3. Burton, Z.F.M.; Moldowan, J.M.; Magoon, L.B.; Sykes, R.; Graham, S.A. Interpretation of source rock depositional environment and age from seep oil, east coast of New Zealand. *Int. J. Earth Sci.* **2019**, *108*, 1079–1091. [CrossRef]
4. Peters, K.E.; Cassa, M. Chapter 5: Applied source rock geochemistry. In *The Petroleum System—From Source to Trap*; Magoon, L.B., Dow, W.G., Eds.; American Association of Petroleum Geologists Memoir 60: Tulsa, OK, USA, 1994; Volume 14, pp. 93–120.
5. Jarvie, D.M.; Hill, R.J.; Ruble, T.E.; Pollastro, R.M. Unconventional shale-gas systems: The Mississippian Barnett Shale of north-central Texas as one model for thermogenic shale-gas assessment. *AAPG Bull.* **2007**, *91*, 475–499. [CrossRef]
6. Johnson, J.E.; Phillips, S.C.; Torres, M.E.; Pinero, E.; Rose, K.K.; Giosan, L. Influence of total organic carbon deposition on the inventory of gas hydrate in the Indian continental margins. *Mar. Pet. Geol.* **2014**, *58*, 406–424. [CrossRef]
7. Burton, Z.F.M. Sediment organic contents required for gas hydrate formation: A survey of published basin and hydrocarbon system models. *Fuels* **2022**, *3*, 280–287. [CrossRef]
8. Zou, C.; Zhu, R.; Chen, Z.Q.; Ogg, J.G.; Wu, S.; Dong, D.; Qiu, Z.; Wang, Y.; Wang, L.; Lin, S.; et al. Organic-matter-rich shales of China. *Earth Sci. Rev.* **2019**, *189*, 51–78. [CrossRef]
9. Lash, G.G.; Blood, D.R. Organic matter accumulation, redox, and diagenetic history of the Marcellus Formation, southwestern Pennsylvania, Appalachian basin. *Mar. Pet. Geol.* **2014**, *57*, 244–263. [CrossRef]
10. Liu, W.; Yao, J.; Tong, J.; Qiao, Y.; Chen, Y. Organic matter accumulation on the Dalong Formation (Upper Permian) in western Hubei, South China: Constraints from multiple geochemical proxies and pyrite morphology. *Palaeogeogr. Palaeoclimatol. Palaeoecol.* **2019**, *514*, 677–689. [CrossRef]

11. He, J.; Ding, W.; Jiang, Z.; Jiu, K.; Li, A.; Sun, Y. Mineralogical and chemical distribution of the Es3L oil shale in the Jiyang Depression, Bohai Bay Basin (E China): Implications for paleoenvironmental reconstruction and organic matter accumulation. *Mar. Pet. Geol.* **2017**, *81*, 196–219. [CrossRef]
12. Harris, N.B.; Freeman, K.H.; Pancost, R.D.; White, T.S.; Mitchell, G.D. The character and origin of lacustrine source rocks in the Lower Cretaceous synrift section, Congo Basin, west Africa. *AAPG Bull.* **2004**, *88*, 1163–1184. [CrossRef]
13. Wu, J.; Liang, C.; Hu, Z.; Yang, R.; Xie, J.; Wang, R.; Zhao, J. Sedimentation mechanisms and enrichment of organic matter in the Ordovician Wufeng Formation-Silurian Longmaxi Formation in the Sichuan Basin. *Mar. Pet. Geol.* **2019**, *101*, 556–565. [CrossRef]
14. Zeng, S.; Wang, J.; Fu, X.; Chen, W.; Feng, X.; Wang, D.; Song, C.; Wang, Z. Geochemical characteristics, redox conditions, and organic matter accumulation of marine oil shale from the Changliang Mountain area, northern Tibet, China. *Mar. Pet. Geol.* **2015**, *64*, 203–221. [CrossRef]
15. Moldowan, J.M.; Sundararaman, P.; Schoell, M. Sensitivity of biomarker properties to depositional environment and/or source input in the Lower Toarcian of SW-Germany. *Org. Geochem.* **1986**, *10*, 915–926. [CrossRef]
16. Ding, J.H.; Zhang, J.C.; Huo, Z.P.; Shen, B.; Shi, G.; Yang, Z.; Li, X.; Li, C. Controlling Factors and Formation Models of Organic Matter Accumulation for the Upper Permian Dalong Formation Black Shale in the Lower Yangtze Region, South China: Constraints from Geochemical Evidence. *ACS Omega* **2021**, *6*, 3681–3692. [CrossRef]
17. Schmitz, B.; Charisi, S.D.; Thompson, E.I.; Speijer, R.P. Barium, SiO_2 (excess), and P_2O_5 as proxies of biological productivity in the Middle East during the Palaeocene and the latest Palaeocene benthic extinction event. *Terra Nova* **1997**, *9*, 95–99. [CrossRef]
18. Abart, F.R.; Wagreich, M.; Gier, S.; Ahmed, M.S.; Sami, M. Late Campanian Climatic-Continental Weathering Assessment and Its Influence on Source Rocks Deposition in Southern Tethys, Egypt. *Minerals* **2023**, *13*, 160.
19. Francois, R.A. Study on the regulation of the concentrations of some trace metals (Rb, Sr, Zn, Pb, Cu, V, Cr, Ni, Mn and Mo) in Saanich Inlet Sediments, British Columbia, Canada. *Mar. Geol.* **1988**, *83*, 285–308. [CrossRef]
20. Nesbitt, H.; Young, G.M. Early Proterozoic climates and plate motions inferred from major element chemistry of lutites. *Nature* **1982**, *299*, 715–717. [CrossRef]
21. Zhang, B.; Cheng, W.; Zhang, Q.; Li, Y.; Sun, P.; Fathy, D. Occurrence Patterns and Enrichment Influencing Factors of Trace Elements in Paleogene Coal in the Fushun Basin, China. *ACS Earth Space Chem.* **2022**, *6*, 3031–3042. [CrossRef]
22. Qiu, X.; Liu, C.; Mao, G.; Deng, Y.; Wang, F.; Wang, J. Major, trace and platinum-group element geochemistry of the Upper Triassic nonmarine hot shales in the Ordos basin, Central China. *Appl. Geochem.* **2015**, *52*, 42–52. [CrossRef]
23. Harris, N.B. *The Deposition of Organic-Carbon-Rich Sediments: Models, Mechanisms, and Consequences*; SEPM Society for Sedimentary Geology: Tulsa, OK, USA, 2005.
24. Murphy, A.E.; Sageman, B.; Hollander, D.J.; Lyons, T.W.; Brett, C.E. Black shale deposition and faunal overturn in the Devonian Appalachian Basin: Clastic starvation, seasonal water-column mixing, and efficient biolimiting nutrient recycling. *Paleoceanography* **2000**, *15*, 280–291. [CrossRef]
25. Teng, J.; Liu, Y. Analysis of distribution, storage potential and prospect for shale oil and gas in China. *Prog. Geophys.* **2013**, *28*, 1083–1108.
26. Dang, W.; Zhang, J.; Tang, X.; Chen, Q.; Han, S.; Li, Z.; Du, X.; Wei, X.; Zhang, M.; Liu, J.; et al. Shale gas potential of Lower Permian marine-continental transitional black shales in the Southern North China Basin, central China: Characterization of organic geochemistry. *J. Nat. Gas Sci. Eng.* **2016**, *28*, 639–650. [CrossRef]
27. Liang, Q.; Zhang, X.K.; Tian, J.; Sun, X.; Chang, H. Geological and geochemical characteristics of marine-continental transitional shale from the Lower Permian Taiyuan Formation, Taikang Uplift, southern North China Basin. *Mar. Pet. Geol.* **2018**, *98*, 229–242. [CrossRef]
28. Peng, Y. *The Shale Gas Accumlation Conditions of Taiyuan Formation in Southern North China Basin*; China University of Geosciences: Beijing, China, 2020.
29. Huo, Z.P.; Zhang, J.C.; Li, P.; Tang, X.; Yang, X.; Qiu, Q.; Dong, Z.; Li, Z. An improved evaluation method for the brittleness index of shale and its application—A case study from the southern north China basin. *J. Nat. Gas Sci. Eng.* **2018**, *59*, 47–55. [CrossRef]
30. Xu, H.L.; Zhao, Z.J.; Lu, F.L.; Yang, Y.; Tang, Z.; Sun, G.; Xu, Y. Tectonic evolution of the Nanhuabei area and analysis about its petroleum potential. *Geotect Metallog.* **2004**, *28*, 450–463.
31. Yu, H.; Lv, F.; Guo, Q.; Lu, W.; Wu, J.; Han, S. Proto-sediment basin types and tectonic evolution in the southern edge of North China Pla. *Earth-Sci. Rev.* **2005**, *27*, 111–117.
32. Zhou, X.; Ni, C.; Yang, F. The Paleozoic prototype basins and their tectonic deformation in North China and their controlling effects upon hydrocarbon accumulation. *Pet. Nat. Gas Geol.* **2010**, *31*, 779–794.
33. Diao, Y.; Wei, J.; Li, Z.; Cao, H.; Li, X. Late carboniferous-early permian sequence stratigraphy and paleogeography in the southern North China Basin. *Acta Geol. Sin.* **2011**, *35*, 88–94.
34. Zhang, M.Q. *Controlling Factor for Shale Enrichment of the Permian Shale in Southern North China Basin*; China University of Geosciences: Beijing, China, 2016.
35. Huo, Z.P.; Gao, J.; Zhang, J.C.; Zhang, D.; Liang, Y. Role of overlying and underlying limestones in the natural hydraulic fracturing of shale sections: The case of marine–continental transitional facies in the Southern North China Basin. *Energy Rep.* **2021**, *7*, 8711–8729. [CrossRef]
36. Haskin, L.A.; Wildeman, T.R.; Haskin, M.A. An accurate procedure for the determination of the rare earths by neutron activation. *J. Radioanal. Chem.* **1968**, *1*, 337–348. [CrossRef]

37. Taylor, S.R.; McLennan, S.M. *The Continental Crust: Its Composition and Evolution*; Blackwell Scientific Publications: Oxford, UK, 1985; pp. 117–140.
38. Bonis, N.R.; Ruhl, M.; Kürschner, W. Climate change driven black shale deposition during the end-Triassic in the western Tethys. *Palaeogeogr. Palaeoclimatol. Palaeoecol.* **2010**, *290*, 151–159. [CrossRef]
39. Burton, Z.F.M.; McHargue, T.; Kremer, C.H.; Bloch, R.B.; Gooley, J.T.; Jaikla, C.; Harrington, J. Peak Cenozoic warmth enabled deep-sea sand deposition. *Sci. Rep.* **2023**, *13*, 1276. [CrossRef] [PubMed]
40. Fedo, C.M.; Nesbitt, H.W.; Young, G.M. Unraveling the effects of potassium metasomatism in sedimentary rocks and paleosols, with implications for paleoweathering conditions and provenance. *Geology* **1995**, *23*, 921–924. [CrossRef]
41. Price, J.R.; Velbel, M.A. Chemical weathering indices applied to weathering profiles developed on heterogeneous felsic metamorphic parent rocks. *Chem. Geol.* **2003**, *202*, 397–416. [CrossRef]
42. McLennan, S.M.; Hemming, S.R.; McDaniel, D.K.; Hanson, G.N. Geochemical approaches to sedimentation, provenance, and tectonics. *Spec. Pap. Geol. Soc. Am.* **1993**, *284*, 21–40.
43. Chen, H.; Tang, D.; Chen, S.; Tang, S. Geochemical characteristics of mudstones from the lower cretaceous strata of the Jixi Basin, NE China: Implications for organic matter enrichment. *Int. J. Coal Geol.* **2021**, *249*, 103904. [CrossRef]
44. Brumsack, H.J. The trace metal content of recent organic carbon-rich sediments; implications for Cretaceous black shale formation. *Palaeogeogr. Palaeoclimatol. Palaeoecol.* **2006**, *232*, 344–361. [CrossRef]
45. Dymond, J.; Suess, E.; Lyle, M. Barium in Deep-Sea Sediment: A Geochemical Proxy for Paleoproductivity. *Paleoceanography* **1992**, *7*, 163–181. [CrossRef]
46. Shen, J.; Schoepfer, S.D.; Feng, Q.; Zhou, L.; Yu, J.; Song, H.; Wei, H.; Algeo, T.J. Marine productivity changes during the end-Permian crisis and Early Triassic recovery. *Earth Sci. Rev.* **2015**, *149*, 136–162. [CrossRef]
47. Schroeder, J.; Murray, R.W.; Leinen, M.S.; Pflaum, R.; Janecek, T.R. Barium in equatorial Pacific carbonate sediment: Terrigenous, oxide, and biogenic associations. *Paleoceanography* **1997**, *12*, 125–146. [CrossRef]
48. Ibach, L.E.J. Relationship between sedimentation rate and total organic carbon content in ancient marine sediments. *AAPG Bull.* **1982**, *66*, 170–188.
49. Algeo, T.J.; Maynard, J.B. Trace-element behavior and redox facies in core shales of Upper Pennsylvanian Kansas-type cyclothems. *Chem. Geol.* **2004**, *206*, 289–318. [CrossRef]
50. Wu, Z.; Zhao, X.; Wang, E.; Pu, X.; Lash, G.; Han, W.; Zhang, W.; Feng, Y. Sedimentary environment and organic enrichment mechanisms of lacustrine shale: A case study of the Paleogene Shahejie Formation, Qikou Sag, Bohai Bay Basin. *Palaeogeogr. Palaeoclimatol. Palaeoecol.* **2021**, *573*, 110404. [CrossRef]
51. Pan, Y.; Huang, Z.; Li, T.; Guo, X.; Xu, X.; Chen, X. Environmental response to volcanic activity and its effect on organic matter enrichment in the Permian Lucaogou Formation of the Malang Sag, Santanghu Basin, Northwest China. *Palaeogeogr. Palaeoclimatol. Palaeoecol.* **2020**, *560*, 110024. [CrossRef]
52. Tyson, R.V. Sedimentation rate, dilution, preservation and total organic carbon: Some results of a modelling study. *Org. Geochem.* **2001**, *32*, 333–339. [CrossRef]
53. Sageman, B.B.; Murphy, A.E.; Werne, J.P.; Ver Straeten, C.A.; Hollander, D.J.; Lyons, T.W. A tale of shales: The relative roles of production, decomposition, and dilution in the accumulation of organic-rich strata, Middle–Upper Devonian, Appalachian basin. *Chem. Geol.* **2003**, *195*, 229–273. [CrossRef]
54. Ruhlin, D.E.; Owen, R.M. The rare earth element geochemistry of hydrothermal sediments from the East Pacific Rise: Examination of a seawater scavenging mechanism. *Geochim. Cosmochim. Acta* **1986**, *50*, 393–400. [CrossRef]
55. Murray, R.W.; Brink, M.B.T.; Gerlach, D.C.; Russ, G.P.; Jones, D. Rare earth, major, and trace elements in chert from the Franciscan Complex and Monterey Group, California: Assessing REE sources to fine-grained marine sediments. *Geochim. Cosmochim. Acta* **1991**, *55*, 1875–1895. [CrossRef]
56. Tenger; Liu, W.; Xu, Y.; Chen, J. Comprehensive geochemical identification of highly evolved marine carbonate rocks as hydrocarbon-source rocks as exemplified by the Ordos Basin. *Sci. China Ser. D* **2006**, *49*, 384–396. [CrossRef]
57. Doner, Z.; Kumral, M.; Demirel, I.H.; Hu, Q. Geochemical characteristics of the Silurian shales from the central Taurides, southern Turkey: Organic matter accumulation, preservation and depositional environment modeling. *Mar. Pet. Geol.* **2019**, *102*, 155–175. [CrossRef]
58. Hatch, J.R.; Leventhal, J.S. Relationship between inferred redox potential of the depositional environment and geochemistry of the Upper Pennsylvanian (Missourian) Stark Shale Member of the Dennis Limestone, Wabaunsee County, Kansas, U.S.A. *Chem. Geol.* **1992**, *99*, 65–82. [CrossRef]
59. Li, W.; Li, X.; Ding, Q.; Chen, S.; Zhang, J. Geochemical characteristics and paleoenvironment of Cretaceous Jiufotang Formation in Xiushui Basin, northern Liaoning. *Glob. Geol.* **2019**, *38*, 154–161.
60. Mei, S. Application of rock chemistry in the study of Presinian sedimentary environment and the source of Uranium mineralization in Hunan Province. *Land Resour. Her.* **1988**, *7*, 25–31.
61. Ye, L.; Qi, T.; Peng, H. Depositional environment analysis of Shanxi formation in Eastern Ordos Basin. *Acta Sedimentol. Sin.* **2008**, *26*, 202–210.
62. Fu, J.; Deng, X.; Wang, Q.; Li, J.; Qiu, J.; Hao, L.; Zhao, Y. Densification and hydrocarbon accumulation of Triassic Yanchang Formation Chang 8 Member, Ordos Basin, NW China: Evidence from geochemistry and fluid inclusions. *Pet. Explor. Dev.* **2017**, *44*, 48–57. [CrossRef]

63. Xiao, J. Geochemical Indicators of Sedimentary Environments—A Summary. *Earth Environ.* **2011**, *39*, 405–414.
64. Wang, P.W.; Chen, Z.L.; Li, X.J. Geochemical characteristics and environmental significance of Dengying formation of Upper Sinian in Qiannan Depression. *Geoscience* **2011**, *25*, 1059–1065.
65. Liu, C.; Liu, K.; Wang, X.; Wu, L.; Fan, Y. Chemostratigraphy and sedimentary facies analysis of the Permian Lucaogou Formation in the Jimusaer Sag, Junggar Basin, NW China: Implications for tight oil exploration. *J. Asian Earth Sci.* **2019**, *178*, 96–111. [CrossRef]
66. Guo, L.; Zhang, S.; Xie, X.; Zhong, S.; Huang, C.; Chen, B. Geochemical characteristics and organic matter enrichment of the Dongyuemiao member mudstone of Lower Jurassic in the Western Hubei-Eastern Chongqing. *Sci. J. Earth Sci.* **2017**, *42*, 1235–1246.
67. Shang, B.H.; Ni, Y.; Song, H.; Liu, S.; Zhang, L. Sequence stratigraphic framework of Late Paleozoic coal measures and their sedimentary evolution in Henan, Yprovince. *Coal Geol. Explor.* **2012**, *40*, 1–5.

Journal of
Marine Science and Engineering

MDPI

Article

Microscopic Analysis of Natural Fracture Properties in Organic-Rich Continental Shale Oil Reservoirs: A Case Study from the Lower Jurassic in the Sichuan Basin, China

Xuefeng Bai [1,2], Saipeng Huang [3,4,*], Xiandong Wang [5,6], Zhiguo Wang [5,6], Youzhi Wang [5,6], Weiqi Ma [5,6], Yanping Zhu [5,6], Mengdi Sun [3], Bo Liu [3], Xiaofei Fu [3], Lijuan Cheng [3], Likai Cui [3] and Yudong Hou [7]

1 School of Geoscience, China University of Petroleum, Qingdao 266580, China
2 Exploration Department, PetroChina Daqing Oilfield Company Limited, Daqing 163000, China
3 Key Laboratory of Continental Shale Hydrocarbon Accumulation and Efficient Development, Ministry of Education, Northeast Petroleum University, Daqing 163318, China
4 Department de Mineralogia, Petrologia i Geologia Aplicada, Facultat de Ciències de la Terra, Universitat de Barcelona (UB), c/Martí i Franquès s/n, 08028 Barcelona, Spain
5 Exploration and Development Research Institute, PetroChina Daqing Oilfield Company Limited, Daqing 163712, China
6 Key Laboratory of Continental Shale Oil of Heilongjiang Province, Daqing 163318, China
7 College of Resources and Safety Engineering, Chongqing University, Chongqing 400030, China
* Correspondence: huangspcugb@hotmail.com

Abstract: Natural fractures are of paramount importance in storing carbon in shale oil reservoirs, where ultra-low porosity and permeability necessitate their essentiality for enhanced oil recovery. Therefore, comprehensively clarifying the characteristics of natural fractures in shale oil reservoirs is imperative. This paper focuses on investigating the microscopic features of natural fractures in organic-rich continental shale oil reservoirs that are commonly found in the Lower Jurassic strata of the Sichuan Basin, employing them as a representative example. Multiple methods were utilized, including mechanical testing, Kaiser testing, multi-scale CT scanning (at 2 mm, 25 mm, and 100 mm scales), and a numerical simulation of fluid seepage in fracture models. The results revealed that the in situ stress of the target seam displays the characteristic of $\sigma_H > \sigma_v > \sigma_h$, with σ_v and σ_h being particularly similar. The relatively high lateral stress coefficient (ranging from 1.020 to 1.037) indicates that the horizontal stresses are higher than the average level. Although the 2 mm CT scan provides a more detailed view of fractures and connected pores, it primarily exhibited more pore information due to the high resolution, which may not fully unveil additional information about the fractures. Thus, the 25 mm shale core is a better option for studying natural fractures. The tortuosity of the different fractures indicated that the morphology of larger fractures is more likely to remain stable, while small-scale fractures tend to exhibit diverse shapes. The simulations demonstrated that the stress sensitivity of fracture permeability is approximately comparable across different fracture scales. Therefore, our research can enhance the understanding of the properties of natural fractures, facilitate predicting favorable areas for shale oil exploration, and aid in evaluating the carbon storage potential of shale oil reservoirs.

Keywords: Sichuan Basin; Lower Jurassic; shale oil reservoir; natural fracture; CT scanning

Citation: Bai, X.; Huang, S.; Wang, X.; Wang, Z.; Wang, Y.; Ma, W.; Zhu, Y.; Sun, M.; Liu, B.; Fu, X.; et al. Microscopic Analysis of Natural Fracture Properties in Organic-Rich Continental Shale Oil Reservoirs: A Case Study from the Lower Jurassic in the Sichuan Basin, China. *J. Mar. Sci. Eng.* **2023**, *11*, 1036. https://doi.org/10.3390/jmse11051036

Academic Editor: George Kontakiotis

Received: 24 April 2023
Revised: 10 May 2023
Accepted: 10 May 2023
Published: 12 May 2023

1. Introduction

Shale oil is one of the most important unconventional resources [1–5], attracting global attention in recent years [6–12]. The use of hydraulic fracturing technology has made it possible to extract oil and gas from previously uneconomical shale formations [13,14], leading to a shale oil production boom, particularly in the United States and Russia [15–17]. As an alternative to conventional energy, shale oil is leading a new energy revolution to address China's energy shortage [18]. Recent research has shown that shale oil also has excellent potential in the Qingshankou Formation (Fm) in the Songliao Basin [19] and the Lianggaoshan Fm and Ziliujing Fms in the Sichuan Basin [20]. Notwithstanding the abundant resources of shale oil, its effective exploitation has been hampered by the extremely low porosity and permeability of the reservoirs [21,22]. As such, the recovery of shale oil remains a pressing research focus.

Natural fractures are the primary channels for oil and gas seepage, making them essential for enhanced recovery [23–26]. In addition, natural fractures in reservoirs are crucial for creating a low-carbon environment, as they serve as significant storage spaces for carbon storage [27–32]. Consequently, many scholars have conducted research on shale natural fractures [21,33–36] and the multi-scale dimensional evaluation of natural fractures in shale has been a hot topic [22,37,38]. Information about natural fractures, such as fracture density, filler, orientation, width, and length, can be obtained from shale outcrops [39] and the natural fracture orientation can serve as a vital indicator of the present-day stress field [40]. Additionally, full-diameter shale drill cores can be used to study natural fractures, and macroscopic fractures can be described by observing fresh, unweathered shale cores [22]. Moreover, the Formation MicroScanner Image (FMI) data of formation resistivity can also identify natural fractures [41], though the accuracy of microfracture observation is insufficient. Further research on fracture observation through scanning electron microscope photographs has also been conducted [42], although thin sections may alter the original state of the fracture.

In recent years, there has been an increasing demand for a more precise observation and description of in situ natural fractures. Non-destructive techniques such as radiation X-ray computed tomography (CT) scanning have become a common means of studying in situ fractures and microscopic pores [43–46]. Gou et al. (2019) [45] used nano-CT and micro-CT to finely study the pore-fracture structure of shales, but the study primarily focused on the pore system. However, numerous advanced representation techniques have been utilized to thoroughly investigate the connectivity of shale pores, as extensively documented in studies by Sun et al. (2016, 2020) [47,48]. On the other hand, the shale fracture shape can be well revealed by using micron-CT [49] and the fracture length, width, connection forms, and tortuosity as well. Furthermore, CT scanning can also be used to monitor fracture propagation in real-time [44]. Additionally, numerical modeling based on the geometries of reservoir rocks can be established using CT scanning, and further coupling studies can be conducted using COMSOL [50].

While CT scanning is a widely used technique for studying natural fractures in shale, there has been limited research on the comparative characterization of these fractures using different core sizes. The investigation of corresponding natural fracture characteristics across different scales remains largely understudied. Therefore, this study investigates the characteristics of natural fractures in continental shale oil reservoirs of the Lower Jurassic in the Sichuan Basin at the microscopic scale using multi-scale cores. To achieve this goal, various methods, such as mechanical testing, Kaiser testing, and multi-scale CT scanning (at 2 mm, 25 mm, and 100 mm scales), were employed. In addition, numerical simulations of the fluid seepage in fracture models were conducted to study the fluid seepage characteristics in fractures at different scales. Thus, this paper presents a novel and comprehensive approach, particularly, by combining the CT scanning technique with the finite element method (FEM) to enable the microscopic observation of fractures and the numerical simulation of fluid flow in fluid–solid coupling within the fracture. This approach allows for an investigation into the characterization of microscopic fractures in shale oil reservoirs. Therefore, the results of this study can contribute significantly to a better understanding of the properties of natural fractures and aid in predicting favorable areas for shale oil exploration.

2. Geological Setting

The Sichuan Basin is a significant resource-based basin in China, known for its abundant shale oil and gas reserves [22]. This basin is a first-order tectonic unit of the Yangzi

Platform that formed during the Indo–Chinese period and was later folded and deformed by the Himalayan movement, resulting in its current tectonic condition. Based on its geomorphological features, the Sichuan Basin can be further classified into six zones: the low and gentle tectonic zone in north Sichuan, the low and steep tectonic zone in west Sichuan, the gentle tectonic zone in central Sichuan, the low and folded tectonic zone in southwest Sichuan, the low and steep tectonic zone in south Sichuan, and the high and steep tectonic zone in east Sichuan (Figure 1a).

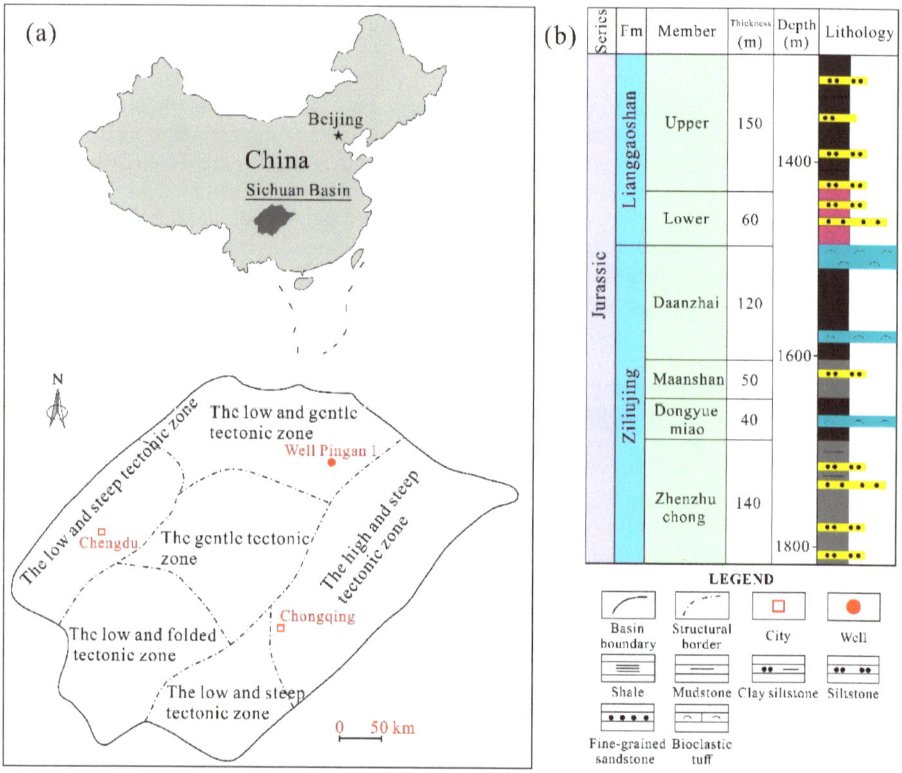

Figure 1. (**a**) The detail location of the Sichuan Basin, and its structural outline. The star in the figure represents the capital of China, the red square-shaped and circle represent the city and well, respectively; (**b**) the comprehensive stratigraphic column of the Jurassic, Sichuan Basin. The different colors in (**b**) indicates the different levels of stratigraphic units, like Series, Fms, and member.

The Jurassic sedimentary assemblage in the Sichuan Basin is primarily composed of clastic rocks interspersed with tuffs, covering an area of approximately 18×10^4 km^2, with thickness varying from 500 to 4500 m. The southern region of the basin presents a relative thinness in comparison to the substantially thicker formations found in the northern basin. The basin's Jurassic formations are divided from bottom to top into the Ziliujing Formation (ZLJ Fm), the Lianggaoshan Formation (LGS Fm), the Shaximiao Formation (SXM Fm), the Suining Formation (SN Fm), and the Penglaizhen Formation (PLZ Fm) (Figure 1b). While the Lower Jurassic is well-preserved in the Sichuan Basin, the SXM Fm of the Middle Jurassic and other formations have suffered various degrees of denudation and they are relatively well-preserved only in the central Sichuan Basin.

This study collected shale samples from the typical exploration well Pingan 1. The upper section of the LGS Fm in Pingan 1 is further divided into three sublayers: the upper LGS-1, the upper LGS-2, and the upper LGS-3. Among them, the percentage of clayey feldspathic quartz shale in the upper LGS-1 is 54.8%. The average content of clay minerals is 44.3% and the quartz content is 46.9%, with small amounts of calcite and pyrite. The porosity of the upper LGS-1 ranges from 0.86% to 4.66%, with an average of 2.84%. The TOC (total organic carbon) content ranges from 0.57% to 2.56%, with an average of 1.35%. The shale of the upper LGS has a high content of brittle minerals; in particular, the brittle mineral content of the upper LGS-1 ranges from 40.6% to 74.7%, with an average of 54.8%. As shown in Figure 1a, Pingan 1 is located in the low and gentle tectonic zone, and its shale oil production has reached 112.8 m^3/d, a high yield [51]. The shale samples were obtained from Pingan 1 at an average depth of 3011.2 m.

3. Experiments and Methodology

The present study incorporates a combination of experimental tests and numerical simulation methods which consequently provide a comprehensive overview of the principles and techniques employed in CT scanning and FEM numerical simulations.

3.1. The Mechanics of the Shale Oil Reservoir

The development of natural fractures in shale oil reservoirs typically involves two processes: fracture initiation and propagation. Type I fractures (tension cracks), which are the most hazardous among type II (shear cracks) and type III (tension–shear cracks), initiate when the tensile stress exceeds the rock's tensile strength. Additionally, fracture toughness is a crucial parameter that indicates the rock's ability to resist fracture propagation. Thus, both of these mechanical parameters have a significant impact on fracture development.

The Brazilian splitting test is a highly effective method for determining both the tensile strength and fracture toughness of rocks [52,53]. To prepare the specimens for the test, samples are taken from vertical strata and cut into three pieces, which are then standardized to ensure a height-to-diameter ratio of 1. The rock samples are then carefully machined to meet the requirements of the International Society for Rock Mechanics' Recommended Test Specification. The size, mass, and density of the specimens are all carefully documented in Table 1. In addition to evaluating the rock's tensile strength, the Brazilian splitting test also assesses its fracture toughness by introducing pre-fabricated cracks into the specimens. Specifically, three semi-disc specimens are extracted from the vertical strata and manual cracks are created to achieve a height-to-crack length ratio of 0.5, as required by the standard. More detailed specifications of the specimens are provided in Table 2.

Table 1. The specimens used for testing tensile strength.

No.	Length (mm)	Diameter (mm)	Weight (g)	Density (g/cm^3)
1-1	26.02	24.68	32.21	2.59
1-2	23.98	24.02	29.58	2.72
1-3	24.55	25.54	32.54	2.59

Table 2. The specimens used for testing fracture toughness.

No.	Height (mm)	Diameter (mm)	Thickness (mm)	Prefabricated Crack Length (mm)
2-1	38.07	76.07	23.13	19.00
2-2	39.51	75.65	23.34	19.00
2-3	36.62	75.48	23.68	19.00

The test apparatus is a rock multi-field coupling triaxial tester with a maximum axial testing force of 500 KN and a deformation measurement error of less than ±0.5%. The apparatus is capable of controlling temperatures ranging from room temperature to 150 °C, with a temperature control accuracy of ±0.5 °C. To determine the tensile strength, the test was performed after axial compression loading at a rate of 0.001 mm/s until the residual strength value was noticeable. During the fracture toughness assessment, the axial loading rate was controlled at a slower pace of 0.0001 mm/s. The highest axial load and the displacement of the specimen at the time of failure were recorded and documented. Finally, the specimen was removed, and a displacement–load curve was plotted to depict the specimen's degradation.

In addition, assessing the present-day in situ stress state can provide insight into the propagation behavior of natural fractures in shale oil reservoirs. The Kaiser effect test, which is an effective experimental method that has been commonly used for determining in situ stress [23,54,55], was carried out to obtain the present-day in situ stress of the Pingan 1 well. The Kaiser effect of rocks can be used to measure the historical maximum stress that the rock has undergone. In general, a uniaxial loading test is performed on rock samples, and the emitted sound signals are detected using an acoustic emission instrument. The stress corresponding to the Kaiser point, which is the abrupt increase in the number of sound emissions observed in the relationship between the emitted sound signals and stress, is considered to be the historical maximum stress that the rock has experienced. Typically, a rock sample is taken vertically from the full-size core and three samples are taken horizontally at an incremental angle of 45°. The Kaiser test was performed on the three group rock cores (a total of 12 specimens) with an average depth of 3011.2 m. After conducting measurements, we were able to determine the in situ stresses of three-layer sections located at depths of 3010.00 m, 3010.60 m, and 3013.00 m.

3.2. The CT Scanning of Multi-Scale Shale Cores

The tests were carried out with a CT scanner apparatus of the GE Phoenix Nanotom S at the Northeast Petroleum University, China. A detailed schematic of the X-ray CT scanning setup is shown in Figure 2. Its maximum resolution is approximately 200 nm, which is sufficiently fine to display the cores' information. The sample diameter that the CT scanner can hold to test ranges from 1 to 120 mm, which fulfills the test requirements. The advantages of using CT scanning include (1) its ability to provide a comprehensive view of very small feature surfaces from a large amount of image data without destroying the sample and (2) the CT images reflect information on the energy decay of the X-rays as they penetrate the object. The pore structure and the relative density value within the core are positively correlated by the grey scale of the 3D CT images. Therefore, the 3D digital core can make a qualitative analysis and quantitative evaluation of pore throat size, connectivity, and morphology.

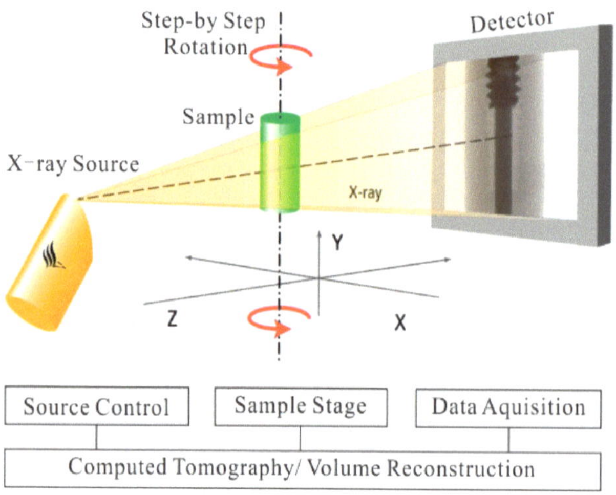

Figure 2. Schematic diagram of the X-ray CT scanning setup. The X-ray source and sample are represented by yellow and green colors, respectively. The coordinate axes show the directions, and red arrows demonstrate the rotation direction during the test.

Compared to experimental analysis methods, such as mercury piezoelectricity and scanning electron microscopy, CT scanning has an advantage in the rapid and non-destructive scanning of rock samples in a wide range and in all directions. The digital cores obtained by CT scanning can be used to study the microscopic pore and fracture characteristics of the reservoir in a more intuitive way, and their application on microscopic pore and fracture structure evaluation is more extensive.

For the purpose of achieving a comparative study on the morphological characteristics of the microfracture at multi-scales, three kinds of core sizes, including 100 mm, 25 mm, and 2 mm, are selected for the scanning test and then analyzed to display the fracture morphology using the program Avizo. The scanning accuracy of the 2 mm core samples ranges from 0.92 to 1.25 μm. Their detailed parameters, including fracture length, width, connection forms, tortuosity, etc., are then displayed and analyzed.

3.3. Natural Fracture Simulation

The use of CT scans can effectively showcase the natural fracture features of shale and enable seepage simulations in typical shale fractures with fluid–solid coupling, allowing for a better understanding of the types of fractures that contribute to the increased seepage capacity of shale reservoirs. To accomplish this, realistic fracture models were created using the CT scans, and seepage simulations were conducted on different natural fracture types. The fluid–solid coupling process was carried out through the use of COMSOL Multiphysics, a simulation software that utilizes various physics interfaces to model fluid–solid coupling processes, such as fluid flow, heat transfer, solid mechanics, and other relevant phenomena. The software uses the FEM to solve the mathematical equations that describe the physical behavior of the system. FEM is a widely used method for modeling fluid–solid interactions [56]. In COMSOL, the fluid–solid coupling is implemented through two physics interfaces: the fluid flow interface and the structural mechanic interface. The fluid flow interface models the behavior of fluids, including velocity, pressure, temperature, and density, while the structural mechanics interface models the behavior of solid materials, including stress, strain, and deformation. The software utilizes boundary conditions to achieve these goals. Additionally, COMSOL allows for non-linear fluid–solid interaction, where the behavior of the fluid and the solid are mutually affected. Overall, COMSOL is a powerful tool for modeling fluid–solid coupling processes, offering a wide range of options for defining the physics, geometry, and boundary conditions of a system.

In this study, we present a fully coupled simulation of the mutual interactions between fluids and solids in coal, implemented using the COMSOL software. The fluid in the channel is assumed to be in a state of fully developed laminar flow, with a constant density and incompressible viscous flow of gas. Thus, the flow of fluid in the channel is described by the incompressible Navier–Stokes equations for the velocity field u (u, v) in the moving spatial coordinate system and pressure p.

$$\rho\frac{\partial u}{\partial t}+\rho((u-u_0)\cdot\nabla)u-\nabla\cdot\left[-pI+\eta\left(\nabla u+(\nabla u)^T\right)\right]=F \tag{1}$$

$$-\nabla\cdot u=0 \tag{2}$$

Here, the I denotes the unit diagonal matrix and F represents the volume force affecting the fluid. Assuming that there is no gravity or other volume force affecting the fluid (i.e., $F = 0$), the velocity of the coordinate system is denoted by u_0 (u_0, v_0).

The fluid exerts a force on the structural wall, generated by viscous resistance and fluid pressure, causing the elastic deformation of the coal skeleton, which is compressed under external loading. The outer boundary of the coal sample is fixed, and all boundaries in contact with the fluid domain are subject to the load imposed by the fluid, except for the outer boundary. The formula is given below:

$$F_T=-n\cdot\left(-pI+\eta\left(\nabla u+(\nabla u)^T\right)\right) \tag{3}$$

Here, n denotes the normal vector of the boundary. This load represents the sum of pressure and viscous forces.

The pressure and velocity of the fluid and solid domains are transmitted through their common boundaries, and the Arbitrary Lagrangian–Eulerian (ALE) method is used to handle geometric deformation and boundary movement through dynamic meshing [57].

$$v_t(x_0, y_0) = \left.\frac{\partial v}{\partial t}\right|_{x_0, y_0} \tag{4}$$

$$v_{Time}(X_n, Y_n) = \left.\frac{\partial v}{\partial t}\right|_{X_n, Y_n} \tag{5}$$

where X and Y are the material frame coordinates and x and y are the coordinates of the spatial frame.

Figure 3 displays the models created from the CT scan data of two different shale core sizes: 100 mm and 25 mm. The models used in the simulations were created using the following process: (1) identifying fractures from scanned sections, as shown in Figure 3a,e, (2) segmenting the data based on specific thresholds, displayed in Figure 3b,f, (3) reconstructing a 3D model structure, as seen in Figure 3c,g, and (4) refining appropriate grids based on the size of the structure, as demonstrated in Figure 3d,h.

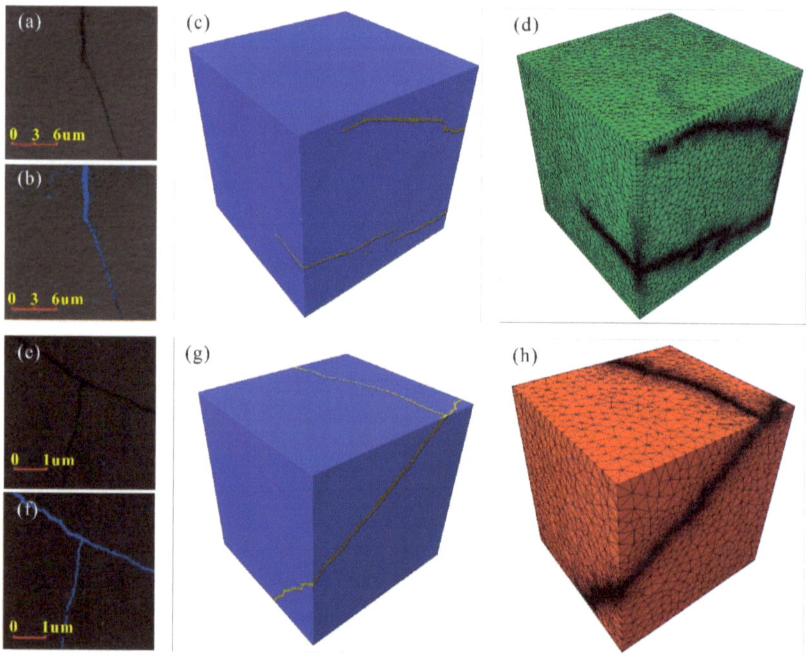

Figure 3. A detailed description of the process for creating models based on the CT scanning data from 100 mm and 25 mm shale cores. The process includes identifying fractures from scanned sections (**a**,**e**), segmenting the data based on specific thresholds (**b**,**f**), reconstructing a 3D model structure (**c**,**g**) displaying with the blue color, and refining appropriate grids based on the size of the structure (**d**,**h**), displaying with green and red colors, respectively.

In addition, the models were established using the following parameters: a density of 2.65 g/cm^3, a tensile strength of 8 MPa, a compressive strength of 46.67 MPa, a Young's modulus of 1.33 GPa, and a Poisson's ratio of 0.21. The model obtained from the 100 mm shale core has a side length of 22,832 μm and a mesh count of 1,342,627, as shown in Figure 3d. In contrast, the model obtained from the 25 mm shale core is much smaller, with a side length of 5576 μm and a mesh count of 492,140. Both models use tetrahedrons to divide the mesh. Within the framework of the fluid–solid coupling simulation, the fractures present in the models serve as the principal conduits for CH_4, the gas under investigation, to permeate through the system.

4. Results

4.1. The Shale Oil Reservoir's Mechanical Data

For the tensile strength test, the shale oil reservoir's breakdown pressures were recorded as 7.26 kN, 10.50 kN, and 3.30 kN (Figure 4a–c) and the corresponding tensile strengths were 7.20 MPa, 11.60 MPa, and 3.35 MPa. In terms of values, there was a significant variation in the tensile strength of the different specimens, even if they were collected from the same depth. When testing the fracture toughness, the breakdown pressures were recorded as 1.55 kN, 2.14 kN, and 1.66 kN (Figure 4d–f). The fracture toughness of the specimen was calculated using the elastic solution of the Brazilian cleavage test for the collected data. The corresponding fracture toughnesses were 0.26 MPa·$m^{1/2}$, 0.36 MPa·$m^{1/2}$, and 0.28 MPa·$m^{1/2}$. Overall, the magnitude of the tensile strengths and fracture toughness varied a little, however, they exhibited similar mechanical characteristics.

The in situ stress obtained by the Kaiser test is displayed in Table 3. The average in situ stresses were determined for the target reservoirs, with the maximum horizontal stress (σ_H), vertical stress (σ_v), and minimum horizontal stress (σ_h) values being 81.21 MPa, 76.83 MPa, and 76.42 MPa, respectively. The stress gradients of σ_H, σ_v, and σ_h varied from 2.69 to 2.71 MPa/100 m, 2.51 to 2.57 MPa/100 m, and 2.50 to 2.57 MPa/100 m, respectively. Moreover, the lateral stress coefficients ranged from 1.020 to 1.037.

Table 3. Present-day in situ stress of the shale oil reservoir of the Pingan 1 well.

No.	σ_H (MPa)	σ_H Stress Gradient (MPa/100 m)	σ_v (MPa)	σ_v Stress Gradient (MPa/100 m)	σ_h (MPa)	σ_h Stress Gradient (MPa/100 m)
3-1	80.99	2.69	77.29	2.57	76.64	2.55
3-2	81.61	2.71	75.64	2.51	75.31	2.50
3-3	81.03	2.69	77.56	2.57	77.32	2.57

4.2. The Microscopic Dimensions of Natural Shale Fractures

The study utilized CT scans to examine the morphological characteristics of natural fractures across multiple scales, specifically focusing on three different core sizes of 100 mm, 25 mm, and 2 mm. Figure 5 displays the CT scanning result of a full-diameter shale core with a diameter of 100 mm. Figure 5a,b shows grayscale images of two different cross-sections, revealing clearly visible natural fractures, one of which measures approximately 1 mm wide. The cross-sectional slicing of the shale core helps to demonstrate the two-dimensional morphological characteristics of natural fractures; however, it is hard to present their three-dimensional (3D) shape. For example, it is apparent that some of the fractures are continuously distributed in the vertical direction when comparing Figure 5a,b, however, it is obvious that the same fracture shows a different shape in different cross-sections. The CT scanning of the full-diameter core in 3D is shown in Figure 5c,d. Depending on the method of presentation, Figure 5c,d displays the CT scanning's grayscale image and pseudo-color image, respectively. The 3D morphology of the fractures is effectively visualized, revealing that the fractures exhibit a vertical orientation, as evidenced by the results obtained. However, in the case of full-diameter shale cores, fractures with dimensions below the millimeter range could not be adequately captured owing to the low precision of the scanning technique employed.

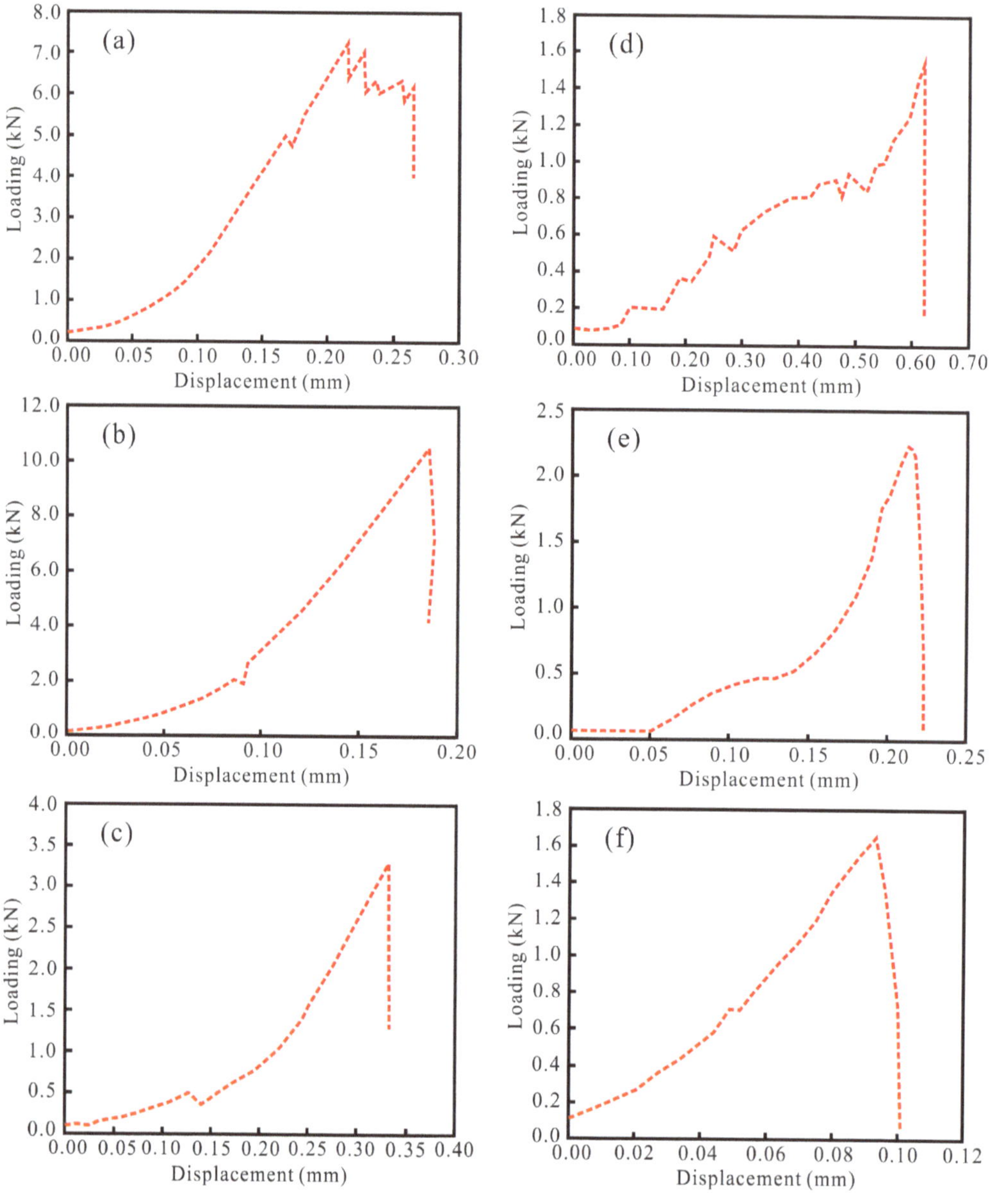

Figure 4. Maximum axial loading and corresponding displacement during tensile strength (**a–c**) and fracture toughness tests (**d–f**). The red dot lines show the displacements along with the variation of the loading.

When scanning a 25 mm shale core, reducing the core size results in a significant improvement in scanning accuracy (Figure 6). The deep black line in Figure 6a demonstrates a fracture with a stronger tortuosity. The two longitudinal cross-sections of the core show at least three fractures that are close to parallel to each other (Figure 6b), indicating that they are most probably bedding planes. The different colors in Figure 6c show connecting pores in the vicinity of the fractures. The volume and shape of the bedding planes are presented by the colored plane (Figure 6d). It can be seen that the fracture planes are not continuously distributed but intermittently. The perspectival modeling of pores and fractures can demonstrate all visible pores and identifiable fractures simultaneously (Figure 6e), though it is worth noting that the pore development between the two fractures is significantly reduced. Moreover, the fractures can be extracted individually to show their detailed information. As shown in Figure 6f, the different colors represent different fracture planes.

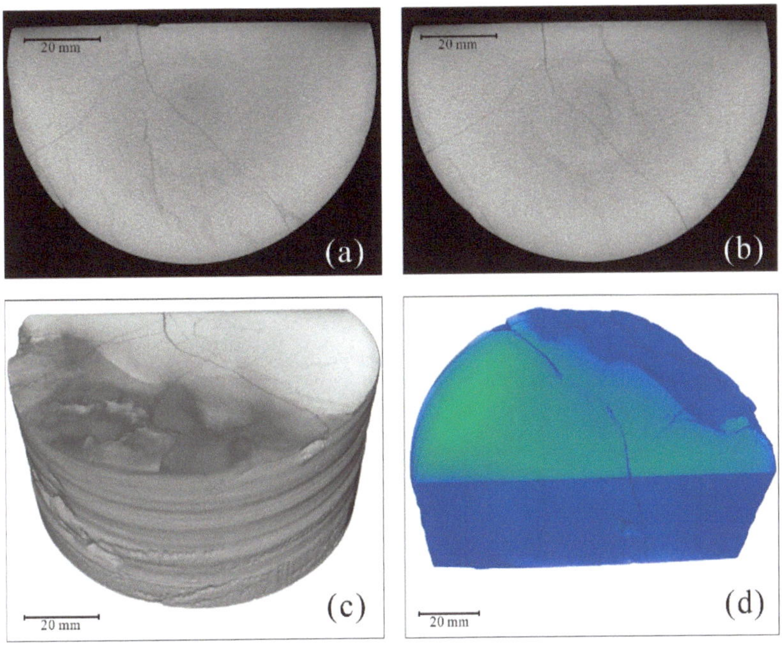

Figure 5. The CT scanning of a 100 mm shale core. (**a**,**b**) Display of the grayscale images of two different cross-sections; (**c**) a 3D representation of the CT scanning's grayscale image; (**d**) a 3D representation of the CT scanning's pseudo-color image. The color in (**d**) is pseudo-color which is a form of expression for the interpretation of CT scanning image data.

By reducing the core scale to 2 mm, the scanning accuracy improves to 0.92 μm. This level of precision allows fractures and pores in the material to be more clearly visible, as shown in Figure 7. Specifically, Figure 7a displays a cross-section that reveals approximately five fractures, one of which measures around 20 μm in width. Different from the 25 mm core, the fractures shown in the 2 mm core are not bedding planes although they come from the same core. Figure 7b shows the enlarged part of Figure 7a; it indicates that the three fractures trend to propagation in the same direction. Figure 7c,d presents the connectivity fracture model and segmental display of the connectivity in the core, respectively. The displayed fractures are much finer than Figure 6f and more disconnected pores appear around the fractures. Based on the inclination angle of the fractures, they appear to belong to an inclined fracture. The pore–fracture model reveals finer details of the micropores, which are distributed randomly (as shown in Figure 7e). Additionally, the segmented model of the pore–fracture body demonstrates that more interconnected pores are present in wider fractures (as shown in Figure 7f).

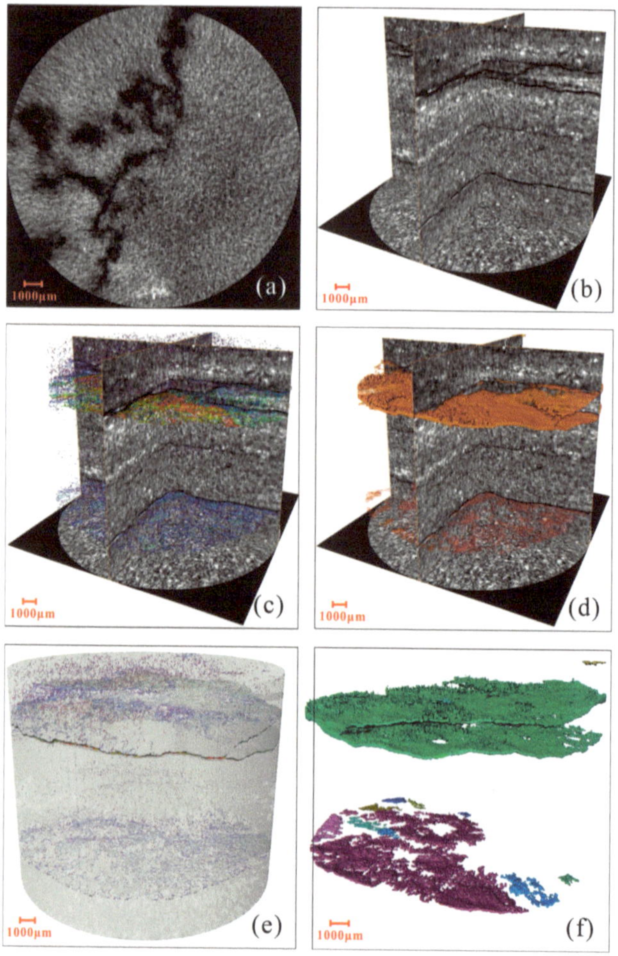

Figure 6. The CT scanning of a 25 mm shale core. (**a**) Display of one of the cross-sections of the core; (**b**) the longitudinal cross-section of the core; (**c**) the longitudinal cross-section of the core showing the pores; (**d**) the longitudinal cross-section of the core showing the natural fractures; (**e**) the perspectival modeling of the pores, and (**f**) the perspectival modeling of the natural fractures. The colorful part in the (**c**–**f**) are mainly applied to exhibit the pore–fracture space.

4.3. The Tortuosity Features at Different Shale's Natural Fracture Scales

The tortuosity of the fractures in the 100 mm and 25 mm shale cores were also obtained and analyzed. Figure 8 shows the statistical fractures' tortuosity of four cross-sections of a 25 mm shale core. This CT scan revealed a total of six visible fractures, identified by serial numbers I, II, III, IV, V, and VI. The fractures can be classified into two groups based on their visible length: the long fractures, consisting of numbers I, II, and III, and the short fractures, consisting of numbers IV, V, and VI. Figure 9 shows the statistical fractures' tortuosity of four cross-sections of a 100 mm shale core. As the fractures did not run through the entire full-diameter shale core, only those parts with fractures were selected for display. There are only two fractures visible which are numbered I and II.

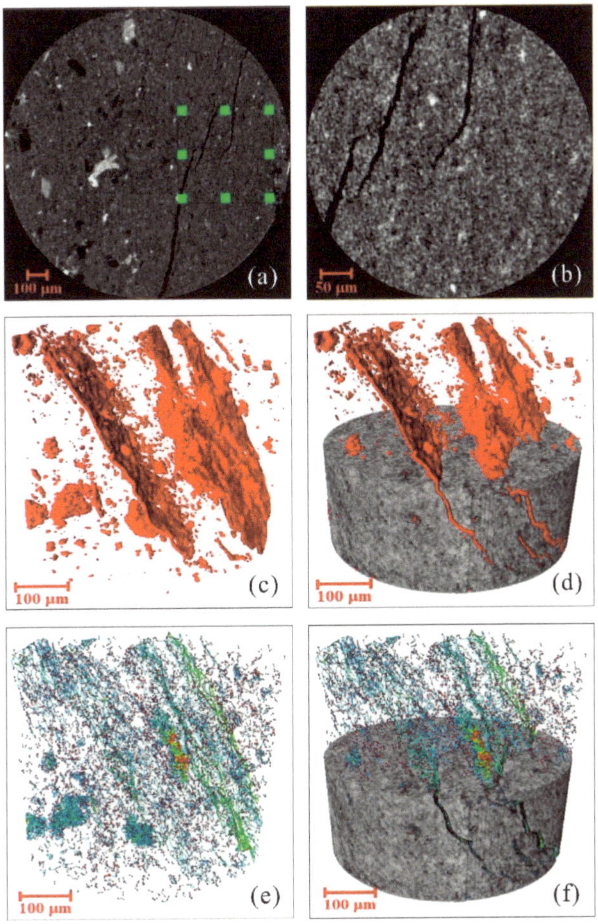

Figure 7. The CT scanning of 2 mm shale core. (**a**) Display of the cross-sections in the core; (**b**) an enlarged portion of (**a**); (**c**) the red color shows the connectivity model; (**d**) segmental display of the connectivity in the core; (**e**) detailed presentation of the pore–fracture model, and (**f**) the segmented display model of the pore–fracture body. The red color in (**c**,**d**) mainly represent the connectivity fracture, while the colorful part in (**e**,**f**) mainly show the pore–fracture space.

The fractures' tortuosity is used to describe the fracture roughness in a cross-section, which is calculated by l/L, where l is the real length of the natural fracture and L is the straight-line distance between the two fracture endpoints [22]. In Figure 10, the tortuosity of each fracture in different cross-sections was calculated and compared for a 25 mm shale core, with the resulting range typically falling between 1.00 and 1.06. For the long fractures of I, II, and III, the tortuosity variation ranged from 0.01 to 0.02, while the range increased to 0.03 for the short fractures of IV, V, and VI. Similarly, in a 100 mm shale core, the fractures' tortuosity exhibited a similar changing trend, with values ranging from 1.002 to 1.016 and a range of tortuosity variation between 0.009 and 0.014 (Figure 11).

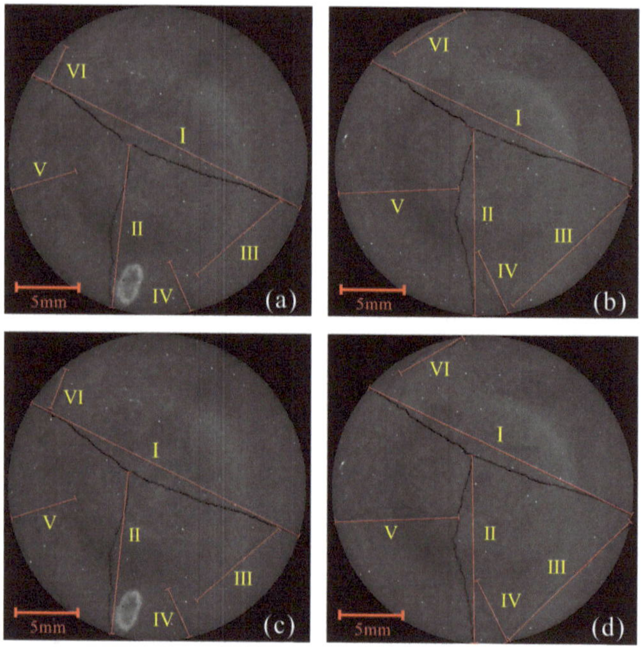

Figure 8. Statistical fractures' tortuosity of four cross-sections of a 25 mm core. The (**a**–**d**) corresponding to the four cross-sections of this 25 mm core. Except for the red scale, the red straight lines in the figure show the distance between the ends of the fractures. The yellow roman numbers of I, II, III, IV, V, and VI in the figure are corresponding to the number of the fractures.

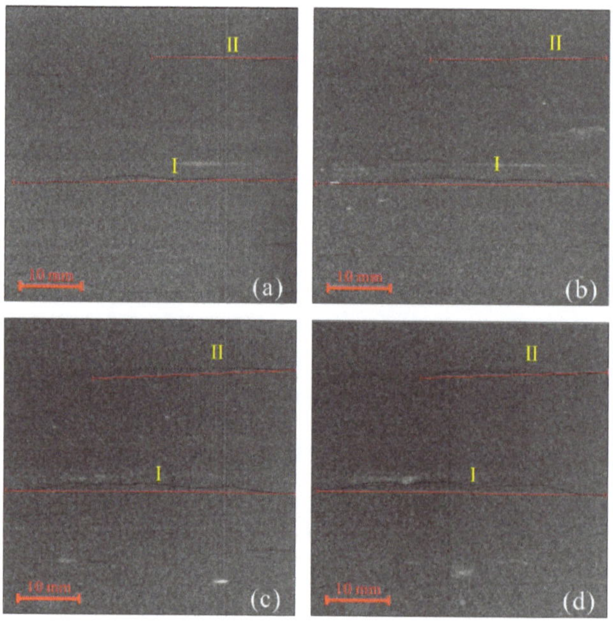

Figure 9. Statistical fractures' tortuosity of four cross-sections of a 100 mm core. The (**a**–**d**) corresponding to the four cross-sections of this 100 mm core. Except for the red scale, the red straight lines in the figure display the distance between the ends of the fractures. The yellow roman numbers of I and II in the figure are the number of the fractures.

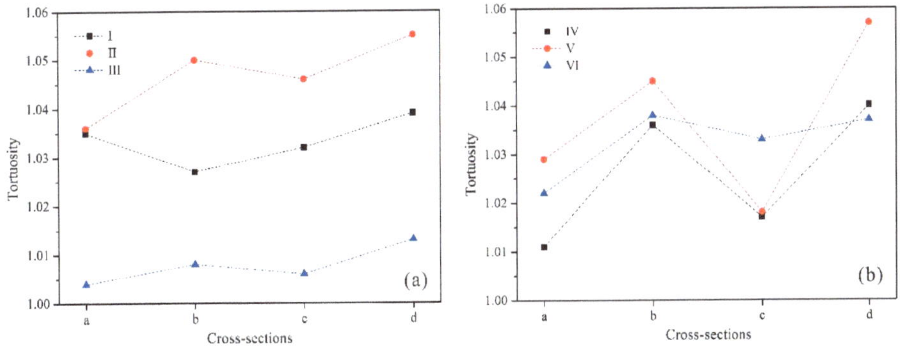

Figure 10. (**a**) The long fractures' tortuosity in four cross-sections of a 25 mm core; (**b**) the short fractures' tortuosity in four cross-sections of a 25 mm shale core. The (a), (b), (c), and (d) represent the four cross-sections in Figure 8, while the I, II, III, IV, V, and VI in the figure are the corresponding fracture number in Figure 8.

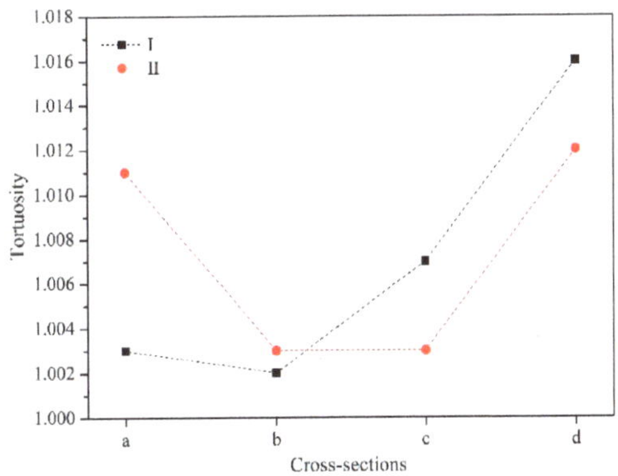

Figure 11. The fractures' tortuosity in four cross-sections of a 100 mm shale core. The (a), (b), (c), and (d) represent the four cross-sections in Figure 9, while the I and II in the figure are the fracture number in Figure 9.

Furthermore, Figure 12 illustrates the probability distribution of the fracture widths in the 25 mm core. The analysis reveals that the majority of the fracture widths fall within the range of 25 μm to 150 μm, with a maximum value of approximately 250 μm. Notably, the frequency distribution displays the most recurrent values to be within the range of 75 μm to 87 μm.

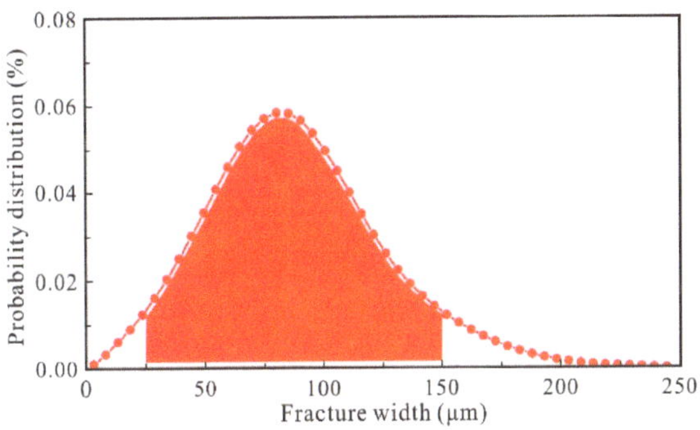

Figure 12. The fracture width probability distribution of the 25 mm core. The red part in the figure shows the main intervals of the fracture width.

4.4. Simulation of Shale Natural Fracture Seepage with Different Sizes

Figure 13a–h presents the simulation results of models obtained from 100 mm and 25 mm shale cores, respectively. Figure 13a,e displays the fracture morphology of the two models. While the fracture in Figure 13e is smaller than in Figure 13a, its morphology is more continuous. The fluid velocity in Figure 13b ranges from 0 to 3 m/s, which is almost 10 times higher than that in Figure 13f, indicating a high degree of permeability. The fracture surface displacement in Figure 13c ranges from 0 to 300 μm, almost three times higher than in Figure 13g. The distribution of fracture surface pressure in Figure 13d,h shows nearly the same pressure range, despite the difference in fracture sizes. Higher pressures only occur at the edge of the fractures in Figure 13d, while they appear throughout the fracture in Figure 13h. The red arrows in Figure 13d,h indicate the direction of fracture surface pressure, which is almost perpendicular to the fracture surface where the fractures bend.

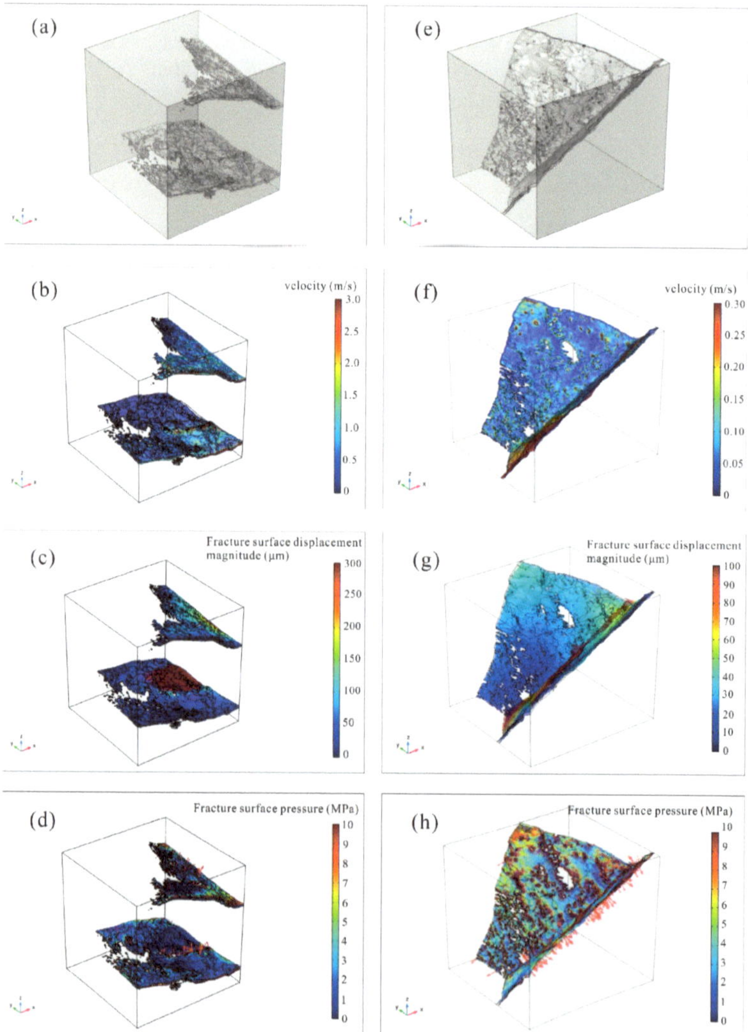

Figure 13. The simulation results show the morphology of fractures (**a**,**e**) and the distribution of fluid velocity (**b**,**f**), fracture surface displacement (**c**,**g**), and fracture surface pressure (**d**,**h**) at 10 MPa surrounding pressure.

5. Discussion

5.1. The In Situ Stress Characteristic of the Shale Oil Reservoir

The in situ stress of the target shale reservoir exhibits the characteristic of $\sigma_H > \sigma_v > \sigma_h$, with σ_v and σ_h being particularly similar. The three kinds of stress conditions, including (1) $\sigma_v > \sigma_H > \sigma_h$, (2) $\sigma_H > \sigma_v > \sigma_h$, and (3) $\sigma_H > \sigma_h > \sigma_v$, correspond to normal faulting, strike-slip, and thrust faulting, respectively [58]. This stress pattern tends to result in fracture patterns that follow a strike-slip mechanism (as depicted in Figure 14). The lateral stress coefficient, which represents the ratio of horizontal stresses to vertical stress [59], is a critical parameter for simulating fracture initiation under in situ stress conditions [23]. According to the regression correlation between the lateral stress coefficient and burial depth proposed by Brown and Hoek (1978) [60], the lateral stress coefficient of the target layers should range from 0.333 to 0.998. Nevertheless, the measured value obtained ranged from 1.020 to 1.037, markedly surpassing the theoretically predicted values. A high lateral stress coefficient implies that the horizontal stresses exceed the average level.

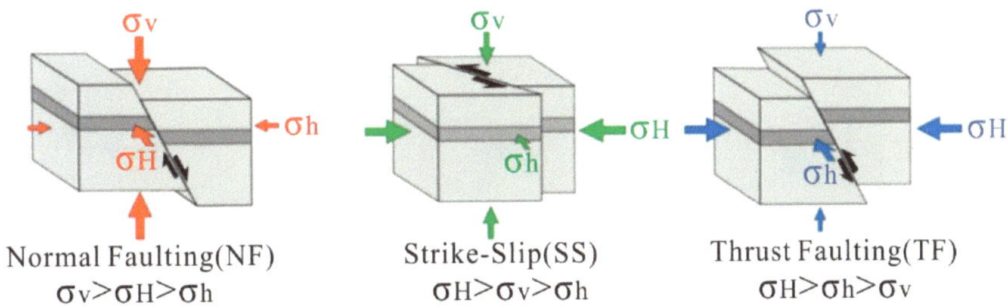

Figure 14. The main tectonic stress patterns. The arrows in the figure demonstrate the direction of the stress, and red, green, and blue colors display different stress patterns.

5.2. Comparison of the Natural Fracture Shale's CT Scanning with Multi-Scales

Three types of shale cores, measuring 100 mm, 25 mm, and 2 mm, were subjected to CT scanning. The 100 mm core revealed fractures with a width at the millimeter level, but due to the low accuracy of the CT scanning, only visible fractures could be detected. Moreover, the pores were almost indiscernible. In contrast, the 25 mm core allowed for a simultaneous display of pores and fractures. Notably, the connecting pores were found to be particularly well-developed near the bedding planes (Figure 6c). Further analysis using the program Avizo revealed that the bedding planes were often discontinuous, and the presence of numerous protrusions significantly increased the volume of fractures. The CT scanning of the 2 mm core yielded a substantially improved resolution, enabling a more comprehensive visualization of the fractures and interconnected pores. However, the higher accuracy of the CT scanning at this scale primarily highlights the pore information, which may not fully capture additional details of the fractures. Thus, in studies focusing on fractures, the 25 mm core size is more appropriate. To summarize, the choice of core size should be guided by the research objectives.

5.3. The Tortuosity Changes in Different Scales

According to Sun et al. (2022) [22] tortuosity is an effective measure of the geometrical complexity of natural fractures. The CT scanning data of the 25 mm shale core indicate that the tortuosity of both long and short fractures exhibited a similar changing trend across different cross-sections. While the tortuosity of an individual long fracture tended to remain stable across different cross-sections, short fractures exhibited more variation in tortuosity, as shown by the data analysis of Figure 15. In particular, the standard deviation of the tortuosity of large fractures was lower than that of small fractures, indicating that the morphology of larger fractures is more likely to remain stable, while small-scale fractures tend to exhibit diverse shapes. Moreover, the standard deviation of tortuosity for large fractures of types I and II from the 100 mm shale core was 0.006 and 0.005, respectively, in agreement with the data presented in Figure 15. This finding indicates that larger fractures exhibit greater shape stability. Additionally, tortuosity is influenced by frequent changes in the width of a fracture, which are determined by irregularities in the fracture wall.

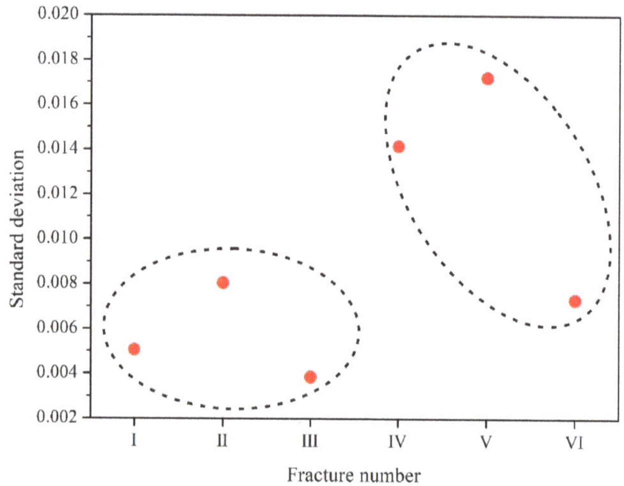

Figure 15. The standard deviation of the fractures of I, II, III, IV, V, and VI. The red dots display the standard deviation of tortuosity, and the dot lines demonstrate the tortuosity of large fractures and small fractures, respectively.

5.4. Sensitivity of Fracture Permeability at Variable Surrounding Pressures

The two fracture models were extracted from the CT results of 100 mm and 25 mm; therefore, the fracture models reflect different sizes at respective scales. The simulation of dynamic fracture permeability response was conducted for shale core sizes of 100 mm and 25 mm while varying the pressure from 2 MPa to 10 MPa, as displayed in Figure 16. The fracture permeability for the 100 mm core varied from 2.04×10^4 μm^2 to 8.67×10^4 μm^2, whereas it decreased from 1.3×10^3 μm^2 to 5.45×10^2 μm^2 for the 25 mm core. Notably, the fracture permeability of the 100 mm core was almost ten times higher than that of the 25 mm core. Despite the significant difference in fracture size between the two models, as depicted in Figure 13, their permeability exhibited a similar changing trend with varying surrounding pressures. Furthermore, the permeability showed a linear decrease with increasing surrounding pressure, as evidenced by the data, showing a reduction of 2.35 and 2.39 times the initial values for the 100 mm and 25 mm shale cores, respectively. The results indicate that the stress sensitivity of fracture permeability remains consistent across various fracture scales. However, it should be noted that fractures with smaller scales may undergo complete closure when subjected to higher surrounding pressures, leading to the complete loss of permeability. As such, our understanding of fracture behavior relies on the assumption that the fractures do not experience total closure.

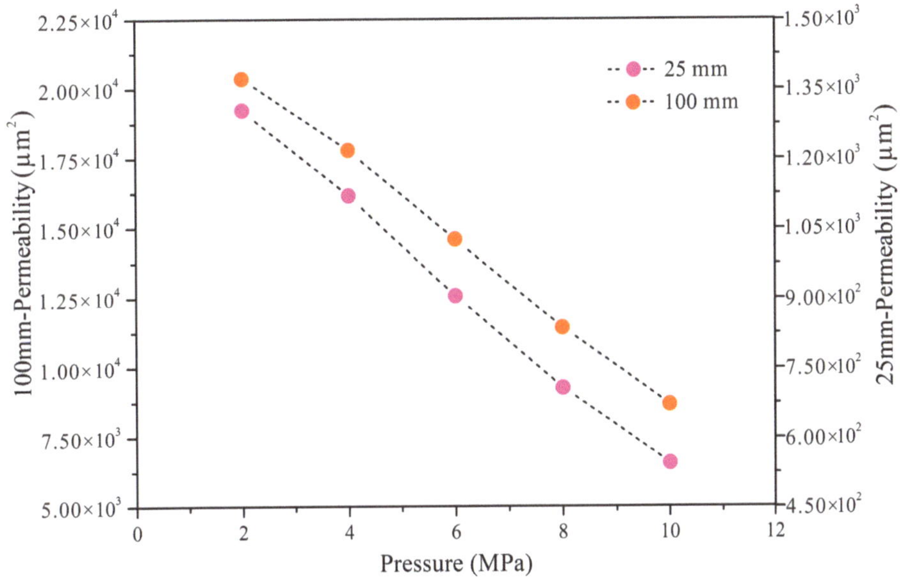

Figure 16. The dynamic permeability changing vs. pressure varies from 2 MPa to 10 MPa. The 100 mm and 25 mm shale cores are simulated.

6. Conclusions

This study utilized a novel and comprehensive approach with the combination of multi-scale CT scanning and numerical simulation of fluid seepage in fracture models to investigate the developmental features of natural fractures in the Lower Jurassic shale oil reservoirs in the Sichuan Basin. The findings of this study can be summarized as follows:

1. The in situ stress of the shale reservoir exhibits the characteristic of $\sigma_H > \sigma_V > \sigma_h$, with σ_V and σ_h being particularly similar, resulting in a strike-slip fracture pattern. The relatively high lateral stress coefficient between 1.020 and 1.037 indicates that the horizontal stresses are higher than the average level. The above understanding helps to predict the natural fracture pattern.
2. While the 2 mm core CT scan offers a more detailed view of the fractures and connected pores, the high-resolution results in more pore information, which may not fully reveal additional information about the fractures. Therefore, if the primary focus of the study is on fractures, then the 25 mm core would be a more suitable option to study the microscopic natural fractures.
3. The tortuosity of various fractures indicates that larger fractures tend to maintain a more stable morphology, whereas smaller-scale fractures tend to exhibit a greater diversity of shapes, which is not conducive to fluid flow.
4. The fracture permeability displays a nearly linear decrease with increasing surrounding pressure, suggesting that the stress sensitivity of fracture permeability is approximately similar across different fracture scales.

Overall, in this research, extensive experimental and numerical simulations are innovatively combined, leading to the comprehensive investigation of microscopic natural fractures in shale oil reservoirs which has significantly improved our understanding of their properties and enabled the identification of favorable areas for shale oil exploration. To advance this research further, we plan to develop numerical fracture models that cater to different natural fracture types in shale oil reservoirs. These models will be integrated with the Thermo-Fluid–Solid coupling method to provide further insights into the evolution mechanism of fracture permeability.

Author Contributions: Conceptualization, S.H.; methodology, S.H., X.B. and X.W.; software, Y.H. and L.C. (Likai Cui); formal analysis, S.H., X.B., W.M., L.C. (Lijuan Cheng) and Z.W.; investigation, S.H., X.B. and Y.W.; resources, S.H., X.B. and W.M.; writing—original draft, S.H. and X.B.; writing—review and editing, S.H., X.B., B.L., M.S. and X.F.; visualization, S.H., X.B. and Y.Z.; supervision, S.H.; funding acquisition, S.H., B.L., M.S. and X.B. All authors have read and agreed to the published version of the manuscript.

Funding: The research work was funded by Heilongjiang Province "Hundred Million" Engineering of Major Projects in Science and Technology (SC2020ZX05A0023) and the Scientific Research and Technological Development Project of China National Petroleum Corporation (2022DJ1809). This study was supported by the National Natural Science Foundation of China (42202155; 42272159; U22A20574), China Postdoctoral Science Foundation (2021MD703807), Heilongjiang Provincial Postdoctoral Science Foundation (LBH-Z20121), China Scholarship Council (202008230018). The authors also acknowledge funding by the Natural Science Foundation of Heilongjiang Province (TD2021D001) and the Key R&D Program of Hainan Province (ZDYF2023GXJS009). The APC was funded by 42202155.

Institutional Review Board Statement: Not applicable.

Informed Consent Statement: Not applicable.

Data Availability Statement: Data are unavailable due to privacy restrictions.

Conflicts of Interest: The authors declare no competing interests.

References

1. Fathoni, A.Z.; Batts, B.D. A literature review of fuel stability studies with a particular emphasis on shale oil. *Energy Fuels* **1991**, *5*, 682–693. [CrossRef]
2. Fu, C.Y. China's shale gas and shale oil resources: Opportunities and challenges. *Energy Explor. Exploit.* **2014**, *32*, 759–769. [CrossRef]
3. Mahzari, P.; Mitchell, T.M.; Jones, A.P.; Oelkers, E.H.; Juri, J.E. Novel laboratory investigation of huff-n-puff gas injection for shale oils under realistic reservoir conditions. *Fuel* **2020**, *285*, 118950. [CrossRef]
4. Liu, B.; Sun, J.; Zhang, Y.; He, J.; Fu, X.; Yang, L.; Xing, J.; Zhao, X. Reservoir space and enrichment model of shale oil in the first member of Cretaceous Qingshankou Formation in the Changling sag, southern Songliao Basin, NE China. *Pet. Explor. Dev.* **2021**, *48*, 608–624. [CrossRef]
5. Wu, T.; Pan, Z.J.; Liu, B.; Connell, L.D.; Fu, X.F. Laboratory characterization of shale oil storage behavior: A comprehensive review. *Energy Fuels* **2021**, *35*, 7305–7318. [CrossRef]
6. Curnow, W.J. The prospects for shale oil in Australia. *Fuel* **1987**, *66*, 292–294. [CrossRef]
7. Zou, C.N.; Yang, Z.; Cui, J.W.; Zhu, R.K.; Hou, L.H.; Tao, S.Z.; Yuan, X.J.; Wu, S.T.; Lin, S.H.; Wang, L.; et al. Formation mechanism, geological characteristics and development strategy of nonmarine shale oil in China. *Pet. Explor. Dev.* **2013**, *40*, 5–27. [CrossRef]
8. Solarin, S.A.; Gil-Alana, L.A.; Lafuente, C. An investigation of long range reliance on shale oil and shale gas production in the U.S. market. *Energy* **2020**, *195*, 116933. [CrossRef]
9. Liu, B.; Yang, Y.; Li, J.; Chi, Y.; Li, J.; Fu, X. Stress sensitivity of tight reservoirs and its effect on oil saturation: A case study of Lower Cretaceous tight clastic reservoirs in the Hailar Basin, Northeast China. *J. Pet. Sci. Eng.* **2020**, *184*, 106484. [CrossRef]
10. Liu, B.; Wang, Y.; Tian, S.; Guo, Y.; Wang, L.; Yasin, Q.; Yang, J. Impact of thermal maturity on the diagenesis and porosity of lacustrine oil-prone shales: Insights from natural shale samples with thermal maturation in the oil generation window. *Int. J. Coal Geol.* **2022**, *261*, 104079. [CrossRef]
11. Bai, L.H.; Liu, B.; Du, Y.J.; Wang, B.Y.; Tian, S.S.; Wang, L.; Xue, Z.Q. Distribution characteristics and oil mobility thresholds in lacustrine shale reservoir: Insights from N_2 adsorption experiments on samples prior to and following hydrocarbon extraction. *Pet. Sci.* **2022**, *19*, 486–497. [CrossRef]
12. Cui, D.; Yin, H.L.; Liu, Y.P.; Li, J.; Pan, S.; Wang, Q. Effect of final pyrolysis temperature on the composition and structure of shale oil: Synergistic use of multiple analysis and testing methods. *Energy* **2022**, *252*, 124062. [CrossRef]
13. Orangi, A.; Nagarajan, N.R.; Honarpour, M.M.; Rosenzweig, J. Unconventional shale oil and gas-condensate reservoir production, impact of rock, fluid, and hydraulic fractures. In Proceedings of the SPE Hydraulic Fracturing Technology Conference, The Woodlands, TX, USA, 24–26 January 2011. [CrossRef]
14. Boak, J.; Kleinberg, R. 4–Shale Gas, Tight Oil, Shale Oil and Hydraulic Fracturing. In *Future Energy*, 3rd ed.; Elsevier: Amsterdam, The Netherlands, 2020; pp. 67–95.
15. Shaldybin, M.V.; Wilson, M.J.; Wilson, L.; Lopushnyak, Y.M.; Brydson, R.; Krupskaya, V.V.; Kondrashova (Deeva), E.S.; Glotov, A.V.; Goncharov, I.V.; Samoilenko, V.V.; et al. The nature, origin and significance of luminescent layers in the Bazhenov Shale Formation of West Siberia, Russia. *Mar. Pet. Geol.* **2019**, *100*, 358–375. [CrossRef]
16. Khakimova, L.; Bondarenko, T.; Cheremisin, A.; Myasnikov, A.; Varfolomeev, M. High pressure air injection kinetic model for Bazhenov Shale Formation based on a set of oxidation studies. *J. Pet. Sci. Eng.* **2019**, *172*, 1120–1132. [CrossRef]

17. Mukhina, E.; Cheremisin, A.; Khakimova, L.; Garipova, A.; Dvoretskaya, E.; Zvada, M.; Kalacheva, D.; Prochukhan, K.; Kasyanenko, A.; Cheremisin, A. Enhanced oil recovery method selection for shale oil based on numerical simulations. *ACS Omega* **2021**, *6*, 23731–23741. [CrossRef]
18. Zou, C.N.; Pan, S.Q.; Jing, Z.H.; Gao, J.L.; Yang, Z.; Wu, S.T.; Zhao, Q. Shale oil and gas revolution and its impact. *Acta Pet. Sin.* **2020**, *41*, 1–12.
19. Jin, Z.J.; Liang, X.P.; Bai, Z.R. Exploration breakthrough and its significance of Gulong lacustrine shale oil in the Songliao Basin, Northeastern China. *Energy Geosci.* **2022**, *3*, 120–125. [CrossRef]
20. Wang, X.; He, S.; Guo, X.W.; Zhang, B.Q.; Chen, X.H. The Resource Evaluation of Jurassic Shale in North Fuling Area, Eastern Sichuan Basin, China. *Energy Fuels* **2018**, *32*, 1213–1222. [CrossRef]
21. Gale, J.F.W.; Laubach, S.E.; Olson, J.E.; Eichhuble, P.; Fall, A. Natural fractures in shale: A review and new observations. *AAPG Bull.* **2014**, *98*, 2165–2216. [CrossRef]
22. Sun, S.S.; Huang, S.P.; Gomez-Rivas, E.; Griera, A.; Liu, B.; Xu, L.L.; Wen, Y.R.; Dong, D.Z.; Shi, Z.S.; Chang, Y.; et al. Characterization of natural fractures in deep-marine shales: A case study of the Wufeng and Longmaxi shale in the Luzhou Block Sichuan Basin, China. *Front. Earth Sci.* **2022**, 1–14. [CrossRef]
23. Huang, S.P.; Liu, D.M.; Yao, Y.B.; Gan, Q.; Cai, Y.D.; Xu, L.L. Natural fractures initiation and fracture type prediction in coal reservoir under different in-situ stresses during hydraulic fracturing. *J. Nat. Gas Sci. Eng.* **2017**, *43*, 69–80. [CrossRef]
24. Zhou, J.; Zhang, L.Q.; Li, X.; Pan, Z.J. Experimental and modeling study of the stress-dependent permeability of a single fracture in shale under high effective stress. *Fuel* **2019**, *257*, 116078. [CrossRef]
25. Huang, S.P.; Liu, D.M.; Cai, Y.D.; Gan, Q. In-situ stress distribution and its impact on CBM reservoir properties in the Zhengzhuang area, southern Qinshui Basin, North China. *J. Nat. Gas Sci. Eng.* **2019**, *61*, 83–96. [CrossRef]
26. Sun, M.D.; Wen, J.J.; Pan, Z.J.; Liu, B.; Blach, T.P.; Ji, Y.P.; Hu, Q.H.; Yu, B.S.; Wu, C.M.; Ke, Y.B. Pore accessibility by wettable fluids in overmature marine shales of China: Investigations from contrast-matching small-angle neutron scattering (CM-SANS). *Int. J. Coal Geol.* **2022**, *255*, 103987. [CrossRef]
27. Hazra, B.; Vishal, V.; Sethi, C.; Chandra, D. Impact of supercritical CO_2 on shale reservoirs and its implication for CO_2 sequestration. *Energy Fuels* **2022**, *36*, 9882–9903. [CrossRef]
28. Jia, B.; Tsau, J.; Barati, R. A review of the current progress of CO_2 injection EOR and carbon storage in shale oil reservoirs. *Fuel* **2019**, *236*, 404–427. [CrossRef]
29. Elenius, M.; Skurtveit, E.; Yarushina, V.; Baig, I.; Sundal, A.; Wangen, M.; Landschulze, K.; Kaufmann, R.; Choi, J.C.; Hellevang, H.; et al. Assessment of CO_2 storage capacity based on sparse data: Skade Formation. *Int. J. Greenh. Gas Control* **2018**, *79*, 252–271. [CrossRef]
30. van Noort, R.; Yarushina, V. Water, CO_2 and argon permeabilities of intact and fractured shale cores under stress. *Rock Mech. Rock Eng.* **2019**, *52*, 299–319. [CrossRef]
31. Peshkov, G.A.; Khakimova, L.A.; Grishko, E.V.; Wangen, M.; Yarushina, V.M. Coupled basin and hydro-mechanical modeling of gas chimney formation: The SW Barents Sea. *Energies* **2021**, *14*, 6345. [CrossRef]
32. Yarushina, V.M.; Wang, L.H.; Connolly, D.; Kocsis, G.; Fæstø, I.; Polteau, S.; Lakhlifi, A. Focused fluid-flow structures potentially caused by solitary porosity waves. *Geology* **2022**, *50*, 179–183. [CrossRef]
33. Gale, J.F.W.; Reed, R.M.; Holder, J. Natural fractures in the Barnett Shale and their importance for hydraulic fracture treatments. *AAPG Bull.* **2007**, *91*, 603–622. [CrossRef]
34. Wei, Y.; Hu, X.; Liu, M.; Wang, W. Investigation of the effect of natural fractures on multiple shale-gas well performance using non-intrusive EDFM technology. *Energies* **2019**, *12*, 932.
35. Pirzada, M.A.; Bahaaddini, M.; Moradian, O.; Roshan, H. Evolution of contact area and aperture during the shearing process of natural rock fractures. *Eng. Geol.* **2021**, *291*, 106236. [CrossRef]
36. Zhang, C.; Liu, D.D.; Jiang, Z.X.; Song, Y.; Luo, Q.; Wang, X. Mechanism for the formation of natural fractures and their effects on shale oil accumulation in Junggar Basin, NW China. *Int. J. Coal Geol.* **2022**, *254*, 103973. [CrossRef]
37. Xian, C.; Zhang, J.; Zhao, C.; Wang, G.; Xing, L. Characterizing and modeling multi-scale natural fractures in the Ordovician-Silurian Wufeng-Longmaxi Shale Formation in South Sichuan Basin. In Proceedings of the Unconventional Resources Technology Conference, Austin, TX, USA, 24–26 July 2017.
38. Micheal, M.; Xu, W.L.; Xu, H.Y.; Zhang, J.N.; Jin, H.J.; Yu, H.; Wu, H.A. Multi-scale modelling of gas transport and production evaluation in shale reservoir considering crisscrossing fractures. *J. Nat. Gas Sci. Eng.* **2021**, *95*, 104156. [CrossRef]
39. Volatili, T.; Agosta, F.; Cardozo, N.; Zambrano, M.; Lecomte, I.; Tondi, E. Outcrop-scale fracture analysis and seismic modelling of a basin-bounding normal fault in platform carbonates, central Italy. *J. Struct. Geol.* **2022**, *155*, 104515. [CrossRef]
40. Laubach, S.E.; Olson, J.E.; Gale, J.F.W. Are open fractures necessarily aligned with maximum horizontal stress? *Earth Planet. Sci. Lett.* **2004**, *222*, 191–195. [CrossRef]
41. Rajabi, M.; Sherkati, S.; Bohloli, B.; Tingay, M. Subsurface fracture analysis and determination of in-situ stress direction using FMI logs: An example from the Santonian carbonates (Ilam Formation) in the Abadan Plain, Iran. *Tectonophysics* **2010**, *492*, 192–200. [CrossRef]
42. Zhang, Y.Y.; He, Z.L.; Jiang, S.; Lu, S.F.; Xiao, D.S.; Chen, G.H.; Li, Y.C. Fracture types in the lower Cambrian shale and their effect on shale gas accumulation, Upper Yangtze. *Mar. Pet. Geol.* **2019**, *99*, 282–291. [CrossRef]

43. Zhang, X.; Lu, Y.; Tang, J.; Zhou, Z.; Liao, Y. Experimental study on fracture initiation and propagation in shale using supercritical carbon dioxide fracturing. *Fuel* **2016**, *190*, 370–378. [CrossRef]
44. Wu, S.T.; Zhai, X.F.; Yang, Z.; Bale, H.; Hong, Y.L.; Cui, J.W.; Pan, S.Q.; Lin, S.H. Characterization of fracture formation in organic-rich shales—An experimental and real time study of the Permian Lucaogou Formation, Junggar Basin, northwestern China. *Mar. Pet. Geol.* **2019**, *107*, 397–406. [CrossRef]
45. Gou, Q.; Xu, S.; Hao, F.; Yang, F.; Zhang, B.; Shu, Z.; Zhang, A.; Wang, Y.; Lu, Y.; Cheng, X.; et al. Full-scale pores and micro-fractures characterization using FE-SEM, gas adsorption, nano-CT and micro-CT: A case study of the Silurian Longmaxi Formation shale in the Fuling area, Sichuan Basin, China. *Fuel* **2019**, *253*, 167–179. [CrossRef]
46. Chen, Y.; Jiang, C.; Leung, J.Y.; Wojtanowicz, A.K.; Zhang, D. Multiscale characterization of shale pore-fracture system: Geological controls on gas transport and pore size classification in shale reservoirs. *J. Pet. Sci. Eng.* **2021**, *202*, 108442. [CrossRef]
47. Sun, M.D.; Yu, B.S.; Hu, Q.H.; Chen, S.; Xia, W.; Ye, R.C. Nanoscale pore characteristics of the Lower Cambrian Niutitang Formation Shale: A case study from Well Yuke #1 in the Southeast of Chongqing, China. *Int. J. Coal Geol.* **2016**, *154–155*, 16–29.
48. Sun, M.D.; Zhang, L.H.; Hu, Q.H.; Pan, Z.J.; Yu, B.S.; Sun, L.W.; Bai, L.F.; Fu, H.J.; Zhang, Y.F.; Zhang, C.; et al. Multiscale connectivity characterization of marine shales in southern China by fluid intrusion, small-angle neutron scattering (SANS), and FIB-SEM. *Mar. Pet. Geol.* **2020**, *112*, 104101. [CrossRef]
49. Tan, Y.; Pan, Z.; Liu, J.; Wu, Y.; Haque, A.; Connell, L.D. Experimental study of permeability and its anisotropy for shale fracture supported with proppant. *J. Nat. Gas Sci. Eng.* **2017**, *44*, 250–264. [CrossRef]
50. Bird, M.B.; Butler, S.L.; Hawkes, C.D.; Kotzer, T. Numerical modeling of fluid and electrical currents through geometries based on synchrotron X-ray tomographic images of reservoir rocks using Avizo and COMSOL. *Comput. Geosci.* **2014**, *73*, 6–16. [CrossRef]
51. He, W.Y.; Bai, X.F.; Meng, Q.A.; Li, J.H.; Zhang, D.Z.; Wang, Y.Z. Accumulation geological characteristics and major discoveries of lacustrine shale oil in Sichuan Basin. *Acta Pet. Sin.* **2022**, *43*, 885–898.
52. Erarslan, N.; Liang, Z.Z.; Williams, D.J. Experimental and numerical studies on determination of indirect tensile strength of rocks. *Rock Mech. Rock Eng.* **2012**, *45*, 739–751. [CrossRef]
53. Dutler, N.; Nejati, M.; Valley, B.; Amann, F.; Molinari, G. On the link between fracture toughness, tensile strength, and fracture process zone in anisotropic rocks. *Eng. Fract. Mech.* **2018**, *201*, 56–79. [CrossRef]
54. Kaiser, J. Erkenntnisse und Folgerungen aus der Messung von Geräuschen bei Zugbeanspruchung von metallischen Werkstoffen. *Arch. Eisenhüttenwesen* **1953**, *24*, 43–45. [CrossRef]
55. Lehtonen, A.; Cosgrove, J.W.; Hudson, J.A.; Johansson, E. An examination of in situ rock stress estimation using the Kaiser effect. *Eng. Geol.* **2012**, *124*, 24–37. [CrossRef]
56. Hou, Y.D.; Huang, S.P.; Han, J.; Liu, X.B.; Han, L.F.; Fu, C.F. Numerical simulation of the effect of injected CO_2 temperature and pressure on CO_2-Enhanced coalbed methane. *Appl. Sci.* **2020**, *10*, 1385.
57. Espino, D.M.; Shepherd, D.E.T.; Hukins, D.W.L. Transient large strain contact modelling: A comparison of contact techniques for simultaneous fluid–structure interaction. *Eur. J. Mech. B-Fluids* **2015**, *51*, 54–60. [CrossRef]
58. Heidbach, O.; Rajabi, M.; Cui, X.; Fuchs, K.; Müller, B.; Reinecker, J.; Reiter, K.; Tingay, M.; Wenzel, F.; Xie, F.; et al. The World Stress Map database release 2016: Crustal stress pattern across scales. *Tectonophysics* **2018**, *744*, 484–498. [CrossRef]
59. Hoek, E.; Brown, E.T. *Underground Excavation in Rock*; CRC Press: Boca Raton, FL, USA, 1980.
60. Brown, E.T.; Hoek, E. Trends in relationships between measured in-situ stresses and depth. *Int. J. Rock Mech. Min. Geomech. Abstr.* **1978**, *15*, 211–215. [CrossRef]

Journal of
Marine Science and Engineering

Article

Experimental Study of Mechanical Properties and Fracture Characteristics of Conglomerates Based on Mohr–Coulomb Criteria

Pengyu Liu [1,2], Yinghao Shen [1,2,*], Mianmo Meng [3], Senlin Luo [1,2], Yi Zhong [1,2] and Qiming Cen [1,2]

1 State Key Laboratory of Petroleum Resources and Prospecting, China University of Petroleum (Beijing), Beijing 102249, China; lpy19950119@163.com (P.L.)
2 The Unconventional Oil and Gas Institute, China University of Petroleum (Beijing), Beijing 102249, China
3 College of Marine Science and Technology, China University of Geosciences, Wuhan 430074, China
* Correspondence: shenyinghao@126.com

Abstract: Gravel is one of the main factors affecting the mechanical properties of conglomerates, which plays a decisive role in crack propagation. In this paper, taking the conglomerate of the Baikouquan Formation in Mahu Sag of Xinjiang as the research object, a three-dimensional model of the conglomerate is constructed by the discrete element numerical simulation method, and the triaxial compression experiment under different confining pressures is simulated. The mechanical properties and fracture morphology of conglomerate are analyzed with gravel content as a variable and verified by laboratory tests. In this simulation, with the increase of gravel content, the compressive strength of the conglomerate decreases, angle of internal friction decreases, and the fractures show different forms. The results show that the gravel morphology, spatial location, and gravel content have an impact on the mechanical properties of the conglomerate. The gravel content affects the formation process of the dominant fracture surface by controlling the distance between gravels so as to control the internal friction angle and it is the main controlling factor for the mechanical properties of the conglomerate. Gravel cracks initiate at the edge of gravel. Stress controls the formation of main cracks under low gravel content, and the influence of gravels under high gravel content makes cracks more discrete and complex.

Keywords: conglomerate; Mahu sag; three-dimensional model; mechanical properties; fracture characteristics

Citation: Liu, P.; Shen, Y.; Meng, M.; Luo, S.; Zhong, Y.; Cen, Q. Experimental Study of Mechanical Properties and Fracture Characteristics of Conglomerates Based on Mohr–Coulomb Criteria. *J. Mar. Sci. Eng.* **2023**, *11*, 1219. https://doi.org/10.3390/jmse11061219

Academic Editor: Assimina Antonarakou

Received: 8 May 2023
Revised: 2 June 2023
Accepted: 8 June 2023
Published: 13 June 2023

1. Introduction

Increasing attention is being paid to the exploitation of unconventional reservoirs, including mainly shale, dense sandstone, and conglomerate reservoirs [1]. Fractures in conglomerates differ markedly from shale fractures in morphology and genesis [2]. The conglomerate reservoirs of the Triassic Baikouquan Formation and Permian Wuerhe Formation are mainly put into development in Mahu Oilfield, Junggar Basin, and Xinjiang [2,3]. The conglomerate reservoirs are characterized by complex lithology, lateral variation blocks, strong heterogeneity, large particle size differences between different lithologies, and obvious differences in mechanical properties due to their fan delta sedimentary environment. Therefore, it is difficult to clarify the mechanical properties of conglomerate reservoirs [4–8].

Acoustic experiments, mechanical experiments, and numerical simulation experiments are currently the main methods used to study the mechanical properties and fracture characteristics of the Mahu aggregate reservoir. The uniaxial compression test results show that the conglomerate has significant brittle failure characteristics, and the failure mode is mainly splitting failure. The compressive strength decreases with the increase in gravel content [9–12]. The indentation hardness test shows that with the increase of gravel particle size, the compressive strength of the conglomerate increases and the elastic modulus decreases logarithmically, and the heterogeneity also increases. Gravel has an effect on the fracture toughness and acoustic characteristics of conglomerates. Under high gravel content, the fracture is complex and bifurcated, the bending degree is high [13], the ultrasonic waveform is seriously disturbed, and the wave velocity increases with the increase in gravel content or gravel size [14,15].

Through the analysis of the fracture morphology of conglomerates, it is concluded that the fracture propagation mode of conglomerate mainly includes the types of crossing gravel, bypassing gravel, embedding, stopping gravel, and being attracted to gravel, mainly expanding around gravel [16–18]. Fracturing fracture morphology is affected by both stress state and gravel characteristics, and cementation surface strength is the dominant factor for the diversion and extension of hydraulic fractures [19–21]. Under the low principal stress difference, the hydraulic fracture is more likely to deflect. With the increase in gravel content, the tortuosity of the fracture also increases. The high-strength gravel can inhibit the fracture from penetrating the gravel and strengthen the overall tensile strength of the conglomerate, which may lead to high fracture pressure. At the same time, the size and arrangement of gravel also affects the expansion of hydraulic fractures. When the size of the gravel is larger, the shielding ability of the fracture is stronger. In the case of a disorderly arrangement of gravel, the fracture deflection is more obvious, and the fracture shape is more complicated [22–26].

The current research shows that the mechanical properties of conglomerates are affected by factors such as gravel particle size, gravel content, cementation between gravel and cement, and the support mode of the conglomerate [27,28]. However, the influence of specific gravel shape and spatial position has not been studied. The specific influence mechanism of gravel content is not clear, and the crack propagation process is not clear. In this paper, based on PFC discrete element numerical simulation software, a three-dimensional mechanical model of conglomerate suitable for the study area is established by correcting the actual mechanical results for the conglomerate of Baikouquan Formation in Mahu Sag, Junggar Basin, Xinjiang. The uniaxial compression test simulation and triaxial compression simulation under different confining pressures are carried out to study the influencing factors and mechanisms of conglomerate mechanical properties and fracture propagation processes, which provide support for reasonable transformation and development countermeasures under different reservoir characteristics.

2. Experimental Methods and Numerical Models

2.1. Triaxial Compression Experiments Based on Moorcullen's Criterion

The study area is the conglomerate reservoir of the Baikouquan Formation in the Mahu Depression of the Junggar Basin, Xinjiang. The target reservoir was used as a specimen to analyze the conglomerate properties, and Figure 1a,b shows that the conglomerate specimens have diverse compositions, interspersed distribution of conglomerates and miscellaneous bases, uneven distribution of conglomerate grain size, and complex compositions. Digitization of the specimens can yield parameters such as conglomerate content, conglomerate grain size distribution, and conglomerate roundness. The stress-strain curves shown in Figure 1c can be obtained by uniaxial and triaxial compression experiments on the specimen conglomerate specimens, and the mechanical parameters such as compressive strength, elastic modulus, and Poisson's ratio can be obtained by analyzing and processing the stress-strain curves.

The Moorcullen criterion can explain the mechanical behavior of a rock during compression. According to the Moorcullen strength criterion, a rock is damaged when it is in a state of ultimate stress, i.e., when the shear stress on one side exceeds the ultimate shear stress it can withstand. When the two principal stresses σ_1 and σ_3 acting at a point in the rock are known, the positive stress σ and the shear stress τ in any plane can be derived from Equation (1). In the plane right-angle coordinate system, the curve of Equation (1) as shown in Figure 1d is a circle, i.e., the Mohr stress circle. A series of Mohr stress circles can be plotted by performing the strength curves of the rock under different stress states as shown in Figure 1e, and the line tangent to these Mohr stress circles is the damage envelope, whose expression is expressed by Equation (2). According to the damaged envelope, parameters such as cohesion C and internal friction angle φ of different conglomerates can be obtained, through which the mechanical properties of conglomerates can be analyzed.

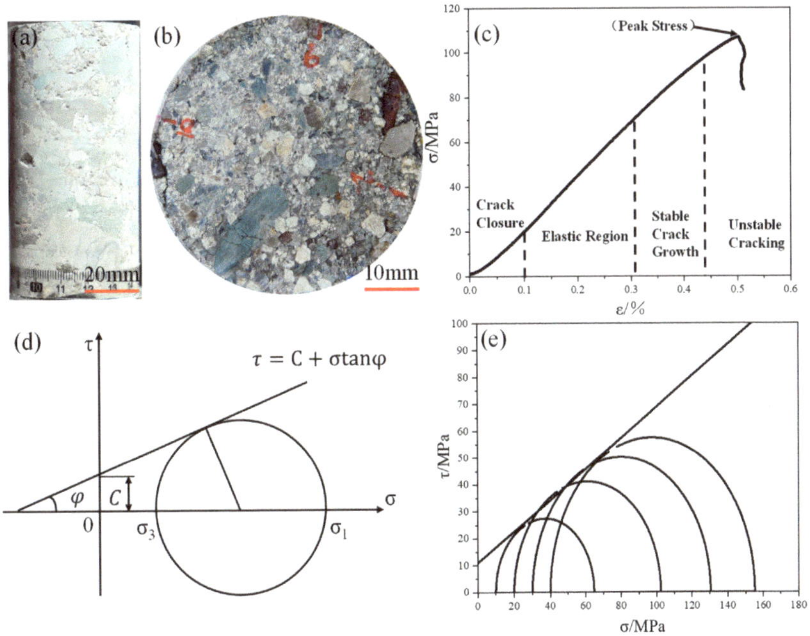

Figure 1. Conglomerate specimens and typical stress-strain curves.(**a**) Side view of conglomerate specimen; (**b**) End view of conglomerate samples; (**c**) Typical stress-strain curve; (**d**) Mohr stress circle; (**e**) Mohr stress circle under different confining pressure.

$$(\sigma - \frac{\sigma_1+\sigma_3}{2})^2 + \tau^2 = \left(\frac{\sigma_1+\sigma_3}{2}\right)^2 \tag{1}$$

$$\tau = C + \sigma \tan \varphi \tag{2}$$

where σ is the normal stress, MPa. τ is the shear stress, MPa. σ_1 and σ_3 are principal stress, MPa. C is cohesion, MPa. φ is the internal friction angle, °.

By the Moorcullen criterion in rock mechanics to know the microscopic rupture conditions for the interface of the shear stress to reach the shear strength of damage, and in the plane of the normal and main stress σ_1 tilt β angle, the normal stress and shear stress as shown in Equations (3) and (4), combined with Equation (2) and the introduction of the Moorcullen rupture criterion to obtain the damage criterion Equation (5), Equation (5) in the plane of the main stress of σ_1 and σ_3 is a straight line, then the uniaxial compressive strength σ_c as shown in Equation (6), that is, in the case of the same cohesion uniaxial compressive strength is proportional to the rupture interface dynamic friction coefficient μ.

$$\sigma = \frac{(\sigma_1+\sigma_3)}{2} + \frac{(\sigma_1-\sigma_3)}{2}\cos 2\beta \tag{3}$$

$$\tau = -\frac{(\sigma_1-\sigma_3)}{2}\sin 2\beta \tag{4}$$

$$\sigma_1\left(\sqrt{\mu^2+1}-\mu\right) - \sigma_3\left(\sqrt{\mu^2+1}+\mu\right) = 2C \tag{5}$$

$$\sigma_c = 2C\left(\sqrt{\mu^2+1}+\mu\right) \tag{6}$$

$$\beta = \frac{\pi}{2} - \frac{1}{2}\tan^{-1}\frac{1}{\mu} \tag{7}$$

$$\beta = \frac{\pi}{4} + \frac{1}{2}\varphi \tag{8}$$

where β is the failure angle, °. μ is the dynamic friction coefficient. σ_c is the uniaxial compressive strength, MPa.

2.2. Linear Parallel Bond Model

Due to the high randomness of conglomerate specimen gravel distribution, variable shape, and uneven particle size distribution, it is not easy to quantify the mechanical properties of conglomerates, and these problems can be effectively solved by means of numerical simulations. The numerical particle flow simulation software PFC is able to characterize the mechanical properties of rocks under the microscopic action of particles and has better results for the simulation of conglomerates, which are multicomponent rocks [29–31].

In the numerical particle flow simulation program, the macroscopic mechanical properties of the model are controlled by the microscopic contact properties. The linear parallel bonding model provides the mechanical behavior of two interfaces, one equivalent to the linear model for infinitesimal interfaces and the other for parallel bonding. The contact force of the linear model consists of a linear force and a damping force, determined by the normal stiffness k_n, the tangential stiffness k_s, and the friction coefficient μ; the damping force is determined by the normal damping ratio β_n, and the tangential damping ratio β_s (Figure 2a), it does not resist relative rotation, and slip is accommodated by imposing a Coulomb limit on the shear force. The parallel bonding model acts in parallel with the linear model to transfer forces and moments between the contacting particles (Figure 2b). Unlike the linear model, the parallel bonding model resists the relative rotation between particles and its behavior is linear elastic until the strength limit is exceeded and the bond breaks, making it unbonded. If the parallel bonding model is activated when the surface gap of the object $g_s \leq 0$, the bond breaks when it exceeds the strength limit, as shown in Figure 2c,d the particles will produce translation or rotation between them, at which point the parallel bonding model is equivalent to the linear model.

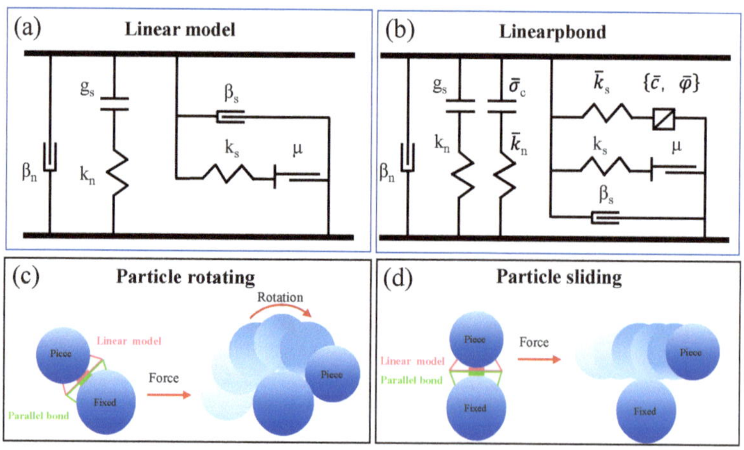

Figure 2. Rheological components (**a**,**b**) and behavior (**c**,**d**) of the linear model and linear parallel bond model.

The forces and displacements are established by Equation (9) for the linear model and by Equation (10) for the linear parallel bond model. In the linear model, the contact force is composed of linear force and dashpot force, and the contact moment is 0. While in the linear parallel model, parallel-bond forces are added, and the parallel-bond forces allow the model to transfer moments. In total, there are four types of energy in the linear parallel bond model: strain energy, slip energy, damping energy, and bond-building strain energy, where strain energy is stored in the linear spring (Equation (11)), slip energy is the total energy consumed by frictional slip (Equation (12)), the dot product of the slip component and the average linear shear force occurring during the timestep gives the increment of slip energy. The slip energy remains equal to zero until the bond breaks. Damping energy is the total energy consumed by the damper (Equation (13)), and strain energy, slip energy, and damping energy are stored in the linear model plane; bond strain energy is stored in the bond parallel bond spring.

$$F_c = F^l + F^d, M_c = 0 \tag{9}$$

$$F_c = F^l + F^d + \overline{F}, M_c = \overline{M} \tag{10}$$

$$E_k = \frac{1}{2}\left(\frac{\left(F_n^l\right)^2}{k_n} + \frac{\left(F_n^l\right)^2}{k_s}\right) \tag{11}$$

$$E_\mu = E_\mu - \frac{1}{2}\left(\left(F_s^l\right)_0 + F_s^l\right)\Delta\delta_s^\mu \tag{12}$$

$$E_\beta = E_\beta - F^d\delta\Delta t \tag{13}$$

where F_c is the contact force, N. F^l is the linear force, N. F^d is the dashpot force, N. M_c is the contact moment, N·m. $\overline{F}$ is the parallel-bond force, N. $\overline{M}$ is the parallel-bond moment, N·m. E_k is the strain energy, N·m/pa. F_n^l is the linear normal force, N. F_s^l is the linear shear force, N. k_n is the normal stiffness, N/m. k_s is the shear stiffness, N/m. E_μ is the slip energy, N·m/pa. $\left(F_s^l\right)_0$ is the linear shear force at the beginning of the timestep, N. $\Delta\delta_s^\mu$ is the slip component of relative shear-displacement increment, m. E_β is the dashpot energy, N·m/pa. δ is the relative translational velocity, m/s. Δt is the time step, s.

2.3. 3D Numerical Model of Conglomerate

Although the mechanical properties of the conglomerate have been studied with the help of numerical simulation methods, they are only limited to the two-dimensional space state, while the distribution of conglomerates in three-dimensional space is more random and the influence on the mechanical properties of conglomerates is more complicated, so it is necessary to construct a three-dimensional numerical model of the conglomerate to study the mechanical properties of the conglomerate.

Before constructing the three-dimensional numerical simulation of the conglomerate, microstructural and elemental analyses of the conglomerate specimens were first performed by scanning electron microscopy to determine the basic compositional form of the conglomerate. The results show that the conglomerate is mainly composed of conglomerate, heterogeneous base, and cement, and different conglomerates are supported in different ways, with heterogeneous base particles supported at low conglomerate content as shown in Figure 3c, and particles supported at high conglomerate content as shown in Figure 3a. The elemental analysis of Figure 3b,d shows that the cement is mainly distributed at the edge of the gravels, with a small amount in the heterogeneous base. The elemental content of the cement is dominated by Al and Ca elements, and some K, Mg, and Na elements are present. The type of cementation is clayey cementation, and the elemental differences between the gravel and the cement are obvious, with clear boundaries. From the scanning electron microscopy results, it can be concluded that the conglomerate is a three-phase composite material composed of a conglomerate, heterogeneous base, and weak cementation surfaces distributed at the edge of the conglomerate.

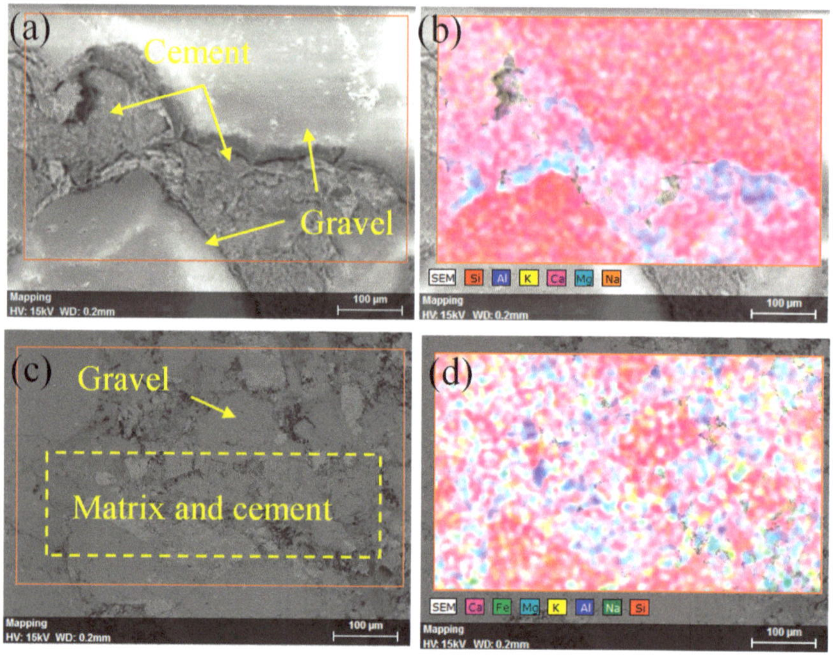

Figure 3. Electron microscope scan of conglomerate (**a**,**c**) and its corresponding elemental analysis (**b**,**d**).

Based on the component analysis of the conglomerate specimens, the conglomerate is modeled in 3D in Figure 4 using the particle flow software PFC. A two-dimensional image of the actual conglomerate (b) was extracted from the actual specimen taken (a) and reconstructed into a cubic conglomerate (c) using small diameter particles for stacking in three dimensions according to (b). To study the effect of different shapes of gravels on the mechanical properties of conglomerates, spherical conglomerate (d) and chain conglomerate (e) were constructed, and finally a conglomerate 3D model (f) with a diameter of 50 mm and a height of 100 mm was generated. The conglomerate model (f) simulates the gravel by particle clusters, the generated spheres simulate the matrix, the contact between gravel and gravel, matrix and matrix is simulated by a linear model, and the cementation surface between gravel and matrix is simulated by a linear parallel bonding model. Using gravel content and morphology as variables, the study generated conglomerate models with different gravel contents. The following issues were considered in the construction of the conglomerate model: (1) the particle clump diameter of the simulated conglomerate is 5 times that of the matrix spheres, considering the size effect and computational efficiency; (2) the sphere accumulation in the restricted space cannot fill the whole space, and it is known that the density of sphere accumulation in the cylinder is related to the space extent, sphere diameter, sphere gradation, and arrangement according to the theory of densest accumulation [32–34]. The gravels were generated by controlling the number of clumps generated to quantify the gravel content; (3) to avoid local stress effects due to gravels generated at the model boundary, the clumps were generated inside the model by numerical modeling to simulate the mechanical behavior of the gravels under load. The model parameters are shown in Table 1.

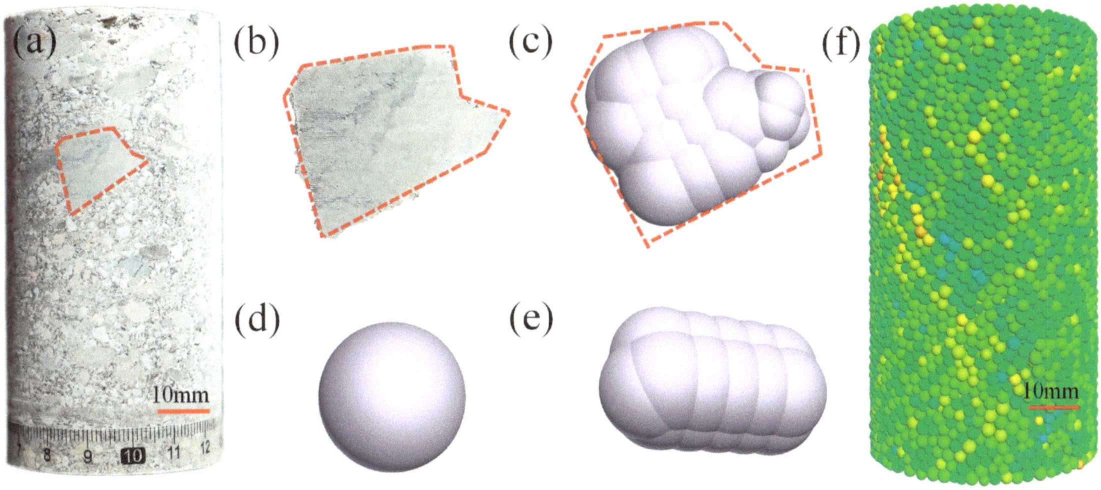

Figure 4. The conglomerate specimen (**a**),real gravel (**b**) and numerical model of conglomerates (**f**) and different shapes of gravel models (**c**–**e**).

Table 1. Table of model parameters.

Contact Property	Effective Young's Modulus (Pa)	Tensile Strength(Pa)	Cohesion (Pa)	Friction Coefficient
Ball to ball	1×10^9	1×10^7	1×10^7	0.57
Ball to clump	1×10^6	1×10^6	5×10^6	0.3

3. Results

3.1. Experimental Results of Conglomerate

The gravel content of the taken conglomerate specimens was statistically analyzed, and then uniaxial compression experiments were performed to obtain the stress-strain curves shown in Figure 5a. The curves show that the elastic phase of the stress-strain curve is longer when the gravel content is 4.23%, and the conglomerate as a whole shows brittle characteristics; when the gravel content is 56.8%, the conglomerate stress-strain curve as a whole is smoother, the crack stabilization growth phase is longer, and the conglomerate as a whole shows plastic characteristics. Figure 5b shows that the compressive strength of the conglomerate decreases with increasing gravel content from 0 to 56.8%, with an overall negative correlation. The results of uniaxial compression experiments with different gravel contents show that the mechanical properties of conglomerates with different gravel contents vary greatly, and the gravel content is the main controlling factor of the mechanical properties of conglomerates.

According to the uniaxial compression test results of conglomerates, specimens with 25–30% conglomerate content were selected for triaxial compression experiments under different confining pressures, and the stress-strain curves shown in Figure 5c were obtained. The curves show that the surrounding pressure obviously changes the morphology of the stress-strain curve, and the plastic characteristics of the conglomerate are more obvious. By linearly solving the data obtained in Figure 5c, the cohesive force of the conglomerate with 25–30% gravel content is 5.39 MPa and the internal friction angle is 29.29°, and the Mohr stress circle and damage envelope, shown in Figure 5d, are made. However, it is difficult to control the gravel content of the specimens under different confining pressures, gravel arrangement, and spatial locations, which leads to some errors in the calculated results in physics experiments. Therefore, the quantitative expression of the gravel quantity and spatial location with the help of the numerical simulation method is needed to specifically study the effect of gravel content on the mechanical properties of conglomerates, which will be analyzed in Section 3.2.2.

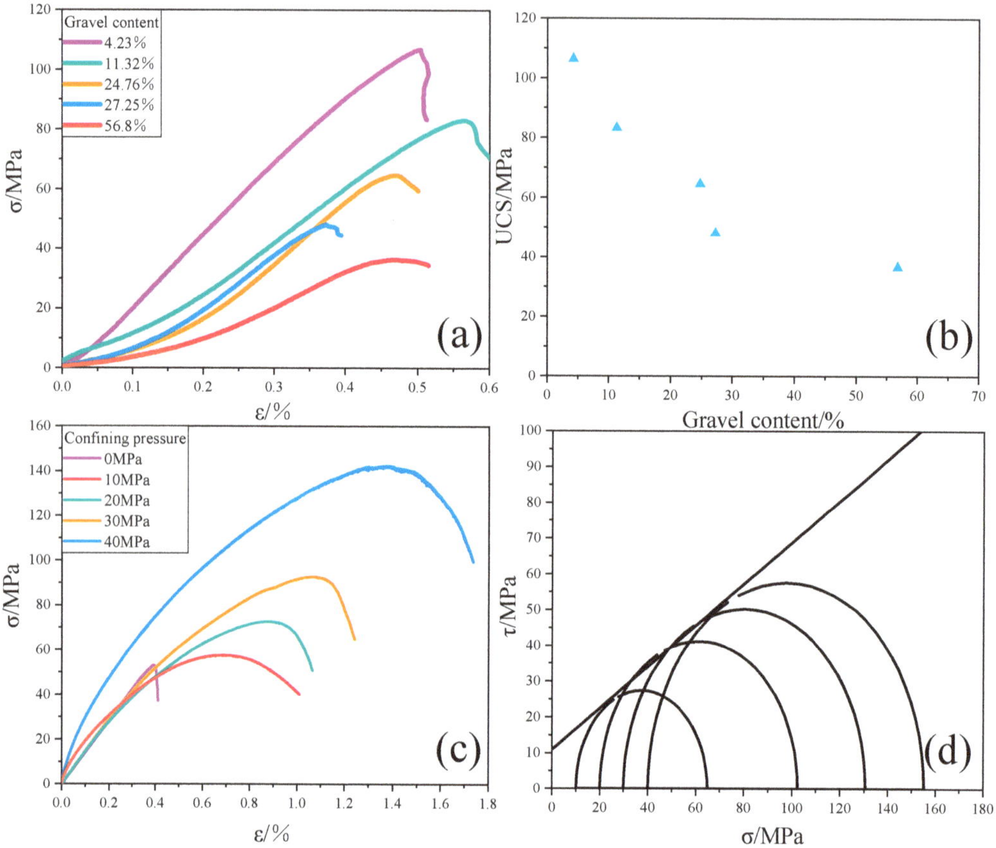

Figure 5. (**a**) Stress-strain curves of uniaxial compression physical experiments on conglomerates; (**b**) Relationship between gravel content and UCS; (**c**) Stress-strain curves of conglomerate triaxial compression physical experiments; (**d**) Mohr stress circle.

In order to analyze the fractures in the conglomerate after the above tests, the conglomerate fractures were carved by microscopy as shown in Figure 6, and reconstructed by CT scanning method. Figure 6a shows that the microfractures extend mainly at the edge of the conglomerate and in the miscellaneous base and that the fractures extending in the miscellaneous base will be affected by other conglomerates and will turn at a large turn of nearly 90°. By performing CT scans after experiments on the above gravels with 25–30% gravel content, followed by noise reduction and 3D reconstruction of the fractures, the results show that the fracture morphology of the gravels with low gravel content is an inclined main fracture. Considering the time cost, the fracture size of the 3D reconstruction is small compared to the core size, but the overall fracture morphology is similar to the reconstructed fracture morphology according to the self-similarity of fractal theory.

3.2. Numerical Simulation Results of Conglomerate

The results of physical experiments indicate that the mechanical properties of the conglomerate vary for different gravel contents, with cracks produced by compression being at the edge of the conglomerate and subject to turning by the spacing of the conglomerate. However, the gravel content of the conglomerate is difficult to standardize in physical experiments and the spatial location and shape of the conglomerate is difficult to characterize. In this section, we investigate the mechanical properties of the conglomerate in three dimensions by quantitatively characterizing the conglomerate content with the help of a three-dimensional conglomerate model established by discrete element numerical simulation methods.

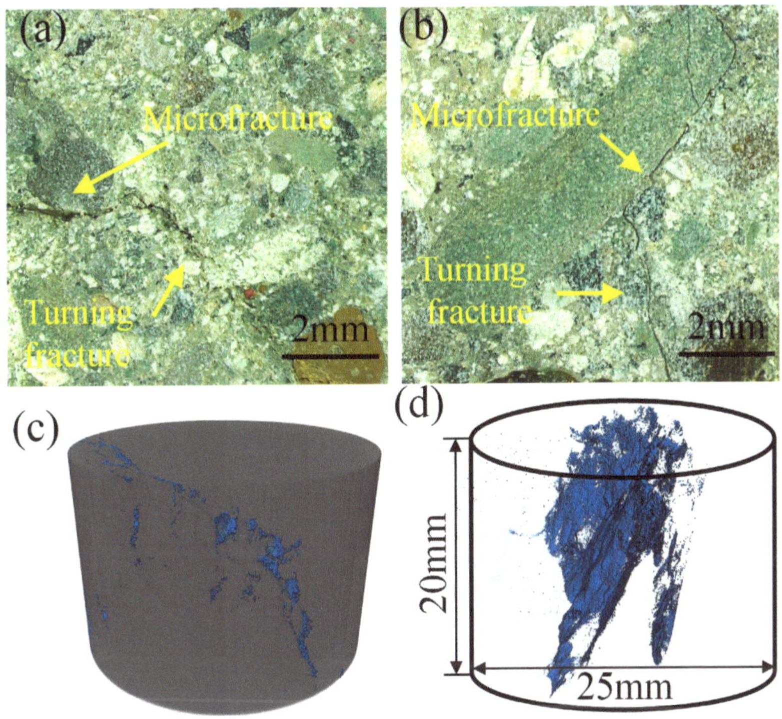

Figure 6. Physical experimental fractures under microscope (**a**,**b**) and CT scan (**c**,**d**) in conglomerate.

3.2.1. Effect of Gravel Morphology and Spatial Location on Mechanical Behavior

To study the effect of gravel morphology on the mechanical properties of conglomerate, uniaxial compression tests were conducted on the spherical, square, and chain conglomerate models shown in Figure 4 to simulate and study the effect of conglomerate morphology on the mechanical properties of conglomerates, and the experimental results showed that the compressive strength of the spherical conglomerate model was 77.75 MPa, the compressive strength of the square conglomerate model was lower at 76.53 MPa, and the compressive strength of the chain conglomerate model was 77.8 MPa. The compressive strength of the chain gravel model is related to the angle between the model axis and the stress direction, with 77.8 MPa for the 90° model, 76.9 MPa for the 45° model, and 74.7 MPa for the 0° model. The compressive strength of the chain gravel model can be equated to a "prefabricated fracture" containing a weak cementation surface, which increases with the increase of the axis angle and stress direction.

The stress state around different conglomerate shapes is demonstrated by slicing the 3D conglomerate model as shown in Figure 7 The chain conglomerate in Figure 7a shows a significant stress concentration at the end of the long axis, making it more susceptible to destabilization and thus cracking at this point. In Figure 7b, the stress state around the block gravel is more uniform than the chain gravel, with stress concentrations at the upper and lower ends of the stress loading direction. In Figure 7c, high stresses are observed around the spherical gravels. The results indicate that the weak cementation surface around the gravel tends to produce stress concentrations in this region, but the shape of the gravel changes the area of stress concentration. In general, the stress concentrations around the less rounded gravel are more directional. Although the 3D gravel model constructed in this paper can provide a more realistic description of the force state of differently shaped gravels, however, compared with the force of the actual gravels in Figure 6, the interaction between gravels of different sizes cannot be effectively described due to the grading problem in the model. Overall, the conglomerate model in this paper is still able to effectively represent the mechanical behavior and fracture extension characteristics of the conglomerate.

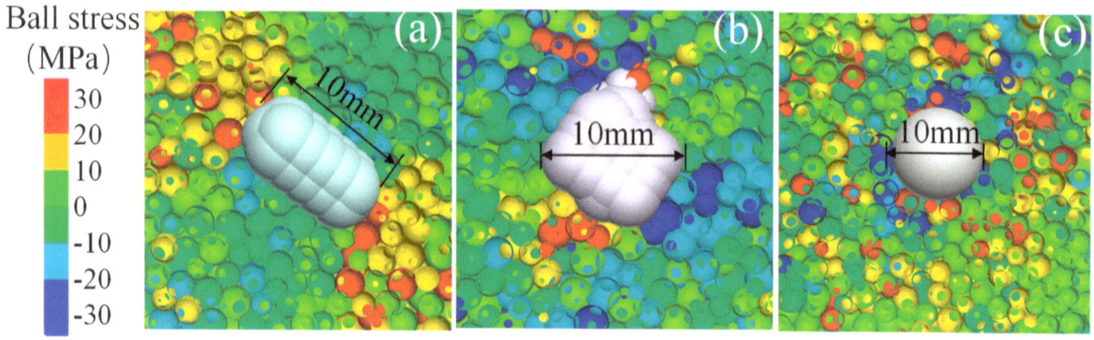

Figure 7. Gravel edge stress diagram with chain gravel (**a**), block gravel (**b**), and spherical gravel (**c**).

3.2.2. Effect of Gravel Content on Mechanical Behavior

From the above study, it is known that the shape of the gravel produces stress concentrations and the gravel spacing affects the expansion of cracks between gravels, both of which affect the mechanical properties of the conglomerate. In practical problems, the gravel shape is mostly polygonal, and the gravel spacing problem is expressed in terms of gravel content. In this section of the study, the conglomerate mechanical properties are investigated using square-shaped conglomerate as the base conglomerate model and conglomerate uniaxial compression tests with conglomerate content as a variable.

Based on the principles of constructing a conglomerate model in Section 2, conglomerate models with the number of conglomerates of 9, 18, 27, 37, 46, 54, 65, 71, and 79 are constructed to represent conglomerate contents of 10%, 20%, 30%, 40%, 50%, 60%, 70%, 80%, and 90%, respectively, and their stress-strain curves are shown in Figure 8a. The results of physical and numerical simulation experiments are presented in Figure 8b, which show that the compressive strength of the conglomerate decreases with increasing gravel content. The results of the physical experiments and the results of the s numerical simulation experiments are in agreement, proving that the model can effectively simulate the mechanical property changes and fracture morphology of the gravels under complex gravel conditions.

By monitoring the formation of cracks during uniaxial compression and recording their spatial locations, the two-dimensional crack slices shown in Figure 9a–d and the three-dimensional crack images shown in Figure 9e–h were obtained. From the two-dimensional fracture slices, it can be seen that the fractures mainly extend along the gravel edge (Figure 9a) and intertwine in the miscellaneous base to form the main fracture (Figure 9b). When the gravel content is high (Figure 9c,d), the fractures intertwine in the miscellaneous base and it is difficult to form the main fracture, but the overall fracture pattern is more complex. The presence of only one gravel but fractures clustered at the upper end of the conglomerate in Figure 9a is due to the fact that only one gravel is contained in this profile in the two-dimensional fracture slice diagram; there are still gravels at other locations in the actual space and the generation of these fractures is influenced by the other gravels. Figure 9b shows that the gravel located at the edges of the model will create cracks and preferentially break, thus affecting the overall strength. From the 3D fracture image, it can be observed more clearly that the fractures extend along the gravel edges and that the fractures are able to form a main fracture pattern at lower gravel contents (Figure 9e,f) and a more complex fracture pattern at higher gravel contents (Figure 9g,h).

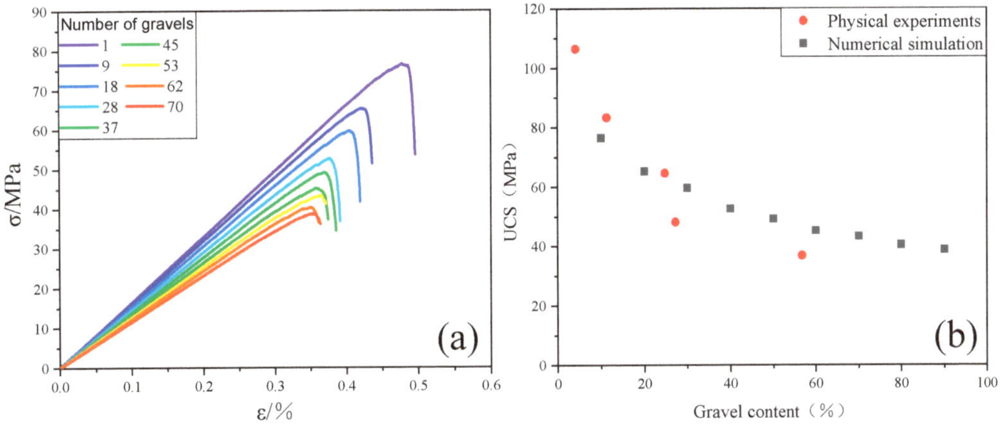

Figure 8. Stress-strain curve of numerical simulation. (**a**) Stress-strain curves at different gravel contents; (**b**) Relationship between gravel content and UCS.

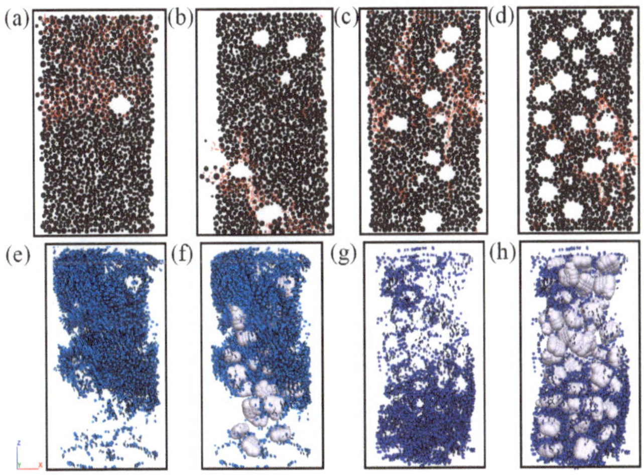

Figure 9. Numerical simulation of 2D crack image (**a–d**) and 3D crack image (**e–h**).

3.2.3. Effect of Surrounding Pressure on the Mechanical Behavior of Gravels

From the above study, the gravel content has a large influence on the mechanical properties of conglomerates. In this section, the conglomerate model constructed in Section 3.2.2 is simulated in triaxial compression tests at 10 MPa, 20 MPa, 30 MPa, and 40 MPa peritectic pressure with the gravel content as the variable, and the stress-strain curves with different gravel contents under different peritectic pressure conditions are obtained as shown in Figure 10. In general, as the surrounding pressure increases, the compressive strength of the conglomerate increases while its peak point strain becomes larger, the initial rupture point appears earlier, the rupture process zone is larger, and the plastic characteristics are more obvious. The compressive strength of the conglomerate decreases with the increase of the number of conglomerates under different confining pressure conditions, which is the same as the results shown in Figures 5 and 8. In the case of high surrounding pressure (Figure 10c,d), the compressive strength of the conglomerate is significantly higher when the gravel content is 10% and 20% than in other cases and decreases steadily with increasing gravel content when the gravel content is greater than 30%.

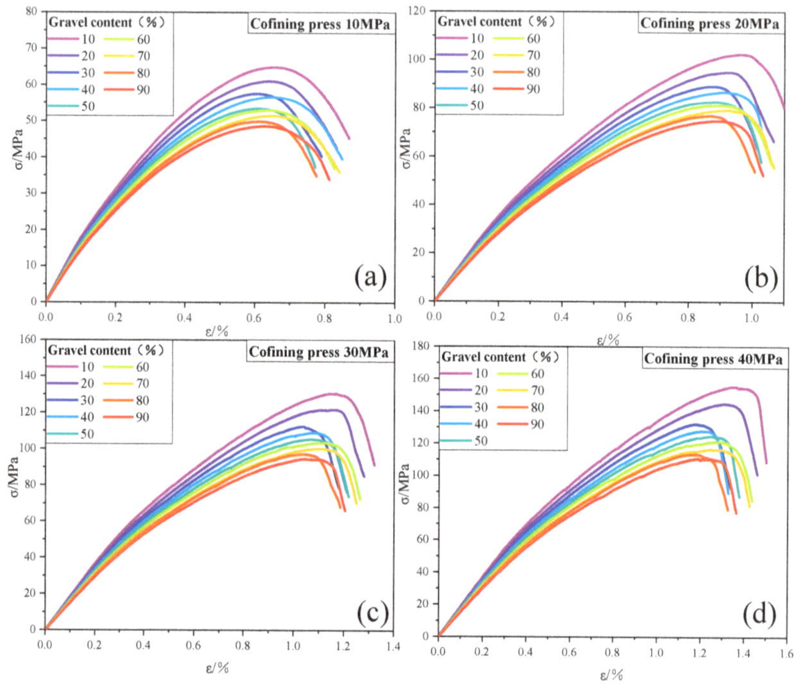

Figure 10. Stress-strain curves for different gravel contents under different confining pressures. (**a**) Stress-strain curve under 10 MPa circumferential pressure; (**b**) Stress-strain curve under 20 MPa circumferential pressure; (**c**) Stress-strain curve under 30 MPa circumferential pressure; (**d**) Stress-strain curve under 40 MPa circumferential pressure.

The results of the above enclosing pressure experiments were solved by planning to obtain the Mohr stress circle and its damage envelope for different gravel contents shown in Figure 11. The mechanical parameters are tabulated as shown in Table 1, and the results show that the cohesive force ranges from 10.41 to 11.52 MPa for different gravel contents, with little change in cohesive force, and the internal friction angle ranges from 20.54° to 30.12°, and the internal friction angle decreases gradually with the increase of gravel content. It can be seen from Figure 11 that when the gravel content is 10% and 20%, the slope of the damaged envelope (internal friction angle) is significantly larger than in the other cases.

The above data showed that the data with gravel content of 10% and 20% deviated from the other data by a large amount. The standard deviation of the data is used as a measure of the amount of deviation, and by processing the data in Table 2, the standard deviation of the data for the angle of internal friction with different gravel contents is 3.01 when the gravel content of 10% and 20% is included, and 1.61 when the two sets of data are not included. The standard deviation of the data for the compressive strength of the conglomerate at different envelope pressures is similar and will not be repeated here. The results indicate that the mechanical properties of the conglomerate at 10% and 20% gravel content are significantly different from the other cases and that the 30% gravel content is the cut-off point for the mechanical properties of the conglomerate.

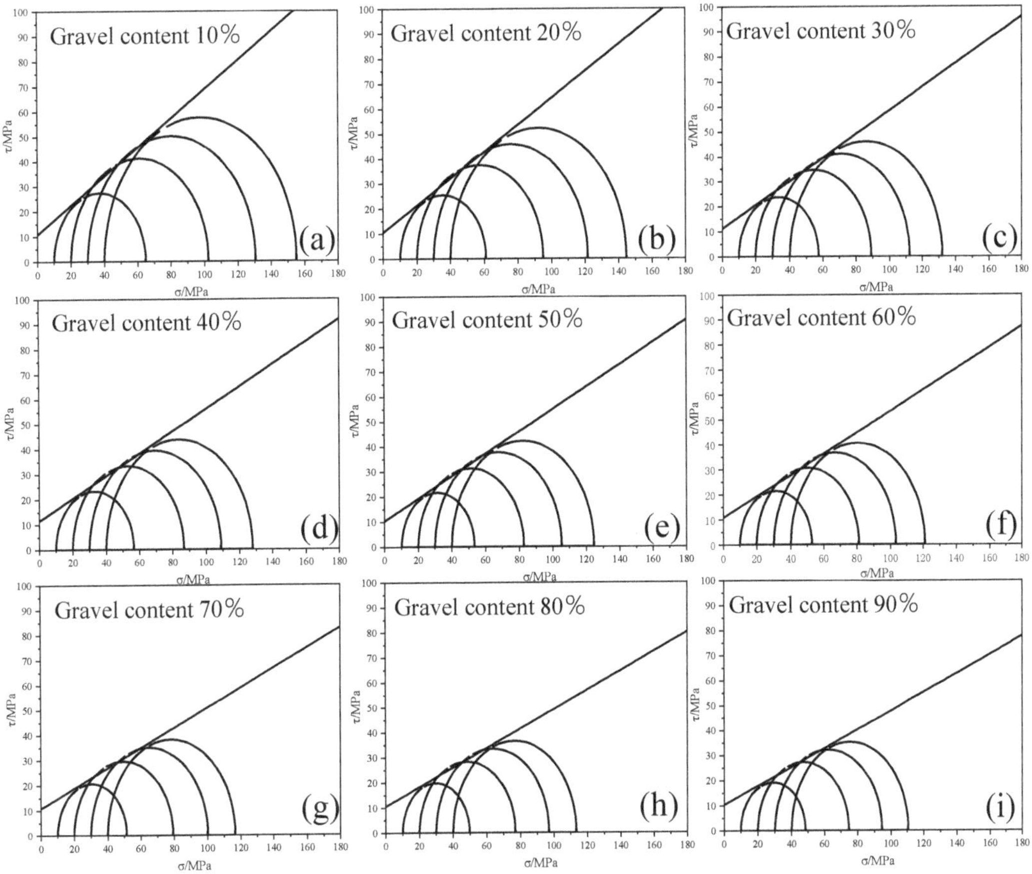

Figure 11. Mohr stress circle and damaged envelope. (**a**) Mohr stress circle with 10% gravel content; (**b**) Mohr stress circle with 20% gravel content; (**c**) Mohr stress circle with 30% gravel content; (**d**) Mohr stress circle with 40% gravel content; (**e**) Mohr stress circle with 50% gravel content; (**f**) Mohr stress circle with 60% gravel content; (**g**) Mohr stress circle with 70% gravel content; (**h**) Mohr stress circle with 80% gravel content; (**i**) Mohr stress circle with 90% gravel content.

Table 2. Table of mechanical parameters for different gravel contents.

Gravel Content (%)			10	20	30	40	50	60	70	80	90
UCS (MPa)	Confining Stress	10 MPa	64.8	61	57.5	56.6	53.4	52.9	51.4	49.8	48.5
		20 MPa	102.4	95	89.2	86.7	82.7	81.3	79.3	76.9	74.8
		30 MPa	130.5	121.7	112.3	109	105.4	103.4	100.2	97.2	94.6
		40 MPa	155.2	144.6	132.1	127.8	124.5	121.2	116.6	113.4	110.7
C (MPa)			10.9	10.7	11.25	11.52	10.41	10.75	10.89	10.64	10.43
Φ (°)			30.121	28.19	25.27	24.08	24.05	23.05	21.85	21.16	20.54

4. Discussion

The conglomerate first passes through the compaction phase during compression, then enters the elastic phase, and next enters the critical microfracture development phase, when damage has occurred at the microscopic level, but the rock still has load-bearing capacity at the macroscopic level. The results of physical experiments (Figure 6) and numerical simulations (Figure 9) demonstrate that microfractures in the three-phase conglomerate model with gravel, heterogeneous base, and weak cementation surface are more likely to be generated in the weak cementation surface, which has the lowest strength, while the microfractures generated in the cementation surface are essentially a frictional behavior. Figure 12a shows the clayey minerals on the surface of the conglomerate being compressed and deformed after the damage, while Figure 12b shows the direction of clay frictional deformation on the gravel surface. As shown in Figures 6 and 9, when microcracks are formed at the edge of the gravel, the further expansion of the cracks is influenced by the gravel, as shown in Figure 12c when microcracks are created at both gravel edges, they will communicate with each other to create new cracks, and when the two gravels are close enough, the priority of the creation of new cracks will be greater than the priority of the stress action, which makes the direction of the connecting cracks deviate from the main crack direction.

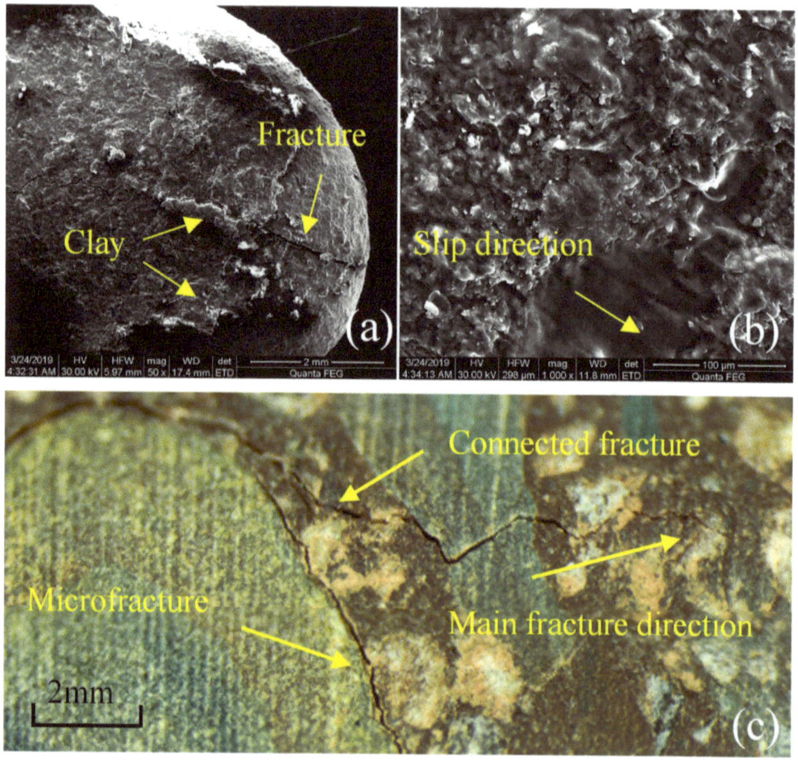

Figure 12. Conglomerate fracture map at different scales. (**a**,**b**) Electron microscope scan of conglomerate; (**c**) Conglomerate diagram under optical microscope.

When the gravel content is high enough, the gravel spacing will be smaller so that the microfractures around different gravels will communicate with each other to produce new microfractures, these microfractures deviate from the main fracture direction thus making the overall fracture pattern more complex, shown in Figure 13, the difference between the maximum and minimum value of the fracture number at 10% gravel content. As shown in Figure 13, the difference between the maximum and the minimum number of cracks at 10% gravel content is 1179.4, and the difference decreases gradually with the increase of gravel content, and the difference is only 645 when the gravel content is 80%, i.e., it is difficult to form the main cracks at low gravel content under high gravel content, but the formed cracks are more uniform and more complex.

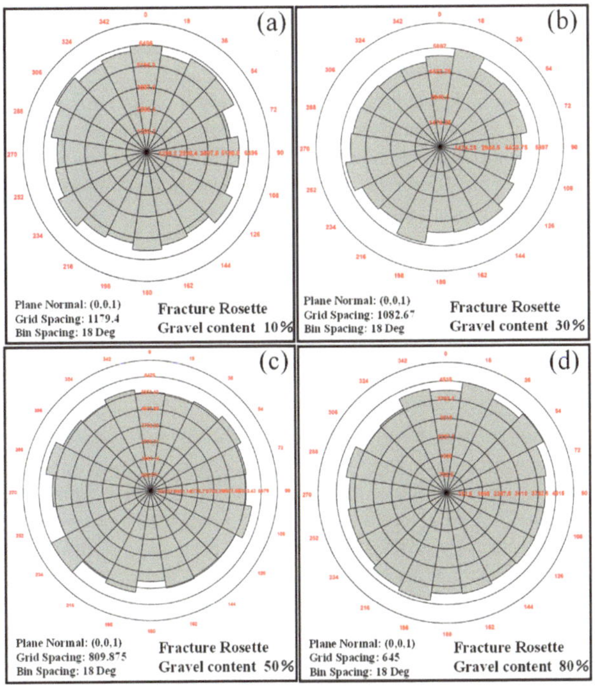

Figure 13. Fracture rosette diagram for different gravel contents. (**a**) Fracture rosette diagram with 10% gravel content; (**b**) Fracture rosette diagram with 20% gravel content; (**c**) Fracture rosette diagram with 30% gravel content; (**d**) Fracture rosette diagram with 40% gravel content.

From the above analysis, it is known that the microfracture starts at the edge of the gravel first, and the fracture is affected by the spacing of the gravel in the process of further expansion. In general, as shown in Figure 14a, when the main fracture with rupture angle β expands along the edge of the gravel, a turn occurs during expansion due to the attraction between the fractures, i.e., the phenomenon shown in Figure 6a; when the main fracture with rupture angle β is created in the miscellaneous base as shown in Figure 14b, the path turns during expansion due to the obstruction of the gravel, i.e., the phenomenon shown in Figure 12c. The angle of the rupture angle β′ of the new fracture formed in both cases is varied by the gravel spacing, and as the gravel content increases, the closer the gravel spacing is, the smaller the newly generated rupture angle β′ is.

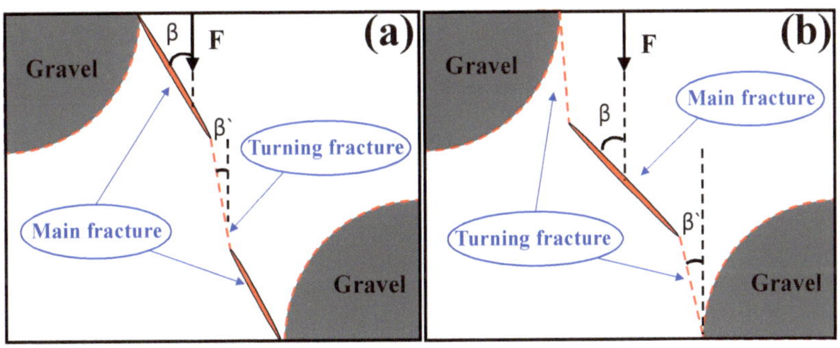

Figure 14. Conglomerate fracture rupture process diagram. (**a**) The first type of fracture rupture process; (**b**) The second type of fracture rupture process.

The rupture angle β when rupture occurs can be obtained from Equations (2) and (3) by associating Equation (7), and the relationship between rupture angle β and internal friction angle φ can be obtained according to the concept of kinetic friction coefficient Equation (8), that is, the rupture angle is proportional to the internal friction angle. From this, it can be seen that as the gravel content increases leading to closer gravel spacing—thus making the rupture angle β′ smaller for the newly generated cracks, leading to a decrease in the overall internal friction angle of the rock—the kinetic friction coefficient decreases and the compressive strength decreases.

5. Conclusions

Conglomerates are affected by complex components such as gravel, and their mechanical behavior is more complex. In this paper, the mechanical properties of conglomerates and their fracture morphology are investigated by indoor physical experiments combined with numerical simulation methods, and the results show that:

(1) The physical experimental results show that the uniaxial compressive strength of conglomerate gradually decreases with increasing gravel content and that gravel content is the main controlling factor of conglomerate mechanical properties;

(2) This paper constructs a fine three-dimensional conglomerate model based on the particle flow discrete element program, which can effectively simulate the conglomerate with complex particle composition and its mechanical behavior;

(3) The numerical simulation results show that the mechanical properties of conglomerates are influenced by the shape of conglomerates and their spatial location. The lower the roundness of the conglomerate, the lower the compressive strength, and the smaller the angle between the long axis of the bar conglomerate and the stress direction, the lower the compressive strength. The lower the roundness of the gravel, the lower the stress concentration at the angular points. As the angle between the long axis of the gravel and the stress direction decreases, the stress concentration at the end face of the gravel becomes more obvious, and the energy accumulated in the stress concentration and the gravel content work together to form a rupture between the gravels more easily, thus lowering the strength of the gravel;

(4) With the increase of gravel content, the internal friction angle of gravel gradually decreases. As gravel content increases, the angle of internal friction decreases. Gravel content influences the formation of the dominant fracture surface by controlling the gravel spacing, which in turn controls the magnitude of the angle of internal friction to influence the compressive strength of the conglomerate;

(5) The conglomerate fracture starts at the edge of the conglomerate, with stress controlling the fracture pattern to form the dominant fracture at low conglomerate content and the influence of the conglomerate at high conglomerate content making the fracture more discrete and complex. With the increase of gravel content, the fractures starting at the edge of the gravel are more likely to communicate with each other during the expansion process, and the high gravel content makes the microfractures at the edge of the gravel develop in large numbers so that it is difficult to form the main fracture, and the fracture pattern is discrete and complex.

Author Contributions: Conceptualization, P.L. and Y.S.; methodology, M.M.; software, S.L.; validation, P.L; formal analysis, M.M.; investigation, Y.Z.; resources, Y.S.; data curation, Q.C.; writing—original draft preparation, P.L.; writing—review and editing, S.L.; visualization, Y.Z. and P.L.; supervision, M.M.; project administration, Y.S. All authors have read and agreed to the published version of the manuscript.

Funding: This research was funded by [The Strategic Cooperation Technology Projects of CNPC and CUPB] grant number [ZLZX2020-01].

Institutional Review Board Statement: Not applicable.

Informed Consent Statement: Not applicable.

Data Availability Statement: The data presented in this study are available on request from the corresponding author.

Conflicts of Interest: The authors declare no conflict of interest.

Nomenclature

σ	normal stress, MPa
τ	shear stress, MPa
σ_1, σ_3	principal stress, MPa
C	cohesion, MPa
φ	internal friction angle, °
F_c	contact force, N
F^l	linear force, N
F^d	dashpot force, N;
M_c	contact moment, N·m
$\overline{F}$	parallel-bond force, N
$\overline{M}$	parallel-bond moment, N·m
E_k	strain energy, N·m/pa
F_n^l	linear normal force, N
F_s^l	linear shear force, N
k_n	normal stiffness, N/m
k_s	shear stiffness, N/m
E_μ	slip energy, N·m/pa
$(F_s^l)_0$	linear shear force at the beginning of the timestep, N
$\Delta\delta_s^\mu$	slip component of relative shear-displacement increment, m
E_β	dashpot energy, N·m/pa
δ	relative translational velocity, m/s
Δt	time step, s
β	failure angle
μ	dynamic friction coefficient
σ_c	uniaxial compressive strength, MPa

References

1. Meng, M.; Ge, H.; Shen, Y.; Ji, W.; Wang, Q. Rock Fabric of Tight Sandstone and Its Influence on Irreducible Water Saturation in Eastern Ordos Basin. *Energy Fuels* **2023**, *37*, 3685–3696. [CrossRef]
2. Zhang, Y.; Ge, H.; Zhao, K.; Liu, S.; Zhou, D. Simulation of pressure response resulted from non-uniform fracture network communication and its application to interwell-fracturing interference in shale oil reservoirs. *Geomech. Geophys. Geo-Energy Geo-Resour.* **2022**, *8*, 114. [CrossRef]
3. Jia, C.Z.; Zou, C.N.; Yang, Z.; Zhu, R.K.; Chen, Z.X.; Zhang, B.; Jiang, L. Significant progress of continental petroleum geology theory in basins of Central and Western China. *Pet. Explor. Dev.* **2018**, *45*, 546–560. [CrossRef]
4. Li, G.X.; Qin, J.H.; Xian, C.G.; Fan, X.; Zhang, J.; Ding, Y. Theoretical understandings, key technologies and practices of tight conglomerate oilfield efficient development: A case study of the Mahu oilfield, Junggar Basin, NW China. *Pet. Explor. Dev.* **2020**, *47*, 1275–1290. [CrossRef]
5. Wei, J.; Liao, H.L.; Wang, H.J.; Chen, J.; Li, N.; Liang, H.; Liu, C.; Zhang, D.; Teng, Z. Experimental investigation on the dynamic tensile characteristics of conglomerate based on 3D SHPB system. *J. Pet. Sci. Eng.* **2022**, *213*, 110350. [CrossRef]
6. Yu, X.H.; Zhai, J.H.; Tan, C.P.; Zhang, L.; Li, X.L.; Gao, Z.P. Conglomerate Lithofacies and Origin Models of Fan Deltas of Baikouquan Formation in Mahu Sag, Junggar Basin. *Xinjiang Pet. Geol.* **2014**, *35*, 619–627.
7. Zhang, C.M.; Wang, X.L.; Zhu, R.; Qu, J.H.; Pan, J.; An, Z.Y. Litho-Facies Classification of Baikouquan Formation in Mahu sag, Junggar Basin. *Xinjiang Pet. Geol.* **2016**, *37*, 606–614.
8. Zhang, C.M.; Song, X.M.; Wang, X.J.; Wang, X.; Zhao, K.; Shuang, Q.; Li, S. Origin and depositional characteristics of supported conglomerates. *Pet. Explor. Dev.* **2020**, *47*, 292–305. [CrossRef]
9. Li, J.R.; Duan, K.; Meng, H.; Wang, J.; Zhang, Q.; Wang, L. On the Mechanical Properties and Failure Mechanism of Conglomerate Specimens Subjected to Triaxial Compression Tests. *Rock Mech. Rock Eng.* **2022**, *56*, 973–995. [CrossRef]
10. Liu, X.J.; Xiong, J.; Liang, L.X.; You, X.C. Rock Mechanical Charactterisitics and Fracture Propagation Mechanism of Sandy Conglomerate Reservoirs in Baikouquan Formation of Mahu Sag. *Xinjiang Pet. Geol.* **2018**, *39*, 83–91.
11. Luo, S.L.; Ge, H.K.; Wang, J.B.; Zhou, W.; Shen, Y.; Liu, P.; Liu, J. Numerical simulation study on the crack propagation of conglomerate. *R. Soc. Open Sci.* **2021**, *8*, 202178. [CrossRef]

12. Wang, J.B.; Ge, H.K.; Liu, J.T.; Shen, Y.; Zhang, Z.; Luo, S.; Liu, D. Effects of Gravel Size and Content on the Mechanical Properties of Conglomerate. *Rock Mech. Rock Eng.* **2022**, *55*, 2493–2502. [CrossRef]
13. Liu, J.T.; Ge, H.K.; Zhang, Z.X.; Wang, X.; Wang, J. Influence of mechanical contrast between the matrix and gravel on fracture propagation of glutenite. *J. Pet. Sci. Eng.* **2022**, *208*, 109639. [CrossRef]
14. Han, X.H.; Zhang, H.; Guo, J.X.; Li, J.; Fang, T. Controlling factors on elastic wave velocities of conglomerate—Experimental and theoretical study. *J. Pet. Sci. Eng.* **2022**, *209*, 109837. [CrossRef]
15. Liu, J.T.; Ge, H.K.; Mou, S.B.; Wang, X.; Wang, J. Characterization of meso-structure of glutenite reservoirs by ultrasonic characteristics and the velocity heterogeneity. *J. Pet. Sci. Eng.* **2022**, *208*, 109436. [CrossRef]
16. Li, N.; Zhang, S.C.; Ma, X.F.; Zou, Y.S.; Chen, M.; Li, S.H.; Zhang, Y.N. Experimental study on the propagation mechanism of hydraulic fracture in glutenite formations. *Chin. J. Rock Mech. Eng.* **2017**, *36*, 2383–2392.
17. Liu, P.; Ju, Y.; Ranjith, P.G.; Zheng, Z.; Chen, J. Experimental investigation of the effects of heterogeneity and geostress difference on the 3D growth and distribution of hydrofracturing cracks in unconventional reservoir rocks. *J. Nat. Gas Sci. Eng.* **2016**, *35*, 541–554. [CrossRef]
18. Ma, X.F.; Zou, Y.S.; Li, N.; Chen, M.; Zhang, Y.; Liu, Z. Experimental study on the mechanism of hydraulic fracture growth in a glutenite reservoir. *J. Struct. Geol.* **2017**, *97*, 37–47. [CrossRef]
19. Chen, C.F.; Wang, J.; Yu, T.X.; Li, Y.; Zou, Y.S.; Ma, X.F.; Liu, L. Proppant Migration Law in Fractures of Conglomerate Reservoirs of Wuerhe Formation in Mahu Sag. *Xinjiang Pet. Geol.* **2021**, *42*, 559–564.
20. Gao, Y.; Zhang, W.; Qin, J.H.; Liang, L.X.; Zhang, J. Numerical Simulation of the Effect of Gravel on the Mechanical Characteristics of Sand Conglomerate. *Sci. Technol. Eng.* **2021**, *21*, 1745–1751.
21. Zhang, W.; Gao, Y.; Liang, L.X. Rock mechanics characteristics of conglomerate reservoir and its effects on fracturing treatment. *Fault-Block Oil Gas Field* **2021**, *28*, 541–545.
22. He, X.D.; Ma, J.X.; Liu, G.; Shi, S.Z.; You, J.; Wu, J.T. Analysis of Rock Mechanics and Assessments of Hydraulic Fracture Network in Conglomerate Reservoirs of Mahu Oilfield. *Xinjiang Pet. Geol.* **2019**, *40*, 701–707.
23. Rui, Z.H.; Guo, T.K.; Feng, Q.; Qu, Z.; Qi, N.; Gong, F. Influence of gravel on the propagation pattern of hydraulic fracture in the glutenite reservoir. *J. Pet. Sci. Eng.* **2018**, *165*, 627–639. [CrossRef]
24. Wang, S.; Qin, J.H.; Yang, X.P.; Li, X.S.; Zhang, Y.; Wang, Y.W. Stress Simulation of Vertical Hydraulic Fracture Propagation Mechanism in Tight Conglomerate Reservoirs of Mahu Area. *Xinjiang Pet. Geol.* **2020**, *41*, 193–198.
25. Yu, Z.C.; Wang, Z.Z.; Jiang, Q.P.; Wang, J.; Feng, Y.; Zheng, J.; Baffour, B.A. Evaluation of low permeability conglomerate reservoirs based on petrophysical facies: A case study from the Triassic Baikouquan Formation, northern Mahu Sag, Junggar Basin, China. *J. Pet. Sci. Eng.* **2022**, *219*, 111082. [CrossRef]
26. Zou, Y.S.; Shi, S.Z.; Zhang, S.C.; Yu, T.; Tian, G.; Ma, X.; Zhang, Z. Experimental modeling of sanding fracturing and conductivity of propped fractures in conglomerate: A case study of tight conglomerate of Mahu sag in Junggar Basin, NW China. *Pet. Explor. Dev.* **2021**, *48*, 1383–1392. [CrossRef]
27. Li, G.; Shu, L.; Jian, X.; Li, Z.; Wang, Z.H. Numerical Simulation for Mechanical Properties of Conglomerate Based on Irregular Particles. *Sci. Technol. Eng.* **2022**, *22*, 8660–8665.
28. Luo, P.; Li, Y.M.; Jiang, Y.S.; Li, Y. Research on micro-morphology characteristics of hydraulic fractures for conglomerate reservoir. *Pet. Geol. Recovery Effic.* **2013**, *20*, 103–106.
29. Li, H.; Yang, C.H.; Ma, H.L.; Shi, X.; Zhang, H.; Dong, Z. A 3D grain-based creep model (3D-GBCM) for simulating long-term mechanical characteristic of rock salt. *J. Pet. Sci. Eng.* **2020**, *185*, 106672. [CrossRef]
30. Li, L.C.; Li, G.; Meng, Q.M.; Wang, H.; Wang, Z. Numerical simulation of propagation of hydraulic fractures in glutenite formation. *Rock Soil Mech.* **2013**, *34*, 1501–1507.
31. Yu, D.H.; Xu, K.T.; Che, H.; Zhang, D.W.; Liu, G.H.; Ma, X.F. Numerical simulation on hydraulic fracture propagation in glutenite reservoir based on microscopic damage multiphase coupling. *Oil Drill. Prod. Technol.* **2016**, *38*, 352–358.
32. Chan, H.-K. Densest columnar structures of hard spheres from sequential deposition. *Phys. Rev. E* **2011**, *84*, 050302. [CrossRef] [PubMed]
33. Mughal, A.; Chan, H.K.; Weaire, D. Phyllotactic Description of Hard Sphere Packing in Cylindrical Channels. *Phys. Rev. Lett.* **2011**, *106*, 115704. [CrossRef] [PubMed]
34. Pickett, G.T.; Gross, M.; Okuyama, H. Spontaneous Chirality in Simple Systems. *Phys. Rev. Lett.* **2000**, *85*, 3652–3655. [CrossRef]

Journal of
*Marine Science
and Engineering*

MDPI

Article

Sequence Stratigraphy, Sedimentology, and Reservoir Characteristics of the Middle Cretaceous Mishrif Formation, South Iraq

Leifu Zhang, Wenqi Zhang *, Yong Li, Benbiao Song, Dawang Liu, Ya Deng, Jiacheng Xu and Yuning Wang

Research Institute of Petroleum Exploration and Development, China National Petroleum Corporation (CNPC), Beijing 100083, China; zhangleifu@petrochina.com.cn (L.Z.); songbenbiao@petrochina.com.cn (B.S.); liudawang@petrochina.com.cn (D.L.); dengya@petrochina.com.cn (Y.D.); xujiacheng@petrochina.com.cn (J.X.); wangyuning@petrochina.com.cn (Y.W.)
* Correspondence: zhangwenqi1982@petrochina.com.cn

Abstract: The Cenomanian–Early Turonian Mishrif Formation is a great contributor to oil production in Iraq. Integrating petrographic, mineralogical, and wireline logging data from 52 wells, this study provides an improved understanding of the sequence stratigraphy, depositional evolution, and reservoir characteristics of the Mishrif Formation in the Mesopotamian Basin, south Iraq. Five types of facies associations are classified: lagoon, shoal, rudist bioherm, shallow marine, and deep marine. Such a classification allows convenient differentiation and interpretation of wireline logs. A sequence stratigraphic framework including five third-order sequences (Mhf 1 to Mhf 5) for the Mishrif Formation is established mainly using wireline logging data of close-distance wells, with the aid of cores and thin sections. Two end-member depositional evolution stages are recognized, from clinoform-like progradational shoal complexes in Mhf 1 within a shallow marine environment, to tidal channels in Mhf 2–3 within a lagoon environment. For Mhf 4–5, abrupt changes in facies associations from north to south indicate the development of an intra-shelf basin where organic-rich mudstones directly overlie the shallow marine grainstone shoals and lagoonal wackestones. Reservoir characteristics and compartmentalization are directly controlled by the sequence stratigraphic framework. Sequence boundaries are featured by wackestones and mudstones overprinted by cementation; they are regionally correlatable and work as regional barriers. Shoal complexes in Mhf 1 and tidal channels in Mhf 2–3 are the main reservoir units. Mudstones and wackestones are intra-reservoir baffles and become more frequently developed towards the south, reflecting the increasing water depth towards south. The characterization of the tidal channels, clinoform-like shoals, and intrashelf basinal deposits in the current study could benefit later development of the Mishrif Formation.

Keywords: Mishrif Formation; Mesopotamian Basin; sequence stratigraphy; reservoir potential; intra-shelf basin; tidal channel

Citation: Zhang, L.; Zhang, W.; Li, Y.; Song, B.; Liu, D.; Deng, Y.; Xu, J.; Wang, Y. Sequence Stratigraphy, Sedimentology, and Reservoir Characteristics of the Middle Cretaceous Mishrif Formation, South Iraq. *J. Mar. Sci. Eng.* **2023**, *11*, 1255. https://doi.org/10.3390/jmse11061255

Academic Editor: George Kontakiotis

Received: 11 May 2023
Revised: 29 May 2023
Accepted: 30 May 2023
Published: 20 June 2023

1. Introduction

Carbonate reservoirs in the Middle East hold considerable oil and gas reserves of the world. In Iraq, the Mishrif Formation of the Cenomanian–Early Turonian in the Mesopotamian Basin contains more than 30% of oil reserves [1], and is regarded as the most important reservoir in several world-class oilfields in southern Iraq, such as the West Qurna and Rumaila oilfields. Characterized by a general shallowing-upward pattern, the Mishrif Formation represents a stable deposition transitioned from an extensive shallow-marine carbonate platform into a lagoonal and supra-tidal environment [2,3]. The deposits are mainly bioclastic limestones, with local build-ups of algae and corals, and rudist bioherms during a ca. 3 million year period [4,5].

Facies associations, reservoir potentials, and depositional settings of the Mishrif Formation have been widely studied [3,6–9]. In addition, equivalent formations to the Mishrif

Formation elsewhere in the Middle East, including the Sarvak Formation of Iran and the Natih Formation of Oman, are also widespread and oil-proliferous in the Arabian Plate. The sequence stratigraphy and sedimentology of these equivalent formations are studied in detail, as is their great hydrocarbon potential [10,11]. However, using different datasets, different depositional models are proposed. While agreeing with the widespread development of rudist buildups, the depositional environment of the Mishrif Formation is still in debate (e.g., ramp, reef, and slope margin) [1,3,12–15].

To date, there are a large number of studies on the sequence stratigraphic frameworks of the Mishrif Formation, many of them focusing on one or a few fields. However, different sequence stratigraphic frameworks are proposed for the Mishrif Formation and its equivalent formations. Early studies with limited data tend to divide the Mishrif Formation into two third-order sequences, separated by an intra-formational unconformity recorded at Amara [15,16]. Using more well data, Refs. [3,8] proposed a scheme that classified the Mishrif Formation into three third-order sequences (S1, S2, and S3). Maximum Flooding Surface (MFS) within S2 and S3 is equivalent to the Middle Cenomanian MFS-130 proposed by [17] and the Late Cenomanian MFS-135 proposed by [1]. Ref. [6] integrated the core and well data of the H field in south Iraq, and classified the Mishrif Formation into six third-order sequences (SB-A to SB-F). For neighboring fields such as the West Qurna and Rumaila oilfields, three and five third-order sequences are proposed for the Mishrif Formation, respectively [18,19]. Different stratigraphy schemes result in uncertainties in terms of regional correlation and comparison of depositional units, also impeding reservoir characterizations and efficient oil and gas development.

Meanwhile, the Mishrif Formation is highly heterogeneous [7,18,20] and depositional facies can change rapidly both vertically and horizontally. For low-gradient carbonate systems such as the Mishrif Formation, a tubular geometry and "layer-cake" horizontal correlation is often applied [10]. For systems analogous to the Mishrif Formation, recent studies employing high-resolution outcrop characterization and seismic data revealed the existence of prograding clinoforms; in addition, that organic-rich intrashelf basins were formed at the time of the Mishrif deposition [10,21–24]. A diagenesis overprint can introduce additional heterogeneity. For example, vugular features can have a strong impact on porosity–permeability characteristics and the flow behavior of carbonate rocks. To accurately represent the flow behavior of vuggy carbonate rocks is essential yet challenging [25,26]. Meteoric, burial, and tectonic-related cementation could significantly decrease the reservoir quality [8,27]. The heterogeneity of the Mishrif Formation in southern Iraq can also be demonstrated by the different production behaviors in neighboring oilfields, for instance, the West Qurna and Rumaila oilfields.

The objectives of this work, therefore, are to (1) investigate the depositional evolution and facies association of the Mishrif Formation in southern Iraq; (2) establish a sequence stratigraphic framework for the Mishrif Formation that allows for regional correlation; (3) unravel the internal heterogeneity such as distribution of clinoforms, tidal channels, and intrashelf basin deposits, as well as reservoir architectures for exploration and production purposes.

2. Geological Setting

The study area in southern Iraq is located in the southern Mesopotamian Basin (Figure 1A), which is an asymmetric foredeep extending from northern Iraq to the Arabian Gulf [1]). The northeastern margin of the basin is bounded by the Zagros Mountains and a series of folded zones, the western margin is bounded by the Mesozoic stable shelf [1]. From the Jurassic to late Cretaceous, along the passive margin of the Neo-Tethys Ocean and the Arabian Plate, the Mesopotamian Basin is characterized by development of a series of shallow intra-shelf basins, for example the Najaf Intra-Shelf Basin/Rumaila Intra-Shelf Basin [28]. In the late Cretaceous, the collision of the Arabian plate and Eurasia Plate resulted in the development of northwest–southeast trending foredeep. Since the late Paleogene, renewed continental convergence formed the Zagros Orogeny characterized by a series of compressional folds and faults in the Zagros fold belt, also resulting in the closure of the Neo-Tethys Ocean [28]. By contrast, the study area is less affected by the compression, but a series of north–south trend folds were developed, perhaps caused by salt diapirism [1,3]. Two oilfields in the study area reflect a present day long-axis north–south trend asymmetrical anticline, with the western flank slightly deeper than the eastern flank and a regional saddle separates the oilfield into north and south regions [2].

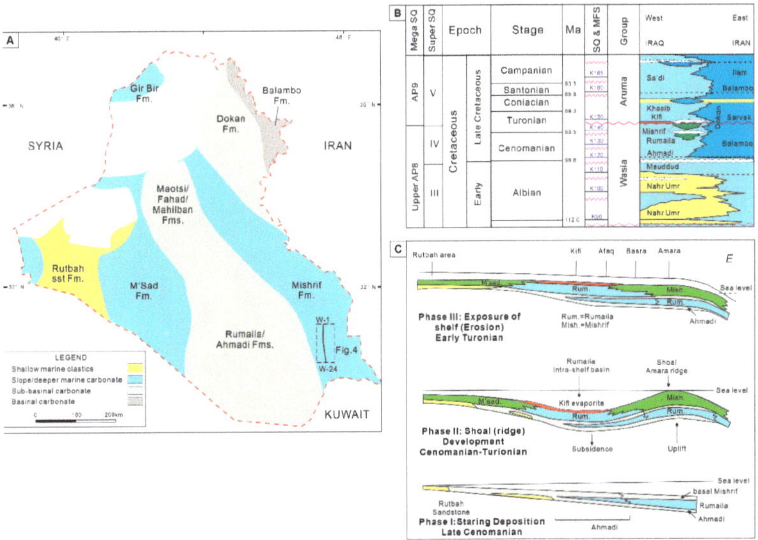

Figure 1. (**A**) Latest Albian–Early Turonian paleogeography map of Iraq, modified from [1]; (**B**) Chronostratigraphy of Iraq and Iran, modified from [1], mega-sequence and super-sequence are from [17]); (**C**) Schematic section showing geological evolution of the Cenomanian–Early Turonian of south Iraq, modified from [1,29].

The Mishrif Formation, firstly described from the Zubair-3 Well in south Iraq, is regarded as being developed in Mid-Cenomanian to Early Turonian based on paleontological studies [3]. It was classified into Supersequence IV within the Megasequence AP8 by [1,17], representing a transition from onlap margin to basin (Figure 1B). The formation is 350 to 400 m thick in the Amara/Buzurgan/Majnoon area and thins towards the southwest, being less than 10 m thick at the Safawi-1 well near the border to Saudi Arabia [1,15].

The evolution of the Mishrif Formation can be divided into three major phases (Figure 1C, [1,29]). The first phase is the start of the deposition: the Rutbah Sandstone in the west and the Ahmadi shales in the east were firstly developed as an east-deepening monocline. Then, shelf carbonates of the M'sad Formation in the west and the Ahmadi–Rumaila formations were developed. Overlying the Rumaila Formation, deposition of the basal Mishrif Formation was initiated. The second phase is the growing shoal/ridges: the Rumaila Formation graded into the Mishrif Formation towards the east and north, while the Mishrif Formation underwent fast deposition due to a combination of sea level fall and tectonic uplift (Figure 1C). Consequently, shoals and ridges were developed and gradually graded to M'sad Formation in the west and Dokan and Balambo Formation in the northeast. Meanwhile, the Rumaila Intra-shelf Basin was initiated, depositing lime mudstones and organic-rich shales of the Rumaila Formation during highstand, and evaporite of the Kifl Formation during lowstand, respectively. The third phase is the exposure of the shelf (erosion): due to the quiescence of tectonic subsidence and uplift as well as sea level fall, the Mishrif Formation underwent subaerial exposure, and graded westwards into shelf carbonates of the Rumaila Formation, evaporite of the Kifl Formation, and the shallow marine carbonates of the M'sad Formation. The termination of the Mishrif deposition associated with subaerial exposure was accompanied by a regional unconformity. The top of the Mishrif Formation can be well differentiated from the overlying

Khasib Formation shales resulting from a rapid flooding (Figure 1B, Figure 2). The Mishrif Formation is time-equivalent and correlatable with the Sarvak Formation of Iran and the Natih Formation of Oman, which have been studied in detail [10,30].

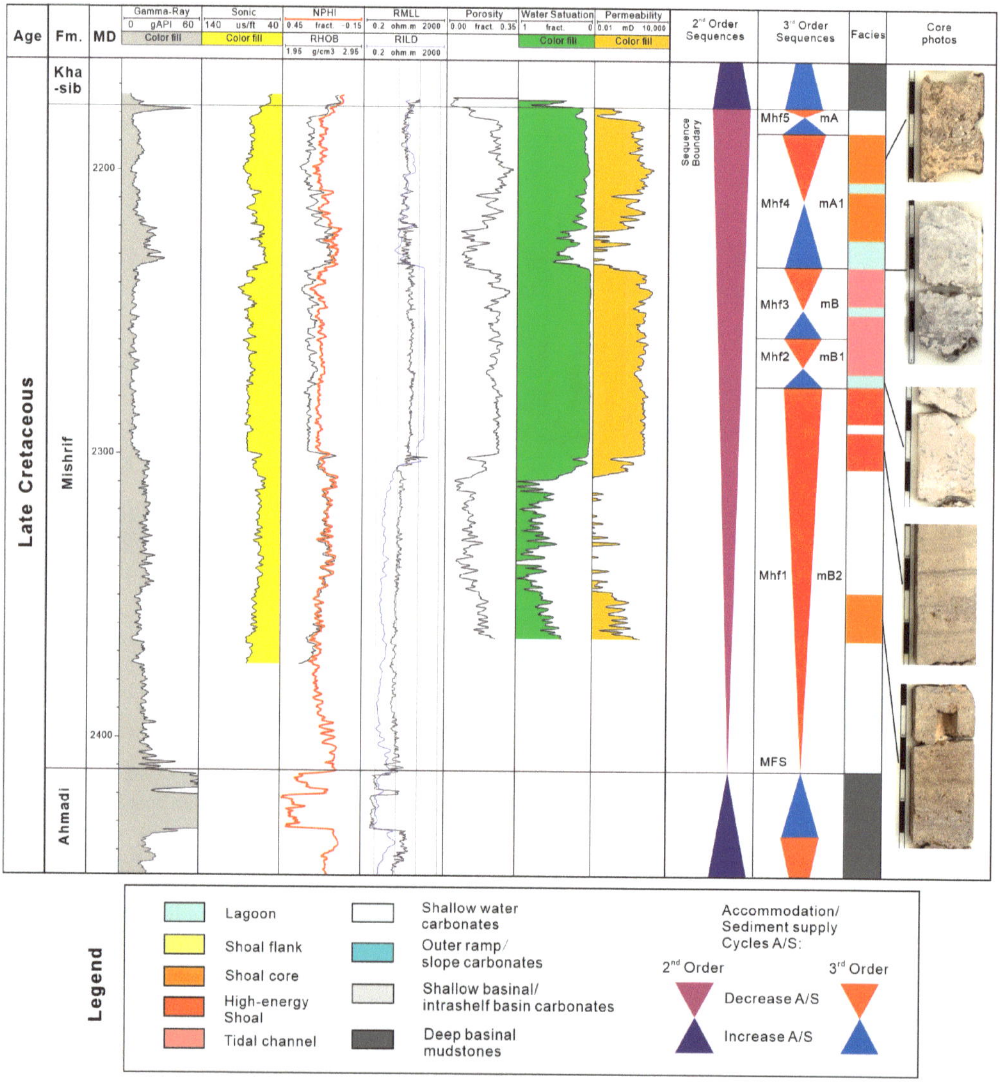

Figure 2. Composite log display of a typical cored well W19, with interpretation of sedimentary facies and sequence stratigraphy.

3. Materials and Methods

More than 2000 thin sections, 200 m thick continuous cores, and wireline logs from more than 50 wells were used in this study. Thin section samples were analyzed using a polarizing microscope, many of them were stained using Alizarin Red S to differentiate the dolomite from calcium carbonate. Petrographic characterizations follow the classification schemes of [31], based on the depositional textures. Sequence stratigraphy framework relies on the identification of the maximum flooding surface (MFS), and transgressive–regressive sequences in the Mishrif Formation were interpreted following the methods and terminology of [32]. They reflect the change in accommodation and sediment supply ratio (A/S ratio). The regional sequence stratigraphy framework for the Arabian Plate provided by [4,17] was also used to support the current correlations.

Lithologies and biofacies generally have a good match to wireline log data of wells, allowing log-based definition of sequence stratigraphic framework. In this study, logs of Gamma-ray (GR), acoustic (DT), thermal neutron porosity (NPHI), bulk density (RHOB), and resistivity (e.g., ILD) were utilized. They are the critical data which support the interpretation of facies association, sequence stratigraphy, and depositional evolution. Along the north–south direction, 52 wells are carefully correlated for this purpose (e.g., Figure 2). Some of the wells are so close and bear similar logging characteristics; thus, only a few representative wells are chosen for the purpose of better showing the internal reservoir characteristics. Consequently, a cross-section comprising 24 wells (W1 to W24 in Figure 1) is established and presented.

4. Results

Sedimentological characterizations based on the integration of wireline logs, cores, and thin sections from 52 wells have led to the identification of 5 facies associations: lagoon, shoal, rudist biostrome, shallow marine, and deep marine. Interpretations of microfacies is beyond the scope of this study, since the main objective is to understand the variations in facies associations under a large-scale sequence stratigraphic framework. Typical characteristics of each facies associations are explained as follows.

4.1. Reservoir Sedimentological Characteristics

4.1.1. Lagoon

Mud-rich wackestones, and rare packstones with abundant benthic foraminifera, ostracods, and gastropods are representative products of this facies association. Miliolids are the most common benthic foraminifera (Figure 3a), whereas rudist debris is scarce. Common bioturbations can be observed on cores (Figure 3a), the abundance of miliolids, muddy texture, and low-diversity of faunas in the thin sections suggest a low energy lagoonal environment. In terms of wireline loggings, high gamma-ray reading and low porosity and sonic motifs (Figure 2) are characteristic of lagoonal facies, yet the gamma ray responses are not as high as the deep marine facies, suggesting the lagoon facies are perhaps not as intensively restricted. The frequent existence of tidal channels that entered the lagoon from the marine-side and deposited high-energy grainstones also demonstrates that the lagoons are not highly restricted. Landwards of the lagoon, tidal flat facies are represented by wackestones, lime mudstones with rare biocomponents. Brecciation and dissolution features observed on cores suggest local exposure, which is characterized by an abrupt peak of gamma ray and sonic logs (Figures 2 and 3g). Tidal flat facies are generally associated with lagoon facies and limited to the top of the Mishrif Formation; therefore, they are not specifically discussed and are incorporated into lagoonal facies.

Lagoon facies association is mainly developed at the top of the Mishrif Formation. From north to south, the distribution of this facies association is less abundant (Figure 4). Tidal channels are frequently observed in the northern area, whereas they abruptly diminish towards the south (near W18 in Figure 4). This perhaps marks the transition of the lagoon environment to shallow marine facies.

4.1.2. Shallow Marine

Massive, thick, fine-sized wackestones to packstones with a variety of bioclasts including rudest, green algae, echinoderms, bivalves, and foraminifera (Figure 3b) represent the shallow marine facies. Shallow marine in this study is the setting between fair-weather wave base and storm wave base. Hence, the mid-ramp, outer ramp, platform slope facies used elsewhere [7] are also considered as a part of shallow marine facies association. Leaching and dissolution are rarely observed for this facies association. Shallow marine facies are predominant in the study area especially during early deposition of the Mishrif Formation. In vertical profile, it works as "background" facies encasing several individual shoal bodies.

Laterally, it is rudist-rich and grain-rich in the north yet becomes wackestone-dominant towards the south (Figure 4).

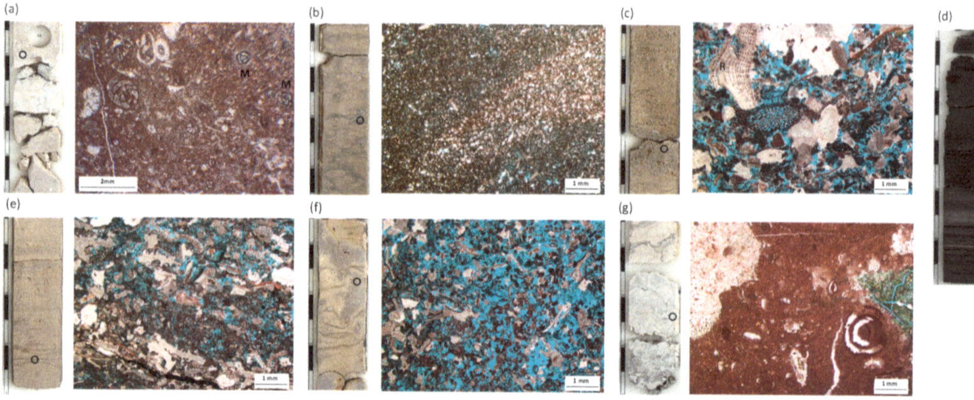

Figure 3. Core photographs and photomicrographs of some of the facies associations of the Mishrif Formation. (**a**) Lagoon facies, W19 Well, 2227.84 m, within Mhf 3; photomicrograph shows typical benthic foraminifera miliolid (M); (**b**) Shallow marine/mid-ramp facies, core shows abundant bioturbation and bivalves fragments and debris, W19 Well, 2303.24 m, within Mhf1; photomicrograph show equant calcite cementation; (**c**) Rudist biostrome facies, W19 Well, 2289.43 m, within Mhf1, core and photomicrograph both show rudist fragments (R); (**d**) Deep-marine facies black shales with good horizontal lamination, W18 Well, 2299.5 m, suggesting restricted, low-energy environment; (**e**) Shoal facies packstone–grainstone, R1 Well, 2292.86 m, within Mhf1, medium-coarse abraded rudist bioclasts, common leaching, and local fracturing; (**f**) Back-shoal facies packstone–grainstone, W19 Well, 2271.83 m, within Mhf2, boring traces on cores and abundant peloids, recrystallized rudist and mollusk bioclasts on photomicrographs; (**g**) Characteristics of the sequence boundary (top Mhf3, mB), W19 Well, 2222.05 m.

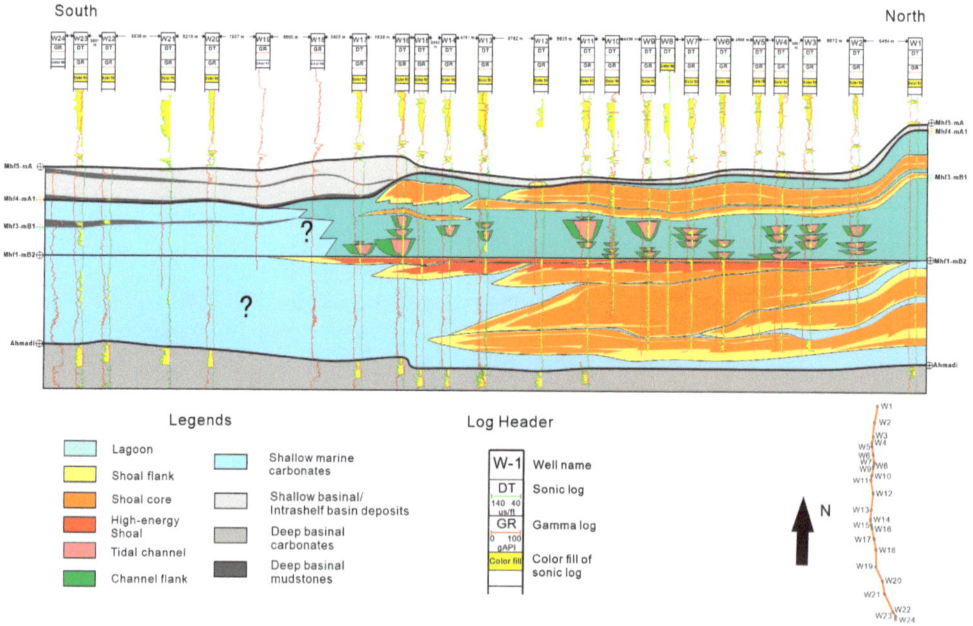

Figure 4. North–south correlation of 24 wells with detailed interpretation of facies associations within a sequence stratigraphic framework. Correlation line as shown in Figure 1A.

4.1.3. Shoal

Grainstones and skeletal–peloidal packstones represent the deposits of shoal facies. On cores, thin bedded laminations can be observed with abundant bioclast and bioturbation; in addition, leaching and local fracturing appear (Figure 3e). Medium-coarse abraded rudist bioclasts, peloids, and skeletal grains of echinoderms, and gastropods are key biocomponents (Figure 3e). The depositional environment of shoals is characterized by relatively shallow and moderate- to high-energy conditions; thus, local shoal build-ups could be re-worked by occasional storms or tidal activities. The co-occurrence of leaching and dissolution, micritization and bioturbation, suggest that shoal facies are developed in a variable environment. Meanwhile, shoal facies are considered as being deposited above a fair-weather wave base, the content of mud is therefore very low, resulting in low-value and blocky gamma-ray logs (Figure 2). In contrast, high, blocky sonic logs suggest good reservoir quality.

Shoals are developed in both the basal and upper part of the Mishrif Formation. At least six rudist-bearing shoals are depicted in the basal part, with thickness ranging from ~5 m to ~40 m, displaying a prograding clinoform feature (Figure 4). Four shoals are depicted at the upper part. The prograding feature is not obvious; one shoal body near W16 is relatively mound-like and has a limited lateral extent (~200 m). They are likely associated with the development of intra-shelf basins.

Behind shoals or rudist build-ups towards the lagoon, back shoal facies are developed and characterized by wackestones and packstones, suggesting a low-moderate environment separating high-energy shoals and low-energy lagoons. Chondrodonta, bivalves, foraminifera, and local rudist fragments are the main biocomponents (Figure 3f). Back-shoal facies are generally associated with shoals; however, it is not easy to differentiate each other on well logs. They are, therefore, not independently discussed in this study. Instead, shoal flank and shoal core are applied, with the latter having higher porosity readings.

4.1.4. Rudist Biostrome

Radiolitid rudist shells with fragments built up an in-situ framework, which can trap sediment and angular bioclasts such as fragmented rudist shells in its interspaces, forming the rudist biostrome in a very shallow, high-energy setting (Figure 3c). Those rudist fragments are generally broken by high-energy water, leaving a poorly sorted texture. Adjacent to shoals, towards the shallow sea, rudist biostrome is locally developed but not as extensive as those rudist-rich build-ups in slope and shoal locations (Mahdi et al., 2013). It is noteworthy that it is difficult to interpret and map the pure rudist biostrome based on well logging data, an accurate identification requires high-resolution three-dimensional seismic characterizations. Because of the erosion and reworking, true rudistid reefs (framestones, floatstones, and boundstones) are rarely encountered [9] and might be re-deposited as "tabular reef" [33].

4.1.5. Deep Marine

This facies association is represented by black shales (Figure 3d). Apparent horizontal laminations suggest a suspension-fallout in a quiet, restricted deep-sea environment without influence of clastic influx or carbonate production. In this study, such black shales only occur at the top of the Mishrif Formation, characterized by abnormally high gamma ray peaks. This shale interval is equivalent with the basinal Khasib Shale as described in [3,8], overlying the Khasib-Mishrif Unconformity. In the southern part of the study area, similar high gamma peaks appear in the upper part of the Mishrif Formation and are considered as sequence boundaries.

The Ahmadi shales and Khasib shales are regionally comparable; they are key surfaces in terms of sequence stratigraphy and hence defined by regional studies such as MFS K120 and K140, respectively [1,17]. Meanwhile, they provide important regional seals for the Mishrif Formation and the Mauddud Formation [34]. The Ahmadi shales and its coeval Shilaif Formation, have decent total organic carbon (TOC) and are an important source of rocks [34,35]. The Khasib shales are perhaps deposited in a restricted environment and grade into marls and marly limestone upwards [34]. Such shales have similar characteristics to shales described elsewhere [36,37]; they perhaps have unconventional reservoir potential but are beyond the scope of the current work.

4.2. Sequence Stratigraphy

The Cenomanian–Turonian Mishrif Formation in Southern Iraq belongs to the upper part of the Arabian Plate megasequence AP8 and supersequence IV as proposed by [1,17]. The unconformity at the top Mishrif Formation forms the boundary of megasequence AP8 and AP7. K140 is the maximum flooding surface (MFS) beneath this unconformity and K120 is the MFS above the Ahmadi Formation [17]. Supersequence IV can be classified into several smaller-order cycles, yet such classifications appear to be a case-by-case scenario varying by different study area and different authors. A chronostratigraphy was established for the study area based on foraminifera, and several different stratigraphic frameworks exist [3,6,8,38].

In this study, five third-order sequences are interpreted for the Mishrif Formation; each third-order sequence is composed of a transgressive and a regressive cycle with the maximum flooding surface (MFS) as the boundary, reflecting the change in accommodation and sediment supply ratio (A/S ratio). From base to top, five third-order sequences are named Mhf 1, Mhf 2, Mhf 3, Mhf 4, and Mhf 5. Within each third-order sequence, the regressive cycle is generally larger than the transgressive cycle. The exception is Mhf 1, only its regressive cycle is within the Mishrif Formation whereas its transgressive cycle is equivalent to the underlying Rumaila Formation (Figure 2) and hence beyond the scope of this study. A high-resolution regional correlation has been carried out in this work based on the integration of well, cores, and petrophysical data, allowing for investigation of the depositional evolution and reservoir architecture.

4.2.1. Sequence Mhf 1

This is the lowermost sequence interpreted in the Mishrif Formation within the study area. The Mhf 1 sequence comprises the lowermost part of the Mishrif Formation and a part of the Rumaila–Ahmadi Formation. Only the regressive cycle of Mhf 1 appears in the Mishrif Formation, the transgressive cycle lies within the underlying Rumaila Formation to Ahmadi Formation [1]. The Rumaila Formation is characterized by “oligosteginal”/sub-basinal carbonates and gradually grades into the underlying Ahmadi Formation; however, the boundary between the Mishrif Formation and the Rumaila Formation cannot be accurately determined in this study. The base of the Mhf 1 is picked at the Ahmadi Formation featured by a transition to a high, blocky gamma-ray interval (Figure 2). The top of Mhf 1 is consistent with the “Disconformity” defined by [3,8]. Additionally, the top Mhf 1 is also consistent with the top of mB2, the lower part of mB, which is widely used by the petroleum industry [33].

The Mhf 1 is becoming thinner from north to south, with an abrupt change in the middle region. Sonic logs (DT) can adequately reflect porosity and reservoir potential, together with gamma-ray logs (GR), could help predict the reservoir distribution. In the north, a majority of the Mhf 1 is stacked shoal complexes with high DT values and low GR values (Figures 2 and 4). By contrast, in the south, few to no shoals appear, while shallow marine facies are predominating. Although a few high-DT value intervals occur locally, the high, serrated to block GR values suggest the shallow marine environment. A low value, blocky motif of GR logged in the north gradually changes to higher value, serrated-blocky motifs in the south, suggesting the water column is thickening towards the south.

A high-resolution well correlation (Figure 4) depicts that at least six rudist-bearing shoals are developed in the Mhf 1. They are likely extensions of the five build-ups interpreted by [33] in this interval. The stacking of grainstone shoals display a north to south progradation pattern, perhaps clinoforms pinching out within a shallow-marine environment [8]. These clinoforms may be comparable with the time-equivalent intervals elsewhere in the Arabian Plate, such as the Savark Formation and the Natih Formation [22,38]. A thin (c.a. 5 m thick) grainstone shoal at the uppermost part of the Mhf 1, featured with low GR and high porosity and sonic values, represents the most favorable reservoir. The top of the Mhf 1 sequence is characterized by subaerial exposure in the north and hardground in the south. Due to exposure, vuggy pores suggesting later dissolution further improve the reservoir quality, making the topmost shoal the highest permeable zones in the Mhf 1.

4.2.2. Sequence Mhf 2

The depositional environment changed drastically after Mhf 1. Overlying the shallow marine-dominated Mhf 1, the Mhf 2 sequence is dominated by peri-tidal lagoonal deposits of wackestones with abundant bioturbation, suggesting a significant sea-level fall compared with Mhf 1. Within the lagoonal background, tidal channels are frequently developed, featured with a channeling-upward pattern (Figure 4): gamma-ray logs display an upward-decrease motif, whereas sonic logs display an upward-increase pattern. This channel-upward pattern is the critical means of interpreting tidal channels in this study since no seismic data are available. Once a tidal channel is interpreted in a certain well, it is carefully correlated with its adjacent wells. Well correlations demonstrate that tidal channels generally cannot extend laterally more than 500 m, which in turn proves that these are meandering tidal channels, rather than the grainstone shoals interpreted by [3,8]. The north–south trend cross-section (Figure 4) is therefore likely perpendicular with the paleoflow direction of the tidal channels. Skeletal shoals or mounds might develop locally (e.g., W14), judging from the occasional shoaling-upward pattern based on logs.

Tidal channels form an important reservoir unit in lagoon-dominated Mhf 2. Well correlation displays a poor inter-channel connectivity. Instead, they form local high-permeability zones and are of great importance in terms of oil development [39]. Towards the south, the Mhf 2 maintains an overall stable thickness but lagoonal facies gradually grade into shallow marine facies/possibly mid-outer ramp facies. This deepening-south trend maintains in the Mhf 3 sequence.

4.2.3. Sequence Mhf 3

The deposition of peritidal lagoonal wackestones and tidal channel grainstones in Mhf 3 demonstrates a similar environment with that of Mhf 2. On the correlation panel based on well logs, tidal channels, featured with a typical channeling-upward/ thinning-upward pattern, are frequently observed within Mhf 3 (Figure 4). Unlike the stable depositional environment of the Mhf 1 in the entire study area, the paleo-environment of Mhf 2 and Mhf 3 clearly changes from the northern to southern region. A small-scale shoal/skeletal build-up is interpreted within the regressive cycle, at the top of Mhf 3, at the middle region of the study area (well 11–12), marking the transition of lagoonal facies to shoal/back-shoal facies, and to mid-outer ramp facies within deeper water depth (Figure 4).

The boundary of Mhf 2 and Mhf 3 appears subtle in the north; however, it becomes apparent by the occurrence of organic-rich mudstones with abnormally high gamma-ray peaks. The top of Mhf 3 is characterized by a series of exposure surfaces in the northern region (Figure 2), supported by the occurrence of microbialite, iron staining in thin sections, and exposure surfaces with karstification features, such as karst pipes in cores (Figure 3g). Additionally, this sequence boundary is highly cemented giving rise to a regional cap-rock.

4.2.4. Sequence Mhf 4

After the exposure of the Mhf 3 sequence boundary, deposition of thick shoals and bioclastic build-ups suggest a flooding and paleo-environment changed from tidal lagoon to shallow marine facies again. In the northern region of the study area, at least two shoal complexes are recognized, featured with upward-increasing high sonic values and low GR values on logs, as well as cross-laminated grainstones with rudist fragments and echinoderms on cores (Figure 2). An abrupt transition occurs at the middle of the study area near well 12 (Figure 4), where large-scale shoals shift to two narrow but thick build-ups, and to shallow marine carbonates to the south (Figure 4). In the southern region of the study area, the transgressive cycle of Mhf 4 is characterized by rapid flooding culminating in the deposition of organic-rich shales, the abrupt change from north to south perhaps marks a transition from a shallower inner ramp, via a rimmed margin to a deeper intra-platform basin.

4.2.5. Sequence Mhf 5

The Mhf 5 sequence is the last third-order sequence within the Mishrif Formation. The top of the Mhf 5 sequence is beneath the unconformity of the top Mishrif Formation (Khasib-Mishrif Unconformity in [8]). Above the unconformity, the Khasib deepwater shales are deposited.

This sequence is generally less than 10m thick in the north of the study area, characterized by low DT values and high GR values. In the northern region of the study area, the transgressive cycle is represented by a gradual increase in GR logs. Towards the southern region, this sequence becomes much thicker (>40 m), the transgressive cycle is reflected by several high GR peaks together with low sonic signatures within this sequence (Figure 4). Organic-rich wackestones and mudstones are deposited, perhaps resulting from deposition within the intra-shelf basin (Rumaila Basin or Najaf Basin, Figure 1B). The regressive cycle is accompanied by a few local build-ups in the north and a decrease in GR values in the south.

A positive carbon excursion is observed in this interval, suggesting that it is a part of the Oceanic Anoxic Event OAE2. The MFS within this sequence is likely consistent with K140 proposed by [4].

5. Discussion

5.1. Depositional Model

The Mishrif Formation in the Arabian Plate has been widely studied, and different depositional models have been proposed. Ref. [12] focused on the importance of rudist build-ups and classified the Mishrif Formation into sub-basinal slope margin, shallow outer reef margin, rudist reef, back-reef, and off-reef. The low-relief feature of the depositional environment was highlighted and was considered similar to ramp. Refs. [1,15] further determined the overall depositional environment as ramp. A shallow marine platform interior and a deep marine mid-outer ramp was identified. Microfacies include shoal, rudist biostrome, lagoon, tidal, mid-ramp, outer-ramp, and deep marine. Refs. [3,8] applied the concept of fair-weather wave base and storm wave base, and used them as criteria to determine the overall shallow open marine environment, rather than ramp. Towards the deep ocean, seven principal facies associations were sequentially recognized, including tidal flat, lagoon, back-shoal, shoal, rudist biostrome, shallow open marine, and deep marine [1,3,8]. In addition, the Natih Formation in Oman, time-equivalent to the Mishrif Formation, was interpreted as inner-ramp, mid-ramp, outer-ramp, and basin interior facies [24,30]. Despite differences in microfacies classification, published studies tend to agree that the Mishrif Formation was deposited within a shallow, low-relief (flat) carbonate platform primarily controlled by relative sea level change.

In this study, wireline logs are the key to unraveling the depositional evolution. Nevertheless, wireline log interpretations are somewhat subjective and, in some cases, cannot differentiate certain microfacies. Therefore, the overall depositional evolution within the entire Mishrif Formation is the focus, rather than the detailed characterization of microfacies. Despite an overall shallowing-upward trend for the Mishrif Formation being commonly agreed [3,8,15], pronounced facies changes resulting from changes in depositional environment are observed in this work. Herein, the well correlation suggests two shallowing phases within an overall shallowing-upward cycle. The first shallowing starts from the basal deep-marine Mhf 1, to shallow marine Mhf 2, to tidal-lagoon facies-dominated Mhf 3. Then, a short-term shallow marine flooding occurs in Mhf 4 but becomes shallower rapidly for the Mhf 5, where subaerial exposure and tidal flat-terrestrial facies occurs. Two end-members represented by Mhf 1 and Mhf 3, respectively, are highlighted according to log-based well correlation (Figure 5): shallow open marine with platform-top grainstone shoals, and shallow open marine with lagoonal tidal channels.

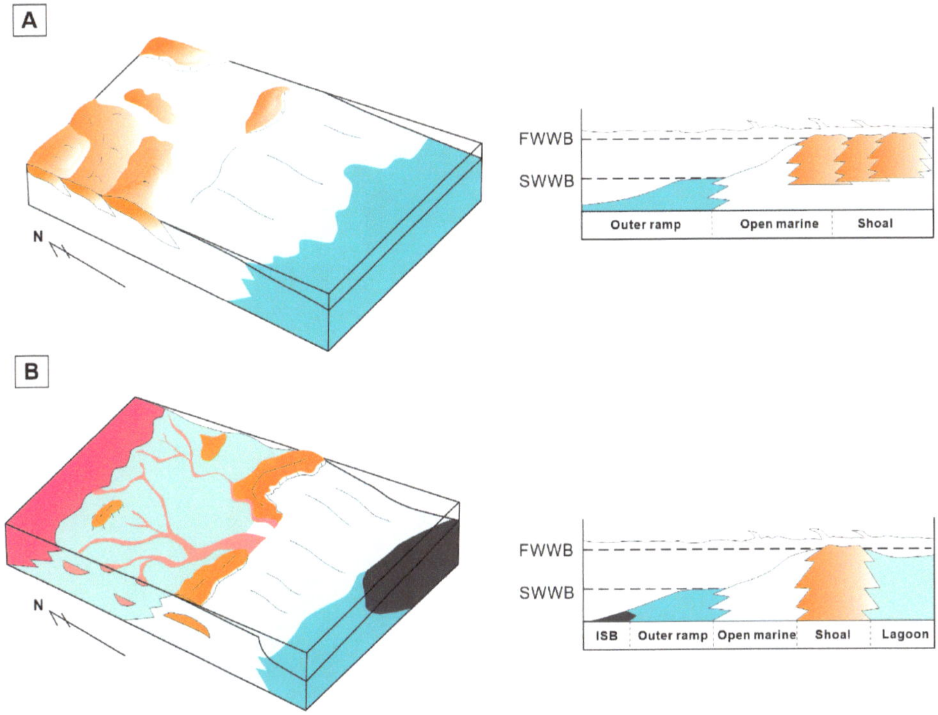

Figure 5. Schematic depositional model for two end-members. (**A**) Shallow open marine with grainstone shoals in Mhf 1; (**B**) Shallow open marine with lagoonal tidal channels in Mhf 3. ISB = Intrashelf basin.

5.1.1. Clinoform-Like Shoal Complex

The overall thickness of the regressive cycle of the Mhf 1 is relatively stable (~100 m, Figure 2). From north to south, however, the sedimentary facies change drastically. Shoals prograde from north to south, a clear progradation pattern appears on the north–south well correlation panel. From the top of Ahmadi Formation to the top of the Mhf 1 sequence boundary, careful correlation using the top of Mhf 1 as datum reveals the stacking of six shoal complexes (Figure 4). The stacking of these shoals displays an apparent prograding pattern, typical characteristic of clinoforms (Figures 5 and 6). Carbonate clinoforms are discovered based on seismic data and the outcrop of Cretaceous epeiric carbonate platform in the Middle East. Examples include outcrop characterization and modeling of Natih E Member in Oman [21], and high-resolution seismic subsurface studies of UAE and Oman [10,22]; they revealed complicated stratigraphic architectures of prograding clinoforms delineating platform margins and stated that clinoform geobodies are more complicated than a simple layer-cake model. Complex clinoform models must be considered in both static and dynamic reservoir modelling workflows so that proper rock properties are reproduced [21]. At the basal part of the Mishrif Formation in south Iraq, clinoform-like shoal complexes are recognized only on the basis of close-distance wells in this study (Figures 4 and 6); hence, unravel the detailed internal stacking characteristics within the fore-reef or shallow marine environments discussed elsewhere [3,15].

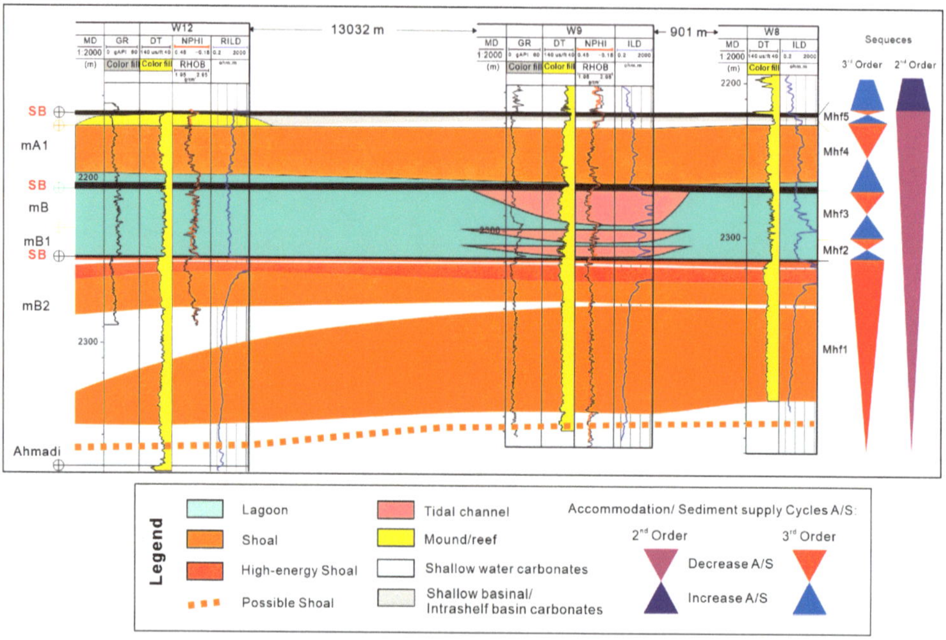

Figure 6. Correlation of W8, W9, and W12 showing prograding clinoform-like shoal complexes in the sequence Mhf 1, tidal channels in Mhf 2 and Mhf 3, isolated shoal complex in Mhf 4, and local mound/reef in Mhf 5. Sequence boundary (SB) is highlighted in black.

Assuming that sonic logs positively correlate with reservoir quality, the closer to the Mhf 1 top surface, the thinner the shoal complex, and the better the reservoir quality. The mechanism for this is perhaps an integrated control of sedimentology and diagenesis [40,41]. First, the continuous sea level fall suggests higher hydrodynamic conditions so that finer grains are elutriated. Second, subaerial exposure of the sequence boundary (Mhf 1 top) made the topmost shoal complex subject to meteoric water dissolution, creating vugs and pipes which improve porosity and permeability. The water depth increases towards the south, and sedimentation is dominated by shallow open marine environment, resulting in deposition of wackestones–packstones. Gamma-ray (GR) values are generally higher in the south (e.g., well 12–15, Figure 4), possibly suggesting a higher proportion of mud and the presence of intercalating mudstone layers; however, further core data are required to assist such an explanation.

The shallowing-upward is depicted by the occurrence of lagoonal facies in Mhf 2 and Mhf 3, suggesting a significant sea-level fall. Large-scale prograding shoal complexes are not developed there, only local rudist/reefal build-ups are developed at the platform margin. Towards the south, lagoonal facies grades into shallow marine facies, with the development of local shoal/build-ups at the margin. At Mhf 3 top, shale beds are developed and become thicker towards the south (Figure 4, Figure 6), reflected by a gamma-ray peak. By contrast, neither shale beds nor local grain-stone build-ups are interpreted on the Mhf 2 top. This implies a possible abrupt lagoon-shallow marine break was developed during the end phase of a regressive cycle of Mhf 2, perhaps related to tectonic movement as Amara paleohigh [1,15]. Tidal channels are frequently developed within the lagoonal facies, based on the typical "channel-upward" motifs on gamma ray and sonic logs.

5.1.2. Tidal Channel

A unique depositional unit in this study is the tidal channels developed within the tidal-lagoon environment. It is of great reservoir potential since grainstones in such channels act as high permeability zones in the upper reservoir unit (Mhf 2–3). A rare high-resolution outcrop characterization of the Natih Formation in Oman suggests that the outcropped tidal channels are generally 0.5–1 km wide, 2–5 m deep, and less than 30 km long (Figure 7A, [30]). They are asymmetric in cross-section and appear as channel belts in the same way as those in siliciclastic depositional systems (e.g., fluvial channel, distributary channel). Channel bases are highly erosive with abundant burrows and few diagenetic modifications. High-energy rudstones, grainstones, and packstones fill the channels interbedding with finer grained carbonates or mudstones. Tidal channels of the Natih Formation outcrops are considered as resulting from tidal backfilling and accretion in tidal creek depositional systems during the transgressive phase [30]. In addition, the three-dimensional seismic azimuth map of the top Natih-e seismic reflector also clearly depicts the channel incisions and channel belt geometry in plane-view with a dimension of more than 1km wide and 30 km long (Figure 7B, [22]). A modern carbonate tidal channel analog of Joulters shoal north of Andros Island on the Great Bahama Bank displays the channel-belt geometry in plane-view (Figure 7C, [42]). For each channel, a main trunk channel is associated with several bifurcating smaller channels, no wash-over fans exit on the seaward side or the lagoonal side. A total of 43 tidal channels are recognized, the mean channel width is c.a. 0.83 km, and the mean channel length is c.a. 5.33 km (calculated based on total channel length of 229 km).

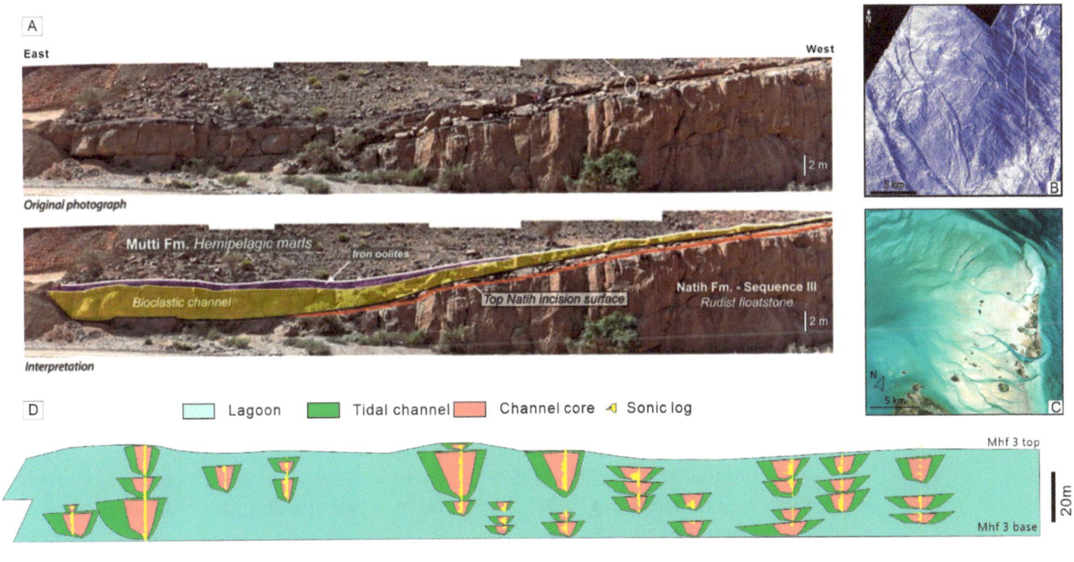

Figure 7. Characteristics of tidal channels. (**A**) Channels at the top of the Natih Formation showing erosion into rudist floatstone, modified from [10]; (**B**) Azimuth map of the top Natih-e seismic reflector showing geometries and incisions of tidal channels, modified from [22]; (**C**) Satellite image of modern analog tidal channels in Great Bahama Bank, taken from [42]; (**D**) Tidal channels interpreted in this study within Mhf 2 to Mhf 3.

In the present study, the tidal channel is identified based on cores and well logs. Within the entire Mishrif Formation, tidal channels are the only depositional unit characterized with apparent cross-lamination (Figures 2 and 8). Wavy, planer muddy laminations appear within grainstone–packstones and faint herringbone bi-directional cross-lamination could be identified in cores. Grains are generally well-sorted, with the appearance of re-worked millimeter-scale rudistid fragments (Figures 2 and 3), suggesting a high-energy condition. On a vertical profile, a "channeling-upward" (Figure 8) pattern is featured by, from base to top, (1) an irregular, erosive base with burrows; (2) grainstone–packstone with abundant bioclastic debris with intercalated thin wackestone–mudstones; (3) a plane, finer grained (wackestone–mudstone) top draping. This pattern is also captured on wireline logs, with a blocky-funnel upward-increase in gamma rays, blocky upward-decrease in sonic logs; the reservoir potential of tidal channels is well reflected by resistivity logs (Figures 4, 6 and 8).

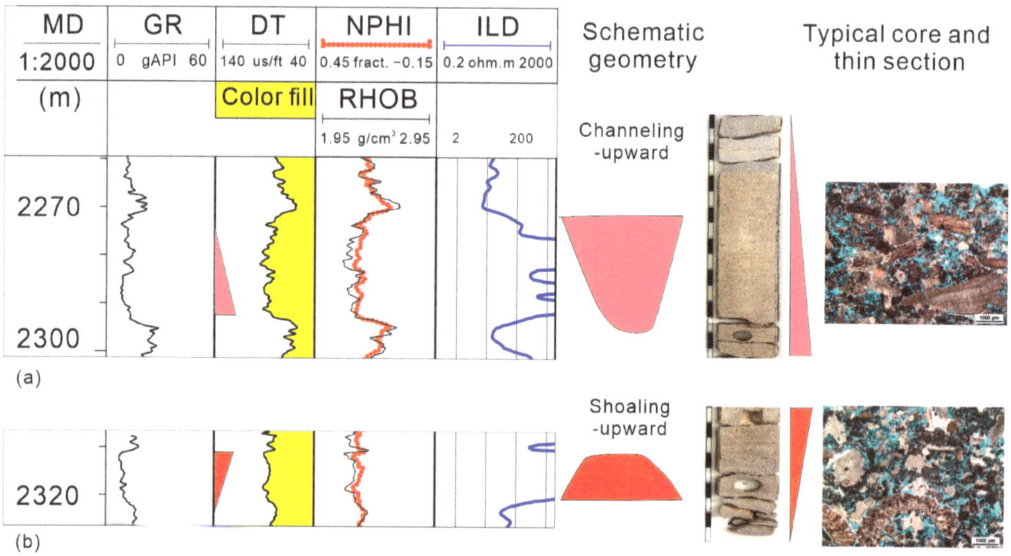

Figure 8. Features of (**a**) channeling-upward pattern and (**b**) shoaling-upward pattern on wireline logs, cores, and thin sections.

It is noteworthy that within a general channel-upward pattern, thin fine-grained intervals with less reservoir potential also occur, separating several thinner channeling-upward cycles. This might be interpreted as hierarchically smaller-scale units. In addition, a proper characterization of the internal channel architecture determines the remaining oil distribution and thus impacts the injection plan [10,22,43]. However, meter-scale heterogeneity is beyond the scope of this study and will be analyzed given more available cores and three-dimensional seismic data.

In the Arabian Plate, lithologies and biofacies generally match well to log data [3]; therefore, the log-based correlation undertaken in this study (Figures 4, 6 and 7D) allows a semi-quantitative characterization of geometrical parameters of tidal channels. In particular, tidal channel depth can be best identified by a "channeling-upward"/decrease-upward pattern based on the sonic logs. Widths of tidal channels cannot be easily measured and thus are not used in this study; three-dimensional seismic surveys are required to do so. In addition, the extension direction of tidal channels is supposed to be southwest–northeast, causing additional difficulty in the accurate measurement of tidal width, given the cross-section is in the north–south direction. A compilation of the 32 tidal channels (Figure 7D) in the study area suggest that minimum depth, maximum depth, and mean depth of tidal channel are 3.27, 22.67, and 8.93, respectively (Figure 9). The mean depth is similar to that of the Natih Formation outcrops as described by [30], with most of the channels having a depth c.a. 5 m. It should be noted that a few tidal channels are deeper than 20 m, this is perhaps due to an amalgamation of multiple smaller tidal channels, as demonstrated by the serrated motif of sonic logs, suggesting high-energy conditions of the tidal channels.

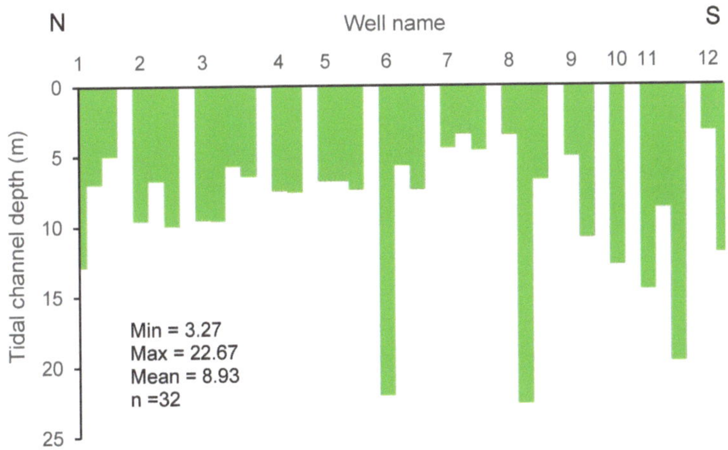

Figure 9. Compilation of tidal channel depth in the study area. X axis is the 12 wells encountering tidal channels from north to south.

In plane-view, tidal channels within carbonate-dominated ramps display a channelized feature similar to those fluvial meandering channels, as demonstrated by the modern sedimentology of the Great Bahama Bank "https://www.beg.utexas.edu/lmod/_IOL-CM01/cm01-step02b.htm (accessed on 1 May 2023)", courtesy of University of Texas at Austin, and the outcrop characterization of carbonate ramps in the Arabian Plate [30]. However, the plane-view of modern analogs and three-dimensional seismic slices both suggest that carbonate tidal channels appear relatively straight and less sinuous than meandering channels of siliciclastic systems (e.g., fluvial, delta).

5.2. Intra-Shelf Basin

Deposition of the Cenomanian–Turonian carbonates has been considered to occur on an extensive platform across the Arabian Plate, associated with an intra-shelf basin (the Najaf basin, or Rumaila basin, by [1]). High-resolution outcrop studies and seismic characterizations both proved that clinoform geobodies prograded towards the intra-shelf basin during relative sea level drops on the Arabian Plate [10,22,24,38]. The Najaf intra-shelf basin started to develop in the Early Cenomanian, southwest of the study area, partly controlled by faulting in the East Baghdad area [1]. Correspondingly, the Mishrif Formation is thinning towards the intra-shelf basin towards the southwest. During the late stage of the Mishrif deposition, the region between the Rumaila field and Zubair field was dominated by restricted lagoonal facies [33], and sub-basinal facies were developed west of the Rumaila field.

The correlation of wells at the middle region of the study area shows that in the sequence Mhf 4, bioclastic build-ups/shoal facies within lagoonal background gradually transits into mid-ramp of shallow marine facies (Figure 10A). Sub-basinal shales started to appear on top of the Mhf 4, between W16 and W18, suggesting the first occurrence of intra-shelf basin deposition (Figure 4). Consequently, it leads to a specific juxtaposition of shoal, lagoon, and sub-basinal shales (Figure 10A) with lagoonal-shoal facies directly grading into intra-shelfal facies, demonstrating an abrupt topographical break. Such an interpretation in this study has been confirmed by high-resolution outcrop characterizations of the Savark Formation (time-equivalent with the Mishrif Formation) in Iran (Figure 10B) by [11,30]. The layered organic-rich mudstones within the intra-shelf basin differentiate themselves from the underlying wackestones of lagoonal facies and the overlying ~80 m thick grainstone shoal (Figure 10B).

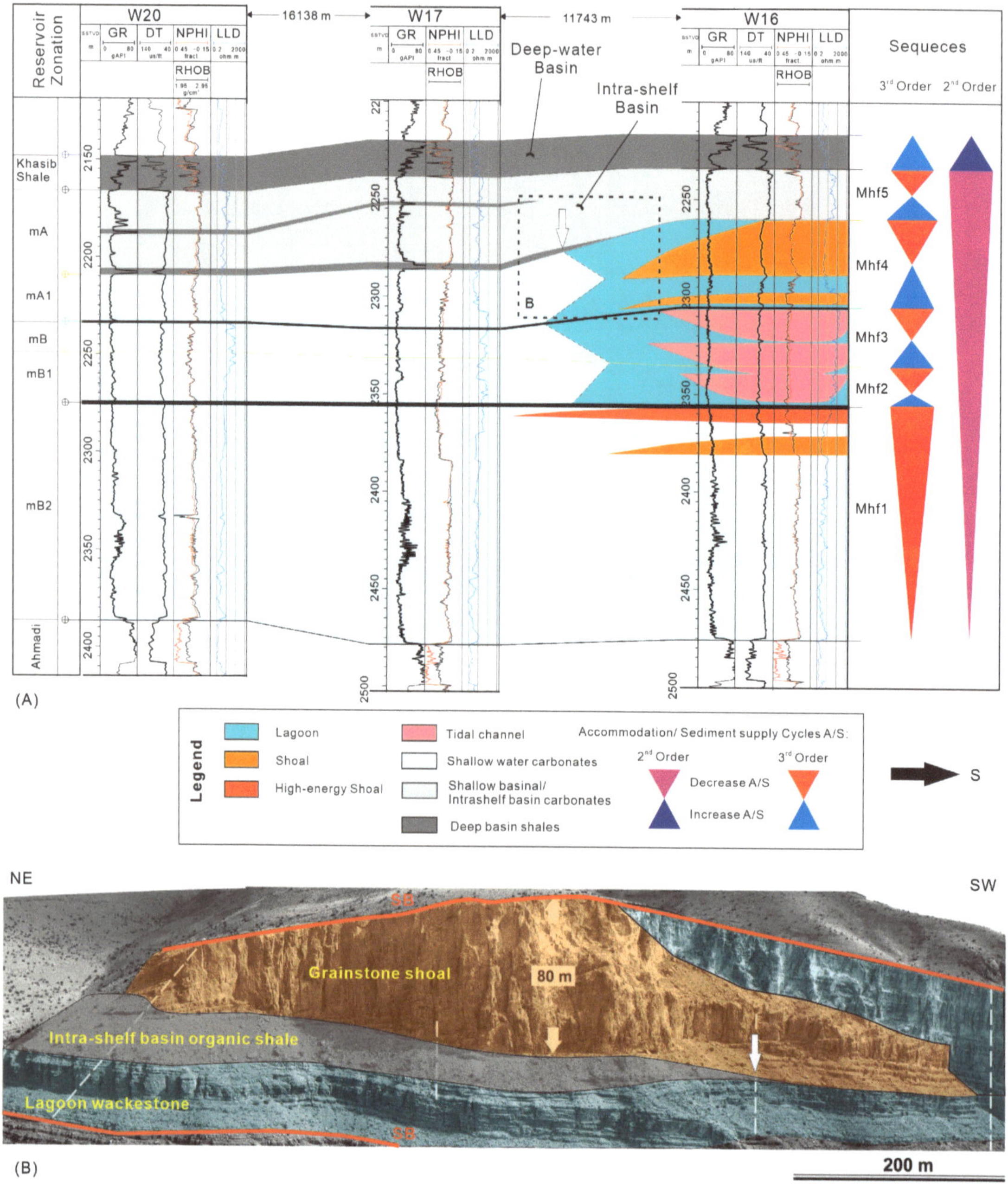

Figure 10. (**A**) Correlation of W16, W17, and W20 showing the stacking characteristics of lagoon, shoal, shallow-marine, and intra-shelf basin facies; see Figure 4 for detailed location of wells; (**B**) Outcrop of the Savark Formation in Iran showing similar stacking of grainstone shoal, lagoon wackestone, and intra-platform basin organic shales, modified from [30].

The development of this intra-shelf basin started in the upper part of the Mishrif Formation (Mhf 4 and Mhf 5) and significantly changed their depositional pattern. The intra-shelf basin changed the paleo-bathymetry such that the increase in accommodation space exceeded carbonate production. Consequently, organic-rich shales were deposited, perhaps consistent with the global anoxic event OAE2 (Oceanic Anoxic Event) at the Cenomanian–Turonian boundary [4], which is time-equivalent with the top Savark Formation within Zagros area [30] and the Natih Formation in Oman [10]. By contrast, the high-energy bioclastic grainstone shoal continued to develop in the platform northwards. When carbonate production exceeds the increasing rate of accommodation space, carbonate clinoforms are estimated to develop along the paleogeography between the platform and the intra-shelf basin. Small-scale bioclastic build-ups are locally developed on the platform margin (Figure 4). The Arabian Plate was characterized by a system whereby carbonate system production generally could not exceed the sea level rise [4,17]. On the flat ramp site, during the transgressive phase, grainstone shoals maintain aggradation, gradually giving rise to a steep topography. The increase in rate of sea-level gradually triggered the change from flat platform to intra-shelf basin via the steepening margin. In fact, sea level change is considered as the main controlling factor for the formation of the ramp–intrashelf depositional system. During different geological periods in the Arabian Plate, other three main intra-shelf basins were developed: the Early Jurassic Marrat intra-shelf basin, the Late Jurassic Hanifa intra-shelf basin, and the Aptian Bab intra-shelf basin; they were all triggered by a similar mechanism [4,17,24].

In the study area, the tectonic movement might also influence the deposition of the Mishrif Formation. A series of north–south trend dome/anticline was formed perhaps due to growing salt structure. Rudistid build-ups aggraded on the crest of dome. High paleotopography could result in erosion and the reworking of the rudistid build-ups which are then transported and dispersed into bioclastic grainstone shoals, similar to "tabular reefs" of several kms wide due to the uplift described by [33]. Such bioclastic build-ups are likely to develop on the flanks of anticlines following the down-dip paleogeography [9] and grade into the intra-shelf basin. In addition, because of the erosion and reworking, true rudistid reefs (framestones, floatstones, and boundstones) are rarely encountered [9].

5.3. Tectonic Influence on Shoal Complexes

Eustatic-level changes clearly exert control on the development of sequence stratigraphy and carbonate facies, as demonstrated earlier. Meanwhile, tectonic movements such as the salt movement and reactivation of basement faults, might happen at the same time with the deposition of rudist build-ups and hence influenced their development and distribution [35]. Being influenced by synchronous basement tectonics, regional studies suggest that the Mishrif Formation has undergone a general change from carbonate ramps in the early Cretaceous into rimmed shelves in middle-late Cretaceous [1,35]. The initiation of the Mishrif Formation was considered as being progradational onto sub-basinal facies towards the northeastern and southwestern direction [1,35]. In the study area, five progradational shoal complexes are interpreted within the Mhf 1 on a north–south cross section which is roughly parallel with the boundary of the Mishrif shallow marine facies and sub-basinal Rumaila/Ahmadi facies (Figure 1A). Consequently, shoals tend to develop in parallel along the platform margins; therefore, this could perhaps interpret why no isolated, localized, small-size build-ups/shoal arc recognized in the cross section. Furthermore, the north–south prograding shoals happen to follow the hinge of a large anticline (Rumaila anticline). Therefore, the influence of the Infracambrian Hormuz Salt and the Samarra-Dujaila-Amara paleohighs also cannot be excluded [3,4]. The growing anticlines are favorable locations for developing rudist build-ups as they kept growing until they reached wave base and underwent erosion and reworking. Reworked rudist fragments were then dispersed into grainstone/packstone and formed "tabular reefs" or banks which are several kilometers wide [9,33]; thus, shoal complexes recognized within the Mhf 1 are likely "tabular reefs". South of the anticline in the study area, shoal complexes grade into deeper marine facies, suggesting the appearance of intra-shelf basin and quiescence of tectonic movements.

5.4. Sequence Stratigraphy within a Regional Framework

There are many previous studies on the sequence stratigraphy of the Mishrif Formation in southern Iraq and division of large-order sequences are widely agreed upon (e.g., megasequence AP8, supersequence IV). Smaller-order sequences, however, may vary depending on different types and availability of data used in their work. The classification of three third-order sequences for the Mishrif Formation has been widely applied regionally [3,24,30]. However, it can vary for different fields, even close-distance fields. For example, three third-order sequences are interpreted for W field [18], five third-order (or equivalent) sequences are interpreted for the R field [19,44], whereas six third-order (or equivalent) sequences are interpreted for the H field [6,18]. For formations equivalent to the Mishrif Formation in the Arabian Plate, different sequence stratigraphic frameworks also exist. For the Savark Formation in Iraq, four sequences, namely Sequence I-IV, are interpreted [30,45]. For the Natih Formation in Oman, five sequences, namely Sequence 1–5, are interpreted [11].

In addition, recognition of maximum flooding surfaces is also subjective according to different authors. MFS K120 and K140 are regionally comparable, represented by Ahmadi shales and Khasib shales, respectively [1,17]. The "disconformity" of the S2 top with significant exposure features by [3] is probably comparable, it is mostly likely the top of Mhf 1 in this study, and it is almost certainly the top of SB-D in the H field [6]. In the study area, K130 lies within Mhf 1, K135 is probably the maximum flooding surface of Mhf 2, and K140 is likely the maximum flooding surface of Mhf 5. Since no biostratigraphic analysis is applied, uncertainties associated with the interpretation of maximum flooding surfaces should be noted. Nevertheless, the stratigraphic framework proposed herein can help unravel reservoir zonation and compartmentalization in the study area, where each sequence is characterized by a specific facies association and is highly recognizable. The sequences are correlatable in the entire study area of several 10 s of km long.

5.5. Reservoir Architecture and Petroleum Potential

Depositional textures of carbonates of the Mishrif Formation have been significantly overprinted by diagenetic processes [3,8,20,39]. Reservoir zonation and heterogeneity are thus an integrated product of deposition and diagenesis. Detailed discussion on the diagenesis is beyond the scope of the current work and will be discussed in another independent study. Within the sequence stratigraphic framework, reservoir zonation is divided into mB2, mB1, mB, mA1, and mA, roughly corresponding to Mhf 1, Mhf 2, Mhf 3, Mhf 4, and Mhf 5, respectively (Figures 2 and 11). Depositional heterogeneities are briefly classified into the following four types.

5.5.1. High-Permeability Zones

Two types of high-permeability zones (permeability of 100 s mD in Figure 11) are recognized: shoal complexes at the top of mB2, and tidal channel within mB1 and mB. The former was developed at the latest regressive stage of the Mhf 1. After their deposition, the sequence boundary of Mhf 1 is accompanied by the first platform exposure of the Mishrif Formation. Diagenesis for this interval is prominent meteoric dissolution, creating dissolution pores and significantly improving reservoir quality. The latter was developed within the Mhf 2 and Mhf 3 (mB 1 and mB). The cross-section shows that grainstone tidal channels are frequently developed but few of them are mutually connected, instead, they form "isolated" high-K zones (Figure 4). Within the Mishrif Formation, these high-permeability zones generally form "thief zones" which represent more rapid water breakthroughs to production wells than predicted by flow simulations. These are of critical importance for reservoir performance; future studies should focus on unraveling their distribution in 3D space.

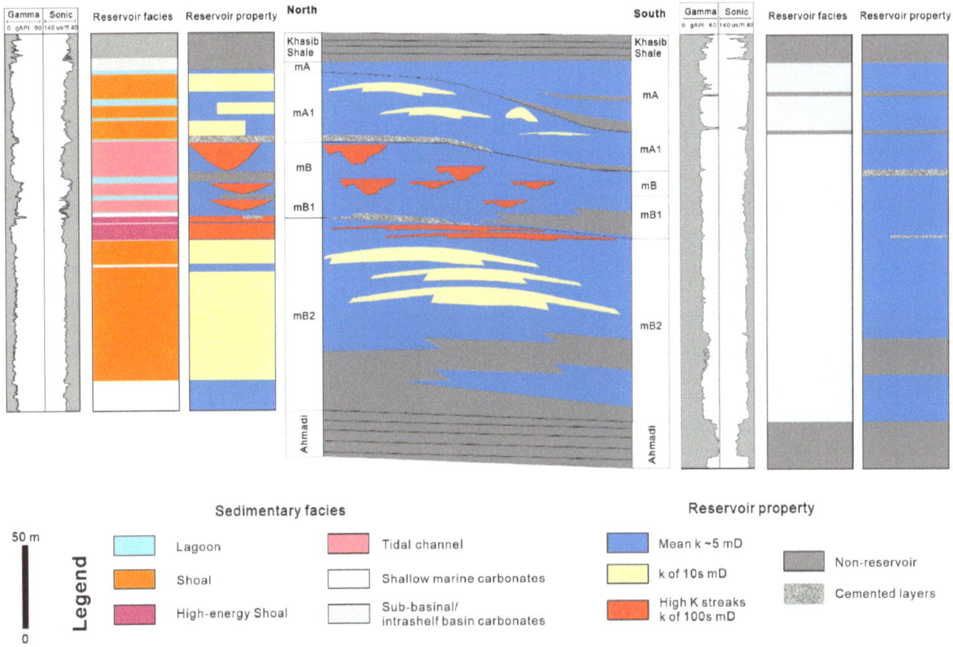

Figure 11. Schematic model showing reservoir zonation of the Mishrif Formation from north to south. Sedimentary facies and reservoir property are highlighted independently.

5.5.2. Medium-Permeability Layers

These layers have permeability of 10 s mD, including prograding shoal complex in mB2, isolated shoals and mounds in mA1. The regressive cycle of the Mhf 1 constitutes several progradational shoal complexes enveloped within a low-K background (Figures 4 and 11), suggesting continuous decrease in A/S ratio. Due to good lateral continuity and connectivity, they also represent decent reservoir potential. Within mA1, grainstones within laterally continuous shoals and more isolated mounds also have reservoir potential.

5.5.3. Intra-Reservoir Baffles

Micro-porous baffles appear within mB2, mB1, and mA (non-reservoir in Figure 11). Baffles in mB2 are likely equivalent to wackestones/marls of the Rumaila Formation, where a single thick layer in the north "splits" into two individual layers southwards. At the basal part of mB1, a thick lagoonal wackestone baffle prominently occurs in the southern region. At the boundary of mA and mB, the sequence boundary of Mhf 4 (mB) was accompanied by heavy cementation in the north and mudstone in the south, reflected by high gamma-ray values. In mA, the thin tidal wackestone layer becomes two thick and laterally extensive sub-basinal mudstones southwards.

5.5.4. Inter-Reservoir/Regional Barriers

The Mishrif Formation was encased by two regional barriers of thick basinal shales with overlying Khasib Formation and underlying Ahmadi Formation; the Khasib shales are seals for the Mishrif Formation. Fabrics of shales might impact the residual oil distribution in the reservoir [35,36], however, it is beyond the scope of the current work.

The study area is characterized by a general deepening-south pattern. Gamma-ray values are generally higher while sonic values are generally lower in south (Figure 11). A generally lower reservoir potential is therefore predicted, despite the development of local high-K build-ups in the south.

6. Conclusions

Five types of facies associations are classified for the Mishrif reservoir in south Iraq: lagoon, shoal, rudist bioherm, shallow marine, and deep marine. Such a classification allows convenient differentiation and interpretation on wireline logs, and could be applied in carbonate reservoir exploration and development for other areas in the Arabian Plate.

The Mishrif reservoir is divided into five third-order sequences from Mhf 1 to Mhf 5. Sea level changes exert clear control on the depositional evolution. When the Mishrif Formation deposition was initiated after transition from Ahmadi shales, a shallow marine environment was prominent. Clinoform-like progradational shoal complexes are identified within the regressive cycle of Mhf 1, in response to continuous sea level fall. Mhf 2 to Mhf 3 are dominated by lagoonal system, encasing tidal channels and subordinate mounds. Sea level rise in Mhf 4 created shallow marine environment allows the final development phase of shoal complexes. Mhf 5 is dominated by marine environment and becomes deeper southwards.

The Rumaila/Najaf intra-shelf basin was initiated in the late stage of Mhf 4. The thickness of Mhf 5 increased significantly towards the south, multiple thick mudstones were developed within the intra-shelf basin, suggesting that the intra-shelf basin was perhaps maturely developed at this stage. Reservoir characteristics of the Mishrif Formation, especially the upper part, were significantly controlled by the development of intra-shelf basin.

The reservoir zonation of the Mishrif Formation is a combined product of deposition and diagenesis. Five reservoir zones including mB2, mB1, mB, mA1, and mA, are recognized, roughly corresponding to five third-order sequences. Two main reservoir units are recognized: shoal complexes at the top of mB2, and tidal channel within mB1 and mB. They are both dominated by grainstones and highly permeable zones, but display different "shoaling-upward" and "channeling-upward" patterns, respectively. The reservoir quality of the former is further enhanced by meteoric dissolution due to subaerial exposure at the sequence boundary. Tidal channels appear "isolated" with limited inter-connectivity, they can form "thief zones" which represent faster water breakthroughs to production wells than predicted by flow simulations and are of critical importance for reservoir performance. High-resolution seismic mapping in the future will help better capture the geometry and distribution of shoal and channel geobodies.

Author Contributions: Conceptualization, W.Z. and Y.L.; methodology, W.Z.; software, D.L., Y.D., J.X., and Y.W.; validation, L.Z.; data curation, Y.W. and B.S.; writing—original draft preparation, L.Z.; writing—review and editing, L.Z. and B.S.; supervision, Y.L.; project administration, B.S.; funding acquisition, B.S. All authors have read and agreed to the published version of the manuscript.

Funding: This research was funded by PETROCHINA, grant number 2017ZX05030-001, 2021DQ0407 and 2021DJ3202.

Institutional Review Board Statement: Not applicable.

Informed Consent Statement: Not applicable.

Data Availability Statement: Not applicable.

Acknowledgments: Jiquan Yin, Mingsheng Feng and Guangya Zhu from RIPED are thanked for their insightful discussions. PETROCHINA is thanked for access to the data and permission to publish.

Conflicts of Interest: The authors declare no conflict of interest.

References

1. Aqrawi, A.; Mahdi, T.; Sherwani, G.; Horbury, A. Characterisation of the Mid-Cretaceous Mishrif reservoir of the southern Mesopotamian Basin, Iraq. In Proceedings of the American Association of Petroleum Geologists Conference and Exhibition, Manama, Bahrain, 7–10 March 2010; pp. 7–10.
2. Al-Mimar, H.S.; Awadh, S.M.; Al-Yaseri, A.A.; Yaseen, Z.M. Sedimentary units-layering system and depositional model of the carbonate Mishrif reservoir in Rumaila oilfield, Southern Iraq. *Model. Earth Syst. Environ.* **2018**, *4*, 1449–1465. [CrossRef]
3. Mahdi, T.A.; Aqrawi, A.A.; Horbury, A.D.; Sherwani, G.H. Sedimentological characterization of the mid-Cretaceous Mishrif reservoir in southern Mesopotamian Basin, Iraq. *Geoarabia* **2013**, *18*, 139–174. [CrossRef]

4. Sharland, P.R.; Casey, D.M.; Davies, R.B.; Simmons, M.D.; Sutcliffe, O.E. Arabian Plate Sequence Stratigraphy—Revisions to SP2. *Geoarabia* **2004**, *9*, 199–214. [CrossRef]
5. Simmons, M.D.; Sharland, P.R.; Casey, D.M.; Davies, R.B.; Sutcliffe, O.E. Arabian Plate sequence stratigraphy: Potential implications for global chronostratigraphy. *Geoarabia* **2007**, *12*, 101–130. [CrossRef]
6. Cantrell, D.; Shah, R.; Ou, J.; Xu, C.; Phillips, C.; Li, X.; Hu, T. Depositional and diagenetic controls on reservoir quality: Example from the upper Cretaceous Mishrif Formation of Iraq. *Mar. Pet. Geol.* **2020**, *118*, 104415. [CrossRef]
7. Ismail, M.J.; Ettensohn, F.R.; Handhal, A.M.; Al-Abadi, A.M. Facies analysis of the Middle Cretaceous Mishrif Formation in southern Iraq borehole image logs and core thin-sections as a tool. *Mar. Pet. Geol.* **2021**, *133*, 105324. [CrossRef]
8. Mahdi, T.; Aqrawi, A. Sequence stratigraphic analysis of the mid-Cretaceous Mishrif Formation, southern Mesopotamian Basin, Iraq. *J. Pet. Geol.* **2014**, *37*, 287–312. [CrossRef]
9. Sadooni, F.N. The nature and origin of Upper Cretaceous basin-margin rudist buildups of the Mesopotamian Basin, southern Iraq, with consideration of possible hydrocarbon stratigraphic entrapment. *Cretac. Res.* **2005**, *26*, 213–224. [CrossRef]
10. Grélaud, C.; Razin, P.; Homewood, P. Channelized systems in an inner carbonate platform setting: Differentiation between incisions and tidal channels (Natih Formation, Late Cretaceous, Oman). *Geol. Soc. Lond. Spéc. Publ.* **2010**, *329*, 163–186. [CrossRef]
11. Razin, P.; Grélaud, C.; van Buchem, F. The mid-Cretaceous Natih Formation in Oman: A model for carbonate platforms and organic-rich intrashelf basins. *AAPG Bull.* **2017**, *101*, 515–522. [CrossRef]
12. Alsharhan, A.S. Facies Variation, Diagenesis, and Exploration Potential of the Cretaceous Rudist-Bearing Carbonates of the Arabian Gulf1. *AAPG Bull.* **1995**, *79*, 531–550.
13. Alsharhan, A.S.; Kendall, C.G.S.C. Paleozoic sequence stratigraphy, depositional systems, and hydrocarbon habitats across the Arabian plate. *AAPG Bull.* **2021**, *105*, 1149–1198. [CrossRef]
14. Mahdi, T.A.; Aqrawi, A.A.M. Role of facies diversity and cyclicity on the reservoir quality of the mid-Cretaceous Mishrif Formation in the southern Mesopotamian Basin, Iraq. *Geol. Soc. Lond. Spéc. Publ.* **2017**, *435*, 85–105. [CrossRef]
15. Aqrawi, A.A.M.; Thehni, G.A.; Sherwani, G.H.; Kareem, B.M.A. Mid-Cretaceous rudist-bearing carbonates of the Mishrif Formation: An important reservoir sequence in the Mesopotamian Basin, Iraq. *J. Pet. Geol.* **1998**, *21*, 57–82. [CrossRef]
16. Reulet, J. Carbonate reservoir in a marine shelf sequence, Mishrif Formation, Cretaceous of the Middle East. In *Exploration for Carbonate Petroleum Reservoirs. Elf Aquitaine*; John Wiley and Sons: New York, NY, USA, 1982; pp. 165–173.
17. Sharland, P.; Archer, R.; Casey, D.; Davies, R.; Hall, S.; Heward, A.; Horbury, A.; Simmons, M. Sequence stratigraphy of the Arabian Plate. *GeoArabia* **2001**, *2*, 1.
18. Deng, H.; Fu, M.; Huang, T.; Gluyas, J.G.; Tong, M.; Wang, X.; Zhou, W.; Liu, F. Ahdeb oil field, Mesopotamian Basin, Iraq: Reservoir architecture and oil charge history. *AAPG Bull.* **2018**, *102*, 2447–2480. [CrossRef]
19. Lehmann, C.; Ryder, K.; Disimone, S.; Perrotta, S. Sedimentology and Sequence Stratigraphy of the Mishrif Formation and it's Impact on Reservoir Heterogeniety, Rumaila Field, Southern Iraq. In Proceedings of the Second EAGE workshop on Iraq, Dead Sea, Jordan, 15–18 September 2013; p. 355. [CrossRef]
20. Zhong, Y.; Zhou, L.; Tan, X.; Guo, R.; Zhao, L.; Li, F.; Jin, Z.; Chen, Y. Characteristics of depositional environment and evolution of Upper Cretaceous Mishrif Formation, Halfaya Oil field, Iraq based on sedimentary microfacies analysis. *J. Afr. Earth Sci.* **2018**, *140*, 151–168. [CrossRef]
21. Adams, E.W.; Grélaud, C.; Pal, M.; Csoma, A.; Al Ja'Aidi, O.S.; Al Hinai, R. Improving reservoir models of Cretaceous carbonates with digital outcrop modelling (Jabal Madmar, Oman): Static modelling and simulating clinoforms. *Pet. Geosci.* **2011**, *17*, 309–332. [CrossRef]
22. Droste, H.; Buchem, F.S.P.V.; Gerdes, K.D.; Esteban, M. High-resolution seismic stratigraphy of the Shu'aiba and Natih formations in the Sultanate of Oman: Implications for Cretaceous epeiric carbonate platform systems. In *Mesozoic and Cenozoic Carbonate Systems of the Mediterranean and the Middle East: Stratigraphic and Diagenetic Reference Models*; Geological Society of London: London, UK, 2010; Volume 329, pp. 145–162.
23. Grélaud, C.; Razin, P.; Homewood, P.W.; Schwab, A.M. Development of Incisions on a Periodically Emergent Carbonate Platform (Natih Formation, Late Cretaceous, Oman). *J. Sediment. Res.* **2006**, *76*, 647–669. [CrossRef]
24. van Buchem, F.S.P.; Razin, P.; Homewood, P.W.; Oterdoom, W.H.; Philip, J. Stratigraphic Organization of Carbonate Ramps and Organic-Rich Intrashelf Basins: Natih Formation (Middle Cretaceous) of Northern Oman. *AAPG Bull.* **2002**, *86*, 21–53.
25. Iraji, S.; Soltanmohammadi, R.; De Almeida, T.R.; Munoz, E.R.; Basso, M.; Vidal, A.C. Laboratory and Numerical Examination of Oil Recovery in Brazilian Pre-salt Analogues Based on CT images. In Proceedings of the Third EAGE Conference on Pre Salt Reservoirs, Rio de Janeiro, Brazil, 23–25 November 2022; Volume 2022, pp. 1–6.
26. Iraji, S.; Soltanmohammadi, R.; Munoz, E.R.; Basso, M.; Vidal, A.C. Core scale investigation of fluid flow in the heterogeneous porous media based on X-ray computed tomography images: Upscaling and history matching approaches. *Geoenergy Sci. Eng.* **2023**, *225*, 211716. [CrossRef]
27. Ahmed, B.; Bakht, S.; Wahid, S.; Hanif, M. Structural Analysis and Reservoir Characterisation of Cretaceous Sequence in Kohala Bala, Khyber Pakhtunkhwa, Pakistan. *Rud. Zb.* **2022**, *37*, 65–81. [CrossRef]
28. Pitman, J.K.; Steinshouer, D.; Lewan, M.D. Petroleum generation and migration in the Mesopotamian Basin and Zagros Fold Belt of Iraq: Results from a basin-modeling study. *Geoarabia* **2004**, *9*, 41–72. [CrossRef]
29. Chatton, M.; Hart, E. *Announcement of a Rock Unit Redefined. Ratawi Formation: Iraq Petroleum Company Unpublished Report*; Oil Exploration Company Archive: Baghdad, Iraq, 1962.

30. Razin, P.; Taati, F.; Van Buchem, F. Sequence stratigraphy of Cenomanian–Turonian carbonate platform margins (Sarvak Formation) in the High Zagros, SW Iran: An outcrop reference model for the Arabian Plate. *Geol. Soc. Lond. Spec. Publ.* **2010**, *329*, 187–218. [CrossRef]
31. Dunham, R.J. *Classification of Carbonate Rocks According to Depositional Textures*; American Association of Petroleum Geologists: Tulsa, OK, USA, 1962; Volume 38, pp. 108–121.
32. Embry, A. Transgressive–regressive (T–R) sequence analysis of the Jurassic succession of the Sverdrup Basin, Canadian Arctic Archipelago. *Can. J. Earth Sci.* **1993**, *30*, 301–320. [CrossRef]
33. Gaddo, J. The Mishrif Formation paleoenvironment in the Rumaila/Tuba/Zubair region of South Iraq. *J. Geol. Soc. Iraq* **1971**, *4*, 1–12.
34. Aqrawi, A.A.; Goff, J.C.; Horbury, A.D.; Sadooni, F.N. *The petroleum Geology of Iraq*; Scientific Press: Beaconsfield, UK, 2010.
35. Sadooni, F.; Aqrawi, A.; Sadooni, F.; Aqrawi, A. Cretaceous Sequence Stratigraphy and Petroleum Potential of the Mesopotamian Basin, Iraq. In *Middle East Models of Jurassic-Cretaceous Carbonate Systems*; Alsharhan, A., Scott, B., Eds.; SEPM Special Publication: Tulsa, OK, USA, 2000.
36. Meng, M.; Ge, H.; Shen, Y.; Ji, W.; Li, Z. Insight into Water Occurrence and Pore Size Distribution by Nuclear Magnetic Resonance in Marine Shale Reservoirs, Southern China. *Energy Fuels* **2022**, *37*, 319–327. [CrossRef]
37. Meng, M.; Peng, J.; Ge, H.; Ji, W.; Li, X.; Wang, Q. Rock Fabric of Lacustrine Shale and Its Influence on Residual Oil Distribution in the Upper Cretaceous Qingshankou Formation, Songliao Basin. *Energy Fuels* **2023**, *37*, 7151–7160. [CrossRef]
38. van Buchem, F.S.P.; Simmons, M.D.; Droste, H.J.; Davies, R.B. Late Aptian to Turonian stratigraphy of the eastern Arabian Plate—Depositional sequences and lithostratigraphic nomenclature. *Pet. Geosci.* **2011**, *17*, 211–222. [CrossRef]
39. Zhang, L.; Li, R.; Deng, H.; Fu, M.; Hu, L.; Guo, X.; Xie, X.; Zhao, C. Identification, distribution characteristics, and effects on production of interlayers in carbonate reservoirs: A case study from the Cretaceous Mishrif Formation in Halfaya Oilfield, Iraq. *J. Pet. Sci. Eng.* **2021**, *202*, 108571. [CrossRef]
40. Hollis, C. Diagenetic controls on reservoir properties of carbonate successions within the Albian–Turonian of the Arabian Plate. *Pet. Geosci.* **2011**, *17*, 223–241. [CrossRef]
41. de Periere, M.D.; Durlet, C.; Vennin, E.; Caline, B.; Boichard, R.; Meyer, A. Influence of a major exposure surface on the development of microporous micritic limestones—Example of the Upper Mishrif Formation (Cenomanian) of the Middle East. *Sediment. Geol.* **2017**, *353*, 96–113. [CrossRef]
42. Purkis, S.J.; Harris, P. Quantitative interrogation of a fossilized carbonate sand body—The Pleistocene Miami oolite of South Florida. *Sedimentology* **2017**, *64*, 1439–1464. [CrossRef]
43. Wei, H.; Zhu, H.; Li, H.; Shi, S.; Gan, S.; Zhai, Z. The Study of the Reservoir Inter-architecture Structure and Remaining Oil in the Eastern Part of Sabei II District. *Rud. Geološko-Naft. Zb.* **2019**, *34*, 13–20. [CrossRef]
44. Al-Ali, M.; Mahdi, M.; Alali, R. Microfacies and depositional environment of Mishrif Formation, north Rumaila Oilfielfd, southern Iraq. *Iraqi Geol. J.* **2019**, *52*, 91–104.
45. Kiani, A.; Saberi, M.H.; ZareNezhad, B.; Mehmandosti, E.A. Reservoir zonation in the framework of sequence stratigraphy: A case study from Sarvak Formation, Abadan Plain, SW Iran. *J. Pet. Sci. Eng.* **2022**, *208*, 109560. [CrossRef]

Journal of
Marine Science and Engineering

MDPI

Article

Effects of Quartz Precipitation on the Abundance and Preservation of Organic Matter Pores in Cambrian Marine Shale in South China

Sile Wei [1,*], Mingyi Hu [1], Sheng He [2], Yanbo Shu [3], Tian Dong [2], Qing He [2], Wei Yang [4] and Quansheng Cai [1]

1 School of Earth Sciences, Yangtze University, Wuhan 430100, China
2 Key Laboratory of Tectonics and Petroleum Resources, Ministry of Education, China University of Geosciences, Wuhan 430074, China; dongtian@cug.edu.cn (T.D.)
3 No. 8 Oil Recovery Plant of PetroChina Changqing Oilfield Company, Xi'An 710021, China; shuyanbo_cq@petrochina.com.cn
4 Guangzhou Marine Geology Survey, China Geological Survey, Ministry of Natural Resources, Guangzhou 510075, China
* Correspondence: weisile@outlook.com

Abstract: To evaluate the effects of quartz precipitation on the abundance and preservation of organic matter pores in marine shale reservoirs, the type of authigenic quartz and the source of silica, as well as the corresponding relation of the Lower Cambrian Shuijingtuo Formation shale in South China were investigated. Quartz in the Shuijingtuo shale occurs as four different types: detrital quartz, replacement of biosiliceous debris, euhedral quartz filled in interparticle pores, and microquartz dispersed in a clay matrix. Euhedral quartz (1–5 μm) and matrix-dispersed microquartz (100–400 nm) are found to be the dominant forms of authigenic quartz. The euhedral quartz accumulates along the interparticle pores, and the porous organic matter fills the interior of the space. Microquartz is mainly wrapped in porous organic matter. Two silica sources were revealed: biogenic silica and clay-derived silica. Biogenic Si is most likely the major source for authigenic quartz in the organic-rich (total organic carbon (TOC) > 2.55 wt.%) samples, which accounts for 23–57 wt.% (average 35 wt.%) of the total Si. Based on petrographic observations, we posit that the precipitation of large-sized euhedral quartz in the interparticle pores most likely originated from biogenic silica in the early stage of diagenesis and that the silica for the clay matrix-dispersed microquartz is provided by biogenic silica and clay-derived silica. The observation of SEM images indicates that the precipitation of early diagenetic euhedral quartz in the interparticle pores enhances rock stiffness, and the buttressing effect can protect the organic matter pores from compaction during the late-stage burial diagenesis. In contrast, the precipitation of late diagenetic microquartz in the clay matrix can lead to a reduction in the capacity of the accommodation space to host retained petroleum, consequently leading to a reduction in the development of organic matter pores and the generation of shale gas.

Keywords: authigenic quartz; silica source; organic matter pores; pore development; marine shale

Citation: Wei, S.; Hu, M.; He, S.; Shu, Y.; Dong, T.; He, Q.; Yang, W.; Cai, Q. Effects of Quartz Precipitation on the Abundance and Preservation of Organic Matter Pores in Cambrian Marine Shale in South China. *J. Mar. Sci. Eng.* **2023**, *11*, 1267. https://doi.org/10.3390/jmse11071267

Academic Editor: Assimina Antonarakou

Received: 18 May 2023
Revised: 4 June 2023
Accepted: 5 June 2023
Published: 21 June 2023

1. Introduction

As an unconventional energy source with huge resource potential, shale gas, which is characterized as self-generating and self-preserving, has been stimulating great exploration in many countries, including America, Canada, China, and others [1]. An organic matter pore, one of the significant pore types in shale reservoirs, serves not only as the primary storage space of adsorbed and free gas, but also as an important micro seepage channel for shale gas [2–4]. Therefore, the abundance and preservation of organic matter pores are of great significance for shale gas exploration and development [2,5–7].

The thermal simulation experiments of sealed gold-tube pyrolysis on shale rocks showed that the generation of a large number of organic matter pores and shale gas is mostly related to the thermal cracking of retained petroleum [8,9]. The higher the amount of retained petroleum or the lower the hydrocarbon expulsion efficiency in shale reservoirs, the more abundantly organic matter pores develop, leading to enrichment of shale gas [10]. For the research of conventional sand reservoirs, petroleum geologists focus on hydrocarbon expulsion efficiency and have carried out significant in-depth research on the controlling factors, such as the organic matter attributes (type, abundance, and thermal maturity), combination structures between the source rock and reservoir rock, and roof/floor sealing conditions of the source rock [7,8,10–13]. However, for unconventional shale reservoirs, the amount of retained petroleum is more significant and plays an important role in shale gas enrichment [8,12]. In addition to the factors that affect the efficiency of hydrocarbon expulsion mentioned above, the changes in pore space before and during the hydrocarbon expulsion period also have an important control effect on the amount of retained petroleum.

Precipitation of authigenic quartz in the shale reservoirs is suggested to rank second to mechanical compaction in affecting the pore space for retained petroleum [14–16]. Nevertheless, research has found that the rigid lattice formed by the self-generated embrittlement of siliceous cement has a strong resistance to compaction [14,17]. Therefore, although siliceous cement occupies some of the primary pores, it holds residual pores, providing favorable space for the migration and filling of residual oil during the oil generation period, and also has a protective effect on the organic matter pores generated by the later thermal cracking of residual oil. However, the siliceous cement observed in shale, especially marine shale, can be of various types and originates from multiple silica sources that formed during different diagenetic periods [18–20]. The complexity of the relationship between the quartz type and the siliceous source brings uncertainties to the assessments of the impacts of siliceous cement on the abundance and preservation of organic matter pores. Therefore, it is necessary to investigate the corresponding relationship between quartz types and silica sources.

The Lower Cambrian Shuijingtuo Formation marine shale in southern China has been recognized as another interval with a tremendous gas exploration potential, after the Upper Ordovician–Lower Silurian Wufeng–Longmaxi Formations shale [21–23]. In this study, a set of marine shale samples collected from the Shuijingtuo shale in the Yichang slope in southern China was investigated to determine the quartz types and silica sources, as well as the corresponding relation between them. More importantly, this study aims to gain insight into the impact of quartz precipitation on the abundance and preservation of organic matter pores.

2. Geological Setting

The Yichang slope zone is located south of the Huangling anticline, the middle of the Yangtze Craton, and is bordered by the Qinling–Dabie orogenic belt to the north and the Xuefeng orogenic belt to the south [24–26] (Figure 1). During the Sinian (Ediacaran) and Cambrian periods, the tectonic movement and sedimentary environment changed drastically in southern China. At the end of the Ediacaran period, the Tongwan orogenic movement formed an alternating structural unit of uplift and depression in the study area [27]. In terms of the sedimentary environment, the transgression from south to north in the Early Cambrian period ended the carbonate deposition of the Late Sinian Dengying Formation and started the deposition of black organic-rich shale in the Early Cambrian Shuijingtuo Formation. The paleogeographical background of uplift and depression restricted the distribution of Early Cambrian sedimentary facies and controlled the thickness of the shale in the Shuijingtuo Formation [22]. The eastern uplift area included shallow-water platform facies with thin shale, and the shale thickness in the local areas was <3 m, similar to that of the YD 3 well (Figure 1). Furthermore, the western depression area mainly belonged to the deep shelf facies, where the shale is thick, such as the 140 m thick shale in the EYY1 well (Figure 1). Overall, the Shuijingtuo shale in the Yichang slope is thick in the west and thin in the east. The lithology column shows that the lower part of the Shuijingtuo shale is dominated by black siliceous organic-rich shale, the middle part is mainly composed of dark gray clay-rich siliceous and calcareous shale, and the upper part mainly consists of gray limestone and dolomite (Figure 1).

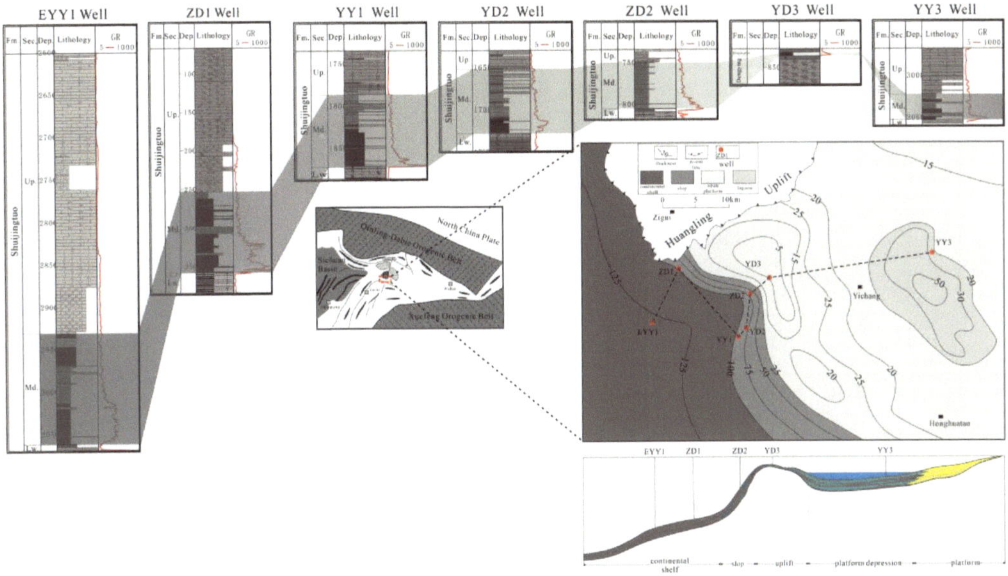

Figure 1. Geological map of the study area and lithology histograms of the Shuijingtuo Formation.

3. Methods

Twenty-nine shale samples were collected from the Shuijingtuo Formation of the EYY1 well in the Yichang slope zone, southern China. All the samples were analyzed for their TOC (total organic carbon), mineral composition, and major elements. Seven shale samples were selected for petrographic observation, including thin section observation, field-emission scanning electron microscopy (FESEM), SEM-based energy dispersive spectroscopy (EDS), and SEM-based cathodoluminescence (CL). The seven samples used for SEM petrographic observations were selected based on their TOC content and mineral compositions, which can essentially cover the overall Shuijingtuo Formation shales.

The TOC content was measured using an Elementar rapid CS cube. The analytical procedure consisted of removing carbonate with 7% HCl, drying at 80 °C for 10 h, and burning with 99.99% oxygen (O2) at 930 °C.

The bulk mineral composition was determined using XRD with an X'Pert PRO diffractometer (Malvern Panalytical B.V., Almelo, The Netherlands). The working voltage, current, radiation, and scanning speed were 40-kV, 40 mA, Cu Kα (λ = 0.15416 nm), and 0.417782° (2θ)/s, respectively, in a continuous mode with a step size of 0.017° in the range of 3–65°. The clay fraction mineralogy was reported by Wei et al. [28].

The major elements, including Si, Al, Fe, Mn, Ti, and Cr, were analyzed using an X-ray fluorescence spectrometer (XRF) (Zsx Primus II, Rigaku, Tokyo, Japan) in Wuhan SampleSolution Analytical Technology Co., Ltd. Powdered samples were dried in an oven for 12 h to remove free and adsorbed water and then heated to determine the loss on ignition. The X-ray tube was a 4.0 kW end window Rh target, and the test conditions comprised a voltage of 50 kV and a current of 60 mA. All the major elemental analysis lines were kα. The data were corrected using the theoretical α coefficient method. The relative standard deviation was <2%.

Thin sections of the shale samples were examined using conventional transmitted polarized light microscopy (Leica DM4500P, Tokyo, Japan). Before SEM imaging, the shale sample was first mechanically cut and the side perpendicular to the bedding plane was polished using Ar ion-beam milling (Leica EM TIC 3X). Subsequently, the samples were inspected using a ZEISS GeminiSEM FESEM instrument coupled with EDS and a cathodoluminescence (CL) detector. SEM imaging, EDS mapping, and CL imaging were performed at an accelerating voltage of 5–10 kV and a working distance of ~5–10 mm.

4. Results

4.1. Bulk Rock Properties

The results of the XRD analysis are depicted in Table 1. The Shuijingtuo shale is dominated by quartz and clay minerals. The quartz content ranges from 14 to 75 wt.%, averaging 53 wt.%. The content of clay minerals ranges from 9 to 53 wt.%, with an average of 20 wt.%. The XRD analysis of the fine powders (<2 μm) shows that the clay minerals are dominated by illite (56–90 wt.%, average 78 wt.%), followed by a mixed layer of illite/smectite (I-S) (10–44 wt.%, average 22 wt.%) [28].

Table 1. Depth, TOC content, mineral compositions, and concentrations of major elements of the Shuijingtuo shale samples.

Depth (m)	Organic	TOC (%)	Quartz (wt.%)	Carbonate (wt.%)	Clay (wt.%)	SiO_2 (wt.%)	Al_2O_3 (wt.%)	Fe_2O_3 (wt.%)	MnO (wt.%)	Zr (ppm)	Al/(Al + Fe + Mn)
2939.08	-lean	1.38	24	16	53	45.21	10.36	3.40	0.03	118	0.70
2975.53	-lean	1.63	28	33	30	31.47	6.21	3.45	0.06	55.6	0.57
2978.08	-lean	1.58	20	44	25	29.35	7.73	4.76	0.07	73.5	0.55
2983.63	-lean	2.47	47	8	30	59.62	12.81	3.32	0.02	122	0.74
2984.98	-lean	2.21	48	8	31	60.80	12.21	3.25	0.02	114	0.74
2989.32	-lean	1.80	34	31	26	40.68	7.00	3.90	0.05	65	0.57
2996.32	-lean	2.27	45	5	34	60.44	13.02	4.85	0.05	119	0.67
3003.50	-rich	2.72	64	5	20	68.67	9.49	2.10	0.02	85.1	0.77
3004.88	-rich	3.08	69	9	12	68.48	7.21	2.27	0.02	60.9	0.70
3007.34	-rich	2.93	58	5	24	65.38	10.71	3.91	0.02	92.9	0.67
3007.75	-rich	3.02	59	5	22	65.46	10.42	2.80	0.03	90.6	0.74
3008.78	-rich	3.47	59	8	21	63.06	9.45	2.75	0.03	83.8	0.72
3011.50	-rich	4.66	62	4	17	65.53	10.11	2.62	0.02	99.3	0.74
3020.54	-rich	4.33	57	8	17	62.42	9.20	3.77	0.02	131	0.65
3023.83	-rich	5.22	62	6	11	63.76	9.40	3.63	0.03	127	0.66
3024.33	-rich	5.24	57	7	20	62.71	9.27	3.55	0.02	123	0.66
3026.38	-rich	6.20	61	8	13	62.86	8.84	2.83	0.03	108	0.70
3030.18	-rich	5.78	60	8	12	62.26	8.68	3.25	0.03	110	0.67
3034.92	-rich	5.76	63	6	18	66.27	8.09	3.00	0.02	98.1	0.67
3037.53	-rich	4.76	55	13	16	57.55	8.54	3.67	0.03	100	0.64
3041.33	-rich	5.88	65	4	18	64.59	7.64	5.96	0.02	93.0	0.49
3043.48	-rich	5.08	63	4	20	69.82	7.22	2.44	0.01	85.8	0.69
3044.76	-rich	6.06	75	5	11	72.98	5.97	2.21	0.01	74.6	0.67
3053.03	-lean	2.54	22	61	11	24.29	3.70	1.58	0.02	48.8	0.64
3053.28	-rich	7.73	68	7	9	65.61	8.13	3.79	0.01	103	0.62
3053.88	-rich	8.25	72	5	10	68.04	6.58	2.96	0.01	83.0	0.63
3054.20	-rich	7.55	56	5	18	64.78	8.57	3.69	0.02	113	0.64
3062.70	-lean	2.34	14	60	16	37.73	8.18	2.77	0.02	112	0.69
3064.45	-lean	2.37	28	42	17	39.67	7.08	2.44	0.01	88.9	0.69

The TOC content of the Shuijingtuo shale varies between 1.38 and 8.25 wt.% (average 4.05 wt.%) (Table 1). The TOC content shows a positive correlation with the quartz content, but exhibits two distinct slopes (Figure 2), implying a difference in the silica source between them. Based on the slope, the samples with TOC < 2.55 wt.% are classified as organic-lean shale, and those with a TOC > 2.55 wt.% are classified as organic-rich shale.

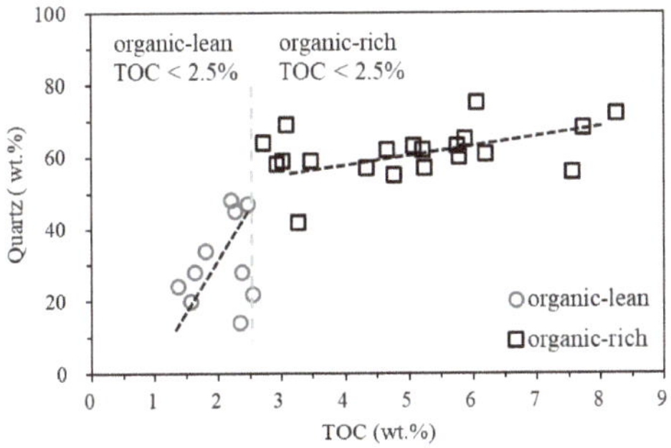

Figure 2. Relationship between TOC and bulk quartz (XRD) content. The organic-lean (TOC < 2.55 wt.%) and organic-rich (TOC > 2.55 wt.%) samples exhibit two distinct positive slopes, implying a difference in silica source between them.

4.2. Quartz Types and Silica Sources

4.2.1. Quartz Types

Compared to sandstone, identifying quartz types in shale reservoirs is more difficult because of the small particle size of the mineral. The widely utilized high-resolution FESEM combined with energy dispersive X-ray (SEM-EDS) and cathodoluminescence (SEM-CL) is a new technique that can differentiate authigenic quartz from detrital quartz at a submicrometer scale, and can identify quartz types based on the information gained from SEM images on crystal shape, spatial distribution, and CL response [29].

Based on the thin section and SEM observations, four quartz types were identified in the Shuijingtuo shale: detrital quartz, replacement of biosiliceous debris, euhedral quartz filling the interparticle pores, and microquartz dispersed in the clay matrix.

Detrital quartz is subangular to angular in shape and ~5–20 μm in size and can be easily identified from the bright CL intensity (Figure 3).

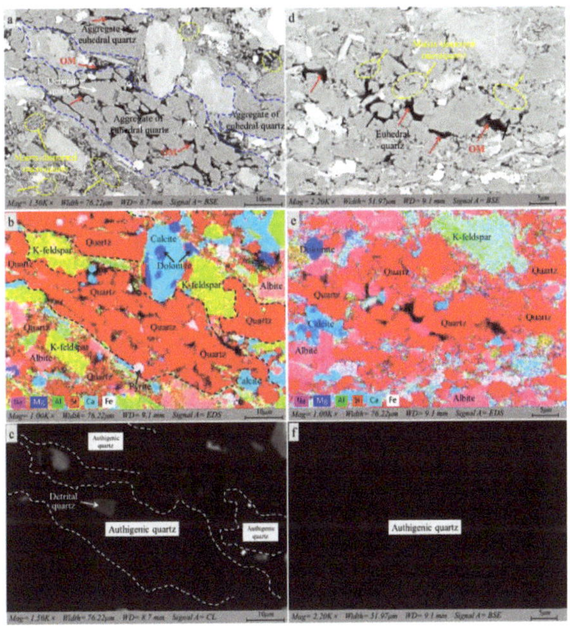

Figure 3. SEM, SEM-CL, and SEM-EDS images showing the petrographic features of authigenic and detrital quartz. (**a**) SEM image from sample 3026.38 m showing that the authigenic euhedral quartz

occurs as aggregates filled in the interparticle pores and the microquartz is dispersed in the claymineral matrix. The euhedral quartz is much larger than the microquartz. (**b**) SEM-EDS images confirming mineral identification, the same area as (**a**). (**c**) SEM-CL images (same area as (**a**)) showing the authigenic quartz has a dark, homogeneous luminescence and detrital quartz has a bright CL intensity. (**d**) SEM image from sample 3007.75 m showing that the authigenic euhedral quartz occurs occasionally as discrete grains. (**e**) SEM-EDS images confirming mineral identification, the same area as (**d**). (**f**) SEM-CL image (same area as (**a**)) showing the authigenic quartz with non-luminescence. Yellow arrows and boxes in (a) represent matrix-dispersed microquartz; red arrows in (**a**,**d**) represent organic matter (OM); blue dashed boxes in (**a**) represent aggregate of euhedral quartz; black arrows in (**d**) represent single euhedral quartz particle.

Radiolarians and sponge spicules are siliceous organisms that are commonly developed in the Lower Cambrian Shale in southern China [30,31], and these organisms were also observed abundantly in this study. As illustrated in Figure 4, siliceous organisms are more abundant in the organic-rich samples than in the organic-lean samples. The biosiliceous debris is poorly preserved and extensively replaced and filled by authigenic minerals. The most abundant replacement and filling of biosiliceous debris are carbonate minerals and pyrite (Figure 5).

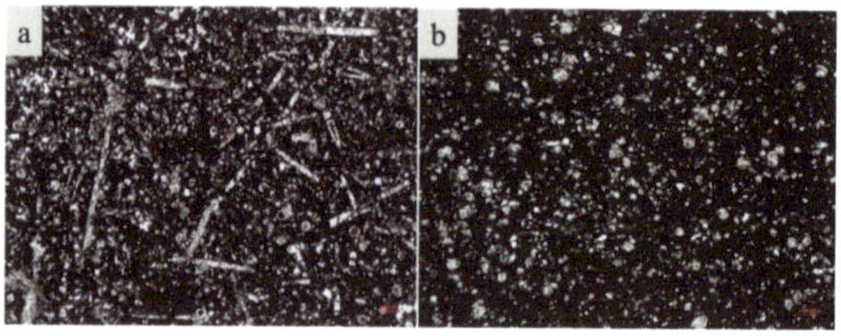

Figure 4. Thin-section images showing difference in the abundance of biosiliceous debris between organic-rich and organ-lean samples. (**a**) Organic-rich sample from 3011.50 m (TOC = 4.66 wt.%) presenting abundant radiolarians and sponge spicules. (**b**) Organic-lean sample from 3064.45 m (TOC = 2.37 wt.%) with essentially nil in the siliceous organism.

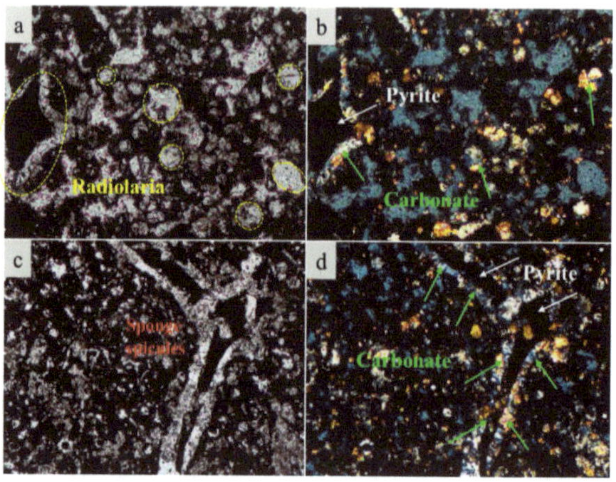

Figure 5. Example showing the replacement and filling of radiolarians and sponge spicules, sample depth = 3011.50 m. The biosiliceous debris is extensively replaced and filled by the carbonate minerals (green arrows) and pyrite (white arrows). Images of (**a**,**c**) were observed with the plane-polarized light. Images of (**b**,**d**) were observed with cross-polarized light. (**a**,**b**) are the same area; (**c**,**d**) are the same area.

Euhedral quartz filling the interparticle pores (Figures 6 and 7) and microquartz dispersed in the clay matrix (Figure 8) are found to be the predominant authigenic quartz types in the Shuijingtuo shale. Both have a dark and homogeneous CL intensity (Figure 3). Clay matrix-dispersed microquartz is a common quartz type that was reported in many other mudrocks [18–20]. Euhedral quartz filling the interparticle pores is a particular quartz type that is rarely reported in shales, except for those from the marine shales in southern China, such as Shuijingtuo shale and Wufeng–Longmaxi shale [17,28]. Euhedral quartz can be easily distinguished from clay matrix-dispersed microquartz by the following three petrographic features. Euhedral quartz occurs commonly as aggregates composed of several macrocrystalline quartz crystals or occasionally as discrete grains. The SEM image statistics from six estimated shale samples show that the size of the macrocrystalline quartz crystals is 1–5 μm (average 3 μm) (Figure 9a). In contrast, matrix-dispersed microquartz is much smaller, primarily 100–400 nm (average 200 nm) (Figure 9b). Different from the euhedral quartz, microquartz typically coexists with clay minerals (Figures 6–8).

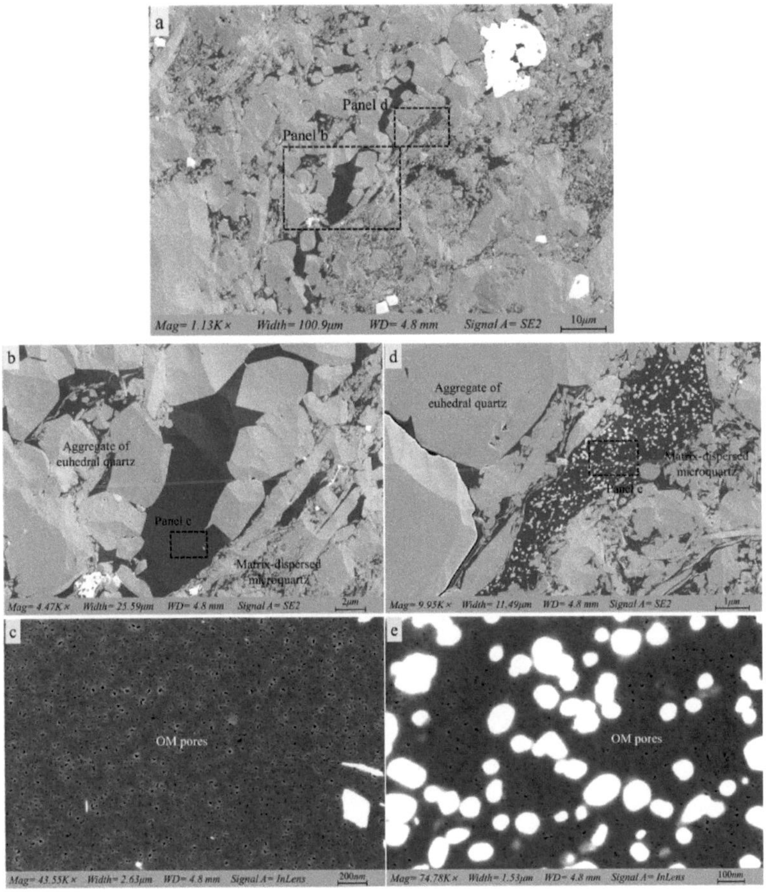

Figure 6. Images showing the euhedral quartz filling the interparticle pores and microquartz dispersed in clay matrix, sample depth = 3026.38 m. (**a**) Abundant euhedral quartz and microquartz dispersed in shale reservoir. (**b**) Enlargement of image (**a**). The euhedral quartz linings along the interparticle pores and the porous organic matter fills the interior of the space. (**c**) Enlargement of image (**b**). Interparticle pore-filled organic matter develops pores with rounded to sub-rounded shape. (**d**) Enlargement of image (**a**). Microquartz dispersed in clay matrix. (**e**) Enlargement of image (**d**) showing abundant organic matter pores.

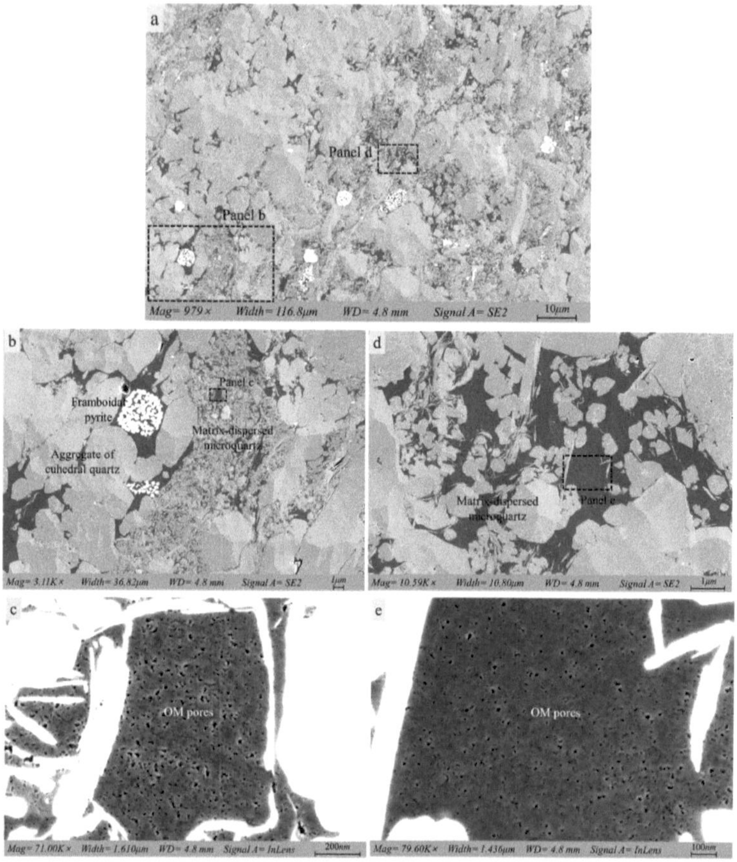

Figure 7. (**a**) SEM images showing the spatial distribution of euhedral quartz and microquartz, sample depth = 3026.38 m. (**b**,**d**) Enlargement of image (**a**) showing that microquartz coexists with the porous organic matter in the clay mineral matrix. (**c**) Enlargement of image (**b**) showing organic matter pores. (**e**) Enlargement of image (**d**) showing organic matter pores.

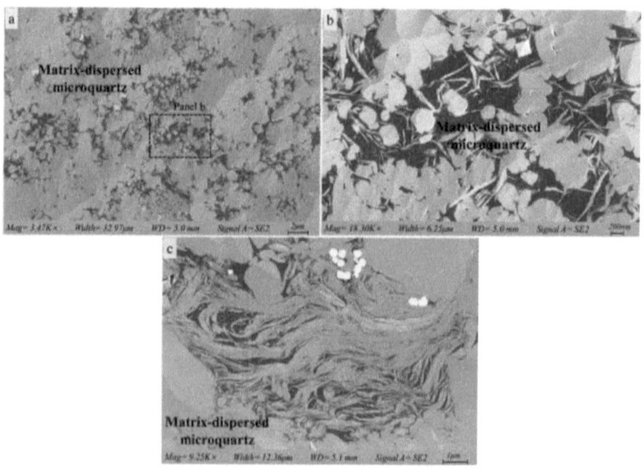

Figure 8. This SEM images showing the microquartz dispersed in clay mineral matrix and is mostly wrapped by the porous organic matter. (**a**) Sample depth = 3041.33 m. (**b**) Enlargement of image (**a**). (**c**) Sample depth = 2996.32 m.

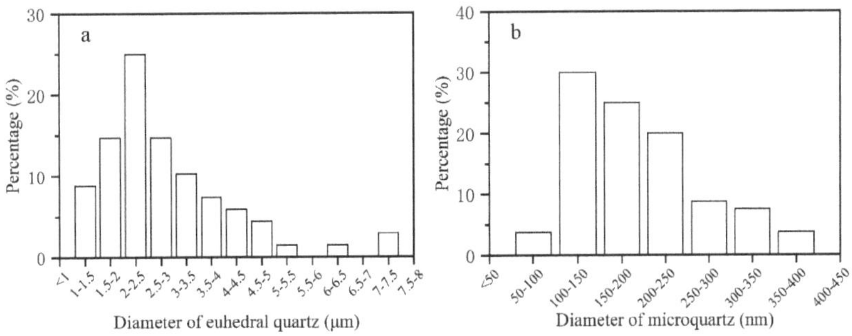

Figure 9. Histogram of diameter of euhedral quartz (**a**) and microquartz (**b**). The size of macrocrystalline quartz crystals is in the range of 1–5 μm (average 3 μm). Matrix-dispersed microquartz is much smaller, mostly between 100 nm and 400 nm (average 200 μm).

The SEM images show that the authigenic quartz typically coexists with porous organic matter in the pore space (Figures 6–8). However, the two types of authigenic quartz have different positional relationships with porous organic matter. In the interparticle pores, the euhedral quartz is observed to be arranged around the pores and the porous organic matter fills the interior of the space composed of euhedral quartz. Conversely, in the clay matrix pores, microquartz is mostly wrapped by porous organic matter.

4.2.2. Silica Sources

Previous studies have suggested that common silica sources include the dissolution of siliceous skeletons (biogenic silica), transition of smectite to illite (clay-derived silica), pressure dissolution of detrital quartz, alteration of feldspar, alteration of volcanic ash, and precipitation of submarine hydrothermal venting [32–34].

Volcanic ash is highly unstable and will dissolve rapidly to release a variety of cations that form bentonite composed of quartz, zircon, apatite, kaolinite, chlorite, zeolite, etc. [35]. In the study area, no volcanic ash layer has been identified in the Shuijingtuo Formation and there are no associated reports about a volcanic eruption during the Early Cambrian period. Furthermore, the mineral compositions detected by XRD analysis reveal that no volcanic ash alteration occurs in the Shuijingtuo shale. Thus, the possibility of volcanic ash being the silica source for authigenic quartz can be excluded. Few intergranular contacts between detrital quartz are observed in the SEM images; therefore, pressure dissolution is less likely to be a silica source. Similarly, no specific petrographic evidence supports feldspar alteration as a potential silica source.

The geochemical proxies (e.g., Al, Fe, Mn, Ti) are effective methods for discriminating the hydrothermal origin from the nonhydrothermal origin of silica [19,20]. The compositions of hydrothermal sedimentary rocks are usually characterized by high Fe and Mn. In contrast, nonhydrothermal sedimentary rocks are enriched in Al and Ti [36]. The ratio of Al/(Al + Fe + Mn), ternary diagram of Al–Fe–Mn, and plot of Fe/Ti–Al/(Al + Fe + Mn) are used to determine the silica input of the hydrothermal source [37,38]. Rocks with a pure hydrothermal origin of silica have a low value of Al/(Al + Fe + Mn) (<0.01), whereas Al/(Al + Fe + Mn) in rocks with pure biogenic silica is >0.60 [39]. The ratio of Al/(Al + Fe + Mn) in the Shuijingtuo shale is 0.49–0.77, with an average value of 0.66 (Table 1), indicating that there is no hydrothermal activity in the study area. Moreover, as illustrated in the Al–Fe–Mn ternary diagram, all the Shuijingtuo samples are plotted in the nonhydrothermal region, which is characterized by enrichment in Al relative to Fe and Mn (Figure 10). Therefore, according to the data for Al, Fe, and Mn, it can be concluded that the Shuijingtuo shale in the study area is related to nonhydrothermal deposition and there is no silica input from the hydrothermal source.

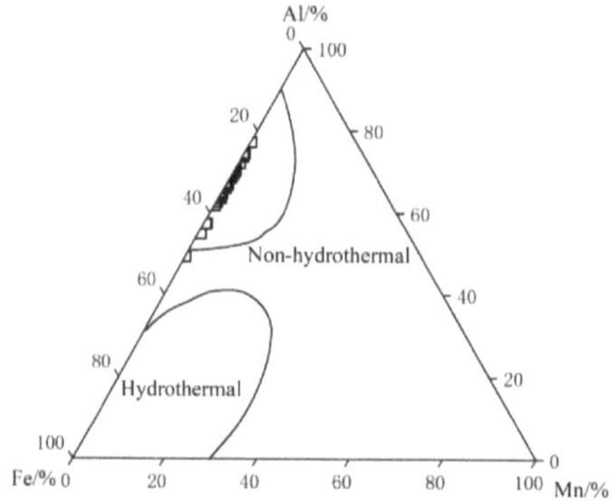

Figure 10. Ternary diagram of Al–Fe–Mn determining the silica input of the hydrothermal and non-hydrothermal source. The examined Shuijingtuo shale samples are plotted in the nonhydrothermal region, indicating that there is no silica input of hydrothermal source of the Shuijingtuo shale in the study area.

The maturity of the examined Shuijingtuo shale is in the stage of dry gas (~1.93–2.20% Ro, [28]), which means that the shale has mostly experienced sufficient temperature to induce substantial illitization of smectite [40]. The clay minerals of the Shuijingtuo shale are mainly composed of illite (78 wt.%, average), followed by a mixed layer of I-S (22 wt.%) [28]. The high ratio of illite to clay minerals implies that the Shuijingtuo shale has possibly completed the conversion from smectite to illite, during which a certain amount of silica was released [41]. Therefore, the transition of smectite to illite is one of the potential sources of silica.

This study suggests the dissolution of siliceous skeletons is the silica source for the authigenic quartz precipitation in the organic-rich samples. The initial silica phase of siliceous skeletons is unstable opal-A, which can quickly transit to metastable opal-CT and ultimately to the stable end product quartz [42,43]. The thin-section images show that many radiolarian and sponge spicules are observed in the organic-rich shales, but are almost absent in organic-lean shales (Figure 4). The substantial replacement of radiolarian and sponge spicules observed within the organic-rich shale samples (Figure 5) confirms that the biosiliceous debris has experienced dissolution and has supplied massive silica for quartz precipitation.

The plot of zirconium (Zr) versus Si is another method that is utilized to examine the biogenic silica source [44–46]. Zr is typically associated with heavy mineral zircon that existed in the detrital fraction and is interpreted to be a proxy for a detrital source of silica [44–46]. A positive correlation between Zr and Si is interpreted to indicate a detrital source for silica, such as in the Jurassic Haynesville-Bossier shale [47], whereas a negative correlation is used to indicate the input of biogenic silica, such as in the Devonian Muskwa shale [45]. In this study, two opposite Zr–Si trend lines were obtained among the Shuijingtuo samples (Figure 11a). Similar results have been reported in the Devonian Horn River Group shale [46] and the Muskwa Formation shale [45]. Coincidentally, the samples in which Zr increases with an increase in Si (detrital trend line) (Figure 11a) are all the organic-lean shales that have a high slope in the TOC–quartz plot (Figure 2). Correspondingly, the samples that lie on the biogenic trend line (Figure 11a) are all the organic-rich shales, which have a low slope in the TOC–quartz plot (Figure 2). Therefore, we infer that the source of silica in the organic-lean samples is major detrital silica, and biogenic silica accounts for at least a certain proportion in the organic-rich samples. The analyses of the silica source in organic-lean and organic-rich shales by Zr–Si plot coincide with the petrographic observations, that is, organic-rich samples are rich in siliceous skeletons, whereas organic-lean samples are essentially nil (Figure 4). The content of biogenic silica will be discussed in the following section.

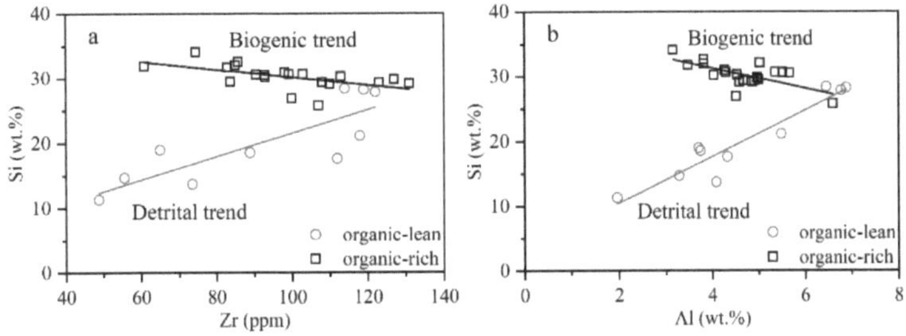

Figure 11. Relationships between Si and Zr (**a**) and Al (**b**) concentration in the Shuijingtuo Shale samples from XRF datasets. The positive lines are termed the "detrital trend", indicating a detrital source for silica. The negative lines are termed the "biogenic trend", indicating the input of biogenic silica. Samples in which Zr and Al increase with the increase of Si are all the organic-lean shales that have a high slope in the TOC–quartz plot (Figure 2). Samples that lie on the biogenic trend line are all the organic-rich shales, which have a low slope in the TOC–quartz plot (Figure 2).

4.3. Content of Biogenic Silica

Excess Si, as proposed by Sholkovitz and Price [48], is defined as the content that exceeds the Si concentration in the terrigenous components (normal detrital background). Excess Si is widely adopted to quantify the degree of enrichment of non-terrigenous silica [49]. Excess (or non-terrigenous) Si is defined as the sum of the biogenic, hydrothermal, volcanic ash alteration silica, and other non-terrigenous input Si [46,50]. Combined with the analyses of silica sources above, the excess Si in the Shuijingtuo shale refers to the biogenic Si content.

Excess Si is calculated as (total Si—detrital Si), which is generally estimated using the following formula: $Si_{excess} = Si_{sample} - [(Si/Al)_{background} \times Al_{sample}]$ [48]. The value of $(Si/Al)_{background}$ is important for excess Si estimation. However, we found that this value varies across different studies. The majority of studies typically use 3.11 as the value of $(Si/Al)_{background}$ [17,20,49,51,52], which is the average of 277 samples from the post-Archaean average shale, also called standard shale [53]. Some studies use the value 2.50 [54], which is the average Si/Al value of samples from the Amazon estuary zone of terrigenous material [48]. The lowest Si/Al value in the estimated samples was designated as the background value [55]. Therefore, the value of $(Si/Al)_{background}$ remains controversial and differs among different studies. Van der Weijden [56] and Tribovillard et al. [57] hold that the composition of the commonly used standard shale and the reference values of normalized elements are not necessarily representative of the local or regional sediments in the study area.

Figure 11b shows the Al–Si plot. Similar to the Zr–Si plot, the organic-lean samples lie on the detrital (positive) trend line, and the organic-rich shales lie on the biogenic (negative) trend line (Figure 11). As analyzed above, silica in organic-lean samples is the major detrital in origin. Therefore, the organic-lean samples can be regarded as normal detrital shales in the EYY1 well. Moreover, samples above the detrital trend line have a higher Si content, which is sourced from the dissolution of siliceous skeletons (radiolarian and sponge spicules). Based on the definition of the excess Si, the excess (biogenic) Si is calculated as the absolute difference between Si measured in a sample and the Si versus Al regression line (detrital trend line) for the organic-lean samples. The determined biogenic Si is shown in Table 2. The results show that the biogenic Si in the organic-rich shales is 7–19 wt.% (average 11 wt.%), and the ratio of biogenic Si to total Si is 23–57 wt.% (average 35 wt.%) (Table 2). If the biogenic Si precipitates in the form of quartz, ~15–42 wt.% biogenic quartz will be formed, accounting for 26–55 wt.% of the total quartz (XRD). Compared to the proportion of clay-derived quartz (discussed below), biogenic silica is the major source of the authigenic quartz in the organic-rich shales. Biogenic silica as the primary source has also been reported in other marine shales, such as the Upper Cretaceous Eagle Ford shale in southern Texas [18], Upper Pennsylvanian Cline shale in the Midland basin [14], Mississippian Bowland shale in the Craven basin [54], and Upper Ordovician Wufeng–Lower Silurian Longmaxi shale in the Sichuan basin [51].

Table 2. Results of the calculated biogenic Si and biogenic quartz of the Shuijingtuo shale samples.

Sample ID	Depth (m)	Organic	TOC (wt.%)	Quartz (XRD wt.%)	Biogenic Si (wt.%)	Biogenic Si /Total Si	Biogenic Quartz (wt.%)	Biogenic Quartz /Quartz (XRD)
EYY24	2939.08	-lean	1.38	24	-	-	-	-
EYY42	2975.53	-lean	1.63	28	-	-	-	-
EYY47	2978.08	-lean	1.58	20	-	-	-	-
EYY61	2983.63	-lean	2.47	47	-	-	-	-
EYY64	2984.98	-lean	2.21	48	-	-	-	-
EYY74	2989.32	-lean	1.80	34	-	-	-	-
EYY90	2996.32	-lean	2.27	45	-	-	-	-
EYY103	3003.50	-rich	2.72	64	10.77	34%	23	36%
EYY106	3004.88	-rich	3.08	69	14.98	47%	32	47%
EYY111	3007.34	-rich	2.93	58	6.92	23%	15	26%
EYY112	3007.75	-rich	3.02	59	7.51	25%	16	27%
EYY114	3008.78	-rich	3.47	59	8.22	28%	18	30%
EYY119	3011.50	-rich	4.66	62	8.12	27%	17	28%
EYY138	3020.54	-rich	4.33	57	8.39	29%	18	32%
EYY147	3023.83	-rich	5.22	62	8.65	29%	19	30%
EYY148	3024.33	-rich	5.24	57	8.40	29%	18	32%
EYY152	3026.38	-rich	6.20	61	9.28	32%	20	33%
EYY159	3030.18	-rich	5.78	60	9.30	32%	20	33%
EYY167	3034.92	-rich	5.76	63	12.29	40%	26	42%
EYY172	3037.53	-rich	4.76	55	7.36	27%	16	29%
EYY180	3041.33	-rich	5.88	65	12.35	41%	26	41%
EYY187	3043.48	-rich	5.08	63	15.58	48%	33	53%
EYY190	3044.76	-rich	6.06	75	19.41	57%	42	55%
EYY204	3053.03	-lean	2.54	22	-	-	-	-
EYY205	3053.28	-rich	7.73	68	11.91	39%	26	38%
EYY206	3053.88	-rich	8.25	72	15.96	50%	34	48%
EYY207	3054.20	-rich	7.55	56	10.70	35%	23	41%
EYY223	3062.70	-lean	2.34	14	-			-
EYY226	3064.45	-lean	2.37	28	-			-

5. Discussions

5.1. Corresponding Relation between Quartz Types and Silica Sources

Two types of authigenic quartz grain are present in the Shuijingtuo shale: one is large-sized euhedral quartz (1–5 μm) filling the interparticle pores and the other is small-sized microquartz dispersed in the clay mineral matrix. The silica for the authigenic quartz is proved to be sourced from the dissolution of siliceous skeletons and the transformation of smectite to illite. These facts raise a question: What is the corresponding relationship between the quartz types and silica sources? Determining the answer to the question is of great significance for the assessment of the impacts of siliceous cement on the abundance and preservation of organic matter pores.

The SEM-CL cannot distinguish between the authigenic quartz formed in the early and late diagenetic stage because both have a dark, non-luminescent CL response. Previous studies have suggested that the dissolution of siliceous skeletons to form the end product quartz occurs at relatively low temperatures [32]. Iijima and Tada [35] reported that the temperature at which opal-A starts to change to opal-CT is within a range between 22 °C and 33 °C (550–1180 m) within Neogene siliceous sediments in northern Japan, and the transformation of opal-CT into quartz starts at a temperature of 28–44 °C (810–1240 m). The reaction of smectite illitization accompanied by silica release occurs at 60–100 °C [58–60].

The comparison of reaction temperature indicates that the precipitation of biogenic quartz is earlier than that of clay-derived quartz.

At shallow depths (lower temperature), the mudrocks have higher porosity (~10–55% at 1 km [61]), which means that there are sufficient primary pores with large pore size for the precipitation of biogenic quartz. Therefore, the large-sized euhedral quartz filling the interparticle pores most likely resulted from the dissolution of siliceous skeletons at an early stage of diagenesis. With an increase in the burial depth and temperature, the transformation of smectite to illite is triggered and the mechanical compaction strength also increases, which leads to a significant loss of porosity and a decrease in pore size. Hence, the quartz sourced from the illitization reaction theoretically has a small size, which was illustrated clearly by the experiments of clay mineral transformation [62,63]. However, we can only conclude that the observed small-sized microquartz is partially derived from the clay transformation because the volume of microquartz presented in the SEM images is much larger than that generated from the transformation of clay minerals, which is discussed below.

The reaction of the illitization of smectite is complex and the amount of released quartz is still disputed. Van de Kamp [64] suggested that, assuming Al conservation, the smectite illitization yields about 17–28 wt.% quartz as a proportion of the original smectite (about 25–41 vol.% of the final illite, the molar volume of clay fraction refers to Środoń et al. [65]). However, Milliken and Olson [15] documented that the illitization based on the Al balance seems untenable because this reaction leads to high volume loss (60%) in the clay fraction. Milliken [16] proposed a 1:1 (smectite:illite) molar balance reaction, which releases 0.47 mol silica without causing an immense volume loss of the clay fraction. If the released silica precipitated in its entirety, it would yield a volume of quartz equivalent to ~6 vol.% of the final clay minerals [16]. However, the observations in high-magnification SEM images differ from the reaction formulas. The point-count volume of the microquartz and clay minerals is listed in Table 3. The results show that the volume ratio of microquartz to clay mineral is 0.67–4.32, which is much higher than that of the aforementioned smectite to illite reactions. Notably, these ratio values do not represent the entire shale sample because there are not enough point-counted SEM images to equate the micro with the macro, which is a common issue for microscopic observations. Nevertheless, these results suggest that the content of microquartz is too high to be obtained fully from the clay mineral transformation. Therefore, the biogenic silica most likely provides mass silica for the microquartz precipitation, which is consistent with the views of Peng et al. [14] in the Cline shale and Milliken and Olson [15] in the Mowry shale. More specifically, the dissolution of siliceous skeletons and transformation of smectite to illite together provide the silica for the precipitation of clay matrix-dispersed microquartz.

Table 3. Point-count volume of clay mineral, microquartz, and organic matter in high-magnification SEM images.

Depth (m)	Organic	TOC (wt.%)	Clay Mineral (vol.%)	Microquartz (vol.%)	Organic Matter (vol.%)	Other Mineral (vol.%)
3053.03	-lean	2.54	41.08	27.5	18.69	12.73
3020.54	-rich	4.33	33.24	40.13	24.58	2.05
3041.33	-rich	5.88	16.67	52.20	31.13	-
3054.20	-rich	7.55	12.99	56.17	29.55	1.29

5.2. Effects on the Abundance and Preservation of Organic Matter Pores

As a significant pore type in the shale pore systems, organic matter pores could supply extensive pore spaces for storing adsorbed and free shale gas [2,27,66,67]. According to the origin, organic matter pores can be divided into primary and secondary (migrated) types [3,9,28]. The primary types are commonly found in kerogen in immature shales and are considered to be inherited from organic macerals [3,68,69]. Formation of the secondary type is stimulated by the evolution of the kerogen and secondary organic matter (solid bitumen or oil) during thermal maturation [28,70–72]. Compared to the primary type, the secondary type is more abundant and well-connected, which is advantageous for the storage and transport of shale gas [12]. Wei et al. [28] reported that most of the porous organic matter in the Shuijingtuo shale is migrated organic matter, which has thermally evolved from the initial liquid petroleum phase of solid bitumen or oil into pyrobitumen. The terms of kerogen, oil, solid bitumen, and pyrobitumen are derived from the work of Mastalerz et al. [73].

The thermal cracking of the retained petroleum was documented to be the main mechanism for the massive shale gas generation and formation of the organic matter pores by the modeling experiments of gas generation and pore evolution by the sealed gold-tube pyrolysis [8,9]. Therefore, the amount of retained petroleum is significantly related to shale gas enrichment and reservoir quality. In addition to well-known factors, such as organic matter attributes (type, abundance, and thermal maturity), combination structures between the source rock and reservoir rock, and roof/floor sealing conditions of the source rock, which are critical for evaluating the amount of retained petroleum, the changes in pore space before and during the hydrocarbon expulsion period have been found to be another important factor that controls the petroleum retention [8,10–12].

The relative time of authigenic quartz precipitation and liquid petroleum migration has an impact on the pore space. The presence of the porous organic matter embayed by euhedral quartz in the interparticle pores (Figures 6b and 7b) suggests that the migration of liquid petroleum occurred later than the precipitation of euhedral quartz. The observations agree with the analysis above that the euhedral quartz with a large grain size resulted from the dissolution of siliceous skeletons in the early diagenetic stage. The precipitation of the early diagenetic euhedral quartz can thus lead to a reduction in primary pores, as it occupies partial primary interparticle pores. It was suggested that the early diagenetic quartz could increase rock mechanical behavior to inhibit mechanical compaction [54,74]. In the primary interparticle pores, part of euhedral quartz exhibits a weak preferred orientation (Figures 3a and 6a), indicating that the quartz has undergone some mechanical compaction at a later stage. Due to the rigidity of the quartz framework, the primary pores retained some space for the subsequent migration of solid bitumen or oil. The rounded to sub-rounded organic matter pores (Figure 6c) indicate that the buttressing effect by the euhedral quartz framework can protect the organic matter pores from compaction during the late-stage burial diagenesis.

The presence of microquartz wrapped by the porous organic matter in the clay matrix pores (Figure 8) suggests a situation where the migration of liquid petroleum is coeval with or later than the precipitation of microquartz. If there is no microquartz precipitation in the clay matrix pores, there would be more pore space for the migration and retention of liquid petroleum, and more shale gas and organic matter pores would be generated.

6. Conclusions

1. In the estimated marine shale reservoir, quartz can be observed as four types, including detrital quartz, replacement of biosiliceous debris, euhedral quartz filled in the interparticle pores, and microquartz dispersed in a clay matrix.
2. The dominant forms of authigenic quartz are euhedral quartz and microquartz. Euhedral quartz is much larger (1–5 μm) than microquartz (100–400 nm). Moreover, they have different positional relationships with porous organic matter. The porous organic matter fills the interior of the space composed of euhedral quartz, whereas microquartz is mostly surrounded by porous organic matter.
3. The source of silica for the precipitation of authigenic quartz includes the dissolution of siliceous skeletons and the transition of smectite to illite. In the organic-rich shale samples, biogenic silica is most likely the major source for the authigenic quartz. The calculated excess Si (biogenic Si) ranges from 7 to 19 wt.% (average 11 wt.%), accounting for 23–57 wt.% (average 35 wt.%) of the total silica.

4. The large-sized euhedral quartz present in the interparticle pores most likely results from the dissolution of siliceous skeletons at an early diagenesis stage. The silica for the clay matrix-dispersed microquartz is jointly provided by biogenic silica and clay-derived silica.
5. The precipitation of early diagenetic euhedral quartz in the interparticle pores enhances rock stiffness, and the buttressing effect can protect the organic matter pores from compaction during the late-stage burial diagenesis. The precipitation of late diagenetic microquartz in the clay matrix can lead to a reduction in the retained petroleum, which would lead to a reduction in the amount of organic matter pores developed.

Author Contributions: Conceptualization, S.W., M.H., S.H., Y.S. and T.D.; methodology, S.W., M.H., S.H. and W.Y.; formal analysis, S.W., Q.H. and Q.C.; investigation, T.D., Q.H., W.Y. and Q.C.; resources, M.H., S.H. and T.D.; data curation, Y.S., Q.H. and W.Y.; writing—original draft preparation, S.W.; writing—review and editing, S.W., M.H., S.H., Y.S., T.D., Q.H., W.Y. and Q.C.; funding acquisition, S.W., S.H. and Q.C. All authors have read and agreed to the published version of the manuscript.

Funding: This research was funded by the Natural Science Foundation of Hubei Province (No. 2022CFB652) and the CNPC Innovation Fund (No. 2021DQ02-0101).

Institutional Review Board Statement: Not applicable.

Informed Consent Statement: Not applicable.

Data Availability Statement: Data supporting the results can be found in the manuscript text, tables, and figures.

Conflicts of Interest: The authors declare no conflict of interest.

References

1. Zou, C.; Yang, Z.; Dong, D.; Zhao, Q.; Chen, Z.; Feng, Y.; Li, J.; Wang, X. Formation, distribution and prospect of unconventional hydrocarbons in source rock strata in China. *Earth Sci.* **2022**, *47*, 1517–1533, (In Chinese with English abstract).
2. Loucks, R.G.; Reed, R.M.; Ruppel, S.C.; Jarvie, D.M. Morphology, genesis, and distribution of nanometer-scale pores in siliceous mudstones of the Mississippian Barnett Shale. *J. Sediment. Res.* **2009**, *79*, 848–861. [CrossRef]
3. Katz, B.J.; Arango, I. Organic porosity: A geochemist's view of the current state of understanding. *Org. Geochem.* **2018**, *123*, 1–16. [CrossRef]
4. Wasaki, A.; Akkutlu, I.Y. Permeability of organic-rich shale. *SPE J.* **2015**, *20*, 1384–1396. [CrossRef]
5. Guo, X.; Li, Y.; Borjigen, T.; Wang, Q.; Yuan, T.; Shen, B.; Ma, Z.; Wei, F. Hydrocarbon generation and storage mechanisms of deep-water shelf shales of Ordovician Wufeng Formation–Silurian Longmaxi Formation in Sichuan Basin, China. *Pet. Explor. Dev.* **2020**, *47*, 193–201, (In Chinese with English abstract). [CrossRef]
6. Zou, C.; Dong, D.; Wang, S.; Li, J.; Li, X.; Wang, Y.; Li, D.; Cheng, K. Geological characteristics, formation mechanism and resource potential of shale gas in China. *Pet. Explor. Dev.* **2010**, *37*, 641–653, (In Chinese with English abstract). [CrossRef]
7. Zhao, W.; Bian, C.; Li, Y.; Liu, W.; Dong, J.; Wang, K.; Zeng, X. Organic matter transformation ratio, hydrocarbon expulsion efficiency and shale oil enrichment type in Chang 73 shale of Upper Triassic Yanchang Formation in Ordos Basin, NW China. *Pet. Explor. Dev.* **2023**, *50*, 12–23, (In Chinese with English abstract). [CrossRef]
8. Hill, R.J.; Zhang, E.; Katz, B.J.; Tang, Y. Modeling of gas generation from the Barnett shale, Fort Worth Basin, Texas. *AAPG Bull.* **2007**, *91*, 501–521. [CrossRef]
9. Ko, L.T.; Loucks, R.G.; Zhang, T.; Ruppel, S.C.; Shao, D. Pore and pore network evolution of Upper Cretaceous Boquillas (Eagle Ford-equivalent) mudrocks: Results from gold tube pyrolysis experiments. *AAPG Bull.* **2016**, *100*, 1693–1722. [CrossRef]
10. Borjigin, T.; Tao, C.; Hu, G.; Shen, B.; Ma, Z.; Pan, A.; Wang, J.; Wang, X.; Xu, E. Effect of hydrocarbon explusion efficiency on shale gas formation and enrichment. *Pet. Geol. Exp.* **2020**, *42*, 325–334+344, (In Chinese with English abstract).
11. Jarvie, D.M.; Hill, R.J.; Ruble, T.E.; Pollastro, R.M. Unconventional shale-gas systems: The Mississippian Barnett Shale of north-central Texas as one model for thermogenic shale-gas assessment. *AAPG Bull.* **2007**, *91*, 475–499. [CrossRef]
12. Ko, L.T.; Loucks, R.G.; Ruppel, S.C.; Zhang, T.; Peng, S. Origin and characterization of Eagle Ford pore networks in the South Texas Upper Cretaceous shelf. *AAPG Bull.* **2017**, *101*, 387–418. [CrossRef]
13. Meng, M.; Peng, J.; Ge, H.; Ji, W.; Li, X.; Wang, Q. Rock fabric of lacustrine shale and its influence on residual oil distribution in the Upper Cretaceous Qingshankou Formation, Songliao Basin. *Energy Fuels* **2023**, *37*, 7151–7160. [CrossRef]
14. Peng, J.; Milliken, K.L.; Fu, Q. Quartz types in the Upper Pennsylvanian organic-rich Cline Shale (Wolfcamp D), Midland Basin, Texas: Implications for silica diagenesis, porosity evolution and rock mechanical properties. *Sedimentology* **2020**, *67*, 2040–2064. [CrossRef]

15. Milliken, K.L.; Olson, T. Silica diagenesis, porosity evolution, and mechanical behavior in siliceous mudstones, Mowry Shale (Cretaceous), Rocky Mountains, USA. *J. Sediment. Res.* **2017**, *87*, 366–387. [CrossRef]
16. Milliken, K. Compactional and mass-balance constraints inferred from the volume of quartz cementation in mudrocks. In *Mudstone Diagenesis; Research Perspectives for Shale Hydrocarbon Reservoirs, Seals, and Source Rocks*; Camp, W.K., Milliken, K.L., Taylor, K., Fishman, N., Hackley, P.C., Macquaker, J.H.S., Eds.; American Association of Petroleum Geologists: Tulsa, OK, USA, 2019; Volume 120, pp. 33–48.
17. Dong, T.; He, Q.; He, S.; Zhai, G.; Zhang, Y.; Wei, S.; Wei, C.; Hou, Y.; Guo, X. Quartz types, origins and organic matter-hosted pore systems in the lower cambrian Niutitang Formation, middle yangtze platform, China. *Mar. Petrol. Geol.* **2021**, *123*, 104739. [CrossRef]
18. Milliken, K.L.; Ergene, S.M.; Ozkan, A. Quartz types, authigenic and detrital, in the Upper Cretaceous Eagle Ford Formation, South Texas, USA. *Sediment. Geol.* **2016**, *339*, 273–288. [CrossRef]
19. Niu, X.; Yan, D.; Zhuang, X.; Liu, Z.; Li, B.; Wei, X.; Xu, H.; Li, D. Origin of quartz in the lower Cambrian Niutitang Formation in south Hubei Province, upper Yangtze platform. *Mar. Petrol. Geol.* **2018**, *96*, 271–287. [CrossRef]
20. Xu, H.; Zhou, W.; Hu, Q.; Yi, T.; Ke, J.; Zhao, A.; Lei, Z.; Yu, Y. Quartz types, silica sources and their implications for porosity evolution and rock mechanics in the Paleozoic Longmaxi Formation shale, Sichuan Basin. *Mar. Petrol. Geol.* **2021**, *128*, 105036. [CrossRef]
21. Liu, Z.; Xu, L.; Wen, Y.; Zhang, Y.; Luo, F.; Duan, K.; Chen, W.; Zhou, X.; Wen, J. Accumulation characteristics and comprehensive evaluation of shale gas in Cambrian Niutitang Formation, Hubei. *Earth Sci.* **2022**, *47*, 1586–1603, (In Chinese with English abstract).
22. Chen, X.; Chen, L.; Jiang, S.; Liu, A.; Luo, S.; Li, H.; Li, P.; Chen, P. Evaluation of shale reservoir quality by geophysical logging for Shuijingtuo Formation of Lower Cambrian in Yichang area, central Yangtze. *J. Earth Sci.* **2021**, *32*, 766–777. [CrossRef]
23. Wang, Y.; Zhai, G.; Liu, G.; Shi, W.; Lu, Y.; Li, J.; Zhang, Y. Geological characteristics of shale gas in different strata of marine facies in South China. *J. Earth Sci.* **2021**, *32*, 725–741. [CrossRef]
24. Gao, J.; Li, Y.; He, S.; He, Z.; Li, S.; Wo, Y.; Li, W.; Zhai, G.; Zhao, J. Exploration discovery of shale gas and its indicative significance to mineralization of MVT Lead-Zinc deposit in Yichang area, West Hubei. *Earth Sci.* **2021**, *46*, 2230–2245, (In Chinese with English abstract).
25. Zhai, G.; Wang, Y.; Bao, S.; Guo, T.; Zhou, Z.; Chen, X.; Wang, J. Major factors controlling the accumulation and high productivity of marine shale gas and prospect forecast in Southern China. *Earth Sci.* **2017**, *42*, 1057–1068, (In Chinese with English abstract).
26. Liu, Y.; Zhai, G.; Xu, X.; Zhang, X.; Bai, L.; Yang, Y.; Cao, S. Logging identification of high quality shale of Niutitang Formation and Doushantuo Formation in western Hubei. *Earth Sci.* **2022**, *47*, 1791–1804, (In Chinese with English abstract).
27. Wei, S.; He, S.; Hu, M.; Yang, W.; Guo, X.; Iglauer, S.; Zhai, G. Supercritical high-pressure methane adsorption on the Lower Cambrian Shuijingtuo shale in the Huangling anticline area, South China: Adsorption behavior, storage characteristics, and geological Implications. *Energy Fuels* **2021**, *35*, 19973–19985. [CrossRef]
28. Wei, S.; He, S.; Pan, Z.; Zhai, G.; Yang, R.; Dong, T.; Yang, W. Characteristics and evolution of pyrobitumen-hosted pores of the overmature Lower Cambrian Shuijingtuo Shale in the south of Huangling anticline, Yichang area, China: Evidence from FE-SEM petrography. *Mar. Petrol. Geol.* **2020**, *116*, 104303. [CrossRef]
29. Milliken, K.L. SEM-Based Cathodoluminescence Imaging for Discriminating Quartz Types in Mudrocks. In Proceedings of the SPE/AAPG/SEG Unconventional Resources Technology Conference, Denver, CO, USA, 12–14 August 2013.
30. Xie, X.; Zhu, G.; Wang, Y. The influence of syngenetic hydrothermal silica fluid on organic matter preservation in lower Cambrian Niutitang Formation, South China. *Mar. Petrol. Geol.* **2021**, *129*, 105098. [CrossRef]
31. Gao, P.; He, Z.; Lash, G.G.; Zhou, Q.; Xiao, X. Controls on silica enrichment of lower cambrian organic-rich shale deposits. *Mar. Petrol. Geol.* **2021**, *130*, 105126. [CrossRef]
32. Worden, R.H.; Morad, S. (Eds.) Quartz Cementation in Oil Field Sandstones: A Review of the Key Controversies. In *Quartz Cementation in Sandstones*; Blackwell Science: Oxford, UK, 2000; pp. 1–20.
33. Rutman, P.; Hoareau, G.; Kluska, J.-M.; Lejay, A.; Fialips, C.; Gelin, F.; Aubourg, C.; Bilbao, E.H. Diagenesis and alteration of subsurface volcanic ash beds of the Vaca Muerta Formation, Argentina. *Mar. Petrol. Geol.* **2021**, *132*, 105220. [CrossRef]
34. Zhao, J.; Jin, Z.; Jin, Z.; Wen, X.; Geng, Y. Origin of authigenic quartz in organic-rich shales of the Wufeng and Longmaxi Formations in the Sichuan Basin, South China: Implications for pore evolution. *J. Nat. Gas. Sci. Eng.* **2017**, *38*, 21–38. [CrossRef]
35. Iijima, A.; Tada, R. Silica diagenesis of Neogene diatomaceous and volcaniclastic sediments in northern Japan. *Sedimentology* **1981**, *28*, 185–200. [CrossRef]
36. Halbach, M.; Halbach, P.; Lüders, V. Sulfide-impregnated and pure silica precipitates of hydrothermal origin from the Central Indian Ocean. *Chem. Geol.* **2002**, *182*, 357–375. [CrossRef]
37. Adachi, M.; Yamamoto, K.; Sugisaki, R. Hydrothermal chert and associated siliceous rocks from the northern Pacific their geological significance as indication of ocean ridge activity. *Sediment. Geol.* **1986**, *47*, 125–148. [CrossRef]
38. Yamamoto, K. Geochemical characteristics and depositional environments of cherts and associated rocks in the Franciscan and Shimanto Terranes. *Sediment. Geol.* **1987**, *52*, 65–108. [CrossRef]
39. Boström, K.; Peterson, M.N.A. The origin of aluminum-poor ferromanganoan sediments in areas of high heat flow on the East Pacific Rise. *Mar. Geol.* **1969**, *7*, 427–447. [CrossRef]

40. Hower, J.; Eslinger, E.V.; Hower, M.E.; Perry, E.A. Mechanism of burial metamorphism of argillaceous sediment: 1. Mineralogical and chemical evidence. *GSA Bull.* **1976**, *87*, 725–737. [CrossRef]
41. Peltonen, C.; Marcussen, Ø.; Bjørlykke, K.; Jahren, J. Clay mineral diagenesis and quartz cementation in mudstones: The effects of smectite to illite reaction on rock properties. *Mar. Petrol. Geol.* **2009**, *26*, 887–898. [CrossRef]
42. Bjørlykke, K.; Egeberg, P.K. Quartz cementation in sedimentary basins. *AAPG Bull.* **1993**, *77*, 1538–1548.
43. Isaacs, C.M. Influence of rock composition on kinetics of silica phase changes in the Monterey Formation, Santa Barbara area, California. *Geology* **1982**, *10*, 304–308. [CrossRef]
44. Blood, R.; Lash, G.; Bridges, L. Biogenic silica in the Devonian shale succession of the Appalachian Basin, USA. In Proceedings of the AAPG 2013 Annual Convention and Exhibition, Pittsburgh, PA, USA, 19–22 May 2013; Volume 50864, pp. 19–22.
45. Wright, A.M.; Spain, D.; Ratcliffe, T.K. Application of inorganic whole rock geochemistry to shale resource plays. In Proceedings of the Canadian Unconventional Resources and International Petroleum Conference, Calgary, AB, Canada, 19–21 October 2010.
46. Dong, T.; Harris, N.B.; Ayranci, K.; Yang, S. The impact of rock composition on geomechanical properties of a shale formation: Middle and Upper Devonian Horn River Group shale, Northeast British Columbia, Canada. *AAPG Bull.* **2017**, *101*, 177–204. [CrossRef]
47. Dowey, P.J.; Taylor, K.G. Extensive authigenic quartz overgrowths in the gas-bearing Haynesville-Bossier Shale, USA. *Sediment. Geol.* **2017**, *356*, 15–25. [CrossRef]
48. Sholkovitz, E.R.; Price, N.B. The major-element chemistry of suspended matter in the Amazon Estuary. *Geochim. Cosmochim. Acta* **1980**, *44*, 163–171. [CrossRef]
49. Ross, D.J.K.; Bustin, R.M. Sediment geochemistry of the Lower Jurassic Gordondale Member, northeastern British Columbia. *Bull. Can. Petrol. Geol.* **2006**, *54*, 337–365. [CrossRef]
50. Ross, D.J.K.; Bustin, R.M. Characterizing the shale gas resource potential of Devonian-Mississippian strata in the Western Canada sedimentary basin: Application of an integrated formation evaluation. *AAPG Bull.* **2008**, *92*, 87–125. [CrossRef]
51. Dong, T.; He, S.; Chen, M.; Hou, Y.; Guo, X.; Wei, C.; Han, Y.; Yang, R. Quartz types and origins in the paleozoic Wufeng-Longmaxi Formations, Eastern Sichuan Basin, China: Implications for porosity preservation in shale reservoirs. *Mar. Petrol. Geol.* **2019**, *106*, 62–73. [CrossRef]
52. Zhang, K.; Jiang, Z.; Yin, L.; Gao, Z.; Wang, P.; Song, Y.; Jia, C.; Liu, W.; Liu, T.; Xie, X.; et al. Controlling functions of hydrothermal activity to shale gas content-taking lower Cambrian in Xiuwu Basin as an example. *Mar. Petrol. Geol.* **2017**, *85*, 177–193. [CrossRef]
53. Wedepohl, K.H. Environmental influences on the chemical composition of shales and clays. *Phys. Chem. Earth* **1971**, *8*, 307–333. [CrossRef]
54. Emmings, J.F.; Dowey, P.J.; Taylor, K.G.; Davies, S.J.; Vane, C.H.; Moss-Hayes, V.; Rushton, J.C. Origin and implications of early diagenetic quartz in the Mississippian Bowland Shale Formation, Craven Basin, UK. *Mar. Petrol. Geol.* **2020**, *120*, 104567. [CrossRef]
55. Qiu, Z.; Liu, B.; Dong, D.Z.; Lu, B.; Yawar, Z.; Chen, Z.H.; Schieber, J. Silica diagenesis in the Lower Paleozoic Wufeng and Longmaxi Formations in the Sichuan Basin, South China: Implications for reservoir properties and paleoproductivity. *Mar. Petrol. Geol.* **2020**, *121*, 104594. [CrossRef]
56. Van der Weijden, C.H. Pitfalls of normalization of marine geochemical data using a common divisor. *Mar. Geol.* **2002**, *184*, 167–187. [CrossRef]
57. Tribovillard, N.; Algeo, T.J.; Lyons, T.; Riboulleau, A. Trace metals as paleoredox and paleoproductivity proxies: An update. *Chem. Geol.* **2006**, *232*, 12–32. [CrossRef]
58. Perry, E.; Hower, J. Burial diagenesis in Gulf Coast Pelitic sediments. *Clays Clay Miner.* **1970**, *18*, 165–177. [CrossRef]
59. Hoffman, J.; Hower, J. Clay mineral assemblages as low grade metamorphic geothermometers: Application to the thrust faulted disturbed belt of Montana, USA. In *Aspects of Diagenesis*; Scholle, P.A., Schluger, P.R., Eds.; SEPM Society for Sedimentary Geology: Tulsa, OK, USA, 1997; pp. 55–79.
60. Bjørlykke, K. Clay mineral diagenesis in sedimentary basins—A key to the prediction of rock properties. Examples from the North Sea Basin. *Clay Miner.* **1998**, *33*, 15–34. [CrossRef]
61. Mondol, N.H.; Bjørlykke, K.; Jahren, J.; Høeg, K. Experimental mechanical compaction of clay mineral aggregates—Changes in physical properties of mudstones during burial. *Mar. Petrol. Geol.* **2007**, *24*, 289–311. [CrossRef]
62. Thyberg, B.; Jahren, J. Quartz cementation in mudstones: Sheet-like quartz cement from clay mineral reactions during burial. *Pet. Geosci.* **2011**, *17*, 53–63. [CrossRef]
63. Metwally, Y.M.; Chesnokov, E.M. Clay mineral transformation as a major source for authigenic quartz in hermos-mature gas shale. *Appl. Clay Sci.* **2012**, *55*, 138–150. [CrossRef]
64. Van de Kamp, P.C. Smectite-illite-muscovite transformations, quartz dissolution, and silica release in shales. *Clays Clay Miner.* **2008**, *56*, 66–81. [CrossRef]
65. Środoń, J.; Zeelmaekers, E.; Derkowski, A. The charge of component layers of illite-smectite in bentonites and the nature of end-member illite. *Clays Clay Miner.* **2009**, *57*, 649–671. [CrossRef]
66. Fu, H.; Yan, D.; Yao, C.; Su, X.; Wang, X.; Wang, H.; Li, Y. Pore structure and multi-scale fractal characteristics of adsorbed pores in marine shale: A case study of the Lower Silurian Longmaxi shale in the Sichuan Basin, China. *J. Earth Sci.* **2022**, *33*, 1278–1290. [CrossRef]

67. Meng, M.; Ge, H.; Shen, Y.; Ji, W.; Li, Z. Insight into water occurrence and pore size distribution by nuclear magnetic resonance in marine shale reservoirs, southern China. *Energy Fuels* **2023**, *37*, 319–327. [CrossRef]
68. Fishman, N.S.; Egenhoff, S.O.; Boehlke, A.R.; Lowers, H.A. Petrology and diagenetic history of the upper shale member of the late Devonian-early Mississippian Bakken Formation, Williston Basin, North Dakota. In *Special Paper of the Geological Society of America*; Geological Society of America: Boulder, CO, USA, 2015; Volume 515, pp. 125–151.
69. Han, Y.; Horsfield, B.; Wirth, R.; Mahlstedt, N.; Bernard, S. Oil retention and porosity evolution in organic rich shales. *AAPG Bull.* **2017**, *101*, 807–827. [CrossRef]
70. Bernard, S.; Wirth, R.; Schreiber, A.; Schulz, H.-M.; Horsfield, B. Formation of nanoporous pyrobitumen residues during maturation of the Barnett Shale (Fort Worth Basin). *Int. J. Coal Geol.* **2012**, *103*, 3–11. [CrossRef]
71. Curtis, M.E.; Cardott, B.J.; Sondergeld, C.H.; Rai, C.S. Development of organic porosity in the Woodford Shale with increasing thermal maturity. *Int. J. Coal Geol.* **2012**, *103*, 26–31. [CrossRef]
72. Yang, W.; He, S.; Zhai, G.; Tao, Z.; Yuan, X.; Wei, S. Maturity assessment of the Lower Cambrian and Sinian shales using multiple technical approaches. *J. Earth Sci.* **2021**, *32*, 1262–1277. [CrossRef]
73. Mastalerz, M.; Drobniak, A.; Stankiewicz, A.B. Origin, properties, and implications of solid bitumen in source-rock reservoirs: A review. *Int. J. Coal Geol.* **2018**, *195*, 14–36. [CrossRef]
74. Milliken, K.L.; Zhang, T.; Chen, J.; Ni, Y. Mineral diagenetic control of expulsion efficiency in organic-rich mudrocks, Bakken Formation (Devonian-Mississippian), Williston Basin, North Dakota, U.S.A. *Mar. Petrol. Geol.* **2021**, *127*, 104869. [CrossRef]

Journal of
Marine Science and Engineering

MDPI

Article

Exploring the Unique Characteristics of High-Pore-Volume Waterflooding and Enhanced Oil Recovery Mechanisms in Offshore Sandstone Reservoirs Using Nuclear Magnetic Resonance Technology

Junrong Liu [1,2,*], Hangyu Li [1,2,*], Jianchun Xu [1,2], Shuyang Liu [1,2], Rongjiang Liu [1,2], Lianjie Hou [1,2] and Qizhi Tan [1,2]

1 School of Petroleum Engineering, China University of Petroleum (East China), Qingdao 266580, China; 20170048@upc.edu.cn (J.X.); shuyang_liu@126.com (S.L.); 1702010228@s.upc.edu.cn (R.L.); s21020144@s.upc.edu.cn (L.H.); hunantanqizhi@163.com (Q.T.)

2 Key Laboratory of Unconventional Oil & Gas Development, China University of Petroleum (East China), Qingdao 266580, China

* Correspondence: liujunrong@upc.edu.cn (J.L.); lihangyu@upc.edu.cn (H.L.)

Simple Summary: The characteristics of high-pore-volume (high-PV) waterflooding and the mechanisms of enhanced oil recovery (EOR) are unclear. an experiment was conducted using nuclear magnetic resonance (NMR) technology-assisted waterflooding to monitor oil migration and pore structure changes during the displacement process. We quantified the ultimate oil displacement efficiency waterflooding and microscopic oil recovery at the pore scale. Additionally, the variations in petrophysical properties seen during this process were analyzed. The experiment revealed that the high-PV displacement process had a high EOR effect, even in the ultrahigh water cut stage. During this period, oil production came mainly from the mesopores. Furthermore, flooding changed the rock structure, increasing the pore volume. However, these changes were inconsistent at both the pore scale and spatial scale. At the pore scale, the volume of macropores decreased, while that of the micropores and mesopores increased. At the spatial scale, the porosity and the average pore size at the front end increased, while those at the back end decreased due to particle migration during flooding. Our findings suggested that pore structure variations at the front ends of experimental cores were representative of those in reservoirs, indicating that high-PV waterflooding could increase the permeabilities of reservoirs.

Abstract: A single paragraph of about 200 words maximum. For research articles, abstracts should give a pertinent overview of the work. We strongly encourage authors to use the following style of structured abstracts, but without headings: (1) Background: Place the question addressed in a broad context and highlight the purpose of the study; (2) Methods: briefly describe the main methods or treatments applied; (3) Results: summarize the article's main findings; (4) Conclusions: indicate the main conclusions or interpretations. The abstract should be an objective representation of the article and it must not contain results that are not presented and substantiated in the main text and should not exaggerate the main conclusions.

Keywords: offshore Sandstone Reservoirs; high-PV waterflooding; EOR; NMR

Citation: Liu, J.; Li, H.; Xu, J.; Liu, S.; Liu, R.; Hou, L.; Tan, Q. Exploring the Unique Characteristics of High-Pore-Volume Waterflooding and Enhanced Oil Recovery Mechanisms in Offshore Sandstone Reservoirs Using Nuclear Magnetic Resonance Technology. *J. Mar. Sci. Eng.* **2023**, *11*, 1296. https://doi.org/10.3390/jmse11071296

Academic Editor: Timothy S. Collett

Received: 30 May 2023
Revised: 21 June 2023
Accepted: 24 June 2023
Published: 26 June 2023

1. Introduction

As the demand for oil continues to grow and onshore oilfields are gradually being depleted, offshore oil and gas resources have become important alternative energy sources [1–4]. Generally, utilizing natural waterflooding is a cost-effective and environmentally sustainable approach for enhancing oil recovery in offshore oilfields, with robust edge-bottom water energy [5–9]. To date, horizontal wells with large-volume water injection are the main development method for such offshore reservoirs, with oil recovery levels reaching 50%. Unlike conventional reservoirs, offshore reservoirs exhibit unique characteristics wherein the water cut increases swiftly after bottom water breakthrough, leading to an extended period of high water cut during the oil production stage.

During the high water cut stage, the characteristic curve of waterflooding differs from the traditional curve, which typically follows a semilogarithmic linear relationship between the oil–water relative permeability ratio (K_{ro}/K_{rw}) and the water saturation (S_W) value [10–13]. Numerous oilfield practices have indicated that the waterflooding characteristic curve tends to warp upward, even during the high water cut stage, suggesting that reservoirs possess significant production potential [7,14–17]. It has been reported that the XJ24-3 offshore oilfield has entered an ultrahigh water cut period, with oil recovery exceeding 60% since 2014. Despite this phenomenon, oil production has continued to increase steadily, as reported by Zhang et al. [16]. Tan et al. [7] found that oil recovery is still increasing by 22.2–34.2% at the high water cut stage (>98%) at the X oilfield in the Bohai Sea. In addition, the curve bends downward when the reservoir enters the ultrahigh water cut stage in indoor experiments [18,19]. Hence, the conventional seepage model cannot describe the oil production patterns of these reservoirs, especially regarding the high water cut stage with high multiple displacements.

In addition, it is difficult to predict the ultimate oil recovery of such oilfields via conventional experimental methods. According to the specification SY/T 5345—2007 [20], the final oil displacement efficiency is the extraction degree of the experimental core after 30 PV displacements or when the water cut is above 99.95%. For offshore reservoirs with strong edge-bottom waters, the displacement multiple is much greater than 30 PV, and considerable oil is still recovered after 30 PV of displacement. Zhang et al. [16] observed that the residual oil saturation decreases from 29.56% to 21.72% when the displacement multiple increases from 30 to 2000 PV, while high-PV water injection can further enhance oil recovery from the Xijiang oil field at the high water cut stage. Ji et al. [21] suggested that the residual oil saturation during the high water cut period is not constant; thus, the oil displacement efficiency measured by conventional core waterflooding experiments cannot represent the ultimate oil recovery rate of high-PV displacement reservoirs.

Thus far, several studies have indicated that wettability alteration and pore structure variation are two important EOR mechanisms of high-PV displacement waterflooding [22–26]. Xiong et al. [22] applied the contact angle to evaluate the wettability of rock during waterflooding and observed a shift in core wettability from oil–wet to intermediate–wet, which is beneficial for enhancing oil displacement efficiency [27–31]. In oil–wet systems, water tends to form continuous channels or 'fingers' through the centers of macropores, displacing oil ahead of the centers [32]. Consequently, oil becomes trapped in the small pores and 'throats'. Conversely, in water–wet systems, the distribution of the two fluids is reversed, relative to the oil–wet system. Residual oil exists in the form of small, spherical globules in the centers of the large pores, and large patches of oil extend over multiple pores that are surrounded by water [29]. These studies suggest that wettability alteration leads to changes in the microscopic distribution of residual oil. To date, most research has focused on macroscopic core oil displacement efficiency, and there have been few studies on microscopic oil recovery at the pore scale.

Additionally, previous research has established that the pore structures of oil reservoirs are significantly changed during waterflooding due to the erosion and migration of mineral particles, resulting in variations in the petrophysical properties (e.g., porosity and permeability) that affect oil displacement efficiency [30–36]. In general, the permeability of a water-driven reservoir increases due to water washing. Li et al. [37] investigated the pore structure distribution during high-PV waterflooding in the Shuanghe Oilfield and found that the pore distribution range widens and many micropores and macropores are produced, while the effect of water injection is not obvious and is hard to predict. Although an increase in macropores is beneficial for increasing permeability, an increase in micropores may not necessarily have the same effect. Hence, the effects of pore structure changes induced by high-PV waterflooding on reservoir permeability are still unclear. Additionally, researchers usually use mercury injection tests to characterize pore structure changes before and after waterflooding. However, the mercury injection experiment is destructive; thus, researchers cannot repeat the waterflooding test using the same core sample. Because of the heterogeneity of the reservoir rock, the test results are not convincing enough.

Nuclear magnetic resonance (NMR), as a non-destructive and efficient method, can be utilized to address the above problems. The transverse relaxation (T_2) spectrum, which is widely used in NMR, has been proposed as an effective approach for characterizing pore structures [37–42] and can be expressed as follows:

$$\frac{1}{T_{2surface}} = \rho_2 \left(\frac{S}{V}\right)_{pore} \tag{1}$$

where ρ_2 represents the surface relaxivity associated with the transverse relaxation process, m/s; $\frac{S}{V}$ denotes the surface-to-volume ratio of the pore, 1/m.

In general, larger pores are associated with longer T_2 values, while smaller pores tend to have shorter T_2 values. However, the pore structure changes seen in different locations are heterogeneous. A new NMR method (the SE–SPI sequence) is applied to scan each layer of the experimental core to further understand the pore structure changes in various parts of the water-driven core.

Moreover, a dynamic T_2 spectrum can accurately characterize oil distribution and oil migration among different pores [43–45]. Despite their potential benefits, few studies have investigated the impacts of high multiple displacements on microscopic oil displacement efficiency using NMR technology.

In this paper, we design an NMR technology-assisted waterflooding experiment to monitor oil migration and pore structure variations during the displacement process. The characteristics of high-PV waterflooding and the relevant EOR mechanisms are investigated. The results of this study provide a theoretical foundation and practical guidance for predicting performance and enhancing productivity in offshore sandstone reservoirs with high water cuts.

2. Methodology

An online NMR-monitored displacement experiment system was set up, as shown in Figure 1, to explore the micro EOR mechanisms and the time-varying law of petrophysical properties during high-PV waterflooding.

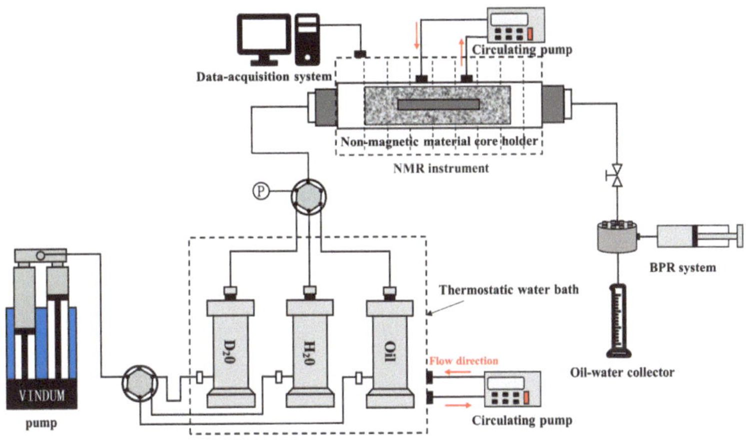

Figure 1. Nuclear magnetic resonance (NMR) technology-assisted experimental system to evaluate waterflooding.

In the experiments, the simulated displacement multiples were greatly improved to accurately reflect the washout strengths of the actual reservoirs. Nuclear magnetic imaging (NMI) technology was utilized to image the sagittal planes of the experimental cores to capture the in situ changes in residual oil during the displacement process. Moreover, the CPMG (Carr–Purcell–Meiboom–Gill) sequence was used to obtain a dynamic T_2 spectrum to quantify the oil displacement efficiency. This method has high test accuracy (1 mg of oil/water could be detected) and could avoid the errors seen in ordinary displacement experiments. Based on the corresponding relationship between relaxation time and pore size, the dynamic T_2 spectrum could be used to analyze microscopic oil recovery at the pore scale. Furthermore, a SE–SPI sequence was applied to obtain a cross-section T_2 spectrum by which to investigate the time-varying laws of petrophysical properties.

2.1. Characterization of Core and Fluid Properties

In this study, we obtained experimental cores from offshore sandstone reservoirs located in Bohai, China. These cores were prepared for waterflooding experiments by cutting into ϕ 25 × 50 mm cylinders. To track the migration of oil within the cores, we followed the processing procedure outlined in Figure 2 and employed NMR instruments. Sand filters were installed at both ends of the core to prevent particle movement from blocking the pipeline during waterflooding. Then, the experimental core was fixed with a nonmagnetic plug and a heat-shrink tube to meet the requirements of subsequent experiments. Prior to waterflooding, the experimental core was subjected to porosity and permeability measurements, which yielded values of 15.4% and 2385 mD, respectively.

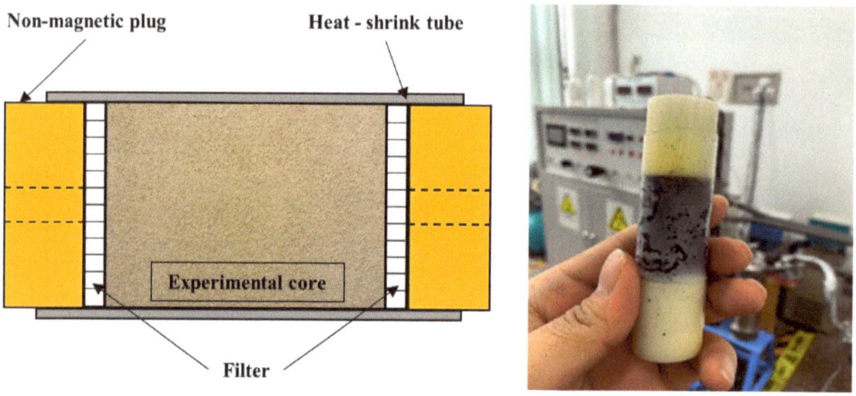

Figure 2. Experimental core treatment.

To avoid the confounding influences of various factors, such as wettability alteration due to asphaltene adsorption, we utilized kerosene and diesel as the hydrocarbon phase instead of crude oil in this study. By selecting these two substances, we could isolate the effects of interest and gain a more accurate understanding of the oil migration behaviors under the specified conditions. The mixing ratio of the two oils was adjusted so that the viscosity of the experimental oil was close to that of crude oil under formation conditions (2.1 mPa·s). Additionally, the oil density was 0.83 g/cm^3 at the experimental temperature.

For the waterflooding experiment, heavy water (D_2O) was chosen as the displacing fluid due to its unique properties. Unlike ordinary water (H_2O), heavy water does not generate NMR signals in a magnetic field. By using heavy water, the NMR signals from the water were eliminated, enabling the observation of oil migration within the pore space using the NMR instrument.

2.2. NMR-Monitored High-PV Waterflooding Experiments

To effectively monitor the movement of oil during waterflooding, we developed a high-PV waterflooding experimental system that utilized NMR technology, as illustrated in Figure 1. The experimental system comprises several essential components, including a piston pump, three accumulators, a core holder, an NMR instrument, a BPR (back pressure regulator) device, and two circulating pumps. The circulating pumps play a crucial role in maintaining the stability of the system temperature. It is worth highlighting that the core of this system features a low-field NMR instrument manufactured by the Niumag Corporation in China. By utilizing the MRI (magnetic resonance imaging) module of the instrument, we could visualize the residual oil and qualitatively assess the effectiveness of oil displacement during waterflooding. Furthermore, the dynamic T_2 spectrum generated by inversion provided a quantitative analysis of microscopic displacement efficiency at the pore scale. The SE–SPI sequence was applied to characterize the pore distributions before and after high-PV waterflooding. Details of the experimental parameters for both modes are provided in Table 1.

Table 1. Key parameters of the NMR instruments.

Function	Pulse Sequences	Field Strength (MHz)	Gradient Strength (T/m)	TW (ms)	TE (ms)	Scanning Times	Temperature (°C)
T_2	CPMG	12	0	1000	0.1	64	32
cross-section T_2	SE-SPI	12	0.0487	1000	0.1	64	32

TW: waiting times; TE: echo spacings.

The experimental protocol followed the industry standard, SY/T 5345—2007, except for an increased displacement multiple of 2000 PV. The specific steps involved in the experiment were as follows. (1) The experimental core was thoroughly cleaned using a toluene–ethanol solution and dried at a temperature of 105 °C. The base NMR signal of the core was then measured. (2) The core was saturated with DI water (H_2O), at an injection rate of 0.05 mL/min, and the resulting water-saturated core was tested using the NMR apparatus to acquire the whole and cross-sectional T_2 spectra. (3) The core was saturated with oil at the same rate, to simulate the initial reservoir, and the NMR signal of the oil-saturated core in its initial state was recorded. (4) High-PV waterflooding experiments were conducted by injecting heavy water at the design speed (10 mL/min) and by recording NMR signals after the 0.5 PV, 1 PV, 30 PV, and 2000 PV waterflooding stages; (5) Following the 2000 PV stage, DI water was used to flood the core for an additional 10 PV to replace the heavy water. Finally, the NMR signals of the core were measured to characterize its pore structure after high-PV waterflooding.

3. Results and Discussion

3.1. Variations in Oil Displacement Efficiency

The NMR instrument captures sagittal images of the experimental cores during waterflooding, as illustrated in Figure 3. Since the NMR signals of the displacement fluid (heavy water) and rock skeleton are very weak, the images primarily reflect the amount of residual oil remaining in the core. The gradual reduction in the NMR signal during waterflooding indicates that this technique effectively reduces the residual oil saturation.

However, after 1 PV of waterflooding, the residual oil reduction is more significant at the injection end than at the exit end, suggesting spatial heterogeneity in the oil displacement effect. After 30 PV, the residual oil at the injection end is further reduced, but some large oil aggregates remain at the exit end due to microscopic heterogeneity. This heterogeneity is due to the presence of local areas with low permeability in the reservoir, which causes high seepage resistance of the crude oil and leads to the formation of microscopic heterogeneous residual oil.

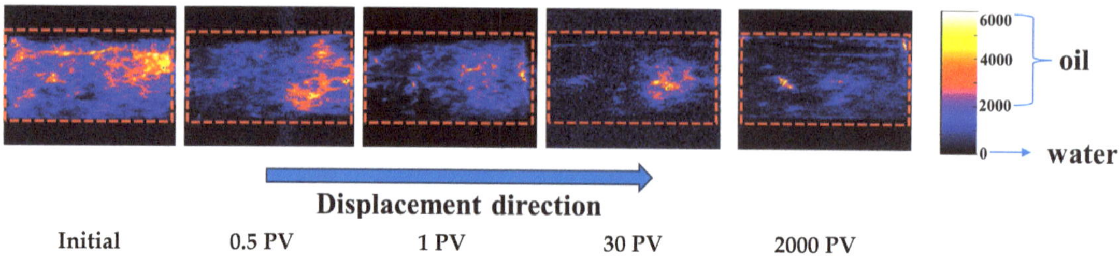

Figure 3. Sagittal images of experimental cores during waterflooding.

As the waterflooding volume increases to 2000 PV, the residual oil greatly decreases, and the oil aggregates disappear, leaving sporadic dot-like residual oil inside the core. These MRI experimental results demonstrate that significant amounts of oil remain in the reservoirs even after conventional waterflooding (<30 PV) and that high-volume waterflooding can further enhance oil displacement efficiency.

To quantify the EOR effect of high-PV waterflooding, we recorded the dynamic T_2 spectrum during the displacement process, as shown in Figure 4. Theoretically, the amount of hydrogen-containing fluid is proportional to its NMR signals (the envelope area of the T_2 spectrum), based on the principles behind NMR technology.

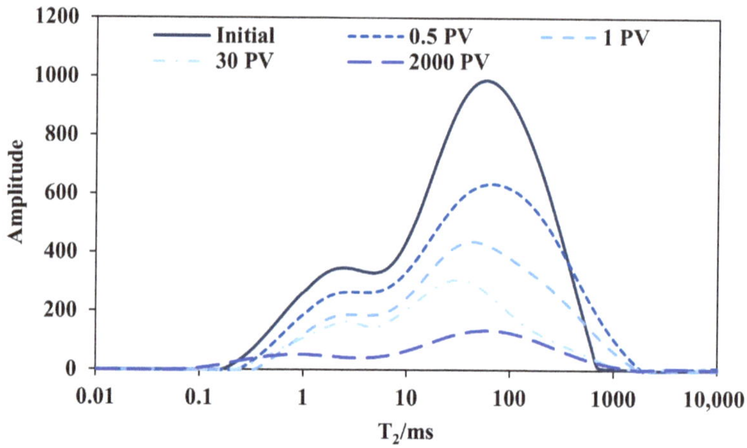

Figure 4. Dynamic T_2 spectrum of the core during waterflooding.

To validate the instrument and confirm its ability to accurately reflect changes in oil volume within the cores, we measured the total NMR signals of cores saturated with varying volumes of oil, as illustrated in Figure 5. The results indicate a strong linear relationship between NMR signals and oil volume, confirming the effectiveness of the instrument and its ability to accurately reflect changes in oil volume within the cores. This finding enables us to derive the NMR signal–oil volume conversion relationship, as presented in Equation (2), and gain a better understanding of the oil migration process within the studied porous media:

$$V = 10^{-4}Q - 0.1133 \quad (2)$$

where Q represents the total signal of NMR; V represents the oil volume in mL.

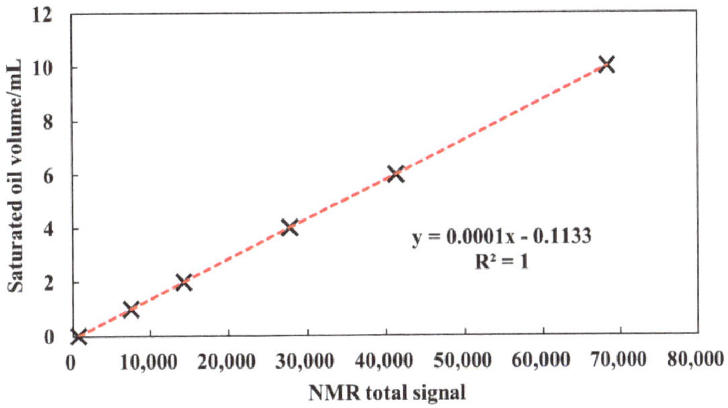

Figure 5. Relationship between the total NMR signals and experimental oil volume.

Then, the oil displacement efficiency can be calculated as follows:

$$R = \frac{V_i - V_r}{V_i} = 1 - \frac{V_r}{V_i} = 1 - \frac{10^{-4}Q_r - 0.1133}{10^{-4}Q_i - 0.1133} \tag{3}$$

where R represents the oil displacement efficiency; V_i represents the initial saturated oil volume in mL; V_r represents the residual oil volume in mL; Q_i represents the total NMR signal of the initial state core; Q_r represents the total NMR signal of the core after waterflooding.

Figure 6 illustrates how waterflooding affects oil displacement efficiency in reservoirs with varying PV values. Moreover, the water cut under different displacement multiples should be measured. In the process of waterflooding, the water cut increases rapidly and reaches 95.1% after water injection at 1 PV. This finding implies that the anhydrous oil production period is short and that the water cut increases rapidly after the water breakthrough. As the displacement multiple reaches 30 PV, the water cut grows to 99.5%, whereas the residual oil saturation is still as high as 33%; thus, the potential of EOR is very high. At the high-PV waterflooding stage (30–200 PV), the oil displacement efficiency increases from 67.0% to 83.7%, with a water cut of more than 99%. These experimental results indicate that the high-PV displacement process still has a good EOR effect at the ultrahigh water cut stage.

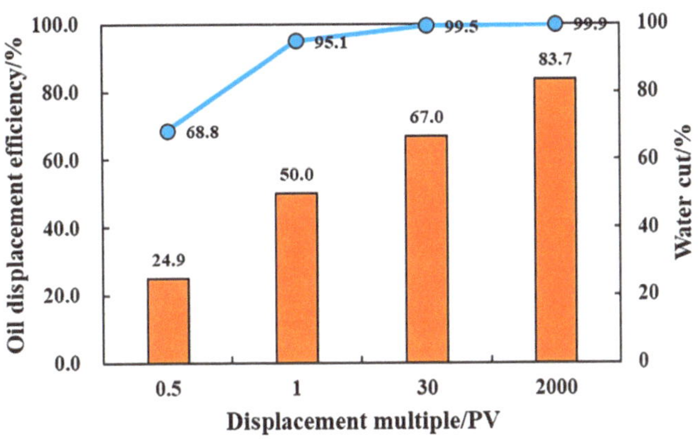

Figure 6. Oil displacement efficiency and water cut values under different displacement multiples.

3.2. Microscopic Oil Displacement Efficiency at the Pore Scale

To further investigate the microscopic EOR mechanisms of high-PV waterflooding, we calculated the oil displacement efficiencies in different pores using dynamic T_2 spectra. This spectrum not only characterizes the oil content of the rock but also reflects the distribution of oil in the pores. To quantify the correspondence between pore size and T_2 relaxation time, the T_2 spectrum (initial oil-saturated core) and mercury injection results were integrated, as demonstrated in Figure 7. We then selected the corresponding peak points (marked with green and black circles) of the two curves to calculate the conversion coefficient with Equation (4) [46]:

$$T_2 = Cr \tag{4}$$

where T_2 is the relaxation time in ms; C is the constant conversion coefficient in ms/μm; and r is the pore radius in μm.

We obtained very similar coefficients of 0.38 and 0.39 from our calculations, indicating a strong correspondence between the T_2 relaxation time and pore radius. The average coefficient, 0.385, was utilized in this experiment. Consequently, the dynamic T_2 spectrum illustrated in Figure 4 was used to characterize the microscopic oil displacement efficiencies of pores with varying radii.

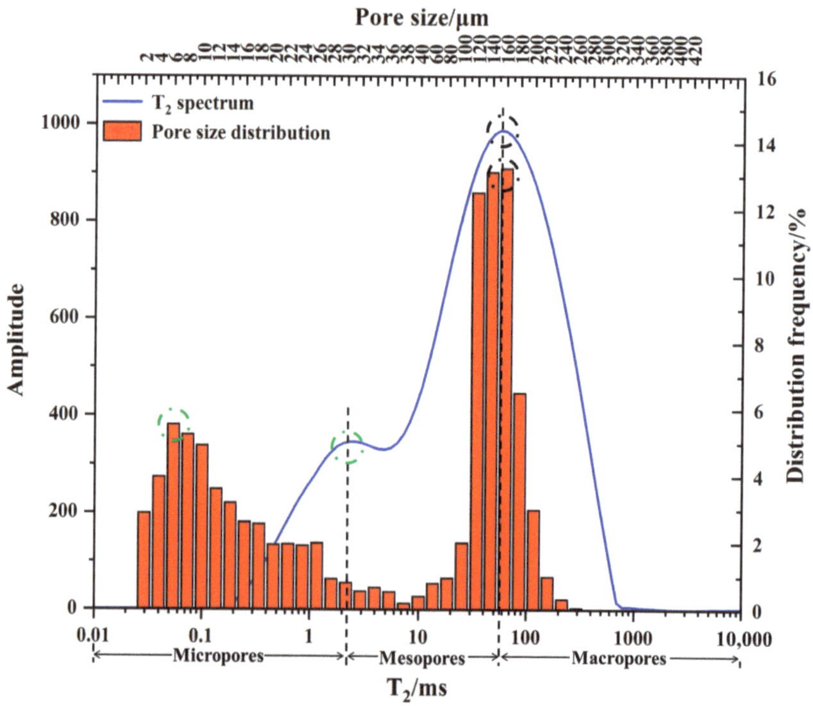

Figure 7. Pore size distribution in the experimental core (NMR and mercury injection test).

Subsequently, we classified the pore types based on their T_2 values, with the peak points being used as the cutoff values to divide the spectra, as described by Dai et al. [47]. Specifically, we classified the pores corresponding to T_2 values ranging from 0–2.3 ms, 2.3–62.4 ms, and 62.4–10,000 ms as micropores, mesopores, and macropores, respectively, as presented in Table 2.

Table 2. Pore size classification.

T_2 Relaxation Time, ms	Pore Radius (R), μm	Pore Type
$T_2 \leq 2.3$	R ≤ 6.0	Micropore
$2.3 < T_2 \leq 62.4$	6.0 < R ≤ 162.1	Mesopore
$T_2 > 62.4$	R > 162.1	Macropore

The microscopic oil displacement efficiencies in different types of pores can be quantified as follows:

$$R_x = 1 - \frac{V_{rx}}{V_{ix}} = 1 - \frac{10^{-4}Q_{rx} - 0.1133}{10^{-4}Q_{ix} - 0.1133} \tag{5}$$

where R_x is the oil displacement efficiency of x-type pores; x refers to different pore types (micropore, mesopore, and macropore); V_{rx} is the residual oil volume in x-type pores, measured in mL; V_{ix} is the initial oil-saturated volume in x-type pores, measured in mL; Q_{ix} is the total NMR signal in the relaxation interval to which the initial x-type pores belong. Q_{xr} is the total NMR signal in the relaxation interval of the x-type pores after waterflooding.

Figure 8 shows the microscopic oil displacement efficiencies in different pore types after various displacement multiples. Due to the greater capillary resistance found in smaller pores, the displacement of stored oil by waterflooding becomes increasingly difficult. As a result, after waterflooding at 1 PV, the oil displacement efficiencies of micropores, mesopores, and macropores are roughly equivalent. During conventional low-multiple waterflooding (1–30 PV), the oil displacement efficiencies in mesopores and macropores increase by 11.8% and 23.1%, respectively, while the increase in micropores is only 1%. However, the residual oil saturation levels in micropores and mesopores remain high, at 41.9% and 38.7%, respectively.

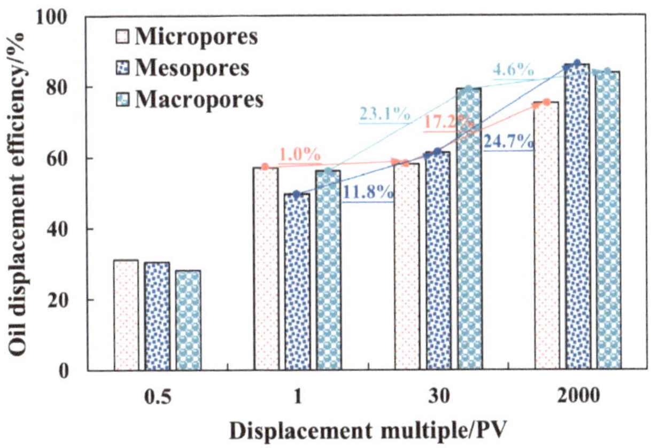

Figure 8. Oil displacement efficiency in different types of pores during waterflooding.

In the high-PV waterflooding stage (30–2000 PV), the oil displacement efficiencies in micropores and mesopores increased by 17.2% and 24.7%, respectively, while the improvement in macropores was 4.6%. The washing action of high-PV waterflooding can alter the wettability of the oil–wet rock surface, enabling the displacement of the oil trapped in micropores or mesopores by capillary force. As most of the oil in macropores is displaced during the low-multiple waterflooding stage (<30 PV), the oil displacement efficiencies in macropores do not improve noticeably during high-PV waterflooding.

In addition, the proportion of oil production from different types of pores during high-PV (30–2000 PV) waterflooding is demonstrated in Figure 9. The results suggest that mesopores are the primary contributors to oil production, accounting for 70.7% of the total production. Although the oil displacement efficiencies in micropores improve significantly, contributing to 14.6% of total production, their contribution is not outstanding due to the tight oil storage space.

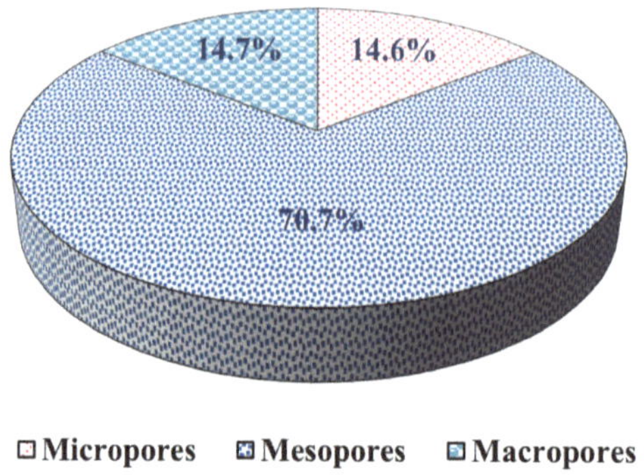

Figure 9. Contributions of the different types of pores to oil production during high-PV (30–2000 PV) waterflooding.

3.3. Time-Varying Petrophysical Properties of High Multiple Waterflooding Cores

3.3.1. Pore Structure Variation

The T_2 spectra of the core that was fully saturated with DI water before and after flooding are presented in Figure 10A. Notably, after heavy waterflooding, the core shows almost no NMR signal. To characterize the pore structure after waterflooding, we injected 10 PV of DI water to displace the heavy water. The relationship between the total NMR signal and the saturated DI water volume is measured and plotted in Figure 10B. The volume of saturated water is equal to the pore volume of the experimental core. Therefore, we calculated the porosities before and after waterflooding to be 15.4% and 17.9%, respectively. This finding indicates that high-PV waterflooding changes the rock structure and increases the rock pore volume.

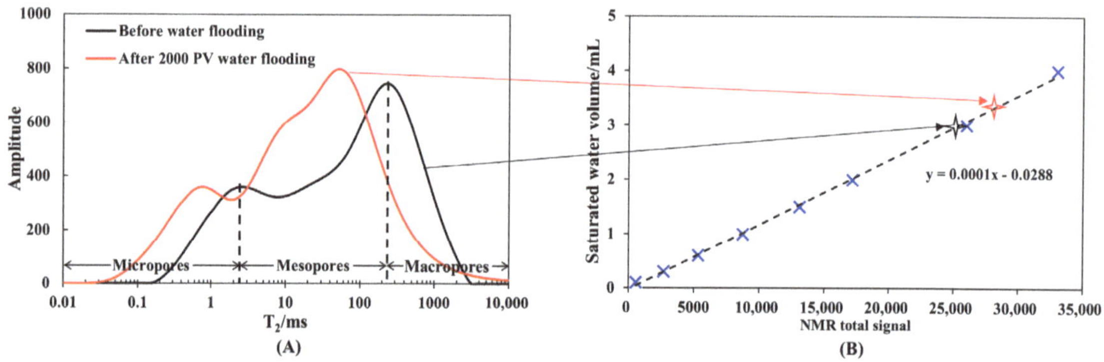

Figure 10. T_2 spectrum of the DI water-saturated core before and after waterflooding (**A**). Relationship between the total NMR signal and saturated DI water volume (**B**).

To investigate the increments in different types of pores, we applied the pore classification method described in Section 3.2 and plotted the results as shown in Figure 11. The volumes of the mesopores and micropores increased by 0.59 mL and 0.32 mL, respectively, while that of the macropores decreased by 0.47 mL. Although the overall porosity of rock increased after high-PV waterflooding, the changes in the different types of pores were not consistent. Among the pore types, the porosity of the mesopores increased the most, whereas that of the macropores decreased.

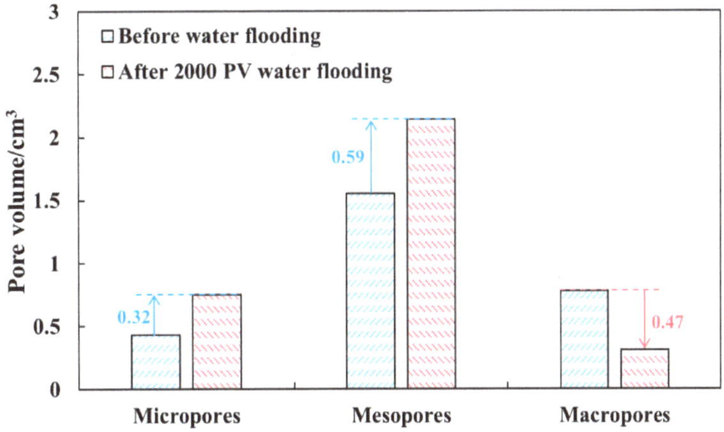

Figure 11. Volumes of the various pores before and after high-PV waterflooding.

To further understand the pore structure variations at different locations of the experimental core, we applied the NMR tomography method to scan each layer, as shown in Figure 12. By comparing the cross-section T_2 spectrum along the water-driving direction before and after waterflooding, we observed different pore structure changes at the front and back of the core. The peak value of the front T_2 spectrum shifted to the right, indicating that the mean pore size increased, whereas the peak value of the back T_2 spectrum shifted to the left, indicating that the mean pore size decreased.

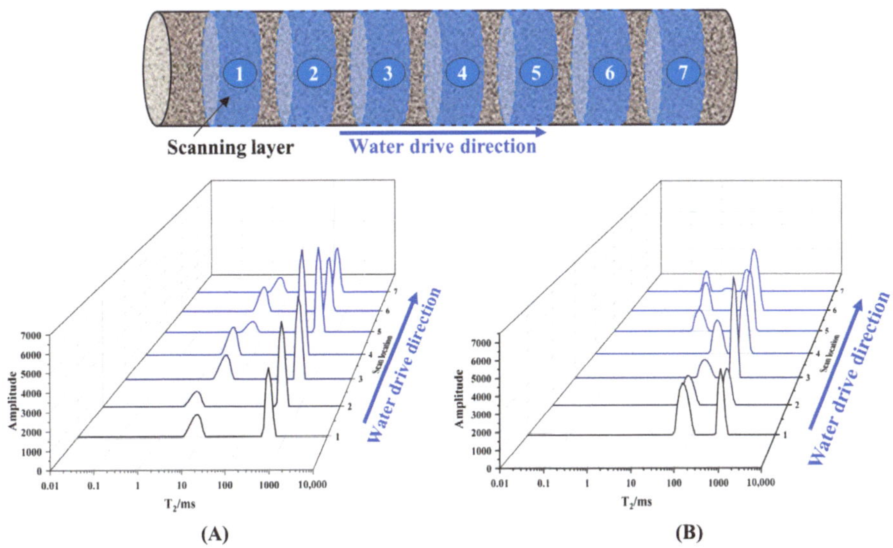

Figure 12. Cross-section T_2 spectrum along the water-driving direction before waterflooding (**A**) and after 2000 PV waterflooding (**B**).

Moreover, we quantified the pore structure variations in the different sections using a similar method to that in Section 3.2, as shown in Figure 13. The pore volumes of each section at the front increased to different degrees, while those at the back section decreased. Recent evidence suggests that the pore structure variation seen during high-PV waterflooding is mainly caused by the particle migration induced by the flushing action of the displacement fluid. Near the injection end, loose rock particles peel off and migrate, increasing the pore volume. As the outlet widens, the flow rate slows, resulting in the accumulation and migration of particles at the outlet, and the pore volume decreases accordingly.

Furthermore, the changes in the different types of pores before and after waterflooding in layers 1 (injection end) and 7 (outlet end) were compared, as shown in Figure 14. In layer 1, the volumes of mesopores and macropores increased by 0.17 mL and 0.01 mL, respectively, while the micropores were basically eliminated. This phenomenon can be attributed to the stripping of particles within a micropore to form a mesopore or macropore, increasing the average pore size at the front end. In contrast, in layer 7, the volume of mesopores decreased sharply by 0.27 mL, while the volume of micropores increased by 0.08 mL, and the volume of macropores remained relatively stable. This result suggests that the migrated particles mainly accumulate in the mesopores, transforming them into micropores, and decreasing the average pore size at the back end. The experimental results reveal that the pore structure variation induced by high-PV waterflooding is inconsistent at the microscale and at the spatial scale.

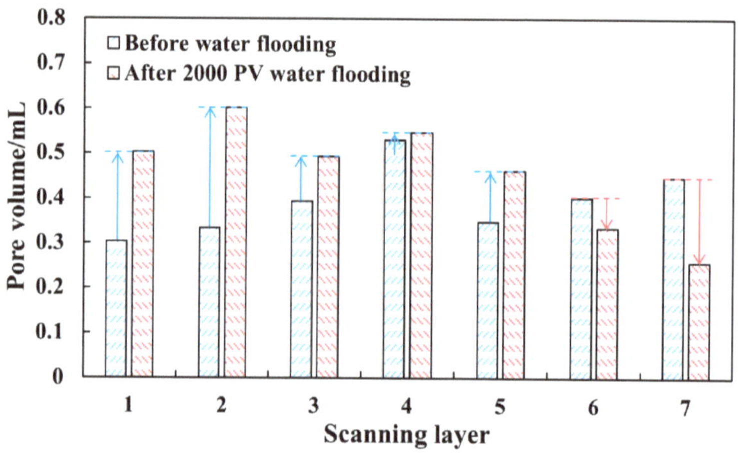

Figure 13. Pore volumes of the different scanning layers before and after waterflooding.

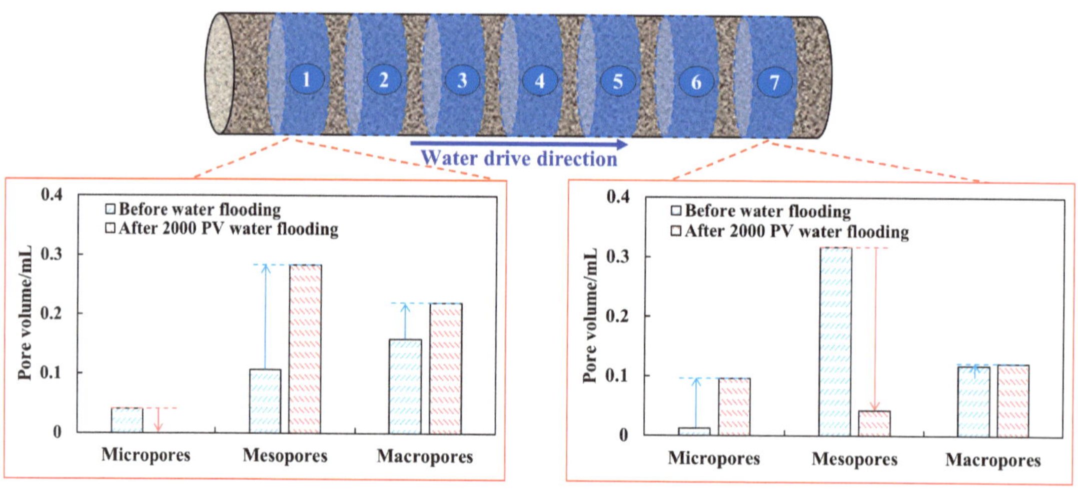

Figure 14. Pore volume variations in the different scanning layers before and after waterflooding.

3.3.2. Permeability Variation

To further analyze the impact of high-PV waterflooding on the reservoir, it is essential to investigate the corresponding permeability variation. The change in pore structure induced by high-PV waterflooding inevitably leads to permeability variation. As the displacement process progresses, the water cut exceeds 95%, and the instantaneous oil/water saturation change is minimal, indicating that the high-PV waterflooding process is a steady-state process. Hence, the water relative permeability can be calculated using Darcy's formula (shown in Equation (6)) by monitoring the pressure change at the injection end during the displacement process, as depicted in Figure 15:

$$K_{rw} = \frac{Q_w \mu_w L}{A \Delta P} \tag{6}$$

where K_{rw} is the water relative permeability, measured in mD; Q_w is the water flux, measured in cm^3/s; μ_w is the viscosity of water, measured in mPa·s; L is the length of the experimental core in cm; A is the cross-section area of the experimental core in cm^2; ΔP is the pressure difference between the two ends of the core, measured in 10^{-1} MPa.

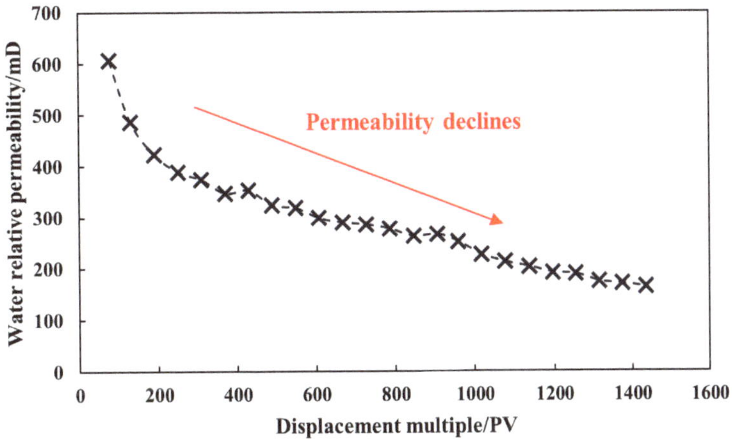

Figure 15. Variations in water relative permeability during waterflooding.

The findings indicate that increasing the displacement multiple decreases water relative permeability. This phenomenon is contradictory to the theoretical prediction that increasing water saturation and porosity increases the water relative permeability [48–50]. This contradiction can be attributed to the inconsistency of rock structure variations, on both the pore scale and the spatial scale. Although the overall porosity of rock increases, at the pore scale, the volume of macropores decreases, while that of micropores and mesopores increases. At the spatial scale, the pore volume at the front increases while that at the back decreases, and the average pore size at the front increases while that at the back decreases. Thus, the permeability at the front increases, while that at the back decreases. The experimental core can be considered to be a series of small cores (scanning layers), and the overall permeability can be calculated as follows:

$$K = \frac{\sum_{i=1}^{n} L_i}{\sum_{i=1}^{n} \frac{L_i}{K_i}} \tag{7}$$

where K represents the overall permeability of the experimental core, measured in mD; i refers to the scanning layers; n refers to the number of scanning layers; L_i represents the length of the i-layer core in cm; and K_i represents the permeability of the i-layer core, measured in mD. According to Equation (7), the overall permeability of the experimental core is governed by the layer with the lowest permeability. According to the experimental results, the particle accumulation at the outlet of the core caused a sharp reduction in the permeability of the entire core. The increase in micropores and the decrease in mesopores seen in layer 7 due to the migration of fine particles decreased the average pore size at the back end, thereby contributing to the permeability reduction. This finding underscores the crucial role of pore structure variations induced by high-PV waterflooding in determining the overall permeability of the rock core.

Notably, particle accumulation at the end of the experimental core during waterflooding may not necessarily occur in actual reservoirs. As shown in Figure 16, reservoirs lack filters that prevent rock particle migration and the capillary end effect, which promotes particle deposition at the outlet of the core samples. Consequently, the pore structure variation at the front of the experimental cores may better reflect the actual changes caused by high-PV waterflooding; the permeability of reservoirs after such flooding is likely to increase, which contrasts with the experimental results. Further studies are needed to explore methods for accurately detecting the change in permeability during high-PV waterflooding in real reservoirs.

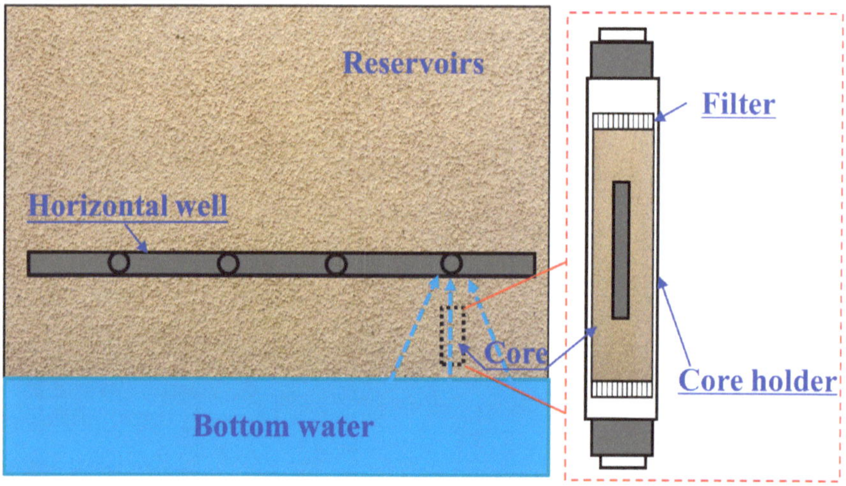

Figure 16. Comparison of the waterflooding process at reservoir scale and laboratory scale.

4. Conclusions

In this study, we present a new NMR-monitored displacement experimental method for investigating microscale EOR mechanisms and time-varying petrophysical properties during high-PV waterflooding. The main findings can be summarized as follows:

(1) Conventional waterflooding ($\leq$30 PV) results in brief oil production without the presence of water. However, it is followed by a swift rise in the water cut reach of up to 99%, while the residual oil saturation is still as high as 33%.

(2) The high-PV displacement process shows a high EOR effect even in the ultrahigh water cut stage, resulting in a notable increase in the recovery rate by 16.7%. Notably, a significant portion of oil production, amounting to 70.7% of the total, primarily originates from the mesopores.

(3) High-PV waterflooding induces changes in the rock structure, increasing porosity. However, the structure variation exhibits inconsistency at the pore scale with a decreased volume of macropores and increased volumes of micropores and mesopores.

(4) The pore structure variation during high-PV waterflooding shows heterogeneity in the spatial scale, with increased pore volume at the front and decreased pore volume at the back. These changes in pore structure are the main reasons for the decline in rock permeability.

(5) The changes in permeability obtained in the core experiment may not reflect those in reservoirs since particle accumulation at the end of the core during waterflooding may not occur in actual reservoirs. The pore structure variation at the front of the experimental cores may truly reflect that found in reservoirs, indicating that the permeability of reservoirs after high-PV waterflooding is likely to be increased.

Overall, the results of this study provide valuable insights into high-PV waterflooding and highlight the need for further research on methods for detecting changes in permeability during the high-PV waterflooding processes of real reservoirs.

Author Contributions: Conceptualization, Methodology, Investigation, Formal analysis, Writing—original draft, Writing—review & editing, J.L.; Supervision, Funding acquisition, Project administration, Writing—review & editing, H.L.; Conceptualization, Supervision, J.X.; Methodology, Investigation, Validation, S.L.; Methodology, Investigation, R.L.; Methodology, Investigation, L.H.; Writing—review & editing, Q.T. All authors have read and agreed to the published version of the manuscript.

Funding: This research was funded by the Shandong Provincial Natural Science Foundation, China (No. ZR2021JQ18) and the National Natural Science Foundation of China (No. 52074337 and No. 52204065, and No. 51904323).

Institutional Review Board Statement: The study did not require ethical approval.

Informed Consent Statement: The study did not involve humans.

Data Availability Statement: The data presented in this study are available on request from the corresponding author.

Conflicts of Interest: The authors declare no conflict of interest.

References

1. Pang, X.; Li, H.; Pang, H. Exploring the mysteries of deep oil and gas formation in the South China Sea to guide Palaeocene exploration in the Pearl River Mouth Basin. *Adv. Geo-Energy Res.* **2022**, *6*, 361–362. [CrossRef]
2. Wang, Y.; Estefen, S.F.; Lourenço, M.I.; Hong, C. Optimal design and scheduling for offshore oil-field development. *Comput. Chem. Eng.* **2019**, *123*, 300–316. [CrossRef]
3. Wang, X.; Liu, W.; Shi, L.; Liang, X.; Wang, X.; Zhang, Y.; Wu, X.; Gong, Y.; Shi, X.; Qin, G. Application of a novel amphiphilic polymer for enhanced offshore heavy oil recovery: Mechanistic study and core displacement test. *J. Pet. Sci. Eng.* **2022**, *215*, 110626. [CrossRef]
4. Gupta, V.; Grossmann, I.E. Offshore oilfield development planning under uncertainty and fiscal considerations. *Optim. Eng.* **2016**, *18*, 3–33. [CrossRef]
5. Zhang, Y.; Li, X.; Sun, X.; Zheng, W.; Tan, X.; Cai, J. Experimental and Numerical Studies of Supercritical Water Flooding for Offshore Heavy Oil Recovery. *Geofluids* **2022**, *2022*, 5362235. [CrossRef]
6. Alvarado, V.; Manrique, E.J. Engineering design challenges and opportunities beyond waterflooding in offshore reservoirs. In Proceedings of the Offshore Technology Conference, Houston, TX, USA, 6–9 May 2013.
7. Tan, J.; Liu, Y.-X.; Li, Y.-L.; Liu, C.-Y.; Mou, S.-R. Study on oil displacement efficiency of offshore sandstone reservoir with big bottom water. *J. Pet. Explor. Prod. Technol.* **2021**, *11*, 3289–3299. [CrossRef]
8. Guo, T.; Su, Y. Current status and technical development direction in heavy oil resrevoir development in Bohai oilfields. *China Offshore Oil Gas* **2013**, *25*, 26–30.
9. Liu, C.; Zhou, W.; Jiang, J.; Shang, F.; He, H.; Wang, S. Remaining Oil Distribution and Development Strategy for Offshore Unconsolidated Sandstone Reservoir at Ultrahigh Water-Cut Stage. *Geofluids* **2022**, *2022*, 6856298. [CrossRef]
10. Xiao, Y.; He, Y.; Zheng, J.; Zhao, J. Modeling of two-phase flow in heterogeneous wet porous media. *Capillarity* **2022**, *5*, 41–50. [CrossRef]
11. Roghanian, R.; Rasaei, M.R.; Haghighi, M. Prediction of Key Points of Water-oil Relative Permeability Curves Using the Linear Regression Technique. *Pet. Sci. Technol.* **2012**, *30*, 518–533. [CrossRef]
12. Can, B.; Kabir, C. Simple tools for forecasting waterflood performance. *J. Pet. Sci. Eng.* **2014**, *120*, 111–118. [CrossRef]
13. Gao, W.; Rui, Y.; Yang, J. Establishment and theoretical basis of a new type of water drive characteristic curve. *Acta Pet. Sin.* **2020**, *41*, 342–347.
14. Song, Z.; Li, Z.; Lai, F.; Liu, G.; Gan, H. Derivation of water flooding characteristic curve for high water-cut oilfields. *Pet. Explor. Dev.* **2013**, *40*, 216–223. [CrossRef]
15. Wang, J.; Shi, C.; Ji, S.; Li, G.; Chen, Y. New water drive characteristic curves at ultra-high water cut stage. *Pet. Explor. Dev.* **2017**, *44*, 1010–1015. [CrossRef]

16. Zhang, W.; Cao, R.; Luo, D.; Sun, C.; Li, W. Displacement characteristics of high-multiple water drive in marine sandstone reservoirs in the Pearl River Mouth Basin, South China Sea. *Pet. Geol. Recovery Effic.* **2018**, *25*, 64–71.
17. Liu, Z.; Yang, Y.; Yao, J.; Zhang, Q.; Ma, J.; Qian, Q. Pore-scale remaining oil distribution under different pore volume water injection based on CT technology. *Adv. Geo-Energy Res.* **2017**, *1*, 171–181. [CrossRef]
18. Bondar, V.V.; Blasingame, T.A. Analysis and interpretation of water-oil-ratio performance. In Proceedings of the SPE Annual Technical Conference and Exhibition, San Antonio, TX, USA, 29 September–2 October 2002.
19. Cao, R.; Jia, Z.; Cheng, L.; Wang, Z.; Ye, T.; Rui, Z. Using high-intensity water flooding relative permeability curve for predicting mature oilfield performance after long-term water flooding in order to realize sustainable development. *J. Pet. Sci. Eng.* **2022**, 110629. [CrossRef]
20. *SY/T 5345-2007*; Test Method for Two Phase Relative Permeability in Rock. Petroleum Industry Press: Beijing, China, 2007.
21. Ji, S.; Tian, C.; Shi, C.; Ye, J.; Zhang, Z.; Fu, X. New understanding on water-oil displacement efficiency in a high water-cut stage. *Pet. Explor. Dev.* **2012**, *39*, 338–345. [CrossRef]
22. Jackson, M.D.; Valvatne, P.H.; Blunt, M.J. Prediction of Wettability Variation within an Oil/Water Transition Zone and Its Impact on Production. *SPE J.* **2005**, *10*, 185–195. [CrossRef]
23. Dai, Z.; Jiang, J.; Li, H.L.; Cao, R.Y.; Xin, J.; Luo, D.H. Wettability experiment of core and mico-mechanism during high-multiple water flooding in heavy oil reservoir. *Sci. Technol. Eng.* **2019**, *19*, 157–163.
24. Xiong, Q.; Yan, Z.; Li, L.; Yang, Y.; Wang, Y.; Chen, X. Study on Ultimate Recovery Factor and Oil Recovery Rate of Water-Flooded Sandstone Oil Fields in Eastern South China Sea. In Proceedings of the International Petroleum Technology Conference, Shanghai, China, 25–27 August 2021.
25. Cao, R.-Y.; Dai, Z.; Wang, Z.-K.; Wang, Y.-H.; Jiang, J.; Li, H.-L.; Jia, Z.-H. Displacement behavior and mechanism of long-term water flooding in sandstone oil reservoirs. *J. Central South Univ.* **2021**, *28*, 834–847. [CrossRef]
26. Wang, S.; Han, X.; Dong, Y.; Shi, H. Mechanisms of reservoir pore/throat characteristics evolution during long-term waterflooding. *Adv. Geo-Energy Res.* **2017**, *1*, 148–157. [CrossRef]
27. Anderson, W. Wettability Literature Survey-Part 6: The Effects of Wettability on Waterflooding. *J. Pet. Technol.* **1987**, *39*, 1605–1622. [CrossRef]
28. Morrow, N.R. Wettability and Its Effect on Oil Recovery. *J. Pet. Technol.* **1990**, *42*, 1476–1484. [CrossRef]
29. Agbalaka, C.C.; Dandekar, A.Y.; Patil, S.L.; Khataniar, S.; Hemsath, J. The effect of wettability on oil recovery: A review. In Proceedings of the SPE Asia Pacific Oil and Gas Conference and Exhibition, Perth, Australia, 20–22 October 2008.
30. Meng, M.; Ge, H.; Shen, Y.; Ji, W.; Wang, Q. Rock Fabric of Tight Sandstone and Its Influence on Irreducible Water Saturation in Eastern Ordos Basin. *Energy Fuels* **2023**, *37*, 3685–3696. [CrossRef]
31. Meng, M.; Peng, J.; Ge, H.; Ji, W.; Li, X.; Wang, Q. Rock Fabric of Lacustrine Shale and Its Influence on Residual Oil Distribution in the Upper Cretaceous Qingshankou Formation, Songliao Basin. *Energy Fuels* **2023**, *37*, 7151–7160. [CrossRef]
32. Raza, S.H.; Treiber, L.E.; Archer, D.L. Wettability of reservoir rocks and its evaluation. *Prod. Mon.* **1968**, *32*.
33. Song, W.; Sun, H.; Sun, G.; Wu, S. Dynamical geologlc process of development liquid–Taking Shengtuo oilfield as an example. *Acta Pet. Sin.* **2002**, *23*, 52–55.
34. Alhuraishawy, A.K.; Bai, B.; Wei, M.; Geng, J.; Pu, J. Mineral dissolution and fine migration effect on oil recovery factor by low-salinity water flooding in low-permeability sandstone reservoir. *Fuel* **2018**, *220*, 898–907. [CrossRef]
35. Afrough, A.; Zamiri, M.S.; Romero-Zerón, L.; Balcom, B.J. Magnetic-Resonance Imaging of Fines Migration in Berea Sandstone. *SPE J.* **2017**, *22*, 1385–1392. [CrossRef]
36. Dan, H.; Yao, G.; Zhou, F. The research of reservoir structural changes after water flooding. *Offshore Oil* **2004**, *24*, 62–66.
37. Li, J.; Cai, Y.; Cui, Y. Variation features of reservoir pore structure after water flush over a long period of time. *Pet. Geol. Recovery Effic.* **2002**, *9*, 68–70.
38. Toumelin, E.; Torres-Verdín, C.; Sun, B.; Dunn, K.-J. Random-walk technique for simulating NMR measurements and 2D NMR maps of porous media with relaxing and permeable boundaries. *J. Magn. Reson.* **2007**, *188*, 83–96. [CrossRef]
39. Meng, M.; Ge, H.; Shen, Y.; Ji, W.; Li, Z. Insight into Water Occurrence and Pore Size Distribution by Nuclear Magnetic Resonance in Marine Shale Reservoirs, Southern China. *Energy Fuels* **2022**, *37*, 319–327. [CrossRef]
40. Liu, J.; Sheng, J.J. Investigation of Countercurrent Imbibition in Oil-Wet Tight Cores Using NMR Technology. *SPE J.* **2020**, *25*, 2601–2614. [CrossRef]
41. Hanafy, A.; Nasr-El-Din, H.A.; Heidari, Z. A Novel Method To Assess Stimulation of Sandstone Cores Damaged by Fines Migration. *SPE J.* **2021**, *27*, 682–704. [CrossRef]
42. Chi, L.; Heidari, Z. Directional-Permeability Assessment in Formations With Complex Pore Geometry With a New Nuclear-Magnetic-Resonance-Based Permeability Model. *SPE J.* **2016**, *21*, 1436–1449. [CrossRef]
43. Liu, J.; Sheng, J.J. Experimental investigation of surfactant enhanced spontaneous imbibition in Chinese shale oil reservoirs using NMR tests. *J. Ind. Eng. Chem.* **2018**, *72*, 414–422. [CrossRef]
44. Yousef, A.A.; Al-Saleh, S.; Al-Kaabi, A.; Al-Jawfi, M. Laboratory investigation of novel oil recovery method for carbonate reservoirs. In Proceedings of the Canadian Unconventional Resources and International Petroleum Conference, Calgary, AB, Canada, 19–21 October 2010.
45. Yang, P.; Guo, H.; Yang, D. Determination of Residual Oil Distribution during Waterflooding in Tight Oil Formations with NMR Relaxometry Measurements. *Energy Fuels* **2013**, *27*, 5750–5756. [CrossRef]

46. Zhao, H.; Ning, Z.; Zhao, T.; Che, F.; Zhang, R.; Hou, T. Applicability comparison of nuclear magnetic resonance and mercury injection capillary pressure in characterisation pore structure of tight oil reservoirs. In Proceedings of the SPE Asia Pacific Unconventional Resources Conference and Exhibition, Brisbane, Australia, 9–11 November 2015.
47. Dai, C.; Cheng, R.; Sun, X.; Liu, Y.; Zhou, H.; Wu, Y.; You, Q.; Zhang, Y.; Sun, Y. Oil migration in nanometer to micrometer sized pores of tight oil sandstone during dynamic surfactant imbibition with online NMR. *Fuel* **2019**, *245*, 544–553. [CrossRef]
48. Geffen, T.; Owens, W.; Parrish, D.; Morse, R. Experimental Investigation of Factors Affecting Laboratory Relative Permeability Measurements. *J. Pet. Technol.* **1951**, *3*, 99–110. [CrossRef]
49. Burdine, N. Relative Permeability Calculations From Pore Size Distribution Data. *J. Pet. Technol.* **1953**, *5*, 71–78. [CrossRef]
50. Honarpour, M.; Mahmood, S.M. Relative-permeability measurements: An overview. *J. Pet. Technol.* **1988**, *40*, 963–966. [CrossRef]

Journal of
Marine Science and Engineering

MDPI

Article

Contribution of Various Shale Components to Pore System: Insights from Attributes Analysis

Lingling Xu [1,2], Renfang Pan [1,2,*], Haiyan Hu [3] and Jianghui Meng [2,4]

1. School of Geosciences, Yangtze University, Wuhan 430100, China; 202073035@yangtzeu.edu.cn
2. Key Laboratory of Exploration Technologies for Oil and Gas Resources, Ministry of Education, Yangtze University, Wuhan 430100, China; mjh@yangtzeu.edu.cn
3. Hubei Key Laboratory of Petroleum Geochemistry and Environment, Yangtze University, Wuhan 430100, China; hyhucom@163.com
4. Cooperative Innovation Center of Unconventional Oil and Gas, Yangtze University, Wuhan 430100, China

* Correspondence: pan@yangtzeu.edu.cn

Abstract: Shale pore systems are the result of the geological evolution of different matrix assemblages, and the composition of gas shale is considered to affect the pore systems in shale reservoirs. This study aimed to investigate the impact of both organic and inorganic constituents on the shale pore system, including specific surface area (SSA) and pore volume in Wufeng–Longmaxi Shale. Multiple linear regression (MLR) was employed to examine the contributions of different components to shale pore structure. The pore structure parameters, including pore SSA and pore volume, were obtained by gas adsorption experiments in 32 Wufeng–Longmaxi Shale (Late Ordovician–Early Silurian) samples. Both pore SSA and pore volume were calculated by the density functional theory (DFT) model on shale samples, and the pore types were determined by high-resolution field emission scanning electron microscopy (FE-SEM). The results of the X-ray diffractometer (XRD) analysis indicate that the Wufeng–Longmaxi Shale is dominated by quartz, clays, carbonates, feldspar, pyrite, and organic matter. Four models were made using SPSS software, all of which showed significant correlation between shale pore size and organic matter (OM) and clays. The content of organic matter played the biggest role in determining the size and structure of the pores. Although the content of quartz is the highest and serves as a rigid skeleton in shale reservoirs, it has complicated effects on the pore structure. In this study, most of the quartz is biogenetic and part of it is transformed from clays in deep shale. Therefore, these two parts of quartz are, respectively, related to organic matter and clays. In essence, the pores related to these two parts of quartz should be attributed to organic matter and clays, which also support the conclusion of the MLR models.

Keywords: pore structure; multiple linear regression; shale composition; diagenesis; overmature shale; attributes analysis

Citation: Xu, L.; Pan, R.; Hu, H.; Meng, J. Contribution of Various Shale Components to Pore System: Insights from Attributes Analysis. *J. Mar. Sci. Eng.* **2023**, *11*, 1327. https://doi.org/10.3390/jmse11071327

Academic Editor: Anabela Oliveira

Received: 14 May 2023
Revised: 20 June 2023
Accepted: 28 June 2023
Published: 29 June 2023

1. Introduction

The breakthrough improvement of shale gas development efficiency [1,2] has inspired the research on shale reservoirs [3–12]. In fact, a large number of pores from nanometers to micrometers are developed in shale reservoirs [7,13,14], which can provide storage space for shale gas, control the behavior of fluid in the pores [5,15,16], and ultimately determine the productivity potential of shale reservoirs. Therefore, the pore system is at the heart of our understanding of shale reservoirs. Dominant factors that control the shale pore systems have long been a question of great interest in a wide range of shale gas fields [12,17–19]. However, the shale pore systems, which are directly carried by the minerals matrix and organic matter (abbreviated as "OM") [4,20], were formed after sedimentary evolution and were controlled by many geological factors, such as shale mineralogy, OM (including abundance and type), burial depth, diagenesis, maturity, and preserved condition [12,19–21]. The constituents of shale consist of two parts: inorganic minerals and OM. Quartz, feldspar, carbonates, and pyrite are generally considered as brittle minerals, while clays and OM as ductile components [22]. It has been demonstrated by many researchers that OM-hosted pores are strongly associated with the OM maturity [5,18,20,23], and the mineral matrix pores are mainly related to burial depth and diagenesis [24]. Recently, there has been renewed interest in the contribution of shale composition to the pore systems [12,25,26], which highly demonstrated the importance of quartz to the shale pore network. Some research reported that the pore specific surface area (abbreviated as "SSA") and pore volume of gas shale increases with the increasing abundance of quartz [13,27–30], and this positive correlation was interpreted to be caused by these rigid minerals that inhibit pore collapse, but other research suggests that quartz does not significantly contribute to the pore network of shales [15,17]. The same phenomenon applies to the study between other shale components and pore structure parameters [21,26,31], which will be well discussed in Section 4. Even though it has been studied in many publications [17,32,33], the effects of shale composition (including organic and inorganic constituents) on pore system are not fully understood. The existing studies fail to resolve the collinearity between independent variables [14,30,34]. So far, only a few studies have noted the collinearity relationship between the content of quartz and OM in the research of the effect of shale components on pore network [13,29,35]. In other words, the highly positive correlation between the content of quartz and shale pore structure parameters may be triggered by high TOC rather than the quartz. Moreover, previous studies often directly use correlation coefficients to represent the degree of correlation between variables when judging the correlation between factors. When the number of samples is small, the correlation coefficient fluctuates greatly and more easily approaches 1 [28,36]. Contrary to previous work, this study introduces test probability P to determine the possibility of correlation between variables using SPSS version 29.0, so as to ensure that the results are statistically significant. In the light of recent research, it is becoming extremely difficult to ignore the contributions of shale composition in shale pore network towards overall and effective porosity. This is particularly relevant in the light of the discovery of various type pores [4,37,38]: To what degree are these pore SSA and pore volume provided by minerals versus OM in the overmature shales? In addition, a systematic understanding of how shale composition contributes to the pore network is still lacking.

This essay investigates the contribution of various compositions to the pore system in overmature shale. The specific objective of this study was to establish MLR models between shale constituents and pore structure parameters to determine the components and contribution proportion that have significant contributions to pore structure. The MLR analysis was conducted by SPSS software to eliminate the collinear relationship between independent variables and quantify the contribution of the independent variables to the dependent variables. Consequently, this research will contribute to a deeper understanding of the pore system in overmature shale reservoirs, clarifying the degree to which different components contribute to shale pore SSA, and that pore volume is also important for shale gas exploration.

2. Materials and Methods

2.1. Samples and Background

Samples used for analysis in this paper were collected from fresh cores of the Wufeng–Longmaxi Shale in South Sichuan shale clays (China) which had experienced similar burial history with a maximum burial depth of 6000–7000 m [39]. Hence, it is approximately believed that all samples had undergone similar diagenetic evolution, and the difference in pore development is mainly caused by their heterogenous composition. Then, the main reason for the differences in pore structure characteristics of the selected samples is their various components. As the Wufeng–Longmaxi Shale is overmature, it is difficult to distinguish the kerogen type by optical observation or Rock-Eval data [40–42]. In this study, the kerogen type was identified by the stable carbon isotopic values (Table 1) according to previous research [40]. The stable carbon isotopic values of selected samples were measured using a FLASH HT EA-MAT 253 IRMS, and the values were calibrated to the PDB standard.

Table 1. Standard of kerogen type for the Wufeng–Longmaxi Shales based on carbon isotopic composition (adopted from [40]).

Kerogen Types	$\delta^{13}C$ (‰)
I	<−29
II_1	−29 to −27
II_2	−27 to −25
III	>−25

The shale samples lack vitrinite particles because the Wufeng–Longmaxi Shale was deposited in the Late Ordovician to Early Silurian, and the reflectance of bitumen (abbreviated as "Rb") was measured to reflect the maturity of shales [43]. In this paper, the Ro* was calculated by the equation: Ro* = 0.938Rb + 0.3145 according to [43], and the random reflection of solid bitumen was measured by a Zeiss Axio Scope A1 reflected-light microscope with a J&M MSP 200 microscope photometer. The mineral components were measured on an X′ Pert3 Powder diffractometer using K-alpha radiation (XRD). Using a scan step of 0.02° and scanning speed at 2°/min, data were obtained at room temperature in 2θ range of 5° to 45°. Quantitative analysis of minerals was calculated by the equation following the Chinese Oil and Gas Industry Standard (SY/T 5163-2018) [12]. An LECO carbon–sulfur analyzer (CS230) was utilized to obtain the TOC content of these samples. Finally, the shale composition results, including organic and inorganic constituents, were normalized to 100% (Table 2).

Table 2. The composition, Ro*, and $\delta^{13}C_{PDB}$ of kerogen of samples from Wufeng–Longmaxi Shales.

Samples ID	Depth (m)	Kerogen $\delta^{13}C_{PDB}$ (‰)	Rb (%)	Ro* (%)	Quartz (%)	Clays * (%)	Carbonates ** (%)	Feldspar † (%)	Pyrite (%)	TOC (%)
Y101H4-4-1	4124.27	−30.1	/	/	37.88	30.26	16.76	11.66	1.10	2.34
Y101H4-4-2	4131.55	−30.6	3.032	3.16	42.62	34.14	11.55	7.33	1.64	2.71
Y101H4-4-3	4140.22	−30.5	2.907	3.04	64.70	18.34	6.84	3.06	2.02	5.04
Y101H4-4-4	4144.14	−30.9	2.88	3.02	56.91	13.36	19.32	2.56	2.05	5.78
Y101H4-4-5	4145.86	/	/	/	58.49	10.45	22.98	2.02	1.12	4.94
Y101H4-4-6	4150.60	/	2.98	3.11	57.59	29.87	4.90	3.69	0.00	3.95
Y101H4-4-7	4154.35	/	/	/	35.55	44.11	15.45	4.39	0.00	0.5
Y101H2-7-1	4102.25	/	/	/	30.16	47.68	11.70	5.72	1.34	3.39
Y101H2-7-2	4109.57	/	/	/	35.47	48.01	5.66	4.72	3.12	3.01
Y101H2-7-3	4139.05	−30.0	3.002	3.13	47.23	35.30	4.77	6.30	1.97	4.44
Z206-1	4244.57	−29.5	2.43	2.6	47.43	32.77	8.05	6.96	1.26	3.53
Z206-2	4254.05	/	/	/	44.95	32.30	7.87	10.11	2.25	2.52
Z206-3	4258.64	/	/	/	43.00	28.39	14.47	9.48	2.11	2.55
Z206-4	4266.76	−30.5	2.471	2.63	72.57	13.56	5.91	2.07	1.35	4.54
Z206-5	4274.36	−30.4	/	/	24.96	32.44	37.92	3.23	0.73	0.71
L204-1	3823.55	/	/	/	34.41	43.78	10.96	4.92	2.40	3.54
L204-2	3835.91	/	/	/	49.56	16.57	24.84	2.68	2.41	3.93
L204-3	3842.21	−30.7	2.698	2.85	69.49	8.53	13.90	1.84	0.68	5.56
L204-4	3844.71	/	/	/	50.26	8.27	33.80	2.18	0.67	4.82
L205-1	4012.06	/	/	/	38.26	33.89	13.50	9.81	1.68	2.86
L205-2	4024.08	−29.7	2.567	2.72	56.67	22.57	11.74	4.31	1.45	3.26
L205-3	4028.30	/	/	/	61.48	7.43	24.01	1.40	1.00	4.68
L205-4	4032.58	/	/	/	57.85	11.96	22.06	2.31	1.74	4.07
L205-5	4034.44	−29.4	2.814	2.95	51.31	15.18	22.98	4.27	2.75	3.51
N213-1	2575.28	−30.9	2.48	2.64	41.67	7.74	43.99	1.47	1.51	3.61
N216-1	2315.30	/	/	/	38.98	9.66	36.00	1.53	10.40	3.42
N216-2	2320.68	−30.4	2.65	2.8	46.67	7.76	38.36	2.83	0.90	3.49
W202-1	2560.20	−29.2	2.53	2.69	76.37	3.89	10.82	0.67	0.37	7.88
W202-2	2573.20	/	/	/	42.06	2.64	50.57	1.46	0.54	2.73

The shale composition results, including organic and inorganic constituents, are given in mass percent normalized to 100% (wt.%). Depth here refers to the depth of the core where the sample is located. TOC = total organic carbon content, wt.%; Ro* = equivalent vitrinite reflectance calculated from the reflectance of bitumen (Rb), Ro* = 0.938Rb + 0.3145 [43]; * Clays = illite + illite/smectite + chlorite; ** Carbonates = calcite + dolomite; † Feldspars = orthoclase + plagioclase; / = samples with no available data.

2.2. FE-SEM

In order to focus on the development of pores related to a certain component and their types depending on their host sites in shales, 28 samples with a relatively high content of a certain component (quartz, clays, carbonates, feldspar, pyrite, TOC) were selected for FE-SEM observation. All samples were prepared by Ar-ion-milling for a much flatter surface [3] before using FE-SEM. The detailed aspects of the FE-SEM observation have been adequately described by previous authors and we recommend readers to [3].

2.3. Low-Pressure Gas Adsorption

Low-pressure gas adsorption is an effective approach for measuring the pore SSA and pore volume of shales [5–8,44], which can effectively reflect capacity of shale gas and heterogeneity of shales [27,45,46].

Before gas adsorption experiments, all samples were ground into about 35 mesh and then degassed under vacuum for 10 h at 120 °C to remove volatiles. The experiments were conducted by a Quantachrome® Autosorb iQ instrument. N_2 physisorption experiments were performed at −195.8 °C. More detailed explanation of the N_2 adsorption experiment is presented in previous studies [8,44]. CO_2 adsorption experiments were carried out at 0 °C with the same instrument. Detailed information about the CO_2 adsorption experiments can be found in previous studies [47,48].

The pore structure was represented by pore SSA and pore volume in this paper. Both the pore SSA and pore volume were calculated by the density functional theory (abbreviated as "DFT") from the adsorption isotherms since DFT models cover the range of micro- to macropores, fit different adsorbates (N_2 and CO_2), and take advantage of reflecting various pore morphologies [44,46,49,50]. More detailed information and explanation of this theory can be found in previous studies [48,49].

2.4. Multiple Linear Regression

MLR analysis is a statistical approach to investigate the interdependent quantitative relationship between two or more variables. In this study, MLR analysis was conducted by using IBM SPSS Statistics 29.0 automatically. In the MLR models, the contents of quartz, clays, carbonates, feldspar, pyrite, and TOC served as independent variables, and the pore SSA and pore volume obtained by CO_2 and N_2 served as dependent variables, respectively. A stepwise method was adopted to select the independent variables that contribute most to the dependent variables and meet the conditions (probability of $F \leq 0.05$) in the model and remove any remaining variables (probability of $F \geq 0.10$) that do not meet the set conditions [51]. Collinearity diagnosis was added to exclude collinearity relationship between the independent variables. In this study, the quality criterion of MLR model includes the following: (1) F-value is used to test the significance of the model (the larger the F-value, the smaller the p-values of significance probability, indicating that the regression model is effective); (2) adjusted R^2, which directly reflects the proportion of the regression value in the model (the closer the adjusted R^2 is to 1, the better the quality of the regression); (3) T-values test the significance of regression parameter for each independent variable (a large T-values implies a small p-value, which means that the independent variable contribute significantly to the equation).

3. Results

3.1. Components, Ro*, and Kerogen Type

The Ro* and $\delta^{13}C_{PDB}$ for the kerogen from all the samples are listed in Table 2, and the main minerals developed in the Wufeng–Longmaxi Shale are summarized in a boxplot (Figure 1). The Wufeng–Longmaxi Shale is mainly composed of variable amounts of quartz, clays, carbonates (calcite and dolomite), feldspar, and TOC, with a small amount of pyrite (Table 2; Figure 1). The inorganic constituents show an inverse relationship in terms of the proportion, that is, the increase of one mineral is compensated by a decrease in the remaining minerals. The content of quartz ranges between 25.14% and 82.91%, with an average of 50.82% (Table 2; Figure 1). The content of clays varies between 1.52% and 49.50%, with an average of 22.96%. The content of carbonates ranges between 4.99% and 51.99%, with an average of 19.72%. The content of feldspar is relatively small, ranging from 0.72% to 11.94%, with a mean value of 4.42%. The content of pyrite is smallest, with an average of 1.81%, the content of all samples, except one sample with 10.77% content, does not exceed 3.2%. The content of TOC ranges between 0.50% to 7.88%, with a mean value of 3.70% (Table 2). In particular, the content of TOC is positively correlated with quartz content, but negatively correlated with clays content. The Ro* ranges from 2.60% to 3.16%, with mean Ro* of 2.88% (Table 2), indicating that all these shale samples are overmature. The Wufeng–Longmaxi Shales have δ^{13}C values of less than −29‰ (Table 2), and are type I [40].

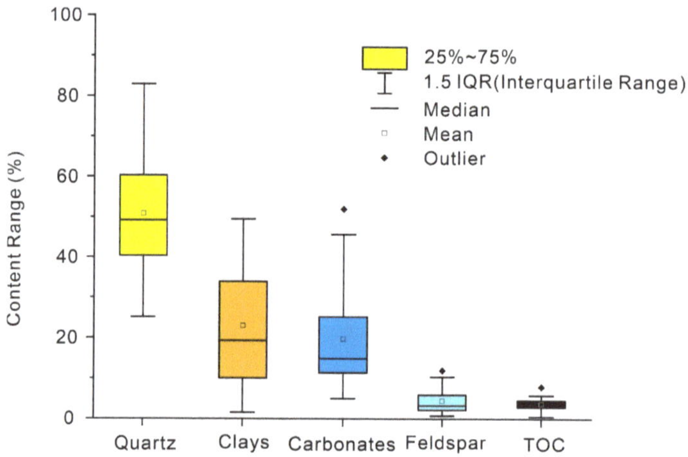

Figure 1. Boxplot for the main composition of the Wufeng–Longmaxi Shale.

3.2. Pores Related to Various Components

Different components of shale develop various types of pores (terminology after [4]). Pores related to quartz include interparticle (abbreviated as "interP") pores (Figure 2a,b) and intraparticle (abbreviated as "intraP") pores (Figure 2c,d), among which the intraP pores are rare; only a very small number of them developed in a few individual grains (Figure 2a,b), with more interP pores developing between quartz individuals or quartz and other minerals such as feldspar, carbonates, or illite (Figure 3b). The number of pores associated with clays is relatively large, and there are a large number of cracks that developed between clays or within them (Figures 3 and 4). A large number of irregular pores developed between clays and other brittle minerals (Figure 3b). However, the degree of pore development in argillaceous shale without OM is extremely low (Figure 4), which only developed few cracks in it. Carbonates developed more intraP pores than quartz grains from the FE-SEM images (Figure 5), which are suspected to be formed from the acid dissolution by circulating acid fluid [6,35]. InterP pores formed at the edge of the grains when along contact with other grains (Figure 5a,b). Feldspar also shows development of developed interP pores and interP pores (Figure 6c,d). Pyrite is the least abundant compared to other components described above in all samples. There are two types of pores related to pyrite, one is intercrystalline pores within pyrite framboids (Figure 6a), and the other is interP pores between pyrites (Figure 6b), which is often associated with OM. Pores associated with OM are the most abundant pores that can be observed in FE-SEM images. At a large scale, OM is dispersed in matrix minerals (Figure 7a,d) and these are connected with each other through the edges of mineral particles. There are nano- to micropores (Figure 7b) that developed within OM. After local amplification, it can be found that there are numerous nanopores that developed within OM, similar to a honeycomb (Figure 7c).

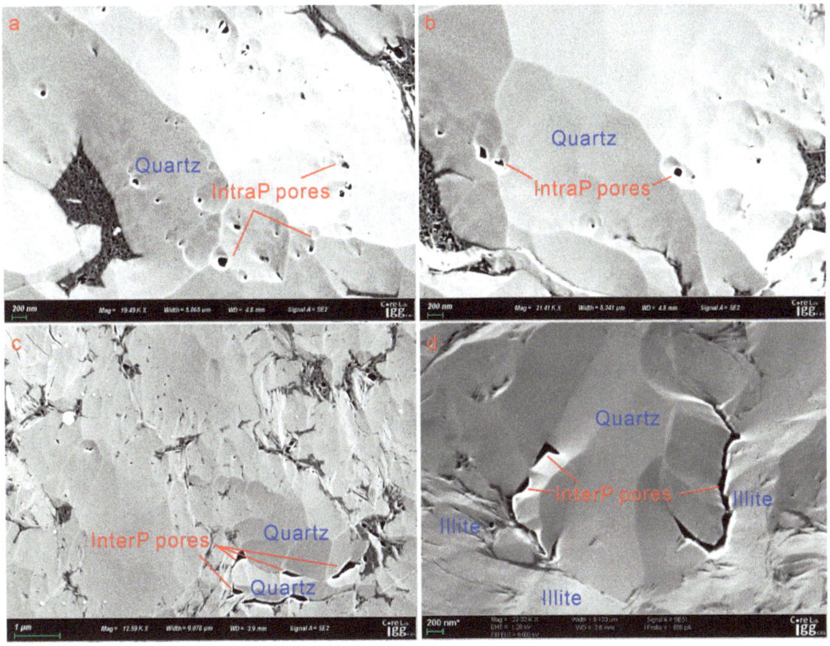

Figure 2. Pores associated with quartz: (**a**,**b**) intraP pores developed within quartz grain; (**c**) interP pores developed between quartz grains; (**d**) interP pores developed between quartz and illite [52].

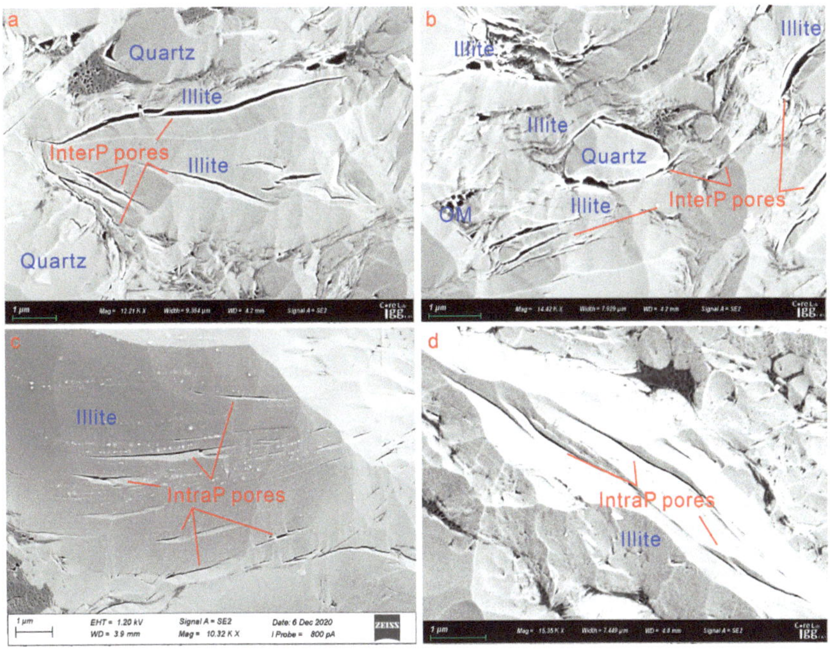

Figure 3. Pores related to clays, presenting as a set of parallel to subparallel cracks/slits generally running horizontally: (**a**,**b**) interP pores developed between clays and other grains; (**c**,**d**) intraP pores developed within clays.

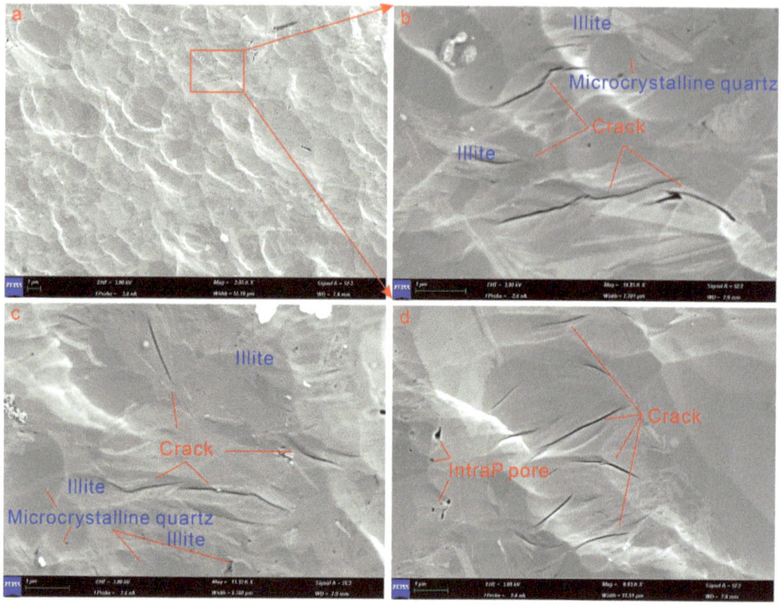

Figure 4. FE-SEM images of argillaceous shale, sample Y101H4-4-7, with 0.5% TOC, which developed few inorganic pores and cracks. (**a**) FE-SEM image of sample Y101H4-4-7 on a large scale; (**b**) partial enlarged image in (**a**); (**c**) cracks and microcrystalline quartz observed in the FE-SEM image of sample Y101H4-4-7; (**d**) cracks and intrP pores developed in the argillaceous shale.

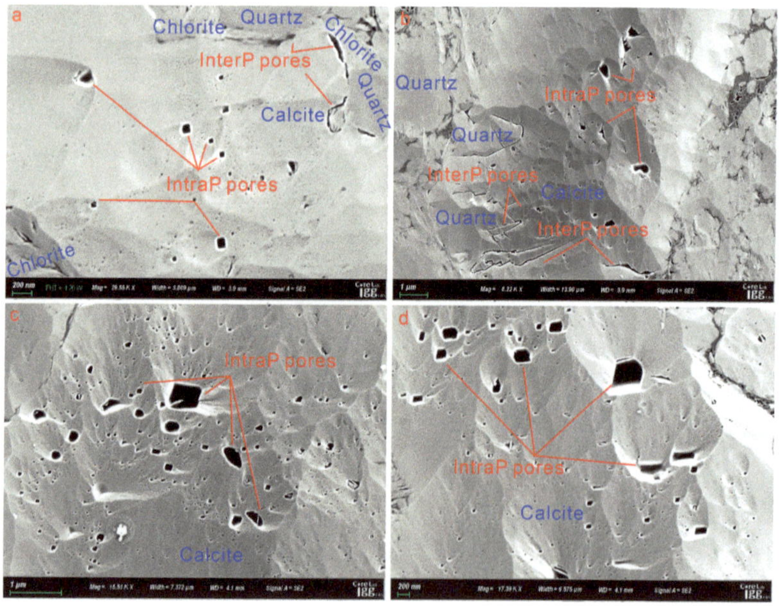

Figure 5. Pores related to carbonates. (**a**) sample ID = Y101H4-4-2, intraP pores developed in calcite particle and interP pores developed between calcite and other grains (quartz or chlorite); (**b**) sample ID = Y101H4-4-3, intraP pores developed in calcite particle and interP pores developed between calcite and quartz; (**c**) sample ID = L204-4, intraP pores developed in calcite; (**d**) sample ID = L204-4, intraP pores developed in calcite.

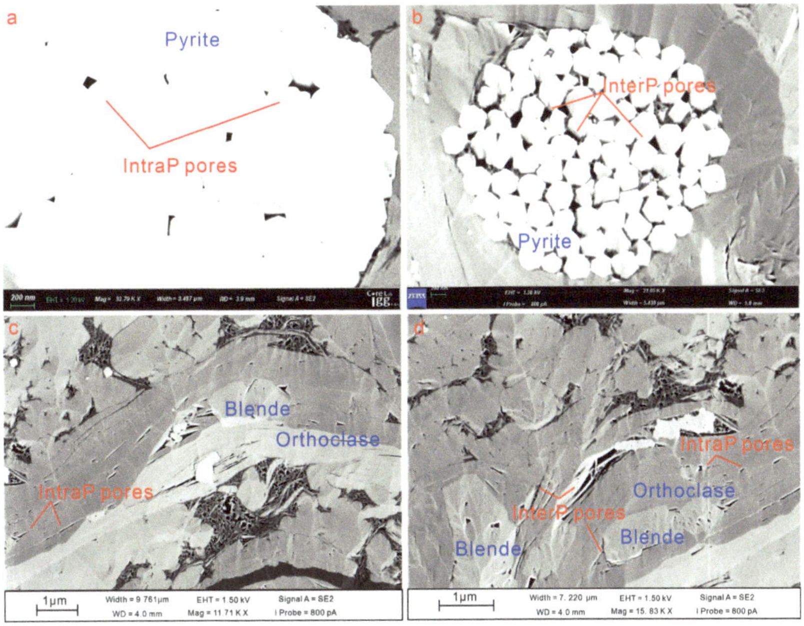

Figure 6. Pores related to pyrite and feldspar: (**a**,**b**) intraP pores developed within pyrite or between pyrite grains; (**c**) intraP pores developed within orthoclase grain; (**d**) interP pores developed between orthoclase and other grains.

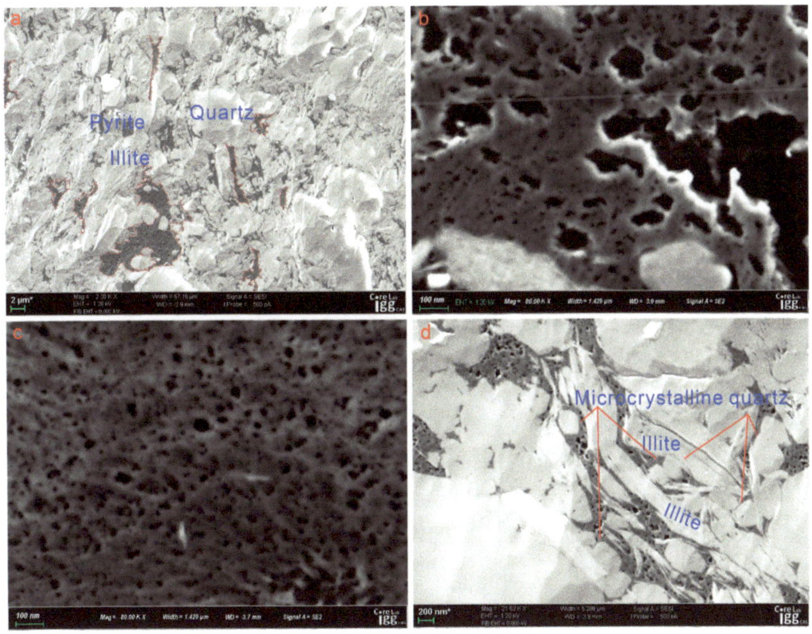

Figure 7. OM-hosted pores: (**a**) sample ID = Y101H2-7-3, OM is dispersed in matrix minerals at a large scale; (**b**) sample ID = Y101H4-4-2, nano- to micropores developed within OM; (**c**) sample ID = Y101H4-4-1, honeycomb-like OM-hosted pores; (**d**) sample ID = Y101H2-7-3, OM-hosted pores developed within clays.

3.3. Pore SSA and Pore Volume

Low-pressure CO_2 adsorption isotherms are shown in Figure 8. The pore SSA and volume obtained by low-pressure CO_2 were calculated through the DFT (Table 3). The CO_2 micropore volumes have the smallest variation range, from 0.003 cm^3/g to 0.009 cm^3/g, with a mean value of 0.006 cm^3/g. The CO_2 SSA ranges from 9.392 m^2/g to 28.227 m^2/g, with an average of 18.903 m^2/g. The low-pressure N_2 adsorption isotherms are shown in Figure 9. The pore SSA and volume obtained by low-pressure N_2 were calculated by the DFT (Table 3). The N_2 SSA varies from 6.14 m^2/g to 34.48 m^2/g, with a mean value of 22.31 m^2/g, and the N_2 volume ranges between 17.64 cm^3/g and 38.56 cm^3/g, with an average of 27.01 cm^3/g.

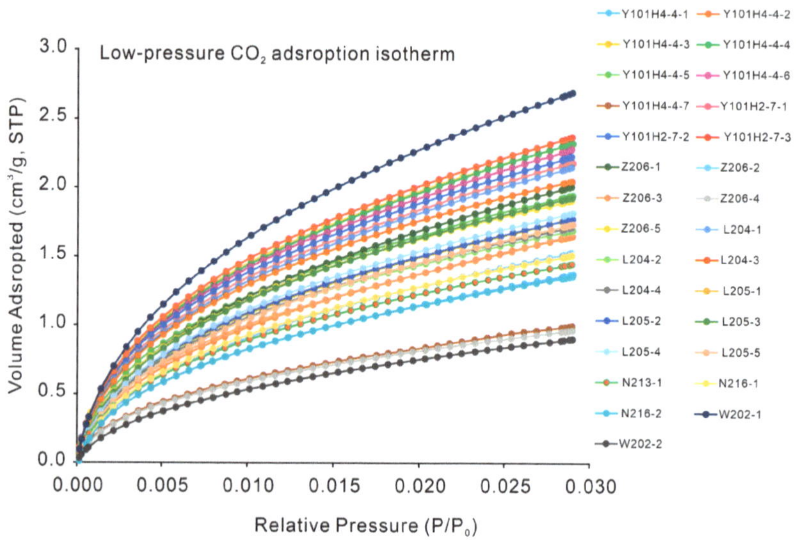

Figure 8. Low-pressure CO_2 adsorption isotherms of the 29 samples from the Wufeng–Longmaxi Shales.

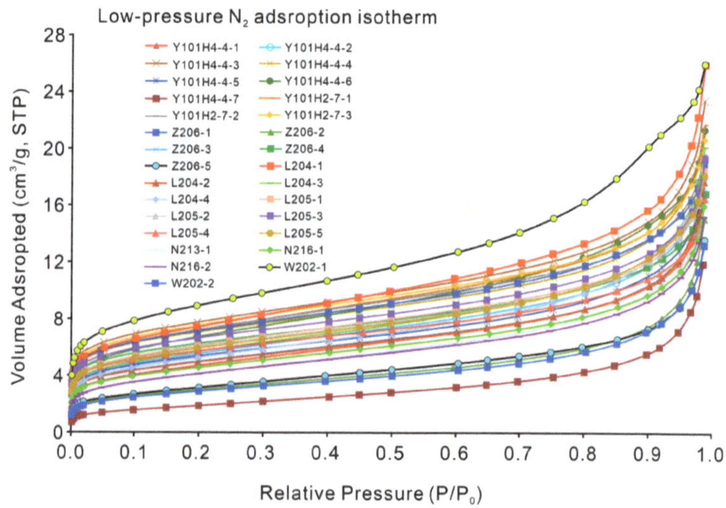

Figure 9. Low-pressure N_2 adsorption isotherms of the 29 samples from the Wufeng–Longmaxi Shales.

Table 3. Pore parameters of samples from Wufeng–Longmaxi Shales.

Samples ID	CO_2		N_2	
	SSA (m^2/g)	V ($\times 10^{-3}$ cm^3/g)	SSA (m^2/g)	V ($\times 10^{-3}$ cm^3/g)
Y101H4-4-1	14.406	4	17.26	24.2
Y101H4-4-2	17.392	5	21.76	28.36
Y101H4-4-3	24.255	7	32.87	34.37
Y101H4-4-4	24.371	7	30.28	31.28
Y101H4-4-5	20.314	6	21.93	26.86
Y101H4-4-6	23.936	7	25.69	31.25
Y101H4-4-7	10.342	3	6.14	17.64
Y101H2-7-1	22.985	7	29.37	31.93
Y101H2-7-2	23.251	7	30.04	29.12
Y101H2-7-3	24.726	7	31.69	30.43
Z206-1	21.098	7	30.08	27.94
Z206-2	16.067	5	10.77	21.69
Z206-3	17.561	5	20.5	25.35
Z206-4	20.049	6	26.8	24.39
Z206-5	10.072	3	11.76	19.37
L204-1	22.495	7	30.01	37.71
L204-2	17.719	5	19.53	25.96
L204-3	21.34	6	22.04	29.39
L204-4	17.956	5	19.86	27.21
L205-1	18.518	6	18.02	27.24
L205-2	18.593	6	24.04	26.57
L205-3	20.266	6	28.42	27.96
L205-4	19.049	6	24.09	23.87
L205-5	18.323	6	23.73	26.6
N213-1	15.139	5	14.12	23.33
N216-1	15.955	5	17.39	23.41
N216-2	14.378	4	14.48	22.26
W202-1	28.227	9	34.48	38.56
W202-2	9.392	3	9.98	19.17

SSA = specific surface area; V = volume.

3.4. Multiple Linear Regression

Four MLR models with corresponding adjusted R^2, F-values, and p-values conducted by SPSS are shown in Table 4. The p-values of the four MLR models were all less than 0.001, indicating these four MLR modes were all statistically significant. The adjusted R^2 ranged from 0.753 to 0.931, showing that there are significant correlations between the residual independent variables and the dependent variables. The factors that contribute to the pore SSA and pore volume are the same. Because there is no statistically significant correlation between quartz, carbonates, feldspar, pyrite, and the CO_2 SSA, CO_2 volume, N_2 SSA, and N_2 volume, the independent variables quartz, carbonates, feldspar, and pyrite were excluded from the MLR models with p-values more than 0.1. The results show that CO_2 SSA, CO_2 volume, N_2 SSA, and N_2 volume were significantly correlated with TOC and clays (Table 4). Furthermore, there are no collinear relationships between clays and TOC in these four MLR models. TOC was the most significant contributor to the total SSA and volume of shale pores since the correlation coefficient is larger, followed by clays (Table 4). The contribution of TOC to the four parameters of shale pore structure is arranged from large to small, as follows: CO_2 SSA, CO_2 volume, N_2 volume, N_2 SSA. The situation for clay minerals is different: N_2 volume, CO_2 SSA, CO_2 volume, N_2 SSA.

Table 4. MLR models with corresponding R, adjusted R^2, and standard error (SE) of the estimate.

Dependent Variable	Model				TOC			Clays		
	Equation	Adjusted R^2	F	p [a]	r	t	p	r	t	p
CO_2 SSA	S = 1.145T + 0.705C	0.931	190.855	<0.001	1.145	19.460	<0.001	0.705	11.974	<0.001
CO_2 volume	V = 1.099T + 0.674C	0.852	81.611	<0.001	1.099	12.728	<0.001	0.674	7.804	<0.001
N_2 SSA	S = 1.037T + 0.657C	0.753	43.766	<0.001	1.037	9.299	<0.001	0.657	5.890	<0.001
N_2 volume	V = 1.071T + 0.708C	0.817	63.452	<0.001	1.071	11.152	<0.001	0.708	7.364	<0.001

Note that S = specific surface area, T = content of TOC, C = content of clays, F = sum of regression square/sum of residual mean square, p [a] = probability of the significance for the model, r = coefficient of the independent variables, T = standardization coefficient/residual, p = test probability value of independent variables.

4. Discussion

According to XRD analysis, the samples are mainly composed of OM, quartz, clays, carbonates, feldspar, and a small amount of pyrite. Although each constituent contributes to the total porosity (Figures 4–7), their contributions to the total pore SSA and volume vary, as indicated by the result of MLR in this paper (Table 4). The constituents of shales may be classified into brittle and ductile in terms of their mechanical behavior. The four MLR models show that the pore SSA and pore volume of the overmature shale samples are mainly contributed by ductile components rather than brittle minerals, in particular OM, which is in agreement with some previous studies [3,5,6,8,20]. In this study, we also see a positive relationship between total porosity and the contents of clays [14,20] and no correlations between total pore volume and the contents of quartz, or the contents of carbonates. The contribution of brittle minerals, clays, and OM to the pore development of overmature shales is discussed as follows, combined with the results of FE-SEM images.

4.1. Contribution of Brittle Minerals to Pore Development

Quartz, the most abundant component in marine shales [53], hardly contributes to the SSA and volume of shale pores, as indicated by the results of MLR (Table 4). However, many publications have reported a positive relationship between the contents of quartz and pore SSA and pore volume, and have concluded that quartz contributes to the development of shale pores [13,14,29,33]. In fact, quartz in marine shales mainly originates from three sources: detrital quartz, biogenic quartz, and clay minerals transformation [54–56]. In most organic-rich marine shales, biogenic quartz accounts for the majority [13,14,25,26,55,57]. As a result, TOC increases with the increasing content of quartz, particularly biogenic quartz [13,25,29,30,58,59]. Indeed, as [35] pointed out, the positive correlations between quartz and pore parameters are attributed to the positive correlation between quartz and OM. Our research sample contains parts of biogenic quartz, and the content of quartz is positively correlated with TOC (Figure 10a). Although we can obtain similar conclusions by using the data from one of the wells as an example and performing a linear regression analysis between the quartz content and pore structure parameters (as shown in Figure 11), these conclusions seem untenable. Maintaining primary pores or generating secondary pores are two ways for shale reservoirs to maintain high porosity [1,20,35,60]. However, on the one hand, we did not observe a significant number of pores associated with quartz in the FE-SEM images, regardless of the type of quartz (as seen in Figure 2). In Section 3.2, we identified three types of pores that are associated with quartz grains. The first type is interP pores located between quartz grains, which are detrital or biogenic quartz, whereas smaller-sized quartz particles transformed by clays typically form within the clays (Figure 12). Under FE-SEM, there are very few pores observed between quartz particles. The second type of pores related to quartz is the dissolution pore [14,57]. The SEM results show that the pores produced by a single particle due to dissolution are limited (Figure 2) compared to carbonates (Figure 5), which is consistent with the previous study [14]. The third type of pores is related to other minerals. The contact between quartz and other minerals (including OM) is more easily compressed due to differences in hardness [17]. Additionally, in this study, there are many quartzes transformed from clay minerals (Figure 12). Since most of this microcrystalline quartz with small size develops within clays [24,54,61,62], it is easy to identify. Nevertheless, numerous pores developed between clays and microcrystalline quartz, and we suggested attributing these pores to clays rather than quartz. Essentially, the higher the content of clays, the more pore spaces are formed in clays. Moreover, the pores related to biogenetic quartz are also linked to the content of OM. Overall, the number of pores contributed by quartz itself is very limited [15]. Although the number of secondary pores generated by quartz is limited, they play a key role in maintaining the primary pore space (Figure 2c) [13,25,29]. However, we found that if there are no secondary pores (mainly OM pores) after deep burial of 6000–7000 m [39], the preserved primary interP pores are very few in overmature shale samples (Figure 4). For example, although the rigid particle content of the Y101H4-4-7 sample is higher than Y101H2-7-1 and the clays content is not significantly different (Table 2), the OM content of Y101H4-4-7 is only 0.50% while that of Y101H2-7-1 is 3.39%. Regardless of the number of pores observed in SEM images (Figures 4 and 13) or the results of gas adsorption experiments (Table 3), the pore development of Y101H2-7-1 is much higher than that of Y101H4-4-7. Due to the strong heterogeneity of shale composition, rigid minerals and ductile ones are mixed together with varied particle size, which cannot form an effective support [63].

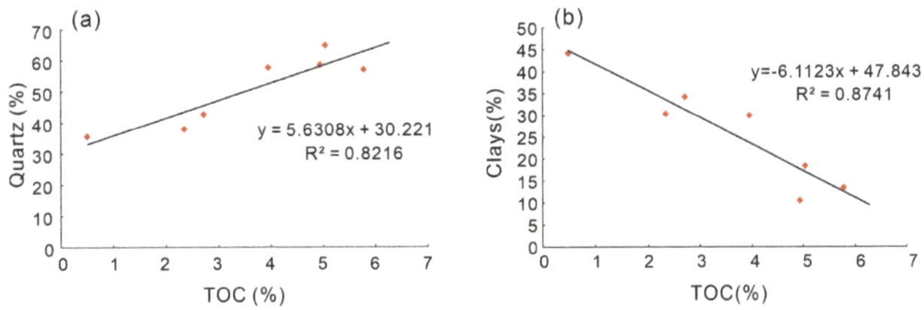

Figure 10. The relationship between the abundance of quartz, clays, and TOC of seven samples selected from well Y101H4-4. (**a**) the relationship between the content of quartz and TOC; (**b**) the relationship between the content of clays and TOC.

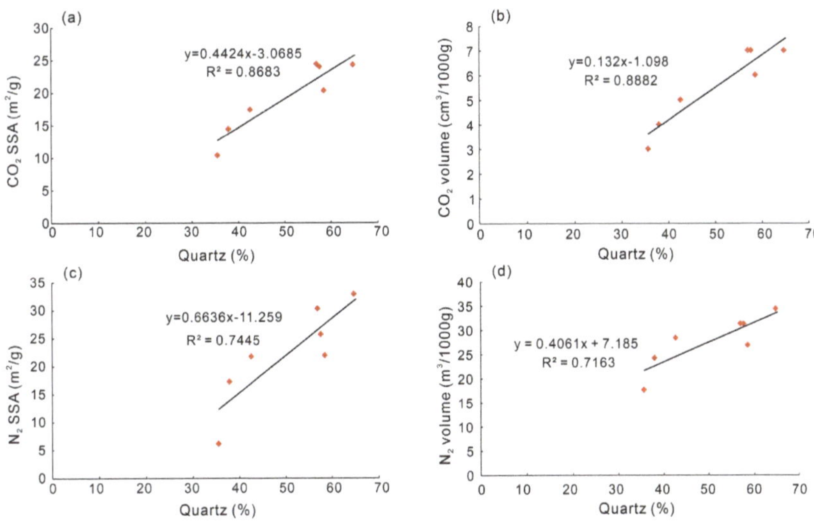

Figure 11. The relationship between the abundance of quartz and pore structure parameters of seven samples from well Y101H4-4. (**a**) the correlation between the quartz content and the value of CO_2 SSA; (**b**) the correlation between the quartz content and CO_2 volume; (**c**) the correlation between the quartz content and N_2 SSA; (**d**) the correlation between the quartz content and N_2 volume.

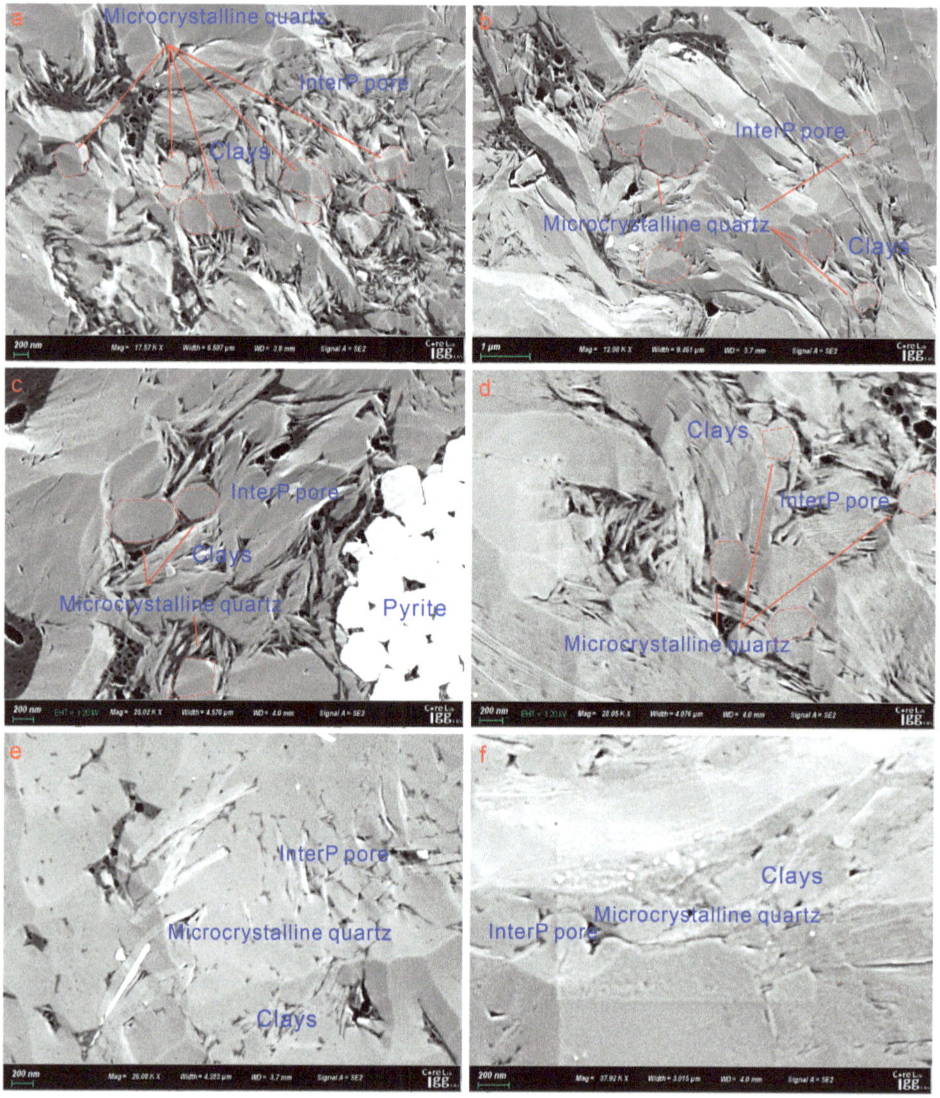

Figure 12. Quartz transformed from clay minerals and considerable pores developed during this process of illitization. (**a**,**b**) sample ID = Y101H4-4-1; (**c**,**d**) sample ID = Y101H4-4-2; (**e**) sample ID = L204-3; (**f**) sample ID = L205-1.

Carbonates, which are abundant in calcareous shales, have a dual effect on shale reservoirs. On one hand, carbonate cementation and precipitates destroy pores and reduce porosity during diagenesis [14,35,62]; on the other hand, they can improve the brittleness of shale reservoirs and play a beneficial role in the later hydraulic fracturing [64,65]. However, their chemical properties are unstable, making them easily dissolved by organic acids generated during the thermal maturation of sedimentary OM, resulting in the formation of dissolution pores [6,66]. Furthermore, the connectivity between pores within carbonates is limited (Figure 6), suggesting that not all pores within carbonates may be equally effective as shale gas storage space. Therefore, the final effect of carbonate minerals on the pore space of a shale reservoir depends on the relative degree of their destructive and constructive effects. The samples in this paper show that the number of constructive pores is almost the same as destructive ones, and as a result, carbonate minerals do not contribute significantly to the total pores volume [14,35].

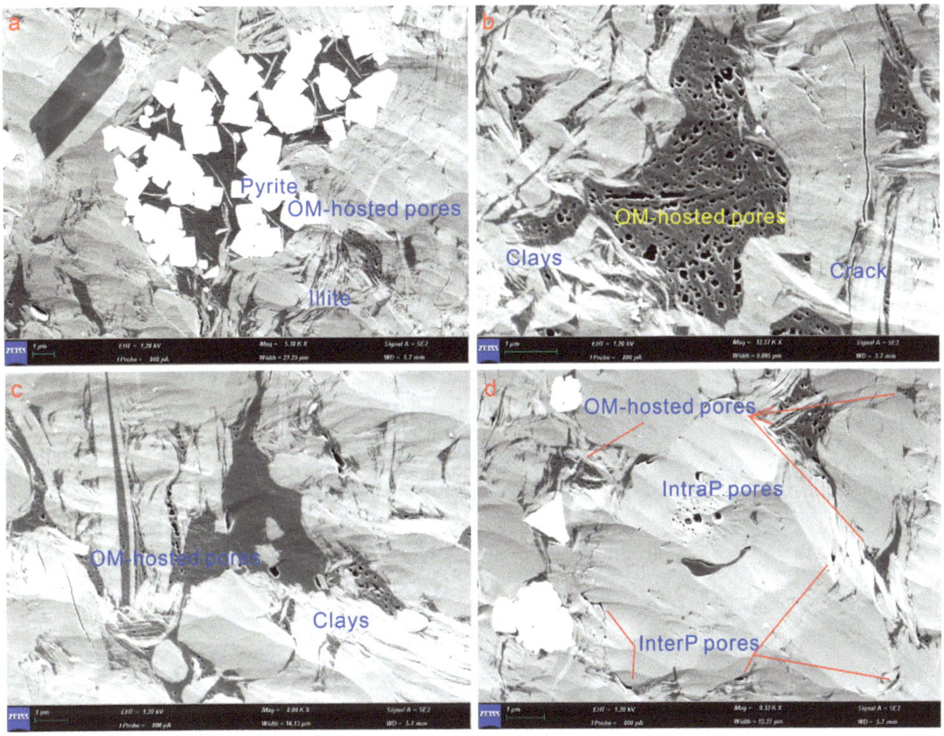

Figure 13. Pores developed in the sample Y101H2-7-1 with 3.39% TOC. (**a**) OM-hosted pores developed in the sample Y101H2-7-1; (**b**) OM-hosted pores and crack developed in the sample Y101H2-7-1; (**c**) OM-hosted pores developed between clays; (**d**) OM-hosted pores and interP pores developed in the sample Y101H2-7-1.

Feldspar, which has a relatively small content in shales, has the same impact on shale pores as carbonates [4,34,62]. Cementation decreases pores while dissolution increases pore space [4,57]. In general, it has no significant contribution to the total pore SSA and volume of shale reservoir due to its small content (Table 4).

Pyrite, though interP pores developed between pyrites or intraP pores developed within it (Figure 6a,b), is too small in content to have affected porosity significantly.

Although rigid particles can improve reservoir brittleness, making it more conducive to later hydraulic fracturing [1,64,65], they generally contribute little to total pore volume and SSA. Because of the different particle sizes, it is difficult to form effective support.

4.2. Contribution of Clays to Pore Development

Clays, the second-most abundant component in the Wufeng–Longmaxi Shale, have been demonstrated to be easily compacted [8,14,63]. The layered structure of clay contributes positively towards porosity [8]. However, if there is no force bearing inside or sheltering outside (Figure 14), these pore spaces may disappear due to compaction [8,67,68]. It has been reported that the porosity of original loose clays decreased by 85% after compaction [62]. Although it is also reported that rigid particles such as quartz, pyrite, and calcite can provide support to inhibit pore reduction [20], the heterogeneity of shale composition makes the pores between particles more easily filled rather than preserved (Figure 4).

In theory, rigid particles with similar size can resist compaction by point contact between rigid grains and form a stiff framework [54], thus preserving interP pore spaces. However, natural shales are a mixture of clays and other minerals that make it difficult to form an effective support and are more easily compacted due to differences in hardness and particle size [62]. Clays can adsorb OM on their surface and play a key role in the burial and preservation of OM [62,69,70].

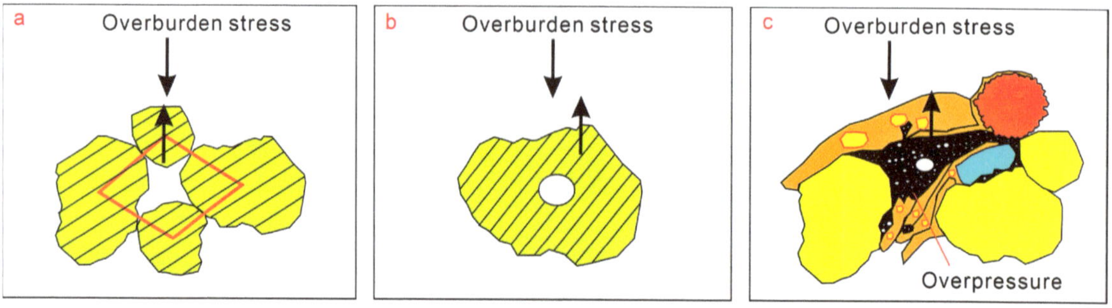

Figure 14. Pores preservation mechanism: (**a**) interP pore supported by rigid minerals, and the overburden stress of the overlying strata acting on the rigid framework; (**b**) intraP pore developed within the rigid minerals, and the overburden stress of the overlying strata acting on the rigid grain; (**c**) pores associated with plastic minerals such as OM or clays, and the pores supported by the overpressure inside.

Consequently, a significant amount of OM can be buried and deposited within clays. During the process of deep burial, typically above 70 °C, clays gradually transform from smectite to illite, leading to the release of significant amounts of silica that precipitates as microcrystalline quartz within the clays [24,54,62]. This process of illitization stiffens the shale framework and increases the brittleness of the reservoir [61], while also reducing the volume of clays and developing considerable pores (Figure 12) [24,54,67]. On the other hand, illitization is generally associated with hydrocarbon generation [71], as OM generates considerable amounts of hydrocarbons during thermal evolution, creating an overpressure environment in the reservoir [72,73]. Under the effect of pressure difference, the fluid may migrate into adjacent clays, becoming the force-bearing body in the pores (Figure 14c) and inhibiting pore collapse [74]. Therefore, we observe a large number of pores developing between the clays and rigid minerals in samples with relatively high TOC content (Figure 12). We infer that this phenomenon occurs because the fluid generated by hydrocarbon generation occupies this part of pore space in the clays, and the overpressure formed by the fluid counteracts the pressure of the overlying strata (Figure 14c), inhibiting the collapse of pores.

Overall, clay minerals can not only produce a large number of secondary pores during burial but can also be easily compacted due to their compressibility. Although some previous studies obtained a negative correlation based on the one-dimensional linear regression analysis of clay mineral content and pore parameters [14,34], in this study, the MLR models pointed out that clay mineral content significantly contributes to the total SSA and total pore volume of shale pores. In fact, simple linear regression analysis ignores the collinearity between multiple factors. For example, clay minerals are negatively correlated with TOC (Figure 10b), which may be the reason why clay minerals are negatively correlated with pore structure parameters. It should be pointed out that clays have no contribution to the maintenance of primary pores due to their ductile nature, but secondary pores may be generated during diagenesis and can be preserved if hydrocarbon generation generates overpressure to provide support (Figure 14c).

4.3. Contribution of OM to Pore Development

OM contributes significantly to the shale reservoir space, particularly in overmature marine shales [20]. However, the presence of OM-hosted pores depends strongly on hydrocarbon generation and thermal evolution, making them the primary pore types in overmature shales [18,20,68,72]. The OM was deposited concurrently with minerals and served as part of the sediments (Figure 15a). As burial depth increased and stress acted on the sediment, ductile minerals such as OM and clays deformed under the effect of in situ overburden stress and occupied the surrounding interP pores (Figure 15b) [68,74]. Meanwhile, as thermal maturity increased, OM evolved further, generating hydrocarbons such as bitumen, oil, gas, and so forth [72]. During this thermal evolution of OM, OM-hosted pores developed in residual OM due to the thermal cracking of in situ OM (Figure 15b) [31,68,74], while overpressure built up due to hydrocarbon generation [72,73]. Driven by the overpressure, parts of the mobile products, including liquid bitumen and oil, migrated to adjacent pores (Figure 15c), including interP pores supported by rigid minerals and pores within clays.

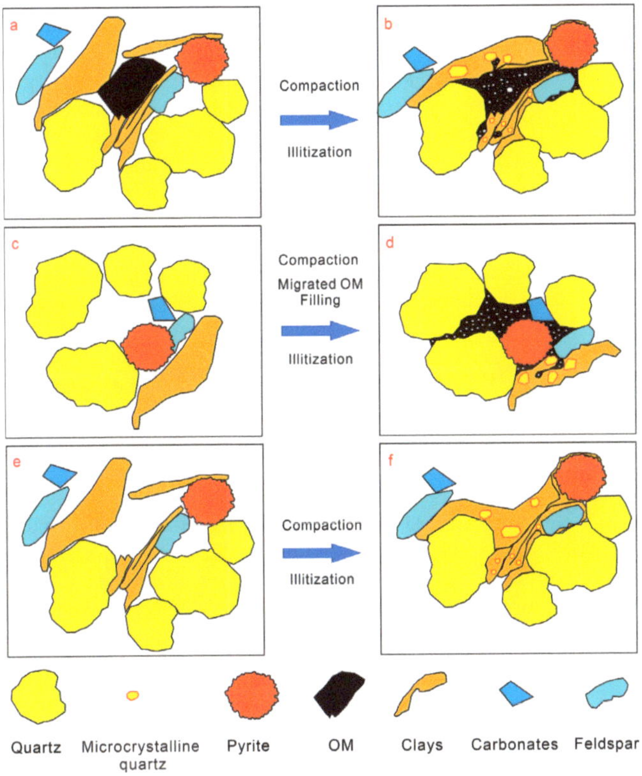

Figure 15. Mechanism of OM on pore preservation: (**a**) OM in situ deposited as part of the sediment framework; (**b**) ductile minerals, such as clays and OM, changed during compaction, secondary pores and microcrystalline quartz developed during diagenesis; (**c**) minerals deposited without OM; (**d**) interP pores associated with OM filling, and secondary pores developed during diagenesis and preserved by OM; (**e**) minerals deposited with higher clays but no OM; (**f**) clays compacted during compaction with pore destruction.

The filling of migrated OM and subsequent products of the further thermal evolution of the OM in these pores prevented the pores from collapsing due to further compaction (Figure 15d). For overmature shales, both residual OM and migrated OM would be further cracked with increasing thermal maturity, and various scales of pores developed within residual OM and migrated OM (Figure 15b,d), while the high cementation of surrounding mineral would form a relatively closed system, thus forming overpressure to preserve pores from compaction. At the same time, illitization of clays generated pores within clays (Figure 15b,d,f). If the sediments were deposited without OM and no OM migrated during burial (Figure 15e), pores would strongly reduce due to compaction (Figures 4 and 15f), but sediments with OM is another case. For both the OM attached to the surface of clays during deposition and the OM that migrated to the clays during burial, after thermal decomposition, its products, which can act as a bearer to inhibit the collapse of pores (Figure 14c), filled the pores generated by the diagenesis of clays (Figure 15b,d). In other words, pores not only can be preserved by rigid minerals (Figure 14a,b) [60] but also by products of hydrocarbon pyrolysis, such as oil and gas, especially under overpressure (Figure 14c). In deep burial shale, OM not only provides pore space to store shale gas, but its products also support pores.

5. Conclusions

The contribution to pore structure of shale components were analyzed by MLR to determine the key factor controlling the gas capacities of marine shales. Our investigation poses a challenge to the previous understanding. Through detailed study on the contribution of shale composition to the pores in Wufeng–Longmaxi Shale, important new information is provided for further understanding of the nature of shale pore systems. Conclusions are as follows: (1) The Wufeng–Longmaxi Shale is dominated by quartz, clays, carbonates, feldspar, and OM, according to XRD analysis. (2) OM and clays are the primary contributors to the storage space in overmature shale, indicated by the MLR, and OM contributes more to the pore structure. Indeed, OM not only provides pore space to store shale gas, but its products also support pores, including OM-hosted pores and pores generated by clay transformation. (3) Total SSA and volume of shale pores increases with the contents of TOC and clays. (4) Quartz, the most abundant component in shale, hardly contributes to the SSA and volume of shale pores. The quartz content of some samples has a positive correlation with the SSA and volume of shale pores because this part of quartz is biogenic, and its abundance is positively related to TOC. Thus, there is a collinear relationship between quartz and TOC. The increment of the pore SSA and pore volume of this part comes from OM rather than quartz. (5) Although carbonates and feldspar are prone to produce dissolution pores during diagenesis, their amount contributes little to the total pore volume and SSA of shale. (6) Shale pores can be preserved in two ways: overpressure existing inside or shelter existing outside the pores.

We acknowledge that we neglected the differences of the thermal maturity and its influence on pore structure. The result of this study explains the contribution of the composition of overmature shale to the SSA and volume of pores. As the experiment was only carried out by gas adsorption without mercury injection capillary pressure or other methods, it is still unclear how shale composition affects the pore structure of macropores with width over 300 nm. A total of 29 samples were analyzed for statistical analysis. However, being a statistical analysis, the correlation models need more data to be more convincing. We hope this paper will be conducive to understanding the mechanism of pore development in overmature shales.

Author Contributions: Conceptualization, L.X. and R.P.; methodology, H.H.; software, J.M.; validation, H.H.; formal analysis, L.X.; investigation, H.H.; resources and data curation, L.X.; writing—original draft preparation, L.X.; writing—review and editing, R.P.; supervision, J.M.; project administration, H.H.; funding acquisition, R.P. All authors have read and agreed to the published version of the manuscript.

Funding: This research was funded by the National Natural Science Foundation of China (grant number 41472123).

Institutional Review Board Statement: Not applicable.

Informed Consent Statement: Not applicable.

Data Availability Statement: Not applicable.

Acknowledgments: The authors would like to thank PetroChina Southwest Oil & Gas field Company for providing core samples. Many thanks also go to the anonymous reviewers for their insightful scientific comments and suggestions that significantly enhanced the quality of the original manuscript. Gratitude should go to Academic Editor Anabela Oliveira and Assistant Editor Sumesa Puangpee for their valuable comments and language improvements that greatly improved the original manuscript.

Conflicts of Interest: The authors declare no conflict of interest.

References

1. Jarvie, D.M.; Hill, R.J.; Ruble, T.E.; Pollastro, R.M. Unconventional shale-gas systems: The Mississippian Barnett Shale of north-central Texas as one model for thermogenic shale-gas assessment. *AAPG Bull.* **2007**, *91*, 475–499. [CrossRef]
2. Muehlenbachs, L.; Spiller, E.; Timmins, C. The housing market impacts of shale gas development. *Am. Econ. Rev.* **2015**, *105*, 3633–3659. [CrossRef]
3. Loucks, R.G.; Reed, R.M.; Ruppel, S.C.; Jarvie, D.M. Morphology, genesis, and distribution of nanometer-scale pores in siliceous mudstones of the Mississippian Barnett Shale. *J. Sed. Res.* **2009**, *79*, 848–861. [CrossRef]
4. Loucks, R.G.; Reed, R.M.; Ruppel, S.C.; Hammes, U. Spectrum of pore types and networks in mudrocks and a descriptive classification for matrix-related mudrock pores. *AAPG Bull.* **2012**, *96*, 1071–1098. [CrossRef]
5. Ross, D.J.K.; Bustin, R.M. The importance of shale composition and pore structure upon gas storage potential of shale gas reservoirs. *Mar. Petrol. Geol.* **2009**, *26*, 916–927. [CrossRef]
6. Chalmers, G.R.; Bustin, R.M.; Power, I.M. Characterization of gas shale pore systems by porosimetry, pycnometry, surface area, and field emission scanning electron microscopy/transmission electron microscopy image analyses: Examples from the Barnett, Woodford, Haynesville, Marcellus, and Doig unitsCharacterization of Gas Shale Pore Systems. *AAPG Bull.* **2012**, *96*, 1099–1119.
7. Clarkson, C.R.; Solano, N.; Bustin, R.M.; Bustin, A.M.M.; Chalmers, G.R.L.; He, L.; Melnichenko, Y.B.; Radliński, A.P.; Blach, T.P. Pore structure characterization of North American shale gas reservoirs using USANS/SANS, gas adsorption, and mercury intrusion. *Fuel* **2013**, *103*, 606–616. [CrossRef]
8. Kuila, U.; Prasad, M. Specific surface area and pore-size distribution in clays and shales. *Geophys. Prospect.* **2013**, *61*, 341–362. [CrossRef]
9. Liu, K.; Ostadhassan, M.; Zhou, J.; Gentzis, T.; Rezaee, R. Nanoscale pore structure characterization of the Bakken shale in the USA. *Fuel* **2017**, *209*, 567–578. [CrossRef]
10. Hazra, B.; Wood, D.A.; Vishal, V.; Varma, A.K.; Sakha, D.; Singh, A.K. Porosity controls and fractal disposition of organic-rich Permian shales using low-pressure adsorption techniques. *Fuel* **2018**, *220*, 837–848. [CrossRef]
11. Ko, L.T.; Ruppel, S.C.; Loucks, R.G.; Hackley, P.C.; Zhang, T.; Shao, D. Pore-types and pore-network evolution in Upper Devonian-Lower Mississippian Woodford and Mississippian Barnett mudstones: Insights from laboratory thermal maturation and organic petrology. *Int. J. Coal Geol.* **2018**, *190*, 3–28. [CrossRef]
12. Xu, S.; Gou, Q.; Hao, F.; Zhang, B.; Shu, Z.; Lu, Y.; Wang, Y. Shale pore structure characteristics of the high and low productivity wells, Jiaoshiba shale gas field, Sichuan Basin, China: Dominated by lithofacies or preservation condition? *Mar. Petrol. Geol.* **2020**, *114*, 104211. [CrossRef]
13. Zhao, J.; Jin, Z.; Jin, Z.; Wen, X.; Geng, Y. Origin of authigenic quartz in organic-rich shales of the Wufeng and Longmaxi Formations in the Sichuan Basin, South China: Implications for pore evolution. *J. Nat. Gas Sci. Eng.* **2017**, *38*, 21–38. [CrossRef]
14. Zhao, J.; Jin, Z.; Jin, Z.; Hu, Q.; Hu, Z.; Du, W.; Yan, C.; Geng, Y. Mineral types and organic matters of the Ordovician-Silurian Wufeng and Longmaxi Shale in the Sichuan Basin, China: Implications for pore systems, diagenetic pathways, and reservoir quality in fine-grained sedimentary rocks. *Mar. Petrol. Geol.* **2017**, *86*, 655–674. [CrossRef]
15. Ross, D.J.K.; Bustin, R.M. Shale gas potential of the lower Jurassic Gordondale member, northeastern British Columbia, Canada. *Bull. Can. Petrol. Geol.* **2007**, *55*, 51–75. [CrossRef]
16. Zhu, X.; Cai, J.; Liu, Q.; Li, Z.; Zhang, X. Thresholds of petroleum content and pore diameter for petroleum mobility in shale. *AAPG Bull.* **2019**, *103*, 605–617. [CrossRef]
17. Wang, X.; Liu, L.; Wang, Y.; Sheng, Y.; Zheng, S.; Wu, W.; Luo, Z. Comparison of the pore structures of Lower Silurian Longmaxi Formation shales with different lithofacies in the southern Sichuan Basin, China. *J. Nat. Gas Sci. Eng.* **2020**, *81*, 103419. [CrossRef]
18. Zhang, Y.; Yu, B.; Pan, Z.; Hou, C.; Zuo, Q.; Sun, M. Effect of thermal maturity on shale pore structure: A combined study using extracted organic matter and bulk shale from Sichuan Basin, China. *J. Nat. Gas Sci. Eng.* **2020**, *74*, 103089. [CrossRef]
19. Gou, Q.; Xu, S.; Hao, F.; Yang, F.; Shu, Z.; Liu, R. The effect of tectonic deformation and preservation condition on the shale pore structure using adsorption-based textural quantification and 3D image observation. *Energy* **2021**, *219*, 119579. [CrossRef]
20. Milliken, K.L.; Rudnicki, M.; Awwiller, D.N.; Zhang, T. Organic matter–hosted pore system, Marcellus formation (Devonian), Pennsylvania. *AAPG Bull.* **2013**, *97*, 177–200. [CrossRef]
21. Liu, R.; Hao, F.; Engelder, T.; Shu, Z.; Yi, J.; Xu, S.; Teng, C. Influence of tectonic exhumation on porosity of Wufeng–Longmaxi shale in the Fuling gas field of the eastern Sichuan Basin, China. *AAPG Bull.* **2020**, *104*, 939–959. [CrossRef]
22. Kang, Y.; Shang, C.; Zhou, H.; Huang, Y.; Zhao, Q.; Deng, Z.; Wang, H.; Ma, Y.Z. Mineralogical brittleness index as a function of weighting brittle minerals—From laboratory tests to case study. *J. Nat. Gas Sci. Eng.* **2020**, *77*, 103278. [CrossRef]

23. Curtis, M.E.; Cardott, B.J.; Sondergeld, C.H.; Rui, C.S. Development of organic porosity in the Woodford Shale with increasing thermal maturity. *Int. J. Coal Geol.* **2012**, *103*, 26–31. [CrossRef]
24. Peltonen, C.; Marcussen, Ø.; Bjørlykke, K.; Jahren, J. Clay mineral diagenesis and quartz cementation in mudstones: The effects of smectite to illite reaction on rock properties. *Mar. Petrol. Geol.* **2009**, *26*, 887–898. [CrossRef]
25. Xu, H.; Zhou, W.; Hu, Q.; Yi, T.; Ke, J.; Zhao, A.; Lei, Z.; Yu, Y. Quartz types, silica sources and their implications for porosity evolution and rock mechanics in the Paleozoic Longmaxi Formation shale, Sichuan Basin. *Mar. Petrol. Geol.* **2021**, *128*, 105036. [CrossRef]
26. Chen, F.; Liu, D.C.; Ding, X.; Zheng, Q.; Lu, S. Pore size distributions contributed by various components in the Upper Ordovician Wufeng Shale from Southeast Chongqing, China. *J. Petrol. Sci. Eng.* **2022**, *208*, 109230. [CrossRef]
27. Han, H.; Zhong, N.; Ma, Y.; Huang, C.; Wang, Q.; Chen, S.; Lu, J. Gas storage and controlling factors in an over-mature marine shale: A case study of the Lower Cambrian Lujiaping shale in the Dabashan arc-like thrust–fold belt, southwestern China. *J. Nat. Gas Sci. Eng.* **2016**, *33*, 839–853. [CrossRef]
28. Gao, Z.; Hu, Q. Pore structure and spontaneous imbibition characteristics of marine and continental shales in China. *AAPG Bull.* **2018**, *102*, 1941–1961. [CrossRef]
29. Dong, T.; He, S.; Chen, M.; Hou, Y.; Guo, X.; Wei, C.; Han, Y.; Yang, R. Quartz types and origins in the paleozoic Wufeng-Longmaxi Formations, Eastern Sichuan Basin, China: Implications for porosity preservation in shale reservoirs. *Mar. Petrol. Geol.* **2019**, *106*, 62–73. [CrossRef]
30. Chen, X.; Shi, W.; Hu, Q.; Hou, Y.; Zhai, G.; Dong, T.; Zhou, Z.; Du, X. Origin of authigenic quartz in organic-rich shales of the Niutitang Formation in the northern margin of Sichuan Basin, South China: Implications for pore network development. *Mar. Petrol. Geol.* **2022**, *138*, 105548. [CrossRef]
31. Ji, W.; Hao, F.; Schulz, H.M.; Song, Y.; Tian, J. The architecture of organic matter and its pores in highly mature gas shales of the lower Silurian Longmaxi Formation in the upper Yangtze platform, south China. *AAPG Bull.* **2019**, *103*, 2909–2942. [CrossRef]
32. Zhao, J.; Jin, Z.; Hu, Q.; Liu, K.; Jin, Z.; Hu, Z.; Nie, H.; Du, W.; Yan, C.; Wang, R. Mineral composition and seal condition implicated in pore structure development of organic-rich Longmaxi shales, Sichuan Basin, China. *Mar. Petrol. Geol.* **2018**, *98*, 507–522. [CrossRef]
33. Hu, H.; Hao, F.; Guo, X.; Yi, J.; Shu, Z.; Bao, H.; Zhu, X. Effect of lithofacies on the pore system of over-mature Longmaxi shale in the Jiaoshiba area, Sichuan Basin, China. *Mar. Petrol. Geol.* **2019**, *109*, 886–898. [CrossRef]
34. Yang, R.; He, S.; Yi, J.; Hu, Q. Nano-scale pore structure and fractal dimension of organic-rich Wufeng-Longmaxi shale from Jiaoshiba area, Sichuan Basin: Investigations using FE-SEM, gas adsorption and helium pycnometry. *Mar. Pet. Geol.* **2016**, *70*, 27–45. [CrossRef]
35. Dong, T.; Harris, N.B.; Ayranci, K.; Twemlow, C.E.; Nassichuk, B.R. Porosity characteristics of the Devonian Horn River shale, Canada: Insights from lithofacies classification and shale composition. *Int. J. Coal Geol.* **2015**, *141*, 74–90. [CrossRef]
36. Gao, Z.; Fan, Y.; Hu, Q.; Jiang, Z.; Cheng, Y. The effects of pore structure on wettability and methane adsorption capability of Longmaxi Formation shale from the southern Sichuan Basin in China. *AAPG Bull.* **2020**, *104*, 1375–1399. [CrossRef]
37. Zhang, Q.; Liu, R.; Pang, Z.; Lin, W.; Bai, W.; Wang, H. Characterization of microscopic pore structures in Lower Silurian black shale (S_1l), southeastern Chongqing, China. *Mar. Pet. Geol.* **2016**, *71*, 250–259. [CrossRef]
38. Yu, B. Classification and characterization of gas shale pore system. *Earth Sci. Front.* **2013**, *20*, 211–220.
39. Guo, X.; Hu, D.; Li, Y.; Wei, X.; Wang, Q.; Zhang, H. Technologies in discovery and exploration of Fuling shale gas field, China. *Nat. Resour.* **2016**, *7*, 271. [CrossRef]
40. Dai, H.M.; Huang, D.; Liu, X.; Yang, Y.; He, X.M.; Bang, H.R.; Tong, D.W. Characteristics and evaluation of marine source rock in Southwestern Shunan. *Nat. Gas Geosci.* **2008**, *19*, 503–508.
41. Hakimi, M.H.; Ahmed, A.F.; Abdullah, W.H. Organic geochemical and petrographic characteristics of the Miocene Salif organic-rich shales in the Tihama Basin, Red Sea of Yemen: Implications for paleoenvironmental conditions and oil-generation potential. *Int. J. Coal Geol.* **2016**, *154*, 193–204. [CrossRef]
42. Wood, D.A.; Hazra, B. Characterization of organic-rich shales for petroleum exploration & exploitation: A review-Part 2: Geochemistry, thermal maturity, isotopes and biomarkers. *J. Earth Sci.-China* **2017**, *28*, 758–778.
43. Schmidt, J.S.; Menezes, T.R.; Souza, I.; Spigolon, A.L.D.; Pestilho, A.L.S.; Coutinho, L.F.C. Comments on empirical conversion of solid bitumen reflectance for thermal maturity evaluation. *Int. J. Coal Geol.* **2019**, *201*, 44–50. [CrossRef]
44. Bardestani, R.; Patience, G.S.; Kaliaguine, S. Experimental methods in chemical engineering: Specific surface area and pore size distribution measurements—BET, BJH, and DFT. *Can. J. Chem. Eng.* **2019**, *97*, 2781–2791. [CrossRef]
45. Strąpoć, D.; Mastalerz, M.; Schimmelmann, A.; Drobniak, A.; Hasenmueller, N.R. Geochemical constraints on the origin and volume of gas in the New Albany Shale (Devonian–Mississippian), eastern Illinois Basin. *AAPG Bull.* **2010**, *94*, 1713–1740.
46. Pang, P.; Han, H.; Hu, L.; Guo, C.; Gao, Y.; Xie, Y. The calculations of pore structure parameters from gas adsorption experiments of shales: Which models are better? *J. Nat. Gas Sci. Eng.* **2021**, *94*, 104060. [CrossRef]
47. Vishnyakov, A.; Ravikovitch, P.I.; Neimark, A.V. Molecular level models for CO_2 sorption in nanopores. *Langmuir* **1999**, *15*, 8736–8742. [CrossRef]
48. Ravikovitch, P.I.; Vishnyakov, A.; Russo, R.; Neimark, A.V. Unified approach to pore size characterization of microporous carbonaceous materials from N_2, Ar, and CO_2 adsorption isotherms. *Langmuir* **2000**, *16*, 2311–2320. [CrossRef]

49. Landers, J.; Gor, G.Y.; Neimark, A.V. Density functional theory methods for characterization of porous materials. *Colloids Surf. A Physicochem. Eng. Asp.* **2013**, *437*, 3–32. [CrossRef]
50. Wei, M.; Zhang, L.; Xiong, Y.; Li, J.; Peng, P. Nanopore structure characterization for organic-rich shale using the non-local-density functional theory by a combination of N_2 and CO_2 adsorption. *Microporous Mesoporous Mater.* **2016**, *227*, 88–94. [CrossRef]
51. Lachniet, M.S.; Patterson, W.P. Use of correlation and stepwise regression to evaluate physical controls on the stable isotope values of Panamanian rain and surface waters. *J. Hydrol.* **2006**, *324*, 115–140. [CrossRef]
52. Zhu, B.; Meng, J.; Pan, R.; Song, C.; Zhu, Z.; Jin, J.; Zhang, C.; Liu, L.; Xu, L.; Zhou, X. Multi-scale characterization of organic matter pore space in deep marine shale combined with mathematical morphology. *Geoenergy Sci. Eng.* **2023**, *223*, 211539. [CrossRef]
53. Fathy, D.; Abart, R.; Wagreich, M.; Gier, S.; Ahmed, M.S.; Sami, M. Late Campanian Climatic-Continental Weathering Assessment and Its Influence on Source Rocks Deposition in Southern Tethys, Egypt. *Minerals* **2023**, *13*, 160. [CrossRef]
54. Metwally, Y.M.; Chesnokov, E.M. Clay mineral transformation as a major source for authigenic quartz in thermo-mature gas shale. *Appl. Clay Sci.* **2012**, *55*, 138–150. [CrossRef]
55. Milliken, K.L.; Ergene, S.M.; Ozkan, A. Quartz types, authigenic and detrital, in the Upper Cretaceous Eagle Ford Formation, south Texas, USA. *Sed. Geol.* **2016**, *339*, 273–288. [CrossRef]
56. Milliken, K.L.; Olson, T. Silica diagenesis, porosity evolution, and mechanical behavior in siliceous mudstones, Mowry Shale (Cretaceous), Rocky Mountains, USA. *J. Sed. Res.* **2017**, *87*, 366–387. [CrossRef]
57. Yang, X.; Yan, D.; Wei, X.; Zhang, L.; Zhang, B.; Xu, H.; Gong, Y.; He, J. Different formation mechanism of quartz in siliceous and argillaceous shales: A case study of Longmaxi Formation in South China. *Mar. Petrol. Geol.* **2018**, *94*, 80–94.
58. Zheng, Y.; Liao, Y.; Wang, Y.; Xiong, Y.; Peng, P. Organic geochemical characteristics, mineralogy, petrophysical properties, and shale gas prospects of the Wufeng–Longmaxi shales in Sanquan Town of the Nanchuan District, Chongqing. *AAPG Bull.* **2018**, *102*, 2239–2265. [CrossRef]
59. Basu, S.; Verchovshy, A.B.; Bogush, A.; Jones, A.P.; Jourdan, A.L. Stability of organic carbon components in shale: Implications for carbon cycle. *Front. Earth Sci.* **2019**, *7*, 297. [CrossRef]
60. Fishman, N.S.; Hackley, P.C.; Lowers, H.A.; Hill, R.J.; Egenhoff, S.O.; Eberl, D.D.; Blum, A.E. The nature of porosity in organic-rich mudstones of the Upper Jurassic Kimmeridge Clay Formation, North Sea, offshore United Kingdom. *Int. J. Coal Geol.* **2012**, *103*, 32–50. [CrossRef]
61. Thyberg, B.; Jahren, J.; Winje, T.; Bjørlykke, K.; Faleide, J.I.; Marcussen, Ø. Quartz cementation in Late Cretaceous mudstones, northern North Sea: Changes in rock properties due to dissolution of smectite and precipitation of micro-quartz crystals. *Mar. Petrol. Geol.* **2010**, *27*, 1752–1764. [CrossRef]
62. Aplin, A.C.; Macquaker, J.H.S. Mudstone diversity: Origin and implications for source, seal, and reservoir properties in petroleum systems. *AAPG Bull.* **2011**, *95*, 2031–2059. [CrossRef]
63. Bjørlykke, K.; Høeg, K. Effects of burial diagenesis on stresses, compaction and fluid flow in sedimentary basins. *Mar. Petrol. Geol.* **1997**, *14*, 267–276. [CrossRef]
64. Jin, X.; Shah, S.N.; Roegiers, J.C.; Zhang, B. An integrated petrophysics and geomechanics approach for fracability evaluation in shale reservoirs. *SPE J.* **2015**, *20*, 518–526. [CrossRef]
65. Zhang, D.; Ranjith, P.G.; Perera, M.S.A. The brittleness indices used in rock mechanics and their application in shale hydraulic fracturing: A review. *J. Petrol. Sci. Eng.* **2016**, *143*, 158–170. [CrossRef]
66. MacGowan, D.B.; Surdam, R.C. Carboxylic acid anions in formation waters, San Joaquin Basin and Louisiana Gulf Coast, USA—Implications for clastic diagenesis. *Appl. Geochem.* **1990**, *5*, 687–701. [CrossRef]
67. Środoń, J. Nature of mixed-layer clays and mechanisms of their formation and alteration. *Annu. Rev. Earth Planet. Sci.* **1999**, 2, 19–53. [CrossRef]
68. Wang, G. Deformation of organic matter and its effect on pores in mud rocks. *AAPG Bull.* **2020**, *104*, 21–36. [CrossRef]
69. Vandenbroucke, M.; Largeau, C. Kerogen origin, evolution and structure. *Org. Geochem.* **2007**, *38*, 719–833. [CrossRef]
70. Kennedy, M.J.; Löhr, S.C.; Fraser, S.A.; Baruch, E.T. Direct evidence for organic carbon preservation as clay-organic nanocomposites in a Devonian black shale; from deposition to diagenesis. *Earth Planet. Sci. Lett.* **2014**, *388*, 59–70. [CrossRef]
71. Abid, I.; Hesse, R. Illitizing fluids as precursors of hydrocarbon migration along transfer and boundary faults of the Jeanne d'Arc Basin offshore Newfoundland, Canada. *Mar. Petrol. Geol.* **2007**, *24*, 237–245. [CrossRef]
72. Hunt, J.M. *Petroleum Geochemistry and Geology*, 2nd ed.; W. H. Freeman & Company: New York, NY, USA, 1996; pp. 265–268.
73. Osborne, M.J.; Swarbrick, R.E. Mechanisms for generating overpressure in sedimentary basins: A reevaluation. *AAPG Bull.* **1997**, *81*, 1023–1041.
74. Loucks, R.G.; Reed, R.M. Scanning-electron-microscope petrographic evidence for distinguishing organic-matter pores associated with depositional organic matter versus migrated organic matter in mudrock. *GCAGS J.* **2014**, 3, 51–60.

Journal of
Marine Science and Engineering

MDPI

Article

Sequence Stratigraphy and Implications for Shale Gas Exploration in the Southern Sichuan Basin, South China

Lingling Xu [1,2], Jianghui Meng [2,3,*], Renfang Pan [1,2], Xue Yang [4], Qimeng Sun [4] and Boyuan Zhu [1,2]

1 School of Geosciences, Yangtze University, Wuhan 430100, China; 202073035@yangtzeu.edu.cn (L.X.); pan@yangtzeu.edu.cn (R.P.); 2021730042@yangtzeu.edu.cn (B.Z.)
2 Key Laboratory of Exploration Technologies for Oil and Gas Resources, Ministry of Education, Yangtze University, Wuhan 430100, China
3 Cooperative Innovation Center of Unconventional Oil and Gas, Yangtze University, Wuhan 430100, China
4 PetroChina Southwest Oil & Gasfield Company, Chengdu 610051, China; yangx63@petrochina.com.cn (X.Y.); sunqimeng2018@petrochina.com.cn (Q.S.)
* Correspondence: mjh@yangtzeu.edu.cn

Abstract: In contrast to the widely used sequence stratigraphic models for passive continental margins, the stacking patterns of strata within epeiric seas, which are influenced by regional tectonic activity, may display opposing characteristics during the same geological period. These variations serve as a record of basin evolution and also affect the accumulation of hydrocarbons within the strata. Our study investigated the development potential of the deep Longmaxi Shale in the southern Sichuan Basin by examining the sequence stratigraphy and sedimentary fill patterns. Using a combination of core observation, well-logging data analysis, and 3D seismic profile interpretation, we aimed to gain an understanding of the sedimentary fill history of the Longmaxi Shale during the Early Silurian. Our analysis revealed that deglaciation and regional tectonic events affected the sequence stratigraphy, resulting in unconformities that were identifiable using seismic data and wireline logs. Through an analysis of thirty wireline logs and two seismic profiles, we identified two third-order sequences suggested in the Lower Longmaxi Formation. Within the two third-order sequences were five systems tracts, with the first exhibiting a complete cycle of sea-level change and the second cycle being incomplete due to regional tectonic events. The graptolite succession on the upper Yangtze Platform provided a temporal view of the sequence stratigraphy and sedimentation rates of the Longmaxi Shale. The thickness trends of the systems tracts reflected the interplay of short-term eustasy fluctuations, subsidence, and uplift. Our analysis suggests that regional subsidence played a significant role in the deposition of the second transgressive systems tract (TST) in the Weiyuan and Luzhou areas, which represents a promising target for shale gas exploration, in addition to the first TST. However, the Changning area experienced a relative sea-level decrease due to the intense uplift of the Qianzhogn Paleo-uplift and the increased supply of sediment and is interpreted as a highstand systems tract (HST); it is not considered to have shale gas exploration potential.

Keywords: sequence stratigraphy; sedimentary evolution; shale gas; Longmaxi Formation; South Sichuan Basin

Citation: Xu, L.; Meng, J.; Pan, R.; Yang, X.; Sun, Q.; Zhu, B. Sequence Stratigraphy and Implications for Shale Gas Exploration in the Southern Sichuan Basin, South China. *J. Mar. Sci. Eng.* **2023**, *11*, 1393. https://doi.org/10.3390/jmse11071393

Academic Editor: George Kontakiotis

Received: 12 June 2023
Revised: 2 July 2023
Accepted: 5 July 2023
Published: 10 July 2023

1. Introduction

The Longmaxi Shale, located in the Upper Yangtze Block of the southern Sichuan Basin, has emerged as the primary target for exploring and developing unconventional shale gas in China [1–3]. Although the Changning, Weiyuan, and Zhaotong shale plays have successfully developed shallow (less than 3500 m deep) shale gas [4], the deep shale in the Luzhou area holds even greater potential for development and is expected to be the next target for shale gas exploration [2]. Studying the strata and fluid properties is crucial for understanding the accumulation of hydrocarbons and accurately predicting their development processes [5–7]. Additionally, a high-resolution sequence stratigraphic framework not only aids in predicting the oil and gas sweet spots and correlating regions but also provides insights into the evolution of the sedimentary basin [1,8–11]. Several studies on the sequence stratigraphy of the Longmaxi Formation in southern Sichuan have been conducted to guide the exploration and development of shale gas resources [1,12–17]. However, due to variations in available data, research methods, and comprehension of sequence stratigraphy, scholars still hold differing opinions on the sequence division and understanding of the Longmaxi Formation in the southern Sichuan Basin. Most researchers have categorized the Lower Longmaxi Formation as a third-order sequence [1,12], while some have divided it into two or more third-order sequences [13].

Nevertheless, they all acknowledge that the Upper Yangtze Sea exhibits consistent variations in relative sea level [1,12–17]. Therefore, the stratigraphic stacking patterns in Weiyuan, Luzhou, and Chongning regions are also consistent. Previous research findings indicated that the transgressive systems tract (TST) represents the most favorable interval for shale gas exploration within a third-order sequence [1]. In other words, based on the correlation between TST and the corresponding strata established through the sequence stratigraphic framework, the shale gas layers with the most potential can be identified. However, recent drilling data demonstrated that the potential distribution of shale gas layers in the Weiyuan, Luzhou, and Chongning regions was inconsistent.

Consequently, the relative sea level changes during the sedimentation of this stratigraphic unit in the Weiyuan, Luzhou, and Changning regions might not be consistent. This could result in a variation of their stratigraphic stacking patterns. Moreover, the migration of the subsidence center, the influence of paleo-uplifts and depressions on the lateral distribution of the sequence stratigraphy, and the evolution of the sequence stratigraphy remain unclear. Identifying the sequence boundary and determining changes in accommodation is critical to establishing the stratigraphic sequence framework of marine shale [18]. Nevertheless, previous studies suggest that the studied strata are controlled not only by eustacy but also by paleogeography and tectonism [19–21], thus making it challenging to determine the ideal sequence framework and evolutionary pattern of the studied area, which is complicated by the interplay of eustacy and regional differential subsidence.

Geological records, including seismic data, lithofacies, sedimentary structures, graptolite abundance and extinction, geochemistry, wireline logs, and organic matter abundance (TOC), are important in analyzing the effects of relative sea-level changes [11,22–27]. Therefore, it is critical to integrate multiple methods to establish a complete and reasonable high-resolution stratigraphic sequence framework for better predicting the "sweet spot". As shale gas exploration and development in the Sichuan Basin moves into the deep field, the deep drilling data in the Luzhou area has been augmented. With this understanding and abundance of data, the objectives of this paper are to reconstruct the evolution of the studied strata under the stratigraphic sequence framework and predict the potential layers for shale gas exploration.

2. Geologic Setting

2.1. Tectonic Setting

The Sichuan Basin, which is located in southwestern China (Figure 1a), is a superimposed basin spanning a vast area measuring 26×10^4 km^2 [28]. The double structures of the basin, consisting of shallow marine deposits and terrestrial strata, were formed during the two evolutionary stages of the basin: (1) the development stage of a marine intracratonic basin and (2) the development phase of an intracontinental composite foreland basin (Figure 2) [22]. South China overlayed the low latitudes of the northern hemisphere in Late Ordovician and Early Silurian times (Figure 3a) and was significantly affected by the global paleoclimate fluctuations and tectonics [29–34]. Tectonic highlands and paleo-uplifts were raised along the eastern margin of the Sichuan Basin during the Late Ordovician to the Early Silurian Yangtze-Cathaysia plate collision and amalgamation [29]. Meanwhile, during the Caledonian Era, the Leshan-Longnvsi highland continued to rise, restricting the upper Yangtze Sea in the north by the Leshan-Longnvsi paleo-high and in the south by Qianzhong Paleo-uplift (Figure 3b) [35,36]. With the increasing extrusion of the South China Plate, the peripheral paleo-uplifts surrounding the upper Yangtze Sea continuously elevated and expanded, leading the Kangdian, Dianqiangui, and Cathaysia lands to become one large land area [19,37,38]. In contrast, the upper Yangtze Sea was open towards the northeast, allowing an influx of open marine waters and preventing complete circulation restriction. While the northern margin of the Yangtze Sea still maintained passive edge characteristics, South China exhibited significant compression and shrinkage of structural features [22,39].

The study area is situated within the southern Sichuan Basin low steep dome belt, which was located at the western edge of the Yangtze Plate. Throughout its geological history, this region has undergone multiple phases of complicated tectonic transformations and superimpositions and has been affected by the tectonic movements of several orogens, including the Caledonian, Hercynian, Indosinian, Yanshanian, and Himalayan [1]. Over a relatively long geological timeframe, this region has formed a distinct geometric shape as a result of its unique structural evolution, leading to the formation of a bowl-shaped syncline structure at the boundary of the Longmaxi Formation.

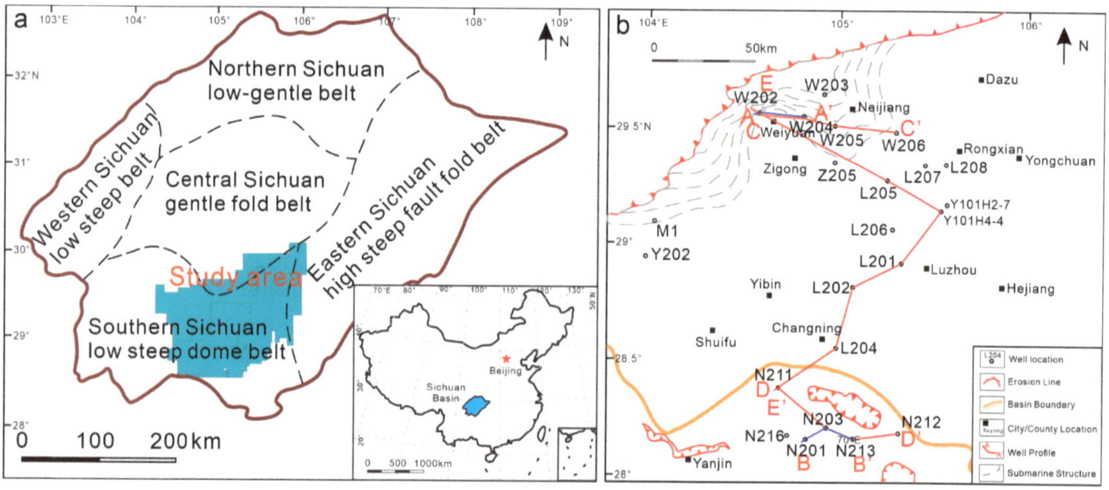

Figure 1. (**a**)Tectonic location map of the study area in the Sichuan Basin, SW China. (**b**) Schematic map of the well locations. Dashed lines in the Weiyuan area delineate the paleo-topography during the Early Silurian period (modified after [21]).

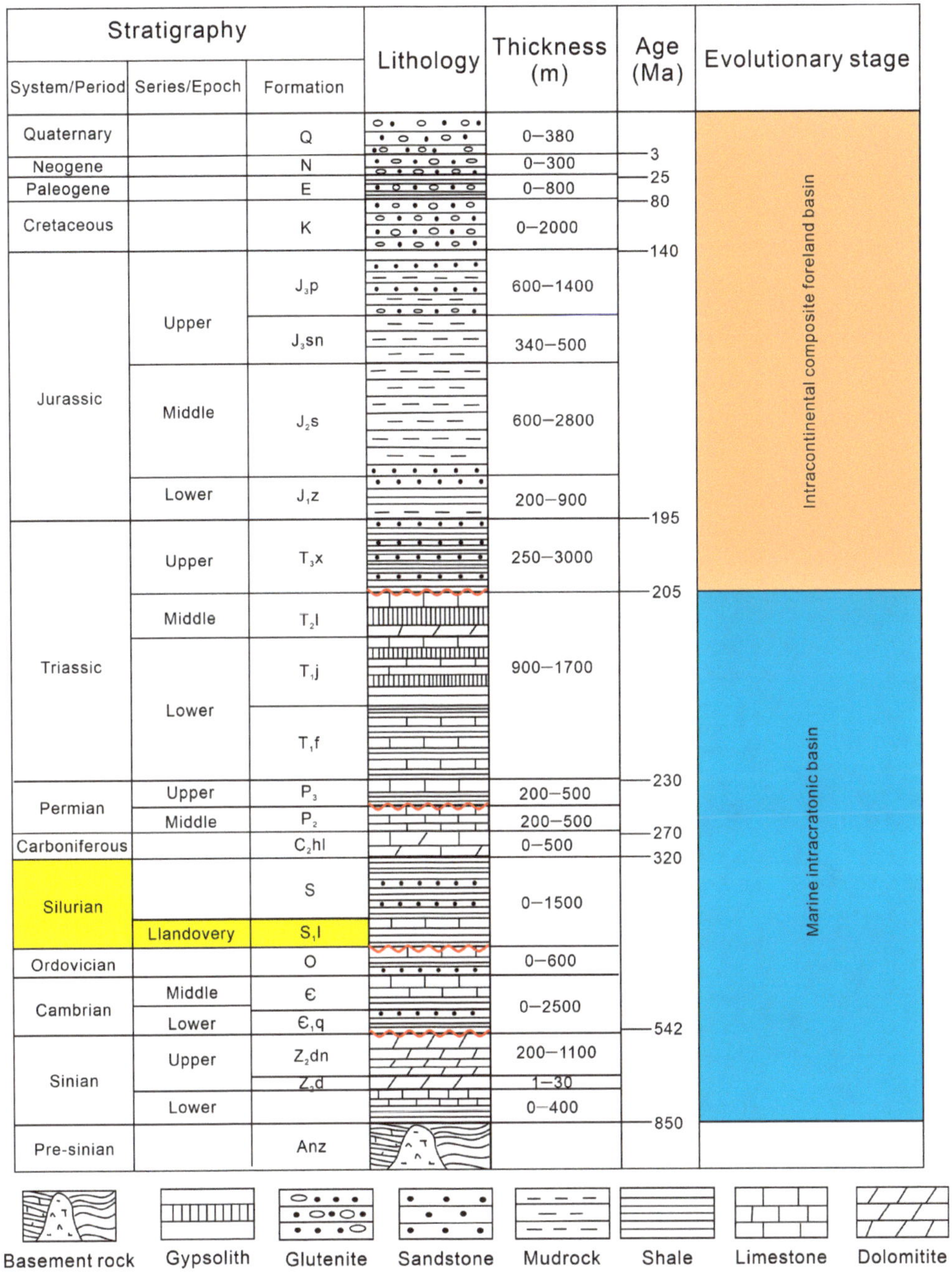

Figure 2. Sedimentary strata and evolutionary stage of the Sichuan Basin (modified from [22]). The highlighted section indicates the location of the research segment. The red wavy lines represent the unconformity surfaces.

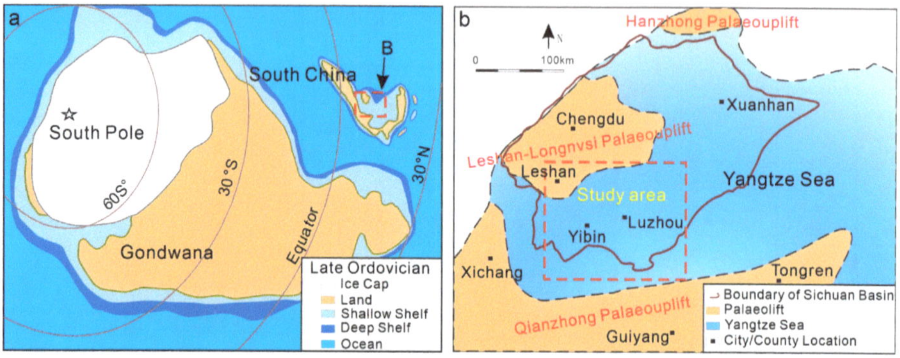

Figure 3. Paleography of the Upper Yangtze Plates. (**a**) Paleography at 445 Ma, Hirnantian, showing the positions of the glacial ice cap and South China [40]. (**b**) Paleogeography of the Early Silurian, showing the positions of the Sichuan Basin and the study area (modified from [41]).

2.2. Regional Stratigraphy

The Longmaxi Formation of the Lower Silurian is divided into two members: The Lower Member and the Upper Member [1]. This article focuses on the Lower Member, which is currently the primary interval for commercial shale gas production. The Longmaxi Formation is widely distributed on the Yangtze Platform [37,42–44]. Below the Longmaxi Formation lies the Wufeng Formation, separated from the underlying black shale of the Longmaxi Formation by a thin layer of interbedded shelly limestone [16]. The underlying unit, called "Guanyinqiao", is known for its thin shelly limestone resulting from Hirnantian glaciation [37,45,46]. The Wufeng Shale and Guanyinqiao together form the Wufeng Formation, consisting of four graptolite zones with a thickness ranging from several meters to thirty meters [47,48]. The Longmaxi Formation comprises the lower organic-rich siliceous shale and the upper silty shale and includes nine graptolite zones with a thickness ranging from several tens of meters to hundreds of meters [49,50].

3. Data and Methods

3.1. Dataset

We used multiple datasets to develop a comprehensive stratigraphic sequence framework for the Lower Longmaxi Formation in the South Sichuan Basin, spanning from the Weiyuan to Changning regions. All data used in this paper were provided by the PetroChina Southwest Oil & Gas Field Company, including a 3D seismic volume with an area of 450 km^2, cores with a total length of 667.18 m from 10 wells, and wireline logs from 32 wells. We also examined one outcrop section located in Nanjiang. The seismic data utilized for analyzing the Longmaxi Formation in the Southern Sichuan Basin was collected before 2011 and reprocessed in 2015 using pre-stack time migration. The frequency of the 3D seismic volume data for the study area ranges from 5 Hz to 65 Hz, with a predominant frequency of 30 Hz. Based on calculations, the maximum vertical resolution attainable from this dataset is approximately 25 m. The logging data used in this article include gamma ray (GR) logs, acoustic (AC) logs, and density (DEN) logs. In addition, 340 core samples were collected for geochemical tests, including determinations for TOC, minerals, major elements, and rare earth elements, which can reflect changes in the sedimentary environment. The TOC content was determined using a LECO carbon-sulfur analyzer (CS230). The samples were crushed into a powder with a particle size of less than 100 mesh. Approximately 1 g of the samples were pretreated with acid for approximately 3 h at temperatures ranging from 60 °C to 80 °C in order to remove carbonate minerals. Subsequently, the samples were subjected to pyrolysis at temperatures up to 550 °C. The experimental procedure for determining TOC was carried out in accordance with the standard [51]. The mineral components were measured using K-alpha radiation (XRD) on an X′ Pert3 Powder diffractometer. Data were obtained at room temperature in the 2θ range of 5° to 45°, with a scanning step of 0.02° and a scanning speed of 2°/min. Mineral quantification analysis was calculated using the equation as per the Chinese Oil and Gas Industry Standard (SY/T 5163-2018) [52]. For the determination of major elements and rare earth elements, we suggest readers refer to previous literature for detailed descriptions [53–55].

3.2. Methodology

The stratigraphic sequence model was based on traditional stratigraphic sequence principles and terminology [9,18,56], and the approach mainly relied upon Rider (1986) [57] and Van Wagoner et al. (1990) [22]. The isochronous framework of twelve wells was identified to constrain the boundary of the sequence stratigraphy based on the standard graptolite zones established by Chen et al. (2015) [1,42,45]. The location of the first appearance of characteristic graptolites served as a representative indicator of the sedimentary age of the corresponding stratigraphic layer. The third-order sequence boundaries were determined by abrupt changes in the seismic profile, wireline logs, outcrop sections, and cores [58]. Due to the relatively low resolution of the seismic data, the maximum flooding surface (MFS) was recognized by the wireline logs, which were characterized by the highest gamma values [22,26,57]. Systems tracts, a subdivision of a sequence, were defined by the stratal stacking patterns, the local stratigraphic relationships, and types of bounding surfaces [18,59].

4. Results

4.1. Stratigraphic Sequence Surfaces

Stratigraphic sequence surfaces mark changes within a lower sequence [18,59]. They are surfaces that appear as unconformities or conformities at the edges and inside of sedimentary basins [1,56]. Based on the available data in this paper, two third-order sequences were defined in the Lower Longmaxi Formation (Figures 4–9). Accordingly, five sequence surfaces were identified, including two maximum flooding surfaces (two MFSs in the Weiyuan and Luzhou areas and one in the Changning area) and three third-order sequence boundaries. The first sequence boundary was named SB1. In the field profile, the lower part of the Longmaxi Formation clearly displayed SB1 in the first third-order sequence stratigraphy as it marked a change of lithology (Figure 4). The lithology under SB1 was composed of shelly limestone from the Wufeng Formation, while above this surface, organic-rich black shale comprised the bottom of the Longmaxi Formation. It was also easy to identify SB1 in the core column by recognizing the shelly limestone and the overlying black shale (Figure 4). This surface, SB1, was characterized by an unconformity in some areas of the Yangtze Plate on the seismic profile [60], especially near the edge of the basin (Figure 5). However, conformity could be observed in deep-water settings, such as the Luzhou and Changning areas (Figure 6), making it impossible to determine the type of sequence boundary solely from local seismic profiles. In contrast, the characteristics of this surface, as observed on wire logs, were obvious and consistent. The gamma value suddenly increased rapidly in a short time (Figures 7 and 9). Consequently, SB1 was interpreted as a third-order sequence boundary as well as a transgressive surface, which coincides with previous research [1,16]. The first MFS served as a globally significant surface mainly controlled by the melting of the Gondwana glacier [16]. Specific, recognizable response characteristics were developed in the field profiles, seismic sections, and well log data that could be used to recognize it (Figures 4–7). Although SB2, the second MFS, and SB3 have not been directly identified and judged through field and seismic profiles, their response characteristics on logging curves were obvious. Specifically, the gamma values of SB2 and SB3 were local minimum values, while the gamma values of the second MFS were local maximum values (Figure 7). The GR logs in the deep depressions (e.g., W202, W204H10-2, and W204) of the Weiyuan area were rather easy to identify, followed by the Luzhou area, and the variations in the Changning area were relatively subtle (Figures 10 and 11). In this article, we have used the boundary of the system tract of the second third-order sequence as the sequence boundary due to data limitations and research purposes.

Figure 4. Field section and core photographs of the Longmaxi Formation, located in Nanjing, are shown on the left with field photographs, the middle, with core photographs from well Y101H4-4, and the right, with digital core photographs from well Y101H4-4. SB1 represents a third-order sequence boundary, while MFS denotes the shift from transgression to highstand regression within the first third-order sequence. Abbreviations: TST-transgressive systems tract; HST-highstand systems tracts.

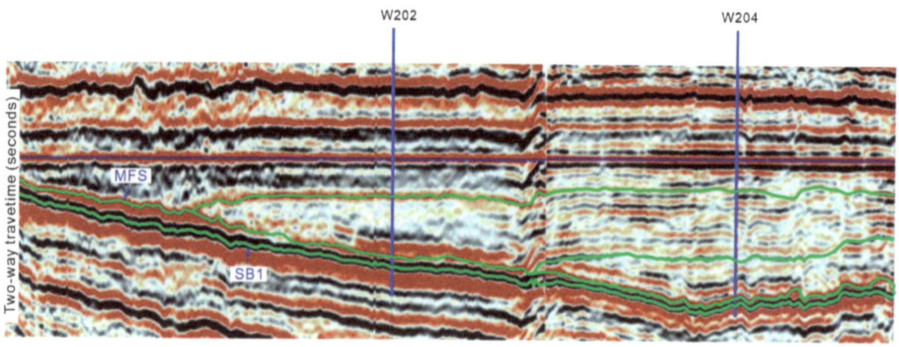

Figure 5. Regional seismic profile from the Weiyuan area (modified from [21]), with SB1 (third-order sequence boundary) and MFS (maximum flooding surface of sequence 1). The location of the seismic profile is shown in Figure 1b. SB = sequence boundary. MFS = maximum flooding surface.

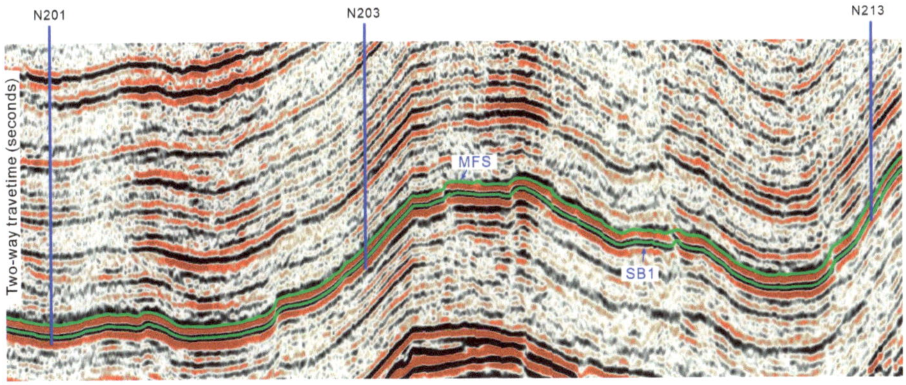

Figure 6. Regional seismic profile from the Changning area, showing the third-order stratigraphic framework. The location of the seismic profile is shown in Figure 1b. SB1 is a third-order sequence boundary. SB = sequence boundary. MFS = maximum flooding surface.

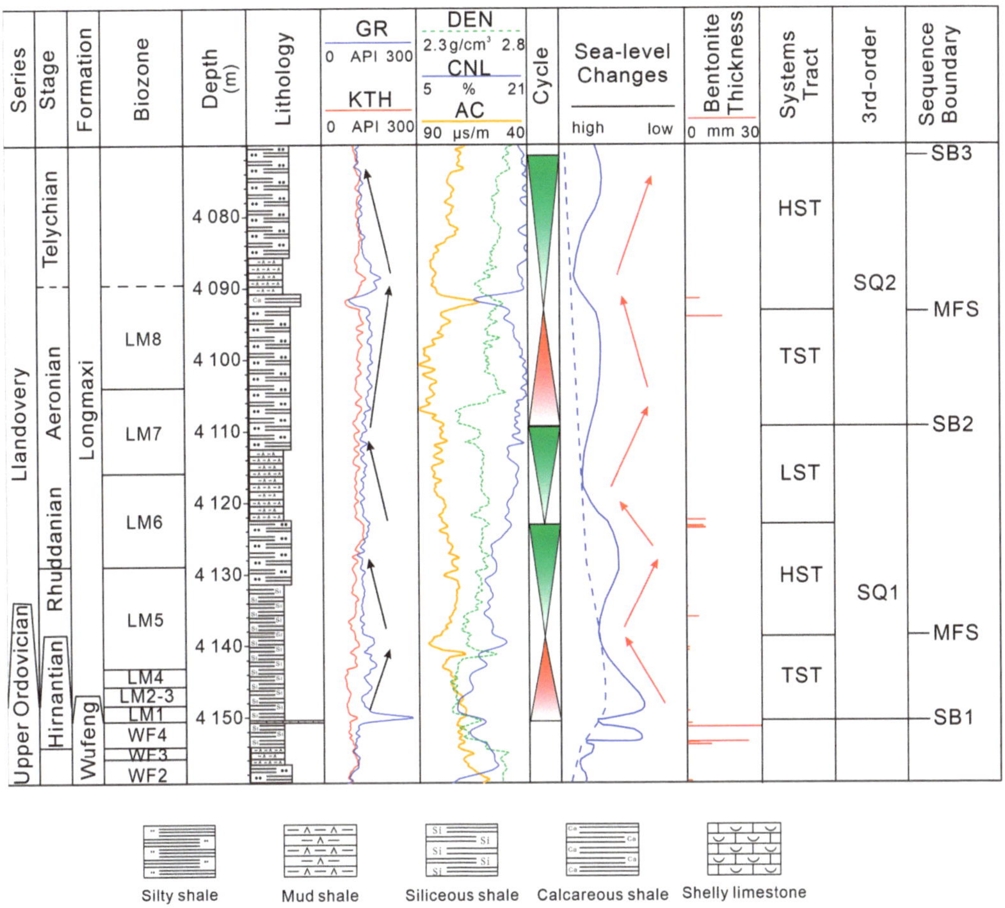

Figure 7. Summary of the stratigraphic sequence column from well Y101H4-4, showcasing the graptolite zone, wireline logs, the bentonite thickness observed from cores, sea-level changes [61],

and the interpretation of sequence stratigraphy. Note the biozone abbreviations are adapted from [45], WF2 = *Dicellograptus complexus*, WF3 = *Paraorthogr. pacificus*, WF4 = *Metabologr. extraordinarius*, LM1 = *Persculptogr. persculptus*, LM2-3 = *Akidograptus ascensus-Parakidogr. Acuminatus*, LM4 = *Cystograptus vesiculosus*, LM5 = *Coronograptus cyphus*, LM6 = *Demirastrites triangulates*, LM7 = *Lituigraptus convolutes*, LM8 = *Stimulograptus sedgwickii*. GR = gamma ray, KTH = gamma ray without uranium, DEN = density, CNL = compensated neutron logging, AC = acoustic, TST = transgressive systems tract, HST = highstand systems tract, SQ = sequence, SB = sequence boundary. A dashed line at biozone boundaries means uncertainty in placement.

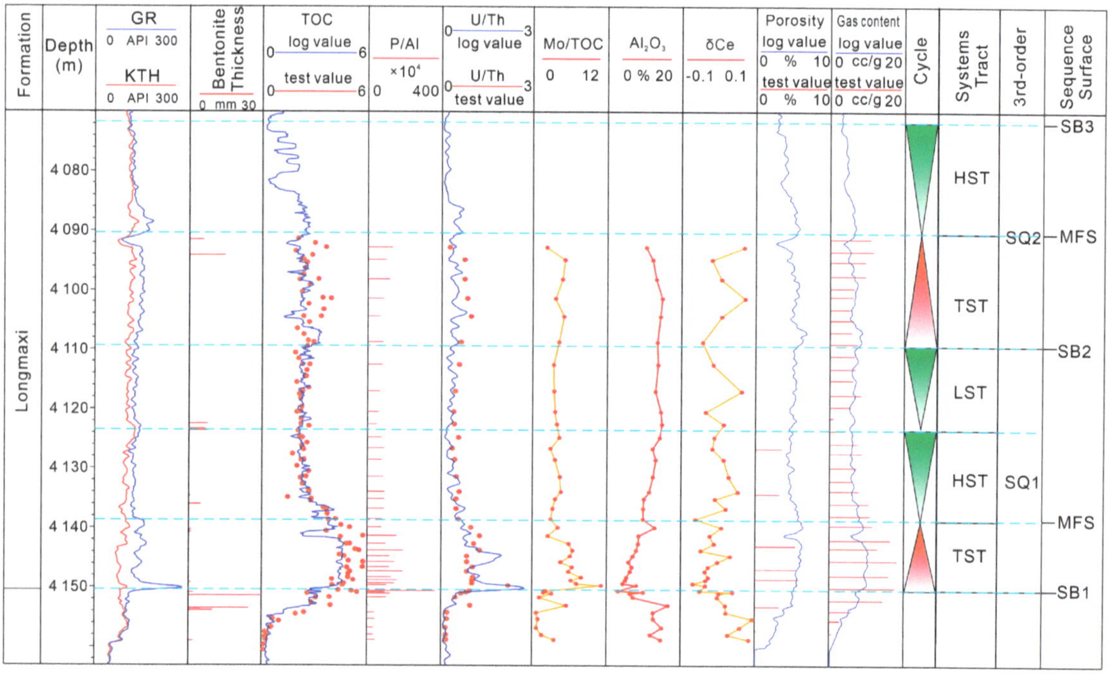

Figure 8. Geological section of the Longmaxi Shale from well Y101H4−4, where the thickness of bentonite was used to indicate the strength of volcanic activity [62]. Paleo-productivity was expressed by the value of P/Al [63], where the paleo-redox state of the depositional environment was expressed by U/Th [16,64]. The degree of water restriction was represented by Mo/TOC [65], and the content of Al_2O_3 served as a proxy for the extent of continental input [66]. Additionally, δCe was used to indicate the trend of relative sea-level changes [66]: $\delta Ce = log\ [3 \times Ce_n/(2 \times La_n + Nd_n)]$. Finally, the gas content was calculated using polynomial regression based on the field test value.

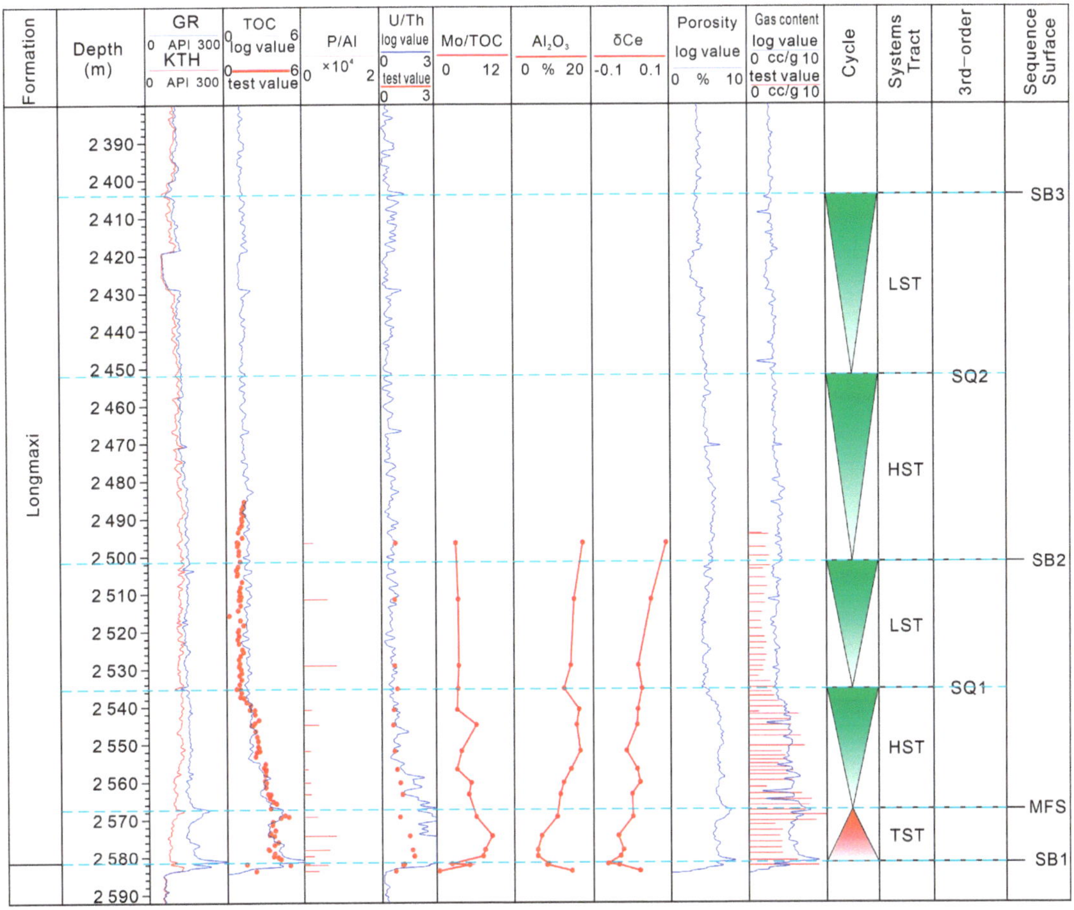

Figure 9. Geological section of the Longmaxi Shale from well N213. Paleo-productivity was expressed by the value of P/Al [63], where the paleo-redox state of the depositional environment was expressed by U/Th [16,64]. The degree of hydrographic restriction was represented by Mo/TOC [65], and the content of Al_2O_3 served as a proxy for the extent of continental input [66]. Additionally, δCe was used to indicate the trend of sea-level changes [66]: $\delta Ce = log\ [3 \times Ce_n/(2 \times La_n + Nd_n)]$. Finally, the gas content was calculated using polynomial regression based on the field test value.

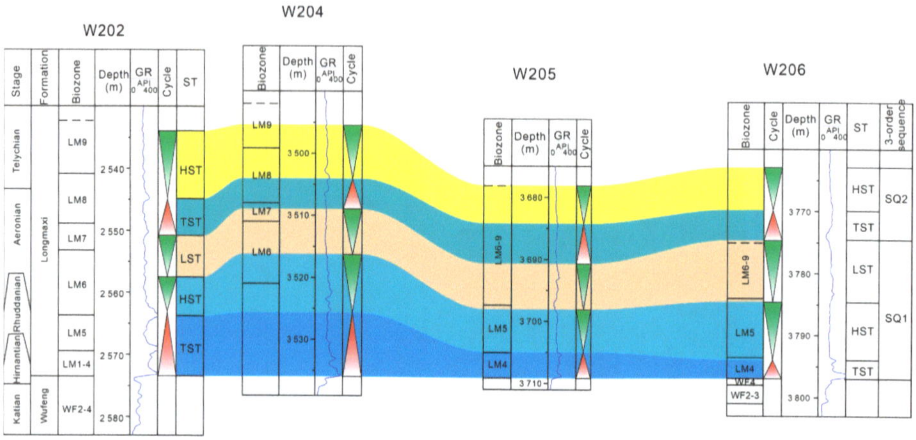

Figure 10. Stratigraphic sequence cross-section of the Wufeng-Longmaxi Formation (A-A′, see Figure 1), showing the stratigraphic sequence correlation in the Weiyuan area. The fourth-order sequences were recognized based on GR under the constraint of biozone. The abbreviations of the biozone were adapted from [45], WF2 = *Dicellograptus complexus*, WF3 = *Paraorthogr. pacificus*, WF4 = *Metabologr. extraordinarius*, LM1 = Persculptogr. persculptus, LM2 = Akidograptus ascensus, LM3 = *Parakidogr. Acuminatus*, LM4 = *Cystograptus vesiculosus*, LM5 = *Coronograptus cyphus*, LM6 = *Demirastrites triangulates*, LM7 = *Lituigraptus convolutes*, LM8 = *Stimulograptus sedgwickii*, LM9 = *Spirograptus guerichi*. GR = gamma ray, SQ = sequence, PSS = parasequence sets. A dashed line at biozone boundaries means uncertainty in placement.

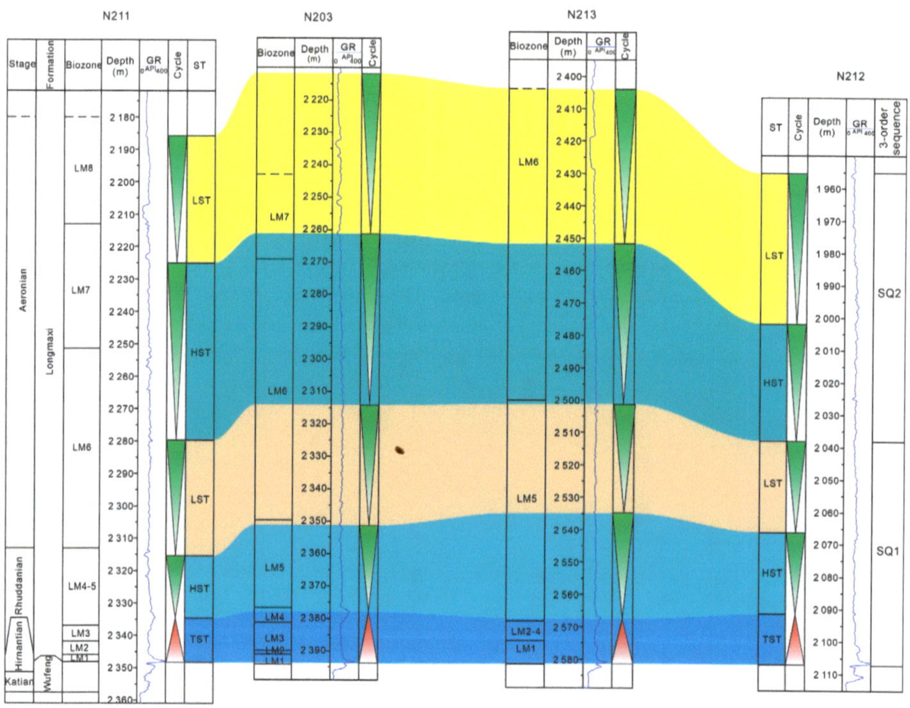

Figure 11. Stratigraphic sequence cross-section of the Wufeng-Longmaxi Formation (C-C′, see Figure 1), showing the sequence stratigraphic correlation in the Changning area. The fourth-order sequences

were recognized based on GR under the constraint of biozone. The abbreviations of the biozone were adapted from [45], WF2 = *Dicellograptus complexus*, WF3 = *Paraorthogr. pacificus*, WF4 = *Metabologr. extraordinarius*, LM1 = *Persculptogr. persculptus*, LM2 = *Akidograptus ascensus*, LM3 = *Parakidogr. Acuminatus*, LM4 = *Cystograptus vesiculosus*, LM5 = *Coronograptus cyphus*, LM6 = *Demirastrites triangulates*, LM7 = *Lituigraptus convolutes*, LM8 = *Stimulograptus sedgwickii*, LM9 = *Spirograptus guerichi*. GR = gamma ray, SQ = sequence, PSS = parasequence sets. A dashed line at biozone boundaries means uncertainty in placement.

4.2. Systems Tracts of the Lower Longmaxi Formation

The systems tracts in this study are further subdivisions of third-order sequences, which can be recognized by their shapes based on GR logs and their positions within a eustatic cycle [9,18,22,56,57]. Taking the Y101H4-4 well as an example for the north area and N213 for the south area, the wireline logs and geochemical data within the sequences are illustrated in Figures 8 and 9, respectively.

The TST of SQ1 represents the organic-rich siliceous shale at the bottom of the Longmaxi Formation, which was deposited from LM1 to LM5 (as shown in Figures 10–12). This particular shale layer can be correlated with the global "hot shale" [67–72] formed during the worldwide transgressive period that resulted from the rapid melting of the Gondwana glacier. The wireline logs of TST in the entire study area exhibited high gamma response characteristics and ended at a maximum flooding surface (MFS), with a local maximum value in GR logs (Figures 8–10). The transgression in the Changning area ended slightly earlier due to the uplift of the Qianzhong Paleo-uplift. In contrast, the transgression in the Luzhou and Weiyuan areas lasted until the *Coronograptus cyphus* (LM5) reached its maximum, submerging the submarine high in the Weiyuan area (Figure 10). However, there was consistency in the geochemical characteristics of the Changning and Luzhou regions (Figures 8 and 9). These include:

- high organic matter abundance with an average value of 3.78% in well N213 and 4.88% in well Y101H4-4;
- strong paleo-productivity with a mean of P/Al of 0.51×10^4 in well N213 and 138.47×10^4 in well Y101H4-4;
- weak oxidizing environment with a mean of U/Th 0.5 in well N213 and 1.12 in well Y101H4-4;
- strong hydrographic restriction with the mean of Mo/TOC of 7.32 from well N213 and 6.23 from Y101H4-4;
- low continental input for the Al_2O_3 content of 7.25% from N213 and 7.21% from Y101H4-4;
- and high gas content with an average value of 5.65 cc/g from well N213 and 15.5 cc/g from Y101H4-4.

The HST of SQ1 was distinguished by the GR logs decreasing upwards, suggesting a decline in the relative sea level during this period. The HST's sedimentation rate (corresponding to LM5 in geological time) began to accelerate (Table 1), leading to a significant increase in thickness and relatively low GR values throughout the study area with a clear funnel shape (Figures 8–10). Subsequently, the HST was covered by the LST, which was deposited during the relative sea-level decline and comprised progradational sets of parasequences with "funnel" or "cylinder" shapes evident in GR logs. Both the HST and LST of SQ1 suffered from poor paleo-productivity and a worse preservation environment for organic matter than the TST in SQ1. Additionally, an increase in δCe denoted a decrease in relative sea level (Figures 8 and 9).

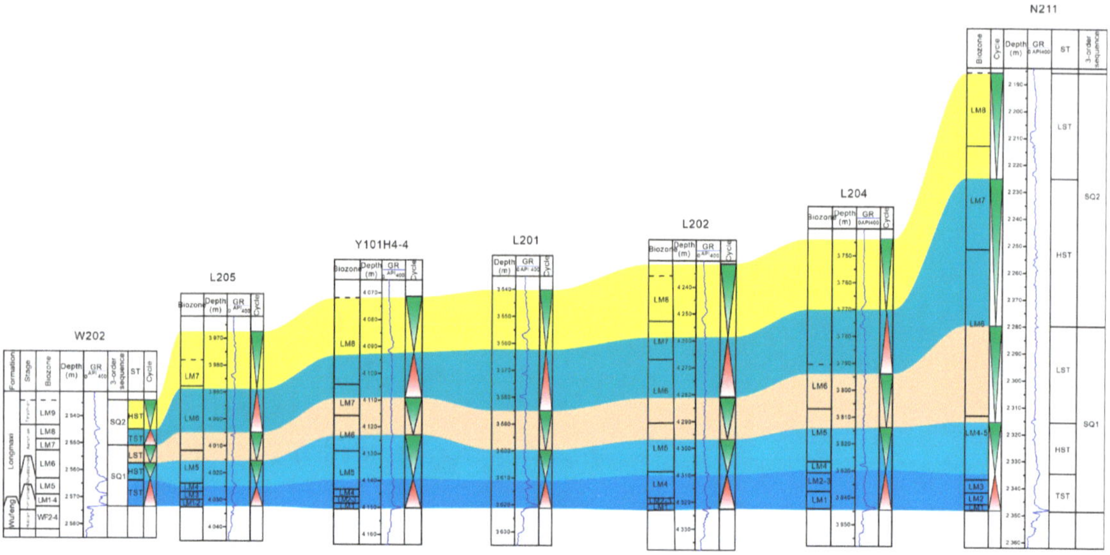

Figure 12. Stratigraphic sequence cross-section of the Wufeng-Longmaxi Formation (B-B′, see Figure 1), showing the stratigraphic sequence correlation in the southern Sichuan Basin. The fourth-order sequences were recognized based on GR under the constraint of biozone. The abbreviations of the biozone are adapted from [45], WF2 = *Dicellograptus complexus*, WF3 = *Paraorthogr. pacificus*, WF4 = *Metabologr. extraordinarius*, LM1 = *Persculptogr. persculptus*, LM2 = *Akidograptus ascensus*, LM3 = *Parakidogr. Acuminatus*, LM4 = *Cystograptus vesiculosus*, LM5 = *Coronograptus cyphus*, LM6 = *Demirastrites triangulates*, LM7 = *Lituigraptus convolutes*, LM8 = *Stimulograptus sedgwickii*, LM9 = *Spirograptus guerichi*. GR = gamma ray, SQ = sequence, PSS = parasequence sets. A dashed line at biozone boundaries means uncertainty in placement.

Table 1. Thickness and sedimentation rates of the stratigraphy from the *Dicellograptus complexus* zone to the *Spirograptus guerichi* zone observed by cores in the southern Sichuan Basin. The ages of the biozone bases are after [45]. "/" means no available data.

Biozone	Age (Ma)	N213		L204		Y101H4-4		L205		W204H10-2	
		Thickness (m)	Rate (m/Ma)	Thickness (m)	Rate (m/Ma)	Thickness (m)	Rate (m/Ma)	Thickness (m)	Rate (m/Ma)	Thickness (m)	Rate (m/Ma)
Spirograptus guerichi (LM9)	438.49	/	/	/	/	/	/	/	/	>29	>80.42
Stimulograptus sedgwickii (LM8)	438.76	/	/	/	/	>14	>51.85	/	/	17.65	24.51
Lituigraptus convolutes (LM7)	439.21	/	/	/	/	11.93	26.51	>27	>61.02		
Demirastrites triangulates (LM6)	440.77	>23.99	>15.38	>35	>22.44	13.04	8.36	23.93	15.34	14.03	8.99
Coronograptus cyphus (LM5)	441.57	68.04	85.05	21.69	27.11	14.24	17.8	12.19	15.24	3.24	4.05
Cystograptus vesiculosus (LM4)	442.47	5.89	2.61	6.77	7.52	2.58	2.87	2.98	3.31	1.82	2.02
Parakidogr. Acuminatus (LM3)	443.40			3.85	2.83	2.65	1.95	3.07	3.30	1.95	2.1
Akidograptus ascensus (LM2)	443.83							2.7	2.62	1.56	3.63
Persculptogr. Persculptus (LM1)	444.43	7.2	12	6.6	11	2.18	3.63			0.6	1

The first systems tract of SQ2 was interpreted as a TST in the Weiyuan and Luzhou areas, but it was interpreted as an HST in the Changning area (Figures 8–11). This systems tract corresponds to the middle part of the Longmaxi Formation. In the study area, the sedimentary thickness was large and decreased laterally towards the north (Figures 8–11). However, there were significant differences in the stratigraphic stacking patterns deposited from SQ2 between the north and south of the study area. The GR logs of wells in the Weiyuan and Luzhou areas clearly exhibited bell-shaped or cylinder-shaped characteristics. Therefore, we interpret this rising gamma value segment as a TST (Figures 6–8). On the contrary, the GR logs in the Changning region showed a funnel shape. Additionally, the GR logs decreased upwards (Figure 9). Thus, we interpret these strata as an HST. It is believed that the second systems tract of SQ2 in the study area exhibited a progradational pattern formed under the control of the global sea-level decline, as evidenced by the funnel-shaped GR logs. Consequently, we defined the second systems tract of SQ2 as an HST in the Weiyuan and Luzhou areas and a LST in the Changning area.

4.3. Spatial Thickness Distribution of Sequences

The connected well profiles and isopach maps of the system tracts can explain the spatial changes of the strata in both horizontal and vertical directions, providing a better understanding of how the South Sichuan Basin was filled with sediment and predicting favorable intervals in both directions. The connected well profiles are presented in Figures 8–10, while the isopach maps are shown in Figures 11 and 12.

During the deposition of the TST of SQ1, thick strata within the Upper Yangtze Plate formed two depocenters—one located in the Weiyuan area to the north and the other in the Changning area to the south (Figure 13). Research has revealed that the paleo-topography of the Weiyuan area contained a structure of landforms, including a ramp, a nose-shaped bulge, a platform, and a depocenter during the Late Ordovician to Early Silurian [21]. Differences in paleo-topography contributed to variations in the thickness of systems tracts, with thicker deposits found in the depocenter and thinner ones in other areas (see Figures 8 and 11). The thickness of HST increased and expanded northward (Figure 13). While the thick strata of the LST of SQ1 in the Changning area have shrunk, those in the Weiyuan area have remained unchanged and were still controlled by paleo-topography (Figure 13). Additionally, the sedimentation rate accelerated during SQ2 deposition, resulting in a generally increased stratigraphic thickness, particularly in the Changning area (Figure 12). The maximum thickness of the HST of SQ2 is 54.7 m, with an average thickness of 48.3 m. As seawater recedes towards the northeast of the Sichuan Basin, the thick stratigraphic range of the entire southern Sichuan region expands (Figure 14).

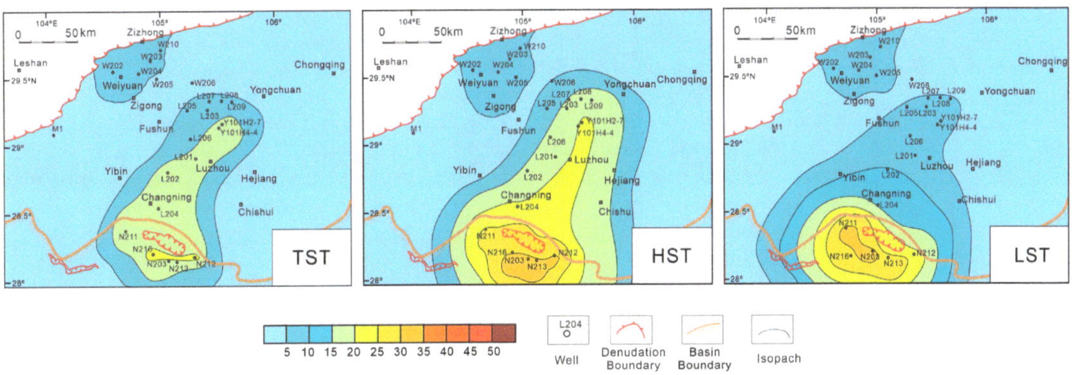

Figure 13. Isopach maps of systems tracts for SQ1.

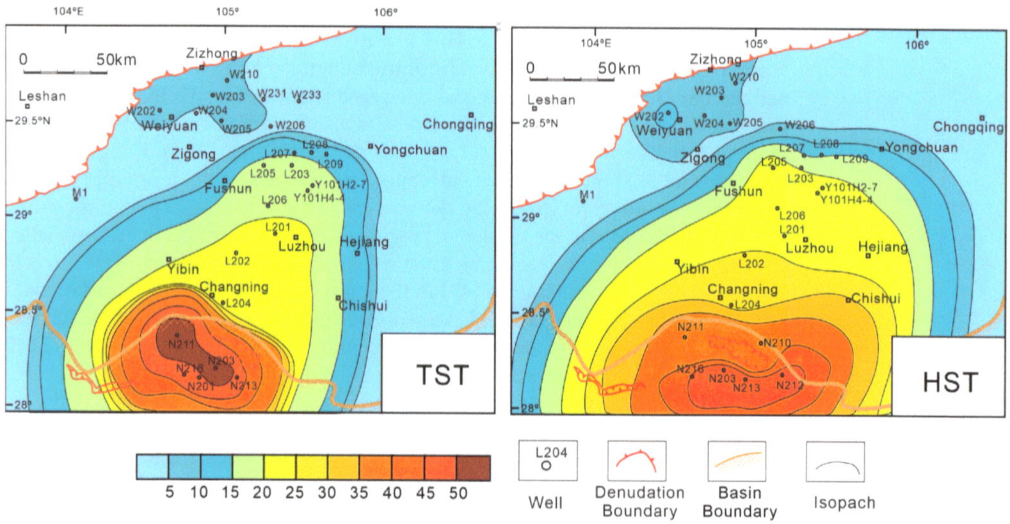

Figure 14. Isopach maps of systems tracts for SQ2.

5. Discussion

5.1. Sedimentary Evolution

We utilized the obtained results and published research to reconstruct the sedimentary history of the study area. During the Late Ordovician and Early Silurian periods, South China was located in the northern hemisphere at low latitudes [40]. The deposition of the sequence stratigraphy within the basin during this period was significantly affected by global sea-level changes, regional tectonic movements, sediment supply, and paleoclimate [1,29–34]. The evolution of the Lower Longmaxi Formation in southern Sichuan was closely related to geological events during this period and consisted of two depositional episodes (Figures 8–10). The TST1 of SQ1 was triggered by deglaciation in the late Hirnantian, resulting in a rapid global sea level rise that led to a second mass extinction event [32,73,74]. This transgressive event can also be compared globally [32,75]. Due to the rapid rise in sea level caused by climate warming, a significant number of cold-water animals went extinct [75]. At the same time, TST1 began to form, and the sea level reached its maximum value in late LM4 (Figures 8–10). This phase was marked by the development of organic-rich siliceous shale with high gamma values across the study area. As a result, the sea area of the Upper Yangtze expanded, depositing parts of the paleo-uplift in the Weiyuan area during LM4 (see Figure 8).

Since then, the global sea level has been declining. At the same time, the Leshan-Longnvsi Paleo-uplift, the Qianzhong Paleo-uplift, and the Jiangnan-Xuefeng Paleo-uplift have continued to uplift and expand into the basin. As a result of these uplifts, the area of the Upper Yangtze Sea has decreased. The strata of HST1 extended into the sea in a progradation pattern with an upward decreasing GR value. Throughout the LST1 sedimentation period, the gamma value remained consistently low and did not show significant changes. However, the center of subsidence had been moved from north to south, while the depocenter migrated from south to north (Figure 13). The first tectonic system of SQ2 differed between the Weiyuan and Changning areas, which was caused by regional compression in South China [1]. We speculate that there is a subsidence center located between the Weiyuan and Luzhou areas. The continuous subsidence of this center had led to a relative sea level rise in sea level in both areas with a characteristic increasing gamma value (Figures 8 and 10). In the late period of TST2, it was believed that there was relatively strong volcanic activity [62] with thick bentonite deposits (as shown in Figure 7).

In contrast, the Changning region situated near the Qianzhong Paleo-uplift experienced a relative sea level decrease, owing to the continuous uplift of the Qianzhong Paleo-uplift. This process also caused an increase in sedimentary thickness in the Changning area due to increased terrestrial input. As a result, we can observe that the thickness of the first systems tract of SQ2 varies significantly along the north-south direction, increasing from less than 5 m in the northwest (Weiyuan) to more than 30 m thick in the south (Changning) (Figure 12). This change is particularly sharp in the southern well, Y101H4-4, which likely resulted from an increased sediment supply due to the rapid uplift of the Qianzhong Paleo-uplift [1]. Thence, the first systems tract of SQ2 in the Weiyuan and Luzhou areas were interpreted as a TST and as an HST in the Changning area. Note that the sedimentation rate in northwest Weiyuan is far lower than that in south Changning (Table 1), indicating increasing sediment input in Changning. The second systems tract of SQ2 was mainly influenced by the global sea level, and the whole Upper Yangtze area was uplifted. It was interpreted as an HST in the Weiyuan and Luzhou areas, considering the previous systems tract, and as an LST in the Changning area.

5.2. *Specificity of Sedimentation in the Study Area*

The sedimentary evolution of the study area during the Early Silurian exhibits unique characteristics. The stratigraphic sequence methods published so far have been mostly applied to passive continental margins, with little discussion on the sedimentation of epeiric seas affected by peripheral basin-mountain activities [76]. During the Early Silurian, the sedimentary filling evolution in the study area was not only related to the global sea level rise but also closely linked to regional tectonic activities [1,77]. As the Yangtze Plate experienced compression and internal deformation, the subsidence center continued to shift northward while the surrounding paleo-uplifts kept rising [77]. This resulted in an increase in the elevation of the strata near the Qianzhong Paleo-uplifts and the corresponding enhancement of sediment supply, leading to an increase in sediment thickness (Figures 9, 12 and 13). Some scholars have described this process as the gradual downward curvature of the Yangtze Platform from the southeast margin to the northwest, forming a deep-water foreland basin, and used the development characteristics of bentonite to explain the intensity of collisions and amalgamation between the Yangtze Plate and surrounding blocks [62,77].

The subsidence center and sedimentary center of the sedimentary basin are inconsistent and constantly changing, which is a unique feature of sedimentary evolution in the process of compression deformation within the plate in the study area. During the SQ1 sedimentary period, the sedimentation of the entire Upper Yangtze Sea was controlled by global sea level changes, forming two sedimentary centers near the ancient uplifts in the north and south (Figures 13 and 14). However, during the SQ2 period, due to the compression deformation inside the Upper Yangtze Plate, the continuous uplift of the Central Guizhou ancient uplift moved northward to the subsidence centers of the Central Sichuan ancient uplift and the Central Guizhou ancient uplift, resulting in a relative sea level decrease in the Changning area near the Central Guizhou ancient uplift. However, the relative sea level of the Weiyuan and Luzhou areas in the north was rising, so the strata in the north and south appeared opposite of the superposition pattern (Figure 15), which also indicates that the developing foreland basin was controlled by peripheral orogeny and intraplate compression [77]. The stratigraphic superposition pattern in different areas of the entire sedimentary basin is not necessarily the same at the same time, which also profoundly affects later oil and gas exploration. The inconsistency and continual change between the subsidence center and depocenter in the Sichuan basins is a characteristic feature of sedimentary filling evolution during the intraplate compression and deformation processes within the study area. During the deposition of the SQ1, sequence stratigraphy in the entire Upper Yangtze Sea was controlled by global sea level changes, resulting in the formation of two sedimentation centers located near the paleo-uplifts in the north and south. However, during the deposition of the SQ2, due to intraplate compression and deformation within the Upper Yangtze Plate, the continued uplift of the Qianzhong Paleo-uplift caused the subsidence center between the Leshan-Longnvsi Paleo-uplift and Qianzhong Paleo-uplift to move northward. This led to a relative drop in sea level in the Changning area near the Qianzhong Paleo-uplift, while the Weiyuan and Luzhou areas to the north experienced a relative sea level rise. As a result, the stratigraphic stacking pattern in the northern and southern regions showed the opposite trend. At the same time, the sedimentation rate accelerated in the Changning area (Table 1). This also indicates that for foreland basins developed under the influence of peripheral orogeny and intraplate compression, the stratigraphic stacking pattern in different regions of the sedimentary basin during the same period may not be consistent, which profoundly affects later oil and gas exploration efforts.

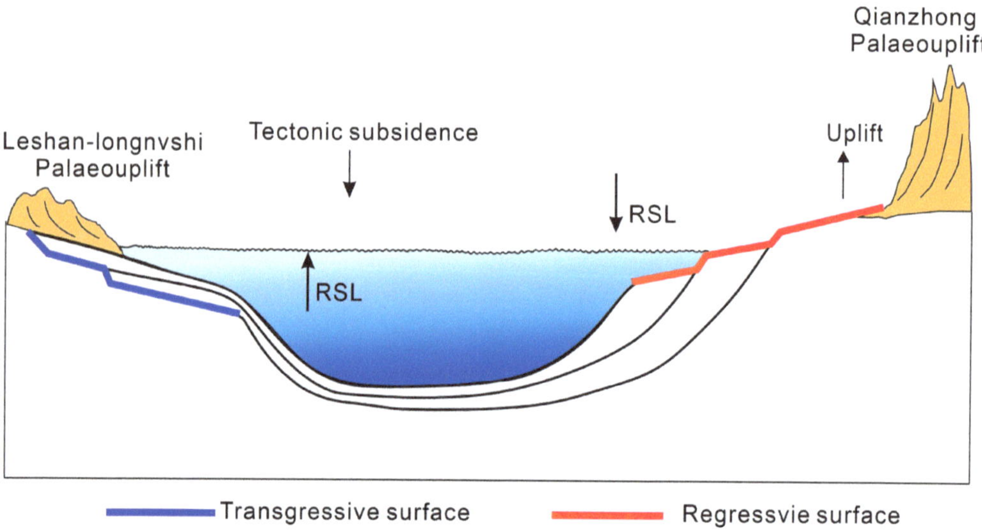

Figure 15. Sedimentary pattern for the first systems tract of SQ2.

5.3. Implications for Shale Gas Exploration

Sequence stratigraphy plays an important role in hydrocarbon evaluation and prediction [1,8,16,58]. Published research has pointed out that the TST is a favorable interval for shale gas [1,30]. This is because TST has a high reducing sedimentary environment, high paleo-productivity, and a sedimentary rate that is favorable for the preservation of organic matter. Therefore, the shale gas exploration potential is also evaluated for TSTs in this paper. The first TST of the Longmaxi Formation has been discussed by numerous publications and proved by the producing wells [16]. The GR values and geochemistry data presented in this paper also support the idea that the first TST of SQ1 is the most promising for shale gas (Figure 7). However, we would like to emphasize the second TST of the Lower Longmaxi Formation, which is referred to as TST2 in the Luzhou and Weiyuan areas for SQ2. Testing data (TOC, P/Al, U/Th, Mo/Toc, porosity, and gas content) from TST2 show that this interval will be the next target for shale gas development in the Luzhou area [2], but there is no shale gas exploration potential in the Changning area (Figure 9). There are also literature explanations for the high paleo-productivity of TST2, stating that the high paleo-productivity of this segment is caused by rising ocean currents [78]. Our research results indicate two favorable intervals for shale gas in the southern Sichuan Basin. The first interval is the black shale deposited throughout the region during the global sea level rise. The second interval is the silty shale deposited in the Weiyuan sag and Luzhou area due to the strong uplift of the Qianzhong Paleo-uplift and compression of the interior of the upper Yangtze Plate, resulting in a subsidence center moving northward.

6. Conclusions

This article presents a series of studies on the sequence stratigraphy, systems tracts, and shale gas exploration potential of the Lower Longmaxi Formation in the southern Sichuan Basin. These studies were conducted through core observation, analysis of wireline logs, description of field sections, as well as interpretation of 3D seismic profiles. The key results that have been identified are as follows.

Based on the sudden logging change, core lithology alteration, and geochemical indicator modification, two third-order sedimentary cycles have been identified. The first cycle showcases a consistent change in the relative sea level of southern Sichuan, while the second cycle differs due to tectonic activity.

The primary cause for the difference in the second sedimentary cycle is the northward displacement of the subsidence center between the ancient lands of central Guizhou and central Sichuan. As the central Guizhou ancient land underwent an uplift, it created an increase in northern accommodation space and a decrease in southern accommodation space. This resulted in a stratigraphic stacking pattern of one advance and one retreat.

By analyzing the sedimentary evolution of the Lower Longmaxi Formation in southern Sichuan, combined with geochemical and logging response characteristics, it can be concluded that the first TST has the best shale gas exploration potential. In addition, for the Weiyuan and Luzhou areas, the second TST also has good shale gas exploration potential.

It should be noted that the data used in this study were primarily obtained from deep water, with limited input from wells in the southern regions of the basin, particularly those located near the basin edge. Therefore, we were unable to observe the presence of onlap in seismic profiles. The stacking patterns of strata were inferred based on the response characteristics of logging curves, which carry a degree of uncertainty. It is necessary to verify the filling model of the basin with wells located closer to the sedimentary basin in the future.

Author Contributions: Conceptualization, L.X. and R.P.; methodology, J.M.; software, B.Z.; validation, Q.S.; investigation, X.Y.; writing—original draft preparation, L.X.; writing—review and editing, J.M. All authors have read and agreed to the published version of the manuscript.

Funding: This research was funded by the National Natural Science Foundation of China (Grant No. 41472123) and the National Natural Science Foundation of China Youth Fund (Gran No. 41402114).

Institutional Review Board Statement: Not applicable.

Informed Consent Statement: Not applicable.

Acknowledgments: The authors would like to thank PetroChina Southwest Oil & Gas field Company for providing core samples and dataset that used in this paper. The authors also want to express gratitude to Zhu (Yiqing Zhu) and Shi (Xuewen Shi) at PetroChina Southwest Oil & Gasfield Company for their provided constructive suggestions to improve this work. In particular, we would like to thank the anonymous reviewers for their patient review and insightful comments. We are also grateful to Academic Editor George Kontakiotis for the valuable comments that greatly improved the original manuscript.

Conflicts of Interest: The authors declare no conflict of interest.

References

1. Chen, L.; Lu, Y.; Jiang, S.; Li, J.; Guo, T.; Luo, C.; Xing, F. Sequence stratigraphy and its application in marine shale gas exploration: A case study of the Lower Silurian Longmaxi Formation in the Jiaoshiba shale gas field and its adjacent area in southeast Sichuan Basin, SW China. *J. Nat. Gas Sci. Eng.* **2015**, *27*, 410–423. [CrossRef]
2. Yang, H.; Zhao, S.; Liu, Y.; Wu, W.; Xia, Z.; Wu, T.; Luo, C.; Fan, T.; Yu, L. Main controlling factors of enrichment and high-yield of deep shale gas in the Luzhou Block, southern Sichuan Basin. *Nat. Gas. Ind.* **2019**, *39*, 55–63.
3. Ma, X.; Xie, J.; Yong, R.; Zhu, Y. Geological characteristics and high production control factors of shale gas reservoirs in Silurian Longmaxi Formation, southern Sichuan Basin, SW China. *Pet. Explor. Dev.* **2020**, *47*, 901–915. [CrossRef]
4. Ma, X. Enrichment laws and scale effective development of shale gas in the southern Sichuan Basin. *Nat. Gas. Ind. B* **2019**, *6*, 240–249. [CrossRef]

5. Mahjour, S.K.; Al-Askari, M.K.G.; Masihi, M. Identification of flow units using methods of Testerman statistical zonation, flow zone index, and cluster analysis in Tabnaak gas field. *J. Petrol. Explor. Prod. Technol.* **2016**, *6*, 577–592. [CrossRef]
6. Ahmadi, S.; Motie, M.; Soltanmohammadi, R. Proposing a modified mechanism for determination of hydrocarbons dynamic viscosity, using artificial neural network. *Pet. Sci. Technol.* **2020**, *38*, 699–705. [CrossRef]
7. Motie, M.; Bemani, A.; Soltanmohammadi, R. On the estimation of phase behavior of CO_2-based binary systems using ANFIS optimized by GA algorithm. In *Fifth CO_2 Geological Storage Workshop*; European Association of Geoscientists & Engineers: Utrecht, The Netherlands, 2018; Volume 1, pp. 1–5.
8. Nadjafi, M.; Mahboubi, A.; Moussavi-Harami, R.; Mirzaee, R. Depositional history and sequence stratigraphy of outcropping Tertiary carbonates in the Jahrum and Asmari formations, Shiraz area (SW Iran). *J. Pet. Geol.* **2004**, *27*, 179–190. [CrossRef]
9. Catuneanu, O. *Principles of Sequence Stratigraphy*; Elsevier: Amsterdam, The Netherlands, 2006; 375p.
10. Lash, G.G.; Engelder, T. Thickness trends and sequence stratigraphy of the Middle Devonian Marcellus Formation, Appalachian Basin: Implications for Acadian foreland basin evolution. *AAPG Bull.* **2011**, *95*, 61–103. [CrossRef]
11. Slatt, R.M.; Rodriguez, N.D. Comparative sequence stratigraphy and organic geochemistry of gas shales: Commonality or coincidence? *J. Nat. Gas Sci. Eng.* **2012**, *8*, 68–84. [CrossRef]
12. Wang, T.; Yang, K.; Xiong, L.; Shi, H.; Zhang, Q.; Wei, L.; He, X. Shale sequence stratigraphy of Wufeng-Longmaxi Formation in southern Sichuan and their control on reservoirs. *Acta Pet. Sin.* **2015**, *36*, 915–925.
13. Wang, Y.; Dong, D.; Li, X.; Huang, J.; Wang, S.; Wu, W. Stratigraphic sequence and sedimentary characteristics of Lower Silurian Longmaxi Formation in Sichuan Basin and its peripheral areas. *Nat. Gas. Ind. B* **2015**, *2*, 222–232. [CrossRef]
14. Guo, X. Sequence stratigraphy and evolution model of the Wufeng-Longmaxi shale in the Upper Yangtze area. *Earth Sci.* **2017**, *2*, 1069–1082.
15. Wang, G.; Zhu, T.; Wang, H.; Wu, J.; Du, W.; Feng, D.; Wang, R. Integrated sequence stratigraphic division and vertical distribution characteristics of marine shale: A case study of the Wufeng Formation-Longmaxi Formation in southeastern Sichuan Basin. *Acta Sedimentol. Sin.* **2019**, *37*, 330–344.
16. Zhu, Y.; Chen, G.; Liu, Y.; Shi, X.; Wu, W.; Luo, C.; Yang, X.; Yang, Y.; Zou, Y. Sequence stratigraphy and lithofacies paleogeographic evolution of Katian Stage—Aeronian Stage in southern Sichuan Basin, SW China. *Pet. Explor. Dev.* **2021**, *48*, 1126–1138. [CrossRef]
17. Cai, Q.; Hu, M.; Kane, Q.I.; Li, M.; Zhang, B.; Hu, Z.; Deng, Q.; Xing, N. Cyclic variations in paleoenvironment and organic matter accumulation of the Upper Ordovician-Lower Silurian black shale in the Middle Yangtze Region, South China: Implications for tectonic setting, paleoclimate, and sea-level change. *Mar. Pet. Geol.* **2022**, *136*, 105777. [CrossRef]
18. Catuneanu, O.; Kendall, C.; Posamentier, H. Sequence Stratigraphy: Methodology and Nomenclature. *Newsl. Stratigr.* **2011**, *44*, 173–245. [CrossRef]
19. Huang, H.; He, D.; Li, Y.; Li, J.; Zhang, L. Silurian tectonic-sedimentary setting and basin evolution in the Sichuan area, southwest China: Implications for palaeogeographic reconstructions. *Mar. Pet. Geol.* **2018**, *92*, 403–423. [CrossRef]
20. Yin, F.; Xu, X.; Wan, F.; Chen, M. Characteristic of sequence and stratigraphical division in evolution of Upper Yangtze region during Caledonian. *J. Stratigr.* **2002**, *26*, 315–319.
21. Shi, Z.; Wang, H.; Lin, C.; Sun, S.; Jin, H.; Hao, C.; Chen, S.; Zhang, R. Paleotopography of Weiyuan-Zigong area in Wufengian-Lungmachian Stages (Ordovician-Silurian transition) and its effect on the quality of shale gas reservoir. *J. Stratigra.* **2020**, *44*, 163–173.
22. Van Wagoner, J.C.; Mitchum, R.M.; Campion, K.M.; Rahmanian, V.D. Siliciclastic Sequence Stratigraphy in Well Logs, Core, and Outcrops: Concepts for High-Resolution Correlation of Time and Facies. In *American Association of Petroleum Geologists Methods in Exploration Series*; AAPG: Tulsa, OK, USA, 1990; 55p.
23. Carter, R.M.; Fulthorpe, C.S.; Naish, T.R. Sequence concepts at seismic and outcrop scale: The distinction between physical and conceptual stratigraphic surfaces. *Sediment. Geol.* **1998**, *122*, 165–179. [CrossRef]
24. Algeo, T.J.; Schwwark, L.; Hoer, J.C. High-resolution geochemistry and sequence stratigraphy of the Hushpuchney Shale (Swope Formation, Eastern Kansas): Implications for climato-environmental dynamics of the Late Pennsylvanian Midcontinent Seaway. *Chem. Geol.* **2004**, *206*, 259–288. [CrossRef]
25. Ver Straeten, C.A.; Brett, C.E.; Sageman, B.B. Mudrock sequence stratigraphy: A multi-proxy (sedimentological, paleobiological and geochemical) approach, Devonian Appalachian Basin. *Palaeogeogr. Palaeoclimatol. Palaeoecol.* **2011**, *304*, 54–73. [CrossRef]
26. Abouelresh, M.O.; Slatt, R.M. Lithofacies and sequence stratigraphy of the Barnett Shale in east central Fort Worth Basin, Texas. *AAPG Bull.* **2012**, *96*, 1–22. [CrossRef]
27. Lagrange, M.T.; Konhauser, K.O.; Catuneanu, O.; Harris, B.S.; Playter, T.L.; Gingras, M.K. Sequence stratigraphy in organic-rich marine mudstone successions using chemostratigraphic datasets. *Earth Sci. Rev.* **2020**, *203*, 103137. [CrossRef]
28. Liu, S.; Yang, Y.; Deng, B.; Zhong, Y.; Wen, L.; Sun, W.; Li, Z.; Jansa, L.; Li, J.; Song, J.; et al. Tectonic evolution of the Sichuan basin, southwest China. *Earth Sci. Rev.* **2021**, *213*, 103470. [CrossRef]
29. He, D.; Li, D.; Zhang, G.; Zhao, L.; Fan, C.; Lu, R.; Wen, Z. Formation and evolution of multi-cycle superposed Sichuan Basin, China. *Sci. Geol. Sin.* **2011**, *46*, 589–606.
30. Lüning, S.; Craig, J.; Loydell, D.K.; Štorch, P.; Fitches, B. Lower Silurian 'hot shales' in North Africa and Arabia: Regional distribution and depositional model. *Earth Sci. Rev.* **2000**, *49*, 121–200. [CrossRef]
31. Feng, H.; Erdtmann, B.D.; Wang, H. Early Paleozoic whole-rock Ce anomalies and secular eustatic changes in the Upper Yangtze region. *China Earth Sci.* **2000**, *3*, 328–336. [CrossRef]

32. Loydell, D.K.; Butcher, A.; Frýda, J. The middle Rhuddanian (lower Silurian) 'hot' shale of North Africa and Arabia: An atypical hydrocarbon source rock. *Palaeogeogr. Palaeoclimatol. Palaeoecol.* **2013**, *386*, 233–256. [CrossRef]
33. Yan, D.; Chen, D.; Wang, Q.; Wang, J. Large-scale climatic fluctuations in the latest Ordovician on the Yangtze block, south China. *Geology* **2010**, *38*, 599–602. [CrossRef]
34. Finnegan, S.; Bergmann, K.; Eiler, J.M.; Jones, D.S.; Fike, D.A.; Eisenman, I.; Hughes, N.C.; Tripati, A.K.; Fischer, W.W. The magnitude and duration of Late Ordovician–Early Silurian glaciation. *Science* **2011**, *331*, 903–906. [CrossRef]
35. Liang, D.G.; Guo, T.L.; Bian, L.Z.; Chen, J.P.; Zhao, Z. Some progress on studies of hydrocarbon generation and accumulation in marine sedimentary regions, Southern China (part 3): Controlling factors on the sedimentary facies and development of Palaeozoic marine source rocks. *Mar. Orig. Pet. Geol.* **2009**, *14*, 1–19.
36. Zeng, X.L.; Liu, S.G.; Huang, W.M.; Zhang, C.J. Comparison of Silurian Longmaxi Formation shale of Sichuan Basin in China and Carboniferous Barnett Formation shale of Fort Worth Basin in United States. *Geol. Bull. China* **2011**, *30*, 372–384.
37. Chen, X.; Rong, J.; Li, Y.; Boucot, A.J. Facies patterns and geography of the Yangtze region, South China, through the Ordovician and Silurian transition. *Palaeogeogr. Palaeoclimatol. Palaeoecol.* **2004**, *204*, 353–372.
38. Mou, C.; Ge, X.; Xu, X.; Zhou, K.; Liang, W.; Wang, X. Lithofacies palaeogeography of the Late Ordovician and its petroleum geological significance in Middle-Upper Yangtze Region. *J. Palaeogeogr.* **2014**, *16*, 427–440.
39. Zhou, M.; Wang, R.; Li, Z. *Ordovician-Silurian Lithofacies Paleogeography and Mineralization in Southern China*; Geological Publishing House: Beijing, China, 1993; 122p.
40. Torsvik, T.H.; Cocks, L.R.M. Gondwana from top to base in space and time. *Gondwana Res.* **2013**, *24*, 999–1030. [CrossRef]
41. Sun, S.; Rui, Y.; Dong, D.; Shi, Z.; Bai, W.; Ma, C.; Zhang, L.; Wu, J.; Chang, Y. Paleogeographic evolution of the Late Ordovician-Early Silurian in Upper and Middle Yangtze regions and depositional model of shale. *Oil Gas Geol.* **2018**, *39*, 1087–1106.
42. Chen, X.; Rong, J.; Mitchell, C.E.; Harper, D.A.T.; Fan, J.; Zhan, R.; Zhang, Y.; Li, R.; Wang, Y. Late Ordovician to earliest Silurian graptolite and brachiopod biozonation from the Yangtze region, South China, with a global correlation. *Geol. Mag.* **2000**, *173*, 623–650.
43. Zhang, K.; Li, Z.; Jiang, S.; Jiang, Z.; Wen, M.; Jia, C.; Song, Y.; Liu, W.; Huang, Y.; Xie, X.; et al. Comparative analysis of the siliceous source and organic matter enrichment mechanism of the Upper Ordovician-Lower Silurian shale in the Upper-Lower Yangtze area. *Minerals* **2018**, *8*, 283. [CrossRef]
44. Li, Y.; Zhang, T.; Ellis, G.S.; Shao, D. Depositional environment and organic matter accumulation of Upper Ordovician-Lower Silurian marine shale in the Upper Yangtze Platform, South China. *Palaeogeogr. Palaeoclimatol. Palaeoecol.* **2017**, *466*, 252–264. [CrossRef]
45. Chen, X.; Fan, J.; Wang, W.; Wang, H.; Nie, H.; Shi, X.; Wen, Z.; Chen, D.; Li, W. Stage-progressive distribution pattern of the Lungmachi black graptolitic shales from Guizhou to Chongqing, Central China. *Sci. China Earth Sci.* **2017**, *60*, 1133–1146. [CrossRef]
46. Fan, J.; Chen, Q.; Melchin, M.J.; David Sheets, H.; Chen, Z.; Zhang, L.; Hou, X. Quantitative stratigraphy of the Wufeng and Lungmachi black shales and graptolite evolution during and after the Late Ordovician mass extinction. *Palaeogeogr. Palaeoclimatol. Palaeoecol.* **2013**, *389*, 96–114. [CrossRef]
47. Wang, H.; Shi, Z.; Zhao, Q.; Liu, D.; Sun, S.; Guo, W.; Liang, F.; Lin, C.; Wang, X. Stratigraphic framework of the Wufeng-Longmaxi shale in and around the Sichuan Basin, China: Implications for targeting shale gas. *Energy Geosci.* **2020**, *1*, 124–133. [CrossRef]
48. Dong, D.; Shi, Z.; Sun, S.; Guo, C.; Zhang, C.; Guo, W.; Guan, Q.; Zhang, M.; Jiang, S.; Zhang, L.; et al. Factors controlling microfractures in black shale: A case study of Ordovician Wufeng Formation-Silurian Longmaxi Formation in Shuanghe Profile, Changning area, Sichuan Basin, SW China. *Pet. Explor. Dev.* **2018**, *45*, 818–829. [CrossRef]
49. Guo, Y.; Li, Z.; Li, D.; Zhang, T.; Wang, Z.; Yu, J.; Xi, Y. Lithofacies palaeogeography of the Early Silurian in Sichuan area. *J. Palaeogeogr.* **2004**, *6*, 20–29.
50. Liang, C.; Jiang, Z.; Yang, Y.; Wei, X. Shale lithofacies and reservoir space of the Wufeng-Longmaxi formation, Sichuan Basin, China. *Pet. Explor. Dev* **2012**, *39*, 736–743. [CrossRef]
51. *GB/T 19145-2022 2022*; Determination for Total Organic Carbon in Sedimentary Rock. China Standardization Administration: Beijing, China, 2022.
52. *SY/T 5163-2018, 2018*; Analysis Method for Clay Minerals and Ordinary Non-Clay Minerals in Sedimentary Rocks by the X-ray Diffraction. National Energy Administration of China: Beijing, China, 2018.
53. Harris, N.B.; Mnich, C.A.; Selby, D.; Korn, D. Minor and trace element and Re–Os chemistry of the Upper Devonian Woodford Shale, Permian Basin, west Texas: Insights into metal abundance and basin processes. *Chem. Geol.* **2013**, *356*, 76–93. [CrossRef]
54. Wei, C.; Dong, T.; He, Z.; He, S.; He, Q.; Yang, R.; Guo, X.; Hou, Y. Major, trace-elemental and sedimentological characterization of the upper Ordovician Wufeng-lower Silurian Longmaxi formations, Sichuan Basin, south China: Insights into the effect of relative sea-level fluctuations on organic matter accumulation in shales. *Mar. Pet. Geol.* **2021**, *126*, 104905. [CrossRef]
55. Deng, Y.; Ren, J.; Guo, Q.; Cao, J.; Wang, H.; Liu, C. Rare earth element geochemistry characteristics of seawater and porewater from deep sea in western Pacific. *Sci. Rep.* **2017**, *7*, 16539.
56. Emery, D.; Myers, K. *Sequence Stratigraphy*; Blackwell: Oxford, UK, 1996; 297p.
57. Rider, M.H. *The Geological Interpretation of Well Logs*; Blackie and Son Limited: Glasgow, UK, 1986.
58. Cai, Q.; Hu, M.; Ngia, N.R.; Hu, Z. Sequence stratigraphy, sedimentary systems and implications for hydrocarbon exploration in the northern Xujiaweizi Fault Depression, Songliao Basin, NE China. *J. Pet. Sci. Eng.* **2017**, *152*, 471–494. [CrossRef]

59. Catuneanu, O. Model-independent sequence stratigraphy. *Earth Sci. Rev.* **2019**, *188*, 312–388. [CrossRef]
60. Wang, Y.; Dong, D.; Huang, J.; Li, X.; Wang, S. Guanyinqiao Member lithofacies of the Upper Ordovician Wufeng Formation around the Sichuan Basin and the significance to shale gas plays, SW China. *Pet. Explor. Dev.* **2016**, *43*, 45–53. [CrossRef]
61. Haq, B.U.; Schutter, S.R. A chronology of Paleozoic sea-level changes. *Science* **2008**, *322*, 64–68. [CrossRef]
62. Wang, Y.; Li, X.; Wang, H.; Jiang, S.; Chen, B.; Ma, J.; Dai, B. Developmental characteristics and geological significance of the bentonite in the Upper Ordovician Wufeng-Lower Silurian Longmaxi Formation in eastern Sichuan Basin, SW China. *Pet. Explor. Dev.* **2019**, *46*, 687–700. [CrossRef]
63. Tribovillard, N.; Algeo, T.J.; Lyons, T.; Riboulleau, A. Trace metals as paleoredox and paleoproductivity proxies: An update. *Chem. Geol.* **2006**, *232*, 12–32. [CrossRef]
64. Jones, B.; Manning, D.A.C. Comparison of geochemical indices used for the interpretation of palaeoredox conditions in ancient mudstones. *Chem. Geol.* **1994**, *111*, 111–129. [CrossRef]
65. Algeo, T.J.; Lyons, T.W. Mo–total organic carbon covariation in modern anoxic marine environments: Implications for analysis of paleoredox and paleohydrographic conditions. *Paleoceanography* **2006**, *21*, PA1016. [CrossRef]
66. Chen, H.F.; Yeh, P.Y.; Song, S.R.; Hsu, S.; Yang, T.; Wang, Y.; Chi, Z.; Lee, T.; Chen, M.; Cheng, C.; et al. The Ti/Al molar ratio as a new proxy for tracing sediment transportation processes and its application in aeolian events and sea level change in East Asia. *J. Asian Earth Sci.* **2013**, *73*, 31–38. [CrossRef]
67. Armstrong, H.A.; Turner, B.R.; Makhlouf, I.M.; Weedon, G.P.; Willianms, M.; Smadi, A.A.; Salah, A.A. Origin, sequence stratigraphy and depositional environment of an upper Ordovician (Hirnantian) deglacial black shale, Jordan. *Palaeogeogr. Palaeoclimatol. Palaeoecol.* **2005**, *220*, 273–289. [CrossRef]
68. Armstrong, H.A.; Abbott, G.D.; Turner, B.R.; Makhlouf, I.M.; Muhammad, A.B.; Pedentchouk, N.; Peters, H. Black shale deposition in an Upper Ordovician-Silurian permanently stratified, peri-glacial basin, southern Jordan. *Palaeogeogr. Palaeoclimatol. Palaeoecol.* **2009**, *273*, 368–377. [CrossRef]
69. Loydell, D.K.; Butcher, A.; Frýda, J.; Lüning, S.; Fowler, M. Lower Silurian "Hot Shales" in Jordan: A new depositional model. *J. Pet. Geol.* **2009**, *32*, 261–270. [CrossRef]
70. Zou, C.; Qiu, Z.; Poulton, S.W.; Dong, D.; Wang, H.; Chen, D.; Lu, B.; Shi, Z.; Tao, H. Ocean euxinia and climate change "double whammy" drove the Late Ordovician mass extinction. *Geology* **2018**, *46*, 535–538. [CrossRef]
71. Yan, D.; Chen, D.; Wang, Z.; Li, J.; Yang, X.; Zhang, B. Climatic and oceanic controlled deposition of Late Ordovician-Early Silurian black shales on the North Yangtze platform, South China. *Mar. Pet. Geol.* **2019**, *110*, 112–121.
72. Li, Y.; Zhang, T.; Shen, B.; Li, Z.; Shao, D.; Lash, G.G. Carbon and sulfur isotope variations through the Upper Ordovician and Lower Silurian of South China linked to volcanism. *Palaeogeogr. Paleoclimatol. Palaeoecol.* **2021**, *567*, 110285. [CrossRef]
73. Finney, S.C.; Berry, W.B.N.; Cooper, J.D.; Ripperdan, R.L.; Sweet, W.C.; Jacobson, S.R.; Soufiane, A.; Achab, A.; Noble, P.J. Late Ordovician mass extinction: A new perspective from stratigraphic sections in central Nevada. *Geology* **1999**, *27*, 215–218. [CrossRef]
74. Trotter, J.A.; Williams, I.; Barnes, C.R.; Lécuyer, C.; Nicoll, R.S. Did cooling oceans trigger Ordovician biodiversification? Evidence from conodont thermometry. *Science* **2008**, *321*, 550–554. [CrossRef]
75. Sheehan, P.M. The Late Ordovician Mass Extinction. *Ann. Rev. Earth Planet. Sci.* **2001**, *29*, 331–364. [CrossRef]
76. Liu, Z.; Algeo, T.J.; Guo, X.; Fan, J.; Du, X.; Lu, Y. Paleo-environmental cyclicity in the Early Silurian Yangtze Sea (South China): Tectonic or glacio-eustatic control? *Palaeogeogr. Palaeoclimatol. Palaeoecol.* **2017**, *466*, 59–76. [CrossRef]
77. Wang, Y.M.; Li, X.; Dong, D.; Zhang, C.; Wang, S. Main factors controlling the sedimentation of high-quality shale in the Wufeng-Longmaxi Fm, Upper Yangtze region. *Nat. Gas Ind. B* **2017**, *4*, 327–339. [CrossRef]
78. Wang, Y.; Chen, B.; Li, X.; Wang, H.; Chang, L.; Jiang, S. Sedimentary characteristics of upwelling facies shale in Lower Silurian Longmaxi Formation, northeast Sichuan area. *Acta Pet. Sin.* **2018**, *39*, 1092–1102.

Journal of
Marine Science and Engineering

MDPI

Article

Properties and Model of a Lacustrine Shale Oil Reservoir: A Case Study of the Upper Fourth Member and Lower Third Member of the Shahejie Formation in Dongying Sag and Zhanhua Sag, Jiyang Depression

Cunfei Ma [1,*], Xianxu Fang [1], Xiaonan Ma [2], Xiantai Liu [3], Bingkun Xu [4] and Xinmin Ge [1]

1 School of Geosciences, China University of Petroleum (East China), Qingdao 266580, China
2 No. 5 Oil Production Plant, Dagang Oilfield Company of CNPC, Tianjin 300280, China
3 Shengli Oilfield Company of SINOPEC, Dongying 257015, China
4 No. 3 Oil Production Plant, Daqing Oilfield CompanyLimited, Daqing 163453, China
* Correspondence: mcf-625@163.com

Abstract: By combining thermal simulation experiments with core data, thin sections, scanning electron microscopy (SEM), source rock, and high-pressure mercury and gas adsorption analysis, this paper evaluates the properties and models of shale reservoirs in the upper fourth member and the lower third member of the Shahejie Formation in Dongying Sag and Zhanhua Sag. The results show that the oil shale is a high-quality source rock with petroleum generation potential. Inorganic mineral pores, organic pores, and fractures have developed with petroleum storage capacity. Clay and organic-rich soft shale lithofacies are interbedded with carbonate-rich hard shale lithofacies to form sandwich-type source–reservoir–caprock assemblages with internal sealing properties. Bitumen occurs mostly in the free state, and to a lesser extent is adsorbed, and shows flow characteristics. The migration resistance of the bitumen is displacement pressure, including capillary force and adsorption resistance, and the main force is pore fluid pressure. The migration modes of the bitumen include both subcritical flow and instantaneous flow, which are controlled by pore fluid pressure, displacement pressure, and the rock's fracture strength. Owing to the multi-scale characteristics of the shale reservoir space, a vein-type model of a multi-scale progressive transport network is developed that obeys Murray's law with a dominant migration pathway in the shale reservoir. The shale oil reservoir is a special lithologic reservoir controlled by the sedimentary environment and has self-generation, self-storage, and self-sealing characteristics with developed pore fluid overpressure and a multi-scale transport network.

Keywords: shale oil; self-generation; self-storage; self-sealing; multi-scale transport; reservoir model

Citation: Ma, C.; Fang, X.; Ma, X.; Liu, X.; Xu, B.; Ge, X. Properties and Model of a Lacustrine Shale Oil Reservoir: A Case Study of the Upper Fourth Member and Lower Third Member of the Shahejie Formation in Dongying Sag and Zhanhua Sag, Jiyang Depression. *J. Mar. Sci. Eng.* **2023**, *11*, 1441. https://doi.org/10.3390/jmse11071441

Academic Editor: George Kontakiotis

Received: 31 May 2023
Revised: 13 July 2023
Accepted: 14 July 2023
Published: 19 July 2023

1. Introduction

Owing to increased difficulties in the exploration and production of conventional reservoirs in China, unconventional reservoirs have gained increasing attention. Abundant lacustrine shales developed in the Cenozoic in China [1,2], and these shales have become an important unconventional oil and gas resource [3–5]. Some oil has been produced from lacustrine shales in the Jiyang Depression and Cangdong Depression of the Bohai Bay Basin [6–8], the Biyang Depression [9,10], the Songliao Basin [11,12], the Santanghu Basin [13–15] and the Jimusar Depression of the Junggar Basin [16,17].

Lacustrine shale is commonly rich in type I kerogen, which makes it an excellent source rock [18–20]. Shales are characterized by abundant organic matter with nano-scale pores and pore throats. Shale oil consists of hydrocarbons that exist in pores in free, adsorbed, and dissolved states. Furthermore, like all fine-grained rocks, lacustrine shales have good sealing capacities for oil and gas [21]. In order to characterize the pore structure of shales, a series of advanced technologies has been introduced based on microscopy, including high-resolution field emission scanning electron microscopy (FE-SEM), atomic force microscopy (AFM), small-angle and ultra-small-angle neutron scattering (SANS and USANS, respectively), and transmission electron microscopy (TEM) [22–27]. For three-dimensional non-destructive observation of shale pores, X-ray microscopy and computed tomography (CT), nano-CT, and focused ion beam (FIB)-SEM have been used [28,29]. In addition, the 3D distribution of the various rock components and pores can be obtained by a digital core technique that combines energy dispersive spectroscopy (EDS) with backscattered electron imaging (BEI) [30–32]. These technologies have shown that the pore structure of lacustrine shale reservoir rock consists of multi-scale pore throats and fractures [33].

Controlled by the geometry and topological characteristics of the pore structure, oil and gas follows the multi-scale seepage law in shale reservoirs [34,35]. The pore throat size is a key factor in controlling the migration of oil and gas. When the pore throat diameter is ~5 nm, the migration of oil and gas is dominated by desorption and diffusion; at ~50 nm, it is dominated by diffusion-slip flow and low-speed non-Darcy flow [36]. Mathematical methods and numerical simulation have been used [37,38] for migration analysis after direct observation using SEM or CT of the generation, migration and accumulation of oil droplets in shale samples in thermal simulation experiments [39,40].

Owing to a lack of economically successful examples of lacustrine shale oil reservoirs, continuous lacustrine reservoir models have been established by analogy with properties of marine shale reservoirs, such as the Bakken Shale, Spraberry–Wolfcamp Shale, Eagle Ford Shale, Niobrara Shale, and Utica Shale in North America [13,41–44]. However, compared with marine shales which cover large areas with relatively consistent properties, lacustrine shales tend to cover small areas and undergo rapid lithofacies change [45,46]. This suggests that a marine shale oil reservoir model is not suitable for most lacustrine shale oil reservoirs because it neglects the strong lateral and vertical heterogeneity of lacustrine shale reservoirs and leads to optimistic resource evaluation and hence exploration failure. In addition, the production characteristics of lacustrine shale reservoirs, such as low initial production, fast production decline, and low recovery, highlight the inadequacy of the marine reservoir model. According to public reports, the recovery of marine shale oil in North America is low, generally at 5–10%, including 2–20% in the Bakken Shale, 8–15% in the Spraberry–Wolfcamp Shale, 6% in the Eagle Ford Shale, and 3% in the Utica Shale. Lacustrine shale oil reservoirs in China have even lower initial production, faster production decline, and lower total recovery. For example, the Paleogene Shahejie Formation in the Bohai Bay Basin and the Triassic Yanchang Formation in the Ordos Basin have less than 5% recovery, which is similar to the Eagle Ford Shale and Utica Shale. Thus, there is a need for further work on lacustrine shales as unconventional reservoirs.

In this study, we used core samples from the upper fourth member and the lower third member of the Shahejie Formation shale oil reservoir from the Dongying Sag and Zhanhua Sag in Jiyang Depression as an example of an unconventional lacustrine system. Core samples were analyzed by systematic testing technology combined with thermal simulation experiments. We report on the properties of the source–reservoir–caprock system and hydrocarbon migration within it, analyze the control of the lacustrine environment on the accumulation of the shale oil and finally build a lacustrine shale oil model. Finally, we offer guidance on the exploration and production of lacustrine shale oil.

2. Overview of Study Area

Jiyang Depression is part of the first-order tectonic unit of the Bohai Bay Basin in East China, which is located west of the Tan-lu fault zone, south of the Chengning uplift and north of the Luxi uplift (Figure 1). The depression is 240 km long and 130 km at its widest point and covers about 26,000 km^2. Jiyang Depression, a typical structure with sags and uplifts, is a Mesozoic–Cenozoic rift–depression composite basin composed of four negative secondary tectonic units, including Dongying, Huimin, Zhanhua, and Chezhen

Sags as well as several positive secondary tectonic units including the Gudao, Yihezhuang, Chenjiazhuang, Wudi, Binxian, and Kendong Qingtuozi uplifts (Figure 1). Dongying and Zhanhua Sags are the richest hydrocarbon accumulation areas in the Jiyang Depression, from which about 30 shale oil wells have obtained good oil production. For example, Y183 well produced more than 11,138.8 t of shale oil from November 2013 to the present.

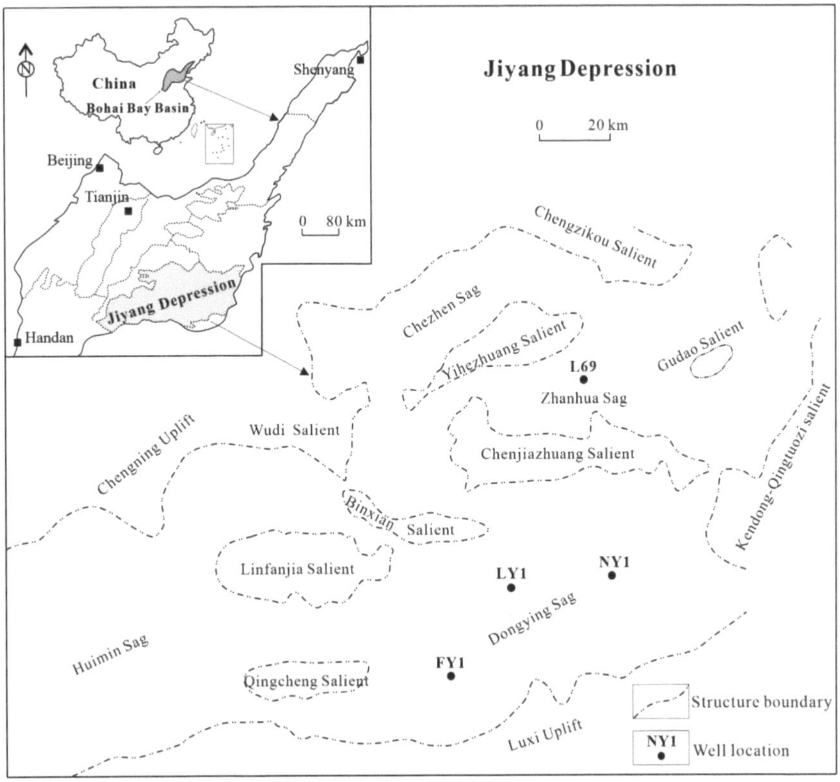

Figure 1. Overview of research area.

The upper fourth member and the lower third member of the Paleocene Shahejie Formation were deposited in a lacustrine transgression systems tract (LTST) or a lacustrine highstand systems tract (LHST) of second-order sequence (Figure 2) [47]. Especially during the active fault depression stage, thick dark shale strata developed under multiple semi-deep or deep lakes, brackish or saline water under alkaline and reducing conditions, and were then transformed in the subsequent early and middle diagenesis stages [1,48,49].

Because shales in the Shahejie Formation in the Jiyang Depression were mainly deposited under brackish-saline water columns [1,48], the lithofacies are mostly clayey limestone and calcareous shale (Table 1). Compared with marine shale reservoirs such as Wolfcamp, Bakken and Eagle Ford in North America, carbonate rocks and shale of Shahejie Formation in Jiyang Depression are rich in clay minerals and develop mud texture, which is similar to Uteland Butte shale reservoirs in North America (Table 1). Shales of the Shahejie Formation in the Jiyang Depression have a present-day depth of 2000–3900 m and a cumulative thickness of 200–400 m. Toc, vitrinite reflectance (Ro) and porosity fall into the range of 1.2–16%, 0.5–1.3%, 3–16%, respectively, with permeability less than 0.1 mD (Table 1). Except for the lithofacies types, reservoir properties are not significantly different from other lacustrine shales in China and North America (Table 1). However, due to the high gas-oil ratio, the crude oil density and viscosity of marine shale oil are smaller than that of lacustrine shale oil (Table 1). The upper fourth member and the lower third member of the Shahejie Formation is the most important shale oil reservoir in the study area.

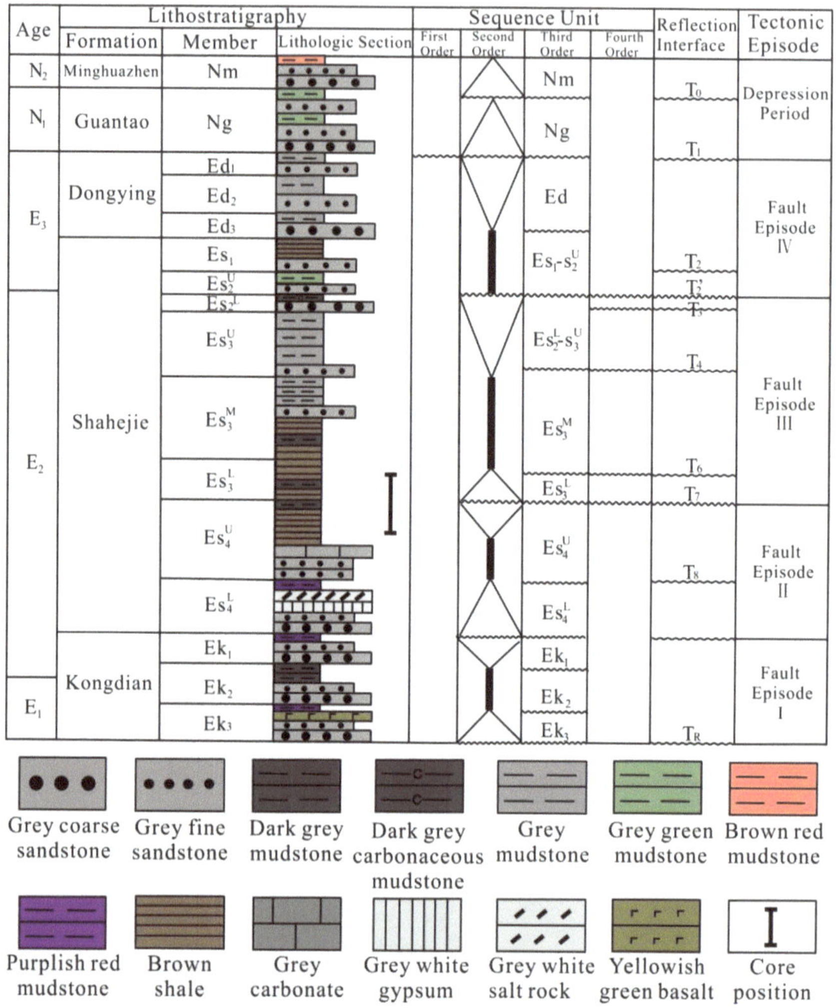

Figure 2. Sequence stratigraphic column for the Dongying Sag (the data was from [1,47–49]).

Table 1. Comparison of geological parameters of main shale oil reservoirs in North America and China [50–55].

Contrast Parameter		North America				China		
Reservoir properties	Shale set	Wolfcamp (Permian Basin)	Bakken	Eagle Ford	Uteland Butte Member, Lower Green River Formation (Uinta Basin)	Shahejie Formation (Jiyang Depression)	Yanchang Formation (Ordos Basin)	Kongdian Formation (Huanghua Depression)
	Sedimentary environment	marine facies	marine facies	marine facies	lacustrine facies	salinized lacustrine facies	lacustrine facies	lacustrine facies

Table 1. *Cont.*

Contrast Parameter		North America				China		
	Lithology	carbonate rocks, shale	carbonate rocks, siltstone, shale	carbonate rocks, calcareous shale	clayey limestone, clayey dolostone, calcareous shale	clayey limestone, calcareous shale	fine sandstone, siltstone, shale	siltstone, shale
	Depth/m	2500–3100	2400–3000	1200–3700	1000–3000	2000–3900	1600–2900	3200–3600
	Thickness/m	20–150	5–12	75	15–130	20–400	5–15	200–400
	Toc/%	2–9	5–10	2–12	2–5	1.2–16	2–10	>2
	Ro/%	0.7–0.9	0.5–1.3	0.45–1.4	0.55–1.10	0.5–1.3	0.9–1.1	0.6–1.1
	Porosity/%	4–12	4–10	8–12	3–20	3–16	2–12	5–10
	Permeability/mD	0.01–1.0	0.01–0.1	0.01–1.0	0.004–0.337	<0.1	0.01–0.2	0.12
Fluid properties	Formation pressure coefficient	1.1–1.2	1.2–1.5	1.35–1.8	1.0–1.8	1.1–2.0	0.77–0.84	0.9–1.2
	Crude oil density/(g/cm^3)	0.71–0.81	0.78–0.83	0.77–0.87	0.82–0.88	0.82–0.92	0.80–0.86	0.86–0.89
	Crude oil viscosity/mPa·s	149.5–666.6	2–10	0.01–1	1–4	4–126	5.5–6.4	10.4–23.5
	Gas-oil ratio (m^3/m^3)	50–140	50–375	90–2100	-	8–97	95–125	68.5–97.5

3. Materials and Methods

Core samples from wells NY1, FY1, LY1, and L69 (Figure 1) were analyzed by a complete set of rock-mineral analysis and thermal simulation experiments. In addition, wells HX4, HX28, and A6 were used for supplementary research. Careful observation and description of the core was carried out to identify the lithofacies, fracture development, and source–reservoir–caprock assemblage of the shale. As a result, 305 representative samples covering the main types of lithofacies were analyzed by thin section, optical microscopy, field emission scanning electron microscope (FE-SEM) with energy dispersion spectrum (EDS), high-pressure mercury, gas adsorption, and source rock pyrolysis analysis, as described in the following section.

Thin sections of shale samples were made to observe the petrological characteristics including the rock component type and distribution, the laminae type and the macro reservoir space type under an optical microscope. Observation and identification of the shale microstructure included mineralogy and reservoir space based on FE-SEM using an FEI Quanta 650. Bitumen in the shale reservoir space was directly observed under low-vacuum FE-SEM. Two-dimensional (2D) scanning of the samples was performed combining FE-SEM with EDS to show pore type distributions, mineral compositions and plane distributions and contact relationships.

The quantitative distribution of macro- to micro-scale reservoir space was determined by the combined application of high-pressure mercury and CO_2 and N_2 gas adsorption methods. The high-pressure mercury intrusion method is used to investigate the distribution of macropores; the N_2 adsorption method is used to analyze the distribution of mesopores and the CO_2 adsorption method is used to quantify the distribution of micropores in the sample.

In this study, the total organic carbon (TOC) content, volatile (free) hydrocarbon content (S_1), pyrolysis yield (S_2) and hydrocarbon generation potential ($S_1 + S_2$) of the shale were obtained using source rock pyrolysis analysis.

Fresh and low-maturity shale samples were selected for thermal simulation experiments. The experimental facilities include an autoclave, a temperature control system and a pressure control system. The autoclave is composed of Hastelloy alloy with a maximum volume of 1000 mL, maximum tolerable temperature of 600 °C and maximum tolerable pressure of 50 MPa. The temperature control system is composed of a thermocouple, a temperature sensor and an AI-518P artificial intelligence temperature controller that precisely controls the heating process. The pressure control system is composed of a pressure sensor and a GBS-STA100 gas-supercharging system.

Prior to the thermal simulation experiment, fresh shale samples were cut into regular lumpy sections along the direction perpendicular to the surface of the laminae; 200 mL distilled water was added to the sample before being placed into the autoclave. Samples were heated to temperatures of 150–600 °C at high pressures of 15–30 MPa for 48 h. Changes occurring in the samples before and after the experiment were evaluated by optical microscopy and FE-SEM to determine the occurrence and migration of the bitumen.

4. Results

4.1. Source–Reservoir–Caprock Characteristics of Shale Oil Reservoir

4.1.1. Oil-Generation Potential

Previous geochemical research has focused on the hydrocarbon potential of the upper fourth member and the lower third member of the Shahejie Formation in the Jiyang Depression [56,57], where the high-quality source rock was deposited mainly in brackish and saline lacustrine environments [58–64]. The kerogen types derived from aquatic plankton are mainly I and II_1 (Table 2). The vitrinite reflectance (Ro) shows that the source rock is located in an oil-generation window (Table 2). The source potential differs considerably owing to the diversity of the lithofacies.

Table 2. Types of organic matter and rock pyrolysis parameters.

Kerogen Type		Ro/%	TOC/%	S_1/(mg/g)	S_2/(mg/g)	$S_1 + S_2$/(mg/g)
I, II_1	minimum	0.5	4.03	1.03	17.59	18.97
	maximum	1.0	12.8	10.0	78.59	82.65
	average	0.8	6.41	4.57	41.28	45.85

The contents of felsic minerals, clay minerals and carbonate minerals show large variation. Generally, carbonate minerals are the most abundant component, at more than 50%, and felsic minerals are 0–33.33%, which is higher than the clay mineral content. Oil shale has a dark color owing to the rich organic matter, displaying lamellar or oriented microscopic stratifications, and always develops calcite veins (Figure 3a,b). The content of carbonate minerals in the oil shale is about 81.4%, with calcite at 77%, followed by felsic minerals at about 10% distributed in the carbonate minerals (Figure 3c). Therefore, oil shale in the reservoir is probably a high-quality source rock because it is rich in organic matter and carbonate minerals, which usually means high oil-generation potential with authigenic attributes.

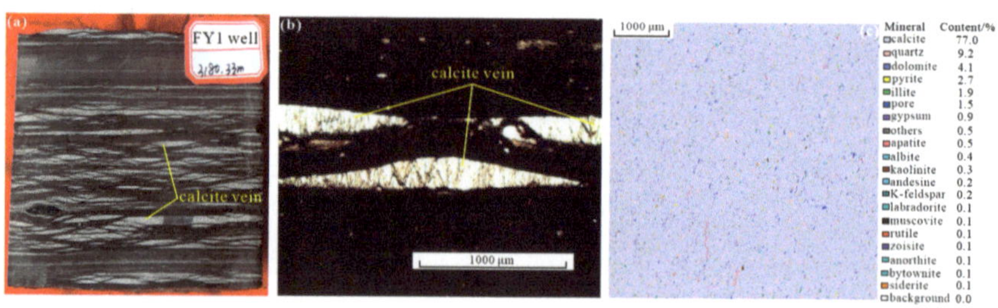

Figure 3. Petrological characteristics of oil shale. (**a**) Parallel calcite veins in core; FY1 well, 3180.33 m. (**b**) Parallel calcite veins under a polarizing microscope, 50×(−); FY1 well, 3181.88 m. (**c**) Calcite (the main mineral) and quartz (second most abundant mineral) in the 2D whole-mineral scan by FE-SEM; L69 well, 3027.30 m.

4.1.2. Reservoir Properties

Shale reservoir space can be divided into pores and fractures (Table 3). Pores were further subdivided into two types: pores in inorganic minerals and pores in organic matter (Table 3). Inorganic mineral pores include intercrystalline (or intergranular) in felsic minerals, carbonates and pyrite, and intracrystalline (or intragranular) pores in felsic and clay minerals, and carbonates (Table 3). These pore types were all further subdivided into primary and secondary pores resulting mainly from the dissolution of felsic minerals, the recrystallization of carbonate minerals and the transformation of clay minerals (Table 3). Organic pores include internal, marginal and boundary pores in organic matter (Table 3) [65]. Fractures include structural, bedding, shrinkage, dissolved and natural fluid pressure fractures based on the origin (Table 3).

Table 3. Classification schemes of shale reservoir space.

<table>
<tr><th></th><th colspan="2">Origin / Location</th><th>Primary Pore</th><th>Secondary Pore</th></tr>
<tr><td rowspan="9">Pore</td><td rowspan="6">Inorganic-mineral pore</td><td rowspan="3">Intercrystalline (intergranular) pore</td><td>Intergranular pore of felsic mineral</td><td>Intergranular dissolved pore of felsic mineral</td></tr>
<tr><td>Intercrystalline pore of carbonate mineral</td><td>Intercrystalline (dissolved) pore of carbonate mineral</td></tr>
<tr><td>Intercrystalline pore of pyrite</td><td>Intercrystalline (dissolved) pore of pyrite</td></tr>
<tr><td rowspan="3">Intracrystalline (intragranular) pore</td><td>Intracrystalline pore of clay mineral</td><td>Intracrystalline (dissolved) pore of clay mineral</td></tr>
<tr><td>Intracrystalline pore of carbonate mineral</td><td>Intracrystalline (dissolved) pore of carbonate mineral</td></tr>
<tr><td>Intragranular pore of felsic mineral</td><td>Intragranular dissolved pore of felsic mineral</td></tr>
<tr><td colspan="2" rowspan="3">Organic pore</td><td>Internal pore in organic matter</td><td></td></tr>
<tr><td>Marginal pore in organic matter</td><td></td></tr>
<tr><td>Boundary pore in organic matter</td><td></td></tr>
<tr><td>Fracture</td><td colspan="4">Structural fracture
Bedding fracture
Shrinkage fracture
Dissolved fracture
Natural fluid pressure fracture</td></tr>
</table>

Pores in the upper fourth member and the lower third member of the Shahejie Formation are mainly discretely distributed in a carbonate matrix based on FE-SEM (Figure 4a). The whole-mineral analysis indicated that carbonate minerals are mostly in contact with quartz, followed by illite (Figure 4b), and the pores are mainly closed to calcite, followed by quartz (Figure 4b). Therefore, the pores are strongly related to the formation of carbonate minerals and quartz, and, to a lesser extent, other minerals. The pore types are mainly primary or secondary intercrystalline (intergranular) pores, including intercrystalline (dissolved) pores of carbonate mineral, intergranular dissolved pores of felsic mineral, intracrystalline (dissolved) pores of clay mineral, and intergranular (dissolved) pores among carbonate, felsic and clay mineral (Figure 4c). Different from Bakken Shale, Barnett Shale and Marcellus Shale, the organic matter in the upper fourth member and lower third member of the Shahejie Formation has developed few internal and marginal pores but abundant boundary pores of organic matter (Figure 4c).

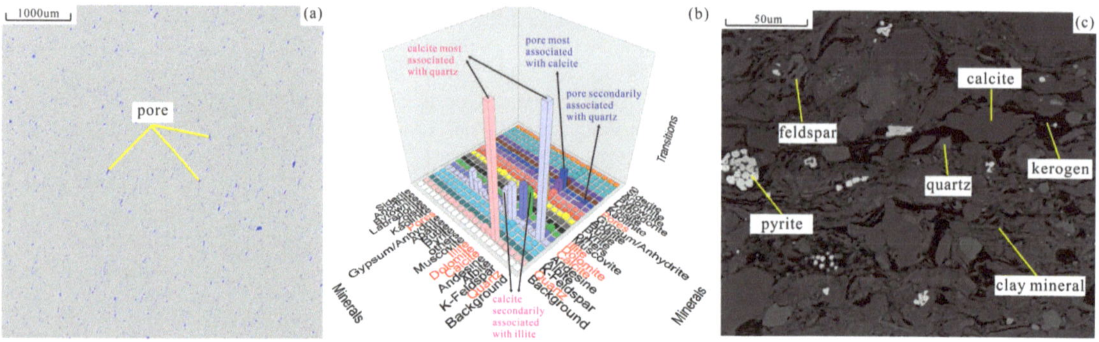

Figure 4. Distribution of rock components in the shale. The images were obtained by FE-SEM. (**a**) Pores distributed in the mineral matrix; L69 well, 3027.30 m. (**b**) Associated minerals and pores in different proportions as shown by the whole-mineral scan; well L69, 3027.30 m. (**c**) Associated minerals and pores; well L69, 3048.10 m.

Despite enormous numbers and various types of shale reservoir spaces, pores in actual shale samples and samples after thermal simulation experiments are not all bitumen-filled. Examples are as follows: Organic pore developed in the kerogen of the sample after performing the thermal simulation experiment (Figure 5a); intracrystalline pore developed in the illite–smectite mixed layer of the sample after the thermal simulation experiment (Figure 5b); intergranular pore developed in the framework of feldspar and the illite–smectite mixed layer of the sample after the thermal simulation experiment (Figure 5c); intergranular dissolved pore developed in the framework of feldspar and calcite of the sample after the thermal simulation experiment (Figure 5d); intergranular pore developed in the framework of minerals (Figure 5e); intercrystalline pore developed in the framework of micritic dolomite (Figure 5f); intercrystalline pore developed in the framework of microfossils (Figure 5g); bedding-parallel vein fracture filled with calcite and bitumen (Figure 5h); bedding-parallel vein fracture filled with calcite and bitumen (Figure 5i); expulsion fracture filled with bitumen (Figure 5j); expulsion fracture filled with bitumen (Figure 5k); bedding and structural fractures filled with bitumen (Figure 5l). As mentioned above, all show filled-in effective reservoir space including organic pores (Figure 5a), large-scale intercrystalline (intergranular) pores close to organic matter (Figure 5b–g) and fractures (Figure 5h–l), especially in hydrocarbon expulsion fractures and bedding-parallel vein fractures (Figure 5h–k). The oil shale contains enormous amounts of intercrystalline (intergranular) pores and boundary pores of organic matter and has developed bedding-parallel vein fractures and hydrocarbon expulsion fractures. Hence, the shale has self-storage properties.

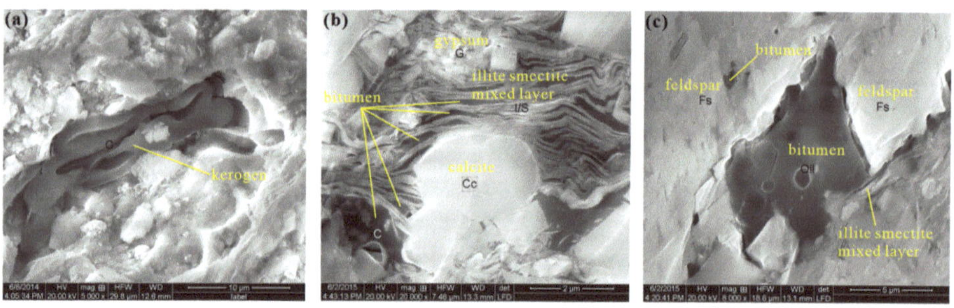

Figure 5. *Cont.*

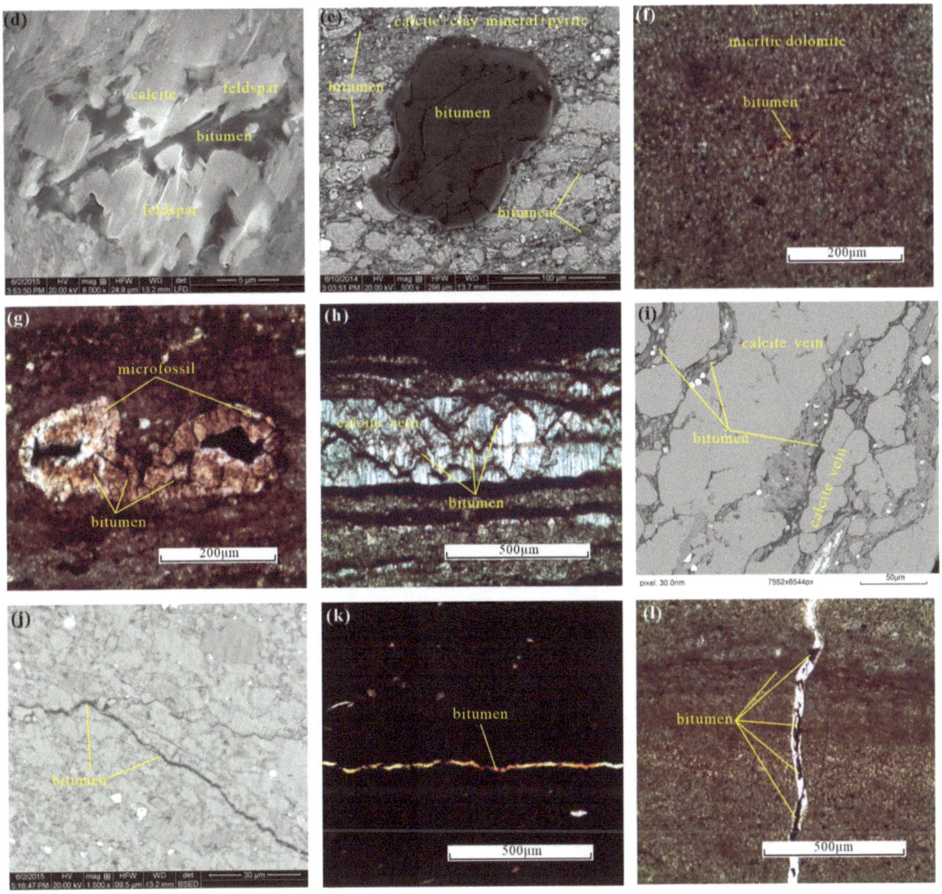

Figure 5. Types of effective reservoir spaces in shale oil reservoir. (**a–e**,**i**,**j**) were obtained by FE-SEM; (**f–h**,**k**,**l**) are polarizing microscope images. (**a**) Organic pore; L69 well, 3059.35 m. The experimental temperature and pressure of the sample were 200 °C and 1 MPa, respectively. (**b**) Intracrystalline pore; L69 well, 3053.20 m. The experimental temperature and pressure of the sample were 200 °C and 20 MPa, respectively. (**c**) Intergranular pore; L69 well, 3053.20 m. The experimental temperature and pressure of the sample were 150 °C and 15 MPa, respectively. (**d**) Intergranular dissolved pore; L69 well, 3053.20 m. The experimental temperature and pressure of the sample were 150 °C and 15 MPa, respectively. (**e**) Intergranular pore developed; L69 well, 3048.10 m. (**f**) Intercrystalline pore, 200×(−); FY1 well, 3168.03 m. (**g**) Intercrystalline pore, 200×(−); FY1 well, 3143.98 m. (**h**) Bedding-parallel vein fracture, 100×(−); NY1 well, 3432.17 m. (**i**) Bedding-parallel vein fracture; L69 well, 3048.10 m. (**j**) Expulsion fracture; L69 well, 3053.20 m. The experimental temperature and pressure of the sample were 200 °C and 20 MPa, respectively. (**k**) Expulsion fracture, 100×(−); FY1 well, 3209.09 m. (**l**) Bedding and structural fractures, 100×(−); FY1 well, 3168.03 m.

4.1.3. Sealing Capacity

Different lithofacies in shale have varying mechanical strengths and physical properties due to differences in the composition of both inorganic minerals and organic matter [66]. The carbonate-rich lithofacies have high mechanical strength, poor physical properties such as low permeability and good sealing ability, whereas low maturity clay-rich or organic-rich lithofacies have weak mechanical strength and can easily be fractured by overpressure fluid [65]. The evolution of the continental lacustrine sedimentary environment in the study area caused the carbonate-rich lithofacies and the clay-rich or organic-rich facies to develop alternately, which formed sets of sandwich-type strata at different scales (Figure 6): Interbedded clayey limestone and oil shale form alternating source rock, reservoir, and caprock (Figure 6a), and clayey limestone acts as caprock, and oil shale acts as source rock and reservoir (Figure 6b), and the calcite veins are limited in the oil shale layer or are distributed on the interface between oil shale and clayey limestone (Figure 6c,d); micritic calcite laminae acts as sealing laminae alternating with organic-rich and terrigenous clastic-rich laminae as oil-generation and reservoir laminae under a polarizing microscope, i.e., sealing laminae alternating with oil-generation and reservoir laminae forms a small-scale source–reservoir–caprock assemblage (Figure 6e,f). The typical pattern has two sets of clayey limestone strata as the roof and the floor with oil shale in the middle (Figure 6a,b). The caprock roof has good mechanical strength and poor physical properties, and oil shale with porosity as the source and reservoir. This type of arrangement can easily form a closed system in which bitumen is continuously generated and accumulated to create pore fluid overpressure [44]. In addition, the calcite veins are limited in the oil shale layer or are distributed on the interface between oil shale and clayey limestone, and are almost absent in the clayey limestone, which means that the clayey limestone (the caprock) is dense enough for fluid to flow. It serves as strong evidence for the sealing ability (Figure 6c,d). Moreover, the sandwich-type strata structure composed of strong and weak lithofacies resists strong external tectonic stress and has self-regulation against deformation to preserve overpressure liquid for long periods, which could help to preserve the shale oil and gas. The production sections of the upper fourth member and the lower third member of the Shahejie Formation in Zhanhua Sag have both a roof and floor. Therefore, the sandwich-type structure of a shale oil reservoir determines its self-sealing ability.

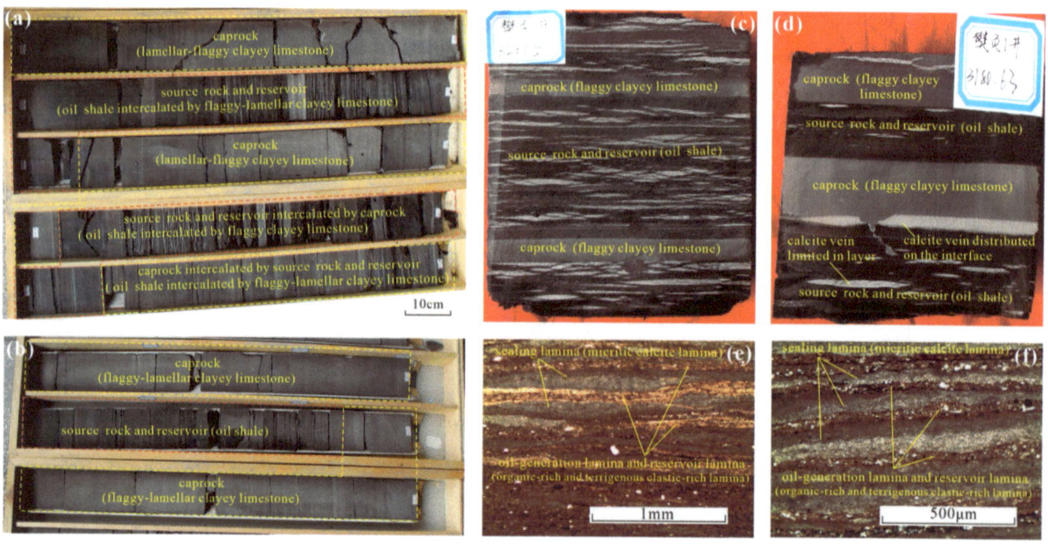

Figure 6. Source–reservoir–caprock assemblage of multi-scale sandwich-type structure in shale oil reservoir. (**a**) Interbedded clayey limestone and oil shale; FY1 well, 3177.03–3182.03 m. (**b**) Clayey limestone intercalated with oil shale interlayer; LY1 well, 3766.61–3769.61 m. (**c**) Flaggy clayey limestone and oil shale; FY1 well, 3201.50 m. (**d**) Calcite veins limited in layers or distributed on the interface in the assemblage; FY1 well, 3180.63 m. (**e**) Micritic calcite laminae alternating with organic-rich and terrigenous clastic-rich laminae under a polarizing microscope, 25×(−); NY1 well, 3345.29 m. (**f**) Sealing laminae alternating with oil-generation and reservoir laminae under a polarizing microscope, 50×(−); NY1 well, 3390.10 m.

4.2. Reservoir Space Structure

The type, size and distribution of reservoir space vary among shales. Conventional qualitative and quantitative methods for measuring reservoir space types and structures include thin-section optical microscopy, SEM, mercury intrusion porosimetry and gas adsorption analyses. The maximum magnification of an optical microscope is usually 500–3000 times, which could be adequate for observing micro-scale pores and fractures. Microfractures are distributed discretely, whilst intercrystalline pores are continuous under the optical microscope. The microfractures effectively connect the intercrystalline micropores. The observation scale of SEM covers the high magnification end of the optical microscope and extends down to the nanopore scale. Hence, SEM is a powerful method for studying shale reservoir pores because it clearly displays their properties at different scales. The fractures are distributed discretely under low magnification, although more hydrocarbon expulsion fractures and intercrystalline pores connected by smaller ones appear as the magnification increases. The intercrystalline pores are connected to form a pore network with the organic matter pores. Pore size is a key parameter regarding the rock's physical properties in shale reservoirs. High-pressure mercury intrusion, N_2 adsorption and CO_2 adsorption methods are used to measure macropores, mesopores and micropores, with pore diameters of >50 nm, 2–50 nm, and <2 nm, respectively [67]. According to the pore volumes of these pores, the shale reservoir space of the upper fourth member and lower third member of the Shahejie Formation is multi-scale and mainly nano–micro scale. The pore volume of the macro pores is larger than that of the mesopores, and that of the mesopores is larger than the pore volume of the micropores.

4.3. Occurrence of Shale Oil

The occurrence of oil and gas in shale reservoirs has been investigated by many groups [68,69]. The three states for bitumen include free, adsorbed and dissolved states [70]. In this study, we utilized ESEM to observe the original shale samples and post-thermal simulation samples and determined that the bitumen occurs mainly in the free state (Figure 7a–i), followed by the adsorbed state (Figure 7j–l). These results are in agreement with those of O'Brien et al. (2002) [40], who also conducted a thermal simulation experiment on organic-rich shale. The bitumen in the free state is distributed mainly in the organic pores, large-scale intercrystalline (intergranular) pores, hydrocarbon expulsion fractures, bedding-parallel vein fractures, bedding fractures and structural fractures, and shows obvious flow traces (Figure 7g–i). The adsorbed bitumen is distributed mainly in the intracrystalline pores of the clay mineral (Figure 7j–l).

(a) HX4 well, 2302.60 m, 500 ℃ (b) HX28 well, 3652.73 m, 600 ℃ (c) HX4 well, 2302.60 m, 600 ℃

Figure 7. *Cont.*

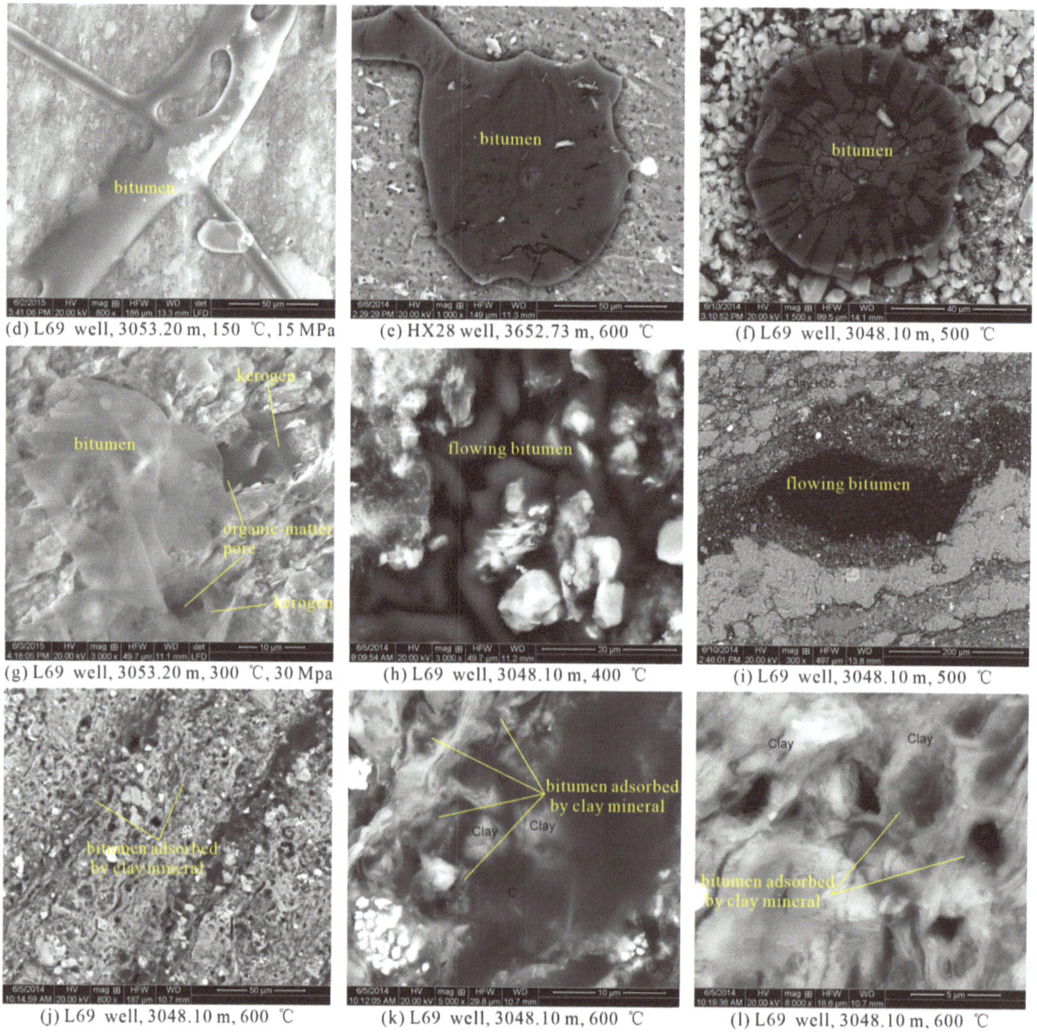

Figure 7. Occurrence of bitumen in the shale samples after thermal simulation experiments shown in FE-SEM images. (**a–c**) Oil drop formed in the sample. (**a**) The experimental temperature and pressure of the sample were 500 °C and 1 MPa, respectively; HX4 well, 2302.60 m. (**b**) The experimental temperature and pressure of the sample were 600 °C and 1 MPa, respectively; X28 well, 3652.73 m. (**c**) The experimental temperature and pressure of the sample were 600 °C and 1 MPa, respectively; HX4 well, 2302.60 m. (**d–f**) Bitumen migrating from the samples. (**d**) The experimental temperature and pressure of the sample were 150 °C and 15 MPa, respectively; L69 well, 3053.20 m. (**e**) The experimental temperature and pressure of the sample were 600 °C and 1 MPa, respectively; HX28 well, 3652.73 m. (**f**) The experimental temperature and pressure of the sample were 500 °C and 1 MPa, respectively; L69 well, 3048.10 m. (**g–i**) Bitumen generated from kerogen and migrating into the pore. (**g**) The experimental temperature and pressure of the sample were 300 °C and 30 MPa, respectively; L69 well, 3053.20 m. (**h**) The experimental temperature and pressure of the sample were 400 °C and 1 MPa; L69 well, 3048.10 m. (**i**) The experimental temperature and pressure of the sample were 500 °C and 1 MPa, respectively; L69 well, 3048.10 m. (**j–l**) Bitumen adsorbed by clay minerals. The experimental temperature and pressure of the sample were 600 °C and 1 MPa, respectively; L69 well, 3048.10 m.

5. Discussion

5.1. Migration Resistance and Dynamics of Shale Oil

Bitumen is displaced from kerogen into nearby reservoir space and migrates in a horizontal direction, which displaces pore water in an oil–water two-phase flow process. Owing to the original hydrophilic conditions of the inorganic mineral surfaces in the shale, capillary pressure creates resistance to water displacement. Felsic and carbonate minerals in shale are extremely small, mainly at the fine-silt scale. However, clay minerals and organic matter are abundant and have large specific surface areas and high adsorption capacities. The shale develops a micro–nano-scale pore and throat network in which the throat is composed of intracrystalline pores of clay mineral or intergranular pores between clay mineral clusters and clastic grains. The fluid in the pores and throats of shale is affected by adsorption resistance owing to a large pore curvature, weak connectivity, high irreducible water saturation and poor physical properties [71–73]. Therefore, the migration of bitumen displacing water overcomes the capillary pressure and adsorption resistance. We thus define the displacement pressure of shale as the sum of capillary pressure and adsorption resistance (Figure 8). According to the equilibrium relationship between the dynamic force and resistance, the fluid pressure of bitumen is the main driving force of the migration, although buoyancy should also be considered for the vertical migration of the bitumen.

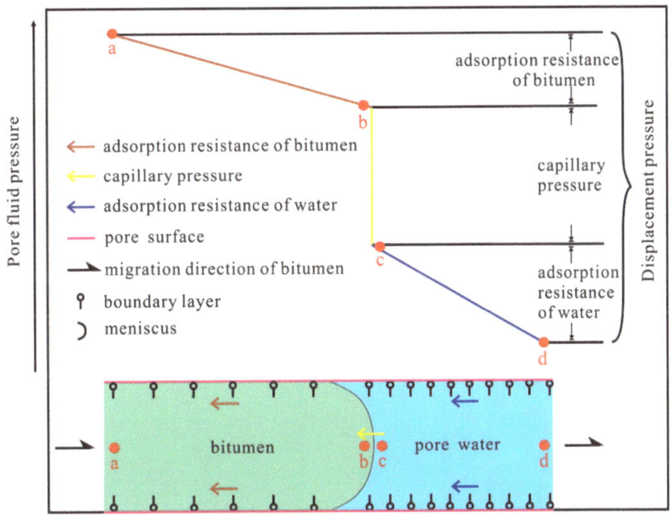

Figure 8. Schematic diagram of dynamic force and resistance in the process of bitumen migration. (a) Bitumen is displaced from kerogen into nearby reservoir space and migrates in a hor-izontal direction. (b) The migration of bitumen displacing water overcomes adsorption resistance of bitumen. (c) The migration of bitumen displacing water overcomes capillary pressure. (d) The migration of bitumen displacing water overcomes adsorption resistance of water.

5.2. Migration Patterns of Shale oil

The main mechanism of pore fluid pressurization is hydrocarbon generation. Fluid displacement is hampered by extremely small pore throats and low permeability. Hence, the volume and, to a greater extent, the rate of hydrocarbon generation are important factors in the buildup of pore fluid pressure [74]. Generally, a rock's fracture strength is larger than its fluid displacement pressure. At low hydrocarbon generation rates, the generated hydrocarbon volume is insufficient to generate a pore fluid pressure that reaches the rock's fracture strength, and the hydrocarbon will be retained in the pore (Figures 9(Aa,Ba) and 10a). If the hydrocarbon generation rate is moderate and the pore fluid pressure is larger than the displacement pressure but smaller than the rock's fracture strength, the hydrocarbon will break through into the adjacent pores according to the migration model of subcritical flow until reaching a balance between hydrocarbon generation and seepage (Figures 9(Ab) and 10b). However, if the hydrocarbon generation rate is high and its quantity is larger than the seepage discharge, hydrocarbons will accumulate and the pore fluid pressure will increase continuously until overcoming the displacement pressure as well as the rock's fracture strength to generate hydrocarbon expulsion fractures; then, the fluid will migrate in the model of instantaneous flow (Figures 9(Ac) and 10c) [75–77]. In particular, when the shale has a high content of organic matter, abundant laminae develop and the rock matrix is very dense but brittle. In this case, the rock's fracture strength is smaller than the displacement pressure and the pore fluid pressure quickly reaches the rock's fracture strength. Then, hydrocarbon expulsion fractures easily form, which enables the migration of instantaneous flow (Figures 9(Bb,Bc) and 10c).

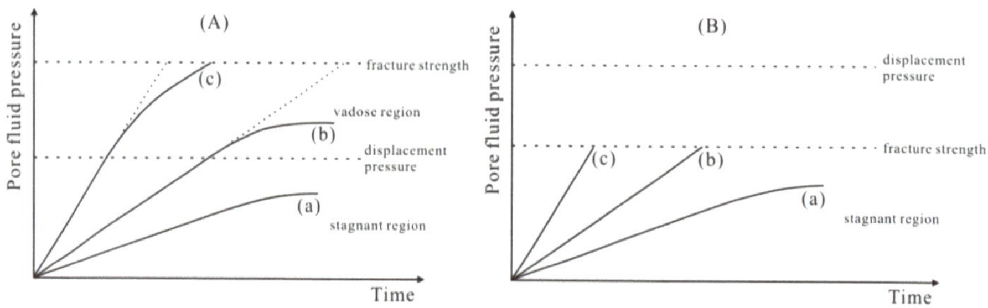

Figure 9. Relationship between the hydrocarbon generation rate and the migration pattern. (**Aa**) The pore fluid pressure is smaller than the displacement pressure and the rock's fracture strength. (**Ab**) The pore fluid pressure is larger than the displacement pressure but smaller than the rock's fracture strength. (**Ac**) The pore fluid pressure is larger than the displacement pressure and the rock's fracture strength. (**Ba**) The pore fluid pressure is smaller than the displacement pressure and the rock's fracture strength. (**Bb**) The pore fluid pressure is larger than the rock's fracture strength. (**Bc**) The pore fluid pressure is larger than the rock's fracture strength.

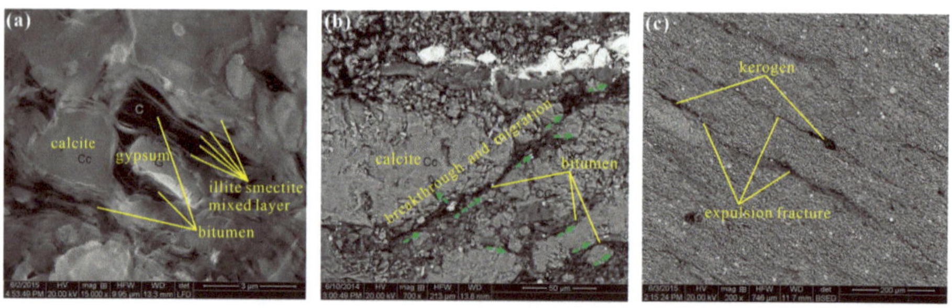

Figure 10. Migration patterns of shale oil in the original or thermal simulated samples by FE-SEM. (**a**) Bitumen detained in an intracrystalline pore of clay mineral; L69 well, 3053.20 m. The experimental temperature and pressure of the sample were 200 °C and 20 MPa, respectively. (**b**) Bitumen breaking through and migrating in an intracrystalline pore of calcite; L69 well, 3048.10 m. (**c**) Expulsion fractures formed with the core of kerogen extending in the parallel direction; L69 well, 3053.20 m. The experimental temperature and pressure of the sample were 200 °C and 20 MPa, respectively.

5.3. Accumulation Model of Shale Oil

The results of shale reservoir pore structure indicate multi-scale properties and a gradually connected network. The large-scale reservoir space is connected with the small-scale one, which determines the multi-scale migration of bitumen and eventually the accumulation in the large-scale reservoir space. The bitumen migrates away from the source kerogen, breaks through into nearby intercrystalline (intergranular) pores and then continues to migrate outside (Figure 11a–c), as indicated by optical microscopy. This phenomenon is observed more clearly under FE-SEM, where free-state bitumen migrates from kerogen organic pores (Figure 11d), enters nearby nano–micro-scale intercrystalline (intergranular) pores with the flow sign (Figure 11e), and accumulates in the micro-scale intergranular dissolved pores (Figure 11f). Especially in organic-rich shale, kerogen gradually accumulates to form an organic-matter network under pressure (Figure 11g–j) [78,79] and is usually connected by hydrocarbon expulsion fractures (Figure 11h–j). Bitumen migrates from the organic-matter network and hydrocarbon expulsion fractures (Figure 11g–k) and eventually accumulates into the larger scale bedding-parallel vein fractures, bedding fractures or structural fractures (Figure 11k,l).

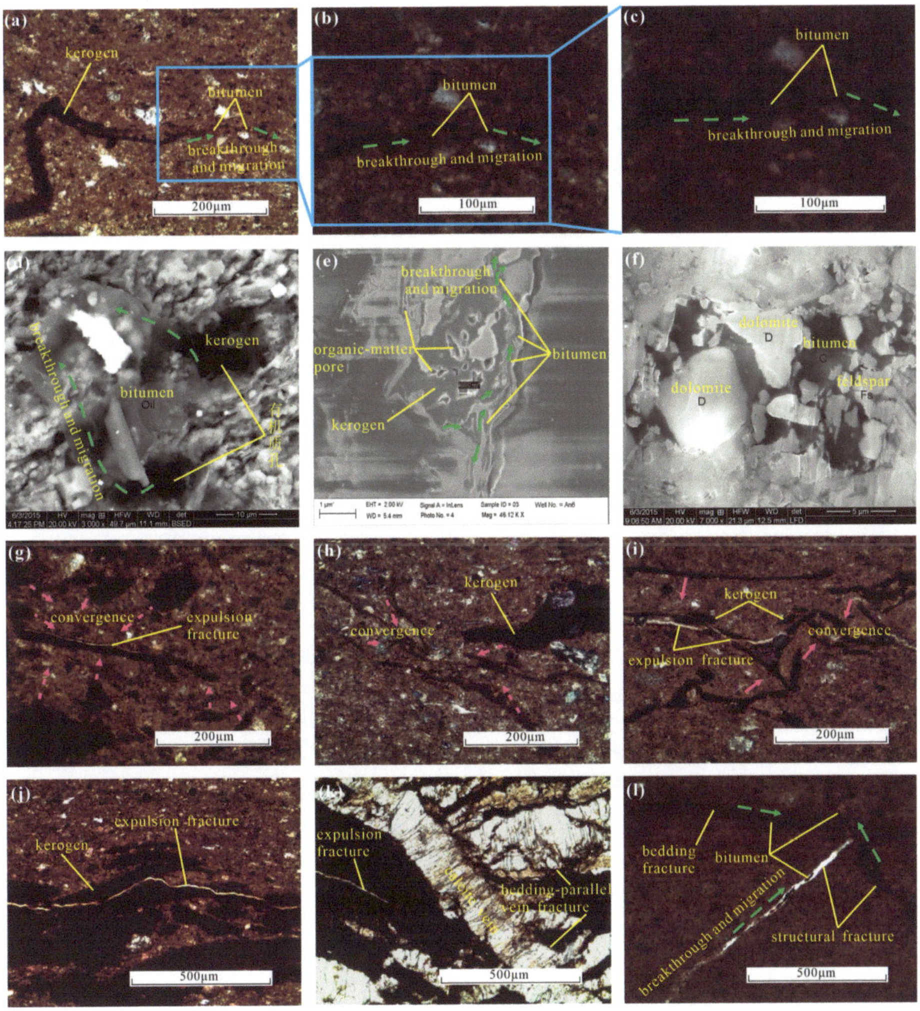

Figure 11. Characteristics of multi-scale migration and accumulation of bitumen in shale reservoir. (**a–c**,**g–l**) are polarizing microscope images; (**d–f**) are FE-SEM images. (**a**) Bitumen migration from kerogen, 100×(−); FY1 well, 3116.48 m. (**b**) Bitumen entering intercrystalline (intergranular) pore next

to kerogen, 200×(−); FY1 well, 3116.48 m. (**c**) Bitumen breaking through intercrystalline (intergranular) pore and migrating farther outside; 500×(−); FY1 well, 3116.48 m. (**d**) Bitumen migrating from kerogen through organic pore; L69 well, 3053.20 m. The experimental temperature and pressure of the sample were 300 °C and 30 MPa, respectively. (**e**) Bitumen migrating from organic pore and breaking through adjacent nano-scale intergranular pore; A6 well, 3060.50 m. (**f**) Bitumen accumulation in macro-scale intercrystalline (intergranular) pore; L69 well, 3053.20 m. The experimental temperature and pressure of the sample were 250 °C and 25 MPa, respectively. (**g**) Kerogen converging to an expulsion fracture to form an organic-matter network, 100×(−); FY1 well, 3040.26 m. (**h**) Kerogen mutual attraction forming an organic-matter network, 100×(−); FY1 well, 3046.53 m. (**i**) Kerogen converging to an expulsion fracture to form an organic-matter network, 100×(−); FY1 well, 3045.23 m. (**j**) Expulsion fracture filled with bitumen in the organic-matter network, 50×(−); FY1 well, 3040.26 m. (**k**) Intercrystalline pore of calcite vein filled with bitumen in bedding-parallel vein fracture, 50×(−); FY1 well, 3183.08 m. (**l**) Bitumen breaking through and migrating in bedding and structural fractures, converging to larger scale of structural fractures, 50×(−); FY1 well, 3168.03 m.

The study of shale gas indicates that the diversity and multiple scales of the reservoir space determine the occurrence, flow pattern and flow mechanism [36,80,81]. The organic pores and intracrystalline (intragranular) pores are mainly at the nanoscale, in which natural gas is in the dissolved and adsorbed states, and the flow type includes Brownian motion, desorption and Knudsen diffusion. The intercrystalline (intergranular) pores are at the micro-scale, in which natural gas is in the adsorbed and free states, and the flow type includes Knudsen diffusion, slippage, bulk diffusion and non-Darcy flow. The scale of natural fractures and hydraulic fractures is larger than that of pores, in which the occurrence of natural gas is in the free state and the flow types are mainly Darcy flow and pipe flow (Figure 12).

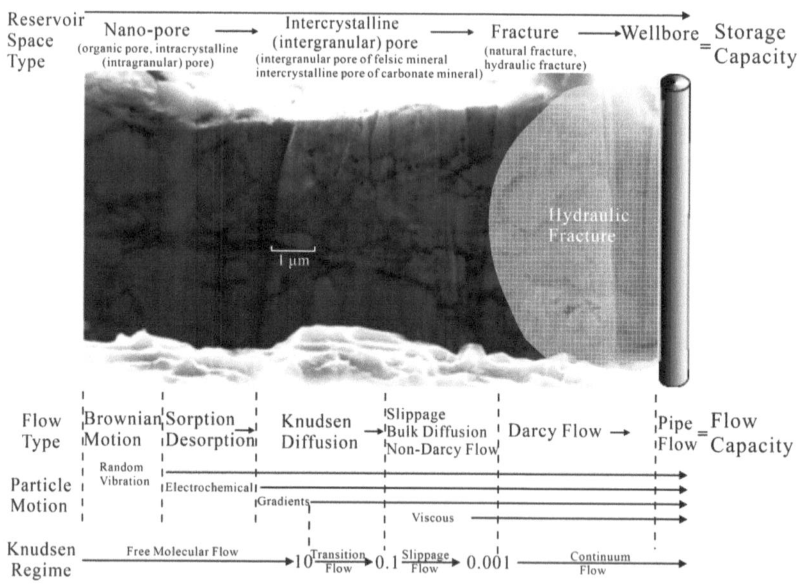

Figure 12. Different reservoir space types and flow characteristics of the shale gas reservoir (the figure was modified after [81]).

Similar to the seepage characteristics of shale gas, shale oil of the upper fourth member and the lower third member of the Shahejie Formation also has modes of multi-scale migration and accumulation (Figure 13). For shale in the oil-generation window, the pyrolysis of kerogen generates bitumen under high temperatures and pressures (Figure 13a), which exists in organic pores in the free state or the dissolved state (Figure 13b). The pore fluid pressure generated by hydrocarbon generation drives bitumen in the free state to migrate into the boundary pores of organic matter and intercrystalline (intergranular) pores mainly according to the migration model of subcritical flow or into hydrocarbon expulsion fractures according to the migration model of instantaneous flow. Secondly, it migrates into nearby reservoir space from the inside of the kerogen by diffusion and is then absorbed on the surfaces of kerogen or inorganic minerals. The bitumen in the free state overcomes the displacement pressure (the sum of capillary pressure and adsorption resistance) as a result of the pore fluid pressure and migrates into the intercrystalline pores by the non-Darcy flow. Intercrystalline pores are found in quartz, calcite and pyrite, clay, feldspar and dolomite (Figure 13c–h). Alternatively, it accumulates in hydrocarbon expulsion fractures by Darcy flow (Figure 13i). The bitumen gradually migrates into the larger-scale intercrystalline pores or fractures and eventually accumulates in bedding and structural fractures (Figure 13j–l); then, it follows the vein-type model of the multi-scale progressive transport network (Figure 13). The displacement pressure of the bitumen shows anisotropy owing to the heterogeneity of the shale fabric. The bitumen always migrates along the path of minimum resistance driven by pore fluid pressure. Although not all of the reservoir space is filled with oil in the pore network, a dominant migration pathway is present with the optimal transport structure obeying Murray's law [82]. That is, at least one pore and throat network at a certain scale forms the most connected pathway. This indicates that the minimum resistance needs to be overcome during the migration and accumulation of bitumen in the optimal pathway, which is key for high-efficiency fluid seepage.

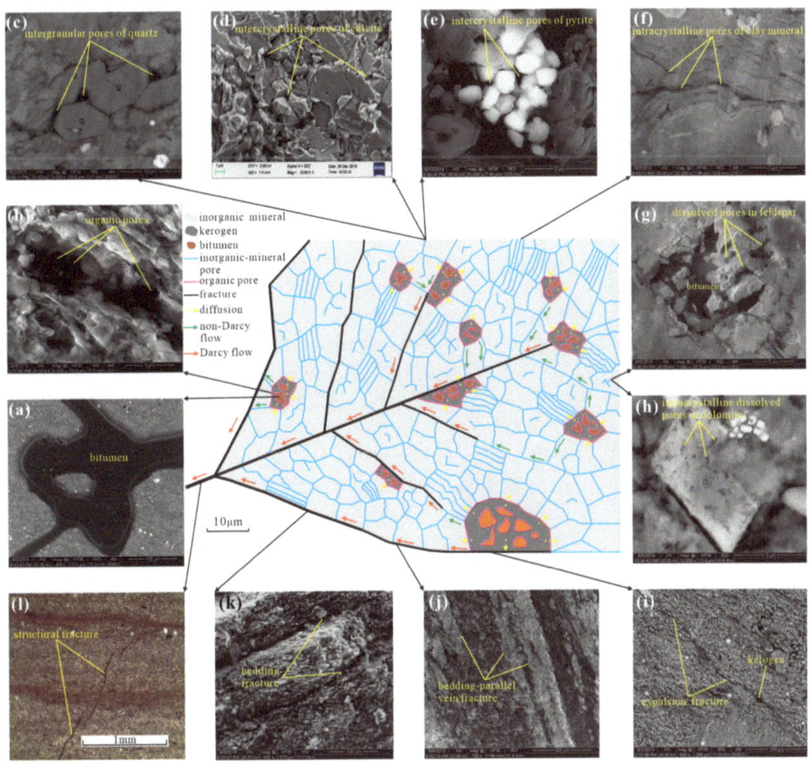

Figure 13. Vein-type model of multi-scale progressive transport network in shale oil reservoir. Except for (**l**), the images were obtained by FE-SEM. (**a**) Bitumen is formed by hydrocarbon generation and

migrates from the sample; L69 well, 3053.20 m. The experimental temperature and pressure of the sample were 200 °C and 20 MPa, respectively. (**b**) Organic pores developed in the kerogen; L69 well, 3053.20 m. (**c**) Intergranular pores located in the framework of quartz; L69 well, 3053.20 m. (**d**) Intercrystalline pores located in the framework of calcite; L69 well, 3694.03.60 m. (**e**) Intercrystalline pores located in the framework of pyrite; L69 well, 3048.10 m. (**f**) Intracrystalline pores developed in the clay minerals; L69 well, 3053.20 m. (**g**) Dissolved pores developed in the feldspar; L69 well, 3053.20 m. The experimental temperature and pressure of the sample were 200 °C and 20 MPa, respectively. (**h**) Intracrystalline dissolved pores developed in the dolomite; L69 well, 3059.35 m. The experimental temperature and pressure of the sample were 600 °C and 1 MPa, respectively. (**i**) Expulsion fracture connected with kerogen; L69 well, 3053.20 m. (**j**) Bedding-parallel vein fractures filled with calcite developed in the shale; L69 well, 3048.10 m. (**k**) Bedding fractures developed in the shale; well L69, 3059.35 m. (**l**) Structural fractures filled with bitumen in the shale in polarizing microscopy image; 25×(−), FY1 well, 3168.99 m.

5.4. Reservoir Model of Shale Oil

Different lithofacies within the studied sections have different potentials for oil generation, reservoir and sealing, and play different roles in the source–reservoir–caprock unit. Lithofacies in the early–middle diagenesis stage are determined mainly by the sedimentary environment; therefore, this is the basis of the shale oil reservoir. Hydrocarbon generation potential is key for the presence of shale oil, for which organic matter quantity, quality and maturity are all requirements. Usually, organic-rich lithofacies develop mainly in the condensed section (CS) of an LTST and an LHST in a humid climate or during transition periods between dry hot and humid climates. Then, they are deposited in still, brackish, strong reducing and stratified water [83–85] and are distributed in low-lying areas of paleotopography adjoining carbonate-rich lithofacies such as clayey limestone. Owing to the uneven paleotopography of the continental lacustrine basin, many local sags and salients have formed several microenvironments, which led to a dispersed distribution of oil shale showing quick lateral phase changes and strong plain heterogeneity.

When the lake level changes, the microenvironments migrate to cause oil shale evolution into carbonate-rich lithofacies. In addition, a lacustrine area with a relatively small area and relatively shallow water depth is sensitively affected by fault activity, climate change and catastrophic flood, and results in thin layers and vertical heterogeneity of the oil shale. Oil shale is spatially surrounded by carbonate-rich lithofacies because of the diversity of the sedimentary environment, which manifests as a sandwich-type strata structure on the profile and a banded shape (or potato type) in the plane view. This structure forms a self-sealing trap environment similar to a lenticular sand trap. Therefore, a shale oil reservoir belongs to a special lithologic trap dominated by shale lithofacies (Figure 14). Because the caprock of the trap is mainly constructed and supported by carbonate-rich lithofacies with high strength and poor physical properties, a shale oil reservoir has strong resistance to tectonic stress and excellent sealing capability, which is favorable for the preservation of shale oil.

Kerogen generates a large amount of organic acid and hydrocarbon from oil shale in the trap, which not only forms organic pores, intercrystalline (intergranular) dissolved pores and intracrystalline (intragranular) dissolved pores but also generates fluid overpressure to form hydrocarbon expulsion fractures or bedding-parallel vein fractures filled with calcite by coupling with diagenesis. Both the pores and fractures can reserve bitumen. Moreover, the sealing trap prevents bitumen in the free state from migrating outside. Then, the bitumen in the trap can choose the pathway with minimum displacement pressure to gradually migrate and accumulate according to the migration pattern of subcritical flow and instantaneous flow driven by the pore fluid pressure (Figure 14).

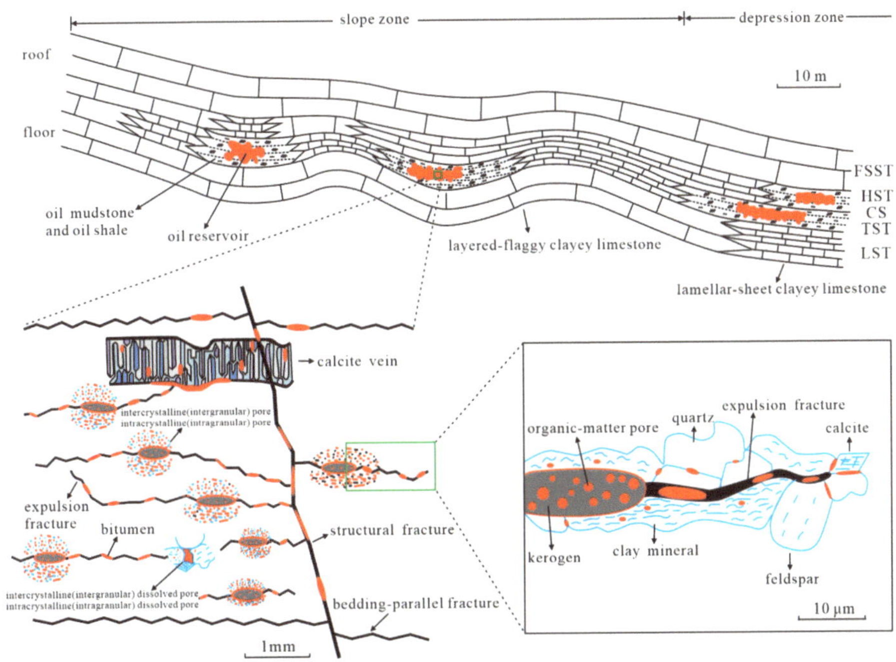

Figure 14. Discrete model of lacustrine shale oil reservoir. FSST: Falling-stage systems tract; HST: Highstand systems tract; CS: Condensed section; TST: Transgressive systems tract; LST: Lowstand systems tract.

Pore fluid overpressure is formed by hydrocarbon generation in the sealing trap, which provides the driving force for bitumen to enter nearby boundary pores of organic matter, intercrystalline (intergranular) pores and hydrocarbon expulsion fractures, and then gradually migrate into the larger scale of the reservoir space including intercrystalline (intergranular) pores, bedding-parallel vein fractures, bedding fractures, and structural fractures. Excessive pore fluid pressure will cause cracking in the sealing caprock, such as clayey limestone. This massive dissipation of bitumen prevents accumulation of the shale oil.

Overall, the shale reservoir is a special lithologic reservoir controlled by a sedimentary environment and has self-generation, self-storage and self-sealing characteristics with developed pore fluid overpressure and a multi-scale transport network (Figure 14). This model presents a positive guide for exploring shale oil in continental lacustrine basins according to five aspects. First, the paleoclimate origin analysis of the oil shale indicates that shale formed mainly in the CS of the LTST and LHST. Second, the sedimentary environment analysis shows that lacustrine shale was deposited in low-lying areas of paleotopography in slope zones and depression zones. Variations in the sedimentary environment led to the development of complex lithofacies and quick phase change with strong macroscopic heterogeneity, which resulted in thin oil layers with quick horizontal change, limited oil accumulation and discrete distribution in the plane. Third, sedimentary evolution created a series of sandwich-type source–reservoir–caprock assemblages with interbedded hard and soft lithofacies. These assemblages have resisted tectonic stress to form a stable sealing trap, which is favorable for the accumulation and storage of bitumen and the development of pore fluid overpressure, which perfectly matches the overpressure characteristics of the shale reservoir of the upper fourth member and the lower third member of the Shahejie Formation in Dongying and Zhanhua Sags. Fourth, the characteristics of the vein-type model of the multi-scale progressive transport network of shale oil indicate strong microscopic heterogeneity and the existence of an optimal migration pathway inside the reservoir. In addition, the maximum connected pore throat radius is key for the seepage of bitumen in the free state in the optimal migration pathway. The large scale of intercrystalline (intergranular) pores and fractures is important for creating reservoir space. In particular, hydrocarbon expulsion fractures formed by hydrocarbon generation and pressurization of kerogen pyrolysis are filled with bitumen near kerogen, which is a particularly favorable reservoir space of continental lacustrine shale. Fifth, based on the self-generation, self-storage and self-sealing characteristics as well as the multi-scale transport network, shale lithofacies are key elements of shale oil reservoirs. When a favorable target evaluation of a shale oil reservoir is conducted, shale lithofacies should be taken as the basic evaluation unit and evaluation of the source qualities, reservoir properties, alteration, fluidity and preservation should be considered to fully reflect the characteristics of the shale oil reservoir. Through systematic analysis of the influencing factors including source quality, reservoir property, alteration, fluidity and preservation in different shale lithofacies, sensitive parameters are optimally selected, and the gray–fuzzy mathematic method is used for comprehensive classification evaluation, which is an effective study method based on the geological understanding of the shale oil reservoir.

A discrete shale oil reservoir model can better explain the single well development characteristics of lacustrine shale oil reservoirs with low initial production, fast decline, and low recovery. The initial monthly oil production of the Y176 vertical well was 24 t. After four months, the production decreased rapidly to 2 t. LYP1-2 horizontal well had similar results, although the decline in the production was lower. Both showed the production characteristics of fractured reservoirs (Figure 15). Hydraulic fracturing can increase the volume of hydraulic fractures in shale reservoirs, connect natural fractures, increase the contact area between fractures and matrix pores, and help the oil in the matrix pores to enter the fractures, thereby increasing production. For example, the initial monthly production of the Y183 vertical well was 7.4 t; after hydraulic fracturing, the monthly production increased to 249.3 t. After fracturing, shale reservoirs showed characteristics of gradual decline as production proceeded, which is similar to that exhibited by fractured reservoirs. For example, the monthly oil production of the Y183 vertical well decreased to 58.9 t after four years of production.

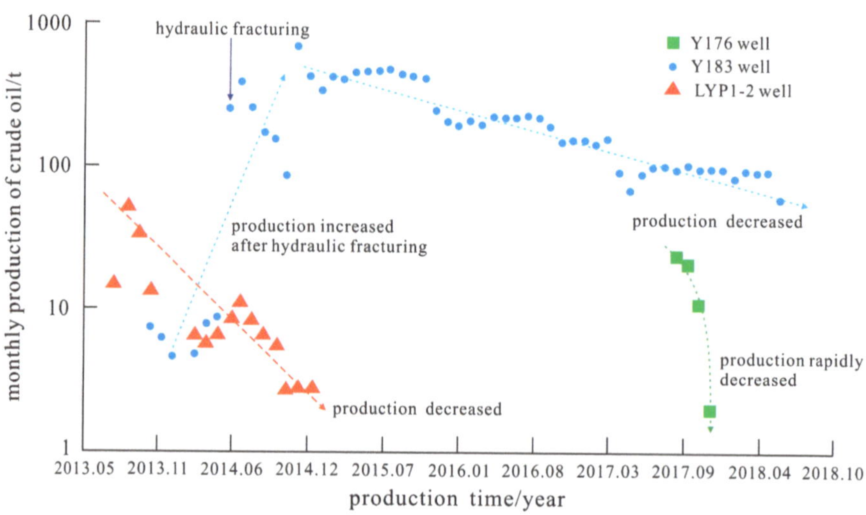

Figure 15. Single well development characteristics of lacustrine shale oil reservoir.

6. Conclusions

(1) The shale oil reservoir has self-generation, self-storage and self-sealing characteristics. The oil shale is rich in organic matter and carbonate minerals, which indicates an excellent hydrocarbon generation index, belongs to high-quality source rock, and has self-generation attributes. Many types of reservoir space are developed including organic pores, inorganic-mineral pores and fractures, which indicate self-storage capacity. Clayey limestone as the roof and floor with interbedded oil shale forms a sandwich-type source–reservoir–caprock assemblage and relatively good sealing space. This assemblage has strong resistance to tectonic stress and is favorable for bitumen storage, which results in the development of pore fluid overpressure.

(2) The shale oil reservoir has characteristics of a multi-scale progressive transport and accumulation network. Bitumen in the free state is stored mainly in organic pores, intercrystalline (intergranular) pores and fractures, and is secondarily distributed in intragranular pores of clay minerals by adsorption. The resistances of bitumen migration include capillary pressure and adsorption resistance, whereas the driving force is mainly pore fluid pressure. The migration patterns of bitumen include subcritical flow and instantaneous flow. Similar to the multi-scale seepage law of shale gas, the vein-type model of the multi-scale progressive transport network is developed because of the multi-scale characteristics of the reservoir space in the shale oil reservoir, where a dominant migration pathway exists.

(3) The shale oil reservoir is a special lithologic reservoir controlled by a sedimentary environment and has self-generation, self-storage and self-sealing characteristics with developed pore fluid overpressure and a multi-scale transport and accumulation network. Shale lithofacies are the main component of the shale oil reservoir, and the sedimentary environment forms the formation foundation. Lacustrine shale lithofacies have different oil potential, reservoir and sealing properties as well as a thin interbedded strata structure caused by sedimentary environment and sedimentary evolution. These conditions generated a sandwich-type source–reservoir–caprock assemblage and formed a special lithologic trap similar to a sandstone lens. In the closed trap environment, a certain pore fluid pressure will drive the multi-scale migration and accumulation of bitumen, but excessive pore fluid pressure will destroy the roof and floor of the shale oil reservoir, which interferes with the bitumen enrichment.

Author Contributions: Conceptualization, C.M. and X.F.; methodology, C.M.; software, X.F.; validation, C.M. and X.F.; formal analysis, X.M.; investigation, X.M.; resources, X.L.; data curation, B.X.; writing—original draft preparation, C.M.; writing—review and editing, X.F.; visualization, X.G.; supervision, C.M.; project administration, C.M.; funding acquisition, C.M. All authors have read and agreed to the published version of the manuscript.

Funding: This research was funded by Natural Science Foundation of China (Grant No. 42172153).

Institutional Review Board Statement: Not applicable.

Informed Consent Statement: Not applicable.

Data Availability Statement: The data presented in this study are available on request from the corresponding author. The data are not publicly available because some of the data involves trade secrets of oilfield companies and cannot be widely disclosed.

Acknowledgments: The authors thank Qiang Jin of China University of Petroleum (East China) for his constructive suggestions on the design of the thermal simulation experiments.

Conflicts of Interest: The authors declare no conflict of interest.

References

1. Jin, Q.; Zhu, G. Progress in research of deposition of oil source rocks in saline lakes and their hydrocarbon generation. *Geol. J. China Univ.* **2006**, *12*, 483–492. [CrossRef]
2. Shi, J.; Jin, Z.; Liu, Q.; Huang, Z. Depositional process and astronomical forcing model of lacustrine fine-grained sedimentary rocks: A case study of the early Paleogene in the Dongying Sag, Bohai Bay Basin. *Mar. Pet. Geol.* **2020**, *113*, 103995. [CrossRef]

3. Zou, C.; Tao, S.; Hou, L. *Unconventional Petroleum Geology*; Geological Publishing House: Beijing, China, 2011; pp. 1–374.
4. Zou, C.; Zhu, R.; Bai, B.; Yang, Z.; Hou, L.; Zha, M.; Fu, J.; Shao, Y.; Liu, K.; Cao, H.; et al. Significance, geologic characteristics, resource potential and future challenges of tight oil and shale oil. *Bull. Mineral. Petrol. Geochem.* **2015**, *34*, 3–17. [CrossRef]
5. Montgomery, S.L.; Jarvie, D.M.; Bowker, K.A.; Pollastro, R.M. Mississippian Barnett Shale, Fort Worth Basin, north-central Texas: Gas-shale play with multi-trillion cubic foot potential. *AAPG Bull.* **2005**, *89*, 155–175. [CrossRef]
6. Liu, H.; Yu, B.; Xie, Z.; Han, S.; Shen, Z.; Bai, C. Characteristics and implications of micro-lithofacies in lacustrine-basin organic-rich shale: A case study of Jiyang Depression, Bohai Bay Basin. *Acta Pet. Sin.* **2018**, *39*, 1328–1343. [CrossRef]
7. Zhao, X.Z.; Zhou, L.H.; Pu, X.G.; Jin, F.M.; Han, W.Z.; Xiao, D.Q.; Chen, S.Y.; Shi, Z.N.; Zhang, W.; Yang, F. Geological characteristics of shale rock system and shale oil exploration breakthrough in a lacustrine basin: A case study from the Paleogene 1st Sub-member of Kong 2 Member in Cangdong Sag, Bohai Bay Basin, China. *Pet. Explor. Dev.* **2018**, *45*, 3–14. [CrossRef]
8. Zhou, L.; Pu, X.; Xiao, D.; Li, H.; Guan, Q.; Lin, L.; Qu, N. Geological conditions for shale oil formation and the main controlling factors for the enrichment of the 2nd Member of Kongdian Formation in the Cangdong Sag, Bohai Bay Basin. *Nat. Gas Geosci.* **2018**, *29*, 1323–1332. [CrossRef]
9. Chen, X.; Yan, Y.; Zhang, X.; Liu, H.; Xue, X.; Luo, X. Generation conditions of continental shale gas in Biyang sag, Nanxiang Basin. *Pet. Geol. Exp.* **2011**, *33*, 137–142. [CrossRef]
10. Li, J.; Shi, Y.; Zhang, X.; Chen, X.; Yan, Y.; Zhu, J.; Lu, S.; Wang, M. Control factors of enrichment and producibility of shale oil: A case study of Biyang Depression. *Earth Sci.* **2014**, *39*, 848–857. [CrossRef]
11. Liu, B.; Lv, Y.; Ran, Q.; Dai, C.; Li, M.; Wang, M. Geological conditions and exploration potential of shale oil in Qingshankou Formation, northern Songliao Basin. *Oil Gas Geol.* **2014**, *35*, 280–285.
12. Huang, W.; Deng, S.; Lu, S.; Yu, L.; Hu, S.; Zhang, J. Shale organic heterogeneity evaluation method and its application to shale oil resource evaluation—A case study from Qingshankou Formation, southern Songliao Basin. *Oil Gas Geol.* **2014**, *35*, 704–711. [CrossRef]
13. Liu, B.; Lv, Y.; Zhao, R.; Tu, X.; Guo, X.; Shen, Y. Formation overpressure and shale oil enrichment in the shale system of Lucaogou Formation, Malang Sag, Santanghu Basin, nw China. *Pet. Explor. Dev.* **2012**, *39*, 744–750. [CrossRef]
14. Liu, B.; Chi, Y.; Huang, Z.; Luo, Q.; Wu, H.; Chen, X.; Shen, Y. Migration mechanism of the Permian hydrocarbon and shale oil accumulation in Malang Sag, the Santanghu Basin. *Oil Gas Geol.* **2013**, *34*, 725–730. [CrossRef]
15. Liu, B.; Lv, Y.; Meng, Y.; Li, X.; Guo, X.; Ma, Q.; Zhao, W. Petrologic characteristics and genetic model of lacustrine lamellar fine-grained rock and its significance for shale oil exploration: A case study of Permian Lucaogou Formation in Malang Sag, Santanghu Basin, nw China. *Pet. Explor. Dev.* **2015**, *42*, 656–666. [CrossRef]
16. Qiu, Z.; Shi, Z.; Dong, D.; Lu, B.; Zhang, C.; Zhou, J.; Wang, H.; Xiong, B.; Pang, Z.; Guo, H. Geological characteristics of source rock and reservoir of tight oil and its accumulation mechanism: A case study of Permian Lucaogou Formation in Jimusar Sag, Junggar Basin. *Pet. Explor. Dev.* **2016**, *43*, 1013–1024. [CrossRef]
17. Song, Y.; Zhou, L.; Guo, X.; Chang, Q.S.; Wang, X. Characteristics and occurrence of lacustrine dolomitic tight-oil reservoir in the middle Permian Lucaogou Formation, Jimusaer Sag, southeastern Junggar Basin. *Acta Petrol. Sin.* **2017**, *33*, 1159–1170.
18. Li, S.; Pang, X.; Li, M.; Jin, Z. Geochemistry of petroleum systems in the Niuzhuang south slope of Bohai Bay Basin—Part 1: Source rock characterization. *Org. Geochem.* **2003**, *34*, 389–412. [CrossRef]
19. Lewan, M.D.; Roy, S. Role of water in hydrocarbon generation from type-I kerogen in mahogany oil shale of the Green River Formation. *Org. Geochem.* **2011**, *42*, 31–41. [CrossRef]
20. Wang, M.; Lu, S.; Xue, H. Kinetic simulation of hydrocarbon generation from lacustrine type I kerogen from the Songliao Basin: Model comparison and geological application. *Mar. Pet. Geol.* **2011**, *28*, 1714–1726.
21. Burton, D.; Woolf, K.; Sullivan, B. Lacustrine depositional environments in the Green River Formation, Uinta Basin: Expression in outcrop and wireline logs lacustrine depositional environments, Uinta Basin. *AAPG Bull.* **2014**, *98*, 1699–1715. [CrossRef]
22. Loucks, R.G.; Reed, R.M.; Ruppel, S.C.; Jarvie, D.M. Morphology, genesis, and distribution of nanometer-scale pores in siliceous mudstones of the Mississippian Barnett Shale. *J. Sediment. Res.* **2009**, *79*, 848–861. [CrossRef]
23. Loucks, R.G.; Reed, R.M.; Ruppel, S.C.; Hammes, U. Spectrum of pore types and networks in mudrocks and a descriptive classification for matrix-related mudrock pores. *AAPG Bull.* **2012**, *96*, 1071–1098. [CrossRef]
24. Slatt, R.M.; O'Brien, N.R. Pore types in the Barnett and Woodford gas shales: Contribution to understanding gas storage and migration pathways in fine-grained rocks. *AAPG Bull.* **2011**, *95*, 2017–2030. [CrossRef]
25. Chalmers, G.R.; Bustin, R.M.; Power, I.M. Characterization of gas shale pore systems by porosimetry, pycnometry, surface area, and field emission scanning electron microscopy/transmission electron microscopy image analyses: Examples from the Barnett, Woodford, Haynesville, Marcellus, and Doig units. *AAPG Bull.* **2012**, *96*, 1099–1119. [CrossRef]
26. Javadpour, F.; Moravvej Farshi, M.; Amrein, M. Atomic-force microscopy: A new tool for gas-shale characterization. *J. Can. Pet. Technol.* **2012**, *51*, 236–243. [CrossRef]
27. Clarkson, C.R.; Solano, N.; Bustin, R.M.; Bustin, A.M.M.; Chalmers, G.R.L.; He, L.; Melnichenko, Y.B.; Radliński, A.P.; Blach, T.P. Pore structure characterization of North American shale gas reservoirs using USANS/SANS, gas adsorption, and mercury intrusion. *Fuel* **2013**, *103*, 606–616. [CrossRef]
28. Wang, Y.; Pu, J.; Wang, L.; Wang, J.; Jiang, Z.; Song, Y.; Wang, C.; Wang, Y.; Jin, C. Characterization of typical 3D pore networks of Jiulaodong formation shale using nano-transmission X-ray microscopy. *Fuel* **2016**, *170*, 84–91. [CrossRef]

29. Liu, S.; Sang, S.; Wang, G.; Ma, J.; Wang, X.; Wang, W.; Du, Y.; Wang, T. FIB-SEM and X-ray CT characterization of interconnected pores in high-rank coal formed from regional metamorphism. *J. Pet. Sci. Eng.* **2017**, *148*, 21–31. [CrossRef]
30. Desbois, G.; Urai, J.L.; Kukla, P.A.; Konstanty, J.; Baerle, C. High-resolution 3D fabric and porosity model in a tight gas sandstone reservoir: A new approach to investigate microstructures from mm-to nm-scale combining argon beam cross-sectioning and SEM imaging. *J. Pet. Sci. Eng.* **2011**, *78*, 243–257. [CrossRef]
31. Golab, A.; Ward, C.R.; Permana, A.; Lennox, P.; Botha, P. High-resolution three-dimensional imaging of coal using microfocus X-ray computed tomography, with special reference to modes of mineral occurrence. *Int. J. Coal Geol.* **2013**, *113*, 97–108. [CrossRef]
32. Tang, X.; Jiang, Z.; Jiang, S.; Li, Z. Heterogeneous nanoporosity of the Silurian Longmaxi Formation shale gas reservoir in the Sichuan Basin using the QEMSCAN, FIB-SEM, and nano-CT methods. *Mar. Pet. Geol.* **2016**, *78*, 99–109. [CrossRef]
33. Takbiri-Borujeni, A.; Fathi, E.; Kazemi, M.; Belyadi, F. An integrated multiscale model for gas storage and transport in shale reservoirs. *Fuel* **2019**, *237*, 1228–1243. [CrossRef]
34. Hu, Q.; Ewing, R.P.; Rowe, H.D. Low nanopore connectivity limits gas production in Barnett Formation. *J. Geophys. Res. Solid Earth* **2015**, *120*, 8073–8087. [CrossRef]
35. Ghanbarian, B.; Javadpour, F. Upscaling pore pressure-dependent gas permeability in shales. *J. Geophys. Res. Solid Earth* **2017**, *122*, 2541–2552. [CrossRef]
36. Javadpour, F.; Fisher, D.; Unsworth, M. Nanoscale gas flow in shale gas sediments. *J. Can. Pet. Technol.* **2007**, *46*, 55–61. [CrossRef]
37. Deng, J.; Zhu, W.; Ma, Q. A new seepage model for shale gas reservoir and productivity analysis of fractured well. *Fuel* **2014**, *124*, 232–240. [CrossRef]
38. Li, Y.; Jiang, Y.; Zhao, J.; Liu, C.; Zhang, L. Extended finite element method for analysis of multi-scale flow in fractured shale gas reservoirs. *Environ. Earth Sci.* **2015**, *73*, 6035–6045. [CrossRef]
39. Figueroa Pilz, F.; Dowey, P.J.; Fauchille, A.L.; Courtois, L.; Bay, B.; Ma, L.; Lee, P.D. Synchrotron tomographic quantification of strain and fracture during simulated thermal maturation of an organic-rich shale UK Kimmeridge Clay: Journal of Geophysical Research. *Solid Earth* **2017**, *122*, 2553–2564. [CrossRef]
40. O'Brien, N.R.; Cremer, M.D.; Canales, D.G. The role of argillaceous rock fabric in primary migration of oil. *Gulf Coast Assoc. Geol. Soc. Trans.* **2002**, *52*, 1103–1112.
41. Martin, R.; Baihly, J.; Malpani, R.; Lindsay, G.; Atwood, W.K. Understanding production from Eagle Ford-Austin chalk system. In Proceedings of the SPE Annual Technical Conference and Exhibition, Denver, CO, USA, 30 October–2 November 2011.
42. Jarvie, D.M. Shale resource systems for oil and gas: Part 2—Shale-oil resource systems. In *Shale Reservoirs–Giant Resources for the 21st Century*; AAPG: Tulsa, OK, USA, 2012; pp. 89–119. [CrossRef]
43. Gong, X.; Tian, Y.; McVay, D.A.; Ayers, W.B.; Lee, W.J. Assessment of Eagle Ford Shale oil and gas resources. In *SPE Unconventional Resources Conference Canada*; Society of Petroleum Engineers: Calgary, AB, Canada, 2013.
44. Ning, F. The main control factors of shale oil enrichment in Jiyang depression. *Acta Pet. Sin.* **2015**, *36*, 905–914. [CrossRef]
45. Shao, L.; Zhang, P.; Hilton, J.; Gayer, R.; Wang, Y.; Zhao, C.; Luo, Z. Paleoenvironments and paleogeography of the Lower and lower Middle Jurassic coal measures in the Turpan-Hami oil-prone coal basin, northwestern China. *AAPG Bull.* **2003**, *87*, 335–355. [CrossRef]
46. Meng, M.; Ge, H.; Shen, Y.; Ji, W.; Li, Z. Insight into water occurrence and pore size distribution by nuclear magnetic resonance in marine shale reservoirs, southern China. *Energy Fuels* **2023**, *37*, 319–327. [CrossRef]
47. Cao, Y.; Jiang, Z. Relationship between hydrocarbon reservoir and genetic types of sequence boundary in rift lacustrine basin. *J. China Univ. Pet.* **2004**, *28*, 1–6.
48. Yang, W.; Jiang, Y.; Wang, Y. Study on shale facies sedimentary environment of lower Es3-upper Es4 in Dongying sag. *J. China Univ. Pet.* **2015**, *39*, 19–26. [CrossRef]
49. Ma, C.; Dong, C.; Luan, G.; Lin, C.; Liu, X.; Elsworth, D. Type, characteristics and effect of natural fluid pressure fractures in shale: A case study of the Paleogene strata in Eastern China. *Pet. Explor. Dev.* **2016**, *44*, 580–589. [CrossRef]
50. Chu, W. Research on the development of shale oil and the fluctuation of international crude oil price. In Proceedings of the 2nd International Conference on Shale Oil Resources and Development Technologies, Beijing, China, 28 April 2019.
51. Li, D.; Liu, Z.; Zhang, G.; Zheng, Z.; Jia, J.; Gao, Y.; Zan, X. Comparison and revelation of tight oil accumulation conditions, distribution characteristics and development status between China and U.S. *Nat. Gas Geosci.* **2017**, *28*, 1126–1138.
52. Birdwell, J.E.; Vanden Berg, M.D.; Johnson, R.C.; Mercier, T.J.; Boehlke, A.; Brownfield, M.E. Geological, geochemical and reservoir characterization of the Uteland Butte member of the Green River Formation, Uinta Basin, Utah. In *Hydrocarbon Source Rocks in Unconventional Plays, Rocky Mountain Region*; Dolan, M.P., Higley, D.K., Lillis, P.G., Eds.; Rocky Mountain Association of Geologists: Denver, CO, USA, 2016; Chapter 12; pp. 352–378.
53. Li, M.Y.; Zhu, R.K.; Hu, S.Y. Geological characteristics and resource potential of overseas terrestrial shale oil. *Lithol. Reserv.* **2022**, *34*, 163–174. [CrossRef]
54. Meng, M.; Peng, J.; Ge, H.; Ji, W.; Li, X.; Wang, Q. Rock fabric of lacustrine shale and its influence on residual oil distribution in the Upper Cretaceous Qingshankou Formation, Songliao Basin. *Energy Fuels* **2023**, *37*, 7151–7160. [CrossRef]
55. Gall, R.D.; Birdwell, J.E.; Brinkerhoff, R.; Vanden Berg, M.D. Geologic characterization and depositional history of the Uteland Butte Member, Green River Formation, southwestern Uinta Basin, Utah in The lacustrine Green River Formation: Hydrocarbon potential and Eocene climate record. *Utah Geol. Assoc.* **2022**, *50*, 37–62. [CrossRef]

56. Hou, D.; Zhang, S.; Xiao, J. The excellent source rocks and accumulation of stratigraphic and lithologic traps in the Jiyang depression, Bohai bay basin, China. *Earth Sci. Front.* **2008**, *15*, 137–146.
57. Zhu, G.; Jin, Q. Geochemical characteristics of two sets of excellent source rocks in Dongying Depression. *Acta Sedimentol. Sinica* **2003**, *21*, 506–512. [CrossRef]
58. Kirkland, D.W.; Evans, R. Source-rock potential of evaporitic environment. *AAPG Bull.* **1981**, *65*, 181–190. [CrossRef]
59. Katz, B.J.; Bissada, K.K.; Wood, J.W. Factors limiting potential of evaporites ashydrocarbon source rocks. *SEG Tech. Program Expand. Abstr.* **1987**, *6*, 923. [CrossRef]
60. Sun, Z.C.; Yang, F.; Zhang, Z.H.; Li, S.J.; Li, D.M.; Peng, L.C.; Zeng, X.L.; Xu, Y.L.; Mao, S.Z.; Wang, Q. *Sedimentary Environment and Oil-Gas Generation of CenozoicSaline Lakes in China*; Petroleum Industry Press: Beijing, China, 1997; pp. 1–338.
61. Jin, Q.; Zhu, G.Y.; Wang, J. Deposition and distribution of high-potential sourcerocks in saline lacustrine environments. *J. China Univ. Pet.* **2008**, *32*, 19–23.
62. Liu, C.Y.; Zhao, J.F.; Ma, Y.P.; Xiong, L.F.; Chen, J.J.; Mao, G.Z.; Zhang, D.D.; Deng, Y. The advances and problems in the study of the characteristics and formation ofhydrocarbon-rich sag. *Earth Sci. Front.* **2014**, *21*, 75–88.
63. Guo, P.; Liu, C.Y.; Gibert, L.; Huang, L.; Zhang, D.W.; Dai, J. How to find high-quality petroleum source rocks in saline lacustrine basins: A case study from the Cenozoic Qaidam Basin, NW China. *Mar. Pet. Geol.* **2020**, *111*, 603–623. [CrossRef]
64. Tang, L.; Song, Y.; Pang, X.P.; Jiang, Z.X.; Guo, Y.C.; Zhang, H.G.; Pan, Z.H.; Jiang, H. Effects of paleo sedimentary environment in saline lacustrine basin on organic matter accumulation and preservation: A case study from the Dongpu Depression, Bohai Bay Basin, China. *J. Pet. Sci. Eng.* **2020**, *185*, 106669. [CrossRef]
65. Ma, C.; Dong, C.; Luan, G.; Lin, C.; Liu, X.; Duan, H.; Liu, S. Characteristics and influencing factors of organic matter pores in paleogene shale, Subei Basin. *J. China Univ. Pet.* **2017**, *41*, 1–13. [CrossRef]
66. Wang, G.; Liu, H.; Xiong, Z.; Wang, J. Control condition of felsic on brittleness of lacustrine shales. *J. China Univ. Pet.* **2016**, *40*, 1–8. [CrossRef]
67. Tian, H.; Zhang, S.; Liu, S.; Zhang, H. Determination of organic-rich shale pore features by mercury injection and gas adsorption methods. *Acta Pet. Sin.* **2012**, *33*, 419–427. [CrossRef]
68. Cardott, B.J.; Landis, C.R.; Curtis, M.E. Post-oil solid bitumen network in the Woodford Shale, USA—A potential primary migration pathway. *Int. J. Coal Geol.* **2015**, *139*, 106–113. [CrossRef]
69. Zhang, X.; Lu, X.; Zhang, L.; Liu, Q. Occurrences of shale gas and their petroleum geological significance. *Adv. Earth Sci.* **2010**, *25*, 597–604. [CrossRef]
70. Zhang, J.; Lin, L.; Li, Y.; Tang, X.; Zhu, L.; Xing, Y.; Jiang, S.; Jing, T.; Yang, S. Classification and evaluation of shale oil. *Earth Sci. Front.* **2012**, *19*, 322–331.
71. Lv, Y.; Zhang, S.; Wang, Y. Research of quantitative relations between sealing ability and thickness of cap rock. *Acta Pet. Sin.* **2000**, *21*, 27–30. [CrossRef]
72. Zhang, L.; Bao, Y.; Liu, Q.; Zhang, S.; Zhu, R.; Zhang, L. Effects of hydrocarbon physical properties on caprock's capillary sealing ability. *Sci. China Earth Sci.* **2010**, *53*, 665–671. [CrossRef]
73. Meng, M.; Ge, H.; Shen, Y.; Ji, W.; Wang, Q. Rock fabric of tight sandstone and its influence on irreducible water saturation in Eastern Ordos Basin. *Energy Fuels* **2023**, *37*, 3685–3696. [CrossRef]
74. Bredehoeft, J.D.; Wesley, J.B.; Fouch, T.D. Simulations of the origin of fluid pressure, fracture generation, and the movement of fluids in the Uinta Basin, Utah. *AAPG Bull.* **1994**, *78*, 1729–1747. [CrossRef]
75. Hao, S.; Liu, G.; Huang, Z.; Gao, Y. Simulation models for primary hydrocarbon migration. *Acta Pet. Sin.* **1994**, *15*, 21–31. [CrossRef]
76. Wang, X.; Zhou, D.; Wang, X. Intermittent fracturing migration of pore fluids-one of the major regime of primary migration of petroleum. *Pet. Explor. Dev.* **1994**, *21*, 20–26.
77. Kobchenko, M.; Panahi, H.; Renard, F.; Dysthe, D.K.; Malthe-Sørenssen, A.; Mazzini, A.; Scheibert, J.; Jamtveit, B.; Meakin, P. 4D imaging of fracturing in organic-rich shales during heating: Journal of Geophysical Research. *Solid Earth* **2011**, *116*, 1–9. [CrossRef]
78. Durand, B. Understanding of HC migration in sedimentary basins (present state of knowledge). *Org. Geochem.* **1988**, *13*, 445–459. [CrossRef]
79. Loucks, R.G.; Ruppel, S.C. Mississippian Barnett Shale: Lithofacies and depositional setting of a deep-water shale-gas succession in the Fort Worth Basin, Texas. *AAPG Bull.* **2007**, *91*, 579–601. [CrossRef]
80. Javadpour, F. Nanopores and apparent permeability of gas flow in mudrocks (shales and siltstone). *J. Can. Pet. Technol.* **2009**, *48*, 16–21. [CrossRef]
81. Sondergeld, C.H.; Newsham, K.E.; Comisky, J.T.; Rice, M.C.; Rai, C.S. Petrophysical considerations in evaluating and producing shale gas resources. In *SPE Unconventional Gas Conference*; SPE: Pittsburgh, PA, USA, 2010; p. 131768.
82. Murray, C.D. The physiological principle of minimum work I. The vascular system and the cost of blood volume. *Proc. Natl. Acad. Sci. USA* **1926**, *12*, 207–214. [CrossRef]
83. Deng, H.; Qian, K. *Analysis on Sedimentary Geochemistry and Environment*; Science Technology Press: Lanzhou, China, 1993; p. 154.

84. Loucks, R.G.; Reed, R.M. Scanning-electron-microscope petrographic evidence for distinguishing organic-matter pores associated with depositional organic matter versus migrated organic matter in mudrock. *Gulf Coast Assoc. Geol. Soc. J.* **2014**, 3, 51–60.
85. Shao, L.; Xiao, Z.; Lu, J.; He, Z.; Wang, H.; Zhang, P. Permo-Carboniferous coal measures in the Qinshui Basin: Lithofacies paleogeography and its control on coal accumulation. *Front. Earth Sci. China* **2007**, 1, 106–115. [CrossRef]

Journal of
Marine Science and Engineering

Article

Marine Controlled-Source Electromagnetic Data Denoising Method Using Symplectic Geometry Mode Decomposition

Yijie Chen [1,2,3,†], Zhenwei Guo [1,2,3,†] and Dawei Gao [1,2,3,*]

1 School of Geosciences and Info-Physics, Central South University, Changsha 410083, China
2 Hunan Key Laboratory of Non-Ferrous Resources and Geological Hazard Detection, Changsha 410083, China
3 Key Laboratory of Metallogenic Prediction of Nonferrous Metals and Geological Environment Monitoring, Ministry of Education, Changsha 410083, China
* Correspondence: 222057@csu.edu.cn
† These authors contributed equally to this work.

Abstract: The marine controlled-source electromagnetic (CSEM) method is an efficient tool for hydrocarbon exploration. The amplitudes of signals decay rapidly with the increasing offset, so signals are easily contaminated by various kinds of noise. A denoising method is critical to improve the data quality, but the diversity of noise makes denoising difficult. Specific frequency signals are transmitted for exploration requirements, and thus traditional filtering methods are not suitable. Symplectic geometry mode decomposition (SGMD), a new method to decompose signals, has an outstanding decomposition performance and noise robustness. Furthermore, it can reduce multiple types of noise by reconstructing the single components. In this study, we introduced SGMD to reduce the noise of marine CSEM data and improved the data quality significantly. The experiments show that SGMD is better than variational mode decomposition and the sym4 wavelet method.

Keywords: marine controlled-source electromagnetic method; denoising method; symplectic geometry mode decomposition

Citation: Chen, Y.; Guo, Z.; Gao, D. Marine Controlled-Source Electromagnetic Data Denoising Method Using Symplectic Geometry Mode Decomposition. *J. Mar. Sci. Eng.* **2023**, *11*, 1578. https://doi.org/10.3390/jmse11081578

Academic Editor: Rouseff Daniel

Received: 6 July 2023
Revised: 2 August 2023
Accepted: 9 August 2023
Published: 11 August 2023

1. Introduction

The marine controlled-source electromagnetic (CSEM) method is a technology applied for hydrocarbon exploration [1,2]. The resistivity information derived from the CSEM data can help distinguish reservoir type [3]. Compared with the marine seismic method, the marine CSEM method is economical and it is particularly sensitive to hydrocarbons below the seafloor [4,5]. As a result, the marine CSEM method has become an effective exploration method for seafloor resources in the past two decades [6].

In marine CSEM data acquisition systems, a horizontal electric dipole (HED) source towed over the seafloor transmits low frequency signals due to the good conductivity of seawater. The receivers are deployed on the seafloor and record the signals. With the increasing offset, the signals decay rapidly and are easily contaminated by noise, reducing the quality of the data, which is essential to inversion and interpretation. Therefore, denoising is an important procedure in the marine CSEM method.

For exploration requirements, specific frequency signals are used as the effective signals in the marine CSEM method. The adaptability of traditional filtering methods is weak, and the noise reduction effect is poor due to the frequency characteristic [7]. Myer proposed short-window stacking to reduce the low-frequency noise, which was verified in the field [8,9], but the spectral leakage from magnetotelluric signals and noise could not be avoided [10]. Methods using signal transformation include variational mode decomposition (VMD) and the wavelet denoising method [11,12]. The advantage of VMD is the noise robustness [13], but the performance is strictly subject to parameters such as the number of the decomposed modes and penalty terms. The wavelet denoising method analyzes the frequency components of signals and avoids the loss of their temporal information [14], but the lack of orthogonality and finite length for some common mother wavelets will affect the denoising results [15,16]. Airwaves dominate in shallow water and lead to misfitting of data; the denoising methods also decompose EM fields and process responses numerically, often dealing with this noise individually [17–20]. Compressive sensing was performed to remove the noise, allowing for a good denoising effect under complex conditions [21], but the strategy to select a suitable dictionary requires further research. To solve this problem, dictionary learning was introduced, which has a stronger correlation with signals compared with a common dictionary, thus achieving better denoising results [22]. Most of the methods mentioned above could achieve good results with a single type of noise. However, the results will be affected when the noise becomes complicated.

Symplectic geometry mode decomposition (SGMD) is a new signal analysis method proposed by Pan et al. [23]. SGMD can decompose strong noisy signals and effectively remove the noise. The core of the method is symplectic geometry similarity transformation, which is applied to solve the eigenvalues of the matrix and reconstruct the single component signals. During the whole decomposition, essential characteristics of the primary component are preserved, which makes it convenient for analyzing the complicated signals. Recently, SGMD has been developed in noise reduction [7,24,25], fault diagnosis [26–28] and modal identification [29]. In order to improve the decomposition results, researchers mainly focus on the criteria of reconstructing single components, including cosine similarity, graph similarity and cosine difference factors, which could enhance the applicability of decomposition algorithms [30–32].

In this paper, we introduced SGMD to marine CSEM data processing, and proposed using power spectral density (PSD) as the criteria for reconstructing single components to depress various types of noise simultaneously. Additionally, we used a Hampel filter to remove impulse noise, which cannot be separated by SGMD. By carrying out synthetic studies with different methods including VMD and the sym4 wavelet denoising method, we utilized forward responses, quantitative indicators and inversion results as the criteria for comparison, demonstrating that the denoising performance of SGMD is better than the other methods.

2. Methodology

2.1. Phase Space Reconstruction

In reality, the measured signal is a nonlinear time series. In order to extract useful information from the time series, two methods were proposed: the coordinate delay reconstruction method and the derivative reconstruction method [33]. Both can theoretically be used for phase-space reconstruction. However, the amplitudes of the signals in the marine CSEM method are weak and the derivatives are sensitive to errors. According to Takens theorem, phase space can be reconstructed by 1-dimensional signals using the coordinate delay reconstruction method, and contains all the dynamic information of the original signal. Therefore, we adopted the coordinate delay reconstruction method in this study.

The reconstruction method works by constructing an m-dimension phase space from different delay times of a 1-dimensional time series. A time series can be discretized as $x = x_1, x_2, \cdots x_n$, where n is the length of the signal. The embedding dimension as well as delay time are critical for reconstruction, and the influence of different parameters also varies. In this study, we adopted the strategy proposed by Bonizz for parameter selection [34]. For a time series x,

$$X = \begin{bmatrix} x_1 & x_{1+\tau} & \cdots & x_{1+(d-1)\tau} \\ \vdots & & \ddots & \vdots \\ x_m & x_{m+\tau} & \cdots & x_{m+(d-1)\tau} \end{bmatrix}, \quad (1)$$

where X is the phase space after reconstruction, d is the embedding dimension, τ is the delay time and $m = n - (d-1)\tau$.

2.2. Algorithm of SGMD

For a time series, the reconstructed phase space is shown as Equation (1).

The covariance symmetric matrix A can be obtained by autocorrelation analysis of the matrix X:

$$A = X^T X, \tag{2}$$

and A is also the Hamilton matrix.

Then, matrix M is constructed by matrix A:

$$M = \begin{bmatrix} A & 0 \\ 0 & -A^T \end{bmatrix}, \tag{3}$$

The square of matrix M is matrix N. According to the definition, both M and N are Hamilton matrices. For a symplectic orthogonal matrix Q, it can be constructed as:

$$Q^T N Q = \begin{bmatrix} B & R \\ 0 & B^T \end{bmatrix}, \tag{4}$$

where matrix B is the upper triangular matrix, transformed by Schmidt orthogonalization, and matrix R is any real matrix. The eigenvalues of B are equal to those of N, which are the square of the eigenvalues of A. The column vectors of matrix Q correspond to the eigenvectors of matrix A.

In fact, the eigenvalues distribution of the matrix A, solved by symplectic geometry similarity transformation, is the symplectic geometry spectra, and the small eigenvalues are often regarded as the noise. The symplectic matrix Q can be replaced with the Householder matrix with properties of protecting the Hamilton matrix from structural destruction. Then, the transformation coefficient matrix S and reconstruction matrix Z is expressed as:

$$S = Q^T X, \tag{5}$$

$$Z = Q\,S, \tag{6}$$

where the column vectors of matrix Z are the initial single components. By using diagonal averaging, these components can be transformed into a new matrix. The procedures are as follows.

Firstly, for each element of matrix Z, which is defined as z_{ij} $(1 \le i \le d, 1 \le j \le m)$, $d^* = \min(m, d)$ and $m^* = \max(m, d)$. For the transformed matrix Z^*, if $m < d$, then $z^*_{ij} = z_{ij}$; otherwise, $z^*_{ij} = z_{ji}$. For each column vector of matrix Z^*, a new series of Y $(y_1, y_2, \cdots y_n)$ can be obtained from:

$$y_k = \begin{cases} \frac{1}{k}\sum\limits_{p=1}^{k} z^*_{p,k-p+1} & 1 \le k < d^* \\ \frac{1}{d^*}\sum\limits_{p=1}^{d^*} z^*_{p,k-p+1} & d^* \le k \le m^* \\ \frac{1}{n-k+1}\sum\limits_{p=k-m^*+1}^{n-m^*+1} z^*_{p,k-p+1} & m^* < k \le n \end{cases}, \tag{7}$$

where n is the length of the original signal. Thus, a series of components Y $(Y_1, Y_2, \cdots Y_d)$ is obtained by diagonal averaging.

However, these components are not completely independent of one another, and the reconstruction of components with similar properties cannot be ignored. In SGMD, the period similarity is used to reconstruct the components, and the reconstruction iteration termination condition is set.

Due to the distribution of the principal components in the front of the matrix, the first component Y1 is compared with the rest of the components. To obtain the first decomposition component SGC_1, highly similar components are added together. Then, as the termination condition, the normalized mean square error (NMSE) between the sum of residuals and the original signal is calculated.

$$\mathrm{NMSE} = \frac{\sum_{i=1}^{n} \mathrm{res_i}}{\sum_{i=1}^{n} x_i} \tag{8}$$

where res is the residuals and x is the original signal. When NMSE is larger than the threshold, the components used in the last iteration are removed from matrix Y and the next iteration is performed on the rest of Y. Otherwise, the iteration ends. After that, the signal can be expressed as:

$$\mathrm{x(n)} = \sum_{i=1}^{N} \mathrm{SGC_i(n)} + \mathrm{res(n)}, \tag{9}$$

where N is the number of decomposition components.

Unfortunately, SGMD cannot remove the impulse noise, especially noise that follows a random distribution, which is common in marine CSEM data. Because the reconstructions are based on period similarity, seriously deviating points cannot be effectively separated. Considering this defect, we utilized a Hampel filter in this study.

2.3. Hampel Filter

Hampel filtering is an outlier removal technique that can be easily applied in practice [35]. It can successfully detect the outliers and protect other parts of the signal.

For a 1-dimensional time series $x = x_1, x_2, \cdots x_n$, assuming that the length of time window is 2k, k elements are placed on each side of the center element. Firstly, the median and standard deviation of the elements in this window are calculated. If the standard deviation is larger than the threshold, the center element will be replaced by the median. The threshold is determined by the amplitude ratio of the outliers to the original signal. A Hampel filter can remove impulse noise rapidly while keeping the characteristics of the signal unchanged.

3. Synthetic Studies

3.1. Synthetic Model

A synthetic model was built to demonstrate the effects of the denoising methods. The canonical reservoir model is shown in Figure 1. The transmitter is towed 50 m above the seafloor and along the y-axis from 0 to 32 km, and the transmitting frequency is 0.25 Hz. The receivers are deployed 0.5 m above the seafloor. We assume that the velocity of transmitter is 1 m/s and the sample interval of the receivers is 1 s.

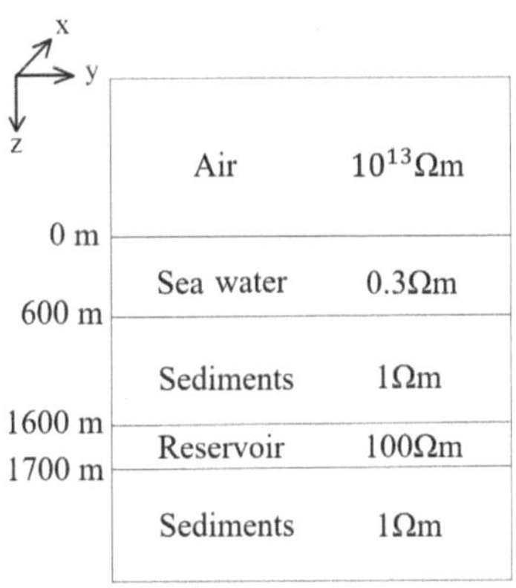

Figure 1. The canonical reservoir model for synthetic data.

In order to calculate the response of the synthetic model, the MARE2DEM code was adopted [36,37]. The electric field E and magnetic field H in the x, y, and z axes can be obtained, but the electric field E in the y axis is used in this study because its amplitude is the largest. Figure 2a shows the magnitude value offset (MVO) plot, which is the forward response, and the signal recorded by the receiver is shown in Figure 2b. It is clear that the simulation signal decayed rapidly and became weak when the offset was large.

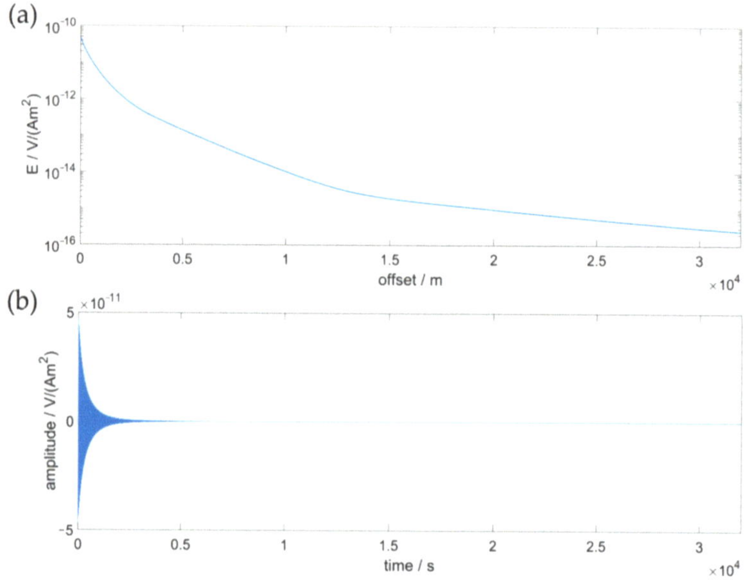

Figure 2. The forward response and the recorded signal. (**a**) Forward response, and (**b**) recorded signal.

3.2. Noise Construction

In this study, four types of noise will be considered, which make the simulation signal closer to that received during real exploration.

(1) Random noise. This kind of noise is not correlated with the signal. Generally, the average magnitude of random noise decreases as the water depth increases. Considering the depth in this study, the magnitude is from 1×10^{-15} V/Am^2 to 2×10^{-15} V/Am^2, and does not follow a Gaussian distribution [38]. This noise is shown in Figure 3a.

(2) Sea water motion noise. This low-frequency noise originates from the seafloor currents, where voltage is induced by the motion of a conductor [39]. The noise is composed of several low-frequency sine signals (0.01 Hz, 0.02 Hz, 0.03 Hz and 0.04 Hz) with a random distribution in the recorded signal. For each sine signal, the magnitude of the noise is fixed within a range of 2×10^{-15}V/Am^2 to 3×10^{-15}V/Am^2. This noise is shown in Figure 3b.

(3) Internal noise. This is from the internal electrode and amplifier noise. This kind of noise is related to the instruments and varies with the magnitude of the original signal; as shown in Figure 3c, it is set as 1% of the original signal [21].

(4) Impulse noise. Due to the complexity of the environment and the error in measurement, impulse noise can be reasonably considered as accidental disturbances. In total, 20 positive impulses are randomly distributed on the signal with a magnitude of 1×10^{-13}V/Am^2 [21]. This noise is shown in Figure 3d.

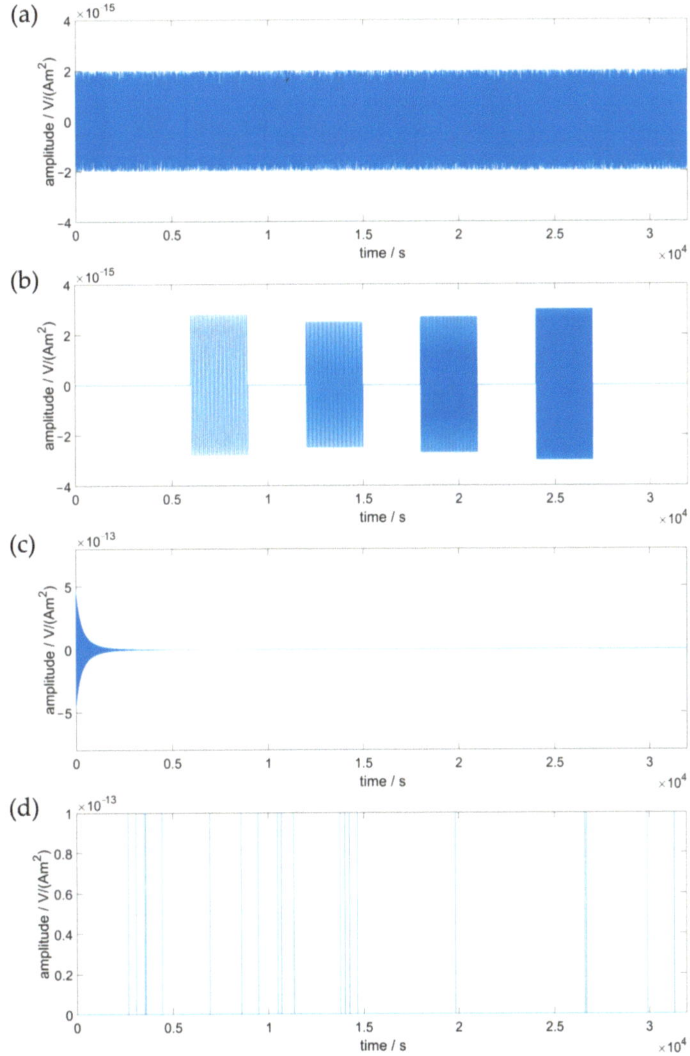

Figure 3. The constructed noise: (**a**) random noise, (**b**) sea water motion noise, (**c**) internal noise, (**d**) impulse noise.

3.3. Synthetic Data

The synthetic data were acquired by adding the four types of noise to the original signal as shown in Figure 2b. The MVO plot, as shown in Figure 4a, was then obtained by solving the system of overdetermined equations, which is composed of synthetic data [40]. The length of the time window was set as 60 s. Figure 4b is the partially enlarged view of the synthetic data, and the distribution of the noise can be observed.

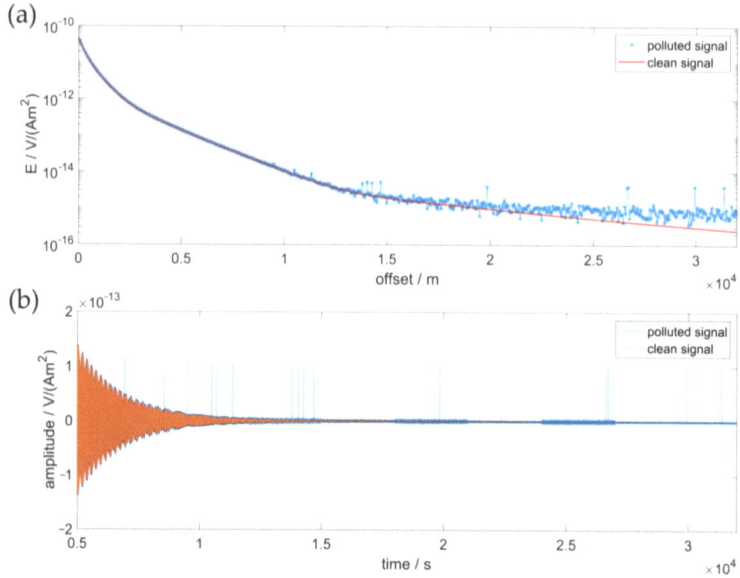

Figure 4. The polluted response and the polluted signal. (**a**) Polluted response and (**b**) polluted signal.

According to Figure 4a, the evident deviations on the polluted MOV plot appeared around 12,000 m, and the impacts became serious as the offset increased. As illustrated in Figure 4b, the polluted signal was heavily influenced by the noise since nearly the same offset and was gradually contaminated over time.

4. Denoising Experiments

4.1. VMD and Wavelet Denoising

VMD begins with the construction of the variational model and separates several intrinsic mode functions (IMFs) based on the central frequency as well as band width [13]. As a frequency domain, non-recursive decomposition method, VMD can avoid mode mixing and edge effects. Meanwhile, it has noise robustness. However, there are six parameters involved when VMD decomposes the signals, and only the appropriate combination of the parameters for each signal allows for the advantages mentioned above. In this study, the strategy for selecting the parameters was from Matlab's Signal Processing Toolbox. After decomposition, the signal was reconstructed by adding the IMFs, and the residual was regarded as the noise [12].

The processed data were acquired by applying VMD to the synthetic data, and the corresponding MVO plot is shown in Figure 5a. Compared with Figure 4a, only a few slight improvements appeared between 18,000 m and 22,000 m. In order to make a clear contrast, the amplitudes of the polluted MVO plot and the processed MVO plot were each normalized by that of the original MVO plot; the result is shown in Figure 5b. After 18,000 m, the blue line is below the red line, indicating that the processed MVO plot is closer to the plot in Figure 2a. However, for the spike pulses caused by impulse noise, the improvement was slight. This proved that VMD can slightly reduce the influences of various noises, except the impulse noise.

The wavelet denoising method is a widely used tool in data processing. It decomposes the signal into multiple components with different frequencies, but unlike the traditional Fourier transform, the representation of the signals in wavelet denoising is in the time-scale domain, allowing the components to be analyzed without losing temporal information. In this study, the mother wavelet was symlet4, and the synthetic data were decomposed into four levels. The wavelet denoising algorithm was minimax threshold with level-dependent soft thresholding [21].

The processed MVO plot and the corresponding normalized amplitude plot are shown in Figure 5c,d. Compared with the results of VMD, there were improvements between 15,000 m and 20,000 m in Figure 5c. In Figure 5d, the blue and red lines separate more clearly. This demonstrates that the wavelet denoising performs better on MVO plots than VMD in terms of the impacts of various noises.

However, the denoising effects of the two methods were insufficient at large offsets where the signals were completely covered by noise. Furthermore, both methods could not effectively remove the impulse noise.

4.2. SGMD Denoising

Before performing SGMD, the termination condition strategy should be discussed. Actually, the termination condition affects the number of decomposition components rather than the first few components. In addition, the principal components are located at the front of matrix transformed by diagonal averaging, which is involved in the matrix decomposition. As a result, the termination condition in the appropriate range will have little effect on the results. The termination condition in this study was set at 0.05.

After that, the effective components were selected and input into the Hampel filter. The PSD illustrates the energy distribution of a signal in the frequency domain and serves as the foundation for selection. We calculate the PSD of each decomposition component, and the component whose frequency had the largest energy that was the same as that of the transmitting signal, was picked out. Considering the time window of the previous procedures and the magnitudes of the signal, the window of the Hampel filter was also 60 s, with a standard deviation of 3.

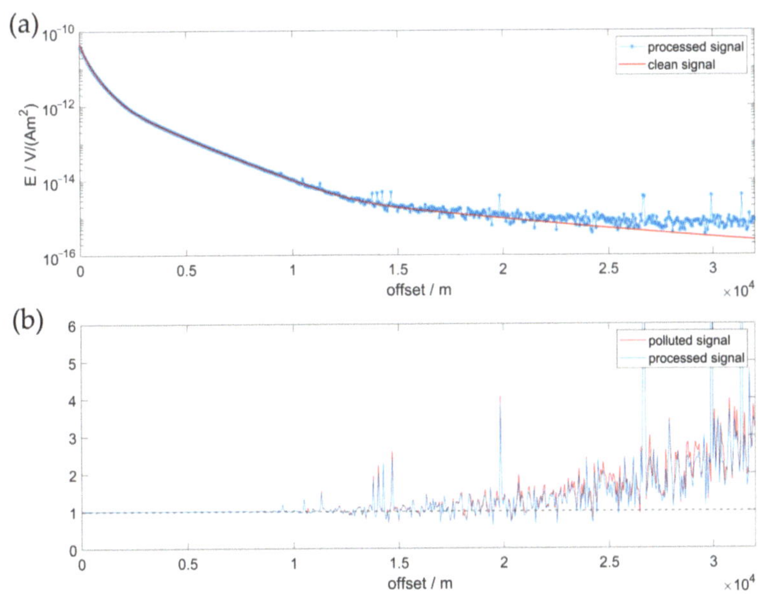

Figure 5. *Cont.*

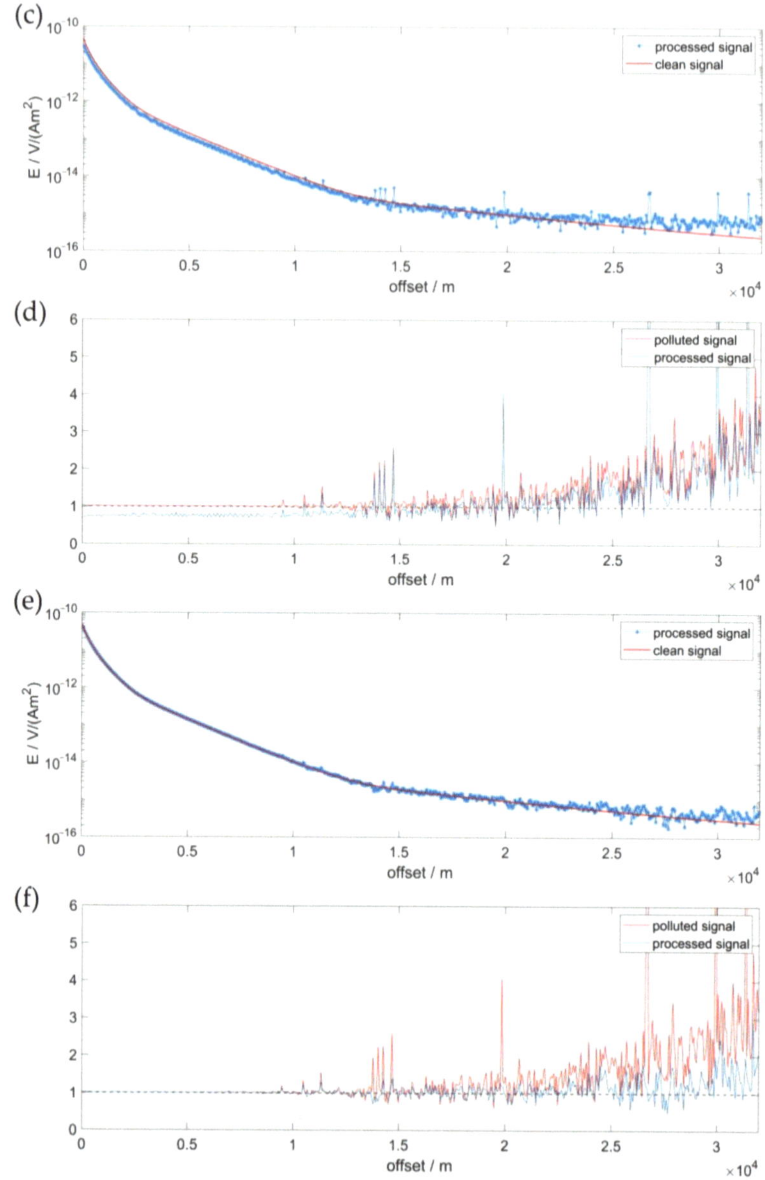

Figure 5. The results of three denoising methods. (**a**) MVO plot processed by VMD; (**b**) normalized amplitude plot of polluted and processed signals by VMD; (**c**) MVO plot processed by wavelet denoising; (**d**) normalized amplitude plot of polluted and processed signals by wavelet denoising; (**e**) MVO plot processed by SGMD; (**f**) normalized amplitude plot of polluted and processed signals by SGMD.

Figure 5e,f show the processed MVO plot and the corresponding normalized amplitude plot. Compared with Figure 5a,c, there were significant improvements in the whole curve, especially at intermediate-to-long offsets. Furthermore, the impulse noise was better suppressed than with the VMD and wavelet denoising methods. In Figure 5f, the blue line clearly separates from the red and is much closer to the standard line, demonstrating that the denoising performance of this method is more complete than those of the two other methods.

4.3. Quantitative Contrast

Generally, the signal-to-noise ratio (SNR) and root-mean-square error (RMSE) are used to evaluate the data quality, which are defined as:

$$\mathrm{SNR} = 10\log_{10}\frac{\sum_{\mathrm{i}=1}^{\mathrm{N}} x(\mathrm{i})^2}{\sum_{\mathrm{i}=1}^{\mathrm{N}} (x(\mathrm{i}) - x'(\mathrm{i}))^2}, \tag{10}$$

$$\mathrm{RMSE} = \sqrt{\frac{1}{\mathrm{N}}\sum_{\mathrm{i}=1}^{\mathrm{N}} (x(\mathrm{i}) - x'(\mathrm{i}))^2}, \tag{11}$$

where $x(\mathrm{i})$ is the clean data and $x'(\mathrm{i})$ is the processed data. However, it is not appropriate to calculate the SNR and RMSE of the whole data set because the magnitudes decay rapidly with increasing offset, resulting in large variations in data quality. Here, we divided the data into 640 segments, and calculated the SNR and RMSE of each segment, which are shown in Figure 6.

As shown in Figure 6a, the black line represents the SNRs of the polluted signal, and it is clear that the hollows were primarily caused by sea water motion noise, while the spike pulses were caused by impulse noise. Random noise and internal noise were distributed throughout the signal, resulting in an overall attenuation trend of the SNRs. Compared with the VMD and wavelet denoising methods, SGMD can clearly depress various types of noise, improving data quality significantly. The denoising performance of VMD on sea water motion noise was better than that of the wavelet denoising method, but for both methods, the effects on random noise and impulse noise were insufficient.

The RMSEs corresponding to the three methods are illustrated in Figure 6b, and the noise distributions were similar to those analyzed above. The RMSEs corresponding to SGMD were much smaller than those of the two other methods, indicating that the processed data were closer to the clean data.

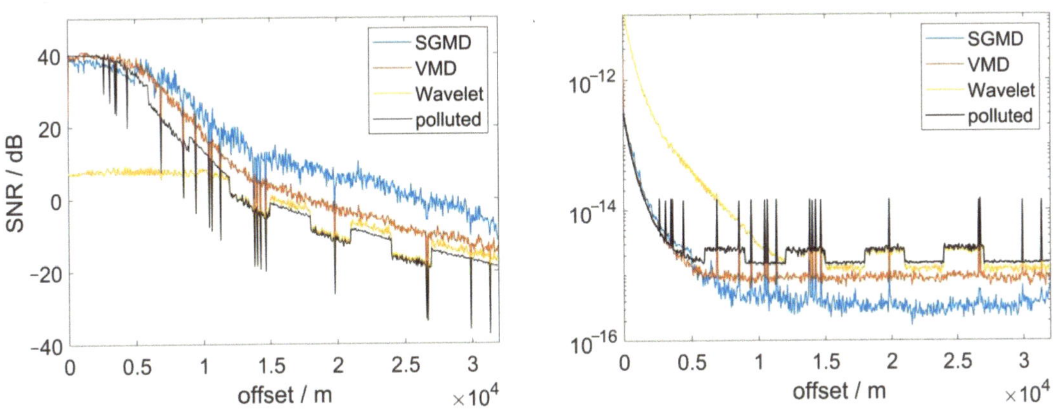

(a) (b)

Figure 6. (**a**) SNRs of the processed and polluted signals, (**b**) RMSEs of the processed and polluted signals.

4.4. Occam Inversion

Another method for assessing data quality is inversion, which reflects the resistivity structure under the seafloor. Occam inversion, proposed by Key, can be divided into two phases [41]. The first phase is to find the model that best fits the observed data and the second phase is to reduce the model roughness as much as possible. MARE2D code was employed to carry out the inversion.

The model used in inversion is slightly different from the forward model: the length of the reservoir in the y axis was fixed at 3000 m, as shown in Figure 7. The noise model is the same as that of the forward simulation and then the polluted and processed data were transferred into E_y, which was utilized as the inversion input.

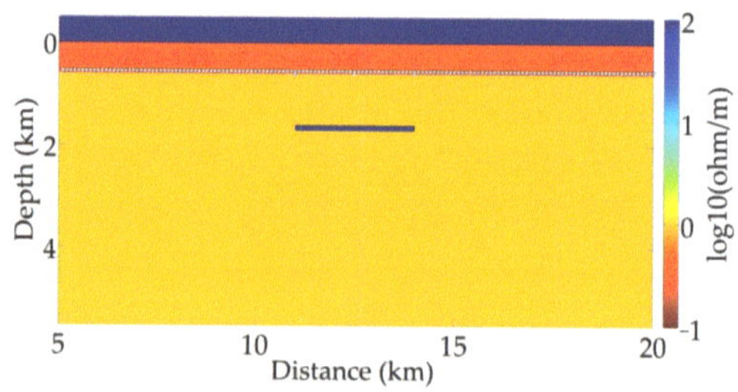

Figure 7. A layered model with a thin and high resistivity reservoir.

Figure 8a is the inversion result of the polluted signal; although the reservoir was detected, the boundaries were unclear and the resistivity was smaller than it was in reality. Figure 8b,c illustrate the results of the VMD and wavelet denoising methods, respectively. Compared with Figure 8a, there was almost no improvement when varieties of noise exist. The result of SGMD is shown in Figure 8d, where the boundaries of resistivity anomaly were smooth and the resistivity was closer to reality, in contrast to the two other methods.

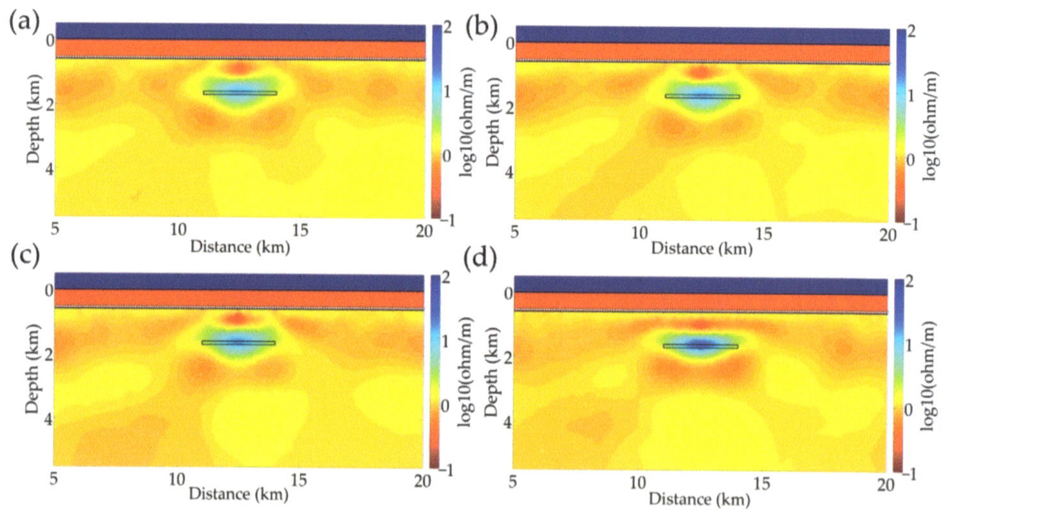

Figure 8. The inversion results of the processed and polluted signals. (**a**) Result of polluted signal; (**b**) result of signal processed by VMD; (**c**) result of signal processed by wavelet denoising; (**d**) result of signal processed by SGMD.

5. Conclusions

The marine CSEM signals are easily affected by various types of noise as the rapid attenuation of amplitudes. Denoising is an indispensable procedure to improve the data quality. In this study, a denoising method using SGMD was proposed to suppress a variety of noises simultaneously. Firstly, a series of single components is obtained by SGMD. Then, we calculate the PSD of each component and pick out the effective signals. Finally, they are put into the Hampel filter and used to reconstruct the processed signal.

We compared the proposed method to the VMD and wavelet denoising methods and verified its superiority in three criteria. The first criterion is the MVO plot, which is the forward response of the marine CSEM method. The MVO plot processed by SGMD was closer to the theoretical plot, especially at intermediate-to-long offsets, and the normalized amplitude plot clearly separated from the noisy result. The second is the quantitative indicator including SNR and RMSE. We plotted the SNRs and RMSEs with varying offsets, and directly showed the improvement of data quality using SGMD. The third is inversion result, which shows that the processed outline of the resistivity anomaly was smooth and the resistivity was closer to reality. All the results demonstrate that the method using SGMD has a significant and comprehensive performance in denoising when various types of noise exist.

In SGMD, the reconstruction of the phase space is the critical step before decomposition, and the combination of delay time and embedding dimension may affect denoising performance; therefore, more research on the parameters may be desired. Furthermore, we used PSD to select the effective components after decomposition, but the relationships between different components require further research.

Author Contributions: Conceptualization, Y.C. and D.G.; methodology, Y.C. and Z.G.; formal analysis, Y.C. and Z.G.; writing—original draft preparation, Y.C.; writing—review and editing, all authors; funding acquisition, Z.G. and D.G. All authors have read and agreed to the published version of the manuscript.

Funding: This research was funded by the National Natural Science Foundation of China (Grant No. 42074169), Major Project of Hunan Province Science and Technology Innovation (2020GK1021), and Open Research Fund Program of Key Laboratory of Metallogenic Prediction of Nonferrous Metals and Geological Environment Monitoring (Central South University), Ministry of Education (Grant No. 2022YSJS16).

Data Availability Statement: The data associated with this research are available and can be obtained by contacting the corresponding author.

Acknowledgments: We thank three anonymous reviewers for their constructive comments.

Conflicts of Interest: The authors declare no conflict of interest.

References

1. Eidesmo, T.; Ellingsrud, S.; MacGregor, L.M.; Constable, S.; Sinha, M.C.; Johansen, S.E.; Kong, F.N.; Westerdahl, H. Sea bed logging (SBL), a new method for remote and direct identification of hydrocarbon filled layers in deepwater areas. *First Break* **2002**, *20*, 144–152.
2. Ellingsrud, S.; Eidesmo, T.; Johansen, S.; Sinha, M.C.; MacGregor, L.M.; Constable, S. Remote sensing of hydrocarbon layers by seabed logging (SBL): Results from a cruise offshore Angola. *Lead. Edge* **2002**, *21*, 972–982. [CrossRef]
3. Mittet, R.; Morten, J.P. Detection and imaging sensitivity of the marine CSEM method. *Geophysics* **2012**, *77*, E411–E425. [CrossRef]
4. Christensen, N.B.; Dodds, K. 1D inversion and resolution analysis of marine CSEM data. *Geophysics* **2007**, *72*, WA27–WA38. [CrossRef]
5. Constable, S. Ten years of marine CSEM for hydrocarbon exploration. *Geophysics* **2010**, *75*, A67–A75. [CrossRef]
6. Key, K. Marine electromagnetic studies of seafloor resources and tectonics. *Surv. Geophys.* **2012**, *33*, 135–167. [CrossRef]
7. Zhu, J.; Wang, G.; Huang, Y.; Zhang, X.; He, R.; Chen, F. A denoising method for ultrasonic testing of rubber composites based on improved symplectic geometric mode decomposition. *Appl. Acoust.* **2023**, *209*, 109428. [CrossRef]
8. Myer, D.; Constable, S.; Key, K. Broad-band waveforms and robust processing for marine CSEM surveys. *Geophys. J. Int.* **2011**, *184*, 689–698. [CrossRef]

9. Attias, E.; Weitemeyer, K.; Minshull, T.A.; Best, A.I.; Sinha, M.; Jegen-Kulcsar, M.; Holz, S.; Berndt, C. Controlled-source electromagnetic and seismic delineation of subseafloor fluid flow structures in a gas hydrate province, offshore Norway. *Geophys. J. Int.* **2016**, *206*, 1093–1110. [CrossRef]
10. Schmidt, O.T.; Colonius, T. Guide to spectral proper orthogonal decomposition. *Aiaa J.* **2020**, *58*, 1023–1033. [CrossRef]
11. Li, S.; Jiang, S.; Wang, Y.; Yu, S. A wavelet correction method for the seawater turbulence noise in marine controlled-source electectromagnetic data. *Geophys. Prospect. Pet.* **2016**, *55*, 657–663.
12. Zhang, J. *Research on Noise Suppression Technology of Marine Controlled-Source Electromagnetic Data Based on VMD*; Jilin University: Changchun, China, 2021.
13. Dragomiretskiy, K.; Zosso, D. Variational mode decomposition. *IEEE Trans. Signal Process.* **2013**, *62*, 531–544. [CrossRef]
14. Baldazzi, G.; Solinas, G.; Del Valle, J.; Barbaro, M.; Micera, S.; Raffo, L.; Pani, D. Systematic analysis of wavelet denoising methods for neural signal processing. *J. Neural Eng.* **2020**, *17*, 066016. [CrossRef] [PubMed]
15. Peng, Z.; Jackson, M.R.; Rongong, J.; Chu, F.; Parkin, R.M. On the energy leakage of discrete wavelet transform. *Mech. Syst. Signal Process.* **2009**, *23*, 330–343. [CrossRef]
16. Xie, H.-B.; Guo, T.; Sivakumar, B.; Liew, A.W.-C.; Dokos, S. Symplectic geometry spectrum analysis of nonlinear time series. *Proc. R. Soc. A Math. Phys. Eng. Sci.* **2014**, *470*, 20140409. [CrossRef]
17. Amundsen, L.; Løseth, L.; Mittet, R.; Ellingsrud, S.; Ursin, B. Decomposition of electromagnetic fields into upgoing and downgoing components. *Geophysics* **2006**, *71*, G211–G223. [CrossRef]
18. Mittet, R.; Gabrielsen, P.T. Decomposition in upgoing and downgoing fields and inversion of marine CSEM data. *Geophysics* **2013**, *78*, E1–E17. [CrossRef]
19. Wright, D. Quantifying the effect of the air/water interface in marine active source EM. *J. Appl. Geophys.* **2015**, *118*, 92–105. [CrossRef]
20. Li, G.; Tang, F.; Li, C.; Lei, W.; Liu, Y. Improved detectivity for detecting gas hydrates using the weighted differential fields of the marine controlled-source electromagnetic data. *J. Mar. Sci. Eng.* **2022**, *10*, 161. [CrossRef]
21. Zhang, P.; Deng, M.; Jing, J.; Chen, K. Marine controlled-source electromagnetic method data de-noising based on compressive sensing. *J. Appl. Geophys.* **2020**, *177*, 104011. [CrossRef]
22. Zhang, P.; Pan, X.; Liu, J. Denoising Marine Controlled Source Electromagnetic Data Based on Dictionary Learning. *Minerals* **2022**, *12*, 682. [CrossRef]
23. Pan, H.; Yang, Y.; Li, X.; Zheng, J.; Cheng, J. Symplectic geometry mode decomposition and its application to rotating machinery compound fault diagnosis. *Mech. Syst. Signal Process.* **2019**, *114*, 189–211. [CrossRef]
24. Pan, H.; Yang, Y.; Zheng, J.; Cheng, J. A noise reduction method of symplectic singular mode decomposition based on Lagrange multiplier. *Mech. Syst. Signal Process.* **2019**, *133*, 106283. [CrossRef]
25. Yu, B.; Cao, N.; Zhang, T. A novel signature extracting approach for inductive oil debris sensors based on symplectic geometry mode decomposition. *Measurement* **2021**, *185*, 110056. [CrossRef]
26. Cheng, J.; Yang, Y.; Li, X.; Pan, H.; Cheng, J. An early fault diagnosis method of gear based on improved symplectic geometry mode decomposition. *Measurement* **2020**, *151*, 107140. [CrossRef]
27. Zhang, G.; Wang, Y.; Li, X.; Tang, B.; Qin, Y. Enhanced symplectic geometry mode decomposition and its application to rotating machinery fault diagnosis under variable speed conditions. *Mech. Syst. Signal Process.* **2022**, *170*, 108841. [CrossRef]
28. Guo, J.; Si, Z.; Xiang, J. Cycle kurtosis entropy guided symplectic geometry mode decomposition for detecting faults in rotating machinery. *ISA Trans.* **2023**, *138*, 546–561. [CrossRef]
29. Zhan, P.; Qin, X.; Zhang, Q.; Sun, Y. Output-Only Modal Identification Based on Auto-regressive Spectrum-Guided Symplectic Geometry Mode Decomposition. *J. Vib. Eng. Technol.* **2023**. [CrossRef]
30. Zhang, X.; Li, C.; Wang, X.; Wu, H. A novel fault diagnosis procedure based on improved symplectic geometry mode decomposition and optimized SVM. *Measurement* **2021**, *173*, 108644. [CrossRef]
31. Guo, J.; Si, Z.; Liu, Y.; Li, J.; Li, Y.; Xiang, J. Dynamic time warping using graph similarity guided symplectic geometry mode decomposition to detect bearing faults. *Reliab. Eng. Syst. Saf.* **2022**, *224*, 108533. [CrossRef]
32. Liu, F.; Wang, H.; Li, W.; Zhang, F.; Zhang, L.; Jiang, M.; Sui, Q. Fault diagnosis of rolling bearing combining improved AWSGMD-CP and ACO-ELM model. *Measurement* **2023**, *209*, 112531. [CrossRef]
33. Packard, N.H.; Crutchfield, J.P.; Farmer, J.D.; Shaw, R.S. Geometry from a time series. *Phys. Rev. Lett.* **1980**, *45*, 712. [CrossRef]
34. Bonizzi, P.; Karel, J.M.; Meste, O.; Peeters, R.L. Singular spectrum decomposition: A new method for time series decomposition. *Adv. Adapt. Data Anal.* **2014**, *6*, 1450011. [CrossRef]
35. Liu, H.; Shah, S.; Jiang, W. On-line outlier detection and data cleaning. *Comput. Chem. Eng.* **2004**, *28*, 1635–1647. [CrossRef]
36. Key, K. Marine EM inversion using unstructured grids: A 2D parallel adaptive finite element algorithm. In Proceedings of the SEG Las Vegas 2012 Annual Meeting, Las Vegas, NV, USA, 4–9 November 2012; pp. 1–5.
37. Key, K. MARE2DEM: A 2-D inversion code for controlled-source electromagnetic and magnetotelluric data. *Geophys. J. Int.* **2016**, *207*, 571–588. [CrossRef]
38. Mittet, R.; Morten, J.P. The marine controlled-source electromagnetic method in shallow water. *Geophysics* **2013**, *78*, E67–E77. [CrossRef]

39. Pethick, A.M. *Multidimensional Computation and Visualisation for Marine Controlled Source Electromagnetic Methods*; Curtin University: Perth, WA, Australia, 2013.
40. Behrens, J.P. *The Detection of Electrical Anisotropy in 35 Ma Pacific Lithosphere: Results from a Marine Controlled-Source Electromagnetic Survey and Implications for Hydration of the Upper Mantle*; University of California: San Diego, CA, USA, 2005.
41. Key, K. 1D inversion of multicomponent, multifrequency marine CSEM data: Methodology and synthetic studies for resolving thin resistive layers. *Geophysics* **2009**, *74*, F9–F20. [CrossRef]

Journal of
Marine Science
and Engineering

MDPI

Article

Complexity and Heterogeneity Evaluation of Pore Structures in the Deep Marine Shale Reservoirs of the Longmaxi Formation, China

Boyuan Zhu [1,2], Jianghui Meng [1,*], Chen Song [3], Renfang Pan [2], Zhengping Zhu [2] and Jineng Jin [2]

[1] Cooperative Innovation Center of Unconventional Oil and Gas, Yangtze University, Wuhan 430100, China; 2021730042@yangtzeu.edu.cn
[2] School of Geosciences, Yangtze University, Wuhan 430100, China; pan@yangtzeu.edu.cn (R.P.); 501096@yangtzeu.edu.cn (Z.Z.); jinjineng@yangtzeu.edu.cn (J.J.)
[3] PetroChina Jidong Oilfield Company, Tangshan 063006, China; songchen2023@petrochina.com.cn
* Correspondence: mjh@yangtzeu.edu.cn

Abstract: The structural evolution and sedimentary differentiation of the Sichuan Basin in China are complex, with intricate reservoir pore structures that significantly impact shale gas production. This study examines the complexity and heterogeneity of the microscopic pore structures in the deep marine shale reservoir in the Longmaxi Formation. Pore structure characterization techniques are used to compare deep and shallow–medium marine shales, and siliceous and silty shales. The results reveal the factors influencing pore structure and their impact on exploration and development. The key points are as follows: (1) The pore structure of deep siliceous shale is the most complex due to its diverse range of pore development patterns, pore types, and sizes. (2) The box dimension of full pore size is about 1.52 for deep marine shale and 1.46 for shallow–medium shale. Organic matter (OM) content, the degree of pore development, and inorganic mineral content all correlate positively with the complexity of the pore structure in deep marine shale, which affects the formation of high-quality reservoirs. (3) Lateral heterogeneity of pore structures shows strong regional variations in the study area. Heterogeneity is more pronounced in the deep marine shale than in the medium and shallow shale formations. OM mesopores significantly influence the overall heterogeneity of the shale pore system. The deep marine shale reservoir is situated in an area with strong regional variations. The pore structure of high-quality reservoirs is more complex than those of shallow–medium marine shales, displaying notable heterogeneity. Pore structures with fractal dimension values close to that of the shallow–medium formations (box dimensions within 1.5) offer promising targets for the exploration and development of deep marine shale gas.

Keywords: Sichuan Basin; Longmaxi Formation; deep marine shale; pore structure; heterogeneity; fractal dimension

Citation: Zhu, B.; Meng, J.; Song, C.; Pan, R.; Zhu, Z.; Jin, J. Complexity and Heterogeneity Evaluation of Pore Structures in the Deep Marine Shale Reservoirs of the Longmaxi Formation, China. *J. Mar. Sci. Eng.* **2023**, *11*, 1613. https://doi.org/10.3390/jmse11081613

Academic Editor: George Kontakiotis

Received: 26 July 2023
Revised: 14 August 2023
Accepted: 16 August 2023
Published: 17 August 2023

1. Introduction

The surface micro-morphology and throat features of dual media pores have a major impact on the migration and accumulation of natural gas. Therefore, pore structures of shale reservoirs are an essential study topic for understanding shale gas enrichment [1,2]. Study of pore systems involves the characterization and analysis of heterogeneity, occurrence characteristics, evolution patterns, and the preservation mechanisms of pores [3–7].

Previous studies have found that organic matter (OM) pores (organic pores) in shale reservoirs are highly developed and complex [8,9]. OM abundance, thermal evolution level, diagenesis, and other factors combine in the creation and development of organic pores [10,11]. Other factors, including steric effects, the presence of hard minerals, and the morphology of the pores themselves, also affect the preservation and connectivity of organic pores [12–14]. Inorganic pores in shale reservoirs are mostly formed by brittle minerals, and inorganic pores with complex structures can be observed under a scanning electron microscope [15,16]. As deeper shales experience more intense thermal evolution, a large number of pores are formed by thermal pyrolysis of OM. At the same time, strong compaction changes both organic and inorganic pores, resulting in relatively lower porosity and alteration of the original pore structure. The development of organic and inorganic pores together forms a complex pore system. The increased complexity and heterogeneity of the pore structures in deep marine shale reservoirs makes reservoir characterization more difficult [17,18]. This, in turn, increases the difficulty of evaluating the petrophysical properties, reservoir capacity, reservoir characteristics, and diageneses of reservoirs. Ultimately, these difficulties with reservoir interpretation can influence the results of shale gas transportation and enrichment. The complexity of pore structures also creates difficulties with reservoir geological modeling and petroleum engineering development [19–21].

The main research objects of this study are the deep and shallow–medium shales in the Luzhou, Weiyuan, and Changning regions. Field emission scanning electron microscopy (FE-SEM) and low-temperature gas adsorption were used to estimate the pore fractal dimension in order to analyze the complexity and heterogeneity of the pore structures. The comparative study highlighted the distinctions between deep marine shale and shallow–medium marine shale, and identified the key elements affecting the complexity and heterogeneity of pore networks in deep marine shale. The findings of this study provide a general theoretical basis for the evaluation and exploration of deep marine shale reservoirs.

2. Geological Setting

The main part of the Luzhou Block in the study area is in the southern Sichuan Basin's low folded structural belt. Its edges are in the gentle structural belt in the central part of the Sichuan Basin and the low folded structural belt in the southwestern part of the basin (Figure 1a) [22]. A comb anticline structure has developed from north to south in the Luzhou area, and several low and steep anticlines have developed in the northeast [23]. The thickness of shale in the Luzhou Block is fairly even across all of southern Sichuan [24].

The Longmaxi Formation lies at the base of the Lower Silurian System. Its lithology is primarily grayish-black and black rocks rich in silty shale, calcareous shale, and siliceous shale (Figure 1b). The formation thickness is between 200 and 500 m [24]. The Long1_1 sub-member is the main formation for shale gas exploration and development, consisting of a set of deep-water shelf facies of organic-rich black shale divided into four sub-layers. The bottom of the Longmaxi Formation is a "sweet spot" formation formed of OM-rich siliceous shale covered with thick silty shale. Shale and tight limestone form the basement of the Wufeng Formation. Overall, the formation has good gas-bearing properties [25].

The deep black shale in the Luzhou area is predominantly silty shale and siliceous shale, with calcareous shale rarely occurring (Figure 1b). Calcareous shales are occasionally found in horizontal layered structures, with carbonatite mineral content generally ranging from 25% to 50% and low total organic carbon (TOC) content. Silty shales generally have a terrestrial brittle mineral content of 60% or less, horizontal layered structures, and low TOC content. The siliceous shale features horizontal bedding and siliceous nodules, with siliceous content exceeding 50% and TOC content generally exceeding 3%. It is rich in biological fossils.

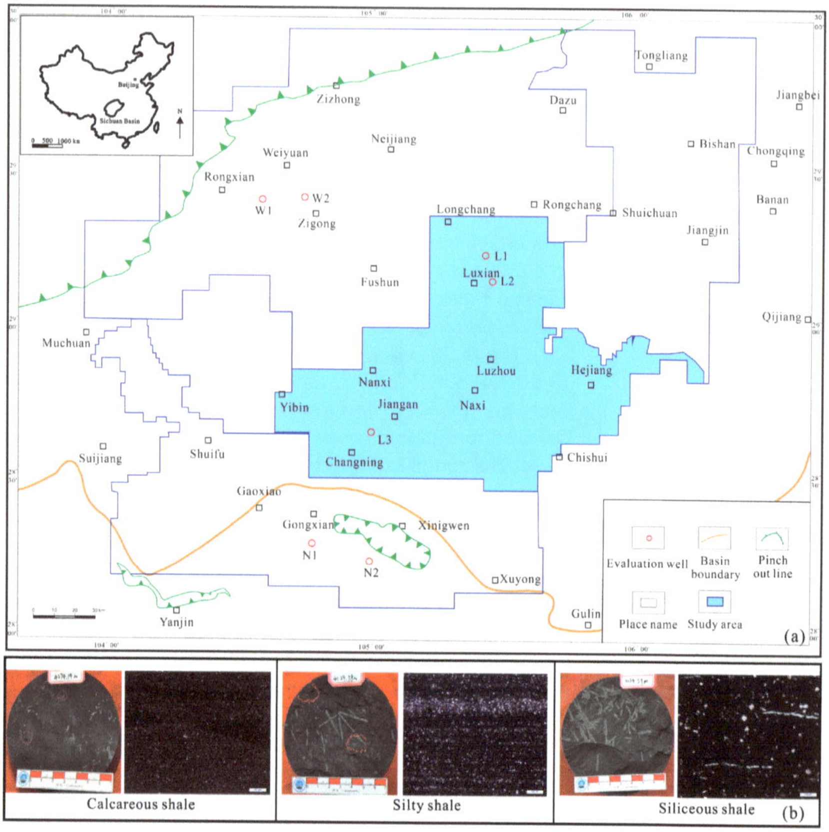

Figure 1. Overview of the study area in the southern Sichuan Basin. (**a**) Overview of the location of the study area; (**b**) common lithofacies of deep marine shale: core and thin rock sections.

3. Samples and Methods

3.1. Sample Data

The scope of current large-scale exploration for shale gas in the Sichuan Basin of China includes shallow shale (burial depth less than 2000 m), shallow–medium shale (burial depth between 2000–3500 m), and deep shale (burial depth between 3500–4500 m). In this study, a total of 16 core samples were analyzed to characterize the pore structure. Eleven samples were obtained from deep formations at approximately 4000 m burial depth in Luzhou and five from shallow–medium formations at around 2000 m burial depth in Changning and Weiyuan (Figure 1a). The sample set comprised 12 deep and shallow–medium siliceous shale formation samples, 3 deep silty shale samples, and 1 deep clay siliceous shale sample. The siliceous shale found in the Longmaxi Formation serves as a high-quality reservoir facies. Compared with other lithologies, it has relatively high TOC, brittle mineral content, porosity, and gas content [26].

SEM observations, nitrogen (N_2) and carbon dioxide (CO_2) adsorption experiments, tests of physical properties, organic geochemistry, and mineral X-ray diffraction were conducted on 16 samples, and model calculation and statistical analysis were carried out.

3.2. FE-SEM Observations

The instruments used for FE-SEM analysis were a Leica RES102 ion polishing instrument and a ZEISS high-resolution FE-SEM. FE-SEM is a kind of high-resolution scanning electron microscope, which is capable of secondary electron imaging, reflective electron imaging, and image processing of the surface morphology of solid samples. The Zeiss FE-SEM produces high-resolution, high-quality images and has a large magnification range and strong stereoscopic imaging capability. Therefore, it is widely used in the characterization of microscopic pore structures. During initial pretreatment, millimeter-scale shale samples were subjected to argon ion polishing of their entire surfaces with an ion polisher. This made it easier to observe the nanoscale pore structure of the shale.

In observing images at different positions along a certain direction using FE-SEM, it is necessary to use an electron gun to emit an electron beam and focus it on the surface of the shale sample while sequentially scanning the sample surface row by row. Secondary electrons emitted from the sample surface are then collected from the monitor. The image signal formed by these electrons is displayed by a cathode ray tube (CRT). In order to keep the position of the sample image displayed by the CRT consistent with the position of the electron beam on the sample surface, the deflection coil of the CRT must be synchronized with the scanning coil.

To accurately quantify the information captured in the SEM images, threshold segmentation of each photo was first performed, converting the entire image to 8-bit grayscale. Subsequently, useful data, such as the OM area, were extracted based on the grayscale values. Finally, 8 to 10 images of each sample with a magnification of around 10 to 20 K were randomly selected for statistical analysis. The statistical results of all images for each individual sample were then collated to obtain data for that sample.

3.3. Low-Temperature Gas Adsorption Method

In N_2 adsorption experiments, based on the BET (Brunauer–Emmett–Teller) theory, the shale specific surface area (SSA) was determined at relative pressures (P/P_0) ranging from 0.05 to 0.35 MPa [27]. Equivalent pore volumes (PVs) were obtained using the BJH (Barrett–Joyner–Halenda) model [28]. N_2 adsorption was used to determine the pore size distribution of mesopores (pore sizes 2–50 nm).

CO_2 molecules have higher adsorption temperatures than N_2, resulting in more vigorous thermal motion. At 0 °C, the saturated vapor pressure of CO_2 reaches 3.495 MPa, making it easier for CO_2 molecules to diffuse into micropores (pore size less than 2 nm) in rock samples. The PV and SSA of micropores can be estimated using the DFT (Density Functional Theory) model [29].

3.4. Estimation of Fractal Dimension

Two-dimensional fractals of pores can be approximated using the SEM-box-counting method, while three-dimensional fractals of pores can be approximated using the Frankel–Halsey–Hill (FHH) theory [30,31].

The box dimension is calculated using the connection between the number of boxes and their sizes [32]. By adjusting the threshold and particle size of the SEM grayscale, the fractal dimensions of different pore sizes can be estimated [8]. The box's dimensions are as follows:

$$D_B = -\lim_{\delta \to 0} \frac{\log N_\delta}{\log_\delta} \tag{1}$$

where N_δ is the number of boxes required to cover the material each time, and δ is the box size (nm). The logarithmic relationship between N_δ and δ is established as follows:

$$\log N_\delta = -D_B \log\delta + c \tag{2}$$

where the absolute value of the slope is the box-counting dimension D_B, and C is a constant. D_{B1} is the box dimension of mesopores, with pore sizes of 10–50 nm; D_{B2} is the box dimension of macropores; D_{B3} is the box dimension of the full range of pore sizes in the FE-SEM images.

The branch of the N_2 adsorption–desorption curve in the low-pressure area (relative pressure less than 0.45) corresponds to pores with diameter less than 4.34 nm and its fractal dimension is D_1 [29]. The branch of the N_2 adsorption–desorption curve in the high-pressure zone corresponds to 4.34–100 nm pores with relative pressure greater than 0.45, and the fractal dimension is D_2 [33]. The FHH fractal dimension is as follows:

$$lnV = (D-3)ln[(P_0/P)] + c^2 \quad (3)$$

where V is the volume of adsorbed gas, m^3; P_0 is saturated vapor pressure, MPa; and p is equilibrium pressure (MPa).

4. Results

4.1. FE-SEM Pore Characteristics

Different lithofacies in shale samples from the deep and shallow–medium marine shale were observed in FE-SEM images (Figure 2). Figure 2a–c show that organic pores are the most developed in deep siliceous shale and that the pores in shallow–medium siliceous shale are the same as those in deep silty shale. Figure 2d–f show that the organic pores of the deep and shallow siliceous shale have high densities of mesopores, while the deep silty shale has a relatively low density of mesopores but a relatively high density of macropores (pore size greater than 50 nm). The organic mesopores and macropores are all irregular; however, macropore structures are more complex. Figure 2g–i show that inorganic pores are not very developed in the deep silty shale, deep siliceous shale, and shallow–medium siliceous shale, as shown by the complex structure of fractures and angular pores. The FE-SEM images reveal that the pore network distribution of the deep siliceous shale has the highest density.

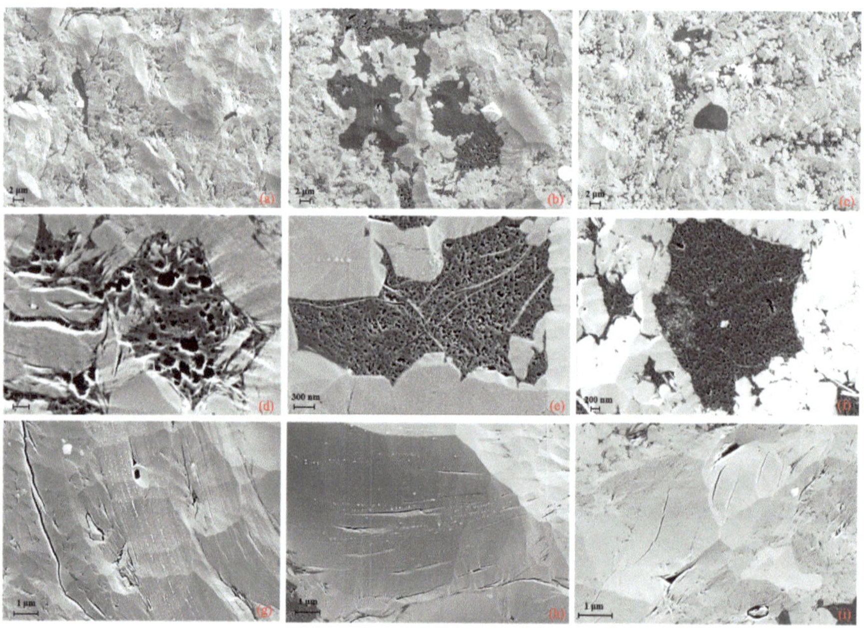

Figure 2. FE-SEM pore characteristics of shale samples in the study area. (**a**) Sample L1-1, 4126 m, pore distribution of silty shale; (**b**) sample L1-4, 4144 m, pore distribution of siliceous shale; (**c**) sample N1-2, 2317 m, pore distribution of siliceous shale; (**d**) sample L1-1, 4126 m, silty shale with developed organic pores; (**e**) sample L1-6, 4149 m, siliceous shale with developed organic pores; (**f**) sample N1-3, 2322 m, siliceous shale with developed organic pores; (**g**) sample L1-1, 4126 m, silty shale with developed inorganic pores; (**h**) sample L1-6, 4149 m, siliceous shale with developed inorganic pores; (**i**) sample N1-1, 2581 m, silty shale with developed inorganic pores.

The pores in the SEM images were statistically analyzed to determine surface porosity. The inorganic surface porosity of deep marine shale is the highest while the formation of organic pores is controlled by lithofacies (Figure 3 gray). The organic surface porosity of the siliceous shale is the highest and that of the deep marine shale is higher than that of the shallow–medium marine shale (Figure 3 yellow). According to surface porosity characteristics, deep siliceous shale has the most developed organic and inorganic pores.

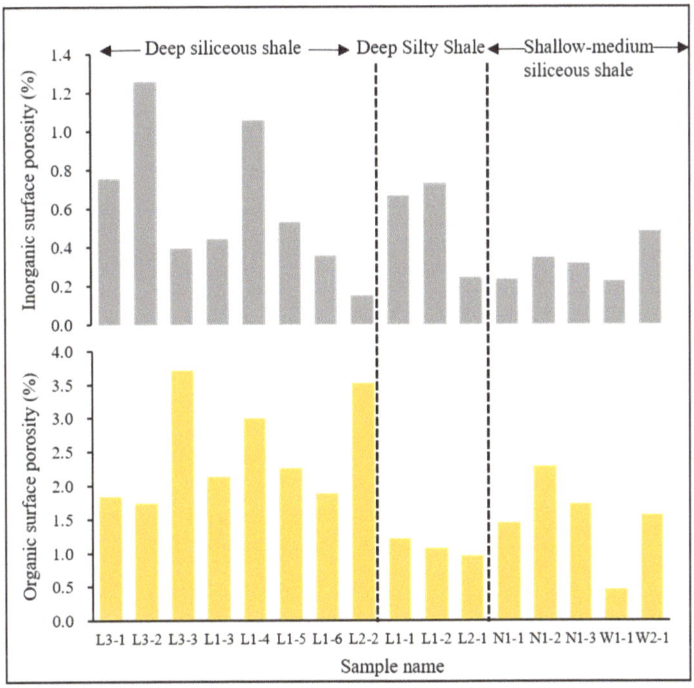

Figure 3. Pore surface porosity based on FE-SEM.

4.2. N_2 Adsorption–Desorption Characteristics

The hysteresis loop characteristics (H1–H4 types) were defined according to IUPAC [34]. Figure 4 shows that the adsorption curves of all samples show a gradual increase in the low-to-medium-pressure range (0–0.8) and a steep incline in the high-pressure range (0.8–1.0), which is typical H3 type behavior. However, deep siliceous shale samples, specifically L-5 and L-6, have wider hysteresis loops with higher adsorption capacities, which is more typical of H2-type behavior. As a result, deep siliceous shale can be classified as a hybrid of H2 and H3 types, referred to here as H2&3 type, while the deep silty shale and shallow–medium marine shale fall within the H3 type.

The presence of the H2&3 type leads to a diverse range of pore characteristics and a decrease in pore connectivity within the rock. On the contrary, the H3-type pore structure appears to be comparatively simple and exhibits comparatively good pore connectivity.

The hysteresis loop of N_2 adsorption–desorption shows that the pores of deep siliceous shale are very developed and their pore structure is the most complex. However, the pore structure complexities of deep silty shale and shallow–middle siliceous shale are similar.

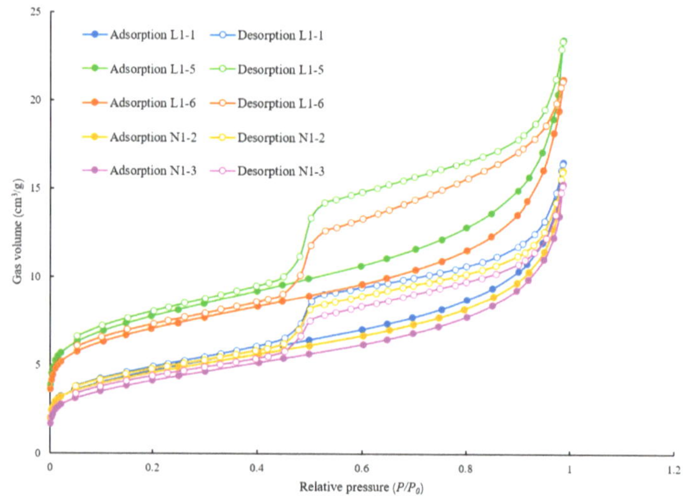

Figure 4. N_2 adsorption–desorption characteristics.

4.3. Pore Size Distribution

The full PV, and the SSA distribution, of each sample's pore sizes can be estimated using the N_2 adsorption BJH model and CO_2 adsorption DFT model.

Figure 5 clearly shows that the mesopores in the shallow–medium siliceous shale contribute more to PV and SSA. In comparison with shallow–medium siliceous shale, the micropores and macropores in deep marine shale greatly increase the PV (Figure 5). Compared with the silty shale, the deep siliceous shale has more macropores (about 14%) (Figure 5).

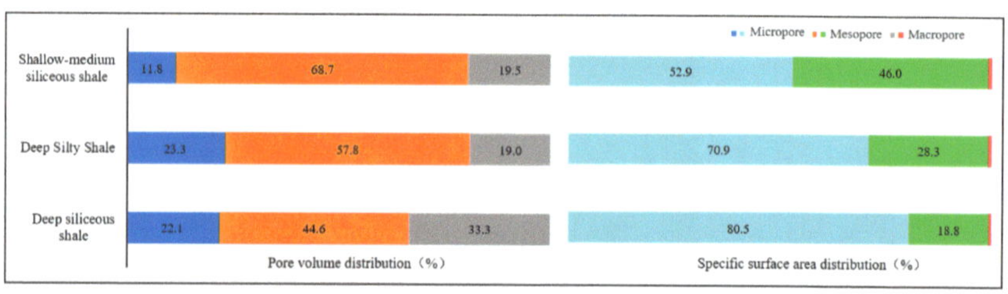

Figure 5. Distribution of PV and SSA proportion.

The variations in PV and SSA with different depths and lithologies reveal that the pore systems in deep marine shale formations are not dominated solely by mesopores. The contributions of micropores and macropores cannot be ignored. The contribution from macropores is more developed.

4.4. Fractal Characteristics of Pores

The complexity of the pore structures increases with the size of the fractal dimension. Box-counting based on SEM images indicates the complexity of the organic pore structure. D_{B1} of wells N1, W1, and W2 in the shallow–medium marine shale is higher than that in the deep marine shale, with an average absolute deviation of +0.16, while D_{B2} and D_{B3} are smaller than that in the deep marine shale (Table 1), with an average absolute deviation of −0.1 and −0.06, respectively. The box-counting dimension relationship between different lithofacies is not significant (Table 1). Analysis of the box-counting dimension reveals that the mesopore structure of the shallow–medium marine shale is more complex than that of the deep marine shale. Conversely, the macropore structure of the deep marine shale is more complex. The organic pore structure of the deep marine shale is evidently more intricate than that of the shallow–medium marine shale.

Table 1. Sample box-counting dimension, FHH fractal dimension, and SEM average pore size.

Sample	Depth	Lithology	D_B			D		SEM Average Pore Size
	m		D_{B1}	D_{B2}	D_{B3}	D_1	D_2	nm
L1-1	4127	Silty shale	1.185	1.647	1.536	2.549	2.766	27.4
L1-2	4134	Silty shale	1.102	1.577	1.514	2.586	2.760	30.1
L1-3	4139	Siliceous shale	1.151	1.566	1.487	2.578	2.758	27.1
L1-4	4143	Siliceous shale	1.170	1.631	1.461	2.634	2.786	24.8
L1-5	4147	Siliceous shale	1.199	1.570	1.489	2.640	2.776	21.6
L1-6	4148	Siliceous shale	1.390	1.475	1.514	2.613	2.761	29.8
L3-1	3825	Siliceous shale	1.139	1.651	1.571	2.596	2.766	64.3
L3-2	3837	Siliceous shale	1.121	1.623	1.580	2.596	2.750	28.6
L3-3	3844	Siliceous shale	1.169	1.620	1.503	2.578	2.758	28.7
L2-1	4113	Silty shale	1.146	1.604	1.512	2.621	2.811	24.1
L2-2	4143	Siliceous shale	1.508	1.754	1.515	2.638	2.809	19.6
N1-1	2580	Siliceous shale	1.422	1.537	1.598	2.508	2.744	19.9
N1-2	2318	Siliceous shale	1.377	1.550	1.478	2.572	2.763	13
N1-3	2323	Siliceous shale	1.406	1.422	1.499	2.526	2.752	22.6
W1-1	1541	Siliceous shale	1.229	1.568	1.249	2.597	2.815	10.4
W2-1	2568	Siliceous shale	1.407	1.475	1.462	2.607	2.733	17.2

Pore structures are characterized based on their FHH fractal dimension. D_1 and D_2 of the shallow–medium marine shale are generally smaller than that of the deep marine shale (Table 1), with an average absolute deviation of −0.04 and −0.012, respectively. Therefore, the pore structure of the deep marine shale is more complex than that of the shallow–medium marine shale. D_1 of the silty shale is small (−0.024 deviation), while the D_2 relationship is not significant (0.008) (Table 1). Therefore, the complexity of the different lithofacies is about the same.

Heterogeneity of Pore Structures

Examination of the fractal dimension across different pore sizes illustrates the heterogeneity of the shale pore system. As shown in Table 1, the fractal dimension varies significantly between different pore sizes. Micropores have a comparatively simple structure, characterized by the smallest fractal dimension [35]. The wide range of variations in the fractal dimension of all the pore sizes in the pore system demonstrates the considerable heterogeneity of the overall pore system.

It is also worth noting that the fractal dimension of full pore size in deep marine shale formations is higher than that in the shallow–medium marine shale (Table 1). This indicates that heterogeneity is more pronounced in the pore system of the deep marine shale of the Longmaxi Formation than in the shallow–medium marine shale.

Variations in the box-counting dimension based on FE-SEM images represent the horizontal and vertical heterogeneity of organic pore structures. Figure 6 shows that the organic pore heterogeneity in Luzhou, Changning, and Weiyuan is strong. The D_{B3} heterogeneity of organic pores in the shallow–medium marine shale is stronger than that in the deep marine shale, and the difference in heterogeneity between siliceous shale and silty shale is not significant (Figure 6). Figure 6 shows that the D_{B3} organic pore heterogeneity over the entire vertical depth range is comparatively weak. The most drastic changes in either the horizontal or vertical directions occur in D_{B1}. Overall heterogeneity of the pore structure is influenced by the heterogeneity of organic mesopores.

Variation in the FHH fractal dimension is used to characterize the horizontal and vertical heterogeneity of pore structures. Figure 7 shows strong horizontal heterogeneity of D_1 and D_2 in the pore structures of Luzhou, Changning, and Weiyuan. The difference in heterogeneity across the shallow–medium marine shale is not obvious, and the difference in heterogeneity between siliceous shale and silty shale is also not obvious (Figure 7). Figure 7 shows that the vertical homogeneity of pore structures at depth D_2 is comparatively good.

The difference in heterogeneity across shallow–medium and deep marine shale formations is not obvious. D_1 changes most drastically in both horizontal and vertical directions.

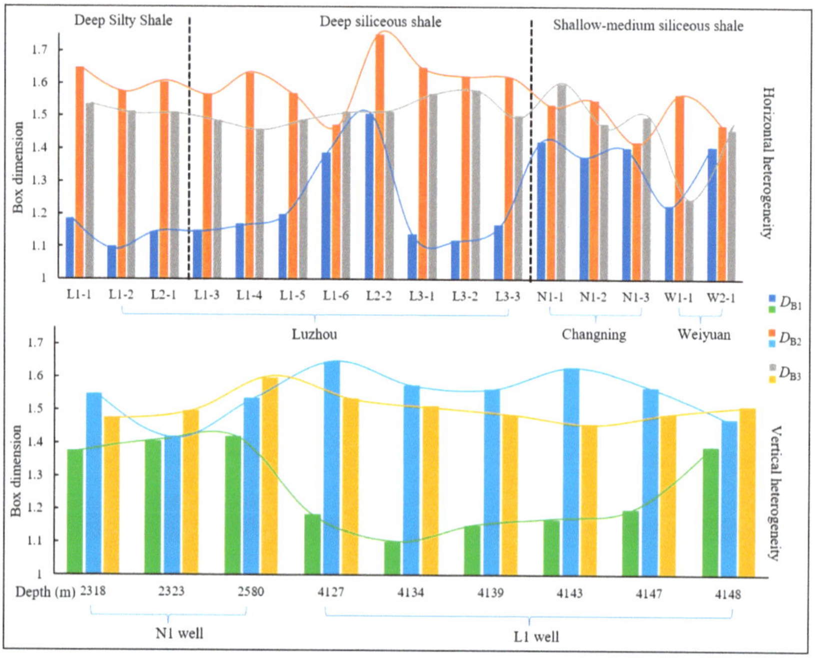

Figure 6. Heterogeneity of organic pore structure based on box dimension.

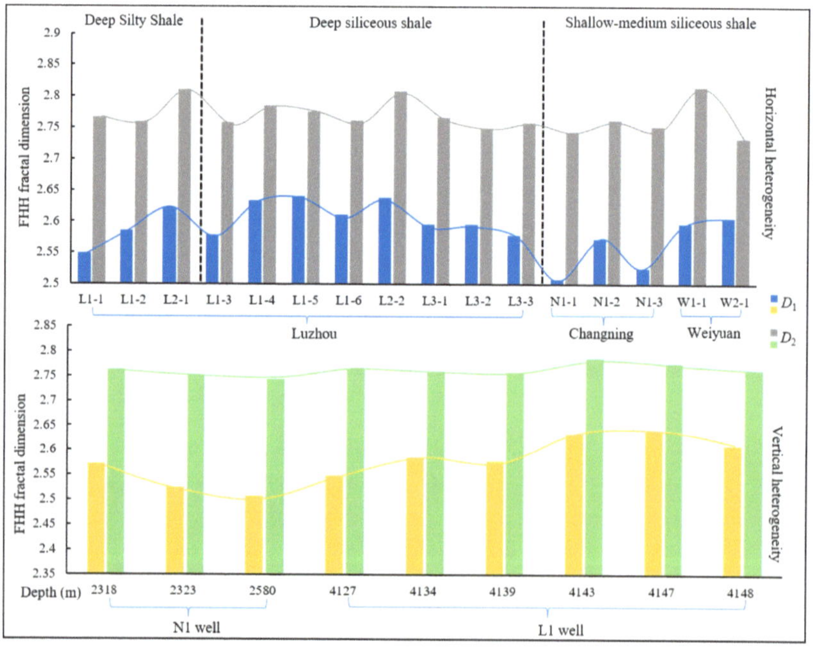

Figure 7. Heterogeneity of pore structure based on FHH fractal dimension.

5. Discussion

5.1. Factors Influencing Pore Structure Complexity

5.1.1. Influencing Factors

The FHH fractal dimension represents the degree of complexity of the pore network. Figure 8 illustrates the relationship between reservoir parameters and fractal dimension after excluding outliers.

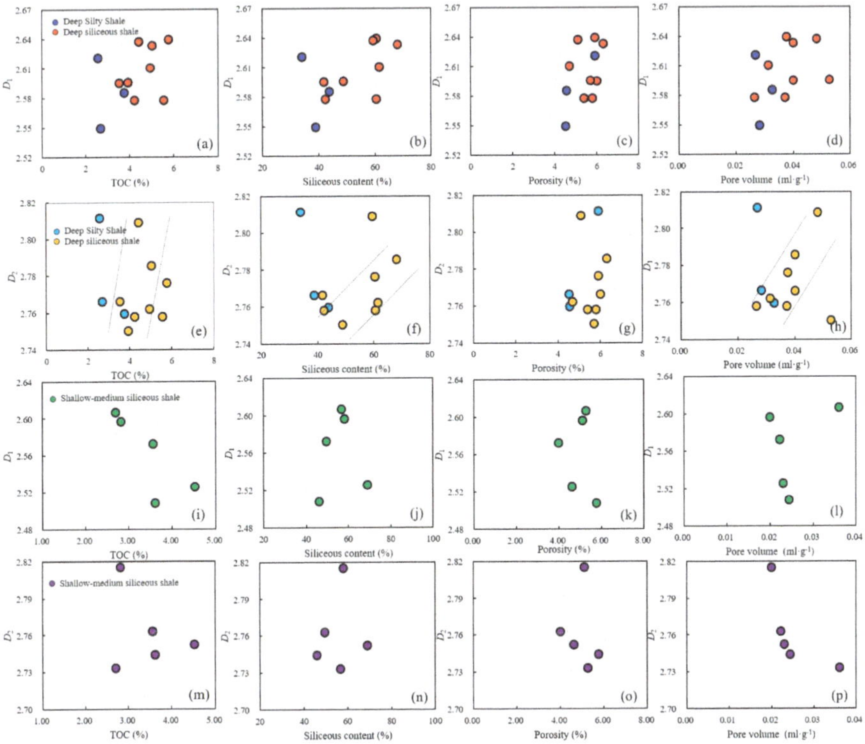

Figure 8. Relationship between fractal dimension and reservoir parameters. (**a–d**) The relationship between deep marine shale D_1 and TOC, siliceous content, porosity, and PV; (**e–h**) the relationship between deep marine shale D_2 and TOC, siliceous content, porosity, and PV; (**i–l**) the relationship between shallow–medium marine shale D_1 and TOC, siliceous content, porosity, and PV; (**m–p**) the relationship between shallow–medium marine shale D_2 and TOC, siliceous content, porosity, and PV.

It can be observed from Figure 8a–d that D_1 of the deep marine shale is positively correlated with TOC, siliceous content, porosity, and PV. Figure 8e–h show that D_2 of the deep siliceous shale is still positively correlated with reservoir parameters. This relationship indicates that the degree of pore development, the amounts of inorganic minerals, and the quantity of OM all have a significant impact on the complexity of the pore structure of deep marine shale.

Figure 8i–l show that D_1 of the shallow–medium siliceous shale is positively correlated with siliceous content to a considerable degree but negatively correlated with TOC and PV. Figure 8m–p show that D_2 of the shallow–medium siliceous shale is positively correlated with TOC and siliceous content but negatively correlated with porosity and PV. This indicates that the complexity of small pores in the shallow–medium marine shale is influenced by the content of inorganic minerals and the degree of pore development. The quantity of OM and the presence of inorganic minerals both affect the complexity of larger pores in the shallow–medium marine shale.

In summary, more factors contribute to the complexity of the pore structure of deep marine shale than to the shallow–medium layers, resulting in a more complex pore structure in the deep marine shale (Table 1). Deep siliceous shale may also be influenced by more factors than silty shale.

5.1.2. Formation of Influencing Factors

The growth of organic pores and their structural characteristics are directly tied to TOC and may be significantly impacted by both the place of origin and quantity of OM [36]. Moreover, the content of siliceous minerals can influence the development and structure of inorganic pores. Also, the interaction between rigid minerals and OM can cause pores to form under high pressure, generating elongated intergranular pores (Figure 2e). The porosity and PV directly reflect the overall degree of pore development.

(1) Organic matter: The overall results show that the complexity of the pore structure of the deep marine shale is controlled by organic matter (Figures 3 and 4; Table 1). The deep marine shale has undergone high levels of thermal evolution and compaction. Hydrocarbon production is beneficial for the formation of organic matter pores, while compaction has a negative effect. The combination of these two processes results in a highly complex structure of organic pores. Additionally, during its sedimentary period, the "sweet spot" interval of the Longmaxi Formation in the Luzhou area was located at the center of deposition, where the water depth was greatest. In contrast, the Changning and Weiyuan areas were located on slopes, leading to differences in OM sources between the two areas [37,38]. Therefore, there is a certain difference in the degree to which the abundance of deep and shallow organic matter controls the complexity of pores.

(2) Inorganic mineral content: Complex inorganic pores (commonly macropores) tend to develop with increasing content of inorganic minerals. If rigid mineral particles are in a dispersed state and do not come into contact with each other, they can compress the organic matter, leading to the destruction of organic pores [39]. Intergranular pores formed by extrusion between inorganic minerals and organic matter have worse storage and connectivity compared with organic pores [40]. Inorganic mineral content and contact relationships of rigid minerals are influential factors in the complexity of inorganic and organic pores in shale reservoirs.

(3) Pore development degree: Compared with the shallow–medium marine shale, the deep marine shale exhibits substantial organic pore development (Figures 2 and 3). The extensive development of organic pores has a substantial impact on the intricate nature of the pore structure seen in deep marine shale formations. The organic pores in the deep marine shale have highly developed micropores, mesopores, and macropores (Figure 5), and both organic and inorganic pores have comparatively high surface porosity (Figure 3). Increased pore development implies more diverse origins for the pore components. In contrast, the shallow–medium marine shale is primarily composed of mesopores. As the degree of pore development in these shales increases, the complexity of the pore structure decreases.

Broadly speaking, organic matter in deep marine shale undergoes substantial thermal evolution and compaction [41,42], leading to an abundance of irregular organic pores. At the same time, destruction of inorganic primary pores and formation of inorganic secondary pores also occur, resulting in a very complex pore structure. As suggested by the distribution of full-size PV contributions and fractal dimension characteristics, this pattern of pore development means that the pore structure of deep marine shale is the most complex. In the Luzhou area, the full-size pores in the deep marine shale are widely dispersed, a phenomenon also observed in the Dazu and Yongchuan areas [43]. In contrast, shallow–medium marine shales typically have widely developed mesopores. The pore structure of deep marine shale is, therefore, more complicated than that of shallow–medium marine shale because pore structures become simpler as pore size decreases.

5.2. Impact of Heterogeneity and Complexity on Exploration

5.2.1. Heterogeneity and Complexity of Reservoir Pore Structures

The Box-counting dimension and FHH fractal dimension serve as crucial parameters for characterizing the heterogeneity of pore structures, as they accurately reflect their complexity and are helpful for identifying factors that contribute to the intricacy of these structures.

The heterogeneity of the deep marine shale pore system is stronger than that of the shallow–medium shale, and the entire shale pore structure has strong lateral heterogeneity. The primary reason for this is the continuous alteration of organic mesopores in shale, as they are transformed into micropores under compaction and into macropores during hydrocarbon generation [44,45]. The heterogeneity of organic mesopores is highly pronounced, leading to a lack of uniformity in pore structures (Figures 6 and 7).

TOC, porosity, brittle mineral content (including siliceous content), and PV are important factors in the classification and evaluation of high-quality shale reservoirs in the Longmaxi Formation. Increases in their values mean that the shale reservoir capacity has improved. The FHH fractal dimension of the deep siliceous shale directly correlates with TOC, porosity, siliceous content, and PV, influenced by a wider range of factors than in the shallow–medium marine shale. Consequently, the better the quality of the deep marine shale reservoir, the more complex the pore structure becomes. Simultaneously, the heterogeneity of the pores, which is influenced by complexity, is also an important factor to be taken into consideration.

5.2.2. Implications for Exploration

A strong shale oil and gas generation base exists in the deep marine shale of the Longmaxi Formation in southern Sichuan [46]. However, our results indicate strong regional specificity of the reservoir. The pore structure of high-quality reservoirs is more complex than in the shallow–medium marine shale, with pronounced heterogeneity in the pore system. A certain degree of heterogeneity in storage space is necessary for effective natural gas storage; however, an overly complex pore structure is not conducive to extraction.

Exploitation of deep marine shale reservoirs is a serious challenge. Clarifying the main influencing factors of pore structure complexity and heterogeneity in deep marine shale will enable us to evaluate and distinguish complex pore space more easily. Further, the reservoir type can be more finely judged. In addition, targeted improvement of organic pore structure may enhance the reservoir's permeability and storage capacity in the future. Therefore, study of the complexity and heterogeneity of pore structures is of great significance for China's drive to commercially exploit deep marine shale gas. For the Longmaxi Formation's deep marine shale reservoir, selecting targets with fractal dimension values close to those of shallow–medium marine shale is a promising direction for deep marine shale gas exploration and development.

6. Conclusions

Deep shale is enriched with a variety of types of organic matter, influenced by thermal evolution, compaction, and mineral contact relationships, leading to development of irregular organic pores. In addition, complex inorganic pore development occurs, resulting in highly complex pore structures in deep marine shale. FE-SEM and fractal dimension do not fully characterize smaller micropores structures; so, the complexity of these pores needs to be further verified by experiments.

Patterns of pore development contribute to the complexity of deep marine shale pore structures. Pores in deep marine shale are widely dispersed and pore sizes are distributed across the entire pore size range, increasing the complexity of both pore types and sizes. However, the shallow–medium marine shale predominantly contains mesopores, so the structure is comparatively simple.

The complexity of deep siliceous shale pore structures is correlated with the abundance of OM, the degree of pore development, and inorganic mineral content. These factors also affect the formation of high-quality reservoirs. Comparatively fewer factors influence the pore structures in deep silty shale and shallow–medium siliceous shale.

Evaluation of lateral heterogeneity in pore structures highlights the strong regional variations in the study area. The heterogeneity of deep marine shale pore systems surpasses that of shallow–medium marine shale. Therefore, the heterogeneity of shale pore systems is largely determined by organic mesopores.

The higher the quality of a deep marine shale reservoir, the more complex the pore structure becomes. At the same time, the pore structures in high-quality reservoirs have strong heterogeneity, which presents challenges for reservoir exploration and development.

Author Contributions: B.Z.: Conceptualization, sorting out ideas, and writing—original draft. J.M.: Writing—review and editing, funding acquisition, and sorting out ideas. R.P.: Writing—review and editing. C.S.: Writing—review and editing, investigation. Z.Z.: Writing—review and editing, investigation. J.J.: Writing—review and editing, investigation. All authors have read and agreed to the published version of the manuscript.

Funding: This research is supported by the Youth Project of the National Natural Science Foundation of China (No. 41402114) and the Outstanding Young and Middle-aged Scientific and Technological Innovation Team Plan Project of Colleges and Universities in Hubei Province (No. t201905).

Institutional Review Board Statement: Not applicable.

Informed Consent Statement: Not applicable.

Data Availability Statement: The data presented in this study are available on request from the corresponding author.

Acknowledgments: The authors would like to thank PetroChina Southwest Oil & Gas field Company.

Conflicts of Interest: The authors declare no conflict of interest.

References

1. Han, K.; Ju, Y.; Wang, G.; Bao, S.; Bu, H.; Neupane, B. Shale composition and pore structure variations in the progradation direction A case study of transitional shales in the Xu-Huai district, Southern North China. *J. Nat. Gas Sci. Eng.* **2016**, *36*, 1178–1187. [CrossRef]
2. Wang, S.; Li, j.; Lu, S.; Zhang, P.; Zhang, J.; Zhang, W. Development characteristics of organic matter pores of marine shale in the Southeastern Chongqing, China. *J. Earth Sci. Environ.* **2019**, *41*, 721–733.
3. Mahjour, S.K.; Correia, M.G.; Santos, A.; Schiozer, D.J. Developing a workflow to represent fractured carbonate reservoirs for simulation models under uncertainties based on flow unit concept. *Oil Gas Sci. Technol.* **2019**, *74*, 15. [CrossRef]
4. Duan, W.; Lin, W.; Tian, J.; Ma, J.; Du, M.; Luo, J. Fractal characteristics and main controlling factors of Wufeng-Longmaxi Formation shale in Luobu Syncline, Southern Sichuan Basin. *Xinjiang Pet. Ceol.* **2022**, *43*, 153–159.
5. Wu, S.; Zhu, R.; Cui, J.; Cui, J.; Bai, B.; Zhang, X.; Jin, X.; Zhu, D.; You, J.; Li, X. Characteristics of lacustrine shale porosity evolution, Triassic Chang 7 Member, Ordos Basin, NW China. *Pet. Explor. Dev.* **2015**, *42*, 167–176. [CrossRef]
6. Zhang, K.; Song, Y.; Jia, C.; Jiang, Z.; Han, F.; Wang, P.; Yuan, X.; Yang, Y.; Zeng, Y.; Li, Y.; et al. Formation mechanism of the sealing capacity of the roof and floor strata of marine organic-rich shale and shale itself, and its influence on the characteristics of shale gas and organic matter pore development. *Mar. Pet. Geol.* **2022**, *140*, 105647. [CrossRef]
7. Liu, S.; Jiao, K.; Zhang, J.; Ye, Y.; Xie, G.; Deng, B.; Ran, B.; Li, Z.; Wu, J.; Li, J.; et al. Research progress on the pore characteristics of deep shale gas reservoirs: An example from the Lower Paleozoic marine shale in the Sichuan Basin. *Nat. Gas Ind.* **2021**, *41*, 29–41.
8. Zhu, B.; Meng, J.; Pan, R.; Hu, H.; Song, C.; Zhu, Z.; Jin, J. New insights into the evaluation criteria for high-quality deep marine shale gas reservoirs in the Longmaxi formation: Evidence from organic matter pore development characteristics. *Front. Ecol. Evol.* **2023**, *11*, 1138991. [CrossRef]
9. Li, Y.; Liu, X.; Cai, C.; Hu, Z.; Wu, B.; Mu, Y.; Duan, X.; Zhang, Q.; Zeng, S.; Guo, J.; et al. Pore Structure Characteristics and Their Controlling Factors of Deep Shale: A Case Study of the Lower Silurian Longmaxi Formation in the Luzhou Area, Southern Sichuan Basin. *ACS Omega* **2022**, *7*, 14591–14610. [CrossRef]
10. Sun, L.; Fu, D.; Wu, Y.; Wang, Z. The current study and development of organic matter pores in organic-rich shale. *Arab. J. Geosci.* **2022**, *15*, 62. [CrossRef]
11. Barry, J.K.; Irene, A. Organic porosity: A geochemist's view of the current state of understanding. *Org. Geochem.* **2018**, *123*, 1–16.
12. Borjigin, T.; Lu, L.; Yu, L.; Zhang, W.; Pan, A.; Shen, B.; Wang, Y.; Yang, Y.; Gao, Z. Formation, preservation and connectivity control of organic pores in the Sichuan Basin. *Pet. Explor. Dev.* **2021**, *48*, 687–699. [CrossRef]

13. Gareth, R.; Chalmers, R.; Marc, B.; Ian, M.P. Characterization of gas shale pore systems by porosimetry, pycnometry, surface area, and field emission scanning electron microscopy/transmission electron microscopy image analyses: Examples from the Barnett, Woodford, Haynesville, Marcellus, and Doig units. *AAPG Bull.* **2012**, *96*, 1099–1119.
14. Wang, Y.; Liu, L.; Cheng, H. Gas adsorption characterization of pore structure of organic-rich shale: Insights into contribution of organic matter to shale pore network. *Nat. Resour. Res.* **2021**, *30*, 2377–2395. [CrossRef]
15. LI, Y.; Hu, Z.; Liu, X.; Cai, Z.; Mu, Y.; Zhang, Q.; Zeng, S.; Guo, J. The pore structure characteristics of deep shale in Longmaxi Formation of Luzhou area. *Fault Block Oil Gas Field.* **2022**, *29*, 584–590.
16. Liu, X.; Zhang, X.; Zeng, X.; Cheng, D.; Ni, H.; Li, C.; Yu, J.; Hu, F.; Li, C.; Wei, B. Pore structure characterization of shales using SEM and machine learning-based segmentation method. *J. China Univ. Pet.* **2022**, *46*, 23–33.
17. Wang, Z.; Jiang, Y.; Fu, Y.; Lei, Z.; Xu, C.; Yuan, J.; Wen, R.; Wang, Z.; Gu, Y.; Yin, X. Characterization of pore structure and heterogeneity of shale reservoir from Wufeng Formation-Sublayers Long-1_1 in Western Chongying based on nuclear magnetic resonance. *Earth Sci.* **2022**, *47*, 490–504.
18. Liu, E.; Liu, C.; Shi, D.; Wang, Y.; Zhu, D.; Xu, Q. Characterization and controlling factors of pore structural heterogeneity of mudstone: A case of upper Paleozoic in northeast Zhoukou Depression, Southern North China Basin. *J. Cent. South Univ.* **2022**, *53*, 3773–3790.
19. Yang, Y.; Wang, K.; Zhang, L.; Sun, H.; Zhang, K.; Ma, J. Pore-scale simulation of shale oil flow based on pore network model. *Fuel* **2019**, *251*, 683–692. [CrossRef]
20. Mahjour, S.K.; Faroughi, S.A. Selecting representative geological realizations to model subsurface CO_2 storage under uncertainty. *Int. J. Greenh. Gas Control* **2023**, *127*, 103920. [CrossRef]
21. Li, D.; Bian, C.; Guo, B.; Zeng, X.; Liang, S. Pore structure and reservoir physical properties for effective development of tight sandstone gas: A case study from the Central Sichuan Basin, China. *Geol. J.* **2022**, *57*, 2497–2510.
22. Li, Y.; Zhou, A.; Xie, W.; Qiu, X.; Dai, Y.; Hu, X.; Cheng, X.; Jiang, Y.; Fu, Y.; Wang, Z. Lithofacies division and main controlling factors of reservoir development in Wufeng Formation-Long'sub-member shale in the Luzhou region, South Sichuan Basin. *Nat. Gas Ind.* **2022**, *42*, 112–123.
23. Zhou, A.; Xie, W.; Qiu, X.; Wu, W.; Jiang, Y.; Dai, Y.; Hu, X.; Yin, X. On Exploration and Development Potential of Shale Gas in Longyi$_1$ ^{4} Sub-bed in Luzhou Block. *Spec. Oil Gas Reserv.* **2022**, *29*, 20–28.
24. Jiang, C.; Zhang, H.; Zhou, Y.; Gan, H.; Pu, J.; Jiang, Y.; Fu, Y.; Gu, Y.; Li, M.; Wang, Z.; et al. Palcogcomorphic characteristics of Wufeng-Longmaxi formation and its influence on development of high-quality shale in Dazu area, Westem Chonging. *J. Cent. South Univ.* **2022**, *53*, 3628–3640.
25. Ma, X. Enrichment laws and scale effective development of shale gas in the southern Sichuan Basin. *Nat. Gas Ind. B* **2019**, *6*, 240–249. [CrossRef]
26. Ma, X.; Xie, J.; Yong, R.; Zhu, Y. Geological characteristics and high production control factors of shale gas reservoirs in Silurian Longmaxi Formation, southern Sichuan Basin, SW China. *Pet. Explor. Dev.* **2020**, *47*, 841–855. [CrossRef]
27. Zhang, L.; Guo, J.; Tang, H.; Liu, J.; Li, Q.; He, J. Pore structure characteristics of Longmaxi shale in the southern Sichuan Basin. *Nat. Gas Ind.* **2015**, *35*, 22–29.
28. Liu, K.; Ostadhassan, M.; Kong, L. Multifractal characteristics of Longmaxi shale pore structures by N_2 adsorption: A model comparison. *J. Pet. Sci. Eng.* **2018**, *168*, 330–341. [CrossRef]
29. Xiong, Y.; Zhou, S.; Jiao, P.; Yang, M.; Zhou, J.; Wei, W.; Cai, J. Fractal analysis of micropore structures in coal and shale based on low-temperature CO_2 adsorption. *Nat. Gas Geosci.* **2020**, *31*, 1028–1040.
30. Panigrahy, C.; Seal, A.; Mahato, N.K.; Bhattacharjee, D. Differential box counting methods for estimating fractal dimension of gray-scale images A survey. *Chaos Solitons Fractals* **2019**, *126*, 178–202. [CrossRef]
31. Pfeifer, P.; Avnir, D. Chemistry in noninteger dimensions between two and three. I. Fractal theory of heterogeneous surfaces. *J. Chem. Phys.* **1983**, *79*, 3558–3565. [CrossRef]
32. Silva, P.M.; Florindo, J.B. A statistical descriptor for texture images based on the box counting fractal dimension. *Phys. A* **2019**, *528*, 121469. [CrossRef]
33. Zhang, Y.; Liu, J.; Xu, H.; Niu, X.; Qin, G.; Cao, D. Comparison between pore structure and fractal characteristics of continental and transitional coal measures shale: A case study of Yan'an and Taiyuan formations at the northeastern margin of Ordos Basin. *Acta Pet. Sin.* **2017**, *38*, 1036–1046.
34. Thommes, M.; Kaneko, K.; Neimark, A.V.; Olivier, J.P.; Reinoso, F.R.; Jean, R.J.; Sing, K.S.W. Physisorption of gases, with special reference to the evaluation of surface area and pore size distribution (IUPAC Technical Report). *Pure Appl. Chem.* **2015**, *87*, 1051–1069. [CrossRef]
35. Zhu, B.; Meng, J.; Pan, R.; Song, C.; Zhu, Z.; Jin, J.; Zhang, C.; Liu, L.; Xu, L.; Zhou, X. Multi-scale characterization of organic matter pore space in deep marine shale combined with mathematical morphology. *Geoenergy Sci. Eng.* **2023**, *223*, 211539. [CrossRef]
36. Zhou, X.; Cuo, W.; Li, X.; Zhang, X.; Liang, P.; Yu, J. Mutual relation between organic matter types and pores with petrological evidence of radiolarian siliceous shale in Wufeng-Longmaxi Formation, Sichuan Basin. *J. China Univ. Pet.* **2022**, *46*, 12–22.
37. Zhu, Y.; Chen, G.; Liu, Y.; Shi, X.; Wu, W.; Luo, C.; Yang, X.; Yang, Y.; Zou, Y. Sequence stratigraphy and lithofacies paleogeographic evolution of Katian Stage-Aeronian Stage in southern Sichuan Basin, SW China. *Pet. Explor. Dev.* **2021**, *48*, 974–985. [CrossRef]
38. Xu, L.; Meng, J.; Pan, R.; Yang, X.; Sun, Q.; Zhu, B. Sequence stratigraphy and implications for shale gas exploration in the Southern Sichuan Basin, South China. *J. Mar. Sci. Eng.* **2023**, *11*, 1393. [CrossRef]

39. Xu, L.; Pan, R.; Hu, H.; Meng, J. Contribution of various shale components to pore system: Insights from attributes analysis. *J. Mar. Sci. Eng.* **2023**, *11*, 1327. [CrossRef]
40. Wang, P.; Jiang, Z.; Ji, W.; Zhang, C.; Yuan, Y.; Chen, L.; Yin, L. Heterogeneity of intergranular, intraparticle and organic pores in Longmaxi shale in Sichuan Basin, South China: Evidence from SEM digital images and fractal and multifractal geometries. *Mar. Pet. Geol.* **2016**, *72*, 122–138. [CrossRef]
41. Liu, S.; Ma, W.; Luba, J.; Huang, W.; Zeng, X.; Zhang, C. Characteristics of the shale gas reservoir rocks in the Lower Silurian Longmaxi Formation, East Sichuan basin, China. *Energy Explor. Exploit.* **2013**, *31*, 187–219. [CrossRef]
42. Lyu, P.; Meng, J.; Pan, R.; Yi, X.; Yue, T.; Zhang, N. Characteristics and differences analysis for thermal evolution of Wufeng-Longmaxi shale, Southern Sichuan Basin, SW China. *Minerals* **2022**, *12*, 906. [CrossRef]
43. Jiang, Z.; Li, X.; Wang, X.; Wang, G.; Qiu, H.; Zhu, D.; Jiang, H. Characteristic differences and controlling factors of pores in typical South China shale. *Oil Gas Geol.* **2021**, *42*, 41–53.
44. Jia, T.; Zhang, S.; Tang, S.; Wang, M.; Xin, D.; Zhang, Q. Characteristics and evolution of low-rank coal pore structure around the first coalification jump: Case study in Southeastern Junggar Basin. *Nat. Resour. Res.* **2022**, *31*, 2769–2786. [CrossRef]
45. Miao, Y.; Li, X.; Lee, J.; Zhou, Y.; Liu, S.; Chang, Y.; Wang, S. Characterization of hydrocarbon/pores generation and methane adsorption in shale organic matter. *Pet. Sci. Technol.* **2018**, *36*, 1187–1193.
46. Guo, X.; Hu, D.; Huang, R.; Wei, Z.; Duan, J.; Wei, X.; Fan, X.; Miao, Z. Deep and ultra-deep natural gas exploration in the Sichuan Basin: Progress and prospect. *Nat. Gas Ind. B* **2020**, *7*, 419–432. [CrossRef]

Journal of
Marine Science and Engineering

MDPI

Article

Estimating Permeability of Porous Media from 2D Digital Images

Gang Lei [1,2], Tianle Liu [1,*], Qinzhuo Liao [3,*] and Xupeng He [4]

1 Faculty of Engineering, China University of Geosciences, Wuhan 430074, China; leigang@cug.edu.cn
2 Shenzhen Research Institute, China University of Geosciences, Shenzhen 518063, China
3 State Key Laboratory of Petroleum Resources and Prospecting, China University of Petroleum-Beijing, Beijing 112249, China
4 EXPEC, Advanced Research Center, Saudi Aramco, Dhahran 34258, Saudi Arabia; xupeng.he@kaust.edu.sa
* Correspondence: liutianle2008@163.com (T.L.); liaoqz@cup.edu.cn (Q.L.)

Abstract: Digital rock physics (DRP) has been widely used as an effective approach for estimating the permeability of porous media. However, conventional implementation of DRP requires the reconstruction of three-dimensional (3D) pore networks, which suffer from intensive memory and underlying uncertainties. Therefore, it is highly significant to develop an approach only based on two-dimensional (2D) cross-sections of parent samples without 3D reconstruction. In this study, we present a novel approach that combines the Kozeny–Carman equation with fractal theory to derive a bridge function that links 2D cross-sectional images and 3D pore structures of parent samples in flow equivalence. Using this bridge function, we predicted the physical properties of the parent samples, including the permeability, bulk porosity, tortuosity, and pore fractal dimension. To validate our model, we performed Lattice Boltzmann (LB) simulations on nine carbonate samples and compared the LB simulation results with our model's predictions. We also compared our predicted results with available data on various porous materials, such as sandstone, glass beads, and carbonate, in the literature. Our findings demonstrate that without reconstructing 3D pore networks, our method provides a reliable estimation of sample permeability using 2D cross-sectional images. This approach not only simplifies the determination of sample permeability in heterogeneous porous media but also sheds new light on the inherent correlations between 2D cross-sectional information and 3D pore structures of parent samples. Moreover, the derived model may be conducible to a better understanding of flow in reservoirs during the extraction of unconventional onshore and offshore oil/gas.

Keywords: porous media; 2D digital image; 3D pore structures; permeability; bridge function

Citation: Lei, G.; Liu, T.; Liao, Q.; He, X. Estimating Permeability of Porous Media from 2D Digital Images. *J. Mar. Sci. Eng.* **2023**, *11*, 1614. https://doi.org/10.3390/jmse11081614

Academic Editor: George Kontakiotis

Received: 28 July 2023
Revised: 14 August 2023
Accepted: 16 August 2023
Published: 17 August 2023

1. Introduction

Permeability, as one of the most fundamental physical properties of porous material [1–5], highly depends on the pore structure [6–9]. Estimating the permeability of porous material from its pore structure has attracted much attention in various environmental and energy applications, such as subsurface water management, geological CO_2 sequestration, geothermal energy extraction, and unconventional onshore and offshore oil and gas recovery processes [10–14].

Generally, there are three approaches for estimating the permeability of porous material, including experimental measurement, empirical models, and digital rock physics (DRP). Experimental measurement is time-consuming and expensive. Moreover, this approach is not repeatable due to its destructive feature. Instead, empirical models, including the well-known Kozeny–Carman (KC) equation and its variations, have been proposed by various researchers [7,10–12,15,16], of which permeability is related by several static parameters, such as porosity, formation factor, specific surface, and other pore structure parameters. Although these empirical models are efficient and cheap [17], they suffer from general applicability due to the requirement of fitting parameters, which are generally rock-type-dependent [12,18].

As an alternative, digital rock physics (DRP) has been widely used [19–28]. However, conventional implementations of DRP require the reconstruction of three-dimensional (3D) pore networks, which suffer from intensive memory and underlying uncertainties. The underlying uncertainties could include resolution loss, structure distortion, non-uniqueness, etc. [26]. Moreover, high-fidelity numerical simulations performed on 3D pore networks are computationally expensive, which limits the general applicability of DRP.

To alleviate such limitations, many efforts have been made to directly estimate permeability from available 2D digital images by investigating the similarities between 2D pore structures and corresponding 3D ones [17,25,26,29,30]. Berryman and Blair [31] predicted the permeability of various porous materials by combining the pore information (porosity and two-point spatial correction function) from 2D SEM images with a form of the KC equation and validated their model with experimental data. However, experimental measurements of the formation factor in their model are missing. Yu and Chen [29] derived a fractal model by incorporating pore structure parameters (e.g., fractal dimension and the pore radius) from 2D images. However, direct extraction pore information from 2D images is different from 3D pore information of parent samples, which is supported by Sisavath et al. [32] and Saxena et al. [26]. Thus, transforming 2D pore information into 3D is necessary. Saxena et al. [26] compared rock permeability in 3D and the permeability computed directly from 2D slices and then proposed a bridge function between the 2D results and the 3D permeability. However, there are two fitting parameters (i.e., the normalized magnitude of variation in the pore radius and the "wavelength" of variation in the pore radius) in their model, which are difficult to be obtained. Recently, by combining the KC equation and Young–Laplace equation, Chen et al. [17] derived a bridge function to link the 2D pore structure and 3D pore structure and predicted the capillary pressure curve (CPC) of porous materials. It was found that their approach is useful to predict the CPC of porous media with a 2D pore structure. However, as they suggested, it was of great practical significance to propose other bridge functions to study the corresponding physical properties (e.g., thermal conductivity coefficient, acoustic conductivity) of porous media. To date, how to effectively transform 2D pore information from 2D digital images into 3D pore information without 3D reconstruction and then predict the 3D permeability of porous material still remains an open question and requires further investigation.

In this paper, we developed a novel, robust, and effective approach to predict the 3D permeability of porous materials from 2D images without 3D reconstruction. Previous studies reveal that interspaces in porous media possess fractal characteristics [12,29,33–37]. Specifically speaking, many scholars [12,29,35,37] suggested that the pore fractal dimension and tortuosity fractal dimension were two key parameters characterizing fluid flow in porous media. Thus, in this paper, these two parameters were applied to characterize the permeability of porous media. As the box-counting algorithm [38] and theoretical models [29,35] can effectively predict the pore fractal dimension [12,39–41], in this paper, the pore fractal dimension was estimated by the box-counting algorithm and theoretical models. In 2015, Wei et al [42]. derived a tortuosity fractal dimension model for fractal porous media that is related to porosity and the pore fractal dimension. As this model is effective at determining the tortuosity fractal dimension [43,44], this model was also used in this paper to model the tortuosity fractal dimension.

Recently, as a promising technique for modeling the fluid flow in porous media at the pore scale, the lattice Boltzmann Method (LBM) has been widely applied by scholars to predict properties (e.g., permeability and relative permeability) of porous media [45–49]. For example, Khodja et al. [46] stated that compared to the traditional methods of computational fluid dynamics, the LBM was ideally suited for massively parallel computation, and it had the advantages of simplicity and flexibility in dealing with complicated geometry. Thus, in this paper, the D3Q19 LBM model was applied to validate our derived model.

In what follows, a new analytical model was derived to transform 2D pore information into a 3D pore structure and give an estimation of the 3D permeability for porous media. Compared with DRP, these models will skip huge workloads of reconstructing 3D pore networks. Subsequently, lattice Boltzmann (LB) simulations were conducted on nine carbonate samples to validate our derived model. In addition, the available experimental data was applied to further validate the developed model and is followed by discussions of this model. Finally, the conclusions are presented.

2. Methodology

In this section, based on the pore morphology of 2D cross-sectional images, an analytical permeability of 3D global matrix pore structure is presented. It should be noted here that, for simplicity, all the 2D cross-sections are perpendicular to the flow direction; thus, the micro-fractures in actual porous media are ignored. The specific steps are as follows (Figure 1): Firstly, pore structure of 2D cross-sectional images will be characterized using fractal theory, and the permeability of 2D pore structure K_{2D} will be derived based on the tube bundle model and fractal geometry [29,39,40]. In the capillary bundle model, the capillary bundles are perpendicular to the images (i.e., capillary bundles run along y direction). Then, by combining the KC equation and fractal geometry, 3D pore structure of porous media in flow equivalence will be estimated, and the permeability of natural samples (in 3D) K_{3D} will be derived. Finally, a bridge function connecting K_{2D} and K_{3D} will be proposed.

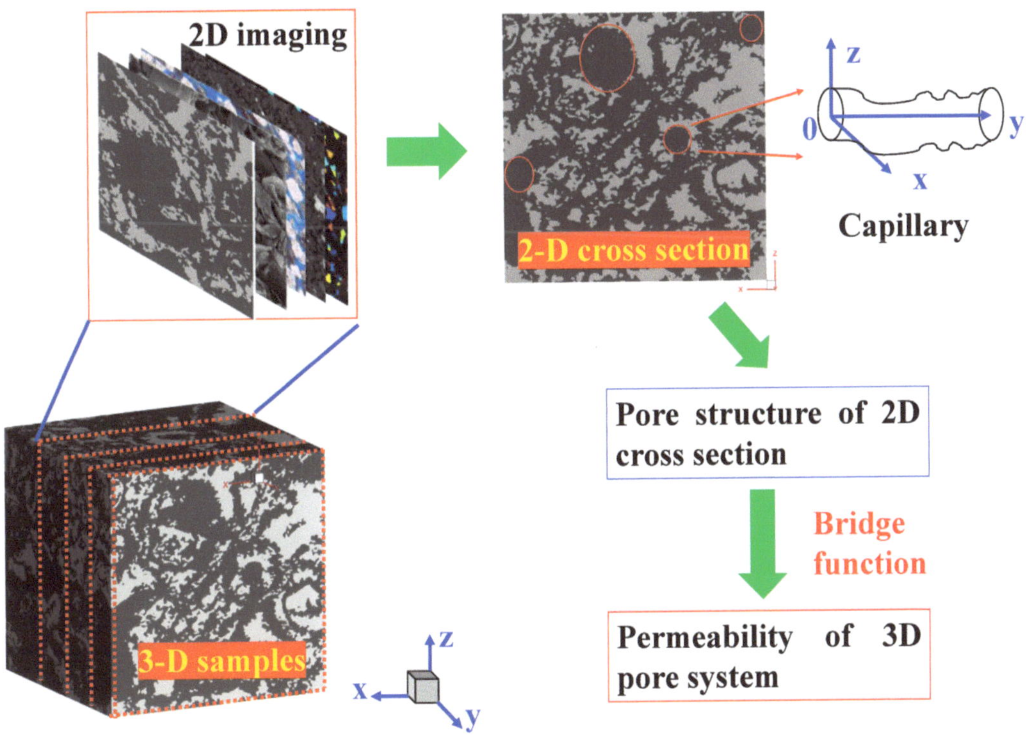

Figure 1. Schematic of transforming the 2D pore information into 3D pore structure and predicting 3D permeability from 2D cross-sections.

Permeability of 2D cross-section: In general, based on 2D imaging (thin section, back scatter electron, micro-CT scanning, scanning electron microscope, scanning electron microscopy, and electron probes, etc.) and digital image processing techniques, pore structure parameters (areal porosity, pore size distribution, average pore radius, pore perimeter, specific surface area, pore fractal dimension, coordination number, and shape factor, etc.) of 2D cross-sections can be determined [26,31,50,51]. For instance, pore fractal dimension of 2D cross-section can be obtained by using the box-counting algorithm [38]. In addition, areal porosity of 2D cross-section can be obtained by thin section analysis or digital image processing techniques. Moreover, by using digital image processing techniques, various pore structure parameters (pore size distribution, the maximum pore size, minimum pore size, average pore radius, pore perimeter, and specific surface area) of 2D cross-sectional images can be easily determined [27,36,46,52,53]. For example, for a given 2D digital image, parameters ($r_{\max}$, $r_{\min}$, and φ_{2D}) can be determined by using the Avizo software, which can effectively determine pore space morphology and extract pore networks [25].

It is well-known that, affected by complex sedimentary and diagenetic processes, pore structure of natural porous media is complicated with obvious heterogeneity, which is difficult to be characterized. Specifically, many scholars suggested that Euclidean mathematics are not suitable to characterize a complex pore structure [29,33,34,54]. With the concept of "Fractal geometry" first proposed by Mandelbrot [33], the problem was well-solved. A large number of studies (experimental and theoretical studies) show that pore structure of most of the sedimentary rocks has obvious fractal phenomena. According to Equation (A4), by combining determined pore structure parameters (e.g., pore radius r, areal porosity φ_{2D}, pore fractal dimension d_f) from 2D imaging and fractal theory, permeability of 2D cross-section (i.e., 2D composite permeated with tortuous capillary tubes that are perpendicular to the 2D cross-section, Figure 1) K_{2D} is given by [12,29,37,40,55]:

$$K_{2D} = \frac{\pi d_f r_{\max}^{d_f}\left(r_{\max}^{3-d_f+d_t} - r_{\min}^{3-d_f+d_t}\right)}{2^{4-d_t}\sqrt{A^{1+d_t}}(3-d_f+d_t)}, \tag{1}$$

where the subscripts max and min denote the maximum and minimum, respectively. More details about Equation (1) can be found in Appendix A. In Equation (1), d_t is tortuosity fractal dimension in 2D space, which is [42]:

$$d_t = (2-d_f+1) + (2-d_f)\frac{\log(d_f) - \log(d_f-1)}{\log(\varphi_{2D})}. \tag{2}$$

Generally speaking, tortuosity fractal dimension d_t ranges from 1 to 2 in 2D space. When the matrix plane is filled with so highly tortuous capillary tubes, d_t is equal to 2; however, for straight capillary tubes, d_t is equal to 1, and on this condition, Equation (1) can be simplified as:

$$K_{2D} = \frac{\pi d_f r_{\max}^{d_f}\left(r_{\max}^{4-d_f} - r_{\min}^{4-d_f}\right)}{8A(4-d_f)}, \tag{3}$$

Furthermore, A in Equations (1) and (3) is the cross-sectional area of the representative elementary volume (REV) in 2D space, which can be calculated by Equation (A3) as:

$$A = \frac{\pi d_f r_{\max}^{d_f}\left(r_{\max}^{2-d_f} - r_{\min}^{2-d_f}\right)}{\varphi_{2D}(2-d_f)} = \frac{\pi d_f r_{\max}^{2}}{\varphi_{2D}(2-d_f)}\left(1 - \frac{r_{\min}^{2-d_f}}{r_{\max}^{2-d_f}}\right). \tag{4}$$

Physically, d_f in Equation (1) is the pore fractal dimension, which ranges from unity to 2 ($1 < d_f < 2$) in 2D pore space. Mathematically, d_f can be determined by box-counting algorithm [38] or by the following equation [29,35]:

$$d_f = d_e - \frac{\ln(\varphi_{2D})}{\ln(r_{\min}/r_{\max})}, \tag{5}$$

wherein the Euclidean dimension d_e is 2 in 2D pore space. Equation (5) reveals that, under a given ratio r_{min}/r_{max}, d_f increases with an increasing of areal porosity φ_{2D}. In addition, by fitting the experimental data, Chen et al. [56] also found a positive correlation between d_f and φ_{2D} and suggested d_f could be estimated by $d_f = 0.974\varphi_{2D}^{0.173}$. Equations (1) through (5) form the basis for analysis of K_{2D}, which were utilized to estimate permeability of 3D parent porous media in flow equivalence.

3D pore structure in flow equivalence: As 2D cross-sections come from 3D parent samples, various studies show that pore structure information (pore morphology) of 2D slice is identical to that of 3D matrix system [17,30,57]. For example, Chen et al. [17] suggested that areal porosity of 2D cross-sections φ_{2D} should be similar to bulk porosity of 3D matrix φ_{3D}. In addition, to estimate permeability in 3D from 2D images, Saxena et al. [26] also assumed that φ_{2D} was similar to φ_{3D}. Figure 2a compares bulk porosity of 3D parent samples versus areal porosity of 2D cross-sections [17,26,31,36,46,53]. Taking the data from Saxena et al. [26], for example, Saxena et al. conducted modeling on 4 samples (i.e., a Fontainebleau sandstone, a Berea sandstone, a Bituminous sand sample, and a Grosmont carbonate). Based on digital rock technology, the bulk porosity φ_{3D} for each sample was determined. Then, multiple 2D cross-sections (lying perpendicular to the flow direction) of a parent cube sample were sampled, and the corresponding areal porosity of each slice was measured using digital image process technology. Similarly, for 3D samples with bulk porosity determined, Wu et al. [36,53] divided the original sample into various layers along vertical direction using low- and high-resolution X-ray CT scanning technologies. Then, areal porosities in each layer of low- and high-resolution digital rock images were measured. Besides digital rock technology, thin section analysis was used by Berryman and Blair [31] and Chen et al. [17] to measure areal porosity of 2D thin. As shown in Figure 2a, porosity of 3D sample φ_{3D} is approximately in the middle between the maximum and minimum values of areal porosities of 2D cross-sections. Specifically, for a 3D porous media with bulk porosity φ_{3D}, most of the areal porosity data φ_{2D} for 2D sectional sections are in the range of $0.9\varphi_{3D}$ and $1.1\varphi_{3D}$. Figure 2b compares bulk porosity of 3D samples versus average porosity of 2D cross-sections. Results suggest that porosity of 3D sample is quite consistent with the average areal porosity of 2D slices. Specifically, the slope of the fitting line is 1.009, which is close to unity. And the intercept of the fitting line is close to 3.38×10^{-4}, which is quite close to 0. As a result, for simplicity of the model, areal porosity of 2D cross-sections φ_{2D} is assumed to be identical to bulk porosity of 3D matrix (parent samples) φ_{3D}, i.e.,

$$\varphi_{3D} = \varphi_{2D}. \tag{6}$$

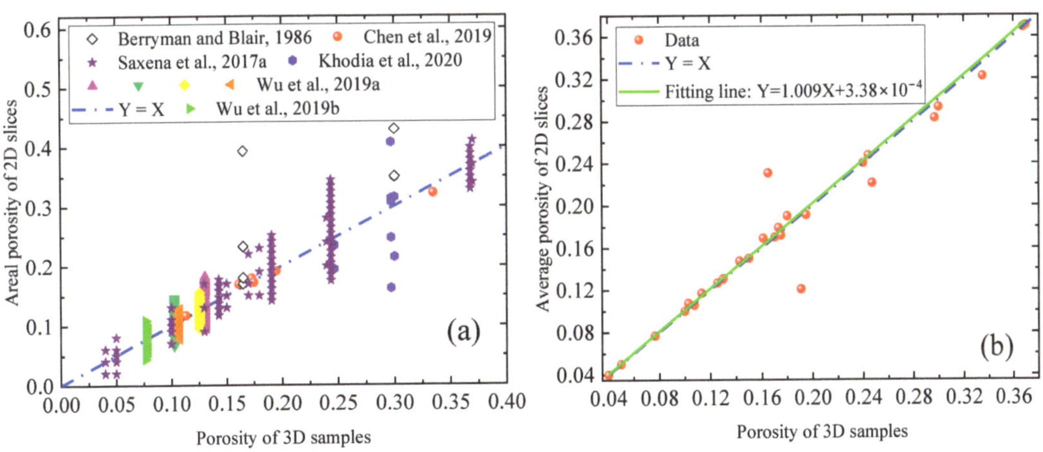

Figure 2. (**a**) Bulk porosity of 3D parent samples versus areal porosity of 2D cross-sections [17,26,31,36,46,53]; (**b**) porosity of 3D samples versus average porosity of 2D slices.

In general, for actual pore structure in 3D space, capillary tube is not straight and tortuosity τ (or tortuosity fractal dimension D_T) is larger than unity. Based on KC equation, Chen et al. [17] developed a bridge function connecting capillary radius in 2D space to that in 3D space. As stated by Chen et al. [17], for a given capillary with pore radius r in 2D space, the corresponding pore radius r_{3D} in 3D space is:

$$r_{3D} = \frac{r}{\tau^4(r_{3D})}. \tag{7}$$

As tortuosity τ is larger than unity, Equation (7) reveals that pore radius r_{3D} in 3D space is less than that in 2D space, and the ratio of r to r_{3D} is tortuosity to the fourth power, i.e., τ^4. With the same method as Chen et al. [17], but assuming that the ratio of specific surface in 3D space to that in 2D space is 3/2 and not the tortuosity, Saxena et al. [26] suggested that the ratio of r to r_{3D} is $9\tau^2/4$. To some extent, $9\tau^2/4$ can be considered a special case of τ^4 (e.g., when τ is assigned as 3/2, $9\tau^2/4$ equals τ^4), so, in this paper, to make our model more general, Equation (7) was used to correlate r and r_{3D}.

According to fractal theory, tortuosity τ of pore radius r_{3D} is [29]:

$$\tau(r_{3D}) = \left(\frac{\sqrt[3]{V}}{2r_{3D}}\right)^{D_T-1}, \tag{8}$$

where V is the bulk volume of the REV. In light of Equation (A10), V can be determined on basis of pore fractal dimension D_f in 3D space as:

$$V = \left\{\frac{\pi D_f r_{3D,max}^{3-D_T}}{4\varphi_{3D}(3-D_T-D_f)}\left[1-\left(\frac{r_{3D,min}}{r_{3D,max}}\right)^{3-D_T-D_f}\right]\right\}^{\frac{3}{3-D_T}}. \tag{9}$$

By combining Equations (7) through (9), we have:

$$\begin{cases} r_{3D,max} = \dfrac{r_{max}}{\left[\sqrt[3]{V}/(2r_{max})\right]^{\frac{4D_T-4}{5-4D_T}}}; & r_{3D,min} = \dfrac{r_{min}}{\left[\sqrt[3]{V}/(2r_{min})\right]^{\frac{4D_T-4}{5-4D_T}}}; \\ \dfrac{r_{3D,max}}{r_{3D,min}} = \left(\dfrac{r_{max}}{r_{min}}\right)^{\frac{1}{5-4D_T}}, & \end{cases} \tag{10}$$

where the subscripts 3D,max and 3D,min denote the maximum and minimum in 3D space, respectively. In what follows, subscript 3D presents in 3D space. Physically, $r_{3D,max}$ should be larger than $r_{3D,min}$, which means D_T in Equation (10) should be less than 1.25. Based on fractal theory, in 3D pore space, D_T is [42]:

$$D_T = (3 - D_f + 1) + (3 - D_f)\frac{\log(D_f) - \log(D_f - 1)}{\log(\varphi_{3D})}, \tag{11}$$

wherein D_f is pore fractal dimension in 3D space, which can be also determined by [29,35]:

$$D_f = 3 - \frac{\ln(\varphi_{3D})}{\ln(r_{3D,min}/r_{3D,max})}. \tag{12}$$

More details about Equation (11) can be found in Equation (A8). By substituting Equation (10) into Equation (12), D_f can be rewritten as:

$$D_f = 3 - \frac{(5-4D_T)\ln(\varphi_{3D})}{\ln(r_{min}/r_{max})}. \tag{13}$$

By combining Equations (11) and (13), we have:

$$D_T = \left[\frac{(5-4D_T)\ln(\varphi_{3D})}{\ln(r_{min}/r_{max})} + 1\right] + \left\langle \frac{(5-4D_T)\ln(\varphi_{3D})}{\ln(r_{min}/r_{max})} \cdot \left\{ \frac{\log\left[3 - \frac{(5-4D_T)\ln(\varphi_{3D})}{\ln(r_{min}/r_{max})}\right]}{\log(\varphi_{3D})} - \frac{\log\left[2 - \frac{(5-4D_T)\ln(\varphi_{3D})}{\ln(r_{min}/r_{max})}\right]}{\log(\varphi_{3D})} \right\} \right\rangle. \quad (14)$$

By solving Equation (14), parameter D_T can be determined. Then, based on Equation (13), D_f can be determined. By substituting Equation (10) into Equation (9), we have:

$$V = \left\{ \frac{\pi D_f r_{max}^{3-D_T} (2r_{max})^{\frac{(4D_T-4)(3-D_T)}{5-4D_T}}}{4\varphi_{3D}(3 - D_T - D_f)} \left[1 - \left(\frac{r_{min}}{r_{max}}\right)^{\frac{3-D_T-D_f}{5-4D_T}}\right] \right\}^{\frac{3(5-4D_T)}{3-D_T}}. \quad (15)$$

By substituting Equation (15) into Equation (10), $r_{3D,max}$ and $r_{3D,min}$ can be determined. Then, according to Equation (A11), 3D permeability K_{3D} of porous media in flow equivalence can be given by:

$$K_{3D} = \frac{\pi D_f 2^{D_T} r_{3D,max}^{D_f} \left(r_{3D,max}^{3-D_f+D_T} - r_{3D,min}^{3-D_f+D_T} \right)}{16 V^{\frac{1+D_T}{3}} (3 - D_f + D_T)}. \quad (16)$$

Figure 3 shows the flow chart for the 3D permeability determination. In light of our derived model, the methodology workflow is summarized as follows:

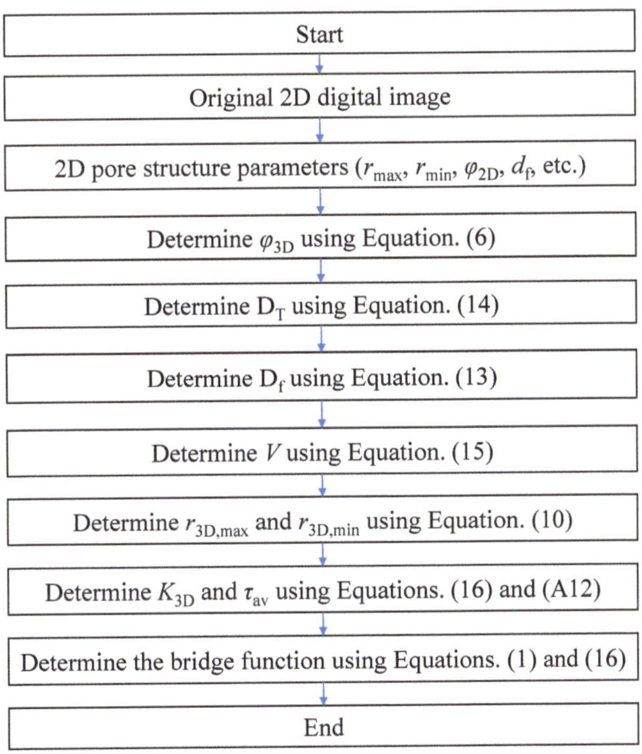

Figure 3. The flow chart for the 3D permeability determination from an original 2D digital image.

Step 1: Determine 2D pore structure parameters (e.g., r_{max}, r_{min}, φ_{2D}, and d_f) from 2D digital image. Specifically, for a given 2D digital image, parameters (r_{max}, r_{min}, and φ_{2D}) can be determined by using the Avizo software [25]. In addition, d_f can be determined with box-counting algorithm [38] or with Equation (5). Subsequently, parameter φ_{3D} can be calculated using Equation (6).

Step 2: Determine parameter D_T by solving Equation (14). Then, determine D_f by using Equation (13). In addition, determine parameter V by using Equation (15). Then, $r_{3D,max}$ and $r_{3D,min}$ can be determined by using Equation (10).

Step 3: Determine K_{3D} and τ_{av} of the corresponding parent sample using Equations (16) and (A12). In addition, the bridge function (i.e., the ratio of K_{2D} to K_{3D}) can be determined by combining Equations (1) and (16). Moreover, other 3D pore structure parameters of the parent sample in flow equivalence can be further estimated.

3. Model Validation

Sampling and Experiments: 3 carbonate core plugs (labeled sample 1, sample 2, sample 3) with a diameter of 25 mm and thickness of 6 mm in the Middle East were selected. For these samples, multiple cubical sites were semi-randomly selected and then used in micro-CT X-ray imaging with the Versa XRM-500 X-ray micro-CT scanner and in simulation with the lattice Boltzmann method (D3Q19 model) to determine the permeability in 3D. Taking sample 1, for example, two cubical sites (site A and site B) were semi-randomly selected to avoid vugs. For site A with a bulk porosity φ_{3D} of 23.5% and a total volume of 1.64^3 mm^3, the volume (voxel) is 648^3, and the voxel volume is 2.53^3 μm^3. With the D3Q19 model stated in Appendix B, the determined permeability in 3D is 27.8×10^{-3} μm^2. In addition, for site B with a bulk porosity φ_{3D} of 23.4%, the volume (voxel) is 681^3, the voxel volume is 1.55^3 μm^3, and the total volume is about 1.06^3 mm^3. Then, based on LBM simulation, the 3D permeability of site B was determined, with the value of 69.4×10^{-3}μm^2. More detailed information about the D3Q19 model can be found in Appendix B. For the sake of simplification, site J (J = A, B, C, ⋯) from the sample I (I = 1, 2, 3) is labeled as sample I–J.

To verify the derived model in Section 2, a cross-section of each sample was sliced and used to estimate the permeability of the corresponding parent sample. Specifically speaking, firstly the 2D pore structure parameters (e.g., r_{max}, r_{min}, φ_{2D}, and d_f) from the 2D digital image were determined; then, based on our derived model, the 3D permeability of the carbonates was predicted. In addition, our predicted 3D permeability was compared to that from D3Q19 LBM to validate the feasibility and effectiveness of our derived model. The experimental data and modeling results are summarized in Table 1. As shown in Table 1 and Figure 4a, the areal porosity (2D porosity) of the cross-section consists of the bulk porosity (3D porosity) of the parent sample. The results (Figure 4a) suggest the slope and the intercept of the fitting line are 1.048 (close to unity) and 7.56×10^{-3} (close to 0), respectively. Similar findings are shown in Figure 2. The results shown in Table 1 suggest that the predicted 3D permeability of each parent sample from the derived model exhibits excellent agreement with the experimental data (the results from the LBM simulations). Taking parent sample 1-A (Figure 4a), for example, in light of the image processing technology, the φ_{2D} of the selected 2D cross-section (Figure 4b) is 31.24%. In addition, with the box-counting algorithm [38], the double logarithmic (Lg) plot of the box size (pixel) versus the total pore number is presented in Figure 4b; then, the pore fractal dimension d_f (the negative of the slope) was determined to be 1.686. As shown in Figure 4b, the pore structure of the 2D cross-section of sample 1-A has a typical fractal scale. In our modeling, d_f and φ_{2D} were assigned as 1.686 and 31.24%, respectively. r_{max} was assigned as 535.06 μm, and φ_{3D} was assigned as 23.5%. Moreover, $r_{3D,max}$ and D_f were assigned as 41.18 μm and 2.81, respectively. Then, based on the derived model, the predicted permeability K_{3D} is 27.8×10^{-3}μm^2, which is in agreement with that determined by the LBM simulation. Figure 4c presents the pore structure parameters (e.g., τ_{av}, $r_{max}/r_{3D,max}$, and D_T) of the samples. As we can see from Figure 4c, the parameter r_{max} is larger than $r_{3D,max}$, which is consistent with the results stated by Chen et al. [17]. For these 9 samples, the ratio

$r_{max}/r_{3D,max}$ ranges from 3.18 to 10.58. In addition, D_T ranges from 1.10 to 1.15, and τ_{av} ranges from 2.19 to 4.69. The results (Figure 4c) also suggest that even from the same parent sample, the different sites behave strongly in heterogeneity. For instance, sample 1-A and sample 1-B come from the same parent sample 1; however, their physical properties vary greatly. Figure 4d presents the average tortuosity versus the ratio $r_{max}/r_{3D,max}$. The results (Figure 4c) show that there exists an approximately linear relationship between the ratio of the maximum pore radius in 2D space to that in 3D space ($r_{max}/r_{3D,max}$) and the average tortuosity of porous media. Physically, when the average tortuosity of porous media is unity, the corresponding ratio $r_{max}/r_{3D,max}$ approximately equals unity. As we can see from the fitting equation, when τ_{av} is assigned as unity, $r_{max}/r_{3D,max}$ is determined to be 0.82, which is close to unity, which verifies our model. The reason for the deviation may be caused by the computational error.

In general, carbonates generally contain clay minerals, and the interaction between clay minerals and fluids (e.g., water, gas, oil) will significantly affect the permeability of porous media [37,40]. For example, Lei et al. [37] derived analytical permeability to study the effect of clay swelling on the permeability of clay-rich argillaceous porous media. In addition, as the interaction between solid minerals (e.g., illite, montmorillonite, and kaolinite) and fluids in porous media (e.g., carbonate media) will change pore structures, the effect of the physical or chemical nature of solid minerals on the fluid flow in porous media is of great significance. However, the derived model in this paper does not consider the interaction between solid minerals and fluids. Thus, in our future work, the interaction between solid minerals and fluids will be taken into account to make our model more reasonable.

Table 1. Experimental data and modeling results.

No.	Samples [a]	LBM		Proposed Model				
		φ_{3D} (%)	K_{3D} (10^{-3} μm^2)	φ_{2D} (%)	d_f	$R_{3D,max}$ (μm)	D_f	K_{3D} (10^{-3} μm^2)
1	1-A	23.5	27.8	31.24	1.686	41.18	2.81	27.8
2	1-B	23.4	69.4	21.35	1.450	43.15	2.79	69.4
3	2-A	31.5	1450	35.71	1.576	64.04	2.78	1450
4	2-B	21.5	216.7	26.44	1.683	70.31	2.79	216.7
5	2-C	34.9	1743.1	38.55	1.611	60.24	2.79	1743.1
6	3-A	30.7	906.9	34.97	1.735	89.07	2.82	906.9
7	3-B	16.2	103.1	10.81	1.415	49.04	2.74	103.1
8	3-C [b]	31.3	3049.4	36.97	1.631	61.83	2.76	3049.4
9	3-D [b]	40.7	2660	36.74	1.636	60.20	2.82	2660

[a] The samples are the sites in the end trims cut off from the original samples in the literature [46]. [b] Samples 3-C and 3-D denote sites D and E of sample 3 in Table 1 of the literature [46].

Dataset of Saxena et al. [26]: To further verify our derived model, the predicted K_{2D} and K_{3D} from the new model were compared with available test data [26]. Saxena et al. conducted numerical experiments on 12 samples from various geologic formations [26]. For these 12 samples, samples 1 to 3 come from the same parent Berea sandstone and represent Berea sandstones A to C, respectively. In addition, samples 4 to 6 represent Bituminous sands A to C, respectively. Samples 7 to 9 represent Fontainebleau (Font) sandstones A to C, respectively. And samples 10 to 12 represent Grosmont carbonates A to C, respectively. During the numerical experiments, Saxena et al. first calculated the K_{3D} of the parent samples based on 3D digital rocks using an LBM simulation; then, they sliced 2D thin sections along the flow direction from the parent samples and performed an LBM simulation to estimate the K_{2D} of the thin section [26]. It should be noted that to calculate the K_{2D} of the thin section, the pore structure was restructured by permeating the pores in the thin section with straight tubes along the flow direction. That is, for the restructured porous media used to determine K_{2D}, the tortuosity fractal dimension d_t was equal to 1. As a result, during our modeling, d_t was assigned as 1, and Equation (3) was used to validate our derived model.

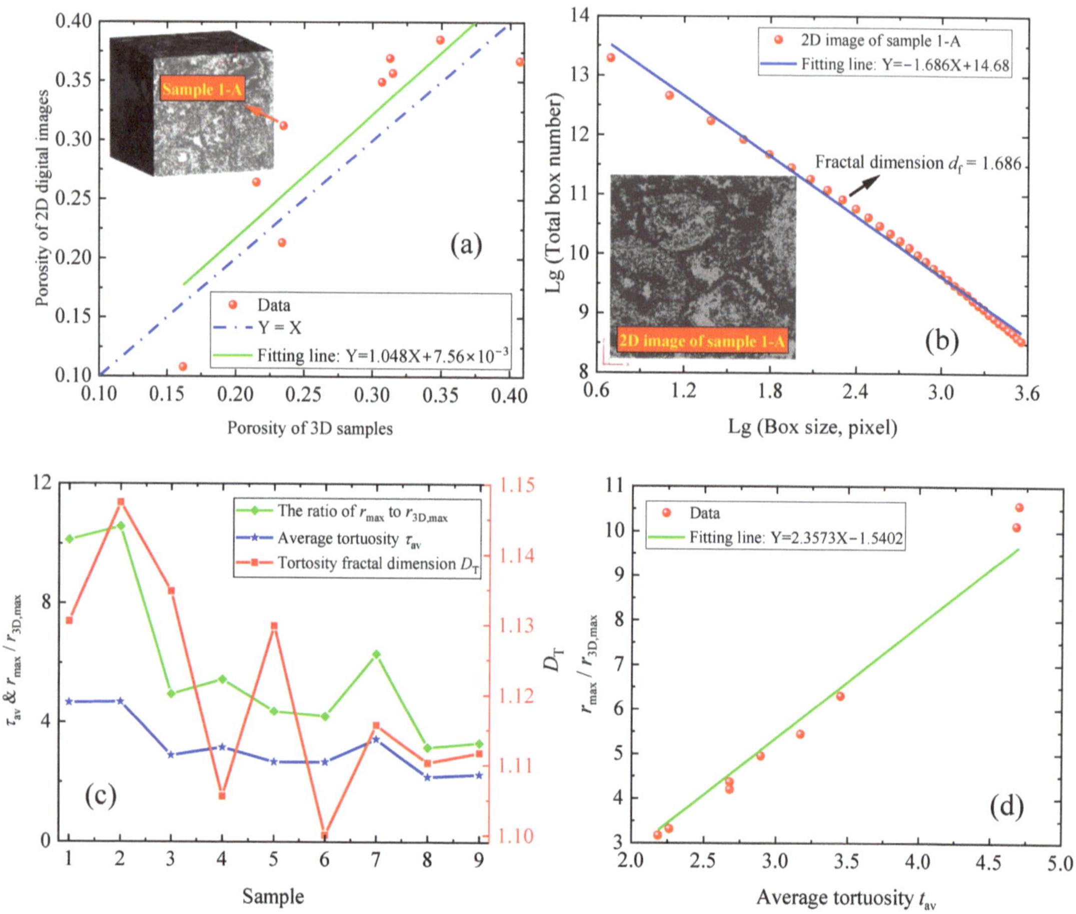

Figure 4. Modeling results of the samples: (**a**) the curves of φ_{2D} versus φ_{3D}; (**b**) d_f of the 2D image of sample 1-A; (**c**) the determined pore structure parameters for various samples; (**d**) τ_{av} versus the ratio $r_{max}/r_{3D,max}$ for various samples.

Taking sample 1 (sample Berea A) with a volume of 800 × 800 × 800 μm^3 and a pixel size of 0.74 μm, for example, the bulk porosity φ_{3D} is 4%, and the areal porosity φ_{2D} of the thin sections ranges from 2% to 6% with an average of 4%. Based on simulations, the LBM 3D permeability of the parent sample was determined as 3×10^{-3}μm^2, and the average LBM 2D permeability of the thin sections is 624×10^{-3}μm^2. In our modeling, both φ_{3D} and φ_{2D} were assigned as 4%. Based on Equations (3) and (13), K_{2D} and K_{3D} can be calculated. The results (Figure 5a–d) suggest that the predictions (K_{2D} and K_{3D}) provide a good match over the test data. Besides K_{2D} and K_{3D}, other parameters (e.g., d_f, τ_{av}, $r_{max}/r_{3D,max}$, D_T, and D_f) of each sample are presented in Figure 5e. As we can see from Figure 5e for these 12 samples, τ_{av} ranges from 1.10 to 1.50, which is smaller than that of the carbonates in Figure 4. This reveals that the carbonates in Figure 4 have stronger heterogeneity than the rocks in Figure 5. The reason may be that the rocks in Figure 4 are carbonate core plugs; however, most of the rocks in Figure 5 are sandstones, which are more homogeneous. In addition, Figure 5e shows that the ratio $r_{max}/r_{3D,max}$ ranges from 1.15 to 1.74, D_T ranges from 1.01 to 1.03, d_f ranges from 1.64 to 1.82, and D_f ranges from 2.69 to 2.83. The results (Figure 5f) also suggest that τ_{av} and the ratio $r_{max}/r_{3D,max}$ are linearly dependent. When τ_{av} is assigned as unity, the ratio $r_{max}/r_{3D,max}$ is estimated to be 0.93, which is approximately equal to unity. Similar findings are shown in Figure 4e.

Dataset of Berryman and Blair [31]: Based on digital image analysis, Berryman and Blair first studied the two-point correlation functions, porosity, and specific surface area of glass beads (55 μm), Ironton Galesville (IG) sandstones, and Berea sandstones [31]. Then, they predicted the permeability of these samples with the following equation:

$$K_{3D} = \frac{\varphi_{2D}^2}{bFS_{2D}^2}, \tag{17}$$

where F is the formation factor, and b is a constant that depends on the cross-section of the tubes. It is recommended that b is assigned as 2 and 3 for circular tubes and flat cracks, respectively.

In order to further verify our derived model, the predictions from our derived model were compared against the available data of Berryman and Blair [31]. Taking sample 3 (sample Berea 1) with an image specific surface of 0.0281 μm^{-1}, an areal porosity φ_{2D} of 17%, and an image permeability K_{2D} of 0.312 μm^2. For example, Berryman and Blair found the bulk porosity (laboratory porosity) φ_{3D} ranged from 15% to 18% with an average value of 16.5%. In addition, based on Equation (14), the laboratory permeability of K_{3D} was determined to be 0.023 μm^2. In our modeling, φ_{3D} and φ_{2D} were assigned as 16.5% and 17%, respectively. The results (Figure 6a) suggest that the predicted results (K_{2D} and K_{3D}) of these samples (i.e., two glass beads, four Berea samples, IG-775 and IG-785) provide a good match over the test data. In addition, the pore structure parameters (e.g., d_f, τ_{av}, $r_{max}/r_{3D,max}$, D_T, and D_f) of each sample are presented in Figure 6b. As pore structures (d_f, τ_{av}, $r_{max}/r_{3D,max}$, D_T, and D_f) of these samples are different, the curves show different behaviors. In addition, the results (Figure 6c) also suggest that τ_{av} is remarkably correlated linearly with the ratio $r_{max}/r_{3D,max}$.

For a given porous medium (bulk porosity φ_{3D}) composed of identical spherical grains, the average particle radius R_{ap} of the porous medium could be written as [26,31,58]:

$$S_{3D} = 4\pi R_{ap}^2 / \left(\frac{4\pi R_{ap}^3/3}{1-\varphi_{3D}} \right) = \frac{3(1-\varphi_{3D})}{R_{ap}} \Rightarrow R_{ap} = \frac{3(1-\varphi_{3D})}{S_{3D}}, \tag{18}$$

where S_{3D} is the specific surface area. Based on the derivation in Appendix C, R_{ap} can be also expressed as:

$$R_{ap} = \lambda R_{avs} + (1-\lambda) R_{avc}, \tag{19}$$

where λ is the weight coefficient of the average spherical particle radius, R_{avs} is the average spherical particle radius, and R_{avc} is the average circular particle radius. More details about R_{avs} and R_{avc} can be found in Equations (A22) and (A24). By combining Equations (15) and (16), the weight coefficient of the average spherical particle radius can be determined. Figure 6d compares the calculated average particle radius for spherical particles and the calculated average particle radius for circular particles with that from the former model. The results from Figure 6d demonstrate that the calculated average particle radius from Equation (15) is approximately in the middle between the calculated values from Equations (A22) and (A24). This indicates that the model gives predicted values that are quite consistent with the results from the former model. In addition, by combining the results from Equations (15) and (16), the weight coefficients for different samples have been determined in Figure 6d. Taking sample 3 for example, when the weight coefficient is unity, the average particle radius can be effectively determined by Equation (A22).

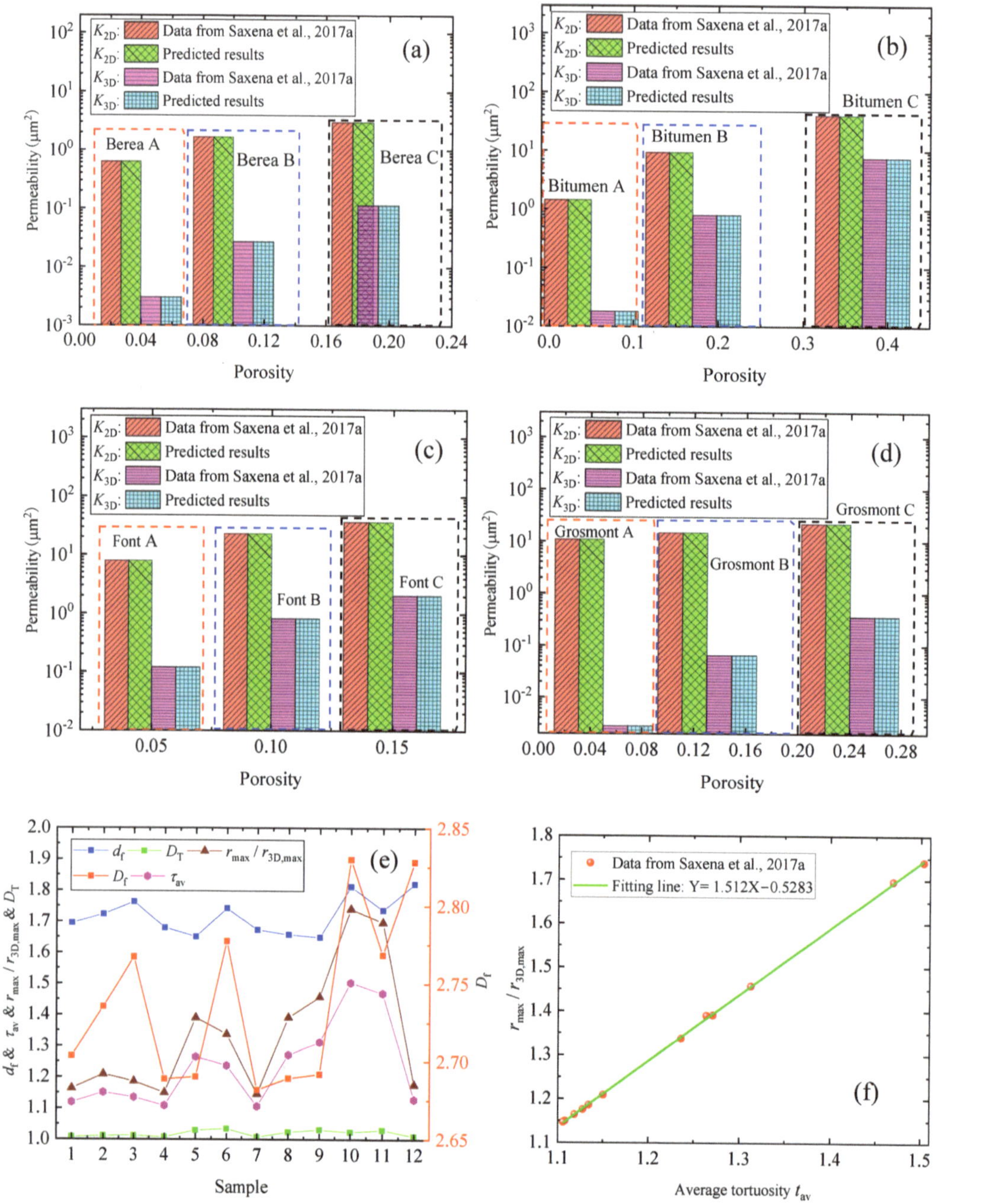

Figure 5. Modeling results of the samples [26]: (**a**) the predicted permeabilities from the derived model versus the data for Berea samples; (**b**) the predicted permeabilities from the derived model versus the data for Bitumen samples; (**c**) the predicted permeabilities from the derived model versus the data for Font samples; (**d**) the predicted permeabilities from the derived model versus the data for Grosmont samples; (**e**) the determined pore structure parameters for various samples; (**f**) τ_{av} versus the ratio $r_{max}/r_{3D,max}$ for various samples.

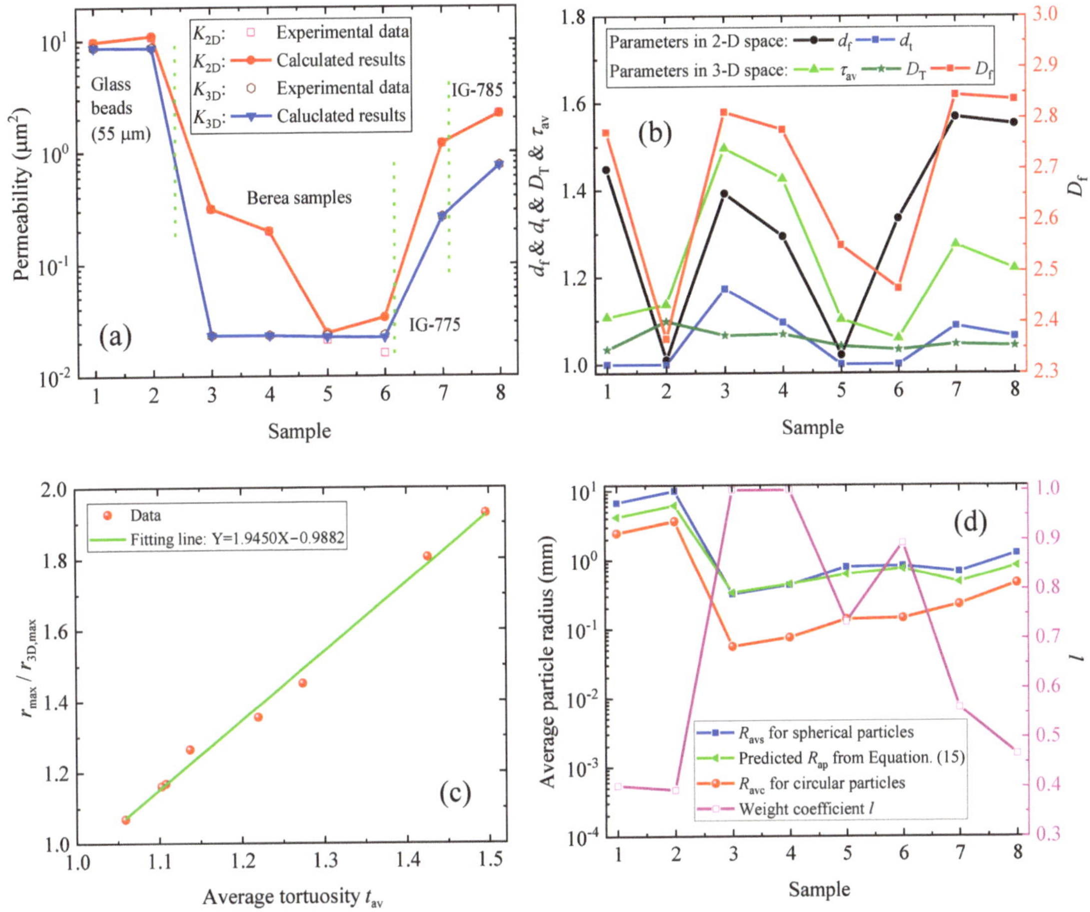

Figure 6. Modeling results of the samples: (**a**) predicted permeability (K_{2D} and K_{3D}) and experimental data; (**b**) the determined pore structure parameters (d_f, d_t, τ_{av}, D_T, and D_f) of various samples; (**c**) τ_{av} versus the ratio $r_{max}/r_{3D,max}$ for various samples; (**d**) the predicted average particle radius for spherical particles and circular particles versus that from Equation (15).

4. Results and Discussions

After validation with exhaustive experimental data, this derived model was utilized for sensitivity analysis of different parameters (D_f and D_T) on the 3D permeability of porous media. Figure 7 shows the influence of pore fractal dimension D_f on the average tortuosity τ_{av} and K_{3D} of porous media in 3D. In this case, the maximum and minimum pore radii in 2D were 3 μm and 0.13 μm, respectively. The parameters φ_{2D} and φ_{3D} were both 0.15, and parameter D_T was 1.1. During the calculation, the parameter D_f ranged from 2.2 to 2.8. Based on our derived model, the maximum pore radius in 3D space $r_{3D,max}$ was determined, which ranged from 1.4 μm to 0.82 μm. In addition, the determined parameter $r_{3D,min}$ ranged from 0.0077 μm to 0.0045 μm. Specifically, when parameter D_f was assigned as 2.2, the corresponding values of $r_{3D,max}$ and $r_{3D,min}$ were 1.4 μm and 0.0077 μm, respectively. As one can see from Figure 7a, there exists a positive relationship between D_f and the average tortuosity τ_{av}. The main reason is that a larger value of D_f means a more complex pore structure of porous materials, leading to a larger value of τ_{av}. Furthermore, the permeability of porous media in 3D decreases with an increase in D_f, which is anticipated. Figure 7b demonstrates that the linear correlativity between τ_{av}

and the ratio $r_{max}/r_{3D,max}$ is very prominent. Moreover, when τ_{av} is assigned as unity, $r_{max}/r_{3D,max}$ is determined to be 0.9951, which is close to unity.

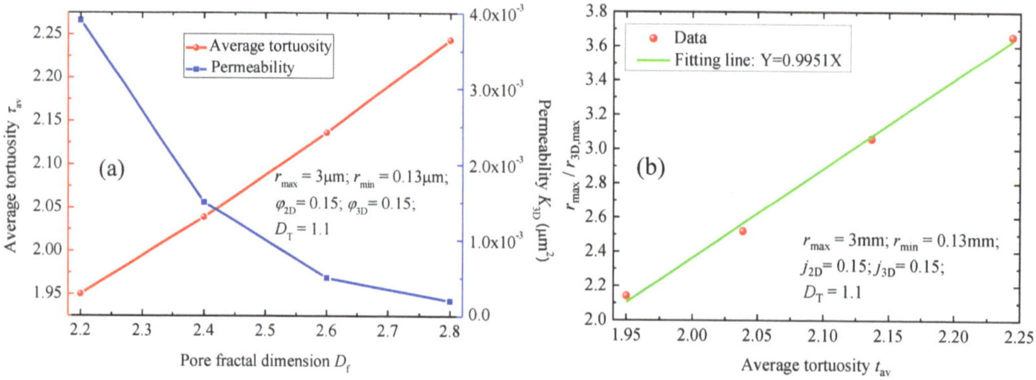

Figure 7. Effect of pore fractal dimension D_f on properties of porous media in 3D: (**a**) curves of D_f versus τ_{av} and K_{3D}; (**b**) τ_{av} versus the ratio $r_{max}/r_{3D,max}$.

The influence of parameter D_T on the τ_{av} and K_{3D} are shown in Figure 8. In this case, the basic parameters applied in the model are summarized in the corresponding figures. During the calculation, the parameter D_f was assigned as 2.2, and the parameter D_T ranged from 1.05 to 1.2. Based on our derived model, the parameter $r_{3D,max}$ ranges from 2.2 μm to 0.07 μm. The results (Figure 8a) suggest that a larger τ_{av} corresponds to a larger D_T. However, K_{3D} decreases as D_T increases. Specifically, for this case, when D_T increases up to a certain value (e.g., $D_T \geq 1.2$), the value of K_{3D} is extremely small. The main reason is that a larger value of D_T means larger seepage resistance, resulting in a small value of K_{3D}. Figure 8b also reveals that τ_{av} has distinct linear correlations with the ratio $r_{max}/r_{3D,max}$. Similar findings have been also demonstrated in Figure 6.

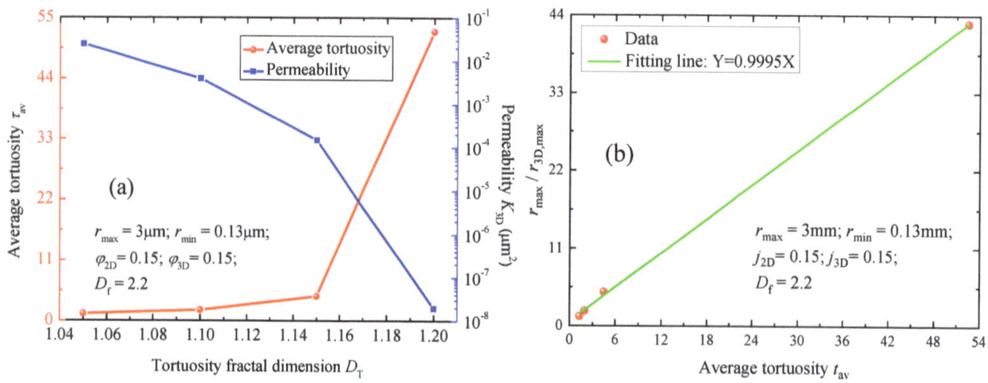

Figure 8. Effect of tortuosity fractal dimension D_T on properties of porous media in 3D: (**a**) curves of D_T versus τ_{av} and K_{3D}; (**b**) τ_{av} versus the ratio $r_{max}/r_{3D,max}$.

Based on the definition of the bridge function f_{bridge} connecting K_{2D} and K_{3D} (i.e., $f_{bridge} = K_{2D}/K_{3D}$), we studied the influences of D_f and D_T on this bridge function f_{bridge}. Figure 9 presents the influences of D_f and D_T on f_{bridge}. For the calculation, the parameters applied in the model are summarized in the corresponding figures, which are identical to those for Figures 7 and 8, respectively. As one can see from Figure 9, the bridge function increases as D_f (or D_T) increases. The main reason is that for a given 2D pore structure, a larger value of D_f (or D_T) corresponds to a smaller value of K_{3D}. Similar findings can be found in Figures 7a and 8a. Moreover, Figure 9 reveals that with an increase of D_f (or D_T), the increase rate of f_{bridge} increases. The main reason is that with an increase of D_f (or D_T), the increase rate of the seepage resistance in the porous media increases.

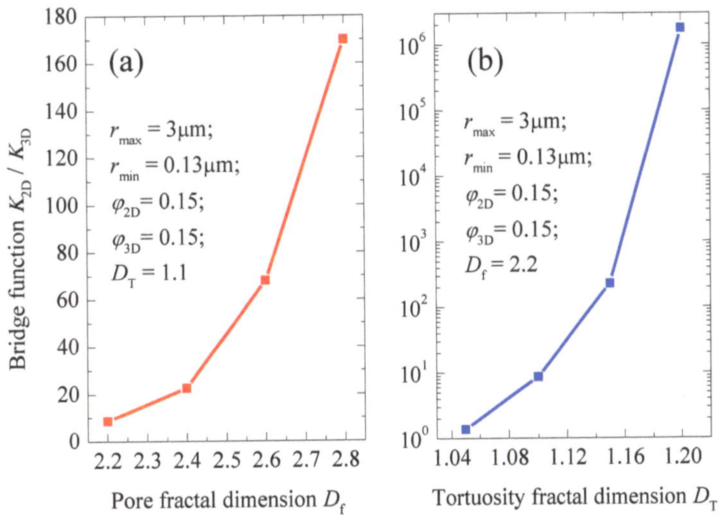

Figure 9. Effects of D_f and D_T on the bridge function f_{bridge}: (**a**) pore fractal dimension D_f versus the bridge function K_{2D}/K_{3D}; (**b**) tortuosity fractal dimension D_T versus the bridge function K_{2D}/K_{3D}.

Model advantages and limitations: The developed model provides a theoretical basis for predicting the 3D permeability of porous media from 2D digital image analysis without the need for 3D reconstruction, achieving high accuracy even with only 2D information. Compared to DRP, our model not only reduces the computational cost of high-resolution scanning, 3D pore network reconstruction, and numerical simulation but also captures the effect of realistic pore shapes on permeability. Furthermore, our model can be used for inverse modeling to estimate relevant parameters, such as tortuosity and the pore fractal dimension in 3D, making it highly practical for estimating permeability in heterogeneous porous media. Overall, our derived model is of great practical significance, as 2D images are easier and cheaper to access than 3D images and reconstructions of 3D pore networks.

However, it is worth noting that our derived model shares the problems of uncertainty and challenge in terms of predicting 3D pore structures and the permeability of porous media. Physically speaking, for a given 2D image, the possible 3D samples are infinite. That is, for a given 2D image, there are infinitely many permeabilities to this problem. As our model is derived based on fractal theory, it is limited to fractal porous media; however, it may not be suitable for some porous media. In addition, in our modeling, the areal porosity of 2D cross-sections φ_{2D} is assumed to be identical to the bulk porosity of the 3D matrix. However, φ_{2D} varies with the slices, and sometimes the difference between the maximum and the minimum 2D porosity is relatively large. Although φ_{3D} is approximately in the middle between the maximum and minimum values of φ_{2D}, 2D porosity is not enough to represent φ_{3D}. Moreover, our developed model is limited to intact porous media and ignores the effect of micro-fractures on the 3D permeability of porous media. Thus, further research is required to reduce the uncertainty in estimating the 3D permeability of porous media from 2D images without reconstruction. Furthermore, in general, the pore surface of porous materials is rough, and the surface fractal dimension is crucial to characterize the fluid flow in porous media [41,59,60]. For example, Lei et al. [59] and Xiao et al. [41] derived theoretical models to study the fluid flow in porous media, and they concluded that the effects of surface fractal dimension on permeability and relative permeability in porous media were significant. Thus, to make our model more reasonable, in our future work, a rough pore surface and surface fractal dimensions will be taken into account. Moreover, as mentioned above, the interaction between solid minerals and fluids will significantly affect the permeability of porous media. To improve the applicability of our derived model, the interaction between solid minerals and fluids will be taken into account. What is more, as the information on fracture systems may not be obtained from 2D cross-sections, in this paper, our derived model focuses on predicting the 3D permeability of intact porous media, and the fractures are ignored. In our future work, we will try to extend our model to study the 3D permeability of fractured porous media from 2D cross-sections of parent samples without 3D reconstruction.

5. Conclusions

In this paper, a novel analytical model was derived to estimate the 3D permeability of porous media from 2D cross-sectional images without reconstruction. Based on fractal theory and the Kozeny–Carman equation, a bridge function was developed to correlate the 2D pore information from the 2D images and the 3D pore structure of the parent samples. The derived model was validated with the results from lattice Boltzmann method (LBM) simulations and various experimental data and is shown to perform well.

The derived model was also used to conduct sensitivity analysis of different parameters (e.g., tortuosity fractal dimension in 3D space D_T, pore fractal dimension in 3D space D_f) on the 3D permeability K_{3D} of porous media. The results indicate that the average tortuosity τ_{av} decreases with an increase of D_f (or D_T). In addition, the average tortuosity τ_{av} is remarkably correlated linearly with the ratio of r_{max} (the maximum pore radius of 2D images) to $r_{3D,max}$ (the maximum pore radius of 3D porous media). Moreover, the permeability K_{3D} decreases as D_f (or D_T) increases. With an increase of D_f (or D_T), the decrease rate of the permeability K_{3D} increases.

The proposed model not only reveals the Intrinsic link of 2D pore information and 3D pore structure in unconventional oil/gas reservoirs but also reduces the computational cost of high-resolution scanning and flow simulation to predict the 3D permeability of unconventional reservoirs. Our proposed model can be applied in many fields, such as CO_2 geology storage, unconventional onshore and offshore oil/gas development, and groundwater seepage.

Although the model focuses on predicting single-phase flow in dry porous media, when the irreducible wetting phase saturation is taken into account, our model can be extended to study the 3D effective permeability of porous media. In addition, it is also available to extend this model to study multi-phase flow in porous media and obtain relative permeability in porous media.

However, it should be noted that our proposed model never considers a rough pore surface (e.g., surface fractal dimension) or interactions in solid minerals–fluid systems. Thus, in our future work, more mechanisms will be taken into account to make our model more reasonable.

Author Contributions: Conceptualization, G.L. and T.L.; methodology, G.L.; software, G.L.; validation, G.L., T.L. and Q.L.; formal analysis, G.L.; investigation, Q.L. and X.H.; resources, Q.L. and X.H.; data curation, G.L.; writing—original draft preparation, G.L.; writing—review and editing, G.L., T.L., Q.L. and X.H.; visualization, G.L. and T.L.; supervision, Q.L. and X.H.; project administration, T.L. and Q.L.; funding acquisition, G.L. and T.L. All authors have read and agreed to the published version of the manuscript.

Funding: This study is jointly supported by the GuangDong Basic and Applied Basic Research Foundation (grant number 2022A1515110376), the Fundamental Research Funds for the Central Universities, China University of Geosciences (Wuhan) (grant number 107-G1323523046) and the National Natural Science Foundation of China (grant number 22208329).

Institutional Review Board Statement: Not applicable.

Informed Consent Statement: Not applicable.

Data Availability Statement: All data have been provided in the paper.

Acknowledgments: The authors are grateful for financial support from the GuangDong Basic and Applied Basic Research Foundation (No. 2022A1515110376), the Fundamental Research Funds for the Central Universities, China University of Geosciences (Wuhan) (No. 107-G1323523046), the National Natural Science Foundation of China (22208329), and the Independent Prospective Basic Project of State Key Laboratory of Offshore Oil Exploitation in 2023.

Conflicts of Interest: We declare that we have no conflict of interest.

Nomenclature

Latin symbols	
A	The cross-sectional area of the representative element (μm^2)
A_{av}	The average pore area (μm^2)
b	A constant that depends on the cross-section of the tubes (dimensionless)
c	Lattice sound speed, which is determined by $\Delta x/\Delta t$ (m/s)
d_e	The Euclidean dimension is 2 in 2D space (dimensionless)
d_f	The pore fractal dimension in 2D space (dimensionless)
d_t	The tortuosity fractal dimension in 2D space (dimensionless)
D_f	The pore fractal dimension in 3D space (dimensionless)
D_T	The tortuosity fractal dimension in 3D space (dimensionless)
$\mathbf{e}_\alpha$	Lattice velocities (m/s)
$\mathbf{e}_y$	The unit vector along the y-axis (m/s)
F	The formation factor (dimensionless)
f	The pore size distribution in 2D space (dimensionless)
f_{3D}	The pore size distribution in 3D space (dimensionless)
$f_\alpha(\mathbf{y}, t)$	The evolution of the density distribution function (kg/m^3)
K_{2D}	Permeability of the 2D cross-section (10^{-3} μm^2)
K_{3D}	3D permeability of porous media in flow equivalence (10^{-3} μm^2)
K_{ij}	Permeability in the i direction when the flow is driven in the j direction (10^{-3} μm^2)
L	The actual streamlined length of a tortuous capillary (μm)
L_j	The size of the computational domain in the j direction (μm)
L_u	The edge length of the cubical unit cell (μm)
N	The total number of pores in the representative element of a 2D cross-section (dimensionless)
N_{3D}	The total number of pores in a representative elementary volume (dimensionless)
N_{total}	The total lattice number (dimensionless)
p	The transient pressure (Pa)
p_{in}	The inlet transient pressure (Pa)
p_{out}	The outlet transient pressure (Pa)
r	The pore radius in 2D space (μm)
r_{av}	The average pore radius in 2D space (μm)
r_{3D}	The pore radius in 3D space (μm)
$r_{3D,av}$	The average pore radius in 3D space (μm)
r_{max}	The maximum pore radius in 2D space (μm)
r_{min}	The minimum pore radius in 2D space (μm)
$r_{3D,max}$	The maximum pore radius in 3D space (μm)
$r_{3D,min}$	The minimum pore radius in 3D space (μm)
R	The particle radius of the porous media (μm)
R_{ap}	The average particle radius of the porous media (μm)
R_{avc}	The average circular particle radius (μm)
R_{avs}	The average spherical particle radius (μm)
R_{max}	The maximum particle radius (μm)
S_{2D}	The specific surface area of the 2D cross-section (μm^{-1})
S_{3D}	The specific surface area of the 3D cross-section (μm^{-1})
$\mathbf{u}_j$	The velocity at the void point j (m/s)
$\mathbf{u}$	The flow velocity (m/s)

V	The pore volume of representative elementary volume (μm^3)
V_{av}	The average pore volume (μm^3)
V_u	The volume of the cubical unit cell (μm^3)
Greek symbols	
τ	Tortuosity (dimensionless)
$\tau(r_{3D})$	Tortuosity of pore radius r_{3D} (dimensionless)
τ_{av}	The average tortuosity (dimensionless)
τ_0	The relaxation time (dimensionless)
φ_{2D}	The areal porosity of 2D cross-sections (dimensionless)
φ_{3D}	The bulk porosity of a 3D matrix (dimensionless)
μ	The dynamic viscosity (mPa·s)
Δy	The lattice distance or the voxel size (μm)
$\mathbf{y}$	The grid location (μm)
Δt	The time step (s)
$\mathbf{u}_i$	The component in the i direction of the volumetric average velocity (m/s)
v	The kinematic viscosity (m^2/s)
ρ	The fluid density in the D3Q19 model (kg/m^3)
ω_α	Fixed weighting factors (dimensionless)
π	Circular constant, which is approximately equal to 3.1415926 (dimensionless)
λ	The weight coefficient of the average spherical particle radius (dimensionless)
Subscript symbols	
α	The direction in the D3Q19 model (dimensionless)
2D	Two-dimensional space
3D	Three-dimensional space
max	Maximum
min	Minimum

Appendix A. Fractal Theory of Porous Media in 2D and 3D Space

As mentioned above, for a 2D cross-section, pore structure parameters can be determined using digital image process technologies and thin section analysis. For example, pore fractal dimension d_f can be calculated with the box-counting algorithm [38]. Moreover, with thin section analysis or digital image processing techniques, areal porosity φ_{2D} and pore size distribution can be easily determined. As the pore structures of most sedimentary porous media follow fractal characteristics, fractal theory has been widely applied to describe the pore structures of porous media [29,34,35]. Based on fractal geometry, with the determined pore fractal dimension d_f, the maximum pore radius r_{max}, and the minimum pore radius r_{min}, the pore size distribution f and the total number of pores in a representative element of 2D cross-section N can be written as [29,35]:

$$\begin{cases} f = d_f r_{min}^{d_f} r^{-(d_f+1)}, \\ N = \left(\frac{r_{max}}{r_{min}}\right)^{d_f}. \end{cases} \tag{A1}$$

Based on Equation (A1), the average pore radius r_{av} and specific surface area S_{2D} of the 2D cross-section are:

$$\begin{cases} r_{av} = \int_{r_{min}}^{r_{max}} rf dr = \frac{d_f r_{min}}{d_f-1}\left[1-\left(\frac{r_{min}}{r_{max}}\right)^{d_f-1}\right]; \\ S_{2D} = \frac{\varphi_{2D}}{A} N \int_{r_{min}}^{r_{max}} 2\pi r f dr = \frac{\varphi_{2D}}{A} 2\pi r_{av}\left(\frac{r_{max}}{r_{min}}\right)^{d_f}, \end{cases} \tag{A2}$$

where A is the cross-sectional area of the representative element (RE), which is:

$$A = \frac{N}{\varphi_{2D}} \int_{r_{min}}^{r_{max}} \pi r^2 f dr = \frac{\pi d_f r_{max}^{d_f}\left(r_{max}^{2-d_f} - r_{min}^{2-d_f}\right)}{\varphi_{2D}(2-d_f)} = \frac{\pi d_f r_{max}^2}{\varphi_{2D}(2-d_f)}\left(1 - \frac{r_{min}^{2-d_f}}{r_{max}^{2-d_f}}\right). \tag{A3}$$

By combining the Hagen–Poiseulle equation, the bundle of the capillary tube model, and Darcy's law, the permeability of the 2D cross-section (i.e., the 2D composite permeated with tortuous capillary tubes that are perpendicular to the 2D cross-section, Figure 1) K_{2D} can be obtained as [12,29,40,54,55]:

$$K_{2D} = \frac{\pi d_f r_{max}^{d_f}\left(r_{max}^{3-d_f+d_t} - r_{min}^{3-d_f+d_t}\right)}{2^{4-d_t}\sqrt{A^{1+d_t}}(3-d_f+d_t)}, \tag{A4}$$

where d_t is the tortuosity fractal dimension in 2D space, which can be determined with [42]:

$$d_t = (2-d_f+1) + (2-d_f)\frac{\log(d_f) - \log(d_f-1)}{\log(\varphi_{2D})}. \tag{A5}$$

when d_t is assigned as unity, K_{2D} can be simplified as:

$$K_{2D} = \frac{\pi d_f r_{max}^{d_f}\left(r_{max}^{4-d_f} - r_{min}^{4-d_f}\right)}{8A(4-d_f)}. \tag{A6}$$

For 3D pore space, with the given pore fractal dimension D_f, the maximum pore radius $r_{3D,max}$, and the minimum pore radius $r_{3D,min}$, the pore size distribution f_{3D} and the total number of pores in a representative elementary volume (SRV) N_{3D} can be written as:

$$\begin{cases} f_{3D} = D_f r_{3D,min}^{D_f} r_{3D}^{-(D_f+1)}; \\ N_{3D} = \left(\frac{r_{3D,max}}{r_{3D,min}}\right)^{D_f}, \end{cases} \tag{A7}$$

where the subscript 3D presents 3D space. Based on fractal theory, in 3D pore space, the actual streamline length L of a tortuous capillary and tortuosity fractal dimension D_T can be determined as [29,42]:

$$\begin{cases} L = (2r_{3D})^{1-D_T} V^{\frac{D_T}{3}}; \\ D_T = (3-D_f+1) + (3-D_f)\frac{\log(D_f)-\log(D_f-1)}{\log(\varphi_{3D})}. \end{cases} \tag{A8}$$

where pore volume V of a representative elementary volume (REV) can be obtained as [61]:

$$V = \frac{N_{3D}}{\varphi_{3D}} \int_{r_{3D,min}}^{r_{3D,max}} \pi r_{3D}^2 L dr_{3D}. \tag{A9}$$

By rewriting Equation (A9), we have:

$$V = \left\{ \frac{\pi D_f r_{3D,max}^{3-D_T}}{4\varphi_{3D}(3-D_T-D_f)} \left[1 - \left(\frac{r_{3D,min}}{r_{3D,max}}\right)^{3-D_T-D_f}\right] \right\}^{\frac{3}{3-D_T}}. \tag{A10}$$

By combining the Hagen–Poiseulle equation, the bundle of the capillary tube model, and Darcy's law, the 3D permeability K_{3D} can be obtained as [12,55,61,62]:

$$K_{3D} = \frac{N_{3D}}{V^{\frac{2}{3}}} \int_{r_{min}}^{r_{max}} \frac{\pi r_{3D}^4}{8(2r_{3D})^{1-D_T} V^{\frac{D_T-1}{3}}} f dr = \frac{\pi D_f 2^{D_T} r_{3D,max}^{D_f}\left(r_{3D,max}^{3-D_f+D_T} - r_{3D,min}^{3-D_f+D_T}\right)}{16V^{\frac{1+D_T}{3}}(3-D_f+D_T)}. \tag{A11}$$

By combining Equation (A7) and Equation (8), the average tortuosity τ_{av} is:

$$\tau_{av} = \int_{r_{3D,min}}^{r_{3D,max}} f_{3D} \left(\frac{\sqrt[3]{V}}{2r_{3D}} \right)^{D_T - 1} dr_{3D} = 2^{1-D_T} D_f r_{3D,min}^{D_f} \left(\sqrt[3]{V} \right)^{D_T - 1} \frac{r_{3D,max}^{1-D_f-D_T} - r_{3D,min}^{1-D_f-D_T}}{1 - D_f - D_T}. \tag{A12}$$

Then, by the definition of specific surface area S_{3D} in 3D space, we have:

$$\begin{aligned} S_{3D} &= \frac{N_{3D}}{V} \int_{r_{3D,min}}^{r_{3D,max}} 2\pi r_{3D} \sqrt[3]{V^{D_T}} (2r_{3D})^{1-D_T} f_{3D} dr_{3D} \\ &= 2^{2-D_T} \pi D_f V^{\frac{D_T - 3}{3}} r_{3D,max}^{D_f} \frac{r_{3D,max}^{2-D_T-D_f} - r_{3D,min}^{2-D_T-D_f}}{2 - D_T - D_f}. \end{aligned} \tag{A13}$$

Appendix B. The Algorithm of the D3Q19 Lattice Boltzmann Method

For single-phase flow simulations, the lattice Boltzmann method (LBM) based on simple BGK relaxation was used here. The fundamental idea of the LBM is to construct a kinetic model simplified from the Boltzmann equation at mesoscale that incorporates the essential physical processes to ensure that the computed macroscopic quantities satisfy the desired governing equation (i.e., Navier–Stokes equation in the current study). The computational grids/lattices are uniformly distributed inside the computational domain. The only unknown $f_\alpha(x,t)$ is the density distribution function and evolves according to the following propagation–collision scheme [63–66]:

$$f_\alpha(\mathbf{y} + \Delta t \mathbf{e}_\alpha,\ t + \Delta t) = f_\alpha(\mathbf{y},\ t) + \frac{f_\alpha^{eq}(\mathbf{y},\ t) - f_\alpha(\mathbf{y},\ t)}{\tau_0}, \tag{A14}$$

and

$$f_\alpha^{eq}(\mathbf{y},\ t) = \rho \omega_\alpha \left[1 + \frac{3}{c^2} \mathbf{e}_\alpha \cdot \mathbf{u} + \frac{9}{2c^4} (\mathbf{e}_\alpha \cdot \mathbf{u})^2 - \frac{3}{2c^2} \mathbf{u} \cdot \mathbf{u} \right], \tag{A15}$$

where Δt is the timestep, $\mathbf{y}$ is the grid location, $c = \Delta y/\Delta t$ is computed with the grid distance Δy (namely the voxel size here), and τ_0 is the dimensionless relaxation time, which is determined by the kinematic viscosity v, namely $\tau_0 = 3v/(c\Delta y) + 0.5$. Additionally, the lattice velocity $\mathbf{e}_\alpha$ and weight ω_α in the moving direction $\alpha \in [0, Q-1]$ were selected according to the D3Q19 model (3-dimension and 19-velocity) [64]. For an arbitrary grid at $\mathbf{y}$, its neighboring grids are exactly located at $\mathbf{y} + \mathbf{e}_\alpha \Delta t$ since the magnitude of $\mathbf{e}_\alpha$ depends on c. The equilibrium distribution function f_α^{eq} was determined from the local fluid density ρ and flow velocity $\mathbf{u}$ that were computed in the absence of internal/external force as follows:

$$\begin{cases} \rho = \sum\limits_{\alpha=0}^{18} f_\alpha; \\ \mathbf{u} = \frac{1}{\rho} \sum\limits_{\alpha=0}^{18} \mathbf{e}_\alpha f_\alpha. \end{cases} \tag{A16}$$

At the initial state, f_α is equal to f_α^{eq}, which was computed with the initial distributions of ρ and $\mathbf{u}$. The bounce-back scheme was used at the complicated solid surface for the non-slip boundary condition and was naturally implemented as the half-way bounce-back scheme with a higher accuracy because the actual solid–fluid interface was exactly located in the middle between the solid and fluid grids in the digital rock simulations. The flows in the current study were driven by a pressure difference $p_{in} - p_{out}$ between the inlet and outlet (equivalent to a density difference due to $p = \rho c^2/3$), and the density constraint was imposed by the robust non-equilibrium extrapolation scheme [67]. The permeability was computed after convergence:

$$K_{ij} = \frac{\rho v L_j}{p_{in} - p_{out}} \langle \mathbf{u}_i \rangle = \frac{\rho v L_j}{p_{in} - p_{out}} \frac{1}{N_{total}} \sum_{j \in void} \mathbf{u}_j, \tag{A17}$$

where $\langle \mathbf{u}_i \rangle$ is the component in the i direction of the volumetric average velocity and computed by a summation of $\mathbf{u}_i$ over the void/fluid grids, N_{total} is the total number of both void and solid grids, K_{ij} is the component in the i direction when the flow is driven in the j direction, and L_j is the size of the computational domain in the j direction. The permeability K_{ij} as a tensor has nine components in total by driving the flow in three different directions using three independent simulations. However, the current simulations in this paper always drive the flow by a pressure difference along the $j = y$ direction and focus on the component in the $i = y$ direction, and K_{yy} is denoted by $K_{3\text{D}}$ as the 3D permeability of the porous media.

It is worth noting that attention has been paid to alleviating the velocity and permeability dependences on the relaxation time due to the nonzero Knudsen effect in the LBM simulations [46,68] where the Knudsen (Kn) number is $\sqrt{\pi/6}(\tau_0 - 0.5)/N_{\text{pore}}$ [69], and N_{pore} is the grid number used to discretize the dominant pore size. However, we could not set $(\tau_0 - 0.5) \to 0$ for $Kn \to 0$; that would result in zero kinematic viscosity and an infinite Reynold (Re) number, leading to the dependence of the permeability on Re. In practical simulations of digital rocks [46,68], using $(\tau_0 - 0.5)$ around 0.2 and a small relative pressure drop of only 1% (i.e., $p_{\text{in}} - p_{\text{out}} = 0.01 p_{\text{in}}$) can make the Kn and Re effects negligible, respectively. On the other hand, Li et al. (2018) found the multi-relaxation time (MRT) model might provide a parameter range that was wider than that of the current BGK model, but the basic mechanism remained the same; i.e., in simulations of intrinsic permeability, one should choose the appropriate relaxation parameter to ensure that Kn is small (but also nonzero as in the current BGK model).

Appendix C. Determination of the Average Particle Size

In this section, the average rock particle radius in porous media with bulk porosity $\varphi_{3\text{D}}$ was estimated with the idealized geometrical model shown in Figure A1, assuming that the unit cell was composed of spherical particles of the same size. Based on Figure A1, the volume of the cubical unit cell V_u and its edge length L_u are:

$$\begin{cases} V_\text{u} = \frac{4\pi R^3}{3(1-\varphi_{3\text{D}})}; \\ L_\text{u} = \sqrt[3]{\frac{4\pi R^3}{3(1-\varphi_{3\text{D}})}} = \sqrt[3]{\frac{4\pi}{3(1-\varphi_{3\text{D}})}} R. \end{cases} \tag{A18}$$

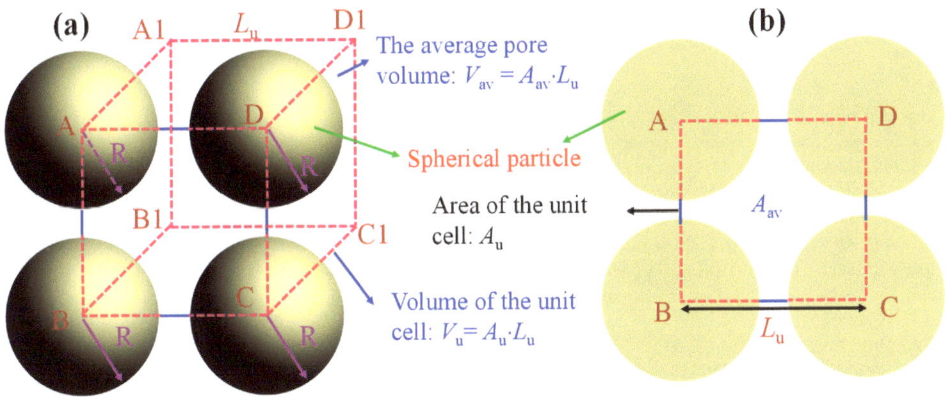

Figure A1. The arrangement of spherical rock particles for the average pore radius of porous media: (**a**) the cubical unit cell; (**b**) front view of the unit cell.

Based on Equation (A18), the average pore volume V_{av} is:

$$V_{av} = V_u - \frac{4\pi R^3}{3} = \frac{4\pi R^3 \varphi_{3D}}{3(1-\varphi_{3D})}. \quad \text{(A19)}$$

Mathematically, the average pore volume V_{av} can be also estimated with:

$$V_{av} = A_{av} L_u = \pi r_{3D,av}^2 \sqrt[3]{\frac{4\pi}{3(1-\varphi_{3D})}} R, \quad \text{(A20)}$$

wherein $r_{3D,av}$ is the average pore radius in 3D space, which is:

$$r_{3D,av} = \int_{r_{3D,min}}^{r_{3D,max}} r_{3D} f_{3D} dr_{3D} = \frac{D_f r_{3D,min}}{D_f - 1} \left[1 - \left(\frac{r_{3D,min}}{r_{3D,max}} \right)^{D_f - 1} \right]. \quad \text{(A21)}$$

By combining Equations (A19) and (A20), the average spherical particle radius R_{avs} is

$$R_{avs} = r_{3D,av} \sqrt{\sqrt[3]{\frac{4\pi}{3(1-\varphi_{3D})}} \frac{3(1-\varphi_{3D})}{4\varphi_{3D}}}. \quad \text{(A22)}$$

By simplifying spherical particles in Figure A1 as circular particles, Xu and Yu derived an analytical model for the maximum particle radius R_{max}, which is [12]:

$$R_{max} = r_{3D,max} \sqrt{\frac{\varphi_{3D}}{1-\varphi_{3D}}}. \quad \text{(A23)}$$

In light of Equation (A23), the average circular particle radius R_{avc} can be also determined as:

$$R_{avc} = r_{3D,av} \sqrt{\frac{\varphi_{3D}}{1-\varphi_{3D}}}. \quad \text{(A24)}$$

Then, by using the average particle radius statistical average method, the average particle radius R_{ap} of the porous media can be expressed as:

$$R_{ap} = \lambda R_{avs} + (1-\lambda) R_{avc}, \quad \text{(A25)}$$

wherein λ is the weight coefficient of the average spherical particle radius.

References

1. Millington, R.J.; Quirk, J.P. Permeability of porous solids. *Trans. Faraday Soc.* **1961**, *57*, 1200–1207. [CrossRef]
2. Whitaker, S. Flow in porous media I: A theoretical derivation of Darcy's law. *Transp. Porous Media* **1986**, *1*, 3–25. [CrossRef]
3. Renard, P.; De Marsily, G. Calculating equivalent permeability: A review. *Adv. Water Eesources* **1997**, *20*, 253–278. [CrossRef]
4. Meng, M.; Ge, H.; Ji, W.; Wang, X.; Chen, L. Investigation on the variation of shale permeability with spontaneous imbibition time: Sand-stones and volcanic rocks as comparative study. *J. Nat. Gas Sci. Eng.* **2015**, *27*, 1546–1554. [CrossRef]
5. Guo, D.; Lv, P.; Zhao, J.; Zhang, C. Research progress on permeability improvement mechanisms and technologies of coalbed deep-hole cumulative blasting. *Int. J. Coal Sci. Technol.* **2020**, *7*, 329–336. [CrossRef]
6. Walsh, J.B.; Brace, W.F. The effect of pressure on porosity and the transport properties of rock. *J. Geophys. Res. Solid Earth* **1984**, *89*, 9425–9431. [CrossRef]
7. Lock, P.A.; Jing, X.; Zimmerman, R.W.; Schlueter, E.M. Predicting the permeability of sandstone from image analysis of pore structure. *J. Appl. Phys.* **2002**, *92*, 6311–6319. [CrossRef]
8. Costa, A. Permeability-porosity relationship: A reexamination of the Kozeny-Carman equation based on a fractal pore-space geometry assumption. *Geophys. Res. Lett.* **2006**, *33*, L02318. [CrossRef]
9. Pan, J.; Zhang, Z.; Li, M.; Wu, Y.; Wang, K. Characteristics of multi-scale pore structure of coal and its influence on permeability. *Nat. Gas Ind. B* **2019**, *6*, 357–365. [CrossRef]
10. Kozeny, J. Uber kapillare leitung der wasser in boden. *R. Acad. Sci. Vienna Proc. Class I* **1927**, *136*, 271–306.

11. Carman, P.C. Fluid flow through granular beds. *Trans. Inst. Chem. Eng.* **1937**, *15*, 150–166. [CrossRef]
12. Xu, P.; Yu, B. Developing a new form of permeability and Kozeny-Carman constant for homogeneous porous media by means of fractal geometry. *Adv. Water Resour.* **2008**, *31*, 74–81. [CrossRef]
13. Nomura, S.; Yamamoto, Y.; Sakaguchi, H. Modified expression of Kozeny-Carman equation based on semilog-sigmoid function. *Soils Found.* **2018**, *58*, 1350–1357. [CrossRef]
14. Liu, C.; Zhang, L.; Li, Y.; Liu, F.; Martyushev, D.A.; Yang, Y. Effects of microfractures on permeability in carbonate rocks based on digital core technology. *Adv. Geo-Energy Res.* **2022**, *6*, 86–90. [CrossRef]
15. Panda, M.N.; Lake, L.W. Estimation of single-phase permeability from parameters of particle-size distribution. *AAPG Bull.* **1994**, *78*, 1028–1039.
16. Rodriguez, E.; Giacomelli, F.; Vazquez, A. Permeability-porosity relationship in RTM for different fiberglass and natural reinforcements. *J. Compos. Mater.* **2004**, *38*, 259–268. [CrossRef]
17. Chen, X.; Yao, G.; Luo, C.; Jiang, P.; Cai, J.; Zhou, K.; Herrero-Bervera, E. Capillary pressure curve determination based on a 2-D cross section analysis via fractal geometry: A bridge between 2D and 3D pore structure of porous media. *J. Geophys. Res. Solid Earth* **2019**, *124*, 2352–2367. [CrossRef]
18. Chen, X.; Yao, G. An improved model for permeability estimation in low permeable porous media based on fractal geometry and modified Hagen-Poiseuille flow. *Fuel* **2017**, *210*, 748–757. [CrossRef]
19. Adler, P.M.; Jacquin, C.G.; Quiblier, J.A. Flow in simulated porous media. *Int. J. Multiph. Flow* **1990**, *16*, 691–712. [CrossRef]
20. Yeong, C.L.Y.; Torquato, S. Reconstructing random media. II. Three-dimensional media from two-dimensional cuts. *Phys. Rev. E* **1998**, *58*, 224–238. [CrossRef]
21. Hilfer, R.; Manwart, C. Permeability and conductivity for reconstruction models of porous media. *Phys. Rev. E* **2001**, *64*, 021304. [CrossRef]
22. Keehm, Y.; Mukerji, T.; Nur, A. Permeability prediction from thin sections: 3D reconstruction and Lattice-Boltzmann flow simulation. Geophys. *Res. Lett.* **2004**, *31*, L04606. [CrossRef]
23. Andrä, H.; Combaret, N.; Dvorkin, J.; Glatt, E.; Han, J.; Kabel, M.; Keehm, Y.; Krzikalla, F.; Lee, M.; Madonna, C.; et al. Digital rock physics benchmarks-part II: Computing effective properties. *Comput. Geosci.* **2013**, *50*, 33–43. [CrossRef]
24. Karimpouli, S.; Tahmasebi, P. Conditional reconstruction: An alternative strategy in digital rock physics. *Geophysics* **2016**, *81*, D465–D477. [CrossRef]
25. Saxena, N.; Mavko, G. Estimating elastic moduli of rocks from thin sections: Digital rock study of 3D properties from 2D images. *Comput. Geosci.* **2016**, *88*, 9–21. [CrossRef]
26. Saxena, N.; Mavko, G.; Hofmann, R.; Srisutthiyakorn, N. Estimating permeability from thin sections without reconstruction: Digital rock study of 3D properties from 2D images. *Comput. Geosci.* **2017**, *102*, 79–99. [CrossRef]
27. Saxena, N.; Hofmann, R.; Alpak, F.O.; Berg, S.; Dietderich, J.; Agarwal, U.; Tandon, K.; Hunter, S.; Freeman, J.; Wilson, O.B. References and benchmarks for pore-scale flow simulated using micro-CT images of porous media and digital rocks. *Adv. Water Resour.* **2017**, *109*, 211–235. [CrossRef]
28. Tahmasebi, P.; Sahimi, M.; Andrade, J.E. Image-based modeling of granular porous media. *Geophys. Res. Lett.* **2017**, *44*, 4738–4746. [CrossRef]
29. Yu, B.; Cheng, P. A fractal permeability model for bi-dispersed porous media. *Int. J. Heat Mass Transf.* **2002**, *45*, 2983–2993. [CrossRef]
30. Rabbani, A.; Ayatollahi, S.; Kharrat, R.; Dashti, N. Estimation of 3-D pore network coordination number of rocks from watershed segmentation of a single 2-D image. *Adv. Water Resour.* **2016**, *94*, 264–277. [CrossRef]
31. Berryman, J.G.; Blair, S.C. Use of digital image analysis to estimate fluid permeability of porous materials: Application of two-point correlation functions. *J. Appl. Phys.* **1986**, *60*, 1930–1938. [CrossRef]
32. Sisavath, S.; Jing, X.; Zimmerman, R.W. Creeping flow through a pipe of varying radius. *Phys. Fluids* **2001**, *13*, 2762–2772. [CrossRef]
33. Mandelbrot, B.B. *The Fractal Geometry of Nature*; WH freeman: New York, NY, USA, 1982.
34. Katz, A.J.; Thompson, A.H. Fractal sandstone pores: Implications for conductivity and pore formation. *Phys. Rev. Lett.* **1985**, *54*, 1325. [CrossRef]
35. Yu, B.; Li, J. Some fractal characters of porous media. *Fractals* **2001**, *9*, 365–372. [CrossRef]
36. Wu, Y.; Tahmasebi, P.; Lin, C.; Zahid, M.A.; Dong, C.; Golab, A.N.; Ren, L. A comprehensive study on geometric, topological and fractal characterizations of pore systems in low-permeability reservoirs based on SEM, MICP, NMR, and X-ray CT experiments. *Mar. Pet. Geol.* **2019**, *103*, 12–28. [CrossRef]
37. Lei, G.; Liao, Q.; Patil, S.; Zhao, Y. A new permeability model for argillaceous porous media under stress dependence with clay swelling. *Int. J. Eng. Sci.* **2021**, *160*, 103452. [CrossRef]
38. Ghiasi-Freez, J.; Kadkhodaie-Ilkhchi, A.; Ziaii, M. Improving the accuracy of flow units prediction through two committee machine models: An example from the South Pars Gas Field, Persian Gulf Basin, Iran. *Comput. Geosci.* **2012**, *46*, 10–23. [CrossRef]
39. Cai, J.; Hu, X.; Standnes, D.C.; You, L. An analytical model for spontaneous imbibition in fractal porous media including gravity. *Colloids Surf. A Physicochem. Eng. Asp.* **2012**, *414*, 228–233. [CrossRef]

40. Lei, G.; Liao, Q.; Patil, S.; Zhao, Y. Effect of clay content on permeability behavior of argillaceous porous media under stress dependence: A theoretical and experimental work. *J. Pet. Sci. Eng.* **2019**, *179*, 787–795. [CrossRef]
41. Xiao, B.; Li, Y.; Long, G.; Yu, B. Fractal permeability model for power-law fluids in fractured porous media with rough surfaces. *Fractals* **2022**, *30*, 2250115. [CrossRef]
42. Wei, W.; Cai, J.; Hu, X.; Han, Q. An electrical conductivity model for fractal porous media. *Geophys. Res. Lett.* **2015**, *42*, 4833–4840. [CrossRef]
43. Liu, L.; Sun, Q.; Wu, N.; Liu, C.; Ning, F.; Cai, J. Fractal analyses of the shape factor in Kozeny-Carman equation for hydraulic permeability in hydrate-bearing sediments. *Fractals* **2021**, *29*, 2150217. [CrossRef]
44. Lei, G.; Qu, J.; Wu, Q.; Pang, J.; Su, D.; Guan, J.; Lu, C. Theoretical analysis of threshold pressure in tight porous media under stress. *Phys. Fluids* **2023**, *35*, 073313.
45. Norouzi, S.; Soleimani, R.; Farahani, M.V.; Rasaei, M.R. Pore-scale simulation of capillary force effect in water-oil immiscible dis-placement process in porous media. In Proceedings of the 81st EAGE Conference and Exhibition, London, UK, 3 June 2019.
46. Khodja, M.R.; Li, J.; Hussaini, S.R.; Ali, A.Z.; Al-Mukainah, H.S.; Jangda, Z.Z. Consistent prediction of absolute permeability in carbonates without upscaling. *Oil Gas Sci. Technol. Rev. D'ifp Energ. Nouv.* **2020**, *75*, 44. [CrossRef]
47. Soleimani, R.; Azaiez, J.; Zargartalebi, M.; Gates, I.D. Analysis of marangoni effects on the Non-isothermal immiscible Ray-leigh-Taylor instability. *Int. J. Multiph. Flow* **2022**, *156*, 104231. [CrossRef]
48. Vasheghani, F.M.; Mousavi, N.M. On the effect of flow regime and pore structure on the flow signatures in porous media. *Phys. Fluids* **2022**, *34*, 115139. [CrossRef]
49. Lei, G.; Xue, L.; Liao, Q.; Li, J.; Zhao, Y.; Zhou, X.; Lu, C. A novel analytical model for porosity-permeability relations of argillaceous porous media under stress conditions. *Geoenergy Sci. Eng.* **2023**, *225*, 211659. [CrossRef]
50. Karimpouli, S.; Tahmasebi, P.; Saenger, E.H. Estimating 3D elastic moduli of rock from 2D thin-section images using dif-ferential effective medium theory3D elastic moduli using 2D images. *Geophysics* **2018**, *83*, MR211–MR219. [CrossRef]
51. Srisutthiyakorn, N.; Hunter, S.; Sarker, R.; Hofmann, R.; Espejo, I. Predicting elastic properties and permeability of rocks from 2D thin sections. *Lead. Edge* **2018**, *37*, 421–427. [CrossRef]
52. Blunt, M.J.; Bijeljic, B.; Dong, H.; Gharbi, O.; Iglauer, S.; Mostaghimi, P.; Paluszny, A.; Pentland, C. Pore-scale imaging and modelling. *Adv. Water Resour.* **2013**, *51*, 197–216. [CrossRef]
53. Wu, Y.; Tahmasebi, P.; Lin, C.; Munawar, M.J.; Cnudde, V. Effects of micropores on geometric, topological and transport properties of pore systems for low-permeability porous media. *J. Hydrol.* **2019**, *575*, 327–342. [CrossRef]
54. Lei, G.; Liao, Q.; Lin, Q.; Zhang, L.; Xue, L.; Chen, W. Stress dependent gas-water relative permeability in gas hydrates: A theoretical model. *Adv. Geo-Energy Res.* **2020**, *4*, 326–338. [CrossRef]
55. Yu, B. Analysis of flow in fractal porous media. *Appl. Mech. Rev.* **2008**, *61*, 050801. [CrossRef]
56. Chen, X.; Yao, G.; Herrero-Bervera, E.; Cai, J.; Zhou, K.; Luo, C.; Jiang, P.; Lu, J. A new model of pore structure typing based on fractal geometry. *Mar. Pet. Geol.* **2018**, *98*, 291–305. [CrossRef]
57. Rabbani, A.; Assadi, A.; Kharrat, R.; Dashti, N.; Ayatollahi, S. Estimation of carbonates permeability using pore network parameters extracted from thin section images and comparison with experimental data. *J. Nat. Gas Sci. Eng.* **2017**, *42*, 85–98. [CrossRef]
58. Comiti, J.; Renaud, M. A new model for determining mean structure parameters of fixed beds from pressure drop meas-urements: Application to beds packed with parallelepipedal particles. *Chem. Eng. Sci.* **1989**, *44*, 1539–1545. [CrossRef]
59. Lei, G.; Wang, C.; Wu, Z.; Wang, H.; Li, W. Theory study of gas-water relative permeability in roughened fractures. Proc. Inst. Mech. Eng. Part C. *J. Mech. Eng. Sci.* **2018**, *232*, 4615–4625. [CrossRef]
60. Suarez-Dominguez, E.J.; Perez-Rivao, A.; Sanchez-Medrano, M.T.; Perez-Sanchez, J.F.; Izquierdo-Kulich, E. Mesoscopic model for the surface fractal dimension estimation of solid-solid and gas-solid dispersed systems. *Surf. Interfaces* **2020**, *18*, 100407. [CrossRef]
61. Cai, J.; Lin, D.; Singh, H.; Wei, W.; Zhou, S. Shale gas transport model in 3D fractal porous media with variable pore sizes. *Mar. Pet. Geol.* **2018**, *98*, 437–447. [CrossRef]
62. Lei, G.; Dong, P.; Wu, Z.; Mo, S.; Gai, S.; Zhao, C.; Liu, Z. A fractal model for the stress-dependent permeability and relative permeability in tight sandstones. *J. Can. Pet. Technol.* **2015**, *54*, 36–48. [CrossRef]
63. Qian, Y.H.; d'Humières, D.; Lallemand, P. Lattice BGK models for Navier-Stokes equation. *EPL* **1992**, *17*, 479–484. [CrossRef]
64. Sukop, M.C.; Thorne, D.T. *Lattice Boltzmann Modeling: An Introduction for Geoscientists and Engineers*; Springer: Berlin, Germany, 2006.
65. Li, J. *Multiscale and Multiphysics Flow Simulations of Using the Boltzmann Equation*; Springer: Cham, Switzerland, 2020.
66. Li, J.; Hussaini, S.R.; Dvorkin, J. Permeability-porosity relations from single image of natural rock: Subsampling approach. *J. Pet. Sci. Eng.* **2020**, *194*, 107541. [CrossRef]
67. Guo, Z.; Zheng, C.; Shi, B. An extrapolation method for boundary conditions in lattice Boltzmann method. *Phys. Fluids* **2002**, *14*, 2007–2010. [CrossRef]

68. Li, J.; Ho, M.T.; Wu, L.; Zhang, Y. On the unintentional rarefaction effect in LBM modeling of intrinsic permeability. *Adv. Geo-Energy Res.* **2018**, *2*, 404–409. [CrossRef]
69. Zhang, Y.; Qin, R.; Emerson, D.R. Lattice Boltzmann simulation of rarefied gas flows in microchannels. *Phys. Rev. E* **2005**, *71*, 047702. [CrossRef]

Journal of
Marine Science and Engineering

MDPI

Article

Experimental Characteristics of Hydrocarbon Generation from Scandinavian Alum Shale Carbonate Nodules: Implications for Hydrocarbon Generation from Majiagou Formation Marine Carbonates in China's Ordos Basin

Yiqing Wang [1], Yaohui Xu [1], Junping Huang [2], Jianglong Shi [2], Heng Zhao [3], Qingtao Wang [4] and Qiang Meng [1,*]

1 Hubei Key Laboratory of Petroleum Geochemistry and Environment, College of Resources and Environment, Yangtze University, Wuhan 430100, China; 2021710462@yangtzeu.edu.cn (Y.W.)
2 Research Institute of Petroleum Exploration & Development-Northwest, PetroChina, Lanzhou 730020, China
3 Jiangsu Design Institute of Geology for Mineral Resources, The Testing Center of China National Administration of Coal Geology, Xuzhou 221006, China
4 Guangzhou Institute of Energy Testing, Guangzhou 511447, China
* Correspondence: mengqiang@yangtzeu.edu.cn

Abstract: The hydrocarbon source rocks of the marine carbonates of the Ordovician Majiagou Formation in the Ordos Basin are generally in the high-overmature stage and are, therefore, not suitable for hydrocarbon thermal simulation experiments. Their hydrocarbon generation potential and hydrocarbon generation characteristics are not clearly understood. Meanwhile, Nordic Cambrian carbonates are similar in lithology, parent material type, and sedimentary age, and are in the low evolution stage, which is suitable for hydrocarbon thermal simulation experiments. Therefore, in this study, we selected the Nordic carbonates for the gold tube thermal simulation experiment to analyze the content and geochemical characteristics of the thermal simulation products. The experimental results are also compared and analyzed with the characteristics of thermal simulation products of Pingliang Formation mud shale (contemporaneous with the Majiagou Formation) and Shanxi Formation coal (in the upper part of the Majiagou Formation), which are similar to the Majiagou Formation in the Ordos Basin. The results showed that the Nordic carbonate has different hydrocarbon production characteristics from the mud shale of the Pingliang Formation of the same parent material type, and although the hydrocarbon production yields of the two are not very different, the carbonate still produces methane at 600 °C. The hydrocarbon production yield of the Nordic carbonates is almost equivalent to that of type-II$_2$ kerogen, indicating that the hydrocarbon production yield is not related to lithology and only to the organic matter type; however, the Nordic carbonate can produce a large amount of H_2S. The alkane carbon isotope changes are mainly controlled by the degree of thermal evolution, showing gradual heaviness with increasing temperature. No carbon isotope sequence reversal occurred during the thermal simulation, and its distribution range is roughly the same as that of the alkane carbon isotope composition of the mud shale of the Pingliang Formation. The ethane carbon isotope composition is as heavy as −21.2‰ at the high-temperature stage, showing similar coal-type gas characteristics. The addition of calcium sulphate ($CaSO_4$) causes the TSR reaction to occur, which has a significant impact on the methane content under high maturity conditions, reducing its content by more than 50% at 600 °C. However, the addition of $CaSO_4$ increases the yield of heavy hydrocarbon gases, such as ethane, and promotes the production of $C_{6\text{-}14}$ hydrocarbons and C_{14+} hydrocarbons at high-temperature stages, and the addition of $CaSO_4$ substantially increases the yield of H_2, CO_2, and H_2S. The thermal simulation results have implications for the hydrocarbon formation mechanism of the early Paleozoic marine carbonate formation system in the stacked basins of the Ordos Basin and the Tarim Basin in China.

Keywords: Ordos Basin; carbonate; thermal simulation; hydrocarbon generation; Cambrian; carbon isotopes; source rock geochemistry

Citation: Wang, Y.; Xu, Y.; Huang, J.; Shi, J.; Zhao, H.; Wang, Q.; Meng, Q. Experimental Characteristics of Hydrocarbon Generation from Scandinavian Alum Shale Carbonate Nodules: Implications for Hydrocarbon Generation from Majiagou Formation Marine Carbonates in China's Ordos Basin. *J. Mar. Sci. Eng.* **2023**, *11*, 1616. https://doi.org/10.3390/jmse11081616

Academic Editor: George Kontakiotis

Received: 3 July 2023
Revised: 10 August 2023
Accepted: 12 August 2023
Published: 18 August 2023

1. Introduction

The carbonate sequence has a pivotal position in the global oil and gas exploration field, with rich hydrocarbon resources and great exploration potential [1–5]. The Ordos Basin, in the central part of the Chinese mainland, is the basin with the largest natural gas distribution area and the largest natural gas accumulation in China, and its Lower Paleozoic Ordovician marine carbonate formation system is an important formation system for natural gas exploration [6,7]. Since the discovery of the Jingbian gas field in the Lower Paleozoic Ordovician in the central part of the basin at the end of the last century, after years of continuous exploration and development, the Jingbian gas field in the central part of the basin and the surrounding areas have been explored to a high degree, but their origin is still controversial [8–10]. It is speculated whether the carbonate rocks of the Ordovician Majiagou Formation can be used as hydrocarbon source rocks to form large gas fields.

The marine carbonate formations of the Majiagou Formation are dominated by sapropel-type kerogen with an average of 0.24 wt.% total organic carbon, which is considered by local scholars [11,12] to be below the lower limit of organic carbon of carbonate hydrocarbon source rocks and not an effective gas source rock with high maturity (vR_o = 2.07–2.68%), having reached the past maturity stage. Kerogen, crude oil, and soluble organic matter are dominated by cracked gas, and therefore, the hydrocarbon products of the lower Paleozoic marine strata are dominated by natural gas [13,14]. The same is true for the Lower Paleozoic marine hydrocarbon source rocks in the Sichuan and Tarim basins, neither of which is suitable for hydrocarbon thermal simulation experiments to reflect their hydrocarbon production characteristics, mechanisms, and hydrocarbon generation potential. Therefore, when studying the hydrocarbon generation mechanism of carbonate rocks, carbonate rock samples from other strata have been chosen as substitutes, such as Jurassic marine carbonates from the Qiangtang Basin of the Qinghai-Tibet Plateau (R_o approximately 0.25–1.05%) [15], shales and marls from the Upper Paleozoic Xiamaling Formation in the Xiahuayuan area of Hebei Province (R_o approximately 0.50–0.65%) [16], Devonian shales and marls from the Luquan area of Yunnan Province (R_o approximately 0.45~0.60%) [17], or mud shales from the Ordovician, which were selected for thermal simulation experiments, such as low-maturity Estonian shales, mature-overmaturity Indogan shales, Sargan shales, and Pingliang shales [18].

However, the above samples differ significantly from the characteristics of Ordovician marine hydrocarbon source rocks in the Ordos Basin, such as inconsistent lithology (carbonates versus mud shales) or large differences in stratigraphic age, causing the hydrocarbon generation parent material to be variable. Several decades of organic geochemical work, including significant advancements in biomarker analysis, have demonstrated organic geochemical indicators of source rock age, terrestrial versus marine source rock organic matter, source rock depositional environments including water column chemistry and toxicity, source rock sedimentary facies, and myriad other paleoclimatic and paleoenvironmental conditions associated with a source rock and its deposition, not to mention indicators of oil thermal maturity, mixing, and biodegradation [19–23]. Therefore, several factors, such as lithology, stratigraphic age, thermal maturity, and hydrocarbon generation parent material, need to be considered when selecting samples for thermal simulation experiments.

In this study, carbonate nodules in the Scandinavian Middle Cambrian–Lower Ordovician Alum Shale of the Baltic Sea were selected for thermal simulation experiments. This carbonate nodule sample has the same lithology and similar parent material type and sedimentary age, and is at a low evolutionary stage compared with the carbonates of the Majiagou Formation in the Ordos Basin. Therefore, we selected it for hydrocarbon generation thermal simulation experiments, and the experimental results are valuable references for the hydrocarbon generation mechanism, characteristics, and hydrocarbon generation potential of the carbonate rocks of the Majiagou Formation in the Ordos Basin.

2. Geological Setting

The Alum Shale Formation in northwestern Europe was deposited from the Middle Cambrian (Miaolingian) to the Early Ordovician (Tremadoc) in the Baltic Basin and surrounding countries. It is a conventional source of oil from Cambrian, Ordovician, and Silurian carbonates and also has potential for shale oil and biogenic or thermogenic shale gas [24] (Figure 1).

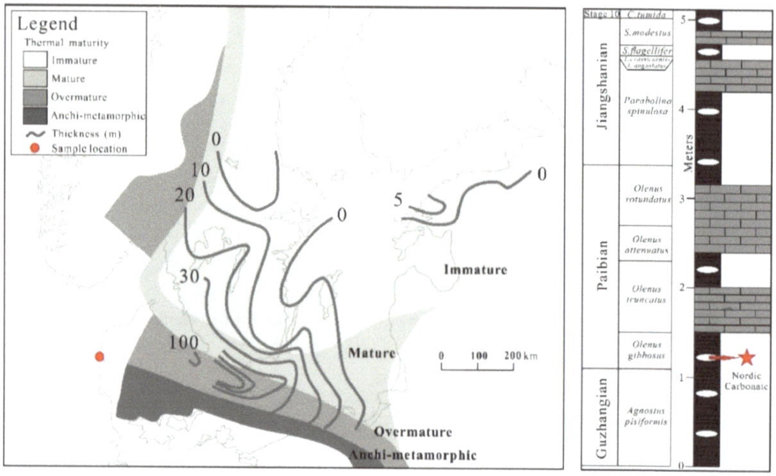

Figure 1. Thickness and thermal maturity of the Alum Shale Formation and stratigraphic column (modified from the reference [23]).

Alum Shale is an important hydrocarbon source rock in the Baltic Basin that varies in maturity from immature to overmature and is dominated by marine organic matter, such as algae and bacteria [24,25]. Approximately 50 small oil fields and 1 medium-sized oil field have been discovered.

3. Samples and Experimental Method

3.1. Samples

The experimental sample was taken from carbonate nodules in the Scandinavian Middle Cambrian–Lower Ordovician Alum Shale in the Baltic Sea (Figure 1), which had a *TOC* content of 1.16 wt.%, with free hydrocarbon (S_1) 0.21 mg/g, oil potential (S_2) 4.08 mg/g, hydrogen index (*HI*) 351 mg/g, T_{max} 429 °C, equivalent vitrinite reflectance 0.57% and organic matter type II_B (Table 1). The sample has a high organic matter content and low thermal maturity, making it suitable for hydrocarbon generation thermal simulation experiments. Whole rock mineral composition analysis showed that the mineral composition of this sample was predominantly calcite at 94.40%, with small amounts of quartz and pyrite and no clay minerals [26], indicating that the sample was a pure carbonate rock with no mud.

The lithology of the Majiagou Formation is carbonate rock–paste salt rock alternately developed, of which the Maj 1 section, Maj 3 section, and Maj 5 section develop the sea recession cyclic table within the evaporite limitation of terrace phase deposition; its lithology is mainly for the salt rock, hard gypsum, sandwiched between thin layers of paste-bearing dolomite, muddy dolomite, and limestone. The Maj 2 section, the Maj 4 section, and the Maj 6 section belong to the sea recession cyclic carbonate terrace phase deposition, and its lithology is mainly for the limestone and dolomite. The commonly called "salt up" and "salt down" are generally distinguished by the boundary of the huge thick saline rocks in the sixth sub-section of the fifth section of the Majiagou Formation ($O_1m_5^6$). The hydrocarbon parent material of the Well Mitan-1 is dominated by sapropel-type kerogen, with an average of 0.4 wt.% total organic carbon, at a high-overmature stage, and the hydrocarbon products are dominated by natural gas [27–30].

Table 1. Thermal simulation of the geochemical basic parameters of the sample.

Lithology	*TOC* (wt.%)	T_{max} (°C)	S_1 (mg/g)	S_2 (mg/g)	S_3 (mg/g)	*HI* (mg/g)	vR_o (%)	Parent Material Types	Source
Nordic carbonate	1.16	429	0.21	4.08	0.68	351	0.57	II_B	this study
mud shale of Pingliang Formation	0.86	443	0.05	1.34	0.77	147	0.80	II	Wang et al., 2014
coal of Shanxi Formation	58.30	424	0.66	97.18	/	167	0.55	III	Peng et al., 2020

The results of the thin section identification showed that the Nordic carbonate sample had a residual structure, with calcite dominating the rock components and generally recrystallized (Figure 2). This formed semi-automorphic powder-fine crystalline calcite of approximately 0.05–0.15 mm grain size, with a small number of self-shaped rhombohedral de-clouded dolomite pseudocrystals of approximately 0.10–0.20 mm grain size; a small amount of residual fossil dragonflies; and residual algal sand flakes. The residual algal sand fragments were generally 0.10–0.30 mm in size and consisted of mud-crystal calcite and mycorrhizal algae. The more complete individual fragments of fossil dragonflies were approximately 1.30 mm long and 0.30 mm wide, pike-shaped or rhombic in shape, 0.70 mm long and 0.45 mm wide, mostly in fragments, and poorly preserved. These contained a large amount of scattered micronized pyrite and hematite.

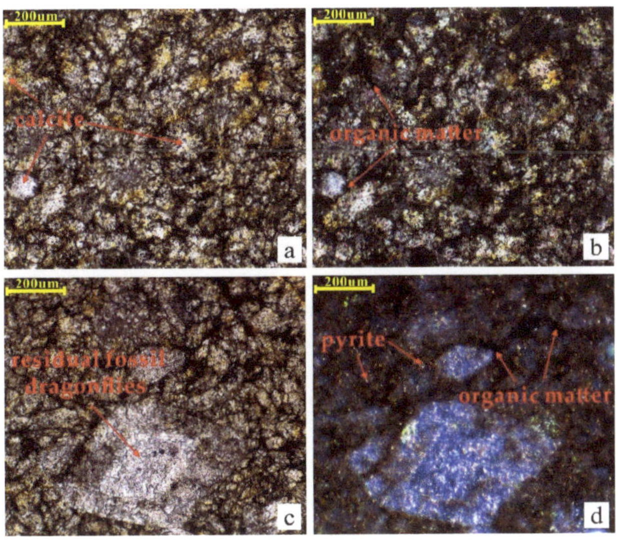

Figure 2. The Nordic carbonate microscopic characteristics. (**a**) Residual structure containing a small amount of residual bacteria algal spherulite, many scattered-shaped or grainy pyrite aggregate-shape distribution, intergranular filling organic matter calcite and pyrite and single polarization; (**b**) a photo with horizon orthogonal polarization, residual structure, calcite recrystallization, generally containing a small amount of residual algal spherulite, containing much grainy pyrite; (**c**) residual fossil dragonflies, including individual and slightly better calcite intergranular residual much of iron and organic matter and single polarization; (**d**) enhancement of (**c**) with sight reflected light photos, where the opaque component is composed of micro granular pyrite and organic matter, and the organic matter is not reflective.

3.2. TOC and Rock Pyrolysis Analysis Methods

The rock samples were crushed and passed through an 0.08 mm (200 mesh) sieve, dried at low temperature, weighed 80–120 mg of samples into the crucible, added with dilute hydrochloric acid (volume ratio of concentrated hydrochloric acid and deionized water is 1:7) for inorganic carbon reaction, and then tested for TOC after removing the inorganic carbon, and the TOC was analyzed with a LECO CS230 Carbon and Sulfur Meter according to the standard method of Determination of Organic Carbon in Sedimentary Rocks (GB/T 19145). The pyrolysis analyses were carried out on the OGE-VI instrument produced by China Petroleum Exploration Institute (CPEI). Before analysis, the sample was crushed and passed through an 0.08 mm (200 mesh) sieve, dried at low temperature, weighed at 80–120 mg into the crucible, and then the sample was analyzed by automatic injection on the instrument. The heating procedure for sample analysis was as follows: during the pyrolysis analysis, the instrument was first rapidly heated to 300 °C at a constant temperature for 3 min to measure the free hydrocarbon S_1 content; then, the temperature was increased to 600 °C at a heating rate of 50 °C/min, and then at a constant temperature for 1 min to detect the content of pyrolysis hydrocarbon S_2; and some of the compounds that can be easily cracked from 300 °C to 392 °C were collected at the same time as the collection of S_2; and the collection of the S_3 was carried out at the same time.

3.3. Thermal Simulation Experiment Method

Hydrocarbon source rock thermal simulation experiments are an important tool to study the hydrocarbon generation process, product characteristics, and hydrocarbon formation evolution of organic matter [31]. Among many thermal simulation instruments, the gold tube thermal simulation system [32,33] in a closed system can simulate the effect of temperature and pressure on the thermal evolution of organic matter simultaneously. Furthermore, gold is chemically inert, which can prevent the catalytic effect of the reaction device on hydrocarbon formation. Gold also has good ductility at high temperature, so it is considered one of the most suitable systems for simulating the gas generation characteristics of organic matter.

The thermal simulation experiment was carried out at the State Key Laboratory of Organic Geochemistry, Guangzhou Institute of Geochemistry, Chinese Academy of Sciences. We used a gold tube–autoclave closed thermal simulation system, and 200–600 mg whole rock samples was individually welded and sealed inside the gold tubes under argon protection (in this experiment, there was also a process of adding calcium sulfate, where the ratio of rock samples to calcium sulfate was approximately 1:1, and the rock samples weighed 120–300 mg). The gold tubes were placed inside the autoclave, and then a high-pressure pump was used to fill the autoclave with high-pressure water, thus exerting pressure on the outer wall of the gold tube, which in turn exerted a certain pressure on the rock sample and the thermal simulation product, using a pressure sensor to regulate and maintain the pressure in each autoclave at 50 MPa. The two sets of samples were warmed using two heating rate programs, slow (2 °C/h) and fast (20 °C/h). Twelve test temperature points were set for each set of samples (with a total of three temperature points set for the series with calcium sulfate addition, at 504 °C, 552 °C, and 600 °C), with a temperature range of 336–600 °C. After the reaction, the autoclave was removed from the heating oven, and the gold tubes were removed, washed, and placed in a fixed-volume vacuum system that was connected to an Agilent model 6890 N gas chromatograph. The gold tube was punctured with a needle under sealed conditions to release pyrolysis gases for analysis using a chromatograph and mass spectrometer.

The quantitative analysis part was carried out with an Agilent 6890 N gas chromatograph to collect the hydrocarbon products in the thermal simulation experiment, and the hydrocarbon components of C_1–C_5 were quantified and analyzed online with the external standard method using the FID detector. After the analysis of the gas components was completed, the appropriate amount of gas was extracted with a transfer device, and the carbon isotope analysis of the gas components was carried out with an Isochrom II GC-IRMS isotope mass spectrometer produced by VG Company of the United Kingdom.

4. Results

Thermal Simulation Experiment Results

Thermal simulation experiments provide the complete process of hydrocarbon generation from hydrocarbon source rocks as the temperature changes, and can reproduce the history of continuous maturation of sedimentary organic matter in the subsurface. However, the most intuitive data from simulation experiments are the composition and yield of hydrocarbons at different temperature points, which differ greatly from the relatively mild and long-lasting geological conditions. Thus, the easyR_o model [34–36], which is commonly used in industry, was used to relate time accumulation at the laboratory scale to maturation under geological conditions (Figure 3).

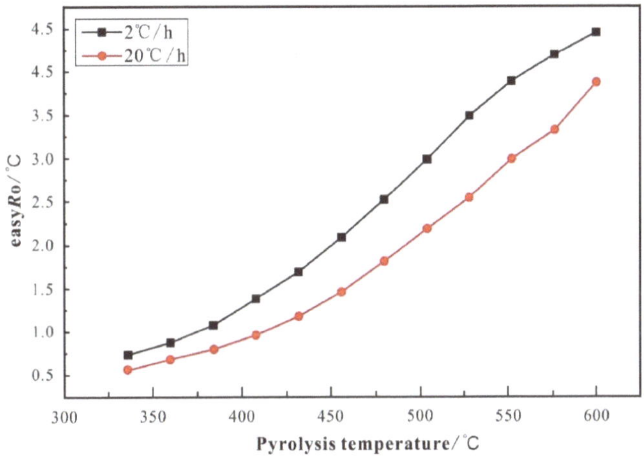

Figure 3. Gold tube thermal simulation experiment of easyR_o values corresponding to different temperature points.

Hydrocarbon generation thermal simulation experiments were carried out under heating rates of 20 °C/h and 2 °C/h. As shown in Figure 2, the easyR_o value continues to increase with increasing pyrolysis temperature. In the experimental temperature range, the corresponding easyR_o value was 0.57% (20 °C/h)/0.74% (2 °C/h) at the lowest temperature of 336 °C, and the easyR_o value was 3.87% (20 °C/h)/4.45% (2 °C/h) at the highest temperature of 600 °C.

During the thermal simulation experiment, gaseous hydrocarbons (C_1–C_5), liquid hydrocarbons (C_{6+}), and non-hydrocarbons (H_2, CO_2, H_2S) were generated. The yield of these components and the carbon isotopic composition of the alkanes varied gradually with increasing thermal simulation temperature (Tables 2 and 3).

Table 2. Thermal simulation experiment of carbonate hydrocarbon production rate under different heating rates with temperature variation characteristics.

Heating Rate	T/ °C	easyr_o/%	Total Amount of G as	CH_4	C_2H_6	C_3H_8	iC_4H_{10}	nC_4H_{10}	iC_5H_{12}	nC_5H_{12}	H_2	CO_2	H_2S	C_{1-5}	C_{6-14}	C_{14+}
			mg/g*TOC*													
20 °C/h	337	0.57	20.73	0.23	0.12	0.08	0.01	0.02	0.00	0.00	0.05	20.22	0.00	0.46	7.28	5.44
	362	0.68	21.79	0.47	0.30	0.21	0.04	0.05	0.01	0.01	0.08	20.63	0.00	1.08	13.38	6.19
	384	0.80	43.12	1.63	1.29	1.07	0.22	0.38	0.08	0.10	0.24	33.36	4.75	4.77	27.22	8.10
	408	0.96	77.08	5.34	4.88	4.29	0.92	1.92	0.59	0.82	0.48	32.73	25.12	18.75	77.88	18.05
	432	1.19	193.00	15.89	14.14	12.89	2.53	5.88	1.84	2.38	0.63	56.40	80.41	55.56	137.49	25.91
	456	1.47	283.17	28.03	23.24	21.94	4.33	10.55	3.47	4.53	0.63	70.87	115.60	96.08	138.00	23.51
	480	1.81	411.44	47.36	35.10	33.03	6.70	14.12	3.60	4.24	0.57	94.86	171.87	144.14	79.20	16.11
	504	2.19	675.04	88.35	51.79	40.66	6.44	7.29	0.37	0.31	0.64	159.02	320.16	195.21	60.41	12.06
	529	2.54	952.04	127.64	45.95	18.20	1.21	0.26	0.01	0.01	0.98	234.81	522.98	193.28	45.34	8.39
	552	2.99	933.33	167.00	22.87	2.57	0.05	0.02	0.00	0.00	1.50	266.33	472.99	192.51	36.94	5.57
	576	3.32	1095.36	192.33	7.54	0.24	0.01	0.01	0.00	0.00	1.87	351.25	542.11	200.12	28.54	2.53
	601	3.87	1161.28	211.68	1.92	0.07	0.00	0.00	0.00	0.00	2.29	415.78	529.53	213.68	19.71	0.36
2 °C/h	336	0.74	34.87	0.82	0.58	0.49	0.14	0.15	0.03	0.04	0.11	32.51	0.00	2.25	17.55	5.13
	360	0.88	56.97	2.68	2.30	2.10	0.52	0.85	0.25	0.28	0.30	40.40	7.29	8.98	40.32	10.78
	384	1.08	114.41	9.40	8.63	7.87	1.65	3.49	1.08	1.31	0.49	53.47	27.02	33.43	90.92	20.73
	408	1.38	192.50	24.17	19.58	18.15	3.47	8.15	2.45	2.99	0.52	65.63	47.39	78.95	158.92	28.71
	432	1.69	293.19	42.95	31.57	30.63	6.34	14.43	4.58	5.68	0.49	88.41	68.12	136.17	138.84	18.18
	457	2.09	468.46	75.07	49.68	46.61	9.62	17.42	3.29	3.60	0.53	136.36	126.28	205.29	81.08	13.29
	481	2.52	630.11	114.62	56.09	37.17	5.37	3.66	0.13	0.10	0.61	196.07	216.30	217.14	57.79	9.32
	504	2.99	953.60	158.21	39.14	6.99	0.32	0.10	0.01	0.01	0.97	309.25	438.61	204.77	38.88	5.39
	529	3.49	1222.12	192.05	4.21	0.31	0.02	0.01	0.00	0.00	1.14	421.81	602.57	196.60	31.08	2.81
	552	3.89	1201.07	217.92	1.21	0.08	0.00	0.00	0.00	0.00	1.50	454.06	526.31	219.20	24.47	1.94
	576	4.19	1236.05	223.84	0.93	0.03	0.00	0.00	0.00	0.00	1.85	516.13	493.26	224.80	13.84	0.88
	600	4.45	1508.19	248.71	0.83	0.01	0.00	0.00	0.00	0.00	2.37	708.38	547.90	249.55	10.25	0.89
2 °C/h adding calcium sulfate	504	2.99	1802.59	148.01	65.92	19.62	0.51	0.05	0.01	0.01	3.66	856.36	708.46	234.12	66.81	19.43
	552	3.89	2341.24	213.34	42.82	0.25	0.01	0.01	0.00	0.00	6.55	1240.90	837.36	256.43	48.12	10.95
	600	4.45	3075.61	103.12	0.75	0.00	0.00	0.00	0.00	0.00	5.52	1592.90	1373.33	103.87	34.50	5.40

Table 3. The carbon isotope composition of the carbonate thermal simulation experiment with temperature variation characteristics.

Heating Rate	T/°C	easyR_o/%	$\delta^{13}C_1$/‰	$\delta^{13}C_{CO2}$/‰	$\delta^{13}C_2$/‰	$\delta^{13}C_3$/‰
20 °C/h	337	0.57	0.0	−10.2	0.0	0.0
	362	0.68	0.0	−9.0	0.0	0.0
	384	0.80	−44.7	−7.5	−38.8	−34.8
	408	0.96	−44.0	−6.7	−36.4	−33.1
	432	1.19	−43.2	−6.7	−35.3	−31.6
	456	1.47	−42.7	−7.2	−35.4	−31.2
	480	1.81	−40.8	−8.9	−34.0	−30.1
	504	2.19	−38.9	−11.3	−31.0	0.0
	529	2.54	−37.4	−15.4	−25.7	0.0
	552	2.99	−35.2	−15.4	0.0	0.0
	576	3.32	−33.0	−14.3	0.0	0.0
	601	3.87	−31.5	−14.1	0.0	0.0
2 °C/h	336	0.74	0.0	−9.0	0.0	0.0
	360	0.88	0.0	−7.5	−38.0	−35.4
	384	1.08	−44.8	−6.9	−36.1	−33.1
	408	1.38	−43.2	−7.1	−34.6	−31.3
	432	1.69	−41.2	−7.1	−33.8	−32.1
	457	2.09	−40.2	−7.8	−32.5	−29.1
	481	2.52	−38.0	−10.0	−29.1	0.0
	504	2.99	−35.0	−12.2	−21.2	0.0
	529	3.49	−32.7	−12.3	0.0	0.0
	552	3.89	−31.6	−12.7	0.0	0.0
	576	4.19	−30.1	−14.6	0.0	0.0
	600	4.45	−28.4	−14.8	0.0	0.0
2 °C/h adding calcium sulfate	504	2.99	−37.3	−6.7	−27.7	0.0
	552	3.89	−34.7	−8.2	−17.4	0.0
	600	4.45	−20.0	−13.8	0.0	0.0

In general, the experimental samples produced more methane and nonhydrocarbon gases at 20 °C/h than at 2 °C/h, and produced or cracked heavy hydrocarbon gases (C_{2+}) and liquid hydrocarbons (C_{6+}) at higher temperatures. This suggested that the slow heating rate during thermal simulation continued the reaction adequately, and therefore, the hydrocarbon yields were higher at lower heating rates, a phenomenon consistent with the results of previous studies [37]. During the thermal simulation, the methane yield increased with increasing thermal simulation temperature and showed an increasing trend at 600 °C. The heavy hydrocarbon gas (C_{2+}) yield increased and then decreased, with its yield approaching zero at 550 °C under slow warming conditions. The trend in the yield of liquid hydrocarbons (C_{6+}) was similar to that of the heavy hydrocarbon gases (C_{2+}). The yields of CO_2 and H_2 of the nonhydrocarbon gases increased with increasing experimental temperature, and the yield of H_2S showed an increasing trend before equilibrium (Table 2). The carbon isotopic compositions of methane, ethane, and propane became progressively heavier with increasing temperature during the thermal simulations, following the kinetic isotope effect [38], with positive carbon isotope sequences ($\delta^{13}C_1 < \delta^{13}C_2 < \delta^{13}C_3$) at all temperature points, and no carbon isotope sequence inversions were observed (Table 3).

5. Comparison of Thermal Simulation Products with Mud Shale of the Pingliang Formation and Coal of the Shanxi Formation

In the thermal simulation experiments, samples from different basins, parent material types, maturities, and other characteristics had different natural gas yields. Considering that natural gas yields are mainly controlled by total organic carbon, we performed *TOC* normalization for gaseous hydrocarbon yields to facilitate the comparison of natural gas yields of various hydrocarbon source rocks. In this study, we compare the thermal modeling products of the Nordic carbonates with those of the Pingliang Formation mud shale [39] (contemporaneous with the Majiagou Formation samples), which is similar to the Majiagou Formation, and the Shanxi Formation coal [40] (in the upper part of the Majiagou Formation). Among them, the Pingliang Formation in the Ordos Basin is an important set of mud shale of the Ordovician, which was deposited in succession with the underlying Majiagou Formation, but its distribution is relatively limited, only in the southwestern margin of the Ordos Basin; the Shanxi Formation is a coal of the Carboniferous Permian, and it is a very important set of hydrocarbon source rocks of the Upper Paleozoic in the Ordos Basin, and it is also the source of hydrocarbon in many atmospheric fields. Both sample groups and the Nordic carbonates in this study used the closed-system gold tube thermal simulation system, and the experimental conditions were mostly the same.

5.1. Gaseous Product Evolutionary Characteristics

5.1.1. Pyrolysis Gaseous Hydrocarbon Characteristics

The experimental results showed that the methane (C_1) yield of the Nordic carbonate increased rapidly with increasing temperature during the thermal simulation and maintained a high growth rate at the highest temperature of 600 °C, indicating that there was still a strong potential for methane production at this time. The C_1 yield of the mud shale of the Pingliang Formation increased continuously from 336 °C to 504–552 °C during the thermal simulation and decreased significantly to almost zero at higher temperatures (easyR_o > 2.99%). The C_1 yields of the coal of the Shanxi Formation were close to those of the Nordic carbonate until 530 °C, with higher C_1 yields for the coal of the Shanxi Formation at higher temperatures (Figure 4a).

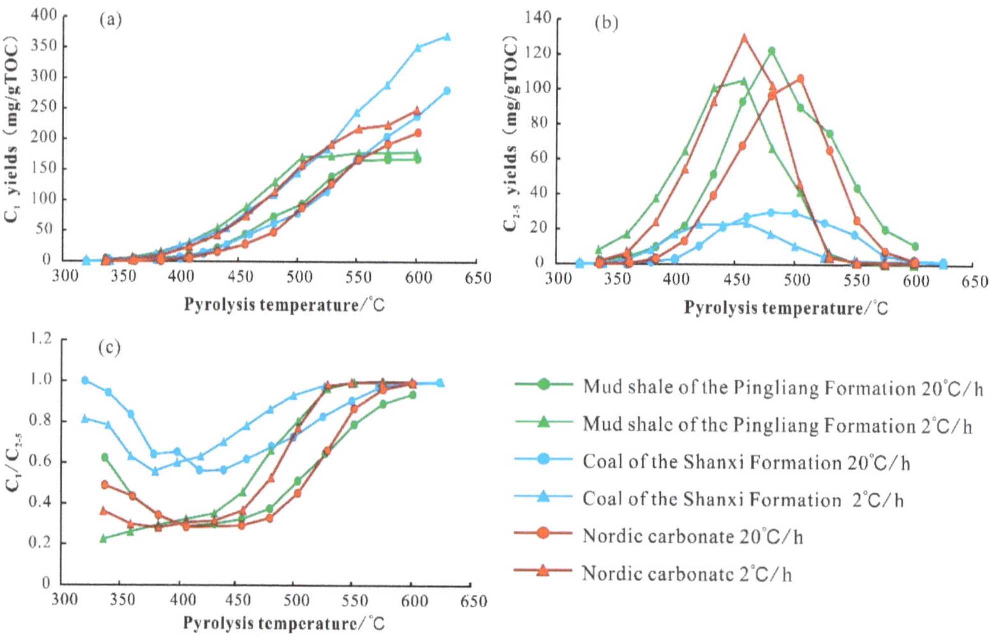

Figure 4. Carbonate thermal simulation of gaseous hydrocarbon production rate with temperature variation characteristics, and out of shale gas state hydrocarbon yield comparison and Shanxi group. (**a**) Carbonate thermal simulation of C_1 yields with temperature variation characteristics, and out of shale gas state hydrocarbon yield comparison and Shanxi group; (**b**) Carbonate thermal simulation of $C_{2\text{-}5}$ yields with temperature variation characteristics, and out of shale gas state hydrocarbon yield comparison and Shanxi group; (**c**) Carbonate thermal simulation of $C_1/C_{2\text{-}5}$ with temperature variation characteristics, and out of shale gas state hydrocarbon yield comparison and Shanxi group.

Unlike methane yield, heavy hydrocarbon gases, such as ethane (C_{2-5}), showed a parabolic pattern of increasing and then decreasing with increasing temperature. The C_{2-5} yields and characteristics of the Nordic carbonate and the mud shale of the Pingliang Formation were relatively similar with a 2 °C/h heating rate, and increased rapidly with increasing pyrolysis temperature from 336 °C, reaching a maximum at 456 °C (easyR_o = 2.09%), and then gradually decreasing, eventually approaching 0 at a pyrolysis temperature of 600 °C (Figure 4b). This indicated that the hydrocarbon gases at the end of the experiment were mainly composed of methane, accounting for more than 99% of the hydrocarbon gases. The predominance of methane during thermal evolution naturally carries implications for the type of petroleum systems (including methane hydrate systems [41–43]) that might develop as a result. In contrast, the coal of the Shanxi Formation had a lower C_{2-5} yield, approximately a quarter of the Nordic carbonate, suggesting a significant difference in the gas compositions generated between the two samples.

The gas drying coefficient (C_1/C_{1-5}) is an important parameter in natural gas geochemical studies and is of great importance for identifying the natural gas genesis, secondary role, and thermal maturity [44–46]. In thermal simulations of Nordic carbonate, C_1/C_{1-5} decreased and then increased with temperature, with a minimum value of 0.5 at 384 °C (easyR_o = 1.08%) for a 2 °C/h heating rate, followed by a relative increase in methane content due to the cracking of C_{2+} hydrocarbons and an increase in the drying coefficient (Figure 4c). Such trends in gas drying coefficients are very common in thermal simulation experiments and are generally thought to be related to the initial cracking of kerogen and the secondary cracking of hydrocarbons [47]. The initial temperature of the thermal simulation experiment was 336 °C, which was difficult to achieve in the geological reality of hydrocarbon formation, and was sufficient to allow rapid hydrocarbon generation from the cracking of kerogen, with short-chain gaseous hydrocarbons and long-chain liquid hydrocarbons forming simultaneously. As the temperature rose, the production of all hydrocarbons increased rapidly, but the rate of increase was relatively higher for the heavy hydrocarbon gases than for methane, causing the proportion of methane in the hydrocarbon gases to gradually decrease. The proportion of methane reached a low point when the secondary cracking threshold was reached. Thereafter, as the temperature increased, the kerogen continued to produce hydrocarbons in addition to C_{2-5} and liquid hydrocarbons, and a large amount of methane was formed by the secondary cracking of hydrocarbons, causing the proportion of methane to increase rapidly and the drying coefficient to increase (Figure 4c).

During the thermal simulation experiments, the C_1/C_{1-5} values of the 2 °C/h series and the 20 °C/h series did not follow the same pattern, with the C_1/C_{1-5} of the 2 °C/h series continuously increasing with the thermal simulation temperature, without a process of decreasing and then increasing. Moreover, the 20 °C/h series showed a process of decreasing and then increasing, and the difference between the C_1/C_{1-5} of the two experiments at the initial temperature point was considerable. At 400–525 °C, the C_1/C_{1-5} values of the Nordic carbonates were slightly less than those of the mud shale of the Pingliang Formation at the same temperature point, while at temperatures greater than 525 °C, the C_1/C_{1-5} values of the Nordic carbonates were slightly more than those of the mud shale of the Pingliang Formation, and relatively speaking, the C_1/C_{1-5} values of both were less different throughout the experiment. Although the C_1/C_{1-5} value of the coal of the Shanxi Formation also underwent a process of first decreasing and then increasing, it had a higher value than that of Nordic carbonate, that is, the content of C_{2-5} in hydrocarbon gases was lower.

The drying coefficients of Pingliang Formation mud shale and Nordic carbonate rocks are relatively lower than those of Shanxi Formation coals. The thermal evolution of the Majiagou Formation in the Ordos Basin (Mitan-1 well, Tong-74 well, and Jingtan-1 well, etc.) [48] has already reached the high-overmature stage, but the drying coefficients (0.947, 0.989, and 0.996) tend to be close to 1, indicating that its natural gas still belongs to the wet gas and not to the dry gas. One of the wells, Mitan-1, is a well that has made a major breakthrough at subsalt, and it is also a more typical well. Therefore, the results of thermal simulation experiments are consistent with the characteristics of the dryness coefficient of the Majiagou Formation.

5.1.2. Pyrolysis Nonhydrocarbon Gas Characteristics

The CO_2 yield of the Nordic carbonate gradually increased with the temperature during the thermal simulation and continued to increase until 600 °C (Figure 5a). In contrast, the mud shale of the Pingliang Formation had higher CO_2 yields at most stages, especially at lower temperatures, and these yields were more than two times higher than those of the Nordic carbonate. At 600 °C, the CO_2 yield of the mud shale of the Pingliang Formation was also more than 30% higher than that of the Nordic carbonate. In contrast, the coal of the Shanxi Formation had a relatively low CO_2 yield, and its content essentially did not increase after temperatures exceeded 450 °C. Compared to the Nordic carbonate CO_2 yield, the CO_2 yield of the coal of the Shanxi Formation was slightly higher in the early part of the thermal simulation, but much lower in the later part. The $CO_2/C_{1\text{-}5}$ ratio decreased rapidly with increasing temperature, indicating that the rate of hydrocarbon gas production was significantly higher than that of CO_2 with increasing temperature. Thus, we suggest that this phenomenon may be related to the lower activation energy of CO_2, which allows for rapid CO_2 production in the first stage and less production in the later stage, resulting in a decreasing ratio.

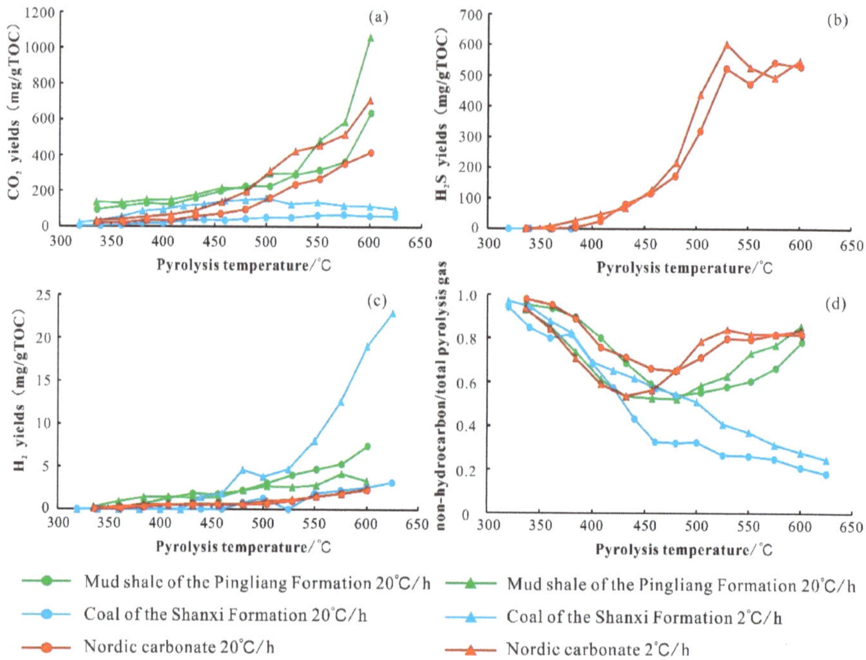

Figure 5. Carbonate thermal simulation of hydrocarbon production rate with temperature variation characteristics, and out of shale and Shanxi coal group of hydrocarbon production rate. (**a**) Carbonate thermal simulation of CO_2 yields with temperature variation characteristics, and out of shale and Shanxi coal group of hydrocarbon production rate; (**b**) Carbonate thermal simulation of H_2S yields with temperature variation characteristics, and out of shale and Shanxi coal group of hydrocarbon production rate; (**c**) Carbonate thermal simulation of H_2 with temperature variation characteristics, and out of shale and Shanxi coal group of hydrocarbon production rate; (**d**) Carbonate thermal simulation of non-hydrocarbon/total pyrolysis gas with temperature variation characteristics, and out of shale and Shanxi coal group of hydrocarbon production rate.

The H_2S yield of the Nordic carbonate increased with increasing pyrolysis temperature during the thermal simulation, reaching a maximum at 528 °C (easyR_o = 3.49%) and gradually decreasing to zero or even negative at higher temperatures (Figure 5b); the mud shale of the Pingliang Formation and the coal of the Shanxi Formation showed almost no H_2S production during the thermal simulation, which may be related to the samples lacking sulfur.

There is no H_2S production in Pingliang Formation mud shale and Shanxi Formation coal, while there is a large amount of H_2S production in Nordic carbonates, and the natural gas of Majiagou Formation in the Ordos Basin, especially most of the subsalt, has H_2S production, for example, the H_2S production of Long-L92 well [49] is as high as 23.23%. From the results of thermal simulation experiments, the Nordic carbonates can also produce a large amount of H_2S, which confirms that the vast majority of the natural gas in the subsalt of the Majiagou Formation contains a higher content of H_2S. Therefore, the H_2S in the Ordos Basin comes from the carbonates, which are the set of hydrocarbon source rocks, instead of the Pingliang Formation mud shale and the Shanxi Formation coal.

During the thermal simulation of the Nordic carbonate and the mud shale of the Pingliang Formation, the H_2 yield increased continuously with increasing pyrolysis temperature, reaching a maximum at the highest temperature of 600 °C. The coal of the Shanxi Formation produced no H_2 gas at the low-temperature stage (T < 440 °C), while at the high-temperature stage (T > 525 °C), the H_2 yield increased rapidly, reaching 22.94 mg/g*TOC* at the highest temperature of 600 °C (Figure 5c).

In the early low-temperature stage, the value of nonhydrocarbon/total pyrolysis gas in the Nordic carbonate and mud shale of the Pingliang Formation was close to 1, that is, nonhydrocarbon gas accounted for the vast majority, and the inflection point appeared in the curve with a further increase in temperature. The non-hydrocarbon/total pyrolysis gas curve showed the characteristics of "first falling and then rising". Of these, changes in CO_2 and H_2S yields played a major role in the inflection point. At 432 °C (easyR_o = 1.69%), for example, the CO_2 and H_2S yields increased sharply, and the nonhydrocarbon/total pyrolysis gas curve appeared to inflect at the low heating rate of the Nordic carbonate. In contrast, the nonhydrocarbon yield of the coal of the Shanxi Formation did not change significantly with temperature, resulting in a continuous decrease in the nonhydrocarbon/total pyrolysis gas ratio with increasing temperature (Figure 5d).

5.2. Liquid Product Evolutionary Characteristics

The characteristics of the liquid hydrocarbon yield with temperature change in the Nordic carbonate during thermal simulation are shown in Figure 6. The oil yield rate showed the parabolic characteristics (first increase and then decrease) under different heating rates, and the oil yield curve change trend of the two heating rates was the same. However, the peak temperature of the oil production rate of the 2 °C/h series was lower than that of the 20 °C/h series, and the difference between the two was approximately 20~50 °C. At 20 °C/h, the oil production rate peaked at 163.40 mg/g*TOC* at 456 °C. At 2 °C/h, the oil production rate peaked at 187.64 mg/g*TOC* at 408 °C, that is, the oil production rate was higher at a low heating rate. Compared with the 2 °C/h series, the oil production peak of the 20 °C/h series lagged behind (Figure 6), and the easyR_o corresponding to the oil production peak was 1.47% and 1.38%, respectively. This was higher than the actual geological conditions, which was determined by the inherent difference between short-term high temperature and long-term low temperature. The oil production rate was higher at the later high heating rate, which fully reflected the mutual compensation relationship between time and temperature in the hydrocarbon production process. Compared to gaseous hydrocarbon yields, liquid hydrocarbon yields reached a maximum at relatively low temperatures, a result that could be attributed to the closed thermal simulation system. In the closed system, organic matter will preferentially crack to form high-carbon-number hydrocarbons, while as the temperature increases to a certain value, the high carbon number in hydrocarbons will also gradually crack to form a low carbon number in hydrocarbons.

The yield of liquid hydrocarbons will gradually decrease after peaking, mainly because the rate of hydrocarbon cleavage is greater than the rate of production. In addition, as thermal evolution proceeds, the high carbon number of hydrocarbons will eventually crack into methane [50,51].

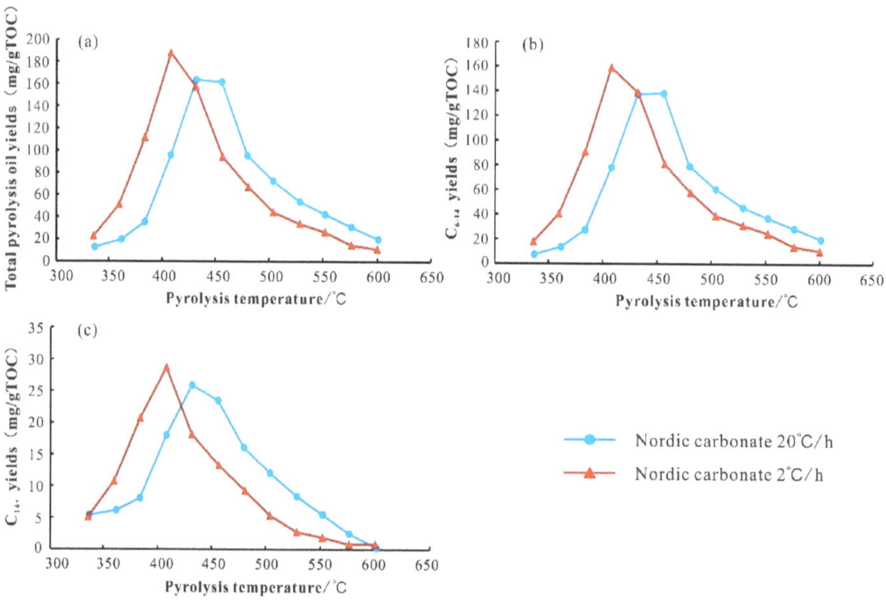

Figure 6. Carbonate thermal simulation in the process of liquid hydrocarbon yield characteristics with the temperature change. (**a**) Carbonate thermal simulation in the process of total pyrolysis oil yields characteristics with the temperature change; (**b**) Carbonate thermal simulation in the process of C_{6-14} yields characteristics with the temperature change; (**c**) Carbonate thermal simulation in the process of C_{14+} yields characteristics with the temperature change.

In the carbonate sequence, organic acid salts are also an important class of organic matter, and their contribution to hydrocarbon generation should not be overlooked [52]. Calcium stearate is a type of organic acid salt, and Wang et al. [53] conducted hydrocarbon generation thermal simulation experiments with calcium stearate. This showed that calcium stearate generated fewer hydrocarbons at the low-temperature stage and began to generate many hydrocarbons at the high-temperature stage; therefore, the contribution of organic acid salt to hydrocarbon generation is considered mainly at the high-overmature stage. In contrast, carbonates have a relatively higher yield of C_{6-14} hydrocarbons and a relatively lower hydrocarbon generation temperature than calcium stearate.

5.3. Carbon Isotope Composition

Carbon isotopic composition is an important parameter for studying the geochemical characteristics of natural gas [54–58]. The experimental results showed that the alkane carbon isotopic compositions of the Nordic carbonate all became progressively heavier with increasing temperature in the thermal simulation experiment, following the kinetic isotope effect. Among them, the methane carbon isotopic composition gradually became heavier from −44.8‰ to −28.4‰, with an overall weight of approximately 16.4‰ (Table 3, Figure 7a). The methane carbon isotope compositions of the Nordic carbonate and the mud shale of the Pingliang Formation had roughly the same distribution range, but differed from the variation pattern in the coal of the Shanxi Formation, which showed an overall lighter and then heavier methane carbon isotope composition. This phenomenon may be due to the nonhomogeneity of the organic matter or the isotopic fractionation effect resulting from changes in the activation energy difference between the early ^{12}C-rich CH_4 and ^{13}C-rich CH_4 [40] (Figure 7a).

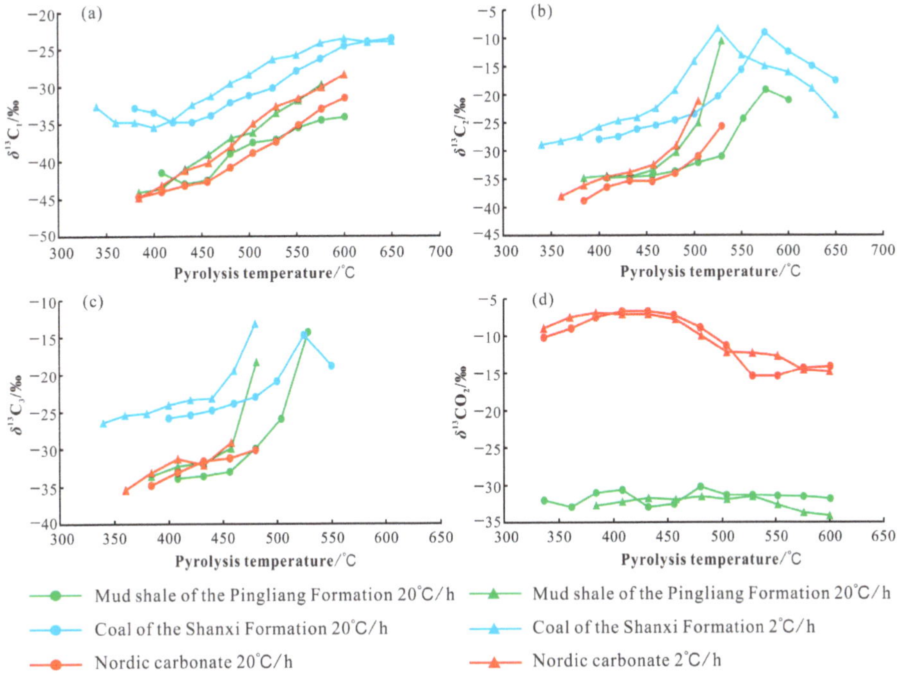

Figure 7. Carbonate rock pyrolysis gas product delta$\delta^{13}C$ characteristics with the temperature and the delta of out shale pyrolysis gas and Shanxi coal products $\delta^{13}C$. (**a**) Carbonate rock pyrolysis gas product delta$\delta^{13}C_1$ characteristics with the temperature and the delta of out shale pyrolysis gas and Shanxi coal products $\delta^{13}C_1$; (**b**) Carbonate rock pyrolysis gas product delta$\delta^{13}C_2$ characteristics with the temperature and the delta of out shale pyrolysis gas and Shanxi coal products $\delta^{13}C_2$; (**c**) Carbonate rock pyrolysis gas product delta$\delta^{13}C_3$ characteristics with the temperature and the delta of out shale pyrolysis gas and Shanxi coal products $\delta^{13}C_3$; (**d**) Carbonate rock pyrolysis gas product delta$\delta^{13}CO_2$ characteristics with the temperature and the delta of out shale pyrolysis gas and Shanxi coal products $\delta^{13}CO_2$.

Methane carbon isotopes are mainly affected by the degree of thermal evolution, and the higher the degree of thermal evolution, the heavier the methane carbon isotopes. The lighter methane carbon isotope composition of natural gas in most parts of the subsalt of the Ordos Basin (Shuang-97 well and Mitan-1 well) [39] is consistent with the methane carbon isotope compositions produced by the thermal simulation experiments.

The ethane carbon isotopic composition gradually increased from −38.0‰ to −21.2‰ (2 °C/h series) with increasing temperature (Figure 7b). The ethane carbon isotopic compositions of the mud shale of the Pingliang Formation were distributed over roughly the same range. At high-temperature stages, the ethane carbon isotopes were heavier than −28‰. It is generally accepted that $\delta^{13}C_2$ values are strongly inherited from the original parent material, and that humic kerogen is more enriched in ^{13}C than humic mud kerogen, so scholars in China often use $\delta^{13}C_2$ values of −28‰ or −29‰ as the boundary between oil- and coal-based gas [54,59], and they play an important role in hydrocarbon genesis discrimination. However, the ethane carbon isotope compositions of the Nordic carbonate and the mud shale of the Pingliang Formation were heavier than −28‰ at high temperatures, and this feature also appears in the thermal simulations of the Luquan marine carbonate source rocks in Yunnan [17], that is, the ethane carbon isotope compositions at high temperatures appeared to be "coal-type gas". This suggested that the ethane carbon isotope composition of oil-type gas may also be heavier than −28‰ under some special circumstances. The ethane carbon isotope compositions of the coal of the Shanxi Formation were all heavier than −28‰, and the ethane carbon isotope compositions gradually became lighter with increasing temperature in the high-temperature section, resulting in a partial inversion of the carbon isotope sequence.

Ethane carbon isotopes, although also affected by the degree of thermal evolution, are most affected by the type of parent material [45]. Oil-type gas (gas generated by sapropel organic matter) is lighter, generally lighter than −28‰, and coal-type gas (gas generated by humic organic matter) is generally heavier than −28‰. The hydrocarbon source rock of Mitan-1 well [39] belongs to carbonate rock in the marine phase, and the organic matter type is II/II_1, and the natural gas generated should be oil-type gas. It also belongs to the self-generated and self-storage type of gas reservoir, but the generated natural gas surprisingly has the characteristics of coal-type gas. And, this feature also appeared in our thermal simulation experiments, which can respond to its consistency with the ethane carbon isotope composition characteristics of many natural gases under the salt of Majiagou Formation (Tao-38 well, Tong-75 well, etc.) [49].

The Nordic carbonate followed a similar pattern of change in propane carbon isotopes during thermal modeling to that of methane and ethane, becoming heavier with increasing temperature (Figure 7c) and following kinetic isotope effects. The mud shale of the Pingliang Formation produced a much heavier propane carbon isotopic composition compared to the Nordic carbonate. The coal of the Shanxi Formation follows the same pattern of variation as the Nordic carbonate, but the propane carbon isotopic composition of the coal of the Shanxi Formation was approximately 8–10‰ heavier.

Notably, the carbon isotope composition sequences of methane, ethane, and propane were all in positive order ($\delta^{13}C_1 < \delta^{13}C_2 < \delta^{13}C_3$) throughout the thermal simulation experiment for the Nordic carbonate, and no partial or complete inversion of the carbon isotope sequences occurred. This suggested that the carbonate rock may not experience carbon isotope sequence inversions during normal thermal evolution, unless they undergo mixing [60] or secondary interactions [46]. In contrast, the inversion of heavy hydrocarbon gas $\delta^{13}C$ in the coal of the Shanxi Formation at the high premature stage indicated the complexity of the heavy hydrocarbon gas sources at the high premature stage, with sequential $\delta^{13}C$ fractionation effects from different sources, which may also be related to the $\delta^{13}C$ fractionation effects from aromatic hydrocarbon demethylation and methyl linkage [40].

The pattern of CO_2 carbon isotope changes during thermal modeling was completely different from that of alkane gases, with the CO_2 carbon isotope compositions generated by thermal modeling of the Nordic carbonate slowly becoming heavier and then gradually lighter. The CO_2 carbon isotopic composition of the mud shale of the Pingliang Formation fluctuated within a certain interval and was of organic origin (Figure 7d). The large difference in CO_2 carbon isotope composition between the two was because the Nordic carbonate used in the thermal simulations was the protolithic sample, whereas the inorganic carbon was removed from the mud shale of the Pingliang Formation with dilute hydrochloric acid prior to the thermal simulations.

6. Organic–Inorganic Interactions: The Effect of Gypsum Rock Presence on Hydrocarbons

The reaction between sulfate and sedimentary organic matter is a typical type of organic–inorganic interaction within sedimentary basins [61]. In our Ordos Basin study, the gypsum rock sequence in a broad sense refers to a set of sedimentary strata dominated by salt rock and gypsum rock or interlayered with sandstone mudstone. As the main evaporite rock type in our study, gypsum rock is mostly composed of gypsum, anhydrite, and variable water plaster, and contains small amounts of rock salt, salt materials, dolomite, clay organic matter, iron oxides, and other minor components. It has the characteristics of strong plasticity, easy flow, and density [62], and is often symbiotic with carbonate rock sequences in marine basins. The sedimentary environment of marine carbonate rocks is often accompanied by the deposition of gypsum rocks [63], and further studies on the influence of gypsum rocks on organic matter hydrocarbon generation are of positive significance in exploring the hydrocarbon generation mechanism of marine hydrocarbon source rocks. Previous studies have explored whether the salt minerals in gypsum rocks can promote the early hydrocarbon production process of hydrocarbon source rocks. Zhao et al. [64] found that the addition of gypsum ($CaSO_4$) increased hydrocarbon production by 50% from the pyrolysis of kerogen in the marine hydrocarbon source rocks of the Ordovician Pingliang Formation in the Ordos Basin, showing a strong catalytic effect and changing the composition of organic matter pyrolysis hydrocarbons. Numerous studies have demonstrated the catalytic effect of sulfate on the thermal evolution of hydrocarbon source rocks [64–66]. To identify the effect of gypsum rock presence on hydrocarbons, $CaSO_4$ was added during the gold tube thermal simulation experiments in this study, and the results are discussed.

The experimental results showed that the Nordic carbonate rock sample with $CaSO_4$ addition had little effect on C_1 yield at 500 °C and 550 °C, while at 600 °C, it caused a sharp decrease in C_1 yield, indicating that gypsum strongly affects C_1 yield at high maturation conditions (Figure 8a). The higher $C_{2\text{-}5}$ yield of the sample with $CaSO_4$ addition at the same temperature suggested that the addition of gypsum may increase the $C_{2\text{-}5}$ content rather than decrease it. However, when the pyrolysis temperature reached 600 °C, the $C_{2\text{-}5}$ yield was almost zero, with or without the addition of $CaSO_4$ (Figure 8b). The addition of $CaSO_4$ significantly increased the yield of H_2, CO_2, and H_2S, and its CO_2 yield increased by more than two times at the same temperature. According to the mechanism of the TSR reaction, although the increase in temperature can promote the TSR (refers to the process in which sulphate minerals are reduced by sulphate and hydrocarbons to produce acid gases, such as H_2S, which is an important mechanism for the formation of natural gas with high H_2S content) reaction when the temperature is too high, it will lead to the decomposition of H_2S, and the content of H_2S will decrease once the production of the TSR reaction is greater than the decomposition [39]. The TSR reaction increased the hydrogen sulfide yield, as well as the carbon dioxide yield, which confirmed the accuracy of the chemical reaction expression $CaSO_4$ + hydrocarbon → H_2S + $CaSO_3$ + hydrocarbon gas + CO_2 of TSR (Figure 8c–e). At 504–600 °C, the addition of $CaSO_4$ resulted in higher $C_{6\text{-}14}$ and C_{14+} hydrocarbon yields, with $C_{6\text{-}14}$ yields increasing by a factor of 1.7–3.4, and C_{14+} yields increasing by a factor of 3.6–6.0 (Table 3), indicating that the addition of $CaSO_4$ promoted the production of $C_{6\text{-}14}$ and C_{14+} hydrocarbons at the high-temperature stage, that is, the addition of $CaSO_4$ promoted the high-temperature stage of crude oil production.

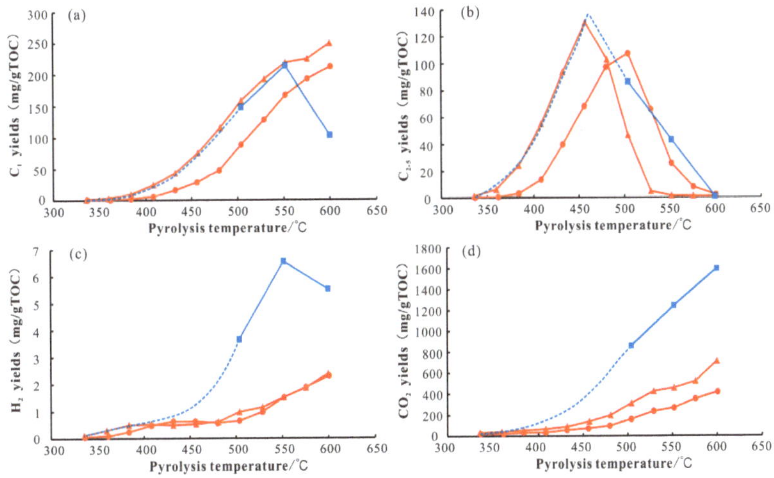

Figure 8. *Cont.*

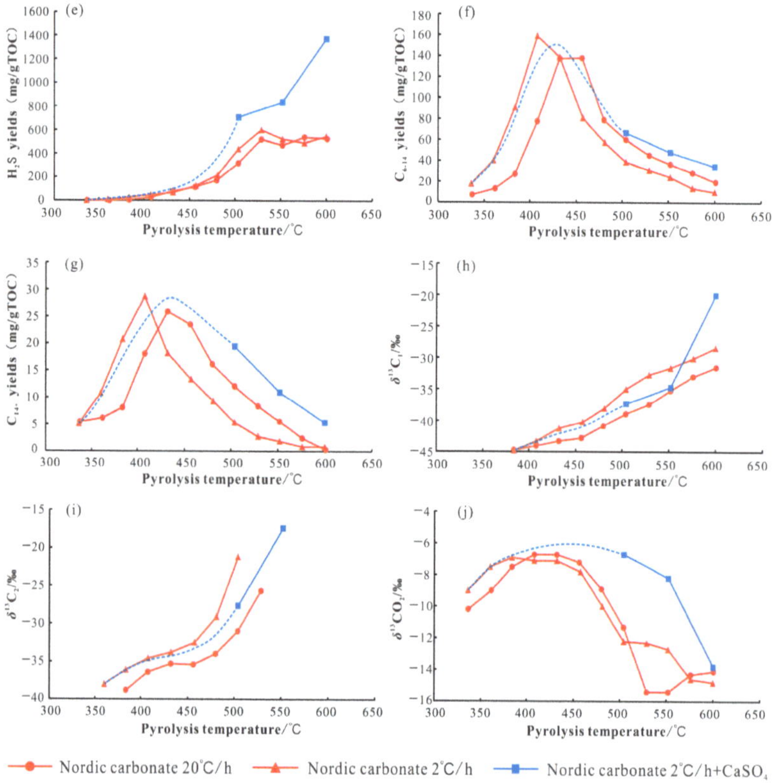

Figure 8. Carbonate thermal simulation results with the temperature change characteristics and product after joining $CaSO_4$ contrast. (**a**) Characterization of the carbonate C_1 yields as a function of temperature versus the temperature change with the addition of $CaSO_4$ and a comparison of the products; (**b**) Characterization of the carbonate $C_{2\text{-}5}$ yields as a function of temperature versus the temperature change with the addition of $CaSO_4$ and a comparison of the products; (**c**) Characterization of the carbonate H_2 yields as a function of temperature versus the temperature change with the addition of $CaSO_4$ and a comparison of the products; (**d**) Characterization of the carbonate CO_2 yields as a function of temperature versus the temperature change with the addition of $CaSO_4$ and a comparison of the products; (**e**) Characterization of the carbonate H_2S yields as a function of temperature versus the temperature change with the addition of $CaSO_4$ and a comparison of the products; (**f**) Characterization of the carbonate $C_{6\text{-}14}$ yields as a function of temperature versus the temperature change with the addition of $CaSO_4$ and a comparison of the products; (**g**) Characterization of the carbonate C_{14+} yields as a function of temperature versus the temperature change with the addition of $CaSO_4$ and a comparison of the products; (**h**) Characterization of the carbonate $\delta^{13}C_1$ as a function of temperature versus the temperature change with the addition of $CaSO_4$ and a comparison of the products; (**i**) Characterization of the carbonate $\delta^{13}C_2$ as a function of temperature versus the temperature change with the addition of $CaSO_4$ and a comparison of the products; (**j**) Characterization of the carbonate $\delta^{13}CO_2$ as a function of temperature versus the temperature change with the addition of $CaSO_4$ and a comparison of the products.

Wang et al. [65] conducted experiments using the addition of salts (pure carbonate, sulfate, and chloride salts) to hydrocarbon source rocks. The results showed that carbonate, sulfate, and chloride salts increased the yield of hydrocarbon pyrolysis of gaseous hydrocarbons from hydrocarbon source rocks with increasing temperature, showing a strong catalytic effect. In particular, the higher the experimental temperature is, the stronger the catalytic effect of the three salts. The catalytic effect of the salts not only enables the salt-lake hydrocarbon source rocks to generate natural gas in large quantities under shallow burial conditions, but also changes the composition of gaseous hydrocarbon products and accelerates the thermal evolution of organic matter. This is consistent with our results.

When $CaSO_4$ was added to the Nordic carbonate sample, the methane carbon isotopic composition became approximately 2–3‰ lighter (504–552 °C) or approximately 8‰ heavier (600 °C), suggesting that the methane may have been oxidized by sulfate at this time (Figure 8h); at 504 °C, the ethane carbon isotopic composition became approximately 6.5‰ lighter, suggesting that the addition of $CaSO_4$ may have facilitated ethane production. When introducing the yield of heavy hydrocarbon gases such as ethane, it was also found that the addition of $CaSO_4$ increased the content of heavy hydrocarbon gases such as ethane and reduced the drying coefficient of the gas, and the $\delta^{13}C_1$ value was still less than the $\delta^{13}C_2$ value after the addition of $CaSO_4$, indicating that it did not lead to carbon isotope inversion (Figure 8i, Table 3). Its propane carbon isotopic composition could not be measured because the amount of propane in the resulting gas was too small; its CO_2 carbon isotopic composition was heavier by approximately 5‰ at 504 °C and 552 °C, and by approximately 1‰ at 600 °C (Figure 8j), suggesting that the addition of $CaSO_4$ led to the presence of carbon isotopic fractionation between CO_2 of organic origin and carbonate rock of inorganic origin, thus leading to a heavier CO_2 carbon isotope or facilitating the decomposition of carbonate minerals. Although the CO_2 carbon isotopes in some highly sulfur-containing natural gases were heavy (approximately 0‰), this did not mean that this CO_2 was not produced due to TSR, as this was the result of TSR-induced CO_2 fractionation with carbon isotopes in reservoir carbonate rocks.

7. Conclusions

The results of the thermal simulation experiments comparing Nordic Cambrian carbonate with mud shale of the Pingliang Formation and coal of the Shanxi Formation showed that the Nordic carbonate and the mud shale of the Pingliang Formation have different hydrocarbon production characteristics, and although the difference in hydrocarbon production rates between the two was not significant, the carbonate continues to produce methane at 600 °C. The difference in hydrocarbon gas production rates between the Nordic carbonate and the coal of the Shanxi Formation was large. The hydrocarbon gas production rate of the Nordic carbonate was approximately equivalent to that of type II_2 kerogen, indicating that the hydrocarbon production rate is not related to lithology and only to the type of organic matter, but the Nordic carbonate can produce much H_2S.

The alkane carbon isotope changes are mainly controlled by the degree of thermal evolution, showing gradual heaviness with increasing temperature. No carbon isotope sequence reversal occurred during the thermal simulation, indicating that the Nordic carbonate may not experience carbon isotope sequence reversal during normal thermal evolution, unless it undergoes mixing or secondary interactions, and its distribution ranges are roughly the same as those of the alkane carbon isotope compositions of the mud shale of the Pingliang Formation. The ethane carbon isotopic composition was as heavy as −21.2‰ in the high-temperature stage, showing similar characteristics to coal-type gas. The large difference in CO_2 isotopic composition between the Nordic Cambrian carbonate and mud shale of the Pingliang Formation was because all CO_2 generated from the mud shale of the Pingliang Formation is derived from the decomposition of organic matter, whereas part of the CO_2 generated from the Nordic carbonate is derived from the decomposition of inorganic minerals, and the carbon isotopic composition of inorganic minerals differs significantly from that of organic matter.

H_2S is also produced from natural gas at the subsalt of the Majiagou Formation in the Ordos Basin. Therefore, we suggest that the H_2S in the Ordos Basin is derived from the carbonate rocks, the set of hydrocarbon source rocks, rather than from the Pingliang Formation mud shale and the Shanxi Formation coal. In addition, many natural gas ethane carbon isotopes under the salt of the Majiagou Formation are also characteristic of coal-type gas, which is consistent with the results of our thermal modeling experiments.

The addition of calcium sulfate did not significantly affect the compositional characteristics of the Nordic carbonate pyrolysis products, but caused the TSR reaction to occur, which significantly affects C_1 yields at the high-maturation stage, reducing their content by more than 50% at 600 °C. However, the addition of $CaSO_4$ increased C_2 yields and promoted the production of $C_{6\text{-}14}$ hydrocarbons and C_{14+} hydrocarbons at the high-temperature stage and substantially increased H_2, CO_2, and H_2S yields. The addition of $CaSO_4$ also caused $\delta^{13}C_1$ values to remain smaller than $\delta^{13}C_2$ values, suggesting no carbon isotope inversion, and the heavier CO_2 carbon isotopic composition suggested that the addition of $CaSO_4$ causes carbon isotope fractionation between CO_2 of organic origin and carbonate rock of inorganic origin, or promotes the decomposition of carbonate minerals.

Author Contributions: Conceptualization, Y.X. and J.H.; methodology, Q.W.; software, Q.M.; validation, J.S., H.Z. and Q.W.; formal analysis, Y.W.; investigation, J.S.; resources, J.H.; data curation, H.Z.; writing—original draft preparation, Y.W.; writing—review and editing, Q.M.; visualization, H.Z.; supervision, Y.X.; project administration, J.S.; funding acquisition, J.H. All authors have read and agreed to the published version of the manuscript.

Funding: This research was funded by [National Natural Science Foundation of China (NSFC)] grant number [41903013], [National Natural Science Foundation of China (NSFC)] grant number [42102180], [Natural Science Foundation of Jiangsu Province] grant number [BK20200171], [PetroChina Exploration and Production Branch Science and Technology Program] grant number [kt2021-04-01], and [PetroChina Exploration and Production Branch Science and Technology Program] grant number [2022KT0103].

Institutional Review Board Statement: Not applicable.

Informed Consent Statement: Informed consent was obtained from all subjects involved in the study.

Data Availability Statement: Data supporting the results can be found in the manuscript text, tables, and figures.

Conflicts of Interest: The authors declare no conflict of interest.

References

1. Liu, H.; Tan, X.; Li, Y.; Cao, J.; Luo, B. Occurrence and conceptual sedimentary model of Cambrian gypsum-bearing evaporites in the Sichuan Basin, SW China. *Geosci. Front.* **2018**, *9*, 1179–1191. [CrossRef]
2. Radwan, A.E.; Husinec, A.; Benjumea, B.; Kassem, A.A.; Aal, A.A.; Hakimi, M.H.; Thanh, H.V.; Abdel-Fattah, M.I.; Shehata, A.A. Diagenetic overprint on porosity and permeability of a combined conventional-unconventional reservoir: Insights from the Eocene pelagic limestones, Gulf of Suez, Egypt. *Mar. Pet. Geol.* **2022**, *146*, 105967. [CrossRef]
3. Jones, R.W. Comparison of carbonate and shale source rocks. In *Petroleum Geochemistry and Source Rock Potential of Carbonate Rocks*; Palacas, J.G., Ed.; American Association of Petroleum Geologists: Tulsa, OK, USA, 1984; Volume 18, pp. 163–180.
4. Palacas, J.G.; Anders, D.E.; King, J.D. South Florida Basin—A prime example of carbonate source rocks of petroleum. In *Petroleum Geochemistry and Source Rock Potential of Carbonate Rocks*; Palacas, J.G., Ed.; American Association of Petroleum Geologists: Tulsa, OK, USA, 1984; Volume 18, pp. 71–96.
5. Katz, B.J.; Dittmar, E.I.; Ehret, G.E. A geochemical review of carbonate source rocks in Italy. *J. Pet. Geol.* **2000**, *23*, 399–424. [CrossRef]
6. Guo, Y.; Fu, J.; Wei, X.; Xu, W.; Sun, L.; Liu, J.; Zhao, Y.; Gao, J.; Zhang, Y. Natural gas accumulation and models in Ordovician carbonates, Ordos Basin, China. *Pet. Explor. Dev.* **2014**, *41*, 393–403. [CrossRef]
7. Meng, M.; Ge, H.; Shen, Y.; Ji, W.; Wang, Q. Rock fabric of tight sandstone and its influence on irreducible water saturation in Eastern Ordos Basin. *Energy Fuels* **2023**, *37*, 3685–3696. [CrossRef]
8. Huang, D.; Xiong, C.; Yang, J.; Xu, Z.; Wang, K. Gas source discrimination and natural gas genetic types of central gas field in Ordos Basin. *Nat. Gas Ind.* **1996**, *16*, 95.
9. Yang, H.; Liu, X.; Zhang, D. Main controlling factors of gas pooling in Ordovician marine carbonate reservoirs in the Ordos Basin and advances in gas exploration. *Nat. Gas Ind.* **2013**, *33*, 1–12.
10. Wang, J.; Jia, H.; Sun, X.; Tao, C.; Zhang, Y.; Ma, L.; Wang, F.; Jiang, H. Comprehensive evaluation on origin and source of natural gas in the Paleozoic in Fuxian Area, Ordos Basin. *Nat. Gas Geosci.* **2022**, *33*, 1476–1484.
11. Dai, J.; Li, J.; Luo, X.; Zhang, W.; Hu, G.; Ma, C.; Guo, J.; Ge, S. Stable carbon isotope compositions and source rock geochemistry of the giant gas accumulations in the Ordos Basin, China. *Org. Geochem.* **2005**, *36*, 1617–1635. [CrossRef]

12. Liang, D.; Guo, T.; Chen, J.; Bian, L.; Zhao, J. Some progresses on studies of hydrocarbon generation and accumulation in marine sedimentary regions, Southern China (Part 1): Distribution of four suits of regional marine source rocks. *Mar. Orig. Pet. Geol.* **2008**, *13*, 1–16. [CrossRef]
13. Li, J.; Li, J.; Li, Z.; Zhang, C.; Cui, H.; Zhu, Z. Characteristics and genetic types of the Lower Paleozoic natural gas, Ordos Basin. *Mar. Pet. Geol.* **2018**, *89*, 106–119. [CrossRef]
14. Meng, M.; Ge, H.; Shen, Y.; Ji, W.; Li, Z. Insight into water occurrence and pore size distribution by nuclear magnetic resonance in marine shale reservoirs, southern China. *Energy Fuels* **2023**, *37*, 319–327. [CrossRef]
15. Ding, W.; Li, C.; Su, A.; He, Z. Study on the comprehensive geochemical cross section of Mesozoic marine source rocks and prediction of favorable hydrocarbon generation area in Qiangtan basin, Tibeta. *Acta Petrol. Sin.* **2011**, *27*, 878–896.
16. Xie, L.; Sun, Y.; Yang, Z.; Chen, J.; Jang, A.; Zhang, Y.; Deng, C. Evaluation of Hydrocarbon Generation of the Xiamaling Formation Shale in Zhangjiakou and Its Significance to the Petroleum Geology in North China. *Sci. China Ser. D Earth Sci.* **2013**, *43*, 1436–1444. [CrossRef]
17. Chen, L.; Zheng, L.; Huang, H.; Ning, C. Carbon isotopic evolution of hydrocarbon gases generated from carbonate source rocks via thermal simulation methods. *Pet. Geol. Exp.* **2022**, *44*, 121–128. [CrossRef]
18. Wang, Q. *The Ordovician Source Rocks Generate Gaseous Hydrocarbon Thermal Simulation Experiment Research*; University of Chinese Academy of Sciences: Beijing, China, 2014.
19. Moldowan, J.M.; Seifert, W.K.; Gallegos, E.J. Relationship between petroleum composition and depositional environment of petroleum source rocks. *AAPG Bull.* **1985**, *69*, 1255–1268.
20. Peters, K.E.; Moldowan, J.M.; Schoell, M.; Hempkins, W.B. Petroleum isotopic and biomarker composition related to source rock organic matter and depositional environment. *Org. Geochem.* **1986**, *10*, 17–27. [CrossRef]
21. Peters, K.E.; Walters, C.C.; Moldowan, J.M. *The Biomarker Guide*; Cambridge University Press: Cambridge, UK, 2005.
22. Burton, Z.F.M.; Moldowan, J.M.; Sykes, R.; Graham, S.A. Unraveling petroleum degradation, maturity, and mixing and addressing impact on petroleum prospectivity: Insights from frontier exploration regions in New Zealand. *Energy Fuels* **2018**, *32*, 1287–1296. [CrossRef]
23. Burton, Z.F.M.; Moldowan, J.M.; Magoon, L.B.; Sykes, R.; Graham, S.A. Interpretation of source rock depositional environment and age from seep oil, east coast of New Zealand. *Int. J. Earth Sci.* **2019**, *108*, 1093. [CrossRef]
24. Schulz, H.M.; Yang, S.; Schovsbo, N.H.; Rybacki, E.; Ghanizadeh, A.; Bernard, S.; Mahlstedt, N.; Krüger, M.; Amann-Hildebrandt, A.; Krooss, B.M.; et al. The Furongian to Lower Ordovician Alum Shale Formation in conventional and unconventional petroleum systems in the Baltic Basin–A review. *Earth-Sci. Rev.* **2021**, *218*, 103674. [CrossRef]
25. Sanei, H.; Petersen, H.; Schovsbo, N.; Jiang, C.; Goodsite, M. Petrographic and geochemical composition of kerogen in the Furongian (U. Cambrian) Alum Shale, central Sweden: Reflections on the petroleum generation potential. *Int. J. Coal Geol.* **2014**, *132*, 158–169. [CrossRef]
26. Meng, M.; Peng, J.; Ge, H.; Ji, W.; Li, X.; Wang, Q. Rock fabric of lacustrine shale and its influence on residual oil distribution in the Upper Cretaceous Qingshankou Formation, Songliao Basin. *Energy Fuels* **2023**, *37*, 7151–7160. [CrossRef]
27. Liu, W.; Wang, X.; Zhang, D.; Cai, Z.; Luo, H.; Chen, X.; Wang, T.; Li, F.; Zhang, W.; Li, F. Restudy on geochemical characteristics and genesis of Jingbian Gas Field in Ordos Basin. *J. Northwest Univ. Nat. Sci. Ed.* **2002**, *52*, 943–956. [CrossRef]
28. Yang, H.; Zhang, W.; Zan, C.; Ma, J. Geochemical characteristics of Ordovician subsalt gas reservoir and their significance for reunderstanding the gas source of Jingbian Gasfield, East Ordos Basin. *Nat. Gas Geosci.* **2009**, *20*, 8–14.
29. Liu, D.; Zhang, W.; Kong, Q.; Feng, Z.; Fang, S.; Peng, W. Lower Paleozoic source rocks and natural gas origins in Ordos Basin, NW China. *Pet. Explor. Dev.* **2016**, *43*, 540–549. [CrossRef]
30. Fu, J.; Yu, Z.; Li, C.; Wang, W.; Huang, Z.; Wu, X.; Wang, S. New discovery and favorable areas of natural gas exploration in the 4th Member of Ordovician Majiagou Formation by Well Mitan 1 in the eastern Ordos Basin. *Nat. Gas Ind.* **2021**, *41*, 17–27. [CrossRef]
31. Schenk, H.; Dieckmann, V. Prediction of petroleum formation: The influence of laboratory heating rates on kinetic parameters and geological extrapolations. *Mar. Pet. Geol.* **2004**, *21*, 79–95. [CrossRef]
32. Zhang, S.; Mi, J.; He, K. Synthesis of hydrocarbon gases from four different carbon sources and hydrogen gas using a gold-tube system by Fischer-Tropsch method. *Chem. Geol.* **2013**, *349–350*, 27–35. [CrossRef]
33. He, K.; Zhang, S.; Mi, J.; Zhang, W. Pyrolysis involving n-hexadecane, water and minerals: Insight into the mechanisms and isotope fractionation for water-hydrocarbon reaction. *J. Anal. Appl. Pyrolysis* **2018**, *130*, 198–208. [CrossRef]
34. Sweeney, J.; Burnham, A. Evaluation of a Simple Model of Vitrinite Reflectance Based on Chemical Kinetics. *AAPG Bull.* **1990**, *74*, 1559–1570.
35. Li, E.; Pan, C.; Yu, S.; Jin, X.; Liu, J. Hydrocarbon generation from coal, extracted coal and bitumen rich coal in confined pyrolysis experiments. *Org. Geochem.* **2013**, *64*, 58–75. [CrossRef]
36. Burnham, A.K. Kinetic models of vitrinite, kerogen, and bitumen reflectance. *Org. Geochem.* **2019**, *131*, 50–59. [CrossRef]
37. Gao, D.; Guo, C.; Jiang, C.; Zhang, L.; Wang, H.; Shi, P.; Chen, Y. Hydrocarbon generation simulation of low-maturity shale in Shanxi Formation, Ordos Basin. *Pet. Geol. Exp.* **2018**, *40*, 454–460. [CrossRef]
38. Xu, Y. *Genetic Theory of Natural Gases and Its Application*; Science Press: Beijing, China, 1994.
39. Wang, Q.; Lu, H.; Greenwood, P.; Shen, C.; Liu, J.; Peng, P. Gas evolution during kerogen pyrolysis of Estonian Kukersite shale in confined gold tube system. *Org. Geochem.* **2013**, *65*, 74–82. [CrossRef]

40. Peng, W.; Liu, Q.; Hu, G.; Lv, Y.; Zhu, D.; Meng, Q.; Guo, F.; Wang, R. Mechanisms of carbon isotope fractionation in the process of natural gas generation: Geochemical evidence from thermal simulation experiment. *Pet. Explor. Dev.* **2020**, *47*, 972–983. [CrossRef]
41. You, K.; Flemings, P.B.; Malinverno, A.; Collett, T.S.; Darnell, K. Mechanisms of methane hydrate formation in geological systems. *Rev. Geophys.* **2019**, *57*, 1146–1196. [CrossRef]
42. Burton, Z.F.M.; Dafov, L.N. Testing the sediment organic contents required for biogenic gas hydrate formation: Insights from synthetic 3-D basin and hydrocarbon system modelling. *Fuels* **2022**, *3*, 555–562. [CrossRef]
43. Burton, Z.F.M.; Dafov, L.N. Salt Diapir-Driven Recycling of Gas Hydrate. *Geochem. Geophys. Geosystems* **2023**, *24*, e2022GC010704. [CrossRef]
44. Wang, X.; Li, X.; Wang, X.; Shi, B.; Luo, X.; Zhang, L.; Lei, Y.; Jiang, C.; Meng, Q. Carbon isotopic fractionation by desorption of shale gases. *Mar. Pet. Geol.* **2015**, *60*, 79–86. [CrossRef]
45. Meng, Q.; Wang, X.; Shi, B.; Lei, Y.; Zhao, H.; Zhao, Q.; Yin, J.; Gao, C. The ^{13}C-depleted methane in terrigenous shale gas: A case study in the Triassic Yanchang Formation, Ordos Basin. *Mar. Pet. Geol.* **2022**, *141*, 105688. [CrossRef]
46. Meng, Q.; Wang, X.; Wang, X.; Shi, B.; Luo, X.; Zhang, L.; Lei, Y.; Jiang, C.; Liu, P. Gas geochemical evidences for biodegradation of shale gases in the Upper Triassic Yanchang Formation, Ordos Basin, China. *Int. J. Coal Geol.* **2017**, *179*, 139–152. [CrossRef]
47. Tian, H.; Xiao, X.; Yang, L.; Xiao, Z.; Guo, L.; Shen, J.; Lu, Y. Pyrolysis of oil at high temperatures: Gas potentials, chemical and carbon isotopic signatures. *Chin. Sci. Bull.* **2009**, *54*, 1217–1224. [CrossRef]
48. Meng, Q.; Shi, J.; Zhao, H.; Huang, J.; Liu, Y.; Wang, Y.; Xie, X.; Xu, Y. Genesis and source of natural gas in Well Mitan-1 of Ordovician Majiagou Formation, mid-eastern Ordos Basin, China. *Nat. Gas Geosci.* **2023**, 1–26.
49. Kong, Q.; Zhang, W.; Li, J.; Zan, C.L. Geochemical characteristics and genesis of Ordovician natural gas under gypsolyte in Ordos Basin. *Nat. Gas Geosci.* **2019**, *30*, 423–432. [CrossRef]
50. Xia, X.; Geoffrey, S.; Ma, Q.; Tang, Y. Compositional and stable carbon isotopic fractionation during non-autocatalytic thermochemical sulfate reduction by gaseous hydrocarbons. *Geochim. Cosmochim. Acta* **2014**, *139*, 472–486. [CrossRef]
51. Wei, Z.; Zou, Y.; Cai, Y.; Wang, L.; Luo, X.; Peng, P. Kinetics of oil group-type generation and expulsion: An integrated application to Dongying Depression. *Bohai Bay Basin. China Org. Geochem.* **2012**, *52*, 1–12. [CrossRef]
52. Liu, W.; Teng, G.; Wang, X.; Li, M.; Hu, G.; Wang, J.; Lu, L.; Zhao, H.; Chen, Q.; Luo, H. New knowledge of hydrocarbon generating theory of organic matter in Chinese marine carbonates. *Pet. Explor. Dev.* **2017**, *44*, 155–164. [CrossRef]
53. Wang, Q.; Liu, W.; Pei, L.; Cai, Z.; Luo, H.; Wang, X.; Zhang, D.; Liu, J. Hydrocarbon generation from calcium stearate: Insights from closed-system pyrolysis. *Mar. Pet. Geol.* **2021**, *126*, 104923. [CrossRef]
54. Liu, Q.; Wu, X.; Wang, X.; Jin, Z.; Zhu, D.; Meng, Q.; Huang, S.; Liu, J.; Fu, Q. Carbon and hydrogen isotopes of methane, ethane, and propane: A review of genetic identification of natural gas. *Earth-Sci. Rev.* **2019**, *190*, 247–272. [CrossRef]
55. Schoell, M. The hydrogen and carbon isotopic composition of methane from natural gases of various origins. *Geochim. Cosmochim. Acta* **1980**, *44*, 649–661. [CrossRef]
56. Schoell, M. Genetic characterization of natural gases. *AAPG Bull.* **1983**, *67*, 2225–2238.
57. James, A.T. Correlation of natural gas by use of carbon isotopic distribution between hydrocarbon components. *AAPG Bull.* **1983**, *67*, 1176–1191.
58. Clayton, C. Carbon isotope fractionation during natural gas generation from kerogen. *Mar. Pet. Geol.* **1991**, *8*, 232–240. [CrossRef]
59. Dai, J. Hydrocarbon isotopic characteristics of natural gas and identification of various natural gases. *Nat. Gas Geosci.* **1993**, *4*, 1–40.
60. Guo, H.; Liu, M.; Wang, Y.; Wang, Q.; Liu, J.; Peng, P. The effect of thermochemical sulfate reduction of n-pentane and condensate on the chemical and isotopic compositions of gases: Implications for the organic reactant and reaction extent. *Mar. Pet. Geol.* **2022**, *145*, 105916. [CrossRef]
61. Wu, J.; Wang, Y.; Qi, W.; Shi, S.; Jiang, F.; Zhang, C. Influence of Gypsum on the Distribution of Aromatic Molecules During Catagenesis and Its Geochemical Significance. *J. Sediment.* **2023**, 1–13. [CrossRef]
62. Zhu, X. *Sedimentary Petrography*; Petroleum Industry Press: Beijing, China, 2008; pp. 214–217.
63. Prothero, D.R.; Schwab, F. *Sedimentary Geology*; Freeman: New York, NY, USA, 2004; p. 557.
64. Zhao, G.; Li, S.; Liu, L. A study on characteristics and kinetics of catalytic degradation from kerogen in carbonate rocks. *Chin. J. Geol.* **2005**, *40*, 47–54. [CrossRef]
65. Wang, J.; Jin, Q.; Ma, G.; Wang, L.; Wang, X. Catalytic action of sulphate evaporates in pyrolytic gas generation of source rocks. *Nat. Gas Geosci.* **2009**, *20*, 26–31.
66. Hu, M.; Cheng, Z.; Zhang, M.; Liu, M.; Song, L.; Zhang, Y.; Li, J. Effect of Calcite, Kaolinite, Gypsum, and Montmorillonite on Huadian Oil Shale Kerogen Pyrolysis. *Energy Fuels* **2014**, *28*, 1860–1867. [CrossRef]

Journal of
Marine Science and Engineering

MDPI

Article

Factors Controlling the Pore Development of Low-Mature Marine–Continental Transitional Shale: A Case Study of the Upper Permian Longtan Shale, Western Guizhou, South China

Manting Zhang [1], Mingyi Hu [1,*], Sile Wei [1], Quansheng Cai [1], Wei Fu [1], Fang Shi [2], Lei Zhang [3] and Haiyan Ding [2]

1 School of Geosciences, Yangtze University, Wuhan 430100, China; 2021730041@yangtzeu.edu.cn (M.Z.)
2 Research Institute of Exploration and Development of Tarim Oilfield Company, PetroChina, Korla 841000, China
3 Lunnan Oil and Gas Production Management Zone of Tarim Oilfield Company, PetroChina, Korla 841000, China
* Correspondence: hmy_1965@163.com

Abstract: The Upper Permian Longtan Shale is a significant reservoir in western Guizhou. To clarify the main factors controlling the low-mature marine–continental transitional shale pore development in western Guizhou, pore types were classified with scanning electron microscopy (SEM), and the pore developmental stages and morphological structures were quantitatively characterized by nitrogen adsorption isotherm analyses. Additionally, the qualitative or semi-quantitative relationships between the pore developmental stages and the main controlling factors were established via geochemical analysis. The results showed that the Longtan Shale pores include intergranular pores, intragranular pores, organic pores, and microfractures. The intergranular pore structures were categorized into ink-bottle, slit, layered, and irregular types. The intragranular pores were found to be of the elliptical, nearly circular, ink-bottle, and irregular varieties. The organic pores were categorized into elliptical, bubble-like, and irregular polygonal variants. The microfractures were only of the elongated type. The clay-mineral-related intergranular pores were the predominant pore type. The organic pores were found to be poorly developed. The mesopores were predominant, followed by macropores. The shale pore diameters ranged between 1 nm and 100 nm, and they are characterized by multiple peaks. The specific surface area (SSA) was primarily provided by nanopores in the range of 5 nm to 10 nm, such that the smaller pores provided a greater contribution to the SSA, and they are more conducive to shale gas adsorption and accumulation. Clay mineral content was the dominant internal factor controlling pore development and the SSA, with the illite–smectite mixed layer being the most obvious controlling factor. While too low or too high clay mineral content is adverse to macropore development, brittle mineral content, carbonate mineral content, and total organic carbon (TOC) content are adverse to pore development and the SSA. Thermal maturity has no remarkable control effect on pore volume and the SSA of non-organic pores.

Keywords: western Guizhou; the Longtan Shale; shale pore; low mature; development characteristics; controlling factors

Citation: Zhang, M.; Hu, M.; Wei, S.; Cai, Q.; Fu, W.; Shi, F.; Zhang, L.; Ding, H. Factors Controlling the Pore Development of Low-Mature Marine–Continental Transitional Shale: A Case Study of the Upper Permian Longtan Shale, Western Guizhou, South China. *J. Mar. Sci. Eng.* **2023**, *11*, 1862. https://doi.org/10.3390/jmse11101862

Academic Editor: Michael Lazar

Received: 1 September 2023
Revised: 21 September 2023
Accepted: 24 September 2023
Published: 26 September 2023

1. Introduction

Shale gas is a clean and efficient energy source; thus, it is highly valued by countries around the world. China has taken a quantum leap in marine-shale commercial exploration, mainly with respect to the marine shale found in the Wufeng–Longmaxi Formation in the Sichuan Basin and its periphery, which has formed the Weiyuan, Changning, Fuling, and other shale gas fields [1–3]. There are multiple types of shale developed in southern China, such as marine shale, marine–continental transitional shale, and continental shale. Unlike marine shale, the cumulative thickness of marine–continental transitional shale in western

Guizhou is large, but the single layer is thin. The single layer is usually between 5 m and 15 m. In the longitudinal direction, the lithology changes rapidly. The shale interval is often associated with tight sandstone or coal seams, with the longitudinal distribution characteristics of gas-bearing coal rock, shale, and gas-bearing tight sandstone interbeds, which have rich shale gas resource potential [4]. However, most of the shale pore research focuses on the marine shale of the Wufeng–Longmaxi Formation, and less attention is paid to the marine–continental transitional shale of the Longtan Formation. The marine-shale exploration practice of the Wufeng–Longmaxi Formation reveals that shale pores play a significant role in controlling shale gas enrichment [5]; thus, it has momentous significance in terms of characterizing pore structure [6–9]. As such, in this study, the main controlling factors of pore development are explored.

Many scholars have proposed various classification schemes for shale pores. The International Union of Pure and Applied Chemistry (IUPAC) classified shale pores, based on pore diameter, into micropores, mesopores, and macropores. This classification scheme is extensively used in the quantitative description and rating of reservoir pores. When studying the siliceous Barnett shale in Fort Worth Basin, North America, Loucks et al. (2012) proposed that matrix pores can be classified into organic pores, intragranular pores, and intergranular pores according to their location relationship between pores and particles [10], but this classification scheme did not include microfractures. Yu Bingsong conducted a comprehensive classification based on two parameters: pore diameter and pore occurrence. In accordance with the relationship between mineral particles and pores, shale pores can be classified into matrix pores and microfractures [11]. The matrix pores can be classified into organic pores, intragranular pores, and intergranular pores. A proposal for the classification of the pore types according to pore morphology, including cylindrical pores, ink-bottle pores, funnel-shaped pores, and slit pores, was put forward by Rouquerol et al. (1994) [12]. The main pore types that develop in different shale types are also different. Marine-shale organic matter is mainly composed of sapropelinite in a large number of organic pores, while the marine–continental transitional shale and continental shale consist chiefly of vitrinite with poor organic pores. The main factors controlling pore development comprise TOC content, thermal evolution degree, kerogen type, mineral composition, etc. It is generally believed that shale porosity increases along with increases in the organic carbon content and maturity, but certain scholars have proposed a different view: When the maturity is low, the liquid hydrocarbons generated from organic matter can block some pores, thus causing a decrease in the effectively connected pores. It has been found that not all organic matter types can form nanoscale organic pores [13–15]. Different from type III kerogen, type I and type II kerogen are more likely to develop organic pores under the same thermal-evolution degrees. This shows that sapropelinite is a favorable factor for organic pore development. In summary, there are various shale pore types and numerous controlling factors, and there are certain differences in the major controlling factors for shale pore structures under different geological backgrounds. Therefore, it is essential to study the shale pore structure and controlling factors under different geological backgrounds.

The Upper Permian Longtan Formation in Guizhou is a potential reservoir for shale gas exploration and development in South China, and it has received widespread attention in recent years. Influenced by multi-stage geological tectonics, the reservoir lithological assemblages and pore structure are complex, and they are characterized by multiple layers, thin layers, large cumulative thicknesses, frequent interbedding, abundant organic matter, low maturity, and high clay mineral content. Though the existing literature has covered the shale pore structure characteristics [6,8,9,16–18], studies exploring areas such as low-mature marine–continental transitional shale pore structure characteristics and the dominant factors controlling shale pore development are lacking. In this research, the Longtan Shale core samples from the HP3 and YL4 wells in western Guizhou were selected. SEM was used to classify the pore types, and nitrogen adsorption isotherms quantified the pore developmental stages and morphological structures. Based on the geochemical test data of the kerogen types, TOC content, maturity, and mineral compositions of the shale samples, the dominant factors controlling shale pore development of the Longtan Formation were defined. These research findings can provide valuable information and references for low-mature marine–continental transitional shale studies on pore characteristics, fine reservoir characterization, and shale gas potential.

2. Geological Setting

Western Guizhou is located in the southwest Guizhou Depression, which belongs to the southwest section of the Upper Yangtze Platform (Figure 1). It has undergone multiple stages and high-intensity tectonic evolution, mainly including the Himalayan Movement, Yanshan Movement, Indosinian Movement, Hercynian Movement, and the Caledonian Movement. Under the influence of these orogenic transformations, folds and faults are developed in western Guizhou, mainly in the form of barrier folds. Anticlines are generally cut by accompanying faults, and the open synclines enable most of the Longtan Formation strata to be preserved.

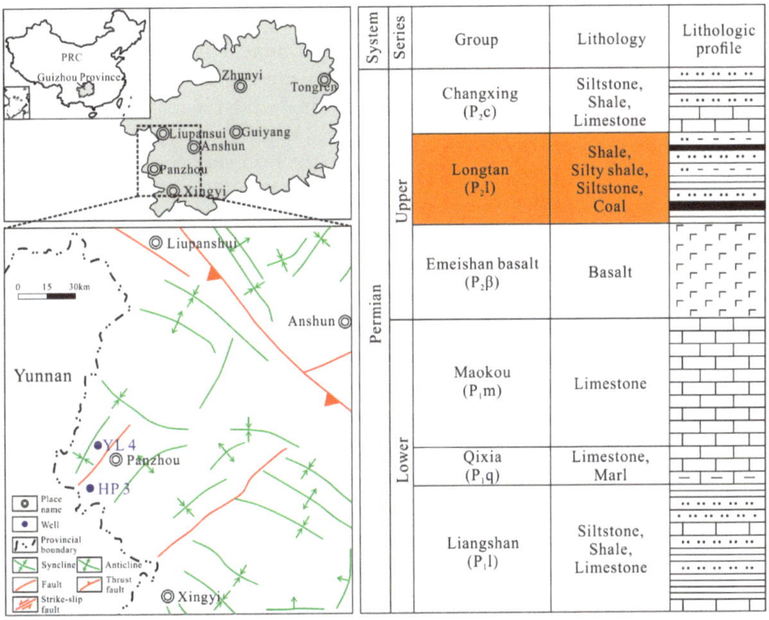

Figure 1. Structure outline map and the stratigraphic system in western Guizhou [19].

The Yangtze Platform experienced several transgression events on a large scale in the Paleozoic. Four shale sets were formed, represented by the Upper Permian, Lower Permian, Upper Ordovician–Lower Silurian, and Lower Cambrian sequences. During the Late Permian, the Longtan Shale was formed when the volcanic activity of the ancient Kangdian land weakened, and when periodical transgression occurred from eastern Guizhou to western Guizhou. A wide range of marine–continental transitional shale in the Longtan Formation was deposited, and the sedimentary environment mainly consisted of delta facies, tidal flat facies, and lagoon facies [19]. The stratum thickness was, generally, relatively large (150 m–300 m), with diverse lithology, which mainly included coal seams, siltstone, argillaceous siltstone, silty mudstone, and shale (Figure 1). Fine sandstone was seen occasionally with pyrite and thin layers of siderite. The Longtan Shale accumulated thickness varies from 30 m to 50 m, and it is characterized by multiple thin single layers and large accumulation thicknesses. It can be seen that the shale frequently alternates with sand and coal, which is the preferred layer series for shale gas exploration. The Longtan Shale is at stage A of the mid-diagenetic stage, and the main types of diagenesis are compaction, cementation, recrystallization, dissolution, and alteration [20].

3. Samples and Methods

Due to the strong weathering of outcrop samples, it is difficult to truly reflect shale reservoir characteristics. The test and analysis samples in this study were collected from drilling cores, which were mainly dark gray silty mudstone, dark gray shale, and gray-black carbonaceous shale. Nineteen Longtan Shale samples were collected from the HP 3 and YL 4 wells in western Guizhou. The samples were studied via the TOC determination test, vitrinite reflectance analysis, kerogen maceral identification and type classification, X-ray diffraction and clay mineral analysis, argon ion polishing SEM, and a low-temperature nitrogen adsorption experiment.

This study used a CS-230 sulfur carbon analyzer for the TOC determination. The experiment was carried out at 25 °C and 60% RH. Vitrinite reflectance analysis, kerogen macerals identification, and type classification were carried out using a Scope.A1 microphotometer. The X-ray diffraction method was used to study the shale mineralogy. The experiment was carried out with an Ultima IV X-ray diffractometer. All the samples were ground into fine powder (i.e., less than 40 μm). Diffraction patterns can help with determining and semi-quantitatively analyzing the relative abundance of minerals. The shale samples were prepared using a Gatan 697 Ilion Type II wide-beam argon ion polishing instrument, observed with ZEISS SIGMA field emission SEM. The low-temperature nitrogen adsorption experiment was conducted with a Quadrascorb SI-specific surface analyzer at 20 °C and a 30% RH. The samples were heated at 90 °C for 1 h and at 350 °C for 5 h.

4. Results

4.1. Organic Geochemical Characteristics

4.1.1. Kerogen Type

There are differences in the hydrocarbon potential among different kerogen types. The maceral identification of the Longtan Shale organic matter shows that sapropelinite ranges between 2.00% and 25.00%, with an average of 8.89%. Exinite ranges from 5.00% to 46.00%, with an average of 20.11%. Vitrinite ranges from 27.00% to 89.00%, with a mean of 61.00%. Inertinite varies between 3.00% and 27.00%, with a mean of 10.00% (Table 1). Overall, vitrinite has the highest values, followed by exinite. The predominant type of kerogen is type III, followed by type II_2 (which has a tendency to generate gas).

4.1.2. TOC Content

Previous exploration practices show that shale with a high TOC content has good hydrocarbon potential. The industrial accumulation of shale gas requires a base of rich gas source material and a certain standard of hydrocarbon-generating organic matter content. Boyer (2006) believed that the shale TOC content should be at least 2% [21]. According to sample tests, the Longtan Shale TOC content varies between 1.10% and 16.70% and averages at 5.43% (Table 1). All the samples showed good hydrocarbon potential.

4.1.3. Thermal Maturity

Organic maturity can reflect the hydrocarbon generation stage and type of kerogen, which can be determined by vitrinite reflectance. Hydrocarbon content and the state generated by organic matter in different mature stages have significant differences. According to the different types of organic-matter gas-generation modes, type III kerogen directly enters the gas-generation stage when the R_o reaches 0.5% [22]. The testing results from the Longtan Shale samples indicate that the thermal maturity is low as it ranges from 0.736% to 1.06% and averages at 0.923% (Table 1). Its evolution stage is mature and at the peak of gas generation.

Table 1. The Longtan Shale mineral composition in western Guizhou.

Well	Sample ID	Depth (m)	Formation	TOC (%)	R_o (%)	Macerals					Mineral Composition (%)			Clay Composition (%)			
						Sapropelinite	Exinite	Vitrinite	Inertinite	Kerogen Type	Brittle	Carbonate	Clay	I/S	I	K	C
HP 3	HP 3 S1	894.07	P_2l	3.22	0.902	7	10	73	10	III	36.30	0	63.70	46.00	2.00	35.00	17.00
HP 3	HP 3 S2	940.45	P_2l	3.79	1.060	23	35	34	8	II_2	25.10	0	74.90	43.00	2.00	31.00	24.00
HP 3	HP 3 S3	962.15	P_2l	6.05	1.024	9	13	51	27	III	38.80	6.10	55.10	62.00	4.00	22.00	12.00
HP 3	HP 3 S4	980.15	P_2l	8.69	0.918	25	34	33	8	II_2	44.10	0	55.90	2.00	1.00	62.00	35.00
HP 3	HP 3 S5	1009.03	P_2l	1.10	0.986	14	10	65	11	III	21.60	1.20	77.20	62.00	2.00	21.00	15.00
YL 4	YL 4 S1	344.88	P_2l	13.70	0.757	13	40	40	7	III	59.60	0.60	39.80	2.00	2.00	49.00	47.00
YL 4	YL 4 S2	346.80	P_2l	1.83	0.736	8	19	70	3	III	30.40	1.40	68.20	26.00	1.00	33.00	40.00
YL 4	YL 4 S3	357.87	P_2l	16.70	0.890	6	16	71	7	III	65.00	25.70	9.30	2.00	2.00	19.00	77.00
YL 4	YL 4 S4	403.71	P_2l	2.46	0.820	14	41	42	3	II_2	45.50	3.90	50.60	20.00	2.00	41.00	37.00
YL 4	YL 4 S5	428.76	P_2l	2.43	0.821	2	11	75	12	III	39.00	12.00	49.00	50.00	5.00	14.00	31.00
YL 4	YL 4 S6	452.46	P_2l	2.65	0.914	2	9	74	15	III	34.30	0.30	65.40	74.00	3.00	12.00	11.00
YL 4	YL 4 S7	488.19	P_2l	3.60	0.920	3	12	75	10	III	27.10	9.90	63.00	78.00	4.00	9.00	9.00
YL 4	YL 4 S8	522.68	P_2l	4.85	0.960	15	46	27	12	II_2	24.90	0.40	74.70	58.00	2.00	24.00	16.00
YL 4	YL 4 S9	548.06	P_2l	8.27	0.961	2	5	89	4	III	30.60	0	69.40	72.00	3.00	14.00	11.00
YL 4	YL 4 S10	561.12	P_2l	3.22	0.961	3	13	77	7	III	39.10	0	60.90	49.00	2.00	16.00	33.00
YL 4	YL 4 S11	573.69	P_2l	4.67	0.966	2	5	81	12	III	41.80	0	58.20	57.00	1.00	22.00	20.00
YL 4	YL 4 S12	585.12	P_2l	7.03	0.975	17	41	32	10	II_2	33.70	0.70	65.60	60.00	2.00	20.00	18.00
YL 4	YL 4 S13	588.98	P_2l	6.69	0.939	2	7	87	4	III	42.20	3.20	54.60	43.00	1.00	25.00	31.00
YL 4	YL 4 S14	612.52	P_2l	2.17	1.023	2	15	63	20	III	21.60	0	78.40	2.00	2.00	49.00	47.00
Average	/	/	/	5.43	0.923	7	10	73	10	III	36.88	3.44	59.68	42.53	2.26	27.26	27.95

4.2. Mineral Composition

The shale mineral composition is rich, and the brittle mineral content greatly affects the fracture generation ability in shale reservoirs. Therefore, the study of mineral composition is particularly important. The Longtan Shale minerals mostly consist of clay minerals (kaolinite, chlorite, illite, and illite–smectite mixed layer), brittle minerals (quartz, feldspar, siderite, and pyrite), and carbonate minerals (calcite and dolomite) (Table 1 and Figure 2). The clay minerals have the highest concentration of 59.68%, ranging from 9.30% to 78.40%. The porosity and microfractures are large, and the pore distribution characteristics are complex when the clay mineral content is high, which leads to a good gas-adsorption capacity. The brittle mineral content is relatively high, ranging between 21.60% and 65.00%, with a mean of 36.88%, which means it is profitable for shale fracturing. The carbonate mineral content ranges between 0% and 25.70% and averages at 3.44% (Table 1).

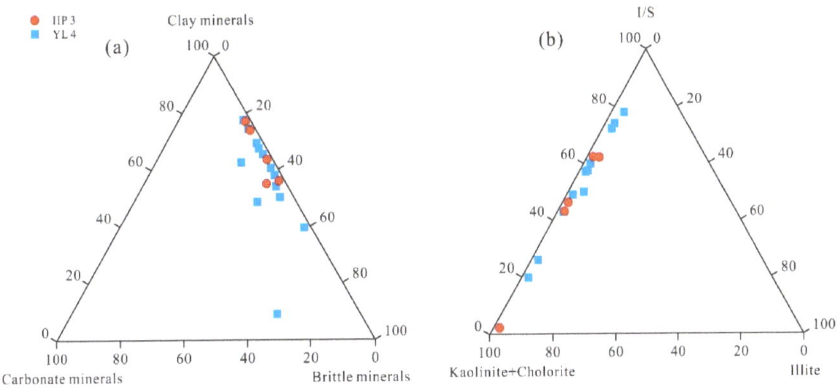

Figure 2. The Longtan Shale mineral distribution: (**a**) triangle diagram of the mineral distribution and (**b**) triangle diagram of the clay mineral distribution. I/S = illite–smectite mixed layer.

The clay minerals consist of illite, illite–smectite mixed layer, and other minerals such as kaolinite and chlorite. The other minerals' content, such as kaolinite and chlorite, is the highest, ranging from 18.00% to 97.00%, with a mean of 55.21%. The illite–smectite mixed layer content ranges between 2.00% and 78.00% and averages at 42.53%. The illite content ranges between 1.00% and 5.00% and averages at 2.26% (Table 1). The high illite–smectite mixed layer content reflects the sedimentary environment, which, at that time, was an oxygen-deficient and relatively reductive environment. It is profitable for organic matter preservation and enrichment, and it provides a suitable environment for hydrocarbon gas generation and enrichment.

4.3. Types of Shale Pores

According to the shale pore structural characteristics and genesis, the Longtan Shale reservoir space consists of intergranular pores, intragranular pores, organic pores, and microfractures (Table 2), and this was determined based on the SLATT method and the O'Brien six-division method of shale pore types [23].

Table 2. The Longtan Shale pore type classification in western Guizhou [23–25].

Pore Types	Genetic Mechanism	Location	Morphological	Pore Diameter
Intergranular pores	Primary sedimentary, cementation, diagenesis	Between clay mineral particles	Silt, layered	50 nm–1 μm
		Between brittle mineral particles, between plastic minerals and brittle mineral particles, between organic matter blocks and brittle minerals	Irregular, ink bottle	100 nm–1 μm
Intragranular pores	Postdiagenesis	Between pyrite crystals	Elliptical, nearly circular, ink bottle	20 nm–500 nm
		Dissolution of soluble minerals	Elliptical, irregular	20 nm–1 μm
Organic pores	Hydrocarbon generation evolution of organic matter	Internal pores of organic matter	Elliptical, bubble-like, irregular polygonal	5 nm–300 nm
Microfractures	Stress effect	Tectonic fractures	Elongated	20 nm–300 nm

4.3.1. Intergranular Pores

Intergranular pores refer to the pores that are supported by brittle mineral particles and which are not filled in particles. Intergranular pores mainly exist between clay mineral floccules or between mineral particles. Clay-mineral-related intergranular pores develop well in the Longtan Shale, which are of the ink-bottle (Figure 3b) and layered (Figure 3a,b) types. The pore diameter ranges between 50 nm and 1 μm. When the clay minerals and brittle minerals are randomly distributed, the clay-mineral-related intergranular pores are usually well preserved because of the support of the brittle mineral. In addition, microfractures are formed within the brittle minerals. Those in contact with plastic clay minerals communicate with clay-mineral-related intergranular pores; therefore, the intergranular pores connectivity was found to be good. The pores between brittle minerals (Figure 3a,g), the pores between plastic mineral particles and brittle mineral particles (Figure 3b), and the intergranular pores between organic matter block edges and mineral particles (Figure 3e) were observed under SEM. They had mostly irregular shapes, with pore diameters varying between 100 nm and 1 μm. Intergranular pores have good internal connectivity and are an important migration channel for shale gas.

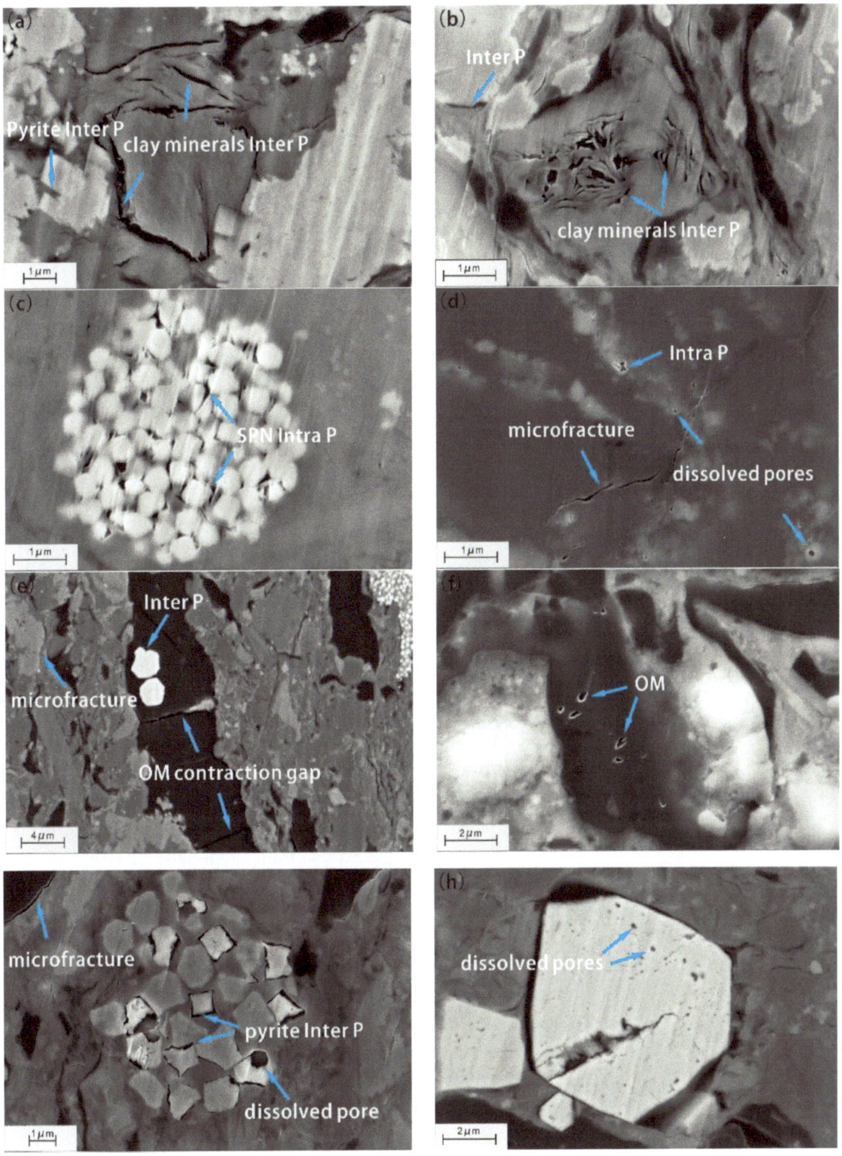

Figure 3. The Longtan Shale pore types and characteristics. SPN = strawberry pyrite nodules; Inter P = intergranular pores; Intra P = intragranular pores; OM = organic matter. (**a**) YL 4 well, 522.68 m; (**b**) YL 4 well, 588.98 m; (**c**) HP 3 well, 894.07 m; (**d**) HP 3 well, 1009.03 m; (**e**) YL 4 well, 403.71 m; (**f**) YL 4 well, 612.52 m; (**g**) YL 4 well, 403.71 m; and (**h**) YL 4 well, 561.12 m.

4.3.2. Intragranular Pores

Intragranular pores are generally formed inside mineral particles, with irregular pore morphology. Strawberry pyrite nodules are relatively common in organic-rich shale. They are formed of numerous small pyrite crystals. There are micropores among these crystals (Figure 3c), which have a pore diameter range of 20 nm to 500 nm. During the diagenesis process, the dissolution of minerals can generate dissolution pores [24] (Figure 3d,g,h), which are mostly elliptical and irregular in shape, albeit with smooth edges. The pore diameter varies between 20 nm and 1 μm, and it is susceptible to compaction.

4.3.3. Organic Pores

The thermal degradation of kerogen toward oil and gas can produce secondary pores and carbon-rich residues during the mature to over-matured stages. When the R_o reaches 0.6% or above, organic pores begin to develop [10,25]. The Longtan Shale R_o is greater than 0.7% (Table 1), which is sufficient for developing organic pores. However, small numbers of organic pores could be observed under a microscope (which is only occasionally seen in a few types of organic matter), and the pore diameter varied between 5 nm and 300 nm. Different from the spongy organic pores found in high- and over-high-mature marine shales, the organic pores that had developed in type III kerogen were mostly bubble-like (Figure 3f), and this may have been caused by the different hydrocarbon generation behaviors of the organic matter [26]. Not all kerogen types are easy-to-form organic pores. Type III kerogen is a type of solid organic matter with a macromolecular cross-linked structure in comparison to type I and II kerogen. Their thermal degradation is a process of "defunctionalization". In this process, it is hard to form organic pores if the macromolecular structure is not depolymerized [27].

4.3.4. Microfractures

A great number of microfractures are formed during the diagenesis process and later the transformation process. Furthermore, microfractures also form at the edge of the organic matter in the hydrocarbon generation process (Figure 3g). These abundant microfractures can provide migration channels for oil and gas, which is beneficial for later transformation. Microfractures in shale are often found between mineral particles (Figure 3e) or between mineral particles and organic matter (Figure 3g). They are usually caused by clay-mineral or organic-matter volume reduction, as well as by uneven internal stress within organic matter [28] (Figure 3d,e). Due to the frequent tectonic activity in western Guizhou, a large number of microfractures with nanoscale width and micro–nanoscale length can be observed under a microscope, most of which are elongated in shape (Figure 3d). Moreover, the brittle mineral content is a significant influence factor for shale porosity, microfracture development degree, gas-bearing capacity, and fracturing treatment methods. The Longtan Shale brittle mineral content is comparatively high, and it averages at 36%, which is more likely to generate fractures and increase the resistance to mechanical compaction. However, it is profitable for primary pore preservation.

4.4. Pore Morphology and Structure Characteristics

Isotherm adsorption curves that are obtained from low-temperature nitrogen adsorption experiments contain information about pore structure and pore type. In 1985, IUPAC introduced the capillary condensation phenomenon and proposed to divide the isothermal adsorption curves of gases into six basic types [29]. In 2015, IUPAC supplemented and improved the scheme proposed in 1985 [30] (Figure 4). The isotherm adsorption curves of the Longtan Shale are mainly IV(a) isotherm (Figure 5) with an inverse "S" shape. This shows that the adsorbent gas undergoes capillary condensation, and that the adsorption curve increases rapidly after reaching a certain pressure. The hysteresis loop is formed due to non-coincidence between the desorption curve and the adsorption curve. It also indicates that the samples are mainly composed of mesoporous pores.

The structure characteristics of the shale pores can be judged based on the morphology of the hysteresis loops on the nitrogen adsorption–desorption curves [31,32]. Various scholars have proposed a number of pore structure classification schemes based on hysteresis loop morphology. According to the differences in hysteresis loop morphology and pore structure, Sing (1985) [29] summarized four types of curves based on the classification of de Boer (1958) [33] and combined pore structure and hysteresis loop morphology; different hysteresis loops correspond to specific pore types (Figure 6). All samples from the Longtan Shale exhibited desorption hysteresis, thereby forming hysteresis loops with the adsorption branches, which means that capillary condensation occurred in the mesoporous samples [29]. According to the desorption hysteresis types (Figure 6), the morphological characteristics of the hysteresis loops in the HP 3 well and YL 4 well belong to type H_2. The interval between the adsorption and desorption curves of the hysteresis loops was relatively large, and the desorption curves showed a feature of rapid decline after reaching a certain relative pressure point. The pores were mainly of the ink-bottle shape with a thin neck and wide body, which act as the "bottleneck" of the ink-bottle type. Most of these pores are clay-mineral-related intergranular pores (Figure 3b). Different from the HP 3 well, some of the hysteresis loops of shale samples in the YL 4 well belong to type H_3. These hysteresis loops show a characteristic of not reaching saturation adsorption in high-pressure zones, which are mainly characterized by slit pores and layered pores (Figure 3a,b). It is noteworthy that most shale samples from the HP 3 well and YL 4 well had unclosed hysteresis loops at low relative pressures, which may be due to the expansion of shale or organic matter during adsorption, incomplete desorption of nitrogen, or the irreversible absorption in pores with similar diameters to nitrogen molecules [15,34].

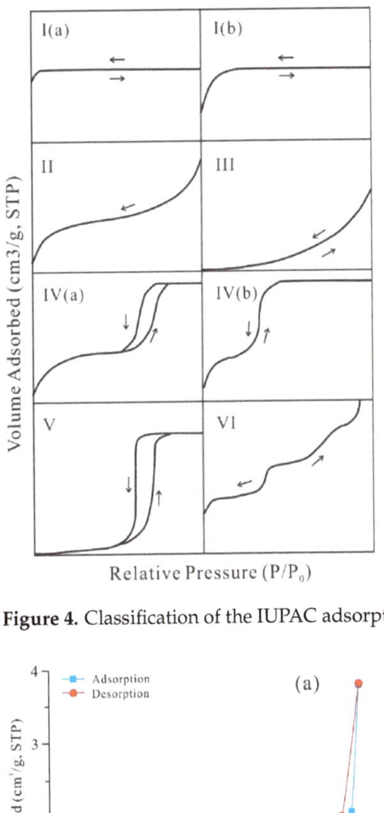

Figure 4. Classification of the IUPAC adsorption isotherm (according to Thommes et al., 2015 [30]).

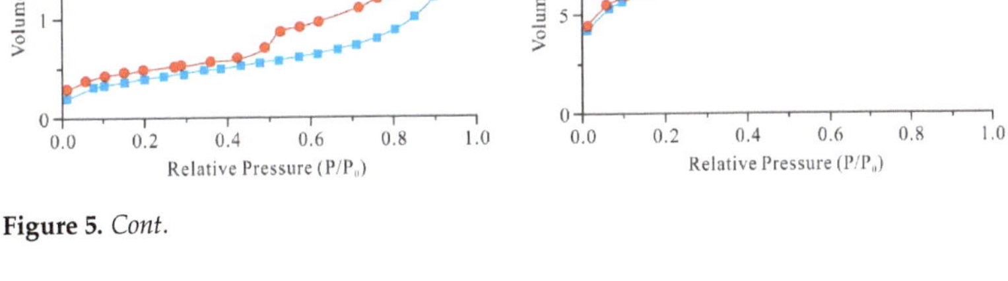

Figure 5. *Cont.*

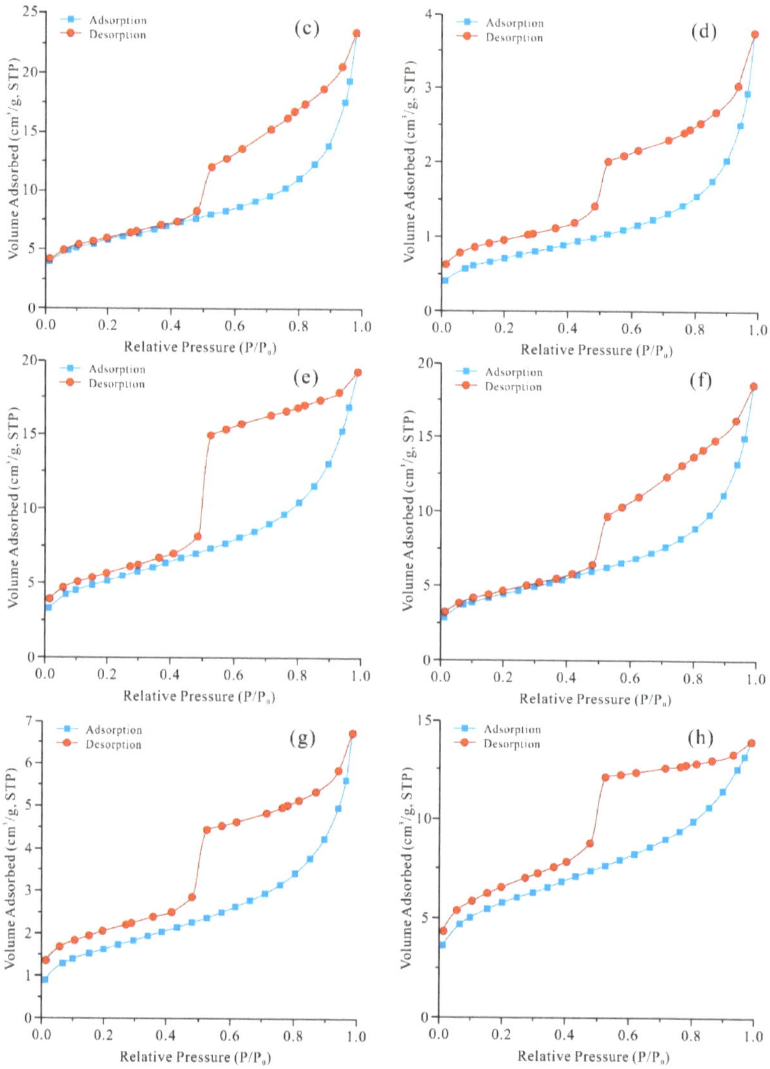

Figure 5. Isotherm adsorption curves of the Longtan Shale in western Guizhou. (**a**) YL 4 well, 357.87 m; (**b**) YL 4 well, 403.71 m; (**c**) YL 4 well, 488.19 m; (**d**) YL 4 well, 573.69 m; (**e**) HP 3 well, 940.45 m; (**f**) HP 3 well, 962.15 m; (**g**) HP 3 well, 980.15 m; and (**h**) HP 3 well, 1009.03 m.

Nitrogen adsorption can be used to characterize pore diameter distribution by differential pore volumes, logarithmic integral pore volumes, the cumulative adsorption amount, and stage adsorption increments [35]. Among them, the differential pore volume can indicate the reservoir space concentration degree in different pore diameter ranges. In this research, the DFT model was applied to describe the shale pore diameter distribution. The pore diameter distribution curves of the Longtan Shale samples all showed multi-peak characteristics (Figure 7), indicating a complex pore structure and a large-scale range in the pore diameter. The main peaks of the HP 3 well samples were concentrated at the 9 nm, 20 nm, and 40 nm range, with the peaks varying from 0.011 cm^3/g to 0.016 cm^3/g. Other peaks appeared at the 1.5 nm, 2 nm, and 4 nm range, with peaks varying between 0.005 cm^3/g and 0.011 cm^3/g. This indicated that the shale samples mostly developed mesopores, followed by micropores. The main peaks of the YL 4 well samples were concentrated at the 20 nm, 40 nm, and 50 nm range, and the peaks varied between 0.0026 cm^3/g and 0.023 cm^3/g. Other peaks appeared at the 1.5 nm, 9 nm, and 70 nm range, with peaks varying from 0.0008 cm^3/g to 0.014 cm^3/g. This showed that the shale samples mostly developed mesopores, followed by macropores.

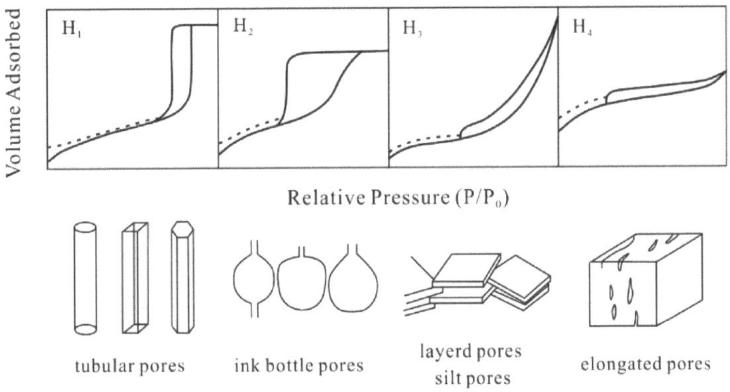

Figure 6. IUPAC hysteresis loop classification [29].

4.5. Pore Volumes and the SSA

The Brunauer–Emmett–Teller (BET) model was used to calculate the pore volume and the SSA [36] (Table 3). The pore volume of the Longtan Shale samples from the HP 3 well ranged between 0.007797 cm^3/g and 0.029955 cm^3/g, averaging at 0.019408 cm^3/g. The SSA varied between 2.90 m^2/g and 20.50 m^2/g, with averages at 12.65 m^2/g. The average pore diameter ranges were between 4.28 nm and 10.40 nm, with averages at 7.21 nm. The pore volume of the YL 4 well shale samples varied between 0.005438 cm^3/g and 0.036434 cm^3/g, averaging at 0.022314 cm^3/g. The SSA varied between 1.20 m^2/g and 27.80 m^2/g, averaging at 14.48 m^2/g. The average pore diameter varied between 4.81 nm and 18.10 nm, with averages at 8.15 nm. By comparison, the pore volume, the SSA, and the average pore diameter of the YL 4 well shale samples were slightly higher than those from the HP 3 well. The Longtan Shale SSA was provided primarily by the pores with a pore diameter varying from 5 nm to 10 nm. The numerous nanoscale pores in the Longtan Formation can greatly increase the SSA and offer plenty of space for shale gas occurrence.

The Longtan Shale total pore volume in western Guizhou varies between 0.005438 cm^3/g and 0.036434 cm^3/g, with averages at 0.021549 cm^3/g. The mesopores (2 nm < Φ < 50 nm) occupy most of the pore volume, varying from 0.002588 cm^3/g to 0.030858 cm^3/g with a mean of 0.017339 cm^3/g. The macropore volume (Φ > 50 nm) ranges between 0.001306 cm^3/g and 0.008306 cm^3/g, with averages at 0.003484 cm^3/g. The micropore volume accounts for a small portion of the total pore volume, varying from 0.000068 cm^3/g to 0.001604 cm^3/g, with a mean of 0.000727 cm^3/g (Table 3). These results indicate that mesopores contribute the most to the Longtan Shale pore volume in western Guizhou, followed by macropores and micropores.

The Longtan Shale SSA was positively correlated with pore volume (Figure 8), indicating that the shale SSA increases along with pore volume. The negative correlation between the average pore diameter and the SSA (Figure 9) indicated that a smaller pore diameter accounts for a larger proportion in the SSA.

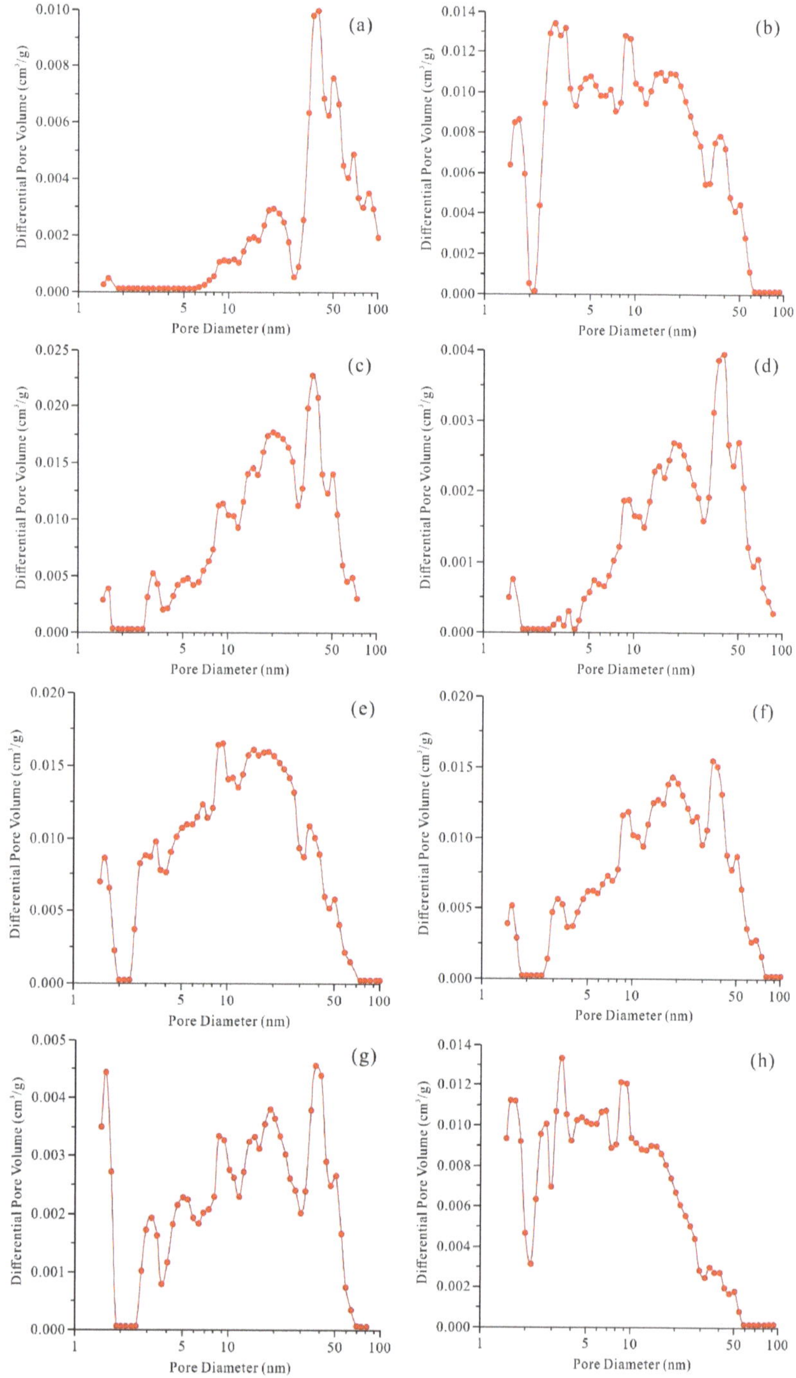

Figure 7. The Longtan Shale pore volume and pore diameter in western Guizhou. (**a**) YL 4 well, 357.87 m; (**b**) YL 4 well, 403.71 m; (**c**) YL 4 well, 488.19 m; (**d**) YL 4 well, 573.69 m; (**e**) HP 3 well, 940.45 m; (**f**) HP 3 well, 962.15 m; (**g**) HP 3 well, 980.15 m; and (**h**) HP 3 well, 1009.03 m.

Table 3. The Longtan Shale pore structure parameters in western Guizhou.

Sample ID	Depth (m)	TOC (%)	R_o (%)	Average Pore Diameter (nm)	Pore Volume (cm^3/g)	Micropore Volume (cm^3/g)	Mesopore Volume (cm^3/g)	Macropore Volume (cm^3/g)	S_{BET} (m^2/g)
HP 3 S1	894.07	3.22	0.902	10.40	0.007797	0.000109	0.006043	0.001645	2.90
HP 3 S2	940.45	3.79	1.060	6.61	0.029955	0.000878	0.024931	0.004146	18.40
HP 3 S3	962.15	6.05	1.024	7.43	0.028353	0.000597	0.021557	0.006199	15.60
HP 3 S4	980.15	8.69	0.918	7.31	0.010682	0.000304	0.008463	0.001915	5.83
HP 3 S5	1009.03	1.10	0.986	4.28	0.020251	0.001344	0.017601	0.001306	20.50
YL 4 S1	344.88	13.70	0.757	18.10	0.005438	0.000068	0.002588	0.002782	1.20
YL 4 S2	346.80	1.83	0.736	5.37	0.036434	0.001561	0.030858	0.004015	27.80
YL 4 S3	357.87	16.70	0.890	16.90	0.006054	0.000088	0.00294	0.003026	1.43
YL 4 S4	403.71	2.46	0.820	5.13	0.028421	0.001372	0.023543	0.003506	23.00
YL 4 S5	428.76	2.43	0.821	5.42	0.018027	0.000729	0.015253	0.002045	13.80
YL 4 S6	452.46	2.65	0.914	6.95	0.035502	0.000856	0.02634	0.008306	21.40
YL 4 S7	488.19	3.60	0.920	7.16	0.03539	0.000781	0.027435	0.007174	20.50
YL 4 S8	522.68	4.85	0.960	7.66	0.031765	0.000742	0.02535	0.005673	16.60
YL 4 S9	548.06	8.27	0.961	6.57	0.023904	0.000814	0.020611	0.002479	14.50
YL 4 S10	561.12	3.22	0.961	4.81	0.029846	0.001604	0.025878	0.002364	26.00
YL 4 S11	573.69	4.67	0.966	9.34	0.0059	0.000078	0.004417	0.001405	2.53
YL 4 S12	585.12	7.03	0.975	7.54	0.025585	0.000763	0.02043	0.004392	13.60
YL 4 S13	588.98	6.69	0.939	7.99	0.012952	0.0003	0.01042	0.002232	6.34
YL 4 S14	612.52	2.17	1.023	5.10	0.017182	0.000816	0.014784	0.001582	14.00
Average	/	5.43	0.923	7.90	0.021549	0.000727	0.017339	0.003484	14.00

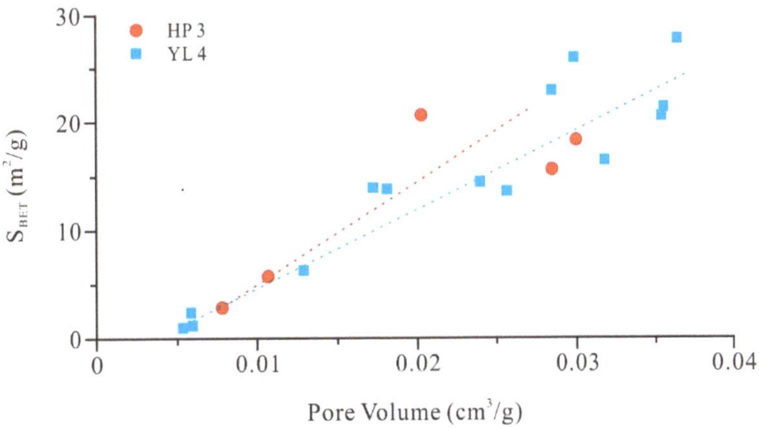

Figure 8. Relationship between pore volume and the SSA.

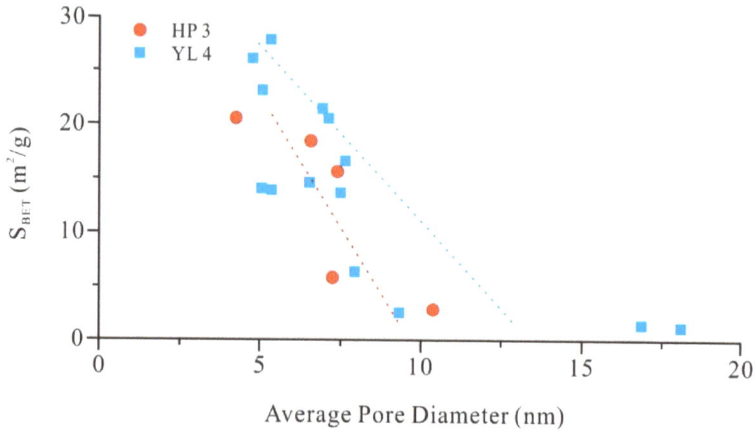

Figure 9. Relationship between the average pore diameter and the SSA.

5. Discussion

5.1. The Influence of Mineral Composition on Pore Development

5.1.1. Relationship between Pore Structure Characteristics and Brittle Minerals

As the supporting minerals in shales, brittle minerals can effectively bear the pressure in diagenesis. However, the brittle mineral content is found to be negatively correlated with the Longtan Shale pore parameters in western Guizhou (Figure 10), indicating that the brittle minerals' content in this area is not conducive to pore development. There is a large amount of terrigenous injection with high quartz and clay mineral content in the Longtan Shale. It is detrimental to quartz framework network development, as well as to organic-matter injection and preservation, thus resulting in the rare development of micropores around brittle minerals. Although brittle minerals have a certain protection on the macropores and mesopores in the periphery of mineral particles, there are a few pores inside brittle minerals, and these are mainly intergranular pores that are formed between brittle mineral particles. The distribution density of these pores is relatively weak. Therefore, there is a negative relationship between brittle mineral content and micropore volume, mesopore volume, macropore volume, and total pore volume. Moreover, the surface area morphology of brittle minerals (e.g., calcite and quartz) related to pores is simple; in addition, it cannot provide adsorption points, and it contributes little to the SSA.

It is a remarkable fact that strawberry pyrite is common in the Longtan Shale, as determined under SEM. The strawberry pyrite is formed in early diagenesis, and it lacks variation in the later diagenetic process, which can represent the earliest shale diagenetic environment. The high feldspar and clay mineral content of the Longtan Shale provides a large number of iron ions, which provides the main growth conditions for strawberry pyrite. According to the correlation between pyrite and reservoir parameters, each pore structure parameter decreases with increasing pyrite content (Figure 11). On the one hand, it shows that the primary reservoir spaces are not pyrite-related inorganic pores, such as pyrite intergranular pores. On the other hand, the euhedral pyrite formed by sulfate reduction in the closed diagenetic stage and overpressure environment occurs in a large amount of plant-residual structural pores. This metasomatizes clay minerals, which compresses the occurrence space of clay mineral pores and is adverse to pore development.

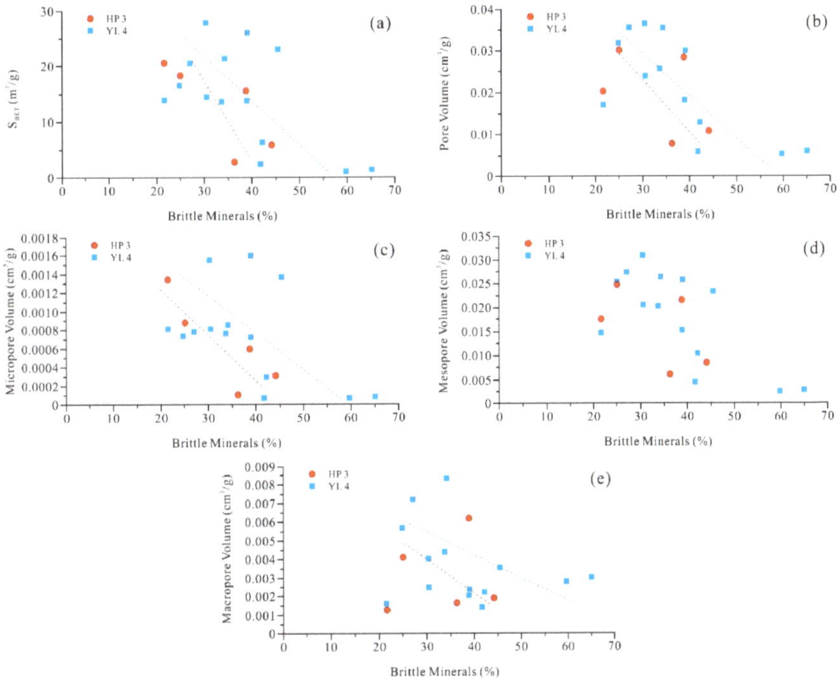

Figure 10. Relationship between pore structure characteristics and brittle minerals.

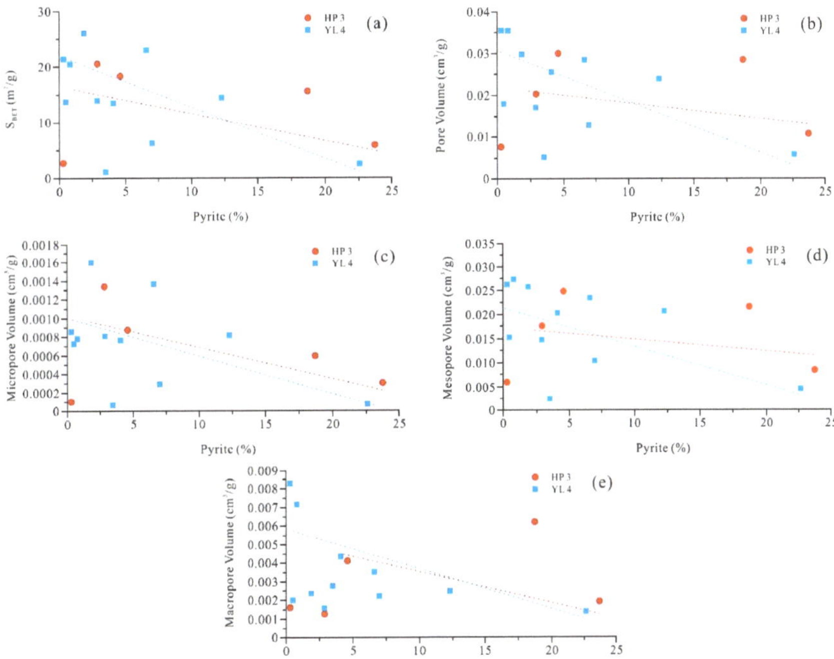

Figure 11. Relationship between pore structure characteristics and pyrite.

5.1.2. Relationship between Pore Structure Characteristics and Carbonate Minerals

The low carbonate mineral content in the Longtan Shale ranges between 0% and 25.70%, with averages at 3.44%. The bedding microfractures and shrinkage microfractures are well-developed, while the dissolution microfractures are rare. Carbonate minerals are soluble minerals that can form secondary dissolution pores under acidic fluids. Most of the secondary dissolution pores exist in isolated point shapes with a small amount of pores, and they do not have reservoir significance. Moreover, carbonate minerals are prone to chemical cementation, thereby forming secondary calcite, enlarging iron dolomite ring edges (achieved by filling in the primary pores or fractures in the form of calcite), or blocking intergranular pores. Consequently, each pore structure parameter is negatively correlated to the carbonate mineral content (Figure 12).

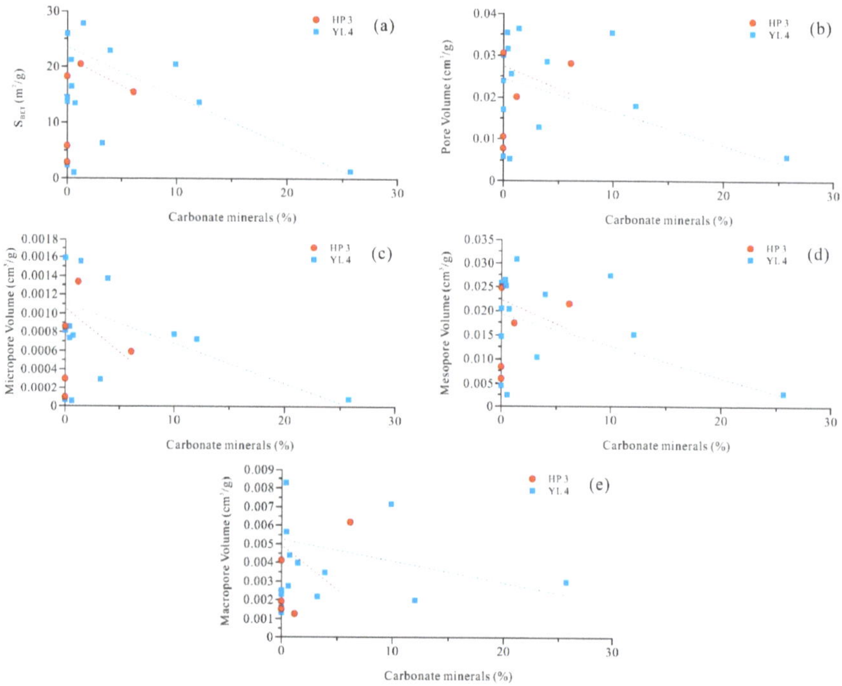

Figure 12. Relationship between pore structure characteristics and carbonate minerals.

5.1.3. Relationship between Pore Structure Characteristics and Clay Minerals

The clay mineral content in the Longtan Shale varies from 9.30% to 78.40%, with averages at 59.68%. The high clay mineral content forms the material foundation of clay-mineral-related pores. Due to the rich content of type III kerogen in the Longtan Shale, it can protect clay-mineral-related pores from the influence of organic matter injection, thus making it the main pore type of this formation. As shown in Figure 13, a positive correlation was found between the clay mineral content and micropore volume, mesopore volume, total pore volume, and the SSA. This indicates that the contribution of clay minerals to shale pores was mainly reflected in the micropores and mesopores, while the control effect on the macropores was not obvious. In the middle diagenetic period of the Longtan Shale, clay minerals undergo compaction and dehydration during the consolidation process, thus leading to volume shrinkage and the formation of new microfractures. These microfractures have a small extension length, but mostly have good connectivity. It is worth noting that the macropore volume shows a tendency to first increase and then decrease with increasing clay mineral content, indicating that too low or too high clay mineral content is unfavorable for macropore development. This is because shales with high clay mineral content are prone to compaction and deformation, leading to a sharp decrease in large primary pores. Moreover, high clay mineral content can cause brittleness reduction and a strengthening of the tensile shear ability of shale, which is adverse to structural fracture development.

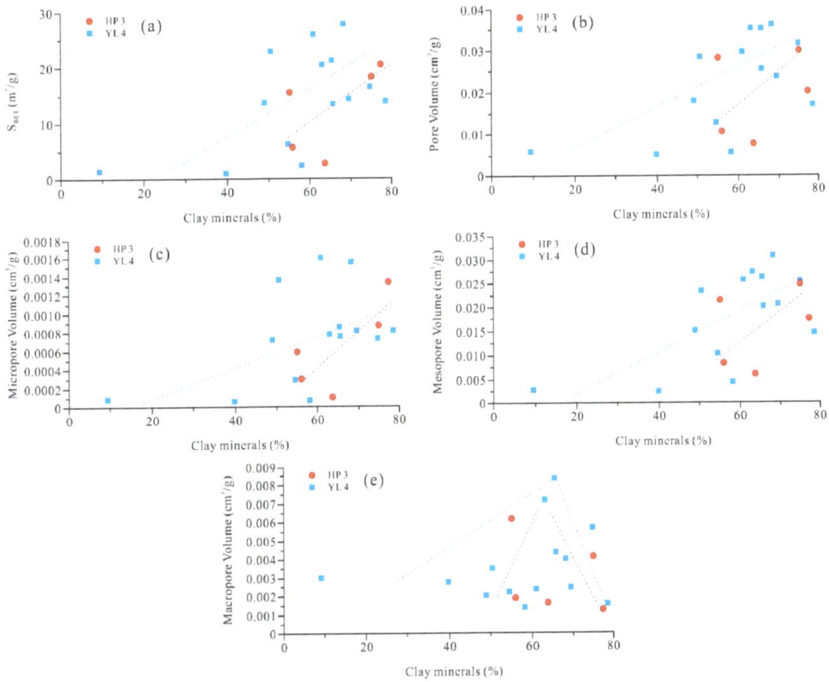

Figure 13. Relationship between pore structure characteristics and clay minerals.

Different clay minerals have great differences in morphology, structure, chemical properties, and physical properties. The effects on shale pore development are also different. X-ray diffraction analysis showed that the Longtan Shale clay minerals consist of illite–smectite mixed layer, chlorite, kaolinite, and slight illite deposits. The Longtan Shale clay minerals do not contain montmorillonite, which indicates that with the increasing depth, along with compaction and dehydration, the increasing temperature and pressure makes montmorillonite gradually transform into illite. In the transitional stage, there is illite–smectite mixed layer. The pores and fractures between clay minerals are the most developed in this stage due to the collapse of montmorillonite.

The pore structure parameters positively correlated with illite–smectite mixed layer content (Figure 14) suggests that illite–smectite mixed layer is profitable for shale pore development. The illite–smectite mixed layer particles have a rough surface, which increases the shale SSA and pore volume, as well as provides more storage and adsorption space for shale gas [37]. The chlorite and kaolinite content negatively correlated with the pore structure parameters (Figures 15 and 16). The negative correlation between the micropore volume and chlorite was found to be significant, which indicated that chlorite has a negative effect on shale pore development, especially on micropore volume. The content of illite was only between 1.00% and 5.00%, with an average of 2.26%, and this was not relative to the pore structure parameters (Figure 17). This indicated that the illite–smectite mixed layer is the main factor controlling clay-mineral-related pores; meanwhile, illite has no obvious control over clay-mineral-related pores and the SSA.

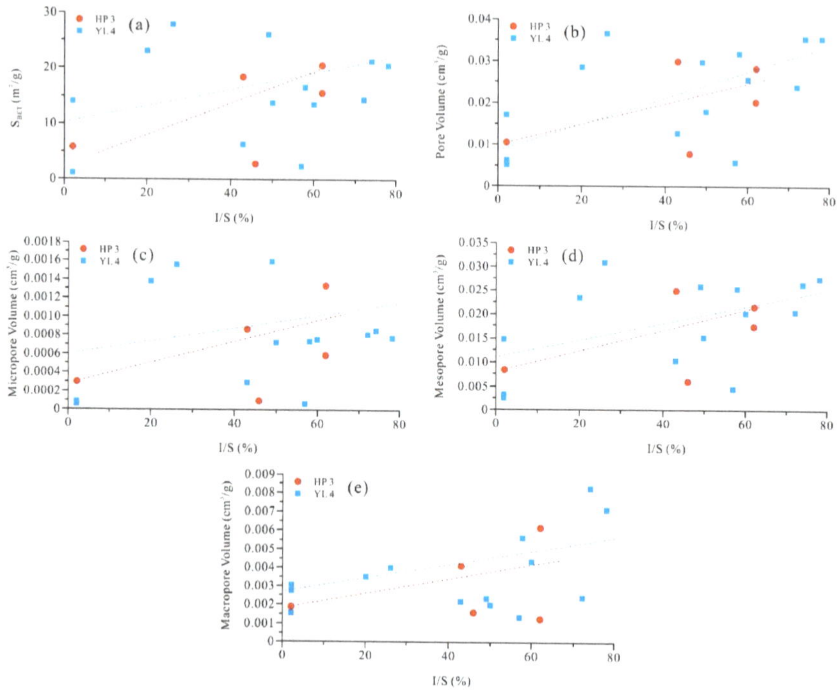

Figure 14. Relationship between pore structure characteristics and I/S.

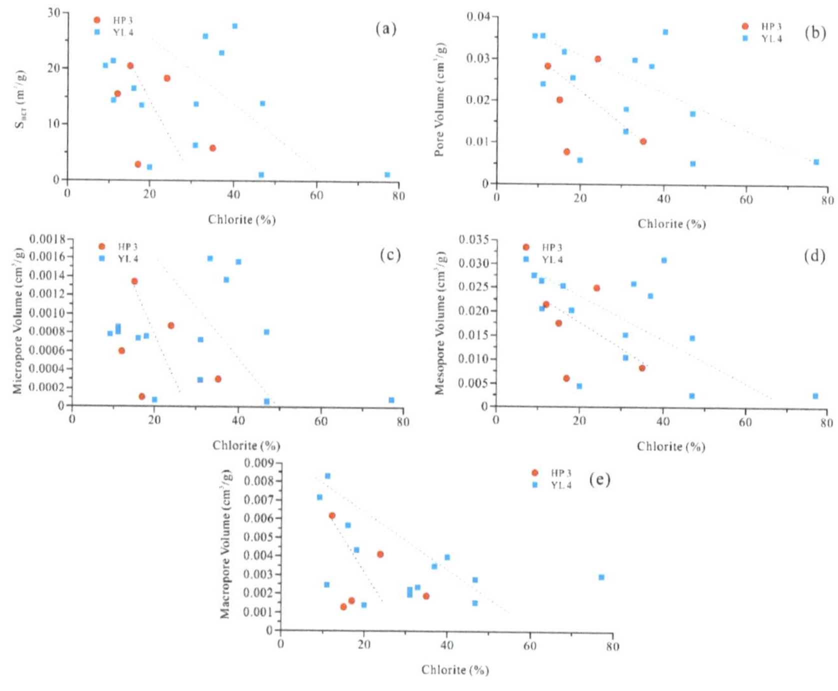

Figure 15. Relationship between pore structure characteristics and chlorite.

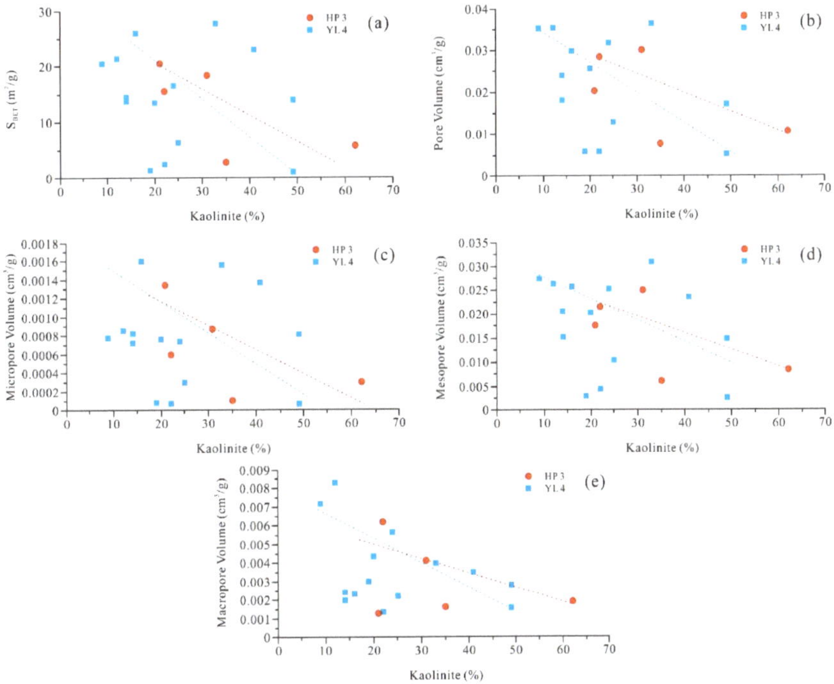

Figure 16. Relationship between pore structure characteristics and kaolinite.

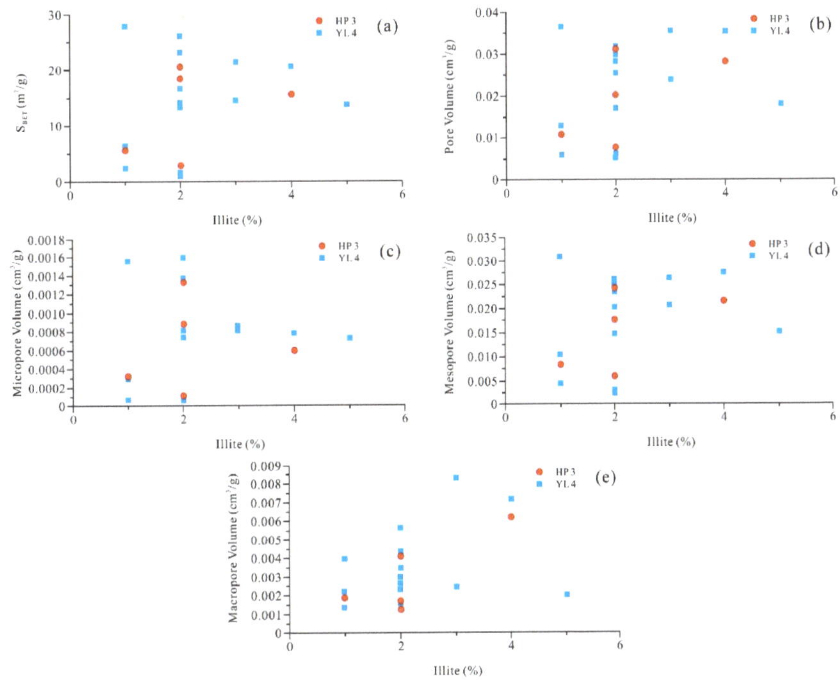

Figure 17. Relationship between pore structure characteristics and illite.

5.2. The Influence of Organic Matter on Pore Development

5.2.1. Relationship between Pore Structure Characteristics and TOC Content

During the process of hydrocarbon generation and expulsion, organic matter can produce large amounts of micro–nanopores, which has important implications for increasing reservoir adsorption and storage space and improving the permeability [38]. The Longtan Shale organic matter content is relatively high in western Guizhou. The organic matter content varies between 1.10% and 16.70% and averages at 5.43%. Unlike marine shale, where the kerogen types are type I and type II_1, and the macerals are mostly sapropelinite. The organic matter of marine and continental transitional shale in western Guizhou mainly originates from higher plant detritus, including type III and type II_2 kerogen; moreover, the maceral components are mainly vitrinite. Ungerer et al. found that the chemical structure of vitrinite has more aromatic and benzene rings, and its molecular structure is stable [39]. It is not easy to phase change and form thermogenic pores; thus, the organic pores are undeveloped. The SSA, micropore volume, mesopore volume, and total pore volume all decrease with the increasing TOC content of the Longtan Shale (Figure 18a–d). There was no correlation found between the macropore volume and TOC content (Figure 18e). This indicates that the contribution of TOC content to the SSA and pore volume is poor. In the Longtan Shale, the organic pores are relatively small in number and size, and the intergranular and intergranular pores are relatively developed (Figure 3f). This may be an important reason as to why no obvious negative correlation was found between the total pore volume and the TOC content [40]. Moreover, the shale plasticity will be strengthened due to the higher organic matter content, and it is more prone to mechanical compaction, which is adverse to pore preservation. In addition, excessive organic matter content will fill some inorganic mineral intergranular pores, thus further reducing the pore volume of the shale.

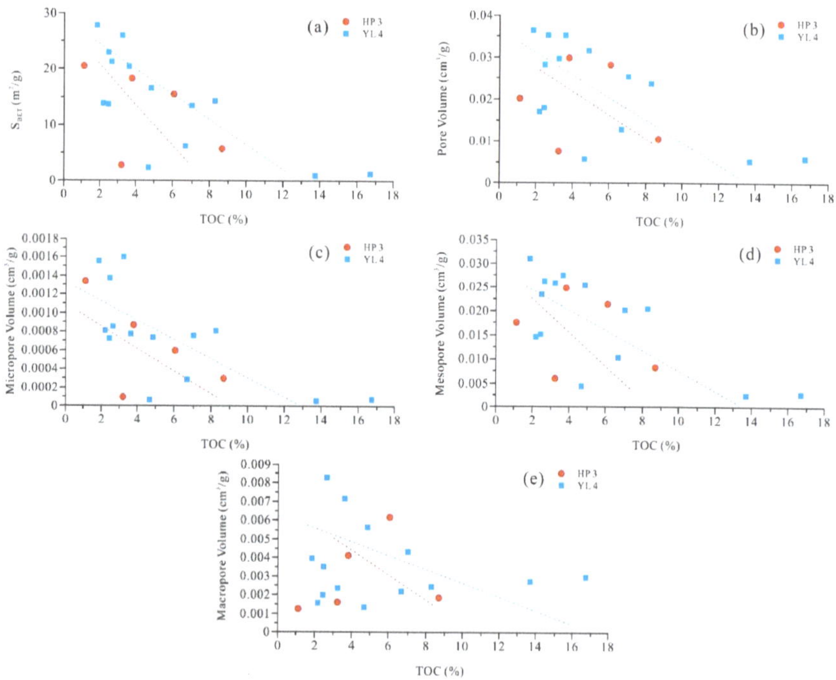

Figure 18. Relationship between the pore structure characteristics and TOC content.

5.2.2. Relationship between Pore Structure Characteristics and the Thermal-Evolution Degree

In shale, organic pore morphology and the degree of development are directly correlated with the hydrocarbon-generation evolution stage [40]. Curtis et al. (2012) believed that kerogen can only develop a great many nanoscale pores when it reaches a certain maturity, and immature-to-low-mature kerogen (R_o < 1.0%) has not undergone the pyrolysis and hydrocarbon expulsion process needed to remain in the original state of no pores or only a few pores [41,42]. Loucks et al. [25] and Curtis et al. [41] also found that when the thermal-evolution degree was less than 0.9%, no organic pores, or particularly few organic pores, were found in the shale SEM images. In accordance with this, undeveloped organic pores were also observed under the microscope in the Longtan Shale (Figure 3e,f).

The Longtan Shale vitrinite reflectance varies from 0.736% to 1.06% and averages at 0.92%. This is a mature evolution and occurs during the middle diagenetic stage; furthermore, this process mainly develops non-organic pores. As shown in Figure 19, the pore structure parameters of the Longtan Shale are irrelevant to thermal maturity (R_o), indicating that thermal maturity has no significant control over the non-organic SSA and pore volume. Loucks et al. and Zhang Y.F. et al. believed that the non-organic pores were mainly formed under the control of diagenesis [10,43], and the gradual increase in burial depth and paleogeotemperature, while the compaction was also more intense, led to the increase in thermal maturity. The strong compaction resulted in the deformation of the plastic minerals and the reduction in mineral pores. However, rigid particles can protect mineral pores from compaction, and mineral composition also has important effects on the non-organic pore evolution [44]. Moreover, cementation and clay mineral transformation also have destructive effects on mineral pores, while dissolution increases the number of mineral pores [43]. In conclusion, thermal maturity has little influence on the non-organic pore-evolution process of the Longtan Shale.

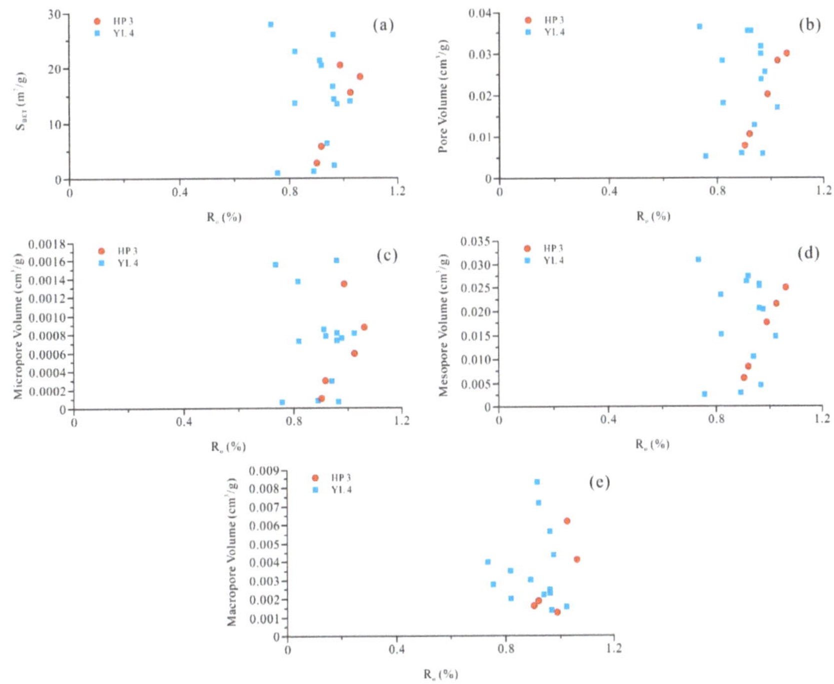

Figure 19. Relationship between the pore structure characteristics and vitrinite reflectance.

6. Conclusions

The physical properties and characteristics of the Upper Permian Longtan Shale pores in western Guizhou were studied, and the relationship between shale pore structure and physical properties was discussed. The following conclusions were made:

(1) The Longtan Shale mineral composition mainly consists of clay minerals and brittle minerals. In addition, the TOC content is rich, and the thermal maturity is low, which is the case due to the processes the shale underwent during its mature evolution stage and the peak of its gas generation period. The shale pore type includes intergranular pores, intragranular pores, organic pores, and microfractures. The intergranular pore structures are of the slit, layered, irregular, and ink-bottle varieties. The intragranular pores range from elliptical, nearly circular, ink-bottle, to irregular types. The organic pores are of the elliptical, bubble-like, and irregular polygonal variants. The microfractures are the elongated type. The clay-mineral-related intergranular pore is the predominant pore type, and it is mainly of the ink-bottle, slit, and layered varieties. The organic pores were found to be poorly developed.

(2) The Longtan Shale pores are mainly mesopores, followed by macropores. The development scale range of the pore diameter is large, ranging from 1 nm to 100 nm, with multiple peak characteristics. The nanopores mainly contributed to the SSA, which varied between 5 nm and 10 nm. The smaller pores provided a greater contribution to the SSA, which is more conducive to shale gas adsorption and accumulation.

(3) Clay mineral content is the dominant internal factor controlling pore development and the SSA, with the most obvious control being due to the illite–smectite mixed layer. However, too low or too high clay mineral content is adverse to macropore development. The macropore volume shows a tendency to first increase and then decrease with increasing clay mineral content. Brittle mineral content, carbonate mineral content, and TOC content are adverse to pore development and the SSA. Thermal maturity has no remarkable control effect on pore volume and the SSA of non-organic pores.

Author Contributions: Writing-original draft, investigation, M.Z.; supervision, M.H.; writing-review & editing, Funding acquisition, S.W.; investigation, Funding acquisition, Q.C.; Data curation, W.F. and F.S.; Visualization, L.Z.; Software, H.D. All authors have read and agreed to the published version of the manuscript.

Funding: This research was funded by Natural Science Foundation of Hubei Province: 2022CFB652 and China National Petroleum Corporation Innovation Found: 2021DQ02-0101.

Data Availability Statement: The data that support the findings of this study are available from the corresponding author upon reasonable request.

Conflicts of Interest: The authors declare no conflict of interest.

References

1. Li, J.; Li, H.; Yang, C.; Wu, Y.; Gao, Z.; Jiang, S. Geological Characteristics and Controlling Factors of Deep Shale Gas Enrichment of the Wufeng-Longmaxi Formation in the Southern Sichuan Basin, China. *Lithosphere* **2022**, *2022*, 4737801. [CrossRef]
2. Xi, Y.; Li, J.; Liu, G.; Cha, C.; Fu, Y. Numerical Investigation for Different Casing Deformation Reasons in Weiyuan-Changning Shale Gas Field during Multistage Hydraulic Fracturing. *J. Pet. Sci. Eng.* **2018**, *163*, 691–702. [CrossRef]
3. Gou, Q.; Xu, S.; Hao, F.; Zhang, B.; Shu, Z.; Yang, F.; Wang, Y.; Li, Q. Quantitative Calculated Shale Gas Contents with Different Lithofacies: A Case Study of Fuling Gas Shale, Sichuan Basin, China. *J. Nat. Gas Sci. Eng.* **2020**, *76*, 103222. [CrossRef]
4. Fu, W.; Hu, W.; Yi, T.; Kane, O.I.; Zhang, M.; Huang, X. Fractal Dimension Analysis of Pores in Coal Reservoir and Their Impact on Petrophysical Properties: A Case Study in the Province of Guizhou, SW China. *Minerals* **2022**, *12*, 1425. [CrossRef]
5. Fan, C.; Li, H.; Qin, Q.; He, S.; Zhong, C. Geological Conditions and Exploration Potential of Shale Gas Reservoir in Wufeng and Longmaxi Formation of Southeastern Sichuan Basin, China. *J. Pet. Sci. Eng.* **2020**, *191*, 107138. [CrossRef]
6. Wu, J.; Yuan, Y.; Niu, S.; Wei, X.; Yang, J. Multiscale Characterization of Pore Structure and Connectivity of Wufeng-Longmaxi Shale in Sichuan Basin, China. *Mar. Pet. Geol.* **2020**, *120*, 104514. [CrossRef]

7. Li, X.; Wang, Y.; Lin, W.; Ma, L.; Liu, D.; Liu, J.; Zhang, Y. Micro-Pore Structure and Fractal Characteristics of Deep Shale from Wufeng Formation to Longmaxi Formation in Jingmen Exploration Area, Hubei Province, China. *J. Nat. Gas Geosci.* **2022**, *7*, 121–132. [CrossRef]
8. Yang, R.; He, S.; Yi, J.; Hu, Q. Nano-Scale Pore Structure and Fractal Dimension of Organic-Rich Wufeng-Longmaxi Shale from Jiaoshiba Area, Sichuan Basin: Investigations Using FE-SEM, Gas Adsorption and Helium Pycnometry. *Mar. Pet. Geol.* **2016**, *70*, 27–45. [CrossRef]
9. Wang, Y.; Cheng, H.; Hu, Q.; Liu, L.; Jia, L.; Gao, S.; Wang, Y. Pore Structure Heterogeneity of Wufeng-Longmaxi Shale, Sichuan Basin, China: Evidence from Gas Physisorption and Multifractal Geometries. *J. Pet. Sci. Eng.* **2022**, *208*, 109313. [CrossRef]
10. Loucks, R.G.; Reed, R.M.; Ruppel, S.C.; Hammes, U. Spectrum of Pore Types and Networks in Mudrocks and a Descriptive Classification for Matrix-Related Mudrock Pores. *AAPG Bull.* **2012**, *96*, 1071–1098. [CrossRef]
11. Bingsong, Y.U. Classification and Characterization of Gas Shale Pore System. *Earth Sci. Front.* **2013**, *20*, 211–220.
12. Rouquerol, J.; Avnir, D.; Fairbridge, C.W.; Everett, D.H.; Haynes, J.H.; Pernicone, N.; Ramsay, J.D.F.; Sing, K.S.W.; Unger, K.K. Recommendations for the Characterization of Porous Solids. *Int. Union Pure Appl. Chem.* **1994**, *66*, 1739–1758. [CrossRef]
13. Kotarba, M.J.; Curtis, J.B.; Lewan, M.D. Comparison of Natural Gases Accumulated in Oligocene Strata with Hydrous Pyrolysis Gases from Menilite Shales of the Polish Outer Carpathians. *Org. Geochem.* **2009**, *40*, 769–783. [CrossRef]
14. Jarvie, D. Evaluation of Hydrocarbon Generation and Storage in the Barnett Shale, Ft. Worth Basin, Texas©. In Proceedings of the Ellison Miles Memorial Symposium, Farmers Branch, TX, USA, 22–23 June 2004.
15. Cao, T.; Deng, M.; Song, Z.; Luo, H.; Hursthouse, A.S. Characteristics and Controlling Factors of Pore Structure of the Permian Shale in Southern Anhui Province, East China. *J. Nat. Gas Sci. Eng.* **2018**, *60*, 228–245. [CrossRef]
16. Sun, W.; Zuo, Y.; Wu, Z.; Liu, H.; Xi, S.; Shui, Y.; Wang, J.; Liu, R.; Lin, J. Fractal Analysis of Pores and the Pore Structure of the Lower Cambrian Niutitang Shale in Northern Guizhou Province: Investigations Using NMR, SEM and Image Analyses. *Mar. Pet. Geol.* **2019**, *99*, 416–428. [CrossRef]
17. Wang, T.; Tian, F.; Deng, Z.; Hu, H. Pore Structure and Fractal Characteristics of Wufeng–Longmaxi Formation Shale in Northern Yunnan–Guizhou, China. *Front. Earth Sci.* **2023**, *10*, 998958. [CrossRef]
18. Yuan, K.; Huang, W.; Chen, X.; Cao, Q.; Fang, X.; Lin, T.; Jin, C.; Li, S.; Wang, C.; Wang, T. The Whole-Aperture Pore Structure Characteristics and Their Controlling Factors of the Dawuba Formation Shale in Western Guizhou. *Processes* **2022**, *10*, 622. [CrossRef]
19. Luo, W.; Hou, M.; Liu, X.; Huang, S.; Chao, H.; Zhang, R.; Deng, X. Geological and Geochemical Characteristics of Marine-Continental Transitional Shale from the Upper Permian Longtan Formation, Northwestern Guizhou, China. *Mar. Pet. Geol.* **2018**, *89*, 58–67. [CrossRef]
20. Ma, X.; Guo, S. Study on Pore Evolution and Diagenesis Division of a Permian Longtan Transitional Shale in Southwest Guizhou, China. *Energy Sci. Eng.* **2021**, *9*, 58–79. [CrossRef]
21. Boyer, C.; Kieschnick, J.; Suarez-Rivera, R.; Lewis, R.E.; Waters, G. Producing Gas from Its Source. *Oilfield Rev.* **2006**, *18*, 36–49.
22. Lin, L.; Zhang, J.; Liu, J.; Long, P.; Tang, X. Favorable Depth Zone Selection for Shale Gas Prospecting. *Earth Sci. Front.* **2012**, *19*, 259–263.
23. Slatt, R.M.; O'Brien, N.R. Pore Types in the Barnett and Woodford Gas Shales: Contribution to Understanding Gas Storage and Migration Pathways in Fine-Grained Rocks. *AAPG Bull.* **2011**, *95*, 2017–2030. [CrossRef]
24. Sun, W.; Zuo, Y.; Wu, Z.; Liu, H.; Zheng, L.; Wang, H.; Shui, Y.; Lou, Y.; Xi, S.; Li, T.; et al. Pore Characteristics and Evolution Mechanism of Shale in a Complex Tectonic Area: Case Study of the Lower Cambrian Niutitang Formation in Northern Guizhou, Southwest China. *J. Pet. Sci. Eng.* **2020**, *193*, 107373. [CrossRef]
25. Loucks, R.G.; Reed, R.M.; Ruppel, S.C.; Jarvie, D.M. Morphology, Genesis, and Distribution of Nanometer-Scale Pores in Siliceous Mudstones of the Mississippian Barnett Shale. *J. Sediment. Res.* **2009**, *79*, 848–861. [CrossRef]
26. Ko, L.T.; Loucks, R.G.; Zhang, T.; Ruppel, S.C.; Shao, D. Pore and Pore Network Evolution of Upper Cretaceous Boquillas (Eagle Ford–Equivalent) Mudrocks: Results from Gold Tube Pyrolysis Experiments. *AAPG Bull.* **2016**, *100*, 1693–1722. [CrossRef]
27. Yang, C.; Xiong, Y.; Zhang, J.; Liu, Y.; Chen, C. Comprehensive Understanding of OM-Hosted Pores in Transitional Shale: A Case Study of Permian Longtan Shale in South China Based on Organic Petrographic Analysis, Gas Adsorption, and X-Ray Diffraction Measurements. *Energy Fuels* **2019**, *33*, 8055–8064. [CrossRef]
28. Liang, J.; Huang, W.; Wang, H.; Blum, M.J.; Chen, J.; Wei, X.; Yang, G. Organic Geochemical and Petrophysical Characteristics of Transitional Coal-Measure Shale Gas Reservoirs and Their Relationships with Sedimentary Environments: A Case Study from the Carboniferous-Permian Qinshui Basin, China. *J. Pet. Sci. Eng.* **2020**, *184*, 106510. [CrossRef]
29. Sing, K.S.W.; Everett, D.H.; Haul, R.A.W.; Moscou, L.; Pierotti, J.H.; Rouquerol, J.; Siemieniewska, T. Reporting Physisorption Data for Gas/Solid Systems with Special Reference to the Determination of Surface Area and Porosity. *Pure Appl. Chem.* **1985**, *57*, 603–619. [CrossRef]
30. Thommes, M.; Kaneko, K.; Neimark, A.V.; Olivier, J.P.; Rodriguez-Reinoso, F.; Rouquerol, J.; Sing, K.S.W. Physisorption of Gases, with Special Reference to the Evaluation of Surface Area and Pore Size Distribution (IUPAC Technical Report). *Pure Appl. Chem.* **2015**, *87*, 1051–1069. [CrossRef]
31. Chen, S.; Zhu, Y.; Wang, H.; Liu, H.; Wei, W.; Fang, J. Shale Gas Reservoir Characterisation: A Typical Case in the Southern Sichuan Basin of China. *Energy* **2011**, *36*, 6609–6616. [CrossRef]

32. Wang, J.; Guo, S. The Whole-Aperture Pore-Structure Characteristics of Marine-Continental Transitional Shale Facies of the Taiyuan and Shanxi Formations in the Qinshui Basin, North China. *Interpretation* **2019**, *7*, T547–T563. [CrossRef]
33. de Boer, J.H.; Everett, D.H.; Stone, F.S. *The Structure and Properties of Porous Materials*; Colston Papers; Academic Press: Cambridge, MA, USA, 1958; Volume 90.
34. Chen, Y.; Qin, Y.; Wei, C.; Huang, L.; Shi, Q.; Wu, C.; Zhang, X. Porosity Changes in Progressively Pulverized Anthracite Subsamples: Implications for the Study of Closed Pore Distribution in Coals. *Fuel* **2018**, *225*, 612–622. [CrossRef]
35. Li, A.; Ding, W.; Jiu, K.; Wang, Z.; Wang, R.; He, J. Investigation of the Pore Structures and Fractal Characteristics of Marine Shale Reservoirs Using NMR Experiments and Image Analyses: A Case Study of the Lower Cambrian Niutitang Formation in Northern Guizhou Province, South China. *Mar. Pet. Geol.* **2018**, *89*, 530–540. [CrossRef]
36. Vishal, V.; Chandra, D.; Bahadur, J.; Sen, D.; Hazra, B.; Mahanta, B.; Mani, D. Interpreting Pore Dimensions in Gas Shales Using a Combination of SEM Imaging, Small-Angle Neutron Scattering, and Low-Pressure Gas Adsorption. *Energy Fuels* **2019**, *33*, 4835–4848. [CrossRef]
37. Hazra, B.; Wood, D.A.; Vishal, V.; Varma, A.K.; Sakha, D.; Singh, A.K. Porosity Controls and Fractal Disposition of Organic-Rich Permian Shales Using Low-Pressure Adsorption Techniques. *Fuel* **2018**, *220*, 837–848. [CrossRef]
38. Xi, Z.; Tang, S.; Li, J.; Zhang, Z.; Xiao, H. Pore Characterization and the Controls of Organic Matter and Quartz on Pore Structure: Case Study of the Niutitang Formation of Northern Guizhou Province, South China. *J. Nat. Gas Sci. Eng.* **2019**, *61*, 18–31. [CrossRef]
39. Ungerer, P.; Collell, J.; Yiannourakou, M. Molecular Modeling of the Volumetric and Thermodynamic Properties of Kerogen: Influence of Organic Type and Maturity. *Energy Fuels* **2015**, *29*, 91–105. [CrossRef]
40. Mastalerz, M.; Schimmelmann, A.; Drobniak, A.; Chen, Y. Porosity of Devonian and Mississippian New Albany Shale across a Maturation Gradient: Insights from Organic Petrology, Gas Adsorption, and Mercury Intrusion. *AAPG Bull.* **2013**, *97*, 1621–1643. [CrossRef]
41. Curtis, M.E.; Cardott, B.J.; Sondergeld, C.H.; Rai, C.S. Development of Organic Porosity in the Woodford Shale with Increasing Thermal Maturity. *Int. J. Coal Geol.* **2012**, *103*, 26–31. [CrossRef]
42. Chandra, D.; Vishal, V.; Bahadur, J.; Agrawal, A.K.; Das, A.; Hazra, B.; Sen, D. Nano-Scale Physicochemical Attributes and Their Impact on Pore Heterogeneity in Shale. *Fuel* **2022**, *314*, 123070. [CrossRef]
43. Zhang, Y.; Yu, B.; Pan, Z.; Hou, C.; Zuo, Q.; Sun, M. Effect of Thermal Maturity on Shale Pore Structure: A Combined Study Using Extracted Organic Matter and Bulk Shale from Sichuan Basin, China. *J. Nat. Gas Sci. Eng.* **2020**, *74*, 103089. [CrossRef]
44. Ross, D.J.K.; Marc Bustin, R. The Importance of Shale Composition and Pore Structure upon Gas Storage Potential of Shale Gas Reservoirs. *Mar. Pet. Geol.* **2009**, *26*, 916–927. [CrossRef]

Journal of
Marine Science and Engineering

MDPI

Article

Multiple Controls on Organic Matter Accumulation in the Intraplatform Basin of the Early Cambrian Yangtze Platform, South China

Qiyang Zhang [1], Entao Liu [1,*], Songqi Pan [2], Hua Wang [1,*], Zhenhua Jing [2], Zhengfu Zhao [2] and Ruiyue Zhu [1]

1 Key Laboratory of Tectonics and Petroleum Resources of Ministry of Education, China University of Geosciences, Wuhan 430074, China; zhangqiyang@cug.edu.cn (Q.Z.); ruiyue.zhu@cug.edu.cn (R.Z.)
2 Sinopec Petroleum Exploration and Production Research Institute, Beijing 100083, China; pansongqi@pku.edu.cn (S.P.); z.jing@pku.edu.cn (Z.J.); zhao_zhengfu@pku.edu.cn (Z.Z.)
* Correspondence: liuentao@cug.edu.cn (E.L.); wanghua@cug.edu.cn (H.W.)

Abstract: Studying the accumulation rules of organic matter (OM) in paleo-ocean sediments can not only enhance our understanding of how OM becomes enriched in ancient oceans but also provide guidance for the exploration of shale gas in unconventional shale strata. A breakthrough has been made in shale gas exploration in the early Cambrian Qiongzhusi Formation in South China. However, less attention has been paid to the intraplatform basin of the Yangtze Platform, and the factors controlling organic matter enrichment in this special region remain unclear. This study focuses on a continuous drilling core across the full well section of the Qiongzhusi Formation in the intraplatform basin of the Yangtze Platform. Through the comprehensive analysis of total organic carbon (TOC), major and trace elements, and Mo isotopes, this study investigates the controlling factors for OM enrichment with $\delta^{98/95}$Mo ratios utilized to identify the existence of euxinic bottom water. The examined 240 m long core can be divided into four units, where the TOC values of the lower Units 1 and 2 (0.2–5.0 wt.%) average higher than the upper Units 3 and 4 (0.2–2.5 wt.%). Redox indicators (U/Th, Ni/Co, EF(Mo)—EF(U)) indicate an increasing oxidation of bottom waters from the bottom upwards. $\delta^{98/95}$Mo data further confirm the presence of weakly euxinic conditions in Units 1 and 2, addressing the ongoing controversy surrounding bottom water redox environments. Primary productivity indicators (Ni/Al, Cu/Al) suggest a relatively low average productivity level within the intraplatform basin. The upwelling indicators EF(Co) * EF(Mn) of different profiles in the Yangtze Platform suggest that low productivity within the intraplatform basin can be mainly attributed to the absence of upwelling. Consequently, this study proposes an organic matter enrichment mechanism for the Qiongzhusi Formation in the intraplatform basin, which emphasizes the significance of the redox environment in the formation of high-quality hydrocarbon source rocks in restricted environments that lack upwelling, setting it apart from the deep ocean. These findings have the potential to provide valuable insights for the exploration of high-quality hydrocarbon source rocks in other similar regions.

Keywords: organic matter; redox conditions; primary productivity; lower Cambrian; Sichuan Basin

Citation: Zhang, Q.; Liu, E.; Pan, S.; Wang, H.; Jing, Z.; Zhao, Z.; Zhu, R. Multiple Controls on Organic Matter Accumulation in the Intraplatform Basin of the Early Cambrian Yangtze Platform, South China. *J. Mar. Sci. Eng.* **2023**, *11*, 1907. https://doi.org/10.3390/jmse11101907

Academic Editor: George Kontakiotis

Received: 4 September 2023
Revised: 28 September 2023
Accepted: 29 September 2023
Published: 2 October 2023

1. Introduction

Significant biological, oceanographic, and geochemical changes occurred during the Ediacaran–Cambrian (E–C) transition [1–6]. Serving as a key record of this interval, lower Cambrian black shales contain valuable information about the marine environment and the evolution of life [7–9]. There are two main modes of organic matter (OM) accumulation: a preservation mode primarily controlled by redox conditions [10,11], with the modern Black Sea being a typical example, and a productivity mode primarily influenced by high productivity [12], such as the modern Arabian Sea. The organic matter enrichment of marine shale is affected by high palaeoproductivity and anoxic environment [13–16]. Additionally, some studies emphasize that OM enrichment is also influenced by factors like water mass restrictions, circulation pattern changes, and terrigenous input [11,17–21]. Recently, an increasing number of studies suggest that the enrichment of OM in shale is likely controlled by multiple factors, including productivity and redox conditions, with potentially different controlling processes in different environmental types [15,20–23].

During the Early Cambrian, widespread deposition of lower Cambrian Qiongzhusi Formation black shale occurred on the Yangtze Block under global sea level rise and transgression [24,25]. During this interval, the Yangtze Block developed various depositional environments, including deep-water slope and basinal facies, transitional facies, carbonate platform facies, and an intraplatform basin [7,26,27] (Figure 1a). Previous studies based on profile sampling and drilling analysis have revealed sea level fluctuations and upwelling events occurring in deep-water slope and basinal facies and carbonate platform facies. These events further influenced the OM enrichment through modifying water column redox conditions and primary productivity within the region [20,28–32]. However, compared to other well-investigated regions of the Yangtze Block [23,33,34], whether the upwelling existed in the region of the intraplatform basin is still unclear.

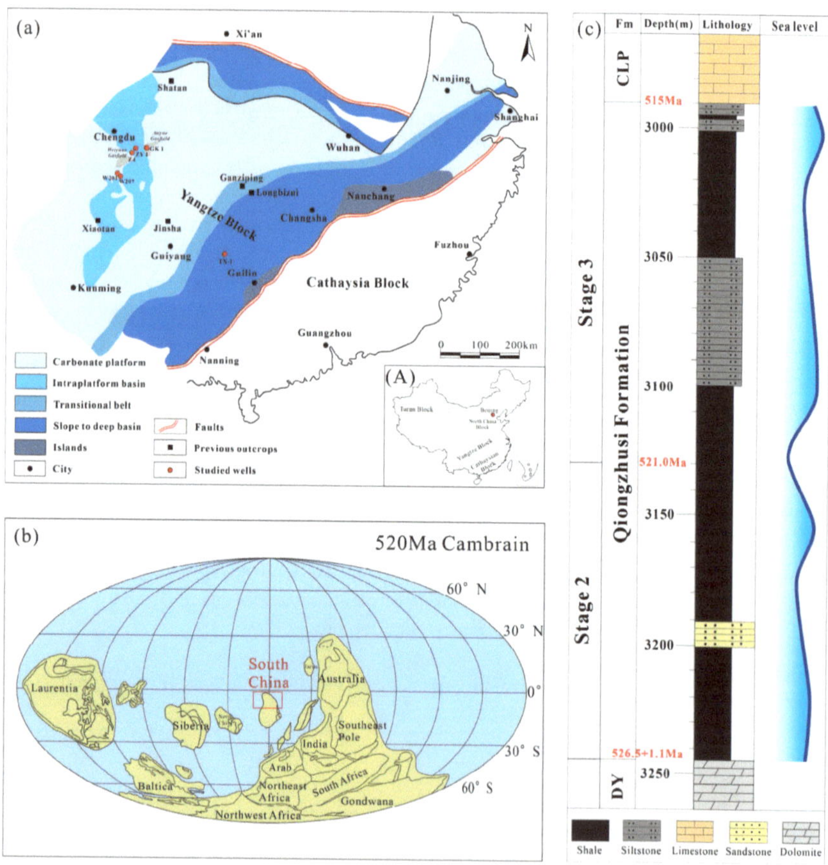

Figure 1. (**a**) Paleogeographic map of the Yangtze Block during the Ediacaran–early Cambrian transition (modified form the References [20,23,24]); (**b**) Early Cambrian global paleogeography (modified form the References [35,36]); (**c**) Stratigraphic column of the Qiongzhusi Formation in the W207 well (the relative sea level changes are modified from Reference [36]).

The facies of the intraplatform basin represent a favorable accumulation zone for high-quality hydrocarbon source rocks during the Early Cambrian [20], with the Anyue and Weiyuan Gas Fields discovered on the two sides [37] (Figure 1a), and possess high hydrocarbon recovery [38]. Compared to the other structural units (e.g., slope to deep ocean and carbonate platform) that have been widely investigated in previous studies (such as in References [23,33,34]), less attention has been paid to the intraplatform basin [20]. Moreover, there exist different points of view on the bottom water redox conditions and OM accumulation mechanisms in the western part of the intraplatform basin [20,39–41]. Previous studies, based on the absence of Ni-Mo sulfide deposits in shale cores and sedimentary iron speciation analysis, suggest that Qiongzhusi Formation shale in the western part of the basin was deposited in an anoxic environment without euxinic bottom waters [39,40] (Figure 1a). Conversely, some studies (such as in References [41,42]) indicate the presence of euxinic bottom waters in the western part of the intraplatform basin (Figure 1a). Consequently, the primary controlling factors for OM enrichment in the lowermost high-total organic carbon (TOC) zone and the mechanism of OM enrichment during the deposition of Qiongzhusi Formation in the western part of intraplatform basin remain unclear.

In this study, we conducted a comprehensive geochemical analysis of the continuous well W207 located in the western part of the intraplatform basin of the Yangtze Platform, South China. This analysis included major elements, trace elements, TOC contents, and Mo isotopes, aiming to reconstruct the bottom water redox conditions and primary productivity conditions during the deposition of lower Cambrian Qiongzhusi Formation shale in the western part of the intraplatform basin, as well as to clarify the primary controlling factors and enrichment mechanisms of OM accumulation at different units.

2. Geological Setting

Around 850–820 Ma [23,43,44], the South China Block gradually separated into the Yangtze Block and the Cathaysia Block under intense rifting processes (Figure 1b). During the E-C transition, the Yangtze Block, evolving from a rift basin into a passive continental margin basin, formed a carbonate platform dominated by carbonate deposits, a narrow marginal transitional zone dominated by black chert and shale series (outer shelf), and a deep-water slope and basinal zone dominated by Liuchapo Formation siliceous rocks [9,20,24,45,46] (Figure 1a), with increasing water depth across the three facies belts.

Recent studies have discovered the presence of an intraplatform basin within the carbonate platform, which developed from the Late Ediacaran to Early Cambrian [47,48] (Figure 1a), with the slopes on both sides of the intraplatform basin showing a steeper gradient in the east and a gentler gradient in the west. The distribution of black shales formed during transgressions may be controlled by the intraplatform basin [20,40,48–50]. A newly drilled evaluation well (W207) near the western side of the intraplatform basin provides an opportunity for this study. In the lower Cambrian, the Maidiping Formation, Qiongzhusi Formation, and Canglangpu Formation developed from bottom to top. However, with the uplift and erosion of Tongwan Movement, the Maidiping Formation only exists in some areas in the middle of Yangtze Platform [51,52] and is missing in the W207 well. The Qiongzhusi Formation unconformably overlies the dolomite of the Dengying Formation (Figure 1c).

Stratigraphic correlation of the Qiongzhusi Formation has been carried out through typical marker beds, biostratigraphy, U-Pb ages, cyclostratigraphy, and geochemistry [36,53]. The black shale at the base of the Shiyantou Formation in the Xiaotan section represents the onset of the last appearance of a small shell fossil assemblage (SSFA-3), which can serve as the boundary between the Maidiping and Qiongzhusi Formation [36]. The CA-ID-TIMS zircon U-Pb ages of 526.86 $\pm$ 0.16 Ma in the Meishucun section reported by Yang [54] provide support for the age of the base of the Qiongzhusi Formation. Lower Cambrian Ni-Mo ore layers discovered at the bottom of the GK1 and Z4 wells can be used as a boundary between the second and third stages of the Cambrian, with a Re—Os isochron age of 521 $\pm$ 5 Ma identified within the layers [53].

Based on biostratigraphic division, Liang et al. [55] found that the appearance of *Tsunyidiscus niutitangensis* in the W207 well corresponds to the extreme values of the GR curve at the boundary of Units 3 and 4. This biozone of trilobite may be associated with *Tsunyidiscus acutus-Tsunyidiscus yanjiazhaiensis* at the GR extreme values (3295.89 m) in the JY1 well, and the age of this interval in the JY1 well has been defined at approximately 519.43 Ma [36,56]. Additionally, existing research has shown that the *Tsunyidiscus acutus-Tsunyidiscus yanjiazhaiensis* found at 3295.89 m in the JY1 well can be fully matched with the fossil assemblage of *Tsunyidiscus niutitangensis* that appears in the Jinsha section, possibly representing the first appearance of trilobites [36,56]. The boundary between the Qiongzhusi and Canglangpu Formation in the Xiaotan section is determined as 515 Ma [31], which may reveal the boundary between the Qiongzhusi and Canglangpu Formation in wells such as ZY1 and GK1 [20].

The Qiongzhusi Formation can be divided into four third-order sequences (SQ1, SQ2, SQ3, SQ4) from bottom to top [20,36,57,58]. Combining the previously defined sequence boundaries with the geochemical data in this study, the Qiongzhusi Formation is divided into four units: Unit 1 (3245–3203 m) primarily consists of black shales; Unit 2 (3203–3111 m) mainly comprises interbedded black shales and dark gray siltstone; Unit 3 (3111–3045 m) is primarily composed of lower black shales and upper dark gray siltstone; and Unit 4 (3045–3000 m) mainly consists of black shales with a top layer of dark gray siltstone (Figure 1c).

3. Samples and Methods

According to the lithology and thickness of the target interval, uniform sampling was carried out for the Qiongzhusi Formation in the W207 well. A total of 90 samples were measured for TOC analysis, major and trace element analysis, and 30 samples for Mo isotope analysis. The location and depth of sample collection are shown in Figure 1 and Table S1. Of course, before all geochemical analysis, samples with veins, nodules, and phosphatic were removed to ensure the authenticity of the data. Then, the remaining samples were ground to 200 mesh size through hand grinding with agate mortar.

3.1. TOC

First, the carbonate in the powder samples (200 mesh size) was removed through adding hydrochloric acid (10%). Then, the residual hydrochloric acid in the sample was removed through washing with distilled water. After the samples were dried overnight, they were analyzed using the CHNOS Elemental Analyzer (Vario EL III) in the Key Laboratory of Geological Processes and Mineral Resources, China University of Geosciences (Wuhan).

3.2. Major and Trace Elements

The analysis of major elements in the samples was performed using the wavelength-dispersive X-ray fluorescence (XRF) spectrometer AXIOSmAX. Mix the processed powder sample, flux ($Li_2B_4O_7$, $LiBO_2$, LiF) and oxidant (NH_3NO_3) in a ratio of 1:10:3. After being heated in a furnace at 1150 °C for 14 min, glass slices were prepared for the XRF test.

Trace elements were analyzed using an Agilent 7700 inductively coupled plasma mass spectrometer (ICP-MS). The volatiles in the samples were removed via burning them in a muffle furnace at 600 °C for 12 h. Then, at a temperature of 190 °C, hydrofluoric acid (HF) and nitric acid (HNO_3) were added into the Teflon beaker containing the powder sample for dissolution treatment. The final solution was then analyzed via ICP-MS with a final analysis accuracy better than ±5%.

The enrichment factors can be calculated as follows:

$$X_{EF} = [(X/Al)_{sample}/(X/Al)_{PAAS}], \tag{1}$$

where the "X" is the selected target trace element. The subscript "sample" refers to the weight ratio of element X to AL in this study sample [59], and "PAAS" denotes the Post-Archean Australian Shale (PAAS) standard [60].

3.3. Mo Isotopic Compositions ($\delta^{98/95}$Mo)

Mo isotopic compositions were analyzed via Thermo-Fisher Scientific Neptune Plus (MC-ICP-MS) in the State Key Laboratory of Isotope Geochemistry, Guangzhou Institute of Geochemistry, Chinese Academy of Sciences, Guangzhou, China. According to the double-spike method [61], an appropriate amount of ^{97}Mo–^{100}Mo double spikes was added to the sample according to the Mo concentration of the samples. Then, at 130 °C, the sample was dissolved and purified with HNO_3, HF, and HCI, and then column separation was performed. The purified Mo was determined via MC–ICP–MS by means of ion-exchange chromatography using BPHA resin [61]. The double-spike and normalizing standard (NIST SRM 3134) was corrected through following the procedure of Siebert et al. [62] and Zhao et al. [63]. At the same time, during the analysis process, the $\delta^{98/95}$Mo of NIST SRM 3134 standard solution and IAPSO seawater was repeatedly measured. The measured values of $\delta^{98/95}$Mo of the two are 0.00 ± 0.05‰ (2SD, n = 15) and 2.06 ± 0.06‰, respectively, which are consistent with the certified values and literature reports [63–65]. Meanwhile, the Mo procedural blank value (<0.1 ng) is much smaller than the total amount of Mo in the samples. Finally, in order to facilitate international comparison and analysis, the $\delta^{98/95}$Mo values in the samples were recalculated based on the NIST SRM 3134 standard solution (0.25‰).

4. Results

As shown in Figures 2 and 3 and Table 1, Unit 1 (3245–3203 m) exhibits relatively high TOC contents (average of 2.5 wt.%), ranging from 0.7 to 4.6 wt.%, with a decreasing trend from bottom to top. The $\delta^{98/95}$Mo values vary between −0.25‰ and 2.23‰, displaying a significant fluctuation range. The ratios of U/Th (0.6–4.4) and Ni/Co (2.6–11.3) (Figure 2), as well as the enrichment levels of EF(U) (3.2–22) and EF(Mo) (9.5–204.2), show relatively high values and exhibit similar variations (Figure 3). Additionally, Ni/Al (6.0×10^{-4} to 28.0×10^{-4}) and Cu/Al (4.1×10^{-4} to 10.0×10^{-4}) ratios are also relatively high (Figure 2).

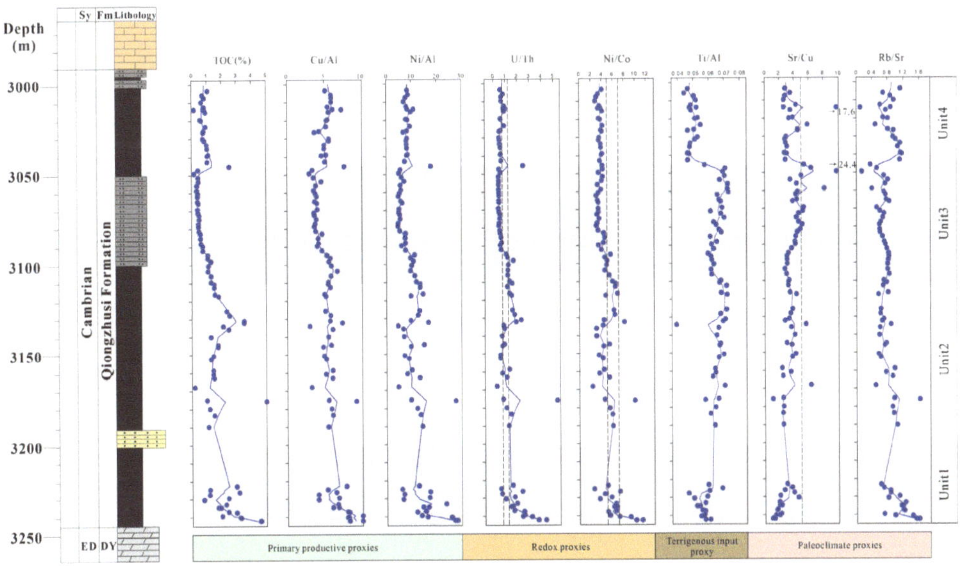

Figure 2. Vertical variations of total organic carbon (TOC) contents, primary productivity proxies (Cu/Al, Ni /Al), redox proxies (U/Th, Ni/Co), and terrigenous input proxy (Ti/Al).

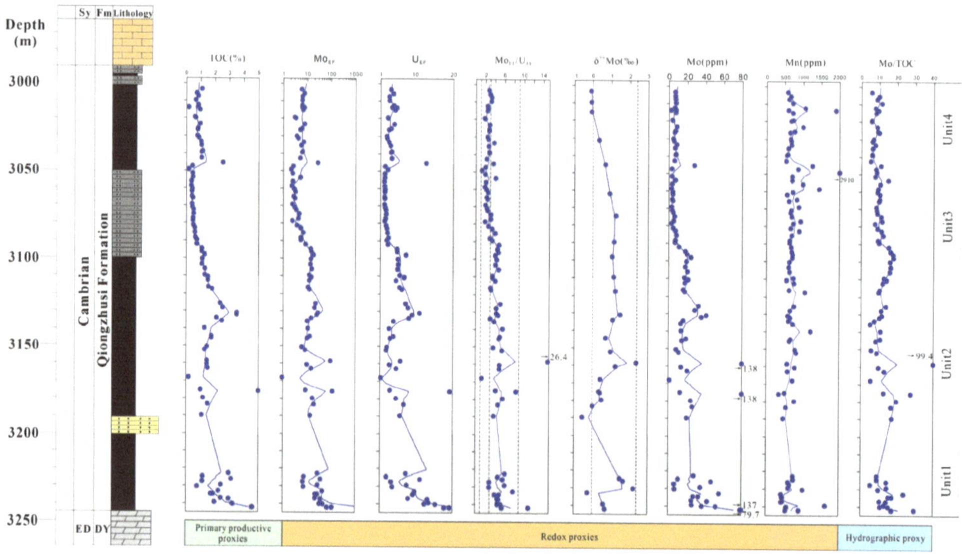

Figure 3. Vertical variations of redox proxies (Mo-EF, U-EF, Mo-EF/U-EF, $\delta^{98/95}$Mo, Mo, and Mn) and hydrographic proxy (Mo/TOC).

Table 1. Mo isotope ($\delta^{98/95}$Mo) values of the lower Cambrian Qiongzhusi Formation.

Samples	Member	Depth	$\delta^{98/95}$Mo	$\delta^{98/95}$Mo(auth) [1]	2SE [2]
		m	‰	‰	
W207-90	Unit 4	3003.41	−0.10	−0.17	0.04
W207-87	Unit 4	3009.68	−0.08	−0.14	0.03
W207-83	Unit 4	3015.33	−0.07	−0.13	0.03
W207-76	Unit 4	3031.33	0.35	0.33	0.02
W207-70	Unit 4	3045.35	0.69	0.70	0.03
W207-62	Unit 3	3061.96	0.92	1.02	0.03
W207-56	Unit 3	3074.73	1.26	1.37	0.03
W207-52	Unit 3	3082.83	1.21	1.29	0.03
W207-49	Unit 3	3089.48	1.17	1.25	0.03
W207-45	Unit 3	3098.12	1.06	1.08	0.04
W207-41	Unit 3	3109.60	1.15	1.18	0.02
W207-37	Unit 2	3117.42	1.24	1.28	0.03
W207-34	Unit 2	3131.03	1.50	1.52	0.03
W207-32	Unit 2	3134.00	1.10	1.14	0.04
W207-29	Unit 2	3144.47	0.72	0.73	0.03
W207-26	Unit 2	3152.30	0.99	1.03	0.02
W207-25	Unit 2	3158.59	2.37	2.38	0.03
W207-24	Unit 2	3160.66	1.25	1.30	0.02
W207-22	Unit 2	3168.00	0.43	0.47	0.03
W207-21	Unit 2	3175.10	0.36	0.35	0.02
W207-20	Unit 2	3175.74	0.42	0.42	0.02
W207-19	Unit 2	3179.90	0.51	0.51	0.03
W207-18	Unit 2	3183.42	0.01	0.00	0.02
W207-17	Unit 2	3189.80	−0.54	−0.58	0.04
W207-15	Unit 1	3224.68	1.50	1.58	0.03
W207-14	Unit 1	3226.13	1.69	1.71	0.03
W207-11	Unit 1	3230.37	2.23	2.41	0.03
W207-10	Unit 1	3232.92	−0.25	−0.25	0.03
W207-05	Unit 1	3239.40	0.62	0.62	0.03
W207-02	Unit 1	3242.10	0.70	0.70	0.03

[1] The corrected $\delta^{98/95}$Mo values were calculated following the formula: $\delta^{98/95}Mo_{auth} = [(\delta^{98/95}Mo_{tot} \times Mo_{tot} - \delta^{98/95}Mo_{det} \times Mo_{det})/Mo_{auth}]$ [66–68], in which the detrital Mo concentration (Mo_{det}) was given by $Mo_{det} = [(Mo/Al)_{PAAS} \times Al_{tot}]$ and the authigenic Mo concentration was calculated by $Mo_{auth} = Mo_{tot} - Mo_{det}$. Mo and Al concentration of PAAS (1.0 ppm, 10%, respectively) [60] and $\delta^{98/95}Mo_{det}$ value of 0.4‰ [69] were used for detritus correction. [2] 2SE means double of Standard Error.

In Unit 2 (3203–3111 m), the TOC content varies from 0.2 to 5.0 wt.%, with an average of 1.9 wt.%. The $\delta^{98/95}$Mo values range from −0.54‰ to 2.37‰, gradually increasing from the bottom to the middle part of the unit (3158.59 m), and then decreasing. The ratios of U/Th (0.3–5.3) and Ni/Co (2.3–10.0) (Figure 2), as well as the enrichment levels of EF(U) (1.5–23.7) and EF(Mo) (1.9–200.1), show relatively consistent values compared to Unit 1 (Figure 3). The ratios of Ni/Al (4.5×10^{-4} to 27.4×10^{-4}) and Cu/Al (3.0×10^{-4} to 9.2×10^{-4}) exhibit similar trends (Figure 2), with higher values at the bottom and lower values in the middle part of Unit 2.

In Unit 3 (3111–3045 m), the TOC content ranges from 0.2 to 1.5 wt.%, with an average of 0.7 wt.%, relatively lower compared to Unit 1 and Unit 2. It gradually decreases in the vertical direction. The $\delta^{98/95}$Mo values are relatively stable, fluctuating between 0.92‰ and 1.26‰, with a decreasing trend in the vertical direction (Figure 3). The ratios of U/Th (0.4–1.7) and Ni/Co (3.0–6.5), as well as the enrichment levels of EF(U) (2.4–8.3) and EF(Mo) (4.3–33.2), are relatively low and show small variations. The ratios of Ni/Al (4.6×10^{-4}–13.3×10^{-4}) and Cu/Al (3.0×10^{-4}–6.7×10^{-4}) gradually decreased from the bottom and then rebounded at the top of the interval, lower than those of Unit 1 and Unit 2 (Figure 2).

In Unit 4 (3045–3000 m), the TOC content varies between 0.2 and 2.5 wt.% (average of 0.9 wt.%), obviously higher than that of Unit 3. The $\delta^{98/95}$Mo values continued to decrease from 0.69‰ to −0.10‰ (Figure 3). The enrichment degree of U/Th (0.5–2.5) and Ni/Co (2.8–4.1) ratios, as well as EF(U) (2.4–12.6) and EF(Mo) (4.6–46.0), was relatively low within this interval. The Ni/Al (6.8×10^{-4}–17.6×10^{-4}) and Cu/Al (3.7×10^{-4}–7.7×10^{-4}) ratios showed a noticeable increase compared to the third stage, steadily rising upwards (Figure 2).

5. Discussion

5.1. The Degree of Oxygenation in the Bottom Water Environment Continues to Increase

The ratios of U/Th and Ni/Co have been widely used to reconstruct paleoceanic redox environments [17,70–74]. The U/Th and Ni/Co ratios in Unit 1 (average of 2.10 and 6.41, respectively) and Unit 2 (average of 1.32 and 5.13, respectively) were relatively high, reflecting an anoxic environment. The relatively low values of these ratios in Unit 3 (average of 0.73 and 4.06) and Unit 4 (average of 0.77 and 3.71) reflect a relatively stable oxic–suboxic environment (Figure 2). Relatively low U/Th and Ni/Co ratios suggest an upward enhancement in the oxidation state of the bottom water environment on the western side of the intraplatform basin [68,73,75,76] (Figure 2). This interpretation is consistent with redox conditions inferred from Mo content, EF(Mo)/EF(U), and Mo/TOC ratios (Figures 3 and 4a).

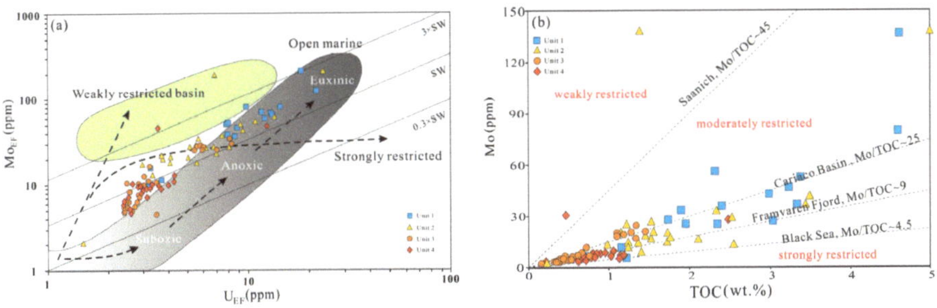

Figure 4. (**a**) EF(Mo) versus EF(U) for Qiongzhusi Formation samples [77,78]. The solid line represents the Mo: U ratio in modern seawater. (**b**) Sediment Mo versus TOC for the Qiongzhusi Formation samples. Four representatives of the modern anoxic basin system (Saanich Inlet, Cariaco Basin, Framvaren Fjord, and Black Sea) are represented by dashed lines in the figure, which reflect different degrees of hydrographic restriction [79].

It should be noted that in restricted environments, Mo concentrations in the water column may decrease due to the lack of open ocean inputs [80–82]. In Unit 1, the Mo content (average of 40.67) and EF(Mo)/EF(U) ratios (5.42 on average) support the anoxic environment indicated by the U/Th and Ni/Co ratios. The extreme point observed in sample W-02 (3242.1 m) (Mo content of 137 ppm, EF(Mo)/EF(U) ratios of 11.06) reflects a possible weakly euxinic environment in the water column (Figure 3). Combining the sea level change curve [36] and various trace elements such as EF(Mo) and EF(U) (Figures 1c and 3), Unit 2 can be divided into two cycles that first increase and then decrease. The relative high points of these cycles (W-20 at 3131 m and W-34 at 3175.7 m) correspond to the occurrence of two transgressive events. The large-scale transgressions connected the intraplatform basin to the open sea, replenishing trace elements (Mo, U, etc.) and increasing the degree of water reduction. A weakly euxinic condition may have occurred at 3175 m (Figure 3), which is consistent with no significant correlation between the TOC and redox-sensitive trace elements such as U/Th, Ni/Al and Cu/Al in this point [83] (Figure 5a–f). Considering the anoxic environment indicated by the U/Th and Ni/Co ratios in Unit 2, the EF(Mo)/EF(U) ratios (average of 5.45) likely indicate dominant anoxic with occasional and weakly euxinic water conditions. This viewpoint is supported by Mo isotope evidence (see details in Section 5.2). Similar short-term euxinic events in anoxic environments have also been observed in the ZY1 well and Jinsha section on the eastern side of the intraplatform basin during the same time [20,31]. In Units 3 and 4, the occurrence of the regression event during Cambrian Age 3 was accompanied by the development of oxidizing bottom water conditions. This interpretation is consistent with the low EF(Mo)/EF(U) ratios (average of 3.09, 3.14) representing an oxic–suboxic environment. From bottom to top, the Qiongzhusi Formation experienced an anoxic–euxinic–oxic–suboxic bottom water environment.

The redox environment of bottom water is closely related to the hydrographic conditions of the basin. The Mo/TOC ratio has been widely used as an indicator of basin water connectivity limitations [81] (Figure 4b). However, due to the possibility of lower Mo concentrations in early Cambrian oceans compared to modern seawater [84–86], Cheng et al. [87] suggested that when the Mo/TOC ratio in sediments exceeds 11 ppm/wt.%, good connectivity between the basin and the open sea can be considered. In Unit 1, the relatively high Mo/TOC values (average of 14.52 ppm/wt.%) indicate low basin water restrictiveness, favorable for the existence of anoxic bottom water environments. The Mo/TOC values in Unit 2 correspond well to the two transgression events, indicating increasing basin openness with the occurrence of transgressions. The oxic–suboxic environment in Unit 3 corresponds to the regression during Cambrian Age 3 (Figures 1c and 3), and the weakening connection between the basin and the open sea prevents the replenishment of Mo consumed in the ocean, resulting in a decrease in Mo and U values (average of 8.78 and 7.86, respectively) [81]. In Unit 4, the level of basin restrictiveness further increased, accompanied by a decrease in Mo/TOC values (Figure 4b).

5.2. $\delta^{98/95}$*Mo Isotopes and Related Redox Conditions*

Mo isotopes in organic-rich sedimentary rocks have been widely used in reconstructing local and global marine redox states because it exhibits high sensitivity to redox conditions [65,67,75,86,88–91]. There are two main modes of Mo output in the ocean. In oxic environments, iron or manganese oxides preferentially adsorb the light Mo isotope, resulting in significant isotope fractionation between sediments and seawater, up to ~3% fractionation [92,93]. In euxinic environments, when the concentration of H_2S in the water solution is high ($[H_2S]aq > 11$ μM), oxyanion molybdate (MoO_4^{2-}), as the main form of modern seawater Mo, is almost completely converted to oxythiomolybdate species ($MoO_{4-x}S_x^{2-}$, x = 1–4), with little isotope fractionation [66,94]. When the concentration of H_2S in the water solution is low ($[H_2S]aq < 11$ μM), the incomplete conversion of molybdate (MoO_4^{2-}) leads to significant isotope fractionation in the sedimentary Mo isotopes [92,95]. This also explains why Mo isotopes can serve as redox proxies. Before analyzing the

Mo isotopes, we eliminated possible detrital effects using the method proposed by Wen et al. [67]. The corrected results show that the measured Mo isotope values ($\delta^{98/95}$Mo) are nearly equivalent to the corrected values ($\delta^{98/95}Mo_{auth}$) (Table 1), indicating a minimal impact of detrital Mo on bulk Mo isotopes. Therefore, we directly discuss the measured values in the following discussion.

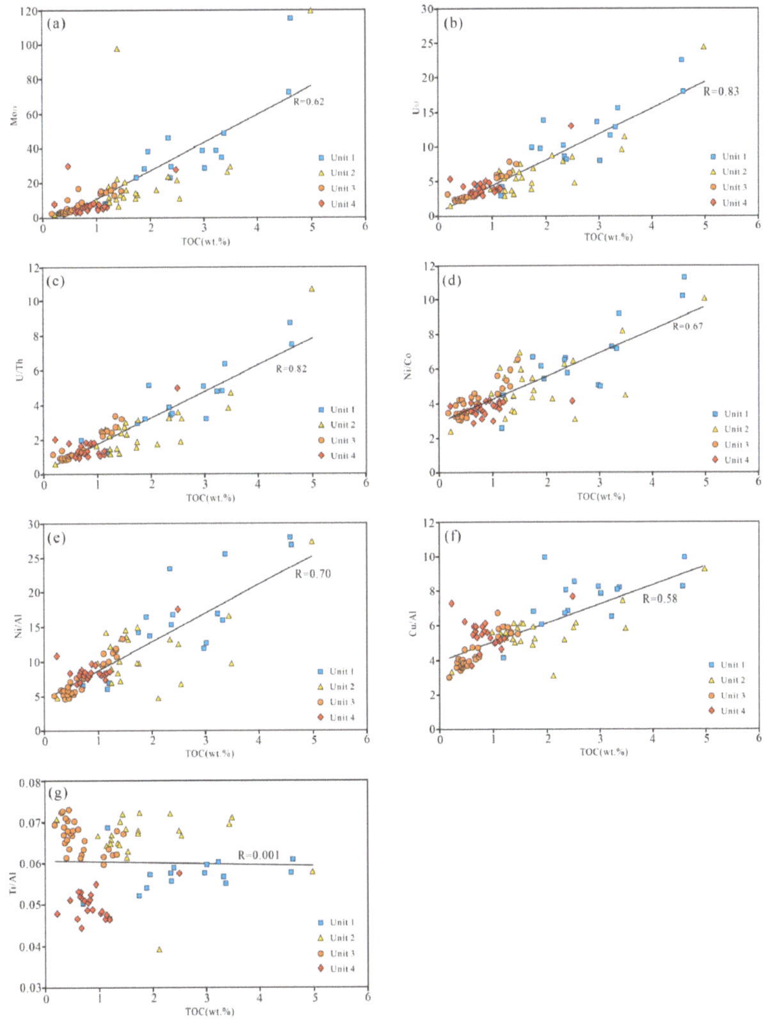

Figure 5. Relationships between TOC contents and **(a)** Mo-EF; **(b)** U-EF; **(c)** U/Th ratios; **(d)** Ni/Co ratios; **(e)** Ni/Al; **(f)** Cu/Al ratios; **(g)** Ti/Al ratios of Qiongzhusi Formation samples.

The presence of weak euxinia in the anoxic bottom water conditions has been confirmed by $\delta^{98/95}$Mo values from the western core of the intraplatform basin. Negative $\delta^{98/95}$Mo values (−0.08‰ to −0.54‰) appear at the bottom of Units 1 and 2 (Figure 3). The occurrence of negative Mo isotopes can usually be attributed to three possible mechanisms: the first possibility is Fe-Mn_{ox} adsorption in the oxic environment [96], but the higher U/Th and Ni/Co ratios in Unit 1 and 2 preclude this possibility. The second possibility is Fe-Mn_{ox} shuttling [68,87,97,98], similar to the modern Baltic Sea and Cariaco Basin [97,99]. Under the shuttle effect of iron and manganese oxide, lighter Mo isotopes enter the sediment, leading to negative $\delta^{98/95}$Mo values. In this case, euxinic conditions in the sediment result in preferential enrichment of Mo relative to U, producing higher EF(Mo)/EF(U) ratios (~3 × SW) and relatively higher Mo concentrations. At the same time, the lack of support for reductive dissolution inhibits Mn accumulation in the sediment, resulting in a relative deficiency of Mn. The EF(Mo)/EF(U) ratios (7.86) and Mo (55.8 ppm) and Mn (424 ppm) values in sample W-10 at the bottom of Unit 2 support the possibility of Fe-Mn_{ox} shuttling. It is worth mentioning that samples W-02 and W-34, mentioned in Section 5.1, also support the possibility of Fe-Mn_{ox} shuttling with their negative $\delta^{98/95}$Mo values (0.7‰ and 0.42‰, respectively). The third possibility is based on the incomplete conversion of molybdate (MoO_4^{2-}) under weakly euxinic water conditions ($[H_2S]aq < 11 \mu M$), causing Mo isotopic fractionation [94,99]. The EF(Mo)/EF(U) ratios of samples with negative $\delta^{98/95}$Mo values in Unit 2 are only about twice that of modern seawater (~1 × SW), which excludes the possibility of Fe-Mn_{ox} shuttling. Therefore, the negative $\delta^{98/95}$Mo values in Unit 2 may be attributed to significant fractionation caused by the incomplete conversion of molybdate (MoO_4^{2-}) under weakly euxinic water conditions. In addition, although Unit 1 is predominantly anoxic, the occurrence of Fe-Mn_{ox} shuttling in samples W-02 and W-10, mentioned above, demonstrates the presence of localized euxinic environments in Unit 1. Thus, the appearance of negative $\delta^{98/95}$Mo values in sample W-02 in Unit 1 may also be due to the incomplete conversion of molybdate (MoO_4^{2-}) under weakly euxinic water conditions or the combined effects of both (such as in Reference [68]).

Except for the negative $\delta^{98/95}$Mo values, samples W-11 (2.23‰) and W-25 (2.37‰) in Units 1 and 2 exhibit relatively high $\delta^{98/95}$Mo values. Similar high values have been reported in profiles from the Yangtze Block and Tarim Basin during the same interval [67,86,87]. The $\delta^{98/95}$Mo values reported by Chen et al. [100] (∼+2.3‰) may reflect the highest $\delta^{98/95}$Mo values of Early Cambrian oceanic seawater, and the two high values are nearly equivalent to contemporaneous seawater $\delta^{98/95}$Mo values (Figure 3), which may indicate smaller isotope fractionation under high-H_2S conditions. The Mo isotopes exhibit significant fluctuations (−0.54‰ to 2.37‰) (Figure 3) in Unit 1 and in the lower part of Unit 2 (before the transgression). Except for samples W-02, W-10, and W-34, mentioned earlier, the other samples did not exhibit Fe-Mn_{ox} shuttling. Such a large range of fluctuations may reflect isotope fractionation under weakly euxinic water conditions [68,95,99]. Overall, whether it is the extreme points that partially appear or the overall fluctuations in Units 1 and 2, these suggest the development of weakly euxinic conditions in the intraplatform basin.

From 3131 m in Unit 2 to 3000 m in Unit 4, Mo isotopes exhibit a decreasing trend, showing a negative bias since the large-scale transgression of Cambrian Stage 2 (Figures 1c and 3). The average $\delta^{98/95}$Mo value of Ni-Mo sulfide ore layers formed during the transgression in the Maluhe and Dazhuliushui profiles is reported as 1.13 ± 0.14‰ (n = 11), which is consistent with the $\delta^{98/95}$Mo values near the location of the Ni-Mo sulfide ore layers in the W207 well (average of 1.18‰, n = 6) [68,101] (Figure 3). Subsequently, with ongoing regression, the confinement of the intraplatform basin gradually increased, possibly leading to a gradual decrease in Mo isotopic values in the intraplatform seawater to values near those of continental input (+0.7‰) [68,101] (Figure 3).

5.3. Relatively Low Primary Productivity May Be Associated with the Absence of Upwelling

Previous studies suggested that primary productivity can be reconstructed through various geochemical indicators, including TOC values, organic biomarkers, C and N isotopes, and other trace elements such as P, Ba, Cu, and Ni [38,102–105]. Cu and Ni elements, which are essential nutrients for the growth of phytoplankton in the photic zone, are often associated with OM in sediments [103]. Considering that the assessment of productivity may not be accurate when relying solely on a single indicator [102,106], we studied variations in primary productivity based on multiple indicators (TOC, Cu, Ni) and eliminated the dilutive effects of detrital fractions on Cu and Ni. Cu/Al and Ni/Al showed a strong correlation with TOC (Figure 5e,f) ($r = 0.58$, $p(\alpha) < 0.05$ for Cu/Al vs.

TOC and r = 0.70, p(α) < 0.05 for Ni/Al vs. TOC), suggesting that most of the Cu and Ni are related to OM and can be used to reconstruct primary productivity in surface waters [20,103]. The shale Cu/Al ratios in the Qiongzhusi Formation minorly decreased from Unit 1 (4.13×10^{-4}–10.0×10^{-4}, average of 7.20×10^{-4}) to Unit 3 (3.0×10^{-4}–6.72×10^{-4}, average of 4.41×10^{-4}), and then increased in Unit 4 (3.71×10^{-4}–7.67×10^{-4}, average of 5.53×10^{-4}), indicating a slight decline in surface water bioproductivity followed by a significant increase upward (Figure 2). A similar trend was also observed in the Ni/Al ratios and TOC values. The Cu/Al (5.47×10^{-4}) and Ni/Al (10.2×10^{-4}) ratios from well W207 are lower than the corresponding ratios observed in the deep-water slope and basinal facies of well TX-1 (Cu/Al = 31.36×10^{-4}, Ni/Al = 29.46×10^{-4}) [34]. This finding is consistent with the conclusions in previous studies [20,107] that the productivity of the intraplatform basin facies is lower than the well-developed upwelling events found in the deep-water slope and basinal facies.

Upwelling events are one of the important factors affecting the accumulation of OM and are widely reported in other facies of the Yangtze Platform [23,31,34] (Figures 6a,b and 7). Nevertheless, within the intraplatform basin, the impact of upwelling on OM accumulation remains unclear. Upwelling from the deep sea is often enriched in cadmium (Cd) elements and relatively depleted in cobalt (Co) and manganese (Mn) elements, so when EF(Co)*EF(Mn) < 0.5 or Co*Mn < 0.4, it is considered to indicate the presence of upwelling. Conversely, when EF(Co)*EF(Mn) > 2 or Co*Mn > 0.4, it indicates a restrictive environment [108] (Figure 6a,c). The EF(Co)*EF(Mn) values (0.36–13.84, average of 1.32) and Co*Mn values (0.64–3.40, average of 1.11) of the Qiongzhusi Formation indicate a relatively restricted environment (Figure 6c,d; Table S2), which is further supported by the relatively low Mo/TOC ratios (Figures 3 and 4b).

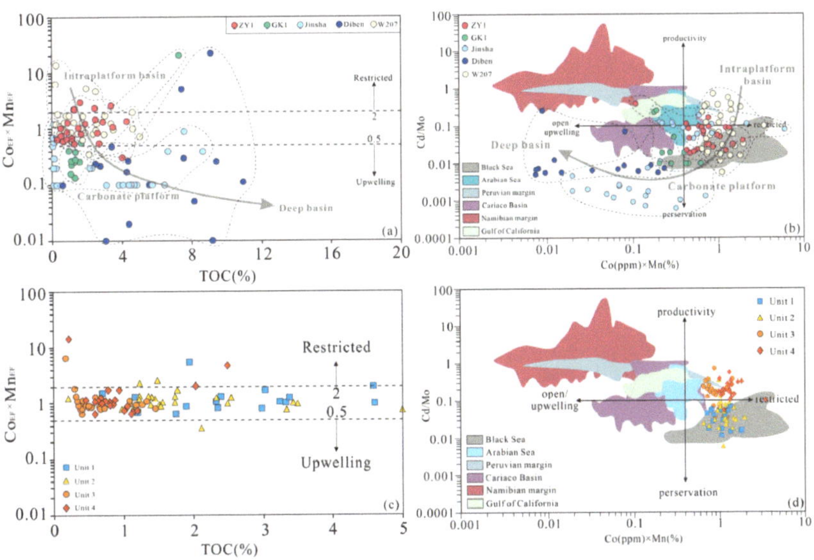

Figure 6. Cross-plots of (**a**,**c**) TOC versus Co-EF × Mn-EF, (**b**,**d**) Co (ppm) × Mn (%) versus Cd/Mo ratios of the Qiongzhusi Formation and its equivalents samples [108]. The sample data in (**a**,**b**) come from published data of different sections and wells in the Yangtze Block. Data sources: ZY1 [20]; GK1 [46]; Jinsha [31]; Diben [20]; W207 (this study, only part of the sample is shown). c and d show the samples of four different units of the Qiongzhusi Formation in Well W207 (this study).

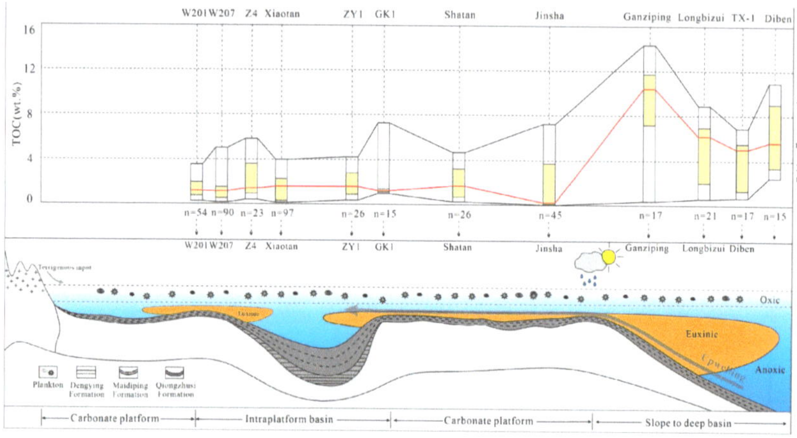

Figure 7. Simplified model of organic matter accumulation control during shale deposition of Qiongzhusi Formation in different facies belts of Yangtze block. TOC data of sections and wells are from Wang et al. [40], Wang et al. [42], Cremonese et al. [109], Gao et al. [20,46], Jin et al. [31,39], Chen et al. [24], Wang et al. [9], Yuan et al. [110].

In the Yangtze Platform, multiple profiles of EF(Co)*EF(Mn) and Co*Mn values within various sedimentary facies reveal a progressive increase in upwelling events from the intraplatform basin towards the deep-water slope and basinal facies. This trend is accompanied by a gradual rise in TOC values (Figures 6a,b and 7). Early findings suggested that upwelling, prevalent within deep-water slope and basinal facies, carried nutrients that effectively enhanced the primary productivity of surface waters, serving as a primary factor for OM accumulation in these regions [23,31,32,34,111]. The EF(Co)*EF(Mn) and Co*Mn values from multiple profiles also corroborate this observation (Figure 6a,b). Therefore, the relatively lower primary productivity in the intraplatform basin facies could potentially be attributed to the absence of upwelling events (Figure 6c,d, and Figure 7). This is in contrast to the deep-water slope and basinal facies, which are characterized by significant upwelling occurrences.

5.4. Mechanisms of OM Enrichment

The enrichment of OM is often influenced by various events, such as sea level changes, upwelling events, and terrigenous input [32,72,86,111–115]. The impact of these events on OM enrichment varies in different geological settings but is ultimately manifested through the control of primary productivity and bottom water preservation conditions [29,72,116].

During the deposition of Unit 1 in Cambrian Stage 2, the strong correlation between redox indices such as U/Th and TOC ($R = 0.82$, $p(\alpha) < 0.05$) suggests that the main bottom water conditions are anoxic, providing favorable conditions for the preservation of OM. The relatively high sea level allowed the intraplatform basin to be connected to the open sea, and the input of nutrients sustained high primary productivity. This is consistent with the correlation exhibited by Cu/Al ($R = 0.52$, $p(\alpha) < 0.05$), Ni/Al ($R = 0.67$, $p(\alpha) < 0.05$), and TOC (Figure 5e,f). Thus, the development of anoxic and high productivity is favorable for increased OM accumulation and preservation (Figure 8A).

Relative sea level changes in the intraplatform basin controlled the redox conditions during the deposition of Unit 2 [117]. Intermittent euxinic and continuous anoxic environments account for the relatively high TOC values. Relatively high primary productivity continuously contributed to OM enrichment with the absence of upwelling events.

During the deposition of Unit 3, as the sea level gradually falls in Cambrian Stage 3, the increasing restriction within the basin led to a reduction in nutrient supply and a decline in primary productivity. At the same time, the predominant bottom water conditions changed from anoxic to suboxic, leading to poor preservation conditions for OM. In addition, the weak negative correlation between Ti/Al and TOC ($R = -0.28$, $p(\alpha) > 0.05$) indicated the dilution effect of increasing terrigenous input on OM [115] (Figure 5g). Thus, the low OM content in Unit 3 should be attributed to low productivity, relatively oxidizing conditions, and the dilution of terrigenous input (Figure 8B).

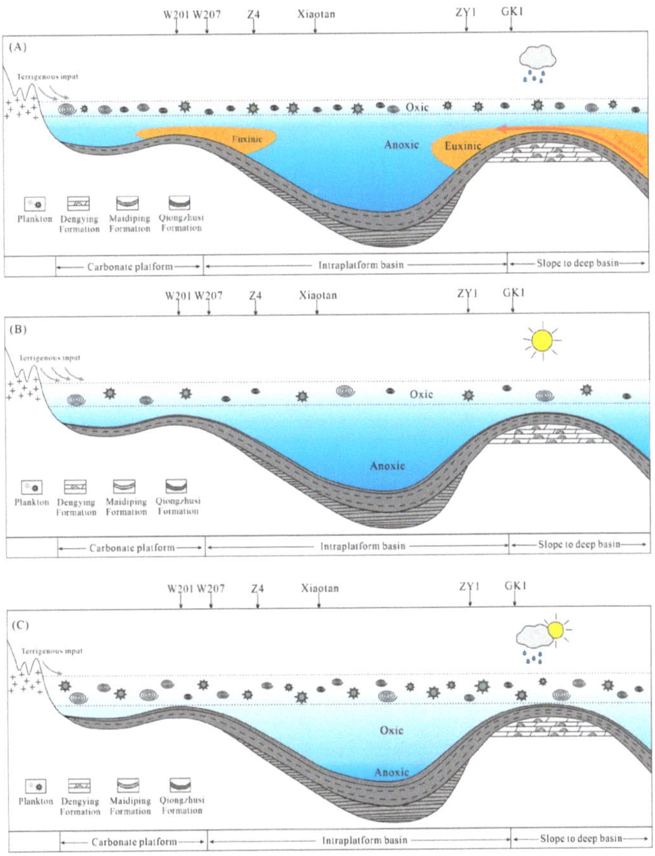

Figure 8. Organic matter accumulation models of Qiongzhusi Formation. (**A**) Units 1 and 2, anoxic bottom water conditions and high primary productivity. (**B**) Unit 3, suboxic bottom water conditions, low primary productivity, and high terrigenous input. (**C**) Unit 4, oxic–suboxic bottom water conditions, relatively high primary productivity, warm and humid climate, and low terrigenous input. Data sources: W201 [40]; W207 (this study); Z4 [42]; Xiaotan [39]; ZY1 [20]; GK1 [46].

Although Unit 4 is still in an oxidizing environment, its TOC value (0.2–2.5 wt.%, average of 0.9 wt.%) showed a recovery compared to Unit 3. A rapid decrease in Ti/Al ratios indicated a weakening of terrigenous input, which may reflect a transition from arid to humid climate in the intraplatform basin [18,23,118,119] (Figure 2). This change is also consistent with the increase in Rb/Sr ratios and decrease in Sr/Cu ratios (Table S2; Figure 2). The relatively warm and humid climate in Unit 4 promoted the bloom of surface water organisms, which is consistent with the first appearance of *Tsunyidiscus niutitangensis* in the lower part of Unit 4 [55], indicating relatively higher primary productivity. Overall, although the widespread oxic–suboxic bottom water conditions in the intraplatform basin are unfavorable for the preservation of OM, the increase in primary productivity and the weakening of terrigenous clastic input have relatively compensated for this deficiency (Figure 8C).

Sweere's analysis of multiple global regions indicates (Figure 6b,d) that Cd/Mo > 0.1 and Co*Mn < 0.4 represent upwelling environments driven by productivity (e.g., Namibian margin, Peruvian margin), while Cd/Mo ratios < 0.1 and Co*Mn > 0.4 reflect restricted basin settings driven by preservation conditions (e.g., the Black Sea). From the Cd/Mo ratios of the four units in this study (Figure 6d), it can be seen that Unit 1 and Unit 2 sample points are mainly located in the "preservation" area, while Unit 3 and Unit 4 samples are mainly located in the "productivity" area.

This is consistent with the TOC values of the shales in Units 1 and 2 being much higher than those in Units 3 and 4. This indicates that preservation conditions are more crucial than productivity conditions in restricted environments that lack upwelling and that they play a critical role in the development of high-quality hydrocarbon source rocks. The enrichment mechanism in the intraplatform basin significantly differs from other facies in the Yangtze Platform, such as the deep-water slope and basinal facies, primarily driven by productivity, or the transitional facies, influenced by both productivity and preservation conditions [20,34,107].

6. Conclusions

(1) During the deposition of the Cambrian Qiongzhusi Formation, the western side of the intraplatform basin within the Yangtze Platform experienced a transition from continuous anoxic conditions with intermittent euxinic to oxic–suboxic bottom water environments. The highly redox-sensitive $\delta^{98/95}$Mo values in Units 1 and 2 support the presence of weakly euxinic conditions in the early Cambrian period, further elucidating the redox environment of the bottom water during this interval.

(2) EF(Co) * EF(Mn) values suggest an absence of upwelling within the intraplatform basin, which diverges from other regions within the Yangtze Platform that showcase substantial upwelling events. The relatively lower primary productivity in the intraplatform basin facies could potentially be attributed to the lack of upwelling events. Primary productivity exhibited a trend of gradual decline from stages 1 to 3, followed by an increase in stage 4.

(3) Qiongzhusi Formation shale exhibits variations in the mechanisms for OM enrichment across distinct units. The lower portion of the Qiongzhusi Formation on the western side of the intraplatform basin (Units 1 and 2) is predominantly influenced by preservation conditions, while the upper portion (Units 3 and 4) is primarily affected by productivity conditions. This indicates that in constrained environments devoid of upwelling, preservation conditions assume a more pivotal role in OM enrichment. This discovery offers valuable insights for the exploration of high-quality hydrocarbon source rocks within the intraplatform basin.

Supplementary Materials: The following supporting information can be downloaded at: https://www.mdpi.com/article/10.3390/jmse11101907/s1, Table S1: Total organic carbon (TOC) and Major elements compositions of the lower Cambrian Qiongzhusi Formation; Table S2: Selected trace elements concentrations and enrichment factors of the lower Cambrian Qiongzhusi Formation.

Author Contributions: Conceptualization, E.L. and Q.Z.; methodology, Q.Z.; software, R.Z.; validation, E.L., S.P. and Z.Z.; formal analysis, E.L.; investigation, Q.Z.; resources, S.P.; data curation, Z.J.; writing—original draft preparation, Q.Z.; writing—review and editing, E.L.; visualization, Q.Z.; supervision, H.W.; project administration, S.P.; funding acquisition, H.W. All authors have read and agreed to the published version of the manuscript.

Funding: This research was funded by Science and Technology Research Project for the Research Institute of Petroleum Exploration and Development, China National Petroleum Corporation (2021YJCQ03).

Institutional Review Board Statement: Not applicable.

Informed Consent Statement: Not applicable.

Data Availability Statement: The data presented in this study are available on request from the corresponding author.

Acknowledgments: We appreciate Research Institute of Petroleum Exploration and Development for providing the data and permission to publish this paper. We are also very grateful to the reviewers and editors for their contributions to improving this paper.

Conflicts of Interest: The authors declare no conflict of interest.

References

1. Knoll, A.H.; Carroll, S.B. Early animal evolution: Emerging views from comparative biology and geology. *Science* **1999**, *284*, 2129–2137. [CrossRef]
2. Kimura, H.; Watanabe, Y. Oceanic anoxia at the Precambrian-Cambrian boundary. *Geology* **2001**, *29*, 995–998. [CrossRef]
3. Amthor, J.E.; Grotzinger, J.P.; Schroder, S.; Bowring, S.A.; Ramezani, J.; Martin, M.W.; Matter, A. Extinction of Cloudina and Namacalathus at the Precambrian-Cambrian boundary in Oman. *Geology* **2003**, *31*, 431–434. [CrossRef]
4. Marshall, C.R. Explaining the Cambrian "Explosion" of Animals. *Annu. Rev. Earth Planet. Sci.* **2006**, *34*, 355–384. [CrossRef]
5. Jiang, S.Y.; Pi, D.H.; Heubeck, C.; Frimmel, H.; Liu, Y.P.; Deng, H.L.; Ling, H.F.; Yang, J.H. Early Cambrian ocean anoxia in South China. *Nature* **2009**, *459*, E5–E6. [CrossRef]
6. Zhang, F.F.; Xiao, S.H.; Romaniello, S.J.; Hardisty, D.; Li, C.; Melezhik, V.; Pokrovsky, B.; Cheng, M.; Shi, W.; Lenton, T.M. Global marine redox changes drove the rise and fall of the Ediacara biota. *Geobiology* **2019**, *17*, 594–610. [CrossRef]
7. Steiner, M.; Wallis, E.; Erdtmann, B.D.; Zhao, Y.; Yang, R. Submarine-hydrothermal exhalative ore layers in black shales from South China and associated fossils—Insights into a Lower Cambrian facies and bio-evolution. *Palaeogeogr. Palaeoclimatol. Palaeoecol.* **2001**, *169*, 165–191. [CrossRef]
8. Zhu, M.; Strauss, H.; Shields, G.A. From snowball earth to the Cambrian bioradiation: Calibration of Ediacaran-Cambrian earth history in South China. *Palaeogeogr. Palaeoclimatol. Palaeoecol.* **2007**, *254*, 1–6. [CrossRef]
9. Wang, J.G.; Chen, D.Z.; Yan, D.T.; Wei, H.Y.; Xiang, L. Evolution from an anoxic to oxic deep ocean during the Ediacaran-Cambrian transition and implications for bioradiation. *Chem. Geol.* **2012**, *306–307*, 129–138. [CrossRef]
10. Demaison, G.; Moore, G. Anoxic environments and oil source bed genesis. *Org. Geochem.* **1980**, *2*, 9–31. [CrossRef]
11. Tyson, R.V. Deposition of Organic-Carbon-Rich Sediments: Models Mechanisms, and Consequences. In *The "Productivity versus Preservation" Controversy: Cause, Flaws, and Resolution*, 1st ed.; Harris, N.B., Ed.; SEPM Special Publication: Tulsa, OK, USA, 2005; Volume 82, pp. 17–33. ISBN 9781565762183.
12. Murphy, A.E.; Sageman, B.B.; Hollander, D.J.; Lyons, T.W.; Brett, C.E. Black shale deposition and faunal overturn in the Devonian Appalachian Basin; clastic starvation, seasonal water–column mixing, and efficient biolimiting nutrient recycling. *Paleoceanography* **2000**, *15*, 280–291. [CrossRef]
13. Abdi, Z.; Rimmer, S.M.; Rowe, H.D.; Nordeng, S. Controls on organic matter accumulation in the bakken formation, williston basin, USA. *Chem. Geol.* **2021**, *586*, 120588. [CrossRef]
14. Si, L.L.; Wei, J.P.; Xi, Y.J.; Wang, H.Y.; Wen, Z.H.; Li, B.; Zhang, H.T. The influence of long-time water intrusion on the mineral and pore structure of coal. *Fuel* **2021**, *290*, 119848. [CrossRef]
15. Chen, L.; Zhang, B.M.; Chen, X.H.; Jiang, S.; Zhang, G.T.; Lin, W.B.; Chen, P.; Liu, Z.H. Depositional environment and organic matter accumulation of the Lower Cambrian Shuijingtuo Formation in the middle Yangtze area, China. *J. Pet. Sci. Eng.* **2022**, *208*, 109339. [CrossRef]
16. Hou, H.H.; Shao, L.Y.; Li, Y.H.; Liu, L.; Liang, G.D.; Zhang, W.L.; Wang, X.T.; Wang, W.C. Effect of paleoclimate and paleoenvironment on organic matter accumulation in lacustrine shale: Constraints from lithofacies and element geochemistry in the northern Qaidam Basin, NW China. *J. Pet. Sci. Eng.* **2022**, *208*, 109350. [CrossRef]
17. Sageman, B.B.; Murphy, A.E.; Werne, J.P.; Ver Straeten, C.A.; Hollander, D.J.; Lyons, T.W. A tale of shales: The relative roles of production, decomposition, and dilution in the accumulation of organic-rich strata, Middle-Upper Devonian, Appalachian basin. *Chem. Geol.* **2003**, *195*, 229–273. [CrossRef]
18. Xu, C.; Shan, X.L.; He, W.T.; Zhang, K.; Rexiti, Y.K.; Su, S.Y.; Liang, C.; Zou, X.T. The influence of paleoclimate and a marine transgression event on organic matter accumulation in lacustrine black shales from the Late Cretaceous, southern Songliao Basin, Northeast China. *Int. J. Coal Geol.* **2021**, *246*, 103842. [CrossRef]
19. Zhao, B.S.; Li, R.X.; Qin, X.L.; Wang, N.; Zhou, W.; Khaled, A.; Zhao, D.; Zhang, Y.N.; Wu, X.L.; Liu, Q. Geochemical characteristics and mechanism of organic matter accumulation of marine-continental transitional shale of the lower permian Shanxi Formation, southeastern Ordos Basin, north China. *J. Pet. Sci. Eng.* **2021**, *205*, 108815. [CrossRef]
20. Gao, P.; Li, S.J.; Lash, G.G.; Yan, D.T.; Zhou, Q.; Xiao, X.M. Stratigraphic framework, redox history, and organic matter accumulation of an Early Cambrian intraplatfrom basin on the Yangtze Platform, South China. *Mar. Pet. Geol.* **2021**, *130*, 105095. [CrossRef]
21. Mansour, A.; Wagreich, M. Earth system changes during the cooling greenhouse phase of the Late Cretaceous: Coniacian-Santonian OAE3 subevents and fundamental variations in organic carbon deposition. *Earth Sci. Rev.* **2022**, *229*, 104022. [CrossRef]
22. Ingall, E.D.; Bustin, R.M.; Van Cappellen, P. Influence of water column anoxia on the burial and preservation of carbon and phosphorus in marine shales. *Geochim. Cosmochim. Acta* **1993**, *57*, 303–316. [CrossRef]
23. Yeasmin, R.; Chen, D.Z.; Fu, Y.; Wang, J.G.; Guo, Z.H.; Guo, C. Climatic-oceanic forcing on the organic accumulation across the shelf during the early cambrian (age 2 through 3) in the mid-upper yangtze block, NE guizhou, south China. *J. Asian Earth Sci.* **2017**, *134*, 365–386. [CrossRef]

24. Chen, D.; Wang, J.; Qing, H.; Yan, D.; Li, R. Hydrothermal venting activities in the Early Cambrian, South China: Petrological, geochronological and stable isotopic constraints. *Chem. Geol.* **2009**, *258*, 168.e181. [CrossRef]
25. Wei, H.; Feng, Q.; Yu, J.; Chang, S. Characteristics and Sources of Organic Matter from the Early Cambrian Niutitang Formtion and Its Preservation Environment in Guizhou. *J. Earth Sci.* **2022**, *33*, 933–944. [CrossRef]
26. Guo, Q.J.; Shields, G.A.; Liu, C.Q.; Strauss, H.; Zhu, M.Y.; Pi, D.H.; Goldberg, T.; Yang, X.L. Trace element chemostratigraphy of two Ediacarane-Cambrian successions in South China: Implications for organosedimentary metal enrichment and silicification in the early Cambrian. *Palaeogeogr. Palaeoclimatol. Palaeoecol.* **2007**, *254*, 194.e216. [CrossRef]
27. Guo, Q.; Strauss, H.; Liu, C.; Goldberg, T.; Zhu, M.; Heubeck, C.; Pi, D.; Vernhet, E.; Yang, X.; Fu, P. Carbon isotopic evolution of the terminal Neoproterozoic and Early Cambrian: Evidence from the Yangtze Platform, South China. *Palaeogeogr. Palaeoclimatol. Palaeoecol.* **2007**, *254*, 140.e157. [CrossRef]
28. Liu, Z.H.; Zhuang, X.G.; Teng, G.E.; Xie, X.M.; Yin, L.M.; Bian, L.Z.; Feng, Q.L.; Algeo, T.J. The Lower Cambrian Niutitang Formation at Yangtiao (Guizhou, SW China): Organic matter enrichment, source rock potential, and hydrothermal influences. *J. Pet. Geol.* **2015**, *38*, 411–432. [CrossRef]
29. Liu, K.; Feng, Q.; Shen, J.; Khan, M.; Planavsky, N.J. Increased productivity as a primary driver of marine anoxia in the Lower Cambrian. *Palaeogeogr. Palaeoclimatol. Palaeoecol.* **2018**, *491*, 1–9. [CrossRef]
30. Zhang, Y.; He, Z.; Jiang, S.; Gao, B.; Liu, Z.; Han, B.; Wang, H. Marine redox stratification during the early Cambrian (ca. 529-509 Ma) and its control on the development of organic-rich shales in Yangtze Platform. *Geochem. Geophys. Geosyst.* **2017**, *18*, 2354–2369. [CrossRef]
31. Jin, C.; Li, C.; Algeo, T.J.; Wu, S.; Cheng, M.; Zhang, Z.; Shi, W. Controls on organic matter accumulation on the early-Cambrian western Yangtze Platform, South China. *Mar. Pet. Geol.* **2020**, *111*, 75–87. [CrossRef]
32. Liu, Z.; Yan, D.; Yuan, D.; Niu, X.; Fu, H. Multiple controls on the organic matter accumulation in early Cambrian marine black shales, middle Yangtze Block, South China. *J. Nat. Gas Sci. Eng.* **2022**, *100*, 104454. [CrossRef]
33. Xia, W.; Yu, B.; Sun, M. Depositional settings and enrichment mechanism of organic matter of the black shales at the bottom of Niutitang Formation, lower Cambrian in Southeast Chongqing: A case study from Well Yuke 1. *Acta Geol. Sin-Engl. Ed.* **2015**, *89*, 287. [CrossRef]
34. Wu, Y.W.; Tian, H.; Gong, D.J.; Li, T.F.; Zhou, Q. Paleo-environmental variation and its control on organic matter enrichment of black shales from shallow shelf to slope regions on the Upper Yangtze Platform during Cambrian Stage 3. *Palaeogeogr. Palaeoclimatol. Palaeoecol.* **2020**, *545*, 109653. [CrossRef]
35. Li, J.H.; Wang, H.H.; Li, W.B.; Zhou, X.B. Discussion on global tectonics evolution from plate reconstruction in Phanerozoic. *Acta Pet Sin.* **2014**, *35*, 207–218.
36. Liu, S.; Jin, S.; Liu, Y.; Chen, A. Astronomical forced sequence infill of Early Cambrian Qiongzhusi organic-rich shale of Sichuan Basin, South China. *Sediment. Geol.* **2022**, *440*, 106261. [CrossRef]
37. Zou, C.; Yang, Z.; Dai, J.; Dong, D.; Zhang, B.; Wang, Y.; Deng, S.; Huang, J.; Liu, K.; Yang, C.; et al. The characteristics and significance of conventional and unconventional Sinian–Silurian gas systems in the Sichuan Basin, central China. *Mar. Pet. Geol.* **2015**, *64*, 386–402. [CrossRef]
38. Kokkinos, N.C.; Nkagbu, D.C.; Marmanis, D.I.; Dermentzis, K.I.; Maliaris, G. Evolution of Unconventional Hydrocarbons: Past, Present, Future and Environmental FootPrint. *J. Eng. Sci. Technol. Rev.* **2022**, *15*, 15–24. [CrossRef]
39. Jin, C.; Li, C.; Algeo, T.J.; Planavsky, N.J.; Cui, H.; Yang, X.; Zhao, Y.; Zhang, X.; Xie, S. A highly redox-heterogeneous ocean in South China during the early Cambrian (~529–514 Ma): Implications for biota-environment co-evolution. *Earth Planet Sci. Lett.* **2016**, *441*, 38–51. [CrossRef]
40. Wang, S.; Zou, C.; Dong, D.; Wang, Y.; Li, X.; Huang, J.; Guan, Q. Multiple controls on the paleoenvironment of the Early Cambrian marine black shales in the Sichuan Basin, SW China: Geochemical and organic carbon isotopic evidence. *Mar. Pet. Geol.* **2015**, *66*, 660–672. [CrossRef]
41. Wang, N.; Wen, L.; Li, M.; Dai, X.; Xu, Y.; Ming, Y.; Yang, S. The origin of abnormally ^{13}C-depleted organic carbon isotope signatures in the early Cambrian Yangtze Platform. *Mar. Pet. Geol.* **2021**, *128*, 105051. [CrossRef]
42. Wang, N.; Li, M.J.; Tian, X.W.; Hong, H.T.; Wen, L.; Wang, W.Z. Climate-ocean control on the depositional watermass conditions and organic matter enrichment in Lower Cambrian black shale in the upper Yangtze Platform. *Mar. Pet. Geol.* **2020**, *120*, 104570. [CrossRef]
43. Wang, J.; Li, Z.X. History of Neoproterozoic rift basins in South China: Implications for Rodinia break-up. *Precambrian Res.* **2003**, *122*, 141–158. [CrossRef]
44. Charvet, J. The Neoproterozoic-Early Paleozoic tectonic evolution of the South China Block: An overview. *J. Asian Earth Sci.* **2013**, *74*, 198–209. [CrossRef]
45. Steiner, M.; Li, G.; Qian, Y.; Zhu, M.; Erdtmann, B.D. Neoproterozoic to Early Cambrian small shelly fossil assemblages and a revised biostratigraphic correlation of the Yangtze Platform (China). *Palaeogeogr. Palaeoclimatol. Palaeoecol.* **2007**, *254*, 67–99. [CrossRef]
46. Gao, P.; Liu, G.; Jia, C.; Young, A.; Wang, Z.; Wang, T.; Zhang, P.; Wang, D. Redox variations and organic matter accumulation on the Yangtze carbonate platform during Late Ediacaran–Early Cambrian: Constraints from petrology and geochemistry. *Palaeogeogr. Palaeoclimatol. Palaeoecol.* **2016**, *450*, 91–110. [CrossRef]
47. Wang, Z.; Jiang, H.; Wang, T.; Lu, W.; Gu, Z.; Xu, A.; Yang, Y.; Xu, Z. Paleo-geomorphology formed during Tongwan tectonization in Sichuan Basin and its significance for hydrocarbon accumulation. *Pet. Explor. Dev.* **2014**, *41*, 338–345. [CrossRef]

48. Li, S.; Gao, P.; Huang, B.; Wang, H.; Wo, Y. Sedimentary constraints on the tectonic evolution of Mianyang-Chang trough in the Sichuan Basin. *Oil Gas Geol.* **2018**, *39*, 889–898.
49. Zhu, M.Y.; Zhang, J.M.; Steiner, M.; Yang, A.H.; Li, G.X.; Erdtmann, B.D. Sinian-Cambrian stratigraphic framework for shallow-to deep-water environments of the Yangtze Platform: An integrated approach. *Prog. Nat. Sci.* **2003**, *13*, 951–960. [CrossRef]
50. Liu, S.; Wang, Y.; Sun, W.; Zhong, Y.; Hong, H.; Deng, B.; Xia, M.; Song, J.; Wen, Y.; Wu, J. Control of intracratonic sags on the hydrocarbon accumulations in the Marine Strata across the Sichuan Basin. *J. Chengdu Univ. of Technol.* **2016**, *43*, 1–23.
51. Du, J.; Wang, Z.; Zou, C.; Xu, C.; Shen, P.; Zhang, B.; Jiang, H.; Huang, S. Discovery of intracratonic rift in the Upper Yangtze and its control effect on the formation of Anyue giant gas field. *Acta Pet Sin.* **2016**, *37*, 1–16.
52. Zhao, J.H.; Jin, Z.J.; Hu, Q.H.; Liu, K.Y.; Liu, G.X.; Gao, B.; Liu, Z.B.; Zhang, Y.Y.; Wang, R.Y. Geological controls on the accumulation of shale gas: A case study of the early Cambrian shale in the Upper Yangtze area. *Mar. Pet. Geol.* **2019**, *107*, 426–437. [CrossRef]
53. Xu, L.; Lehmann, B.; Mao, J.; Qu, W.; Du, A. Re-Os age of polymetallic Ni-Mo-PGE-Au mineralization in early Cambrian black shales of South China—A reassessment. *Econ. Geol.* **2011**, *106*, 511–522. [CrossRef]
54. Yang, C. High Precision Geochronology of the Upper Ediacaran-Lower Cambrian in South China. Ph.D. Thesis, University of Chinese Academy of Sciences, Beijing, China, 2017. (In Chinese).
55. Liang, F.; Jiang, W.; Dai, B.; Chen, Y.; Luo, C.; Zhang, Q.; Tong, K.L.; Hu, X.; Lu, B. Enrichment law and resource potential of shale gas of Qiongzhusi Formation in Weiyuan-Ziyang area. *Nat. Gas. Geosci.* **2022**, *33*, 755–763.
56. Pan, X.Q.; Dong, X.X.; Dai, Q.K.; Luo, J.Z.; Liu, Z.W.; Wang, T.; Hua, H. Fossil assemblages and stratigraphic correlation of the Early Cambrian Jiulaodong Formation in Jinyan-Qianwei Area, Sichuan Basin. *Acta Micropaleontol. Sin.* **2021**, *38*, 241–256. [CrossRef]
57. Liu, J.; Zhang, K.Y. Characteristics of high-resolution stratigraphic sequence of Maidiping Formation and Qiongzhusi Formation in Jingyan area, Sichuan, China. *J. Chengdu Univ. Technol.* **2018**, *45*, 585–593.
58. Zhao, L.K.; Li, W.H.; He, Y.; Zhou, H.F.; Liu, R.; Li, Y.; Wang, W.; Zhong, Y.; Wang, Y.Y. Sedimentation and filling laws of Maidiping-Qiongzhusi formations in Sichuan Basin and their significance of oil and gas geological exploration. *Nat. Gas. Explor. Dev.* **2020**, *43*, 30–38.
59. Lawrence, M.G.; Greig, A.; Collerson, K.D.; Kamber, B.S. Rare earth element and yttrium variability in South East Queensland waterways. *Aquat. Geochem.* **2006**, *12*, 39–72. [CrossRef]
60. Taylor, S.R.; McLennan, S.M. *The Continental Crust: Its Composition and Evolution*; Blackwell Scientific Publications: Oxford, UK, 1985; p. 311.
61. Li, J.; Liang, X.R.; Zhong, L.F.; Wang, X.C.; Ren, Z.Y.; Sun, S.L.; Zhang, Z.F.; Xu, J.F. Measurement of the isotopic composition of molybdenum in geological samples by MC-ICP-MS using a novel chromatographic extraction technique. *Geostand. Geoanal. Res.* **2014**, *38*, 345–354. [CrossRef]
62. Siebert, C.; Nägler, T.F.; Kramers, J.D. Determination of molybdenum isotope fractionation by double-spike multicollector inductively coupled plasma mass spectrometry. *Geochem. Geophys. Geosyst.* **2001**, *2*, 1032. [CrossRef]
63. Zhao, P.P.; Li, J.; Zhang, L.; Wang, Z.B.; Kong, D.X.; Ma, J.L. Molybdenum mass fractions and isotopic compositions of International Geological Reference Materials. *Geostand. Geoanal. Res.* **2016**, *40*, 217–226. [CrossRef]
64. Greber, N.D.; Siebert, C.; Nägler, T.F.; Pettke, T. $\delta^{98/95}$Mo values and molybdenum concentration data for NIST SRM 610, 612 and 3134: Towards a common protocol for reporting Mo data. *Geostand. Geoanal. Res.* **2012**, *36*, 291–300. [CrossRef]
65. Yin, L.; Li, J.; Tian, H.; Long, X. Rhenium-osmium and molybdenum isotope systematics of black shales from the Lower Cambrian Niutitang Formation, SW China: Evidence of a well oxygenated ocean at ca. 520 Ma. *Chem. Geol.* **2018**, *499*, 26–42. [CrossRef]
66. Poulson Brucker, R.L.; McManus, J.; Severmann, S.; Berelson, W.M. Molybdenum behavior during early diagenesis: Insights from Mo isotopes. *Geochem. Geophys. Geosyst.* **2009**, *10*, Q06010. [CrossRef]
67. Wen, H.; Fan, H.; Zhang, Y.; Cloquet, C.; Carignan, J. Reconstruction of early Cambrian ocean chemistry from Mo isotopes. *Geochim. Cosmochim. Acta* **2015**, *164*, 1–16. [CrossRef]
68. Wu, Y.W.; Tian, H.; Li, J.; Li, T.F.; Ji, S. Reconstruction of oceanic redox structures during the Ediacaran-Cambrian transition in the Yangtze Block of South China: Implications from Mo isotopes and trace elements. *Precambrian Res.* **2021**, *359*, 106181. [CrossRef]
69. Dahl, T.W.; Canfield, D.E.; Rosing, M.T.; Frei, R.E.; Gordon, G.W.; Knoll, A.H.; Anbar, A.D. Molybdenum evidence for expansive sulfidic water masses in similar to 750 Ma oceans. *Earth Planet Sci. Lett.* **2011**, *311*, 264–274. [CrossRef]
70. Hatch, J.R.; Leventhal, J.S. Relationship between inferred redox potential of the depositional environment and geochemistry of the Upper Pennsylvanian (Missourian) stark shale member of the Dennis Limestone, Wabaunsee County, Kansas, USA. *Chem. Geol.* **1992**, *99*, 65.e82. [CrossRef]
71. Jones, B.; Manning, D.A. Comparison of geochemical indices used for the interpretation of palaeoredox conditions in ancient mudstones. *Chem. Geol.* **1994**, *111*, 111–129. [CrossRef]
72. Rimmer, S.M.; Thompson, J.A.; Goodnight, S.A.; Robl, T.L. Multiple controls on the preservation of organic matter in Devonian–Mississippian marine black shales: Geochemical and petrographic evidence. *Paleogeogr. Paleoclimatol. Paleoecol.* **2004**, *215*, 125–154. [CrossRef]
73. Algeo, T.J.; Li, C. Redox classification and calibration of redox thresholds in sedimentary systems. *Geochim. Cosmochim. Acta* **2020**, *287*, 8–26. [CrossRef]
74. Qiu, Z.; Tao, H.; Lu, B.; Chen, Z.; Wu, S.; Liu, H.; Qiu, J. Controlling Factors on Organic Matter Accumulation of Marine Shale across the Ordovician-Silurian Transition in South China: Constraints from Trace-Element Geochemistry. *J. Earth Sci.* **2021**, *32*, 887–900. [CrossRef]

75. Ruebsam, W.; Dickson, A.J.; Hoyer, E.M.; Schwark, L. Multiproxy reconstruction of oceanographic conditions in the southern epeiric Kupferschiefer Sea (Late Permian) based on redox-sensitive trace elements, molybdenum isotopes and biomarkers. *Gondwana Res.* **2017**, *44*, 205–218. [CrossRef]
76. Liu, J.S.; Algeo, T.J. Beyond redox: Control of trace-metal enrichment in anoxic marine facies by watermass chemistry and sedimentation rate. *Geochim. Cosmochim. Acta* **2020**, *7*, 296–317. [CrossRef]
77. Algeo, T.J.; Tribovillard, N. Environmental analysis of paleoceanographic systems based on molybdenum–uranium covariation. *Chem. Geol.* **2009**, *268*, 211–225. [CrossRef]
78. Tribovillard, N.; Algeo, T.J.; Baudin, F.; Riboulleau, A. Analysis of marine environmental conditions based on molybdenum–uranium covariation—Applications to Mesozoic paleoceanography. *Chem. Geol.* **2012**, *324–325*, 46–58. [CrossRef]
79. Algeo, T.J.; Lyons, T.W.; Blakey, R.C.; Over, D.J. Hydrographic conditions of the Devono–Carboniferous North American Seaway inferred from sedimentary Mo–TOC relationships. *Palaeogeogr. Palaeoclimatol. Palaeoecol.* **2007**, *256*, 204–230. [CrossRef]
80. Calvert, S.; Pedersen, T. Geochemistry of recent oxic and anoxic marine sediments: Implications for the geological record. *Mar. Geol.* **1993**, *113*, 67–88. [CrossRef]
81. Algeo, T.J.; Lyons, T.W. Mo–total organic carbon covariation in modern anoxic marine environments: Implications for analysis of paleoredox and paleohydrographic conditions. *Paleoceanogr. Paleoclimatol.* **2006**, *21*, PA1016. [CrossRef]
82. Algeo, T.J.; Maynard, J.B. Trace-metal covariation as a guide to water-mass conditions in ancient anoxic marine environments. *Geosphere* **2008**, *4*, 872–887. [CrossRef]
83. Algeo, T.J.; Maynard, J.B. Trace-element behavior and redox facies in core shales of Upper Pennsylvanian Kansas-type cyclothems. *Chem. Geol.* **2004**, *206*, 289–318. [CrossRef]
84. Wang, X.; Shi, X.; Zhao, X.; Tang, D. Increase of seawater Mo inventory and ocean oxygenation during the early Cambrian. *Palaeogeogr. Palaeoclimatol. Palaeoecol.* **2015**, *440*, 621–631. [CrossRef]
85. Xiang, L.; Schoepfer, S.D.; Shen, S.Z.; Cao, C.Q.; Zhang, H. Evolution of oceanic molybdenum and uranium reservoir size around the Ediacaran-Cambrian transition: Evidence from western Zhejiang, South China. *Earth Planet. Sci. Lett.* **2017**, *464*, 84–94. [CrossRef]
86. Zhu, G.; Li, T.; Zhao, K.; Li, C.; Cheng, M.; Chen, W.; Yan, H.; Zhang, Z.; Algeo, T.J. Mo isotope records from Lower Cambrian black shales, northwestern Tarim Basin (China): Implications for the early Cambrian ocean. *Geol. Soc. Am. Bull.* **2022**, *134*, 3–14. [CrossRef]
87. Cheng, M.; Li, C.; Zhou, L.; Algeo, T.J.; Zhang, F.; Romaniello, S.; Jin, C.S.; Lei, L.D.; Feng, L.J.; Jiang, S.Y. Marine Mo biogeochemistry in the context of dynamically euxinic mid-depth waters: A case study of the lower Cambrian Niutitang shales, South China. *Geochim. Cosmochim. Acta* **2016**, *183*, 79–93. [CrossRef]
88. Siebert, C.; Nägler, T.F.; von Blanckenburg, F.; Kramers, J.D. Molybdenum isotope records as a potential new proxy for paleoceanography. *Earth Planet. Sci. Lett.* **2003**, *211*, 159–171. [CrossRef]
89. Wille, M.; Kramers, J.D.; Nägler, T.F.; Beukes, N.J.; Schröder, S.; Meisel, T.; Lacassie, J.P.; Voegelin, A.R. Evidence for a gradual rise of oxygen between 2.6 and 2.5 Ga from Mo isotopes and Re-PGE signatures in shales. *Geochim. Cosmochim. Acta* **2007**, *71*, 2417–2435. [CrossRef]
90. Dahl, T.W.; Anbar, A.D.; Gordon, G.W.; Rosing, M.T.; Frei, R.; Canfield, D.E. The behavior of molybdenum and its isotopes across the chemocline and in the sediments of sulfidic Lake Cadagno, Switzerland. *Geochim. Cosmochim. Acta* **2010**, *74*, 144–163. [CrossRef]
91. Kendall, B.; Komiya, T.; Lyons, T.W.; Bates, S.M.; Gordon, G.W.; Romaniello, S.J.; Jiang, G.; Creaser, R.A.; Xiao, S.; McFadden, K.; et al. Uranium and molybdenum isotope evidence for an episode of widespread ocean oxygenation during the late Ediacaran Period. *Geochim. Cosmochim. Acta* **2015**, *156*, 173–193. [CrossRef]
92. Barling, J.; Arnold, G.L.; Anbar, A.D. Natural mass-dependent variations in the isotopic composition of molybdenum. *Earth Planet. Sci. Lett.* **2001**, *193*, 447–457. [CrossRef]
93. Goldberg, T.; Archer, C.; Vance, D.; Poulton, S.W. Mo isotope fractionation during adsorption to Fe (oxyhydr) oxides. *Geochim. Cosmochim. Acta* **2009**, *73*, 6502–6516. [CrossRef]
94. Neubert, N.; Nägler, T.F.; Böttcher, M.E. Sulfidity controls molybdenum isotope fractionation into euxinic sediments: Evidence from the modern Black Sea. *Geology* **2008**, *36*, 775–778. [CrossRef]
95. Helz, G.R.; Miller, C.V.; Charnock, J.M.; Mosselmans, J.F.W.; Pattrick, R.A.D.; Garner, C.D.; Vaughan, D.J. Mechanism of molybdenum removal from the sea and its concentration in black shales: EXAFS evidence. *Geochim. Cosmochim. Acta* **1996**, *60*, 3631–3642. [CrossRef]
96. Barling, J.; Anbar, A.D. Molybdenum isotope fractionation during adsorption by manganese oxides. *Earth Planet. Sci. Lett.* **2004**, *217*, 315–329. [CrossRef]
97. Scholz, F.; McManus, J.; Sommer, S. The manganese and iron shuttle in a modern euxinic basin and implications for molybdenum cycling at euxinic ocean margins. *Chem. Geol.* **2013**, *355*, 56–68. [CrossRef]
98. Ostrander, C.M.; Kendall, B.; Olson, S.L.; Lyons, T.W.; Gordon, G.W.; Romaniello, S.J.; Zheng, W.; Reinhard, C.T.; Roy, M.; Anbar, A.D. An expanded shale δ^{98}Mo record permits recurrent shallow marine oxygenation during the Neoarchean. *Chem. Geol.* **2020**, *532*, 119391. [CrossRef]
99. Arnold, G.L.; Anbar, A.D.; Barling, J.; Lyons, T.W. Molybdenum isotope evidence for widespread anoxia in mid-proterozoic oceans. *Science* **2004**, *304*, 87–90. [CrossRef] [PubMed]
100. Chen, X.; Ling, H.F.; Vance, D.; Shields-Zhou, G.A.; Zhu, M.; Poulton, S.W.; Och, L.M.; Jiang, S.Y.; Li, D.; Cremonese, L.; et al. Rise to modern levels of ocean oxygenation coincided with the Cambrian radiation of animals. *Nat. Commun.* **2015**, *6*, 7142. [CrossRef]

101. Xu, L.; Lehmann, B.; Mao, J.; Nägler, T.F.; Neubert, N.; Böttcher, M.E.; Escher, P. Mo isotope and trace element patterns of Lower Cambrian black shales in South China: Multi-proxy constraints on the paleoenvironment. *Chem. Geol.* **2012**, *318–319*, 45–59. [CrossRef]
102. Paytan, A.; Griffith, E.M. Marine barite: Recorder of variations in ocean export productivity. *Deep Sea Res. Part II Top. Stud. Oceanogr.* **2007**, *54*, 687–705. [CrossRef]
103. Tribovillard, N.; Algeo, T.J.; Lyons, T.; Riboulleau, A. Trace metals as paleoredox and paleoproductivity proxies: An update. *Chem. Geol.* **2006**, *232*, 12–32. [CrossRef]
104. Little, S.; Vance, D.; Lyons, T.; McManus, J. Controls on trace metal authigenic enrichment in reducing sediments: Insights from modern oxygen-deficient settings. *Am. J. Sci.* **2015**, *315*, 77–119. [CrossRef]
105. Xu, D.T.; Wang, X.Q.; Shi, X.Y.; Tang, D.J.; Zhao, X.K.; Feng, L.J.; Song, H.Y. Nitrogen cycle perturbations linked to metazoan diversification during the early Cambrian. *Palaeogeogr. Palaeoclimatol. Palaeoecol.* **2020**, *538*, 109392. [CrossRef]
106. Schoepfer, S.D.; Shen, J.; Wei, H.; Tyson, R.V.; Ingall, E.; Algeo, T.J. Total organic carbon, organic phosphorus, and biogenic barium fluxes as proxies for paleomarine productivity. *Earth Sci. Rev.* **2015**, *149*, 23–52. [CrossRef]
107. Xia, P.; Hao, F.; Tian, J.; Zhou, W.; Fu, Y.; Guo, C.; Yang, Z.; Li, K.; Wang, K. Depositional Environment and Organic Matter Enrichment of Early Cambrian Niutitang Black Shales in the Upper Yangtze Region, China. *Energies* **2022**, *15*, 4551. [CrossRef]
108. Sweere, T.; van den Boorn, S.; Dickson, A.J.; Reichart, G.J. Definition of new trace-metal proxies for the controls on organic matter enrichment in marine sediments based on Mn, Co, Mo and Cd concentrations. *Chem. Geol.* **2016**, *441*, 235–245. [CrossRef]
109. Cremonese, L.; Shields-Zhou, G.; Struck, U.; Ling, H.F.; Och, L.; Chen, X.; Li, D. Marine biogeochemical cycling during the early Cambrian constrained by a nitrogen and organic carbon isotope study of the Xiaotan section, South China. *Precambrian Res.* **2013**, *225*, 148–165. [CrossRef]
110. Yuan, Y.; Cai, C.; Wang, T.; Xiang, L.; Jia, L.; Chen, Y. Redox condition during Ediacaran-Cambrian transition in the Lower Yangtze deep water basin, South China: Constraints from iron speciation and $\delta^{13}C_{org}$ in the Diben section, Zhejiang. *Chin. Sci. Bull.* **2014**, *59*, 3638–3649. [CrossRef]
111. Gao, P.; He, Z.L.; Li, S.J.; Lash, G.G.; Li, B.; Huang, B.Y.; Yan, D.T. Volcanic and hydrothermal activities recorded in phosphate nodules from the Lower Cambrian Niutitang Formation black shales in South China. *Palaeogeogr. Palaeoclimatol. Palaeoecol.* **2018**, *505*, 381–397. [CrossRef]
112. He, T.H.; Lu, S.F.; Li, W.H.; Sun, D.Q.; Pan, W.Q.; Zhang, B.S.; Tan, Z.Z.; Ying, J.F. Paleoweathering, hydrothermal activity and organic matter enrichment during the formation of earliest Cambrian black strata in the Paleo-environmental variation and its control on organic matter enrichment of black shales from shallow shelf to slope regions on the Upper Yangtze Platform during Cambrian Stage 3e northwest Tarim Basin, China. *J. Pet. Sci. Eng.* **2020**, *545*, 106987. [CrossRef]
113. Ma, Y.Q.; Fan, M.J.; Lu, Y.C.; Liu, H.M.; Hao, Y.Q.; Xie, Z.H.; Liu, Z.H.; Peng, L.; Du, X.B.; Hu, H.Y. Climate-driven paleolimnological change controls lacustrine mudstone depositional process and organic matter accumulation: Constraints from lithofacies and geochemical studies in the Zhanhua Depression, eastern China. *Int. J. Coal Geol.* **2016**, *167*, 103–118. [CrossRef]
114. Tyson, R.V. Sedimentation rate, dilution, preservation and total organic carbon: Some results of a modelling study. *Org. Geochem.* **2001**, *32*, 333–339. [CrossRef]
115. Mansour, A.; Wagreich, M.; Gentzis, T.; Ocubalidet, S.; Tahoun, S.S.; Elewa, A.M. Depositional and organic carbon-controlled regimes during the Coniacian-Santonian event: First results from the southern Tethys (Egypt). *Mar. Pet. Geol.* **2020**, *115*, 104285. [CrossRef]
116. Pedersen, T.; Calvert, S. Anoxia vs. productivity: What controls the formation of organic-carbon-rich sediments and sedimentary Rocks. *AAPG Bull.* **1990**, *74*, 454–466.
117. Dong, T.; Harris, N.B.; Ayranci, K. Relative sea-level cycles and organic matter accumulation in shales of the Middle and Upper Devonian Horn River Group, northeastern British Columbia, Canada: Insights into sediment flux, redox conditions, and bioproductivity. *Geol. Soc. Am. Bull.* **2018**, *130*, 859–880. [CrossRef]
118. Yarincik, K.M.; Murray, R.W.; Peterson, L.C. Climatically sensitive eolian and hemipelagic deposits in the Cariaco Basin, Venezuela, over past 578,000 years: Results from Al/Ti and K/Al. *Paleoceanography* **2000**, *15*, 210–228. [CrossRef]
119. Beckmann, B.; Flögel, S.; Hofmann, P.; Schulz, M.; Wagner, T. Orbital forcing of Cretaceous river discharge in tropical Africa and ocean response. *Nature* **2005**, *437*, 241–244. [CrossRef] [PubMed]

Journal of
Marine Science and Engineering

MDPI

Article

Shale Gas Exploration and Development Potential Analysis of Lower Cambrian Niutitang Formation and Lower Silurian Longmaxi Formation in Northwestern Hunan, South China, Based on Organic Matter Pore Evolution Characteristics

Yanan Miao [1,2], Pengfei Wang [3,*], Xin Li [3], Haiping Huang [1], Can Jin [4] and Wei Gao [5]

1 School of Energy Resource, China University of Geosciences (Beijing), Beijing 100083, China; huah@ucalgary.ca (H.H.)
2 China National Oil and Gas Exploration and Development Company, Ltd., Beijing 100034, China
3 China National Offshore Oil Corporation Research Institute, Beijing 100028, China
4 Kunlun Digital Technology Co., Ltd., PetroChina, Beijing 100007, China
5 No. 4 Oil Production Plant of Daqing Oil Co., Ltd., Daqing 163511, China; 18811368076@163.com
* Correspondence: pengfeijincan@163.com; Tel.: +86-13717925779; Fax: +86-01066554775

Abstract: Shale gas production is obviously higher within the Silurian Longmaxi Formation than that of the Cambrian Niutitang Formation according to the drilling test results in the northwest Hunan area. To clarify the reasons behind this variation, core samples from the two sets of shales were studied for a comprehensive comparison and analysis of their organic matter (OM) pore structure. Methods were used, including the total organic carbon content test, the vitrinite reflectance test, X-ray diffraction, and focused ion-beam scanning electron microscopy (FIB-SEM). The results show that these two shales have similar reservoir characteristics, both with abundant organic matter and high content of brittle minerals. However, the Longmaxi shale with 2.3% to 3.0%Ro presents lower thermal maturity than the Niutitang shale with over 3.0%Ro. In the case of pore structure associated with OM, a huge difference exists between the two shales. The OM pore shape of the Longmaxi shale is very regular, being mostly round and oval, while the OM pore shape of the Niutitang shale is irregular, being flat with a thin middle and thick and elongated ends. An important factor affecting OM pore evolution is thermal maturity. In turn, the thermal maturity is controlled by the tectonic evolution process, especially the maximum paleo-burial depth. In conclusion, the paleo-burial depth of the Lower Cambrian Niutitang shale in northwest Hunan is too large, which leads to the excessive evolution of organic matter in the shale, and the physical and chemical properties are similar to graphite, which leads to the disappearance of OM pores. Shale gas has no effective reservoir space and is largely dispersed in geological history. At the same time, due to the insufficient hydrocarbon-generation evolution time of the Lower Silurian Longmaxi Formation shale in this area, the shale could not form enough gaseous hydrocarbons and a large amount of effective reservoir space (OM pores with regular shape and large pore size), and finally failed to become a gas reservoir. Therefore, the exploration and development potential of the marine shale gas of the Lower Cambrian Niutitang Formation and the Lower Silurian Longmaxi Formation in northwest Hunan is poor.

Keywords: marine shale; tectonic evolution; OM pore; thermal maturity; northwest Hunan

Citation: Miao, Y.; Wang, P.; Li, X.; Huang, H.; Jin, C.; Gao, W. Shale Gas Exploration and Development Potential Analysis of Lower Cambrian Niutitang Formation and Lower Silurian Longmaxi Formation in Northwestern Hunan, South China, Based on Organic Matter Pore Evolution Characteristics. *J. Mar. Sci. Eng.* **2023**, *11*, 1910. https://doi.org/10.3390/jmse11101910

Academic Editor: George Kontakiotis

Received: 13 August 2023
Revised: 19 September 2023
Accepted: 27 September 2023
Published: 2 October 2023

1. Introduction

South China has had remarkable success with the exploration and production of marine shale gas [1,2]. For instance, the China National Petroleum Corporation (CNPC) has worked hard to discover and develop the Lower Silurian Longmaxi shale in the Sichuan Basin. Additionally, it built China's first shale gas exploration and development demonstration site in Fuling, Chongqing, which has generated a total amount of 3×10^{10} m^3 of shale gas [3–5]. Meanwhile, the Lower Cambrian Niutitang shale in the Yichang, Hubei, region has been explored and developed with great success by the China Geological Survey. Here, 7.74 × 10^4 m^3 of gas was produced per day by the successfully installed Well EYYI [6–9]. In order to make additional advancements, shales of the Longmaxi Formation and the Niutitang Formation were drilled by the China Geological Survey in western Hunan. The production data do, however, clearly demonstrate that the Longmaxi Formation produces more shale gas than the Niutitang Formation. The gas output of the Longmaxi shale from Well SY1 is 2.0 × 10^4 m^3, compared to the Niutitang shale's 0.3 × 10^4 m^3 [10–12].

Southern China's deep-water shelf setting is where the Lower Silurian Longmaxi shale and Lower Cambrian Niutitang shale were deposited. These shales present an excellent material basis for reservoir development and hydrocarbon enrichment, because they share similar sedimentary and reservoir characteristics, such as high TOC content, high thermal maturity, high brittle mineral content, and similar types of kerogen [13–16]. However, there is a giant distinction between the gas production of the two shale sets in western Hunan. Several researchers recommend that the tectonic evolution method ought to play a major role in gas reservoir formation [17,18], controlling the organic matter generation and pore evolution of shales. Good tectonic preservation conditions, such as early shallow burial and late rapid settlement, give the mandatory caps and traps for the buildup of shale gas [5].

Previous studies have stated that the Longmaxi and Niutitang shales in western Hunan present similar depositional settings and mineral components while being different in thermal maturities [19–22]. The thermal maturity of the Longmaxi shale is between 2.2% and 3.0%, while that of the Niutitang shale is over 3.0%. In particular, the OM pore structure, which has been shown to be the most efficient storage for hydrocarbon gas in shale reservoirs, will be destroyed by excessively high thermal maturity (>3.5%) (Table 1), according to previous studies [23,24]. Additionally, the thermal maturity of the reservoir can reveal the degree of tectonic evolution. For example, in tectonic regions with comparable heat flow values, an excessively high thermal maturity reveals a deeper paleo-burial depth and relatively intense thermal evolution [6,10].

Table 1. Reservoir parameters and gas content of the Lower Silurian Longmaxi and Lower Cambrian Niutitang shale samples in Northwest Hunan, South China.

Formation	Depth (m)	Average TOC (%)	Average R_o (%)	Average Gas Content (m^3/t)	Straight well Gas Production (10^4 m^3/d)	Porosity (%)
Longmaxi	1024–1083	2.5	2.68	1.5	2.0	1.50
Niutitang	2600–2749	4.83	3.28	0.86	0.3	1.66

Previous studies have stated that Longmaxi and Niutitang shales in western Hunan present similar depositional settings and mineral components but differing tectonic evolution processes, according to a comparison of their tectonic properties [19,20]. It is therefore hypothesized that the two sets of shales have distinct thermal maturities as a result of their various tectonic evolution processes [25,26], which could have an impact on the pore structure and gas storage capability of the shales. Likewise, in northwest Hunan, the Longmaxi shale produces more gas than the Niutitang shale. The Longmaxi Formation in northwest Hunan still produces substantially less gas than the Longmaxi Formation in the Changning–Weiyuan region of the Sichuan Basin, south China (Figure 1). Therefore, this investigation contrasts the Changning–Weiyuan region of northwest Hunan with the Niuchantang Formation and the Longmaxi Formation shale (Figure 2). To assess and evaluate the exploration and development prospects of the shale formations in northwest Hunan, the Longmaxi Formation and Niutitang Formation are explored.

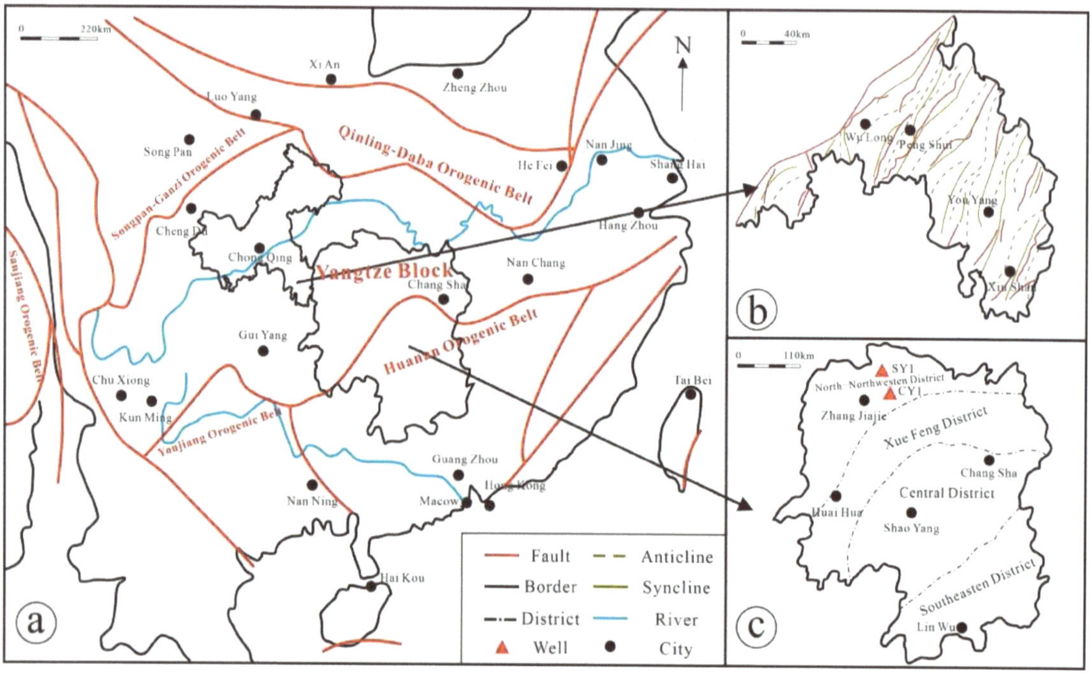

Figure 1. The Longmaxi and Niutitang shales' well locations and study area, in Northwest Hunan, South China. (**a**) Geological map of the Yangtze Block in South China; (**b**) geological map of southeast Chongqing Province, where samples of marine shale were selected; and (**c**) geological map of Hunan Province, where samples of transitional shale were selected. (Well locations are marked by red points).

In this paper, core samples of the Longmaxi and Niutitang shales in western Hunan were examined for a thorough investigation of the reservoir features, utilizing spectromicroscopy analysis, whole-rock mineral assemblage analysis, and a carbon–sulfur analytical tester. In addition, the organic matter pore structure was seen and identified using FIB-SEM. The evolution of the pore structure between the two sets of shales was compared with the regional tectonic evolution characteristics, which can shed light on how the tectonic evolution has an impact on the pore structure and storage capacity of the shales, ultimately contributing to the significant difference in gas production between the two sets of shales.

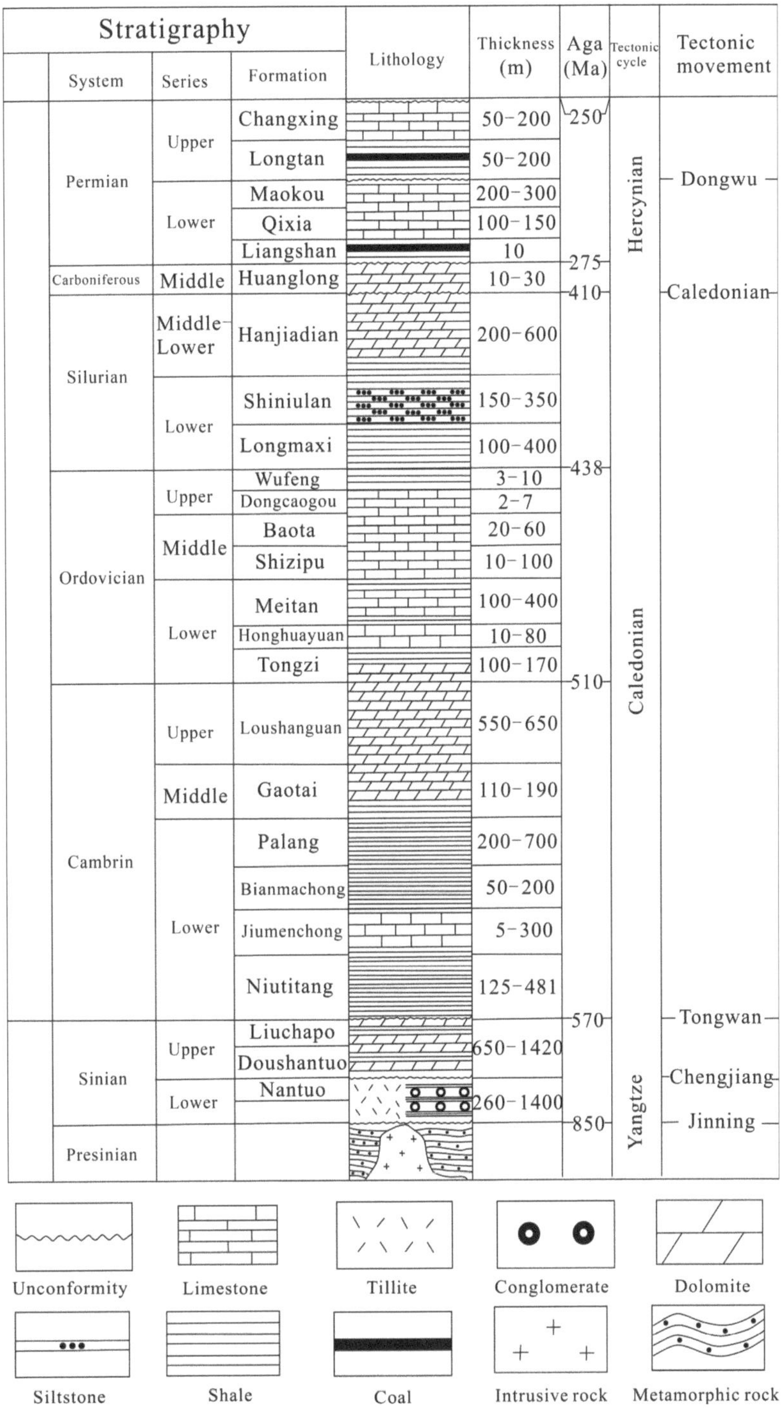

Figure 2. Detailed stratigraphic log graph of the Paleozoic in Hunan, southern China (showing Longmaxi and Niutitang formations).

2. Geological Setting

The upper Yangtze Plate includes the northwest region of Hunan and the Sichuan basin (Figure 3) [19,20]. From the Sinian Period to the Devonian, this plate primarily exhibited cratonic marine basin behavior. Subsequently, the Yangtze landmass underwent rapid transgression and created a broad shelf facies depositional environment [23]. At the end of the Ordovician, there was a local depositional gap caused by the Yichang Movement. Finally, another crust sank with extensive transgression and deposited a series of marine strata in the Early Silurian [27,28]. Geotechnically, the northwest region of Hunan lies between the Sichuan Basin and the Xuefeng Mountain uplift; the latter is located at the southern edge of the upper Yangtze Plate and is a part of the folded belt that resembles a trough in western Hunan and Hubei [29,30]. Since the Triassic, the basin has undergone tectonic inversion, giving rise to a number of thrust-nappe formations and foreland-like basins. The Yanshan Movement, which transformed the area after the Jurassic and Cretaceous, can be divided into three major stages: the stable subsidence and intermittent uplift stage before the early Yanshan Movement, the thrust-nappe stage during the early Yanshan Movement, and the extensional deformation stage after the early Yanshan Movement [29,30].

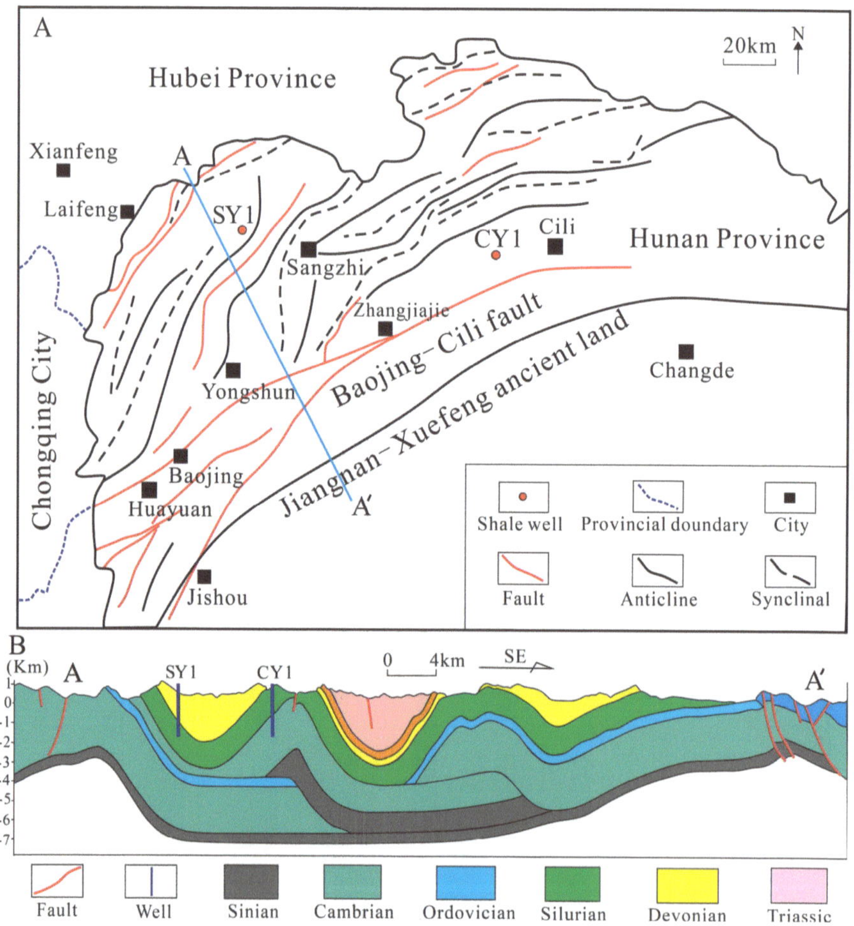

Figure 3. (**A**) Tectonic plan of northwest Hunan, southern China; (**B**) profile map of northwest Hunan, southern China from A to A′ wiih the blue solid line in (**A**).

In building the fundamental tectonic framework in northwest Hunan, the early stage of the Yanshan Movement was the dominant time period among them [31]. The research area has changed into the current NNE or NE-trending fold and fault system after the Jurassic and Himalayan migrations, and is primarily governed by the Cili–Baojing Fault [19,27]. From north to south, the research region is divided into two anticlines: the Yidu–Hefeng anticlinorium and the Sangzhi–Shimen synclinorium. The early-middle Triassic strata make up the majority of the synclinal core whereas the Sinian and Ordovician strata are predominantly exposed in the anticlinal core of the Sangzhi–Shimen synclinorium, which is generally stable and has a NE-NNE tendency [27,29]. The Jiangnan–Xuefeng uplift's western part includes the study region, where the major structure is an NNE-trending fold with significant changes in dip angle on both flanks and substantial folding deformation. The syncline largely preserves the advantageous shale parts [32,33].

2.1. The Longmaxi Shale

The paleogeographic pattern of the middle Yangtze Plate experienced significant changes in the Late Ordovician–Early Silurian: the southeastern margin of the upper Yangtze Plate converged with the Cathaysia Plate in the Late Ordovician, causing fold orogeny to form underwater uplift, which controlled the Longmaxi Formation in the northwestern Hunan are deposited during a transition from deep-water shelf to shallow-water shelf environment and dominated by deep-water shelf facies with organic-rich shale section concentrated in the lower deep-water shelf facies [34–36].

At the beginning of the Early Silurian Longmaxi, the area inherited the relatively closed deep-water stagnant basin environment from the Wufeng period in the Late Ordovician, forming a set of graptolite shale-dominated deposits, which are conformable or disconformable contact with the underlying Ordovician [18,24,37]. After the middle Longmaxi period, due to the rapid dumping of large amounts of fine terrigenous debris into the basin, silty sediments increase, with gray-green to gray shale, silty shale, and sandy shale dominating, and biological fossils are common with graptolite, indicating a shallow marine shelf environment with terrestrial provenance [38,39].

The thickness gradually decreases from the Sangzhi–Shimen area to the north and south, respectively. The changes in lithology and thickness indicate that the sedimentary center is located in the Sangzhi–Shimen area [40,41]. The lithology of the Lower Silurian Longmaxi Formation is mainly composed of black and gray-black mud shale, with common pyrite nodules and carbonate mineral strips, rich in graptolite, mostly deposited in the deep-water shelf and widely distributed, and is the main hydrocarbon generating strata in the upper Yangtze Plate. Drilling data reveals that the shale with gas indication is 50~90 m thick, and the organic-rich section is mainly concentrated in the lower deep-water shelf facies [40,41].

2.2. The Niutitang Shale

The black shales of the Lower Cambrian Niutitang Formation in the northwestern Hunan area are widely distributed with large thicknesses, and the thickness variation is controlled by sedimentary facies [12,32]. In the Early Cambrian, South China Pangea was disintegrated, and the regional tectonic subsidence was intensified. The Niutitang Formation is characterized by deep-water, low-energy, anoxic, and reducing marine sedimentation, depositing deep-water shelf facies gradually transitioning to bathyal sea-deep sea facies [20,30,42].

The rock type can be divided into two parts, The lithology of the lower part is black, grayish black, dark gray shale, siliceous shale interbedded with thin carbonaceous shale, gray limestone, stone coal, and rich in pyrite, indicating deep-water shelf sedimentation in the early stage. Upward the color gradually becomes lighter, and the shale grain size has a tendency to increase [29]. The bottom lithology of the upper part is a thin layer of black siliceous rock, and the middle and upper part is dominated by silty shale, where there locally exists a thin layer of sandstone. The lithology of the Lower Cambrian

Niutitang Formation in northwestern Hunan is mainly composed of black carbonaceous shale interbedded with fine sandstone and silty shale [20,30,42]. The overall lithology variation of the Niutitang Formation in the study area is not significant, and the thickness of organic-rich shale revealed by drilling is 50 m [20,42].

3. Data and Methods

In northwest Hunan, samples of the Longmaxi shale (SY1 well) and Niutitang shale (CY1 well) are examined to better understand their reservoir, geochemical, and OM pore properties via the total organic carbon content (TOC) test, the vitrinite reflectance (R_o) test, X-ray diffraction, and focused ion-beam scanning electron microscopy (FIB-SEM).

3.1. Total Organic Carbon Content (TOC) Test

A LECO CS-200 carbon–sulfur analyzer was used to determine the TOC content of the shale samples. Prior to testing, samples were grounded into fragments with sizes of 80 mesh, and hydrochloric acid was used to remove carbonate composition. Afterward, samples were washed with distilled water and pyrolyzed at 540 °C to measure TOC [15,34,35].

3.2. Vitrinite Reflectance (R_o) Test

Vitrinite is typically absent in the reservoirs in the Lower Cambrian Niutitang Formation and the Lower Silurian Longmaxi Formation in southern China, though, as a result of the high degree of thermal evolution of these shales. As a result, it cannot be used to directly measure the maturity index. Based on a conversion equation between bitumen reflectance (R_b) and R_o: Eq-R_o = $0.618R_b + 0.4$ [43]. This study employed the Eq-Ro to assess the thermal maturity of the target shales. By using a Leica DM4500P polarizing microscope and a CRAIC microspectrophotometer, the vitrinite reflectance was determined [15].

3.3. X-ray Diffraction

Mineral compositions were measured using a Bruker D8 DISCOVER diffractometer equipped with a Cu X-ray tube and a Sol-X solid-state detector. The laboratory's environment had a 24 °C temperature and a 35% humidity level. Five grams of each sample were powdered, screened, and chosen with a size of approximately 300 mesh. For the experiment, slides that had been coated with a solution of the sample mixed with ethanol were necessary. The machinery was run at 45 kV. The 20 s/0.02° step scanning frequency was maintained. Finally, using the K-radiation principle, mineral components can be produced.

3.4. Focused Ion-Beam Scanning Electron Microscopy (FIB-SEM)

Shale samples must be ground, polished with argon ions, and carbon sprayed to improve conductivity before being ready for FIB-SEM examination. OM pores are observed using a FIB-SEM device (FEI HELIOS NanoLab 650; FEI business) (Table 2). High-resolution imaging, energy-dispersive X-ray spectroscopy, secondary electron morphology, and backscattered electron contrast are all functions of SEM imaging.

Table 2. Reservoir geological parameters of Longmaxi and Niutitang shale samples used for FIB-SEM.

Sample	Formation	Depth (m)	TOC (%)	R_o (%)	Solid Kerogen	Mineral (%)				
						Quartz	Feldspar	Clay	Carbonate	Pyrite
#1	Longmaxi	1569.2	2.08	2.56	I	54.3	7.4	31.2	2.7	3.3
#2	Niutitang	2712.4	4.12	3.42	I	52.1	5.6	27.3	8.7	4.1

The matrix minerals, such as quartz and calcite, are light gray under backscattering conditions, while pores are black. OM is dark gray. The color contrast makes it simple to recognize OM pores. The secondary electron morphology was used to better investigate the microscopic properties of porous media. The surface morphology of the OM pores can be highlighted by the secondary electron morphology, which can produce bright light around pores and make the pores easier to see even at low resolutions.

3.5. N_2 Adsorption and Mercury Intrusion Porosimetry (MIP)

A Micromeritics Tristar II 3020 surface area analyzer was used to assess the pore size and test the mesopore (2 nm–300 nm) dispersion properties of shale under low pressure (0.127 Mpa). Prior to measurement, samples were crushed into 60–80 mesh particles and vacuum-degassed for 24 h at 383.15 K to remove adsorbed moisture and volatile materials. The amount of adsorption and desorption can be measured while changing relative pressure by placing the particles in liquid nitrogen at a constant temperature of 77 K. The Barrett–Joyner–Halenda method was used to compute the volumes of the pores.

The macropore (300–1000 nm) information was gathered using MIP analysis. After the pressure rises, mercury molecules progressively enter minuscule pores. The quantity of mercury molecular infiltration was used to determine the pore volume and pore size distribution, and the Young-Dupre equation was used to get the surface area. An automated mercury porosimeter (Auto Pore IV 9500 Porosimeter) was used to measure the sample. Prior to testing, samples with a diameter of 2.5 cm were dried at 110 °C for 24 h to get rid of all adsorbed and free water. In order to prevent microfractures, extra caution was also used when collecting from the core.

3.6. Organic Matter Isolation

Inorganic minerals and OM that were soluble in chloroform were eliminated using chemical and physical techniques from samples. In order to further remove silicate minerals, samples were first crushed and ground to a fine particle size (0.18 mm). To eliminate pyrite from leftovers, hydrochloric acid and zinc particles free of arsenic were applied for 12 h. The leftovers were then rinsed once more with distilled water. In order to effectively separate impurities, samples were put in a centrifuge tube and a heavy liquid with a relative density of 2.0 g/mL to 2.1 g/mL was added. For 20 min, particles were centrifuged at a speed between 2000 and 3000 rpm. After the ice melted, the separated OM was freeze-dried (at 5 °C) for 6 h, and after that, it was placed in a drying oven (at 60 °C). Finally, the dried samples were Soxhlet extracted with chloroform to remove soluble OM and dried in a drying oven (<60 °C).

4. Results

4.1. Geological Characteristics of Shales Reservoir

4.1.1. TOC Contents of the Two Shales

According to test results, shale samples from the Lower Silurian Longmaxi Formation and the Lower Cambrian Niutitang Formation in northwest Hunan had an average TOC concentration of more than 2.0%, which is over the assessment standard for organic-rich shale. The Longmaxi shale samples have an average TOC content of 2.7% (Figure 4), a maximum TOC content of 4.25%, and a lowest TOC content of only 1.58%. Shale samples from the Niutitang Formation had an average TOC level of 2.3% (Figure 4), with the highest TOC contents of 6.87% and minimum TOC contents of only 0.45%.

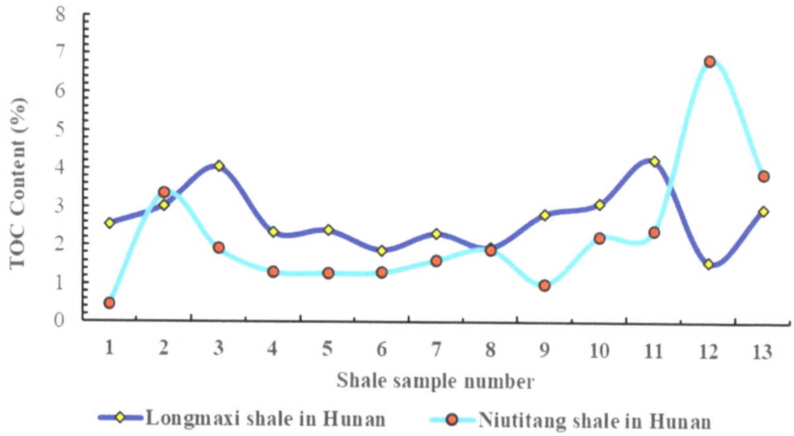

Figure 4. TOC content distribution of Longmaxi and Niutitang shale samples in northwestern Hunan in this study.

4.1.2. Thermal Maturities of the Two Shales

The test findings demonstrate a significant variation between the thermal maturity of shales from the Lower Cambrian Niutitang Formation and the Lower Silurian Longmaxi Formation in northwest Hunan. The distribution of the Longmaxi shale samples is rather even, with an average Ro of just 2.5%, a maximum Ro of 2.84%, and a minimum Ro of only 2.14% (Figure 5). Shale samples from the Niutitang Formation, however, have an average Ro of 4.0%, a maximum Ro of 4.23%, and a minimum Ro of 3.77% (Figure 5). In comparison to Longmaxi Formation shale, Niutitang Formation shale's organic matter evolution has reached the metamorphic stage.

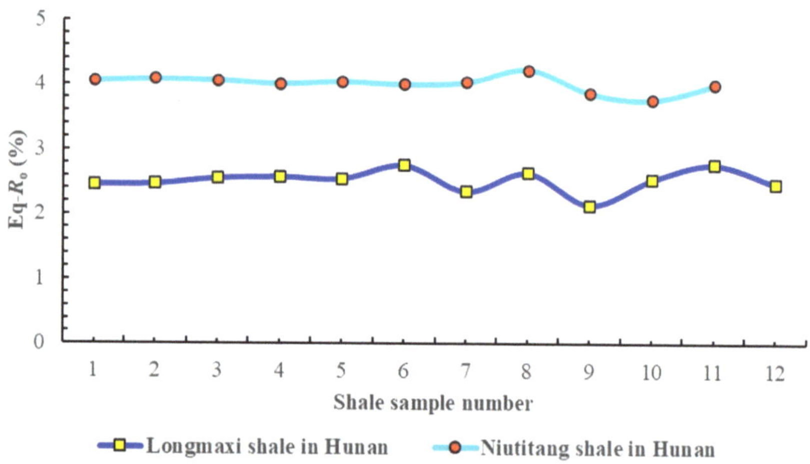

Figure 5. The thermal maturity distribution of shale samples from Longmaxi and Niutitang in northwest Hunan.

4.1.3. Mineral Components of the Two Shales

In northwest Hunan, there is little distinction between the mineral makeup of the Lower Silurian Longmaxi Formation and the Lower Cambrian Niutitang Formation shale. With an average proportion of close to 50%, quartz, feldspar, and other brittle minerals predominate in both shales (Figure 6). But the high proportion of brittle minerals makes it easy to fracture shale using hydraulic pressure. A similar sedimentary environment can be seen in the clay's low brittle mineral content and the limited number of carbonate minerals, which are present in other minerals.

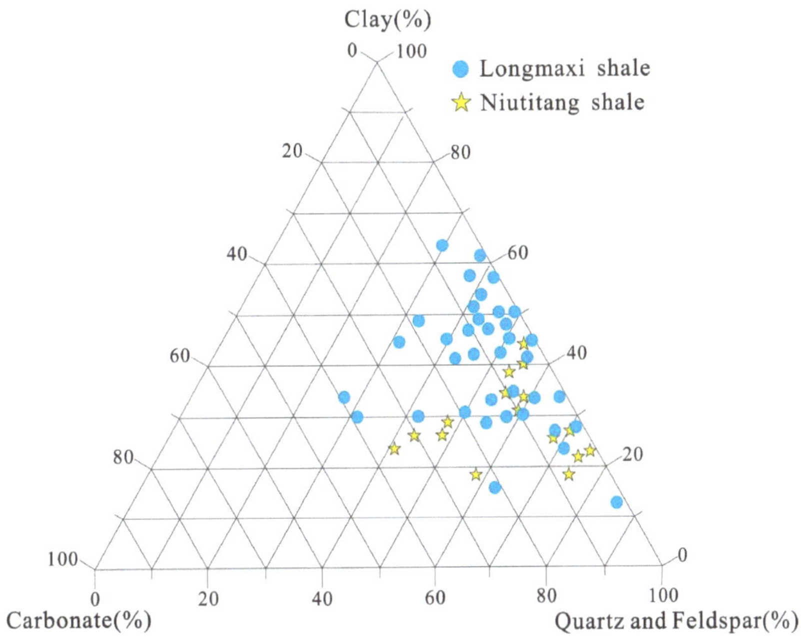

Figure 6. Triangular plot of mineral composition of Longmaxi shale and Niutitang shale in northwestern Hunan in this study.

4.2. OM Pore Properties of the Longmaxi Shale

Between pyrite grains, pyrobitumen forms in shale. The liquid hydrocarbon was produced during the early thermal evolution of the solid kerogen and injected between the pyrite grains, giving rise to the pyrobitumen in its free state, which was then transformed into solid bitumen during the thermal evolution process. The pyrobitumen between the pyrite grains in the Longmaxi shale reservoir contains a significant number of OM pores (Figure 7A). When combined with the secondary electronic function and zoomed into in Figure 7A, the OM pores are revealed to have a regular morphology (usually round and oval) (Figure 7B,E,F). The secondary electron technique is used to observe and characterize the OM pores in dispersed pyrobitumen, and it is discovered that the pyrobitumen itself has formed a significant number of OM pores (Figures 7C,D and 8A–F).

4.3. OM Pore Properties of the Niutitang Shale

The same observation field and secondary electron technologies are used to view and distinguish the OM pores of two different forms of pyrobitumen in the Niutitang shale. According to the findings (Figures 9A,B and 10E,F), there are only a few sporadic OM pores in the pyrobitumen in the Niutitang shale compared to the Longmaxi shale at the same magnification and per unit observation field of view. Additionally, the OM pore diameter in the free pyrobitumen of the Niutitang shale is substantially smaller than that of the Longmaxi shale, measuring only a few tens of nanometers. Additionally, the Niutitang shale's free pyrobitumen contains OM holes that are uneven in shape, primarily elongated and flattened, with fine middles and thick ends (Figures 9C,D and 10C,D). These pores also have shallow pore depths and uneven distribution. The weak pore structure of OM often makes gas occurrence difficult.

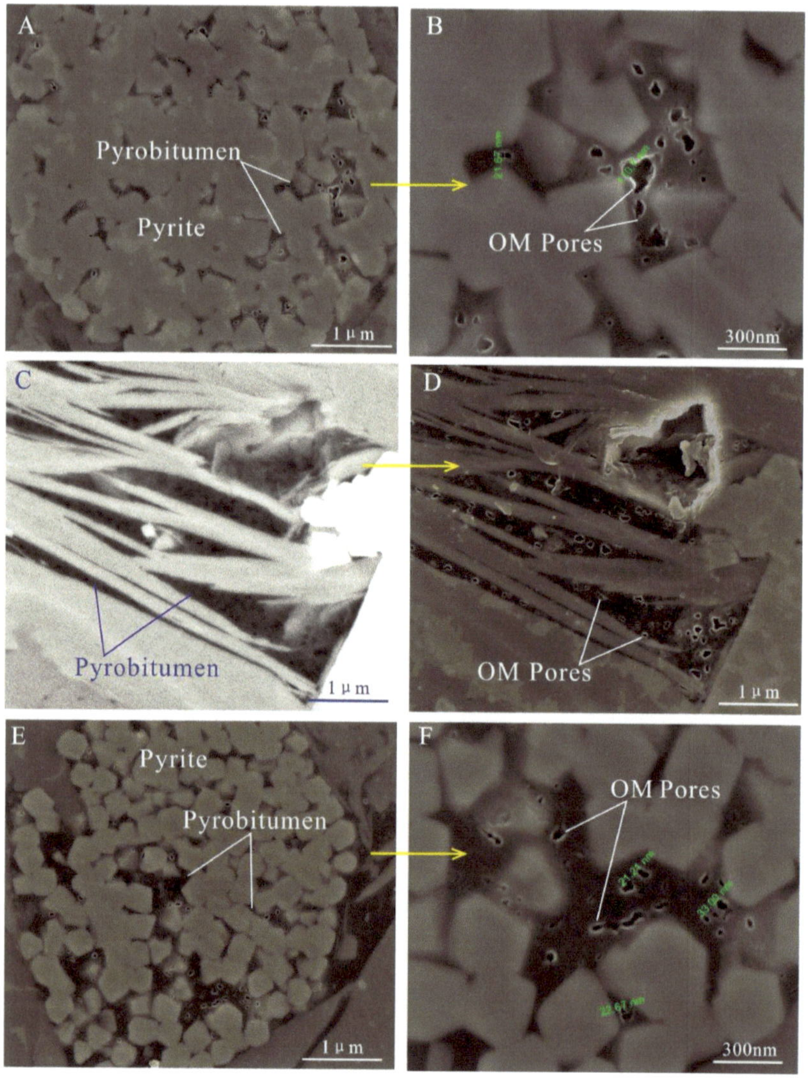

Figure 7. Characteristics of pyrobitumen and organic matter pore development in the Longmaxi Shale sample in northwestern Hunan under FIB-SM microscope: (**A**,**B**) distribution characteristics of pyrobitumen among pyrite grains; (**C**,**D**) large number of organic matter pores in pyrobitumen; (**E**,**F**) the banded pyrobitumen in Longmaxi Shale samples in northwestern Hunan has a large number of pores, with large pore size and uniform morphological distribution, mostly round and oval shape, which has excellent reservoir structure for effective storage of large amount of gas.

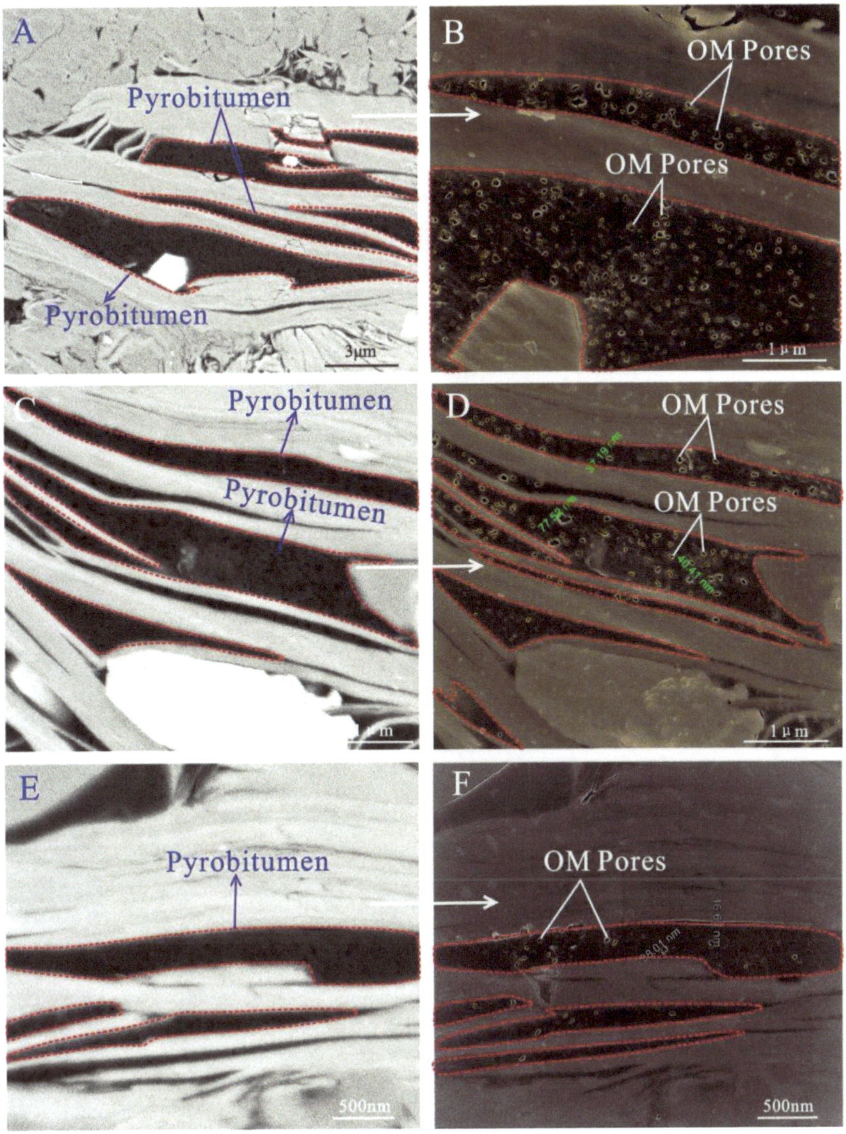

Figure 8. FIB-SM microscopy of dispersed pyrobitumen and its internal organic matter pore development in the Longmaxi Shale sample in northwestern Hunan: (**A**,**C**,**E**) distribution characteristics of dispersed pyrobitumen in the mineral matrix; (**B**,**D**,**F**) large amounts of organic pores developed in the dispersed bitumen are observed when zooming in.

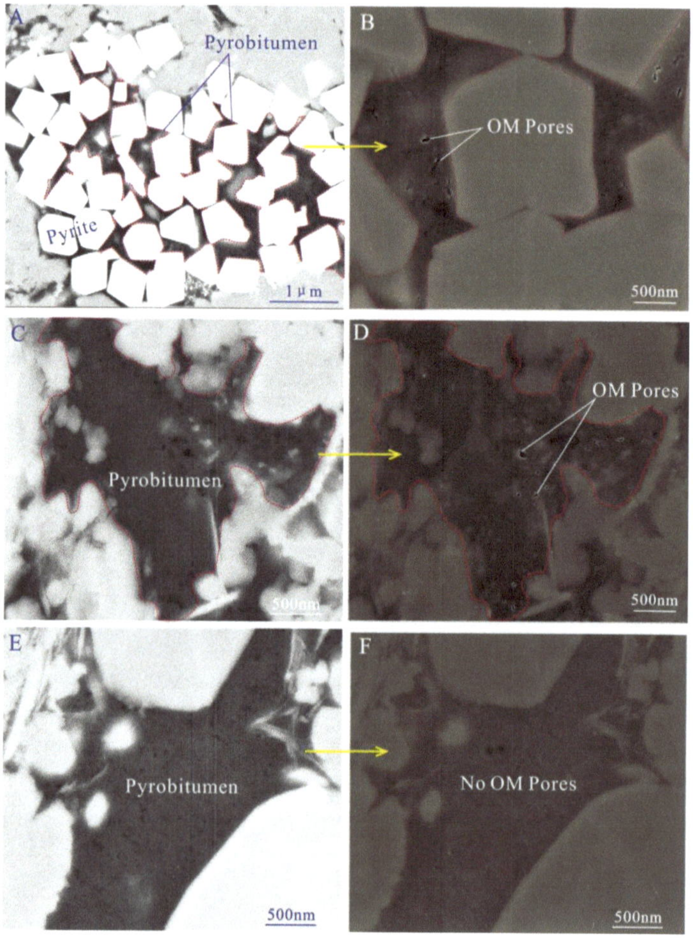

Figure 9. FIB-SM microscopic characteristics of pyrobitumen and organic matter pore development in the shale sample from Niutitang, northwestern Hunan: (**A**,**B**) distribution characteristics of pyrobitumen among pyrite grains; (**C**,**D**) very few organic matter pores developed in pyrobitumen; (**E**,**F**) the banded pyrobitumen in the Niutitang Shale samples in northwestern Hunan has few pores, small pore size, irregular shape, mostly flat and uneven distribution, and poor pore structure of organic matter, which is not conducive to the effective storage of a large amount of gas.

The results show that the pyrobitumen has very few OM pores, and the number of OM pores per unit field of view is significantly lower than that of the Longmaxi shale. The same observation technique (the same magnification, observation field of view, and secondary electron technique) is applied to the dispersed pyrobitumen in the Niutitang shale. Some pyrobitumen exhibits no light even when secondary electron technology is used, suggesting the absence of OM pores within the pyrobitumen in the associated observing field of view (Figure 9E,F). Additionally, the OM pores in the Longmaxi shale have substantially smaller sizes than those in the OM pores. The majority of the pores had an irregular morphology, an elongated, flattened form, and a shallow pore depth (Figure 10A–D). Poor pore structure makes it difficult for gas to occur.

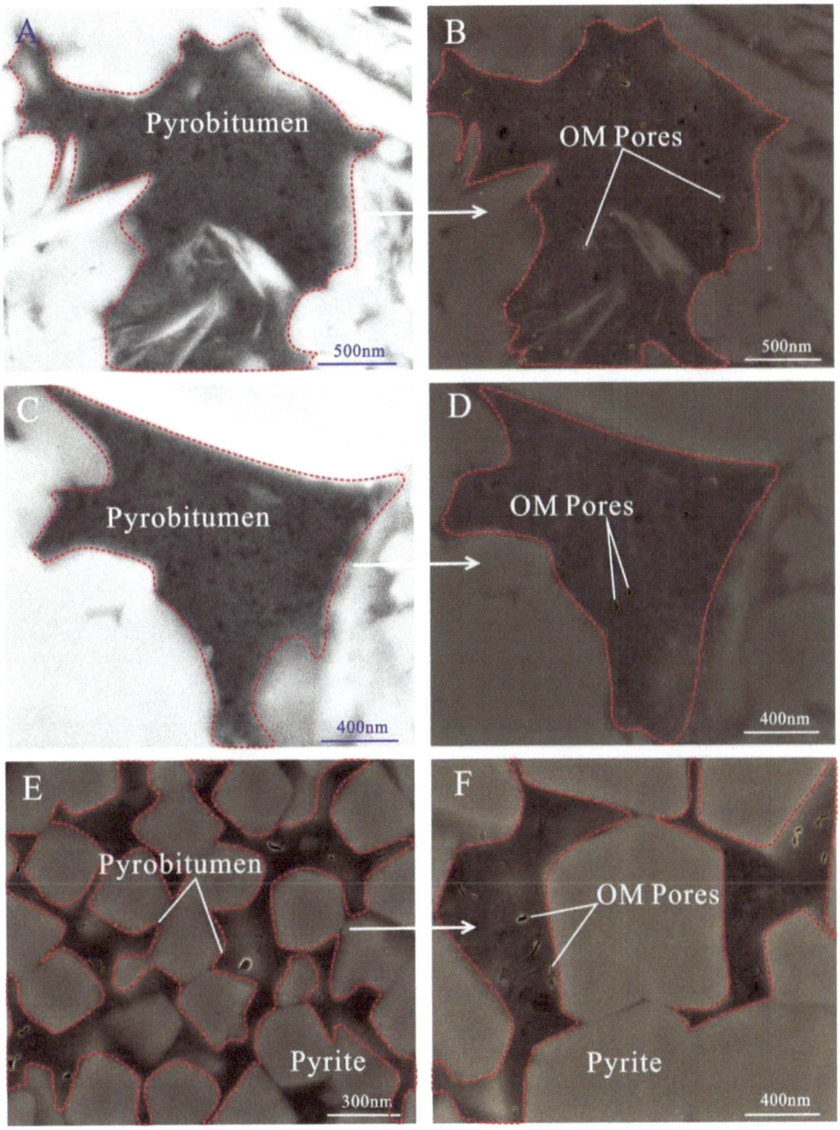

Figure 10. FIB-SM microscopy of dispersed pyrobitumen and its internal organic matter pore development in the Niutitang Shale sample: (**A**,**C**,**E**) distribution characteristics of dispersed pyrobitumen in the mineral matrix; (**B**,**D**,**F**) with magnification, a very small amount of organic matter pores are developed inside the dispersed pyrobitumen, which also has structural characteristics as poor pore size, irregular morphology, mostly flat and uneven distribution.

5. Discussion

5.1. Comparison of the Longmaxi and Niutitang Shales' OM Pore Structures

Solid kerogen and pyrobitumen are the two main types of organic matter (OM) found in shale. The initial organic material (OM) deposited during the strata formation is known as solid kerogen, and it is primarily composed of the dead biological skeletons of microbes and algae [44–46]. A type of solid bitumen is pyrobitumen. The mineral matrix pores (mostly the InterP pores and IntraP pores) created by diagenesis during the initial deposition process are injected with liquid hydrocarbon during the early stage of thermal development [47]. Pyrobitumen is a liquid hydrocarbon that turns into solid bitumen at a later stage of thermal evolution. According to Loucks et al. (2014) [48], pyrobitumen is where OM pores in shale are primarily produced (Figure 11).

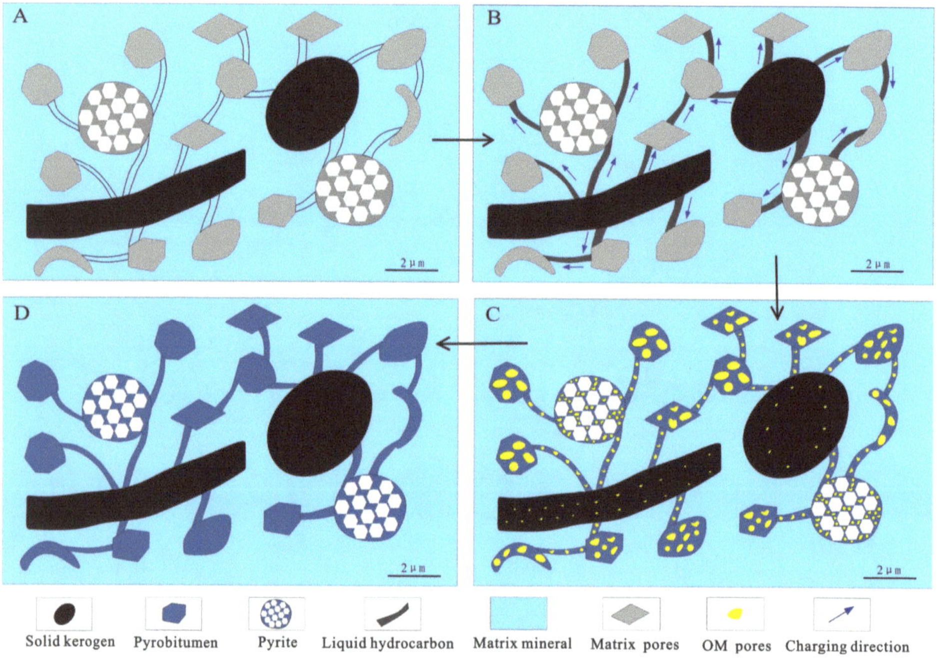

Figure 11. OM pores evolution of marine shales with high evolutionary maturity in southern China: (**A**) primitive diagenetic stage; (**B**) liquid hydrocarbon charging period; (**C**) pore-forming period; (**D**) pore extinction period.

Based on their distinct locations in the reservoir mineral matrix during diagenetic evolution, the pyrobitumen is divided into two primary categories in this study: the free state, which is found between pyrite grains, and the scattered state, which is found in the intergranular pores of feldspar, quartz, and calcite. Under the microscope, dispersed pyrobitumen has enormous and banded properties and covers a much bigger area than free pyrobitumen. As a result, the scattered pyrobitumen—the subject of the microscope analysis—develops more OM pores.

Both the Longmaxi shale and the Niutitang shale include organic matter (OM) pores of various pyrobitumen kinds, and notable distinctions in the OM pore structure can be detected between the two sets of shales (Figure 12A). Overall, the Longmaxi shale in northwest Hunan has a substantially superior OM pore structure than the Niutitang shale. Thus, it can be concluded that the vast difference in gas production between the Longmaxi and Niutitang shales in northwest Hunan is probably caused by their various OM pore shapes (Figure 12B).

(1) According to electron microscopy, the number of OM pores in the Longmaxi shale is significantly higher than that in the Niutitang shale. The Niutitang shale also contains scattered pyrobitumen that lacks OM pores.
(2) The shallow OM pores of the Niutitang shale make it unsuitable for adsorption and preservation because gas can quickly escape.
(3) The OM pore shape of the Longmaxi shale is very regular, being mostly round and oval, while the OM pore shape of the Niutitang shale is irregular, being flat with a

thin middle and thick and elongated ends, which is not favorable for gas adsorption and preservation.

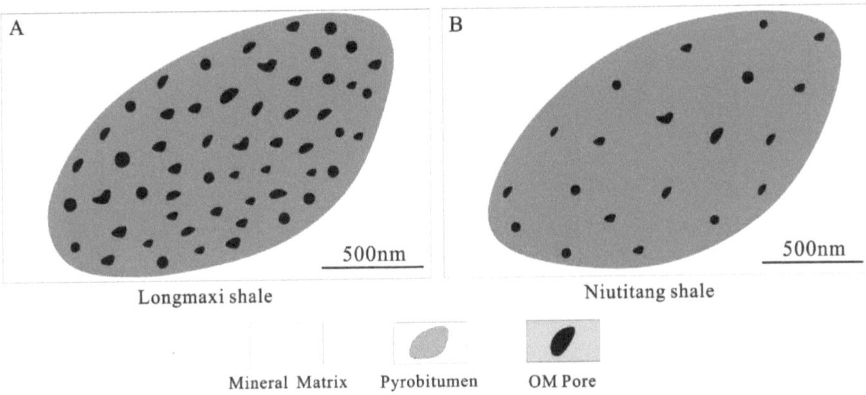

Figure 12. Comparison of organic matter pore structure patterns of the Lower Silurian Longmaxi and Niutitang shales in northwestern Hunan: (**A**) organic matter pore structure pattern of Longmaxi Shale: organic matter pores in the pyrobitumen are numerous, large in pore size, and uniform in morphology, mostly round and elliptical in SEM unit observation field of view; (**B**) organic matter pore structure pattern of Niutitang Shale: organic matter pores in the pyrobitumen are few in number, small in pore size, irregular in morphology, mostly flattened, and unevenly distributed.

5.2. Tectonic Evolution on OM Pore Structure

The strata uplift and subsidence controlled by regional tectonic evolution determine the hydrocarbon-generation evolution process of shale, which will have the following two effects on the evolution of the OM pore structure of shale [49–51].

(1) If the strata subside quickly and are deep buried, the hydrocarbon-generation evolution of shale will intensify and the OM pore generation will increase accordingly. Oppositely, if the strata undergo substantial uplift during the regional tectonic evolution, the hydrocarbon-generation evolution of shale will weaken and the OM pore generation will decrease accordingly.
(2) If the hydrocarbon-generation evolution of shale strata lasts for a long time, the organic matter can be fully evolved, which will generate a sufficient number of OM pores with large diameter and such OM pore structure is conducive to the gas occurrence. Oppositely, if the hydrocarbon-generation evolution of shale strata lasts for a short time, the organic matter cannot be fully evolved, which will result in poor OM structure, a forming limited number of OM pores with a small diameter.

5.2.1. Hydrocarbon-Generation Evolution Process

In northwest Hunan, the stratum burial histories of Well SY1 (drilled into the Lower Silurian Longmaxi shale) and Well CY1 (drilled into the Lower Cambrian Niutitang shale) are compared (Figures 13 and 14). The two wells are close together and are located in the same tectonic region, and consequently, the Longmaxi shale and the Niutitang shale share a similar strata burying history and hydrocarbon-generation evolution process. Both of the shales experienced a similar tendency of subsidence since the strata formation, with extensive subsidence from the end of the Carboniferous (implying the same beginning time of massive hydrocarbon generation). Moreover, the mass uplift of the strata began in the mid-Jurassic (indicating the same ending time of hydrocarbon generation). It indicates that the Longmaxi shale and Niutitang shale in northwest Hunan have undergone a similar length of time for the evolution of hydrocarbons [9,19,20,27]. In other words, there are significant changes in the OM pore structure between the two groups of shales that are not related to the duration of the hydrocarbon-generation evolution.

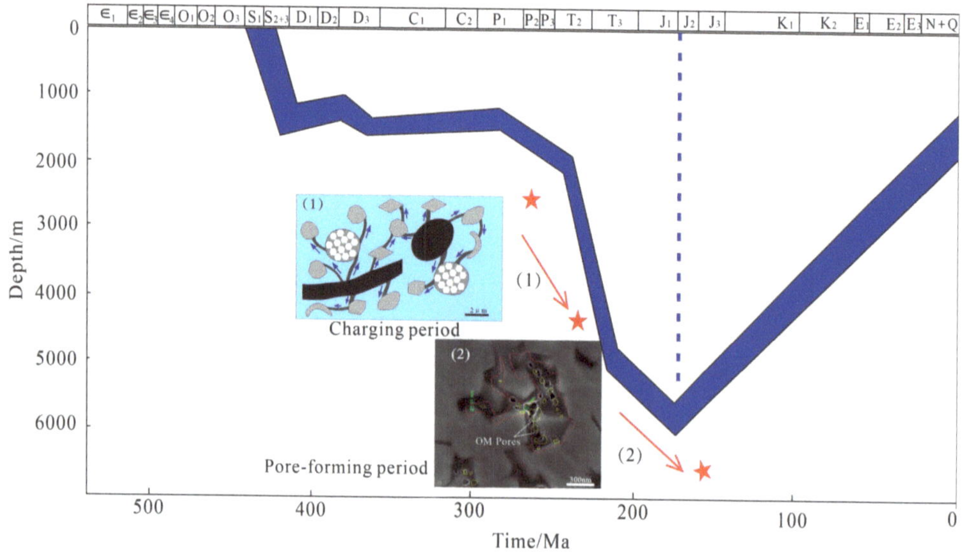

Figure 13. OM pores evolution of the Silurian Longmaxi Formation shale in Well SY1, Northwest Hunan.

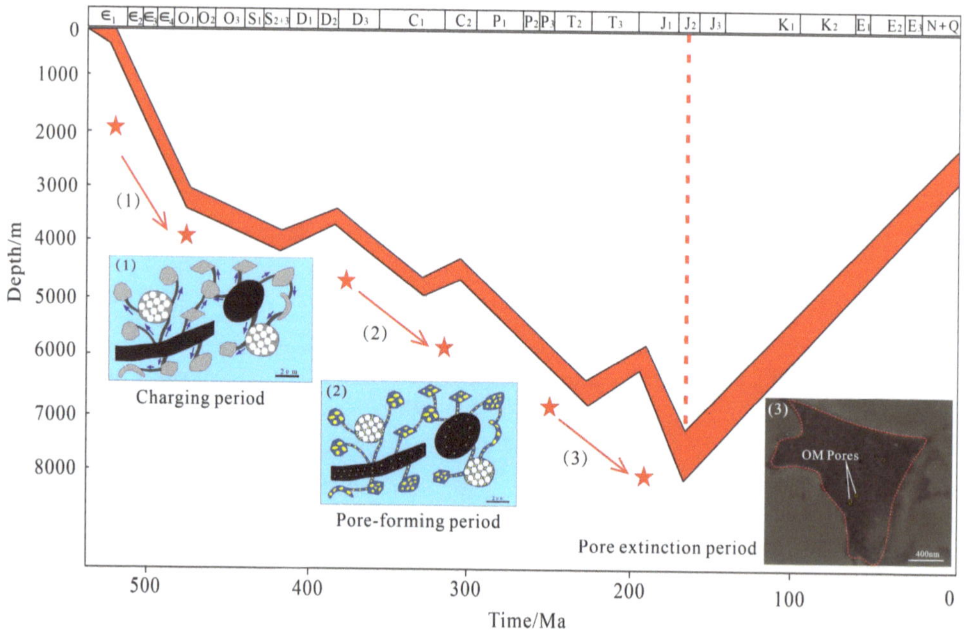

Figure 14. OM pores evolution of the Cambrian Niutitang Formation shale in Well CY1, Northwest Hunan.

5.2.2. Hydrocarbon-Generation Evolution Degree

The OM pore structure of shale can be destroyed by excessive thermal evolution [34,35,44,45]. Due to their shared tectonic affiliation with the northwest Hunan region, the Lower Silurian Longmaxi shale and the Lower Cambrian Niutitang shale both had similar heat flow histories at a particular regional scale [9,19,20,27]. As a result, the primary element in determining the thermal maturity of the two sets of shale reservoirs is the maximum paleo-burial depth of the strata. The maximum paleo-burial depth of the Longmaxi Formation and the Niutitang Formation are very different, which causes the degree of thermal evolution of shales to vary [34,35,51,52].

The greatest paleo-burial depth of the Longmaxi Formation is barely 6000 m (Figure 13), and thermal maturity reaches in the range of 2.3–3.0%. The OM pore structure is not damaged by excessive evolution, which is favorable for the gas occurrence [12,24]. The OM has undergone sufficient but not excessive thermal evolution, preventing OM from becoming graphitized and preserving a significant amount of OM pores. However, the Niutitang shale has experienced excessive thermal evolution and can be buried as deep as 8000 m (Figure 14). The thermal maturity even exceeds 3.5%, and the OM is prone to graphitization. As a result, the OM pore structure is disrupted and is no longer capable of efficiently storing huge amounts of gas.

5.3. Shale Gas Exploration Potential Analysis

In northwest Hunan, the Longmaxi Formation shale has more OM pores than the Niutitang Formation shale, according to the electron microscope. Clearly, the Niutitang Formation shale in northwest Hunan has less promise for exploration and development than the Longmaxi Formation shale. Nevertheless, how does the Longmaxi Formation shale in northwest Hunan differ from that in the Changning–Weiyuan region? A demonstration zone for shale gas development and exploration has been established in the Changning–Weiyuan region of southern China (Figure 15). The Lower Silurian Longmaxi deposit shale is the primary shale gas-producing deposit in the Changning–Weiyuan region. Numerous specialists and academics have established that the Longmaxi Formation shale in the Changning–Weiyuan area has acquired a significant number of OM pores [24,53]. These OM pores are crucial for the efficient storage of shale gas because they have a consistent form and a big pore diameter. For this reason, the Changning–Weiyuan region has the potential to serve as a demonstration area for China's shale gas production and exploration.

The exploration and development potential of the Lower Silurian Longmaxi and Niutitang Formation shales in northwest Hunan is illustrated, and the distinctions between them and the Longmaxi Formation shale in the Changning–Weiyuan area are underlined in order to emphasize the comparative research. The Longmaxi Formation shale samples from Well N215 in the Changning region were chosen for this study in order to determine the development features of OM pores using scanning electron microscopy. The reservoir formation and development features of the Longmaxi Formation shale in the Changning–Weiyuan area are examined in conjunction with the burial history of local layers (Figure 16). TOC of the Lower Silurian Longmaxi Formation shale in the Changning area of the Sichuan Basin is above 2.0%, and the equivalent vitrinite reflectance (Equal-R_o) is mainly between 2.5–3.0%, which is similar to the reservoir characteristics of Longmaxi Formation shale in northwestern Hunan [53,54].

Firstly, the Longmaxi shale in the Changning–Weiyuan area has a high density of OM pores, a large pore diameter, and regular form compared to the Lower Cambrian Niutitang shale generated in northwest Hunan (Figures 7 and 8). The burial depth of shale is modest, which prevents the excessive evolution of organic matter and preserves a significant number of OM pores in the shale, and the greatest ancient burial depth of Longmaxi shale is only 7000 m, as shown by the burial history map of this region (Figure 17). As a result, it is clear that the Lower Cambrian Niutitang Formation shale in Hunan's northwest is primarily the result of a significant ancient burial depth of more than 8000 m (Figure 14). Since the organic matter in shale has experienced excessive evolution, its physical and chemical characteristics are similar to those of graphite, which has caused OM pores to vanish. The Niutitang Formation had significant losses of shale gas as a result of the hydrocarbon gas' loss of useful storage space, which ultimately prevented the formation of a reservoir.

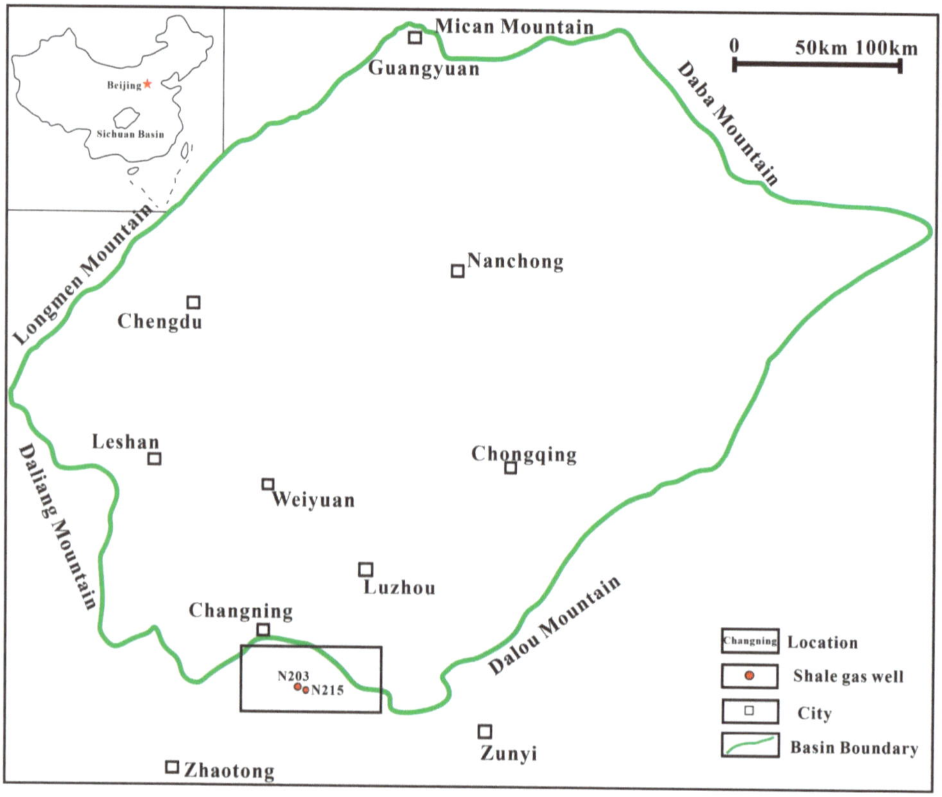

Figure 15. Location of Changning–Weiyuan area and shale drilling of Lower Silurian Longmaxi Formation in the periphery of Sichuan Basin, South China (Revised from [53,54]).

The Longmaxi Formation shale in the Changning–Weiyuan area has a regular pore structure and a larger pore diameter than the Longmaxi Formation shale in northwest Hunan, which is more beneficial for the preservation of a significant amount of hydrocarbon gases. However, under a microscope, the OM pores of the Longmaxi Formation shale in northwest Hunan are smaller and irregular (Figures 7 and 8), and the shale's reservoir capacity is visibly worse than that of the Longmaxi Formation shale in the Changning–Weiyuan area. In terms of exploration and development outcomes, the Longmaxi shale in northwest Hunan has a clearly lower hydrocarbon accumulation potential than the Changning–Weiyuan. Shale from the same formation, Longmaxi, is used. Why does the northwest of Hunan have a distinct exploration and development potential than the Changning–Weiyuan region?

According to related research, the length of time that shale spends producing hydrocarbons has an impact on both the quantity of gaseous hydrocarbons produced and the structural integrity of the OM pores. The amount of gaseous hydrocarbons, the number of OM pores, the pore diameter, and the regularity of the form increase with the length of the hydrocarbon-generation development. The integrated reservoir is enriched by a high number of gaseous hydrocarbons that have a lot of efficient storage space (OM pores) [55,56]. On the other hand, if the shale does not have enough time for the creation of hydrocarbons, not enough gaseous hydrocarbons, and not enough OM pores can be produced, preventing the large-scale accumulation of shale gas. Shale's process of generating hydrocarbons is governed by the formation's elevation and subsidence. Shale will enter the process of continuous hydrocarbon generation when the formation is continuously subsiding. However, the production of shale's hydrocarbons will slow down or stop when the formation is raised significantly or continually to a particular height.

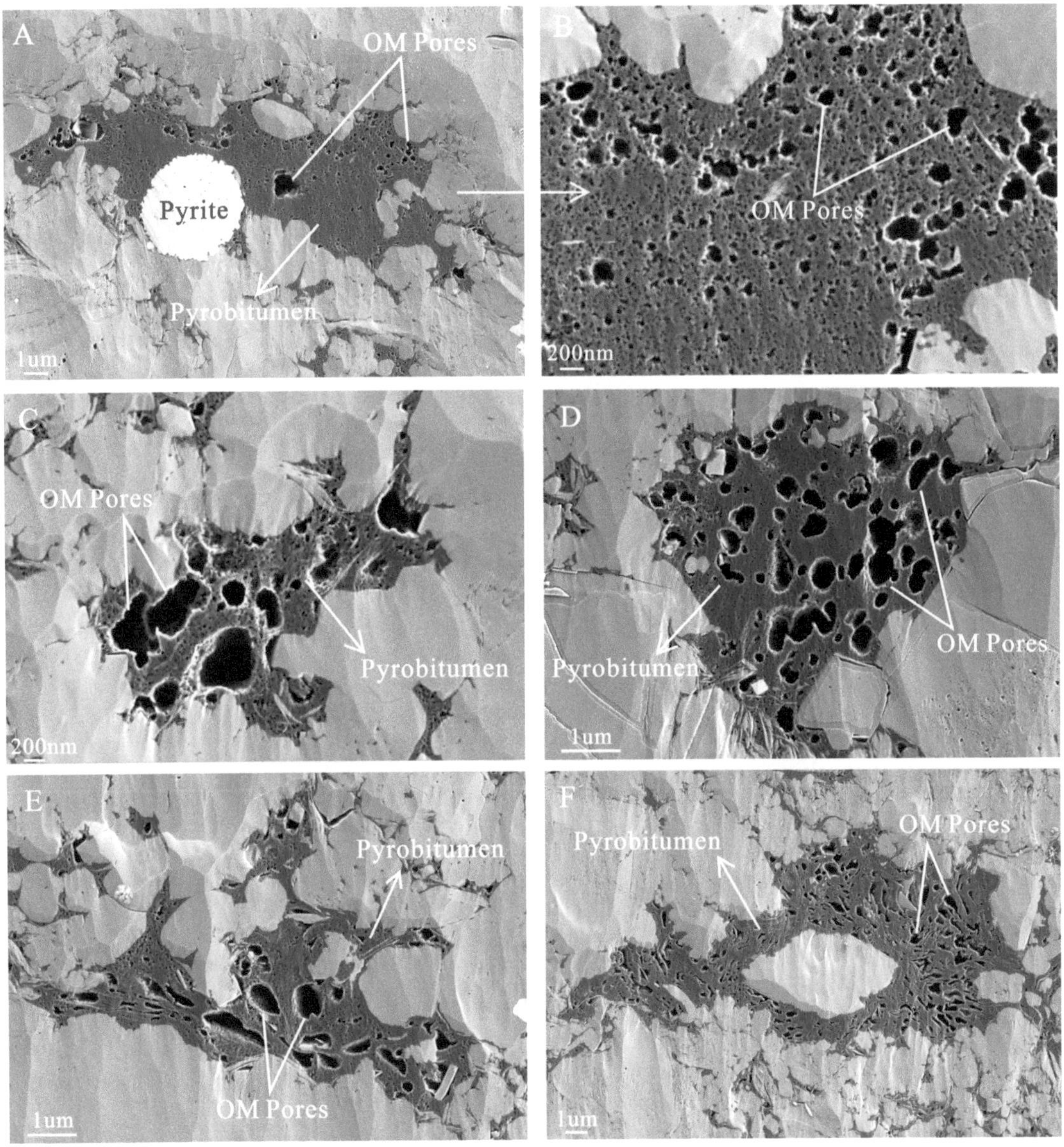

Figure 16. Organic matter pore structure characteristics of Lower Silurian Longmaxi shale in Weiyuan area, Changning: (**A–F**) The shale of Longmaxi Formation has a large number of organic pores under the microscope. Compared with the shale of Longmaxi Formation in northwest Hunan, the pore size of Longmaxi Formation is larger, the number is more in unit view, the shape is more regular, and it has a greater potential to store gas.

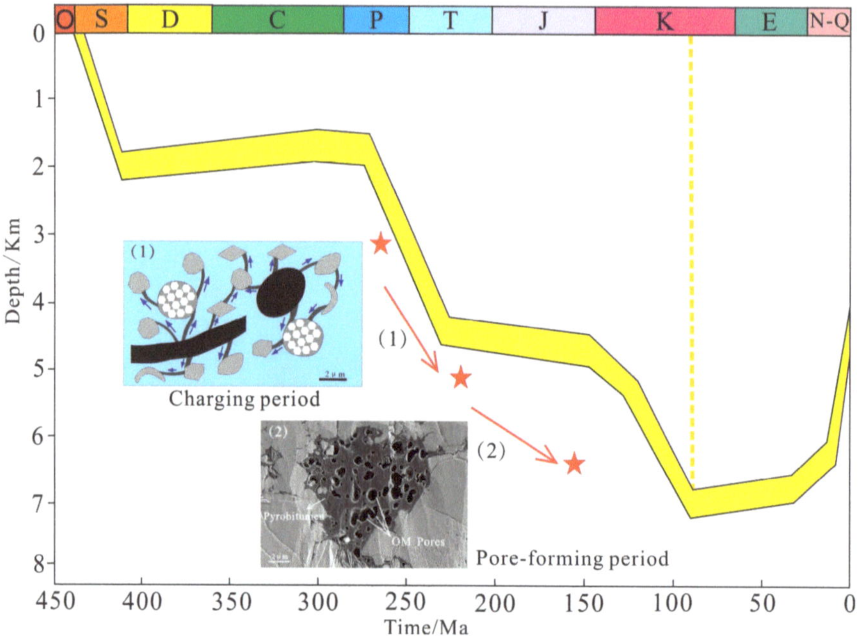

Figure 17. OM pores evolution of the Silurian Longmaxi Formation shale in Well N215, Changning area, Sichuan Basin.

From the stratigraphic burial history of the Changning–Weiyuan region and northwest Hunan. The Longmaxi Formation shale in northwest Hunan was exposed by the SY1 well, which revealed that the formation started to settle extensively in the early Triassic and reached its maximum subsidence depth in the late Early Jurassic. The Longmaxi Formation shale has now reached its maximum degree of thermal development, and its rate of gas generation has also reached its maximum. The formation was subsequently raised, and shale's ability to produce hydrocarbons was diminished until it eventually stopped at the start of the Paleogene. The Longmaxi Formation shale in northwest Hunan's hydrocarbon-generation evolution time only lasted 150 Ma from the beginning of subsidence to the end of the evolution.

The Changning–Weiyuan region's burial history indicates that the Longmaxi Formation shale started to settle in the middle of the Permian and reached its maximum burial depth in the middle of the Cretaceous. The intensity of shale gas hydrocarbon is at its peak at this time. The formation was then raised, which caused shale to produce less hydrocarbons until the end of the Paleogene. From the beginning of subsidence until the end of hydrocarbon generation, the Longmaxi Formation shale in the Changning–Weiyuan area saw an evolution time of almost 250 Ma.

In this study, using ImageJ digital analysis software, OM pores of Lower Silurian Longmaxi Formation shale in the Changning area, Sichuan Basin, Lower Silurian Longmaxi Formation shale, and Lower Cambrian Niuitang Formation shale were extracted digitally. The number of OM pores and the maximum pore diameter were statistically analyzed, and the results showed that: The Longmaxi Formation shale in the Changning area of the Sichuan Basin has the largest number of OM pores under the microscope, and the maximum pore diameter is more than 900 nm (Figure 18A–D), while the Longmaxi Formation shale in northwestern Hunan has a significantly lower number of OM pores under the microscope (Figure 18E–H), and the maximum pore diameter of organic matter is only 157 nm (Figure 18F). On the other hand, the Lower Cambrian Niutitang Formation shale in northwestern Hunan has the smallest number of OM pores under the microscope (Figure 18I–L), and the largest pore diameter is only 76 nm (Figure 18I).

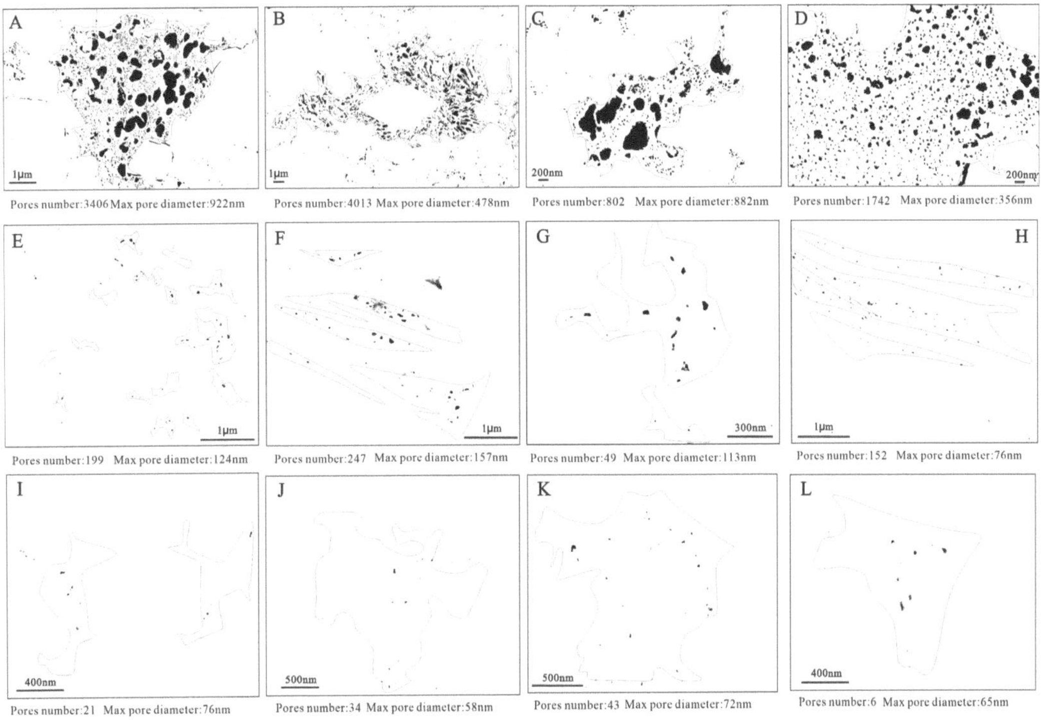

Figure 18. Digital extraction of OM pores in Lower Silurian Longmaxi Formation shale and Lower Cambrian Niutitang Formation shale in Changning area of Sichuan Basin and northwest of Hunan Province by ImageJ digital analysis software: (**A–D**) Longmaxi Formation shale in Changning, Sichuan Basin; (**E–H**) Longmaxi Formation shale in Northwestern Hunan; (**I–L**) Niutitang Formation shale in Northwestern Hunan.

The quantitative comparison can be realized by the pore volume (PV) test of isolated OM after extraction, and the interference of pores associated with mineral components can be avoided. It can be seen, in Figure 19, that the PVs of OM pores are similar at the closed thermal degrees and burial conditions, while are various at the different thermal degrees and burial conditions. In the Hunan area, the PVs of the OM pore in the Niutitang Formation shale are the smallest, with an average value of 0.09 mL/g. The PVs of OM pores in the Longmaxi shale are 1.5 times that in the Niutitang Formation shale, with an average value of 0.15 mL/g. In contrast, OM pores of the Longmaxi Formation shale in the Changning area of Sichuan are the most developed, with the average PV reaching 0.34 mL/g, which is equivalent to 4 times that of the Formation Niutitang Formation shale in the Hunan area. Meanwhile, it can also be seen from Figure 19 that OM pores with small apertures (<300 nm) share high percentages in both shales in the Hunan area, while OM pores with small apertures (300 nm~1000 nm) share high percentages in the Longmaxi Formation shale in the Changning area of Sichuan. This indicates that with the decrease in thermal evolution and the delay of uplift time, the small OM pores are enlarged and gradually merge into large pores. The quantitative comparison of the pore volume of OM pores after extraction is in high agreement with the results of semi-quantitative statistics under the SEM observation (Figure 18).

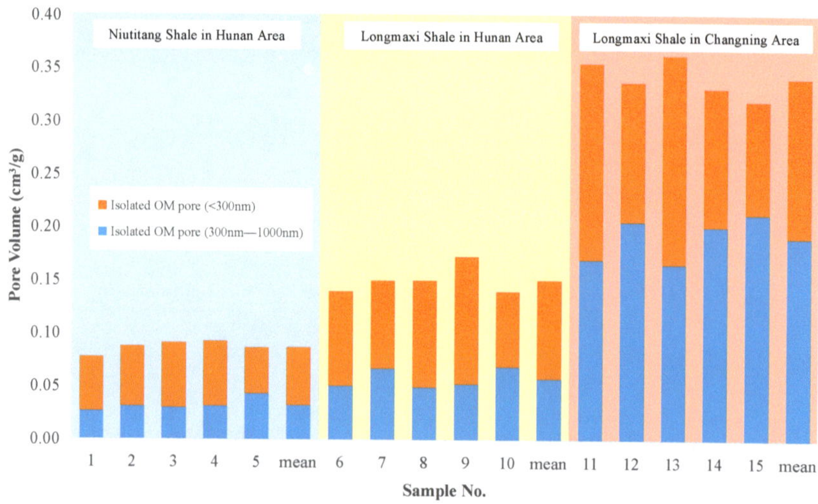

Figure 19. Pore volumes of isolated OM measured via N_2 adsorption and mercury intrusion porosimetry (MIP), samples selected from Niutitang shale in Hunan Area, Longmaxi shale in Hunan Area, and Longmaxi shale in Changning Area of Sichuan.

In conclusion, the excessive evolution of organic matter in the Lower Cambrian Niutitang shale in northwest Hunan is caused by the excessive paleo-burial depth of the rock. Additionally, the rock's physical and chemical characteristics are comparable to those of graphite, which causes the organic matter pores to vanish. Shale gas is widely scattered throughout geological time and lacks an adequate reservoir space. The Lower Silurian Longmaxi Formation shale in this area was unable to generate enough gaseous hydrocarbons or a significant amount of effective reservoir space (OM pores with regular shape and large pore size), and as a result, it was unable to develop into a gas reservoir. As a result, the Lower Cambrian Niutitang Formation and the Lower Silurian Longmaxi Formation maritime shale gas resources in northwest Hunan have weak exploration and development potential.

6. Conclusions

Utilizing FIB-SEM, this study investigated the variations in organic matter pore structure between the Lower Silurian Longmaxi and the Lower Cambrian Niutitang shale in northwest Hunan and examined the contribution of regional tectonic evolution in those variations. The following are the primary conclusions:

(1) The Lower Cambrian Niutitang shale and the Lower Silurian Longmaxi shale both exhibit high TOC concentration and high brittle mineral content, which are similar reservoir features. The thermal maturity of the Longmaxi shale, however, is much lower than that of the Niutitang shale, the latter having a thermal maturity of greater than 3.0% while the former ranges from 2.3% to 3.0%.

(2) The Lower Cambrian Niutitang shale and the Lower Silurian Longmaxi shale have substantially different organic matter pore structures. The Niutitang shale features few irregularly sized organic matter pores that mostly have an extended flat appearance, making it undesirable for gas occurrence.

(3) The thermal maturity of shale reservoirs, which in turn impacts the variability of organic matter pore structure, can be affected by the maximum paleo-burial depth of the strata governed by tectonic evolution. As with the Niutitang shale in our study, excessive shale development and the degradation of organic matter pore structure will result if the maximum paleo-burial depth reaches 8000 m or higher.

(4) The Longmaxi Formation shale in northwest Hunan failed to develop enough gaseous hydrocarbons and a significant quantity of effective reservoir space (OM pores with big pore size) as a result of insufficient hydrocarbon-generation evolution time. The Lower Silurian Longmaxi Formation and the Lower Cambrian Niutitang Formation in northwest Hunan have limited exploration and development potential for marine shale gas.

Author Contributions: Y.M.: Literature search, chart making, research design, data collection, data analysis, manuscriptwriting. P.W.: Research design, data collection and manuscript revision. X.L.: Chart making, research design, data collection. H.H.: Supervision, review & editing. C.J.: Data collection, data analysis. W.G.: Literature search, making charts. All authors have read and agreed to the published version of the manuscript.

Funding: This work is financially supported by grants from the National Natural Science Foundation of China (No. 42002050 and No. 42202157), the Open Project of the Key Laboratory of Structure and Petroleum Resources, Ministry of Education (No. TPR-2021-19), and the China Postdoctoral Science Foundation Funded Project (No. 2020M680815).

Institutional Review Board Statement: Not applicable.

Informed Consent Statement: Not applicable.

Data Availability Statement: Data will be made available on request to the corresponding author.

Conflicts of Interest: The authors declare no conflict of interest.

References

1. Meng, M.; Peng, J.; Ge, H.; Ji, W.; Li, X.; Wang, Q. Rock fabric of lacustrine shale and its influence on residual oil distribution in the Upper Cretaceous Qingshankou Formation, Songliao Basin. *Energy Fuels* **2023**, *37*, 7151–7160. [CrossRef]
2. Meng, M.; Ge, H.; Shen, Y.; Ji, W.; Li, Z. Insight into water occurrence and pore size distribution by nuclear magnetic resonance in marine shale reservoirs, southern China. *Energy Fuels* **2023**, *37*, 319–327. [CrossRef]
3. Chen, Q.; Kang, Y.; You, L.; Yang, P.; Zhang, X.; Cheng, Q. Change in composition and pore structure of Longmaxi black shale during oxidative dissolution. *Int. J. Coal Geol.* **2017**, *172*, 95–111. [CrossRef]
4. Gou, Q.; Xu, S.; Hao, F.; Yang, F.; Zhang, B.; Shu, Z.; Zhang, A.; Wang, Y. Full-scale pores and micro-fractures characterization using FE-SEM, gas adsorption, nano-CT and micro-CT: A case study of the Silurian Longmaxi Formation shale in the Fuling area, Sichuan Basin, China. *Fuel* **2019**, *253*, 167–179. [CrossRef]
5. Guo, X.; Hu, D.; Li, Y.; Wei, Z.; Wei, X.; Liu, Z. Geological factors controlling shale gas enrichment and high production in Fuling shale gas field. *Pet. Explor. Dev.* **2017**, *44*, 513–523. [CrossRef]
6. He, Z.; Li, S.; Nie, H.; Yuan, Y.; Wang, H. The shale gas "sweet window": "The cracked and unbroken" state of shale and its depth range. *Mar. Pet. Geol.* **2018**, *101*, 334–342. [CrossRef]
7. He, Z.; Nie, H.; Li, S.; Luo, J.; Wang, H.; Zhang, G. Differential enrichment of shale gas in upper Ordovician and lower Silurian controlled by the plate tectonics of the Middle-Upper Yangtze, south China. *Mar. Pet. Geol.* **2020**, *118*, 104357. [CrossRef]
8. Yang, W.; He, S.; Iglauer, S.; Guo, X.; Zhai, G.; Zhou, Z.; Dong, T.; Tao, Z.; Wei, S. Porosity characteristics of different lithofacies in marine shale: A case study of Neoproterozoic Sinian Doushantuo formation in Yichang area, China. *J. Pet. Sci. Eng.* **2020**, *187*, 106856. [CrossRef]
9. Wang, P.; Yao, S.; Jin, C.; Li, X.; Zhang, K.; Liu, G.; Zang, X.; Liu, S.; Jiang, Z. Key reservoir parameter for effective exploration and development of high-over matured marine shales: A case study from the cambrian Niutitang formation and the silurian Longmaxi formation, south China. *Mar. Pet. Geol.* **2020**, *121*, 104619. [CrossRef]
10. Hu, H.; Hao, F.; Lin, J.; Lu, Y.; Ma, Y.; Li, Q. Organic matter-hosted pore system in the Wufeng-Longmaxi (O_3w-S_1l) shale, Jiaoshiba area, Eastern Sichuan Basin, China. *Int. J. Coal Geol.* **2017**, *173*, 40–50. [CrossRef]
11. Hou, Y.; Zhang, K.; Wang, F.; He, S.; Dong, T.; Wang, C.; Qin, W.; Xiao, Y.; Tang, B.; Yu, R.; et al. Structural evolution of organic matter and implications for graphitization in over-mature marine shales, south China. *Mar. Pet. Geol.* **2019**, *109*, 304–316. [CrossRef]
12. Zhao, J.; Jin, Z.; Hu, Q.; Liu, K.; Liu, G.; Gao, B.; Liu, Z.; Zhang, Y.; Wang, R. Geological controls on the accumulation of shale gas: A case study of the early Cambrian shale in the Upper Yangtze area. *Mar. Pet. Geol.* **2019**, *107*, 423–437. [CrossRef]
13. Jiao, K.; Yao, S.; Liu, C.; Gao, Y.; Wu, H.; Li, M.; Tang, Z. The characterization and quantitative analysis of nanopores in unconventional gas reservoirs utilizing FESEM–FIB and image processing: An example from the lower Silurian Longmaxi Shale, upper Yangtze region, China. *Int. J. Coal Geol.* **2014**, *128*, 1–11. [CrossRef]

14. Ji, W.; Song, Y.; Jiang, Z.; Chen, L.; Li, Z.; Yang, X.; Meng, M. Estimation of marine shale methane adsorption capacity based on experimental investigations of Lower Silurian Longmaxi formation in the Upper Yangtze Platform, south China. *Mar. Pet. Geol.* **2015**, *68*, 94–106. [CrossRef]
15. Ji, W.; Song, Y.; Rui, Z.; Meng, M.; Huang, H. Pore characterization of isolated organic matter from high matured gas shale reservoir. *Int. J. Coal Geol.* **2017**, *174*, 31–40. [CrossRef]
16. Meng, M.; Ge, H.; Shen, Y.; Ji, W.; Wang, Q. Rock fabric of tight sandstone and its influence on irreducible water saturation in Eastern Ordos Basin. *Energy Fuels* **2023**, *37*, 3685–3696. [CrossRef]
17. Lan, Z.; Li, X.; Chu, X.; Tang, G.; Yang, S.; Yang, H.; Liu, H.; Jiang, T.; Wang, T. SIMS U-Pb zircon ages and Ni-Mo-PGE geochemistry of the lower Cambrian Niutitang Formation in South China: Constraints on Ni-Mo-PGE mineralization and stratigraphic correlations. *J. Asian Earth Sci.* **2017**, *137*, 141–162. [CrossRef]
18. Jin, Z.; Nie, H.; Liu, Q.; Zhao, J.; Jiang, T. Source and seal coupling mechanism for shale gas enrichment in upper Ordovician Wufeng Formation—Lower Silurian Longmaxi Formation in Sichuan Basin and its periphery. *Mar. Pet. Geol.* **2018**, *97*, 78–93. [CrossRef]
19. Qin, M.; Guo, J.; He, H.; Huang, Y.; Jiao, P.; Liu, C.; Zheng, Z.; Guo, J.; Cao, Z.; Wang, C.; et al. Geological conditions and gas-bearing characteristics of shale gas in complex structure area out of Sichuan basin: A case of Wufeng—Longmaxi formation in Northwestern Hunan, China. *J. Central South Univ. (Sci. Tech.)* **2018**, *49*, 1979–1990.
20. Liu, A.; Ou, W.; Huang, H.; Wei, K.; Li, H.; Chen, X. Significance of paleo-fluid in the Ordovician–Silurian detachment zone to the preservation of shale gas in western Hunan–Hubei area. *Nat. Gas Ind. B* **2018**, *5*, 565–574. [CrossRef]
21. Li, F.; Wang, M.; Liu, S.; Hao, Y. Pore characteristics and influencing factors of different types of shales. *Mar. Pet. Geol.* **2019**, *102*, 391–401. [CrossRef]
22. Liu, G.; Zhai, G.; Zou, C.; Cheng, L.; Guo, X.; Xia, X.; Shi, D.; Yang, Y. A comparative discussion of the evidence for biogenic silica in Wufeng-Longmaxi siliceous shale reservoir in the Sichuan basin, China. *Mar. Pet. Geol.* **2019**, *109*, 70–87. [CrossRef]
23. Liu, Z.; Gao, B.; Zhang, Y.; Du, W.; Feng, D.; Nie, H. Types and distribution of the shale sedimentary facies of the Lower Cambrian in Upper Yangtze area, South China. *Pet. Explor. Dev.* **2017**, *44*, 20–31. [CrossRef]
24. Nie, H.; Sun, C.; Liu, G.; Du, W.; He, Z. Dissolution pore types of the wufeng formation and the longmaxi formation in the sichuan basin, south china: Implications for shale gas enrichment. *Mar. Pet. Geol.* **2019**, *101*, 243–251. [CrossRef]
25. Hassan, M.; Leila, M.; Ahmed, M.; Issa, G.; Hegab, O. Geochemical characteristics of natural gases and source rocks in Obayied sub-basin, north Western Desert, Egypt: Implications for gas-source correlation. *Acta Geochim.* **2023**, *42*, 241–255. [CrossRef]
26. Leila, M.; Loiseau, K.; Moretti, I. Controls on generation and accumulation of blended gases (CH4/H2/He) in the Neoproterozoic Amadeus Basin, Australia. *Mar. Pet. Geol.* **2022**, *140*, 105643. [CrossRef]
27. Xu, Z.; Wang, Y.; Jiang, S.; Fang, C.; Liu, L.; Wu, K.; Luo, Q.; Li, X.; Chen, Y. Impact of input, preservation and dilution on organic matter enrichment in lacustrine rift basin: A case study of lacustrine shale in Dehui Depression of Songliao Basin, NE China. *Mar. Pet. Geol.* **2022**, *135*, 105386. [CrossRef]
28. Wang, Q.; Wang, T.; Liu, W.; Zhang, J.; Feng, Q.; Lu, H.; Peng, P. Relationships among composition, porosity and permeability of Longmaxi Shale reservoir in the Weiyuan Block, Sichuan Basin, China. *Mar. Pet. Geol.* **2019**, *102*, 33–47. [CrossRef]
29. Wu, J.; Zhang, S.; Cao, H.; Sun, P.; Luo, X. Fracability evaluation of shale gas reservoir-A case study in the Lower Cambrian Niutitang formation, northwestern Hunan, China. *J. Pet. Sci. Eng.* **2017**, *164*, 675–684. [CrossRef]
30. Xi, Z.; Tang, S.; Wang, J. The reservoir characterization and shale gas potential of the Niutitang formation: Case study of the SY well in northwest Hunan area, South China. *J. Pet. Sci. Eng.* **2018**, *171*, 687–703. [CrossRef]
31. Zhou, L.; Kang, Z.; Wang, Z.; Peng, Y.; Xiao, H. Sedimentary geochemical investigation for paleoenvironment of the Lower Cambrian Niutitang Formation shales in the Yangtze Platform. *J. Pet. Sci. Eng.* **2017**, *159*, 376–386. [CrossRef]
32. Yang, X.; Fan, T.; Wu, Y. Lithofacies and cyclicity of the lower cambrian niutitang shale in the mayang basin of western hunan, south china. *J. Nat. Gas Sci. Eng.* **2016**, *28*, 74–86. [CrossRef]
33. Zhang, Q.; Littke, R.; Zieger, L.; Shabani, M.; Tang, X.; Zhang, J. Ediacaran, Cambrian, Ordovician, Silurian and Permian shales of the Upper Yangtze Platform, South China: Deposition, thermal maturity and shale gas potential. *Int. J. Coal Geol.* **2019**, *216*, 103281. [CrossRef]
34. Wang, P.; Jiang, Z.; Chen, L.; Yin, L.; Li, Z.; Zhang, C.; Wang, G. Pore structure characterization for the Longmaxi and Niutitang shales in the Upper Yangtze Platform, South China: Evidence from focused ion beam–He ion microscopy, nano-computerized tomography and gas adsorption analysis. *Mar. Pet. Geol.* **2016**, *77*, 1323–1337. [CrossRef]
35. Wang, P.; Jiang, Z.; Ji, W.; Zhang, C.; Yuan, Y.; Chen, L.; Yin, L. Heterogeneity of intergranular, intraparticle and organic pores in Longmaxi shale in Sichuan Basin, South China: Evidence from SEM digital images and fractal and multifractal geometries. *Mar. Pet. Geol.* **2016**, *72*, 122–138. [CrossRef]
36. Wang, Y.; Wang, L.; Wang, J.; Jiang, Z.; Jin, C.; Wang, Y. Characterization of organic matter pores in typical marine and terrestrial shales, China. *J. Nat. Gas Sci. Eng.* **2018**, *49*, 56–65. [CrossRef]
37. Xu, H.; Zhou, W.; Zhang, R.; Liu, S.; Zhou, Q. Characterizations of pore, mineral and petrographic properties of marine shale using multiple techniques and their implications on gas storage capability for Sichuan Longmaxi gas shale field in China. *Fuel* **2019**, *241*, 360–371. [CrossRef]

38. Xu, S.; Liu, R.; Hao, F.; Engelder, T.; Yi, J.; Zhang, B.; Shu, Z. Complex rotation of maximum horizontal stress in the Wufeng-Longmaxi Shale on the eastern margin of the Sichuan Basin, China: Implications for predicting natural fractures. *Mar. Pet. Geol.* **2019**, *109*, 519–529. [CrossRef]
39. Xu, S.; Gou, Q.; Hao, F.; Zhang, B.; Shu, Z.; Lu, Y.; Wang, Y. Shale pore structure characteristics of the high and low productivity wells, Jiaoshiba shale gas field, Sichuan Basin, China: Dominated by lithofacies or preservation condition? *Mar. Pet. Geol.* **2020**, *114*, 104211. [CrossRef]
40. Yang, R.; Hao, F.; He, S.; He, C.; Guo, X.; Yi, J.; Hu, H.; Zhang, S.; Hu, Q. Experimental investigations on the geometry and connectivity of pore space in organic-rich Wufeng and Longmaxi shales. *Mar. Pet. Geol.* **2017**, *84*, 225–242. [CrossRef]
41. Zhao, J.; Jin, Z.; Jin, Z.; Hu, Q.; Hu, Z.; Du, W.; Yan, C.; Geng, Y. Mineral types and organic matters of the Ordovician-Silurian Wufeng and Longmaxi Shale in the Sichuan Basin, China: Implications for pore systems, diagenetic pathways, and reservoir quality in fine-grained sedimentary rocks. *Mar. Pet. Geol.* **2017**, *86*, 655–674. [CrossRef]
42. Xi, Z.; Tang, S.; Zhang, S.; Yi, Y.; Dang, F.; Ye, Y. Characterization of quartz in the Wufeng Formation in northwest Hunan area, south China and its implications for reservoir quality. *J. Pet. Sci. Eng.* **2019**, *179*, 979–996. [CrossRef]
43. Jacob, H. Classification, structure, genesis and practical importance of natural solid oil bitumen ("migrabitumen"). *Int. J. Coal Geol.* **1989**, *11*, 65–79. [CrossRef]
44. Loucks, R.G.; Reed, R.M.; Ruppel, S.C.; Jarvie, D.M. Morphology, genesis, and distribution of nanometer-scale pores in siliceous mudstones of the Mississippian Barnett Shale. *J. Sediment. Res.* **2009**, *79*, 848–861. [CrossRef]
45. Loucks, R.G.; Reed, R.M.; Ruppel, S.C.; Hammes, U. Spectrum of pore types and networks in mudrocks and a descriptive classification for matrix-related mudrock pores. *AAPG Bull.* **2012**, *96*, 1071–1098. [CrossRef]
46. Cao, T.; Liu, H.; Pan, A.; Fu, Y.; Deng, M.; Cao, Q.; Huang, Y.; Yu, Y. Pore evolution in siliceous shales and its influence on shale gas-bearing capacity in eastern Sichuan-western Hubei, China. *J. Pet. Sci. Eng.* **2022**, *208*, 109597. [CrossRef]
47. Milliken, K.L.; Rudnicki, M.; Awwiller, D.N.; Zhang, T. Organic matterehosted pore system, Marcellus formation (Devonian), Pennsylvania. *AAPG Bull.* **2013**, *97*, 177–200. [CrossRef]
48. Loucks, R.G.; Reed, R.M. Scanning-electron-microscope petrographic evidence for distinguishing organic-matter pores associated with depositional organic matter versus migrated organic matter in mudrock. *Gulf Coast Assoc. Geol. Soc. J.* **2014**, *3*, 51–60.
49. Wu, S.; Yang, Z.; Zhai, X.; Cui, J.; Bai, L.; Pan, S.; Cui, J. An experimental study of organic matter, minerals and porosity evolution in shales within high-temperature and high-pressure constraints. *Mar. Pet. Geol.* **2019**, *102*, 377–390. [CrossRef]
50. Zhang, K.; Jiang, Z.; Yin, L.; Gao, Z.; Wang, P.; Song, Y.; Jia, C.; Liu, W.; Liu, T.; Xie, X.; et al. Controlling functions of hydrothermal activity to shale gas content-taking lower Cambrian in Xiuwu Basin as an example. *Mar. Pet. Geol.* **2017**, *85*, 177–193. [CrossRef]
51. Yu, Y.; Wang, Z.; Zhang, K.; Feng, X.; Cheng, M. Effects of shear on development characteristics of organic matter pores in shale: A case study of shale in the Niutitang Formation of the well XAD1. *J. Pet. Sci. Eng.* **2022**, *211*, 110166. [CrossRef]
52. Zhang, K.; Jia, C.; Song, Y.; Jiang, S.; Jiang, Z.; Wen, M.; Huang, Y.; Liu, X.; Jiang, T.; Peng, J.; et al. Analysis of Lower Cambrian shale gas composition, source and accumulation pattern in different tectonic backgrounds: A case study of Weiyuan Block in the Upper Yangtze region and Xiuwu Basin in the Lower Yangtze region. *Fuel* **2020**, *263*, 115978. [CrossRef]
53. He, X.; Wu, J.; Yong, R.; Zhao, S. Hydrocarbon accumulation and key exploration & development technologies of Changning-Weiyuan marine shale gas field, southern Sichuan. *Pet. Res.* **2021**, *6*, 1–15.
54. Shi, X.; Kang, S.; Luo, C.; Wu, W.; Zhao, S.; Zhu, D.; Zhang, H.; Yang, Y.; Xiao, Z.; Li, Y. Shale gas exploration potential and reservoir conditions of the Longmaxi Formation in the Changning area, Sichuan Basin, SW China: Evidence from mud gas isotope logging. *J. Asian Earth Sci.* **2022**, *233*, 105239. [CrossRef]
55. Li, X.; Wang, P.; Wang, S.; Jin, C.; Zhang, C.; Zhao, R.; Jiang, Z. Influence of the duration of tectonic evolution on organic matter pores structure and gas enrichment in marine shale: A case study of the Lower Silurian Longmaxi shale in southern China. *Front. Earth Sci.* **2022**, *9*, 787697. [CrossRef]
56. Wang, P.; Zhang, C.; Liu, A.; Zhang, P.; Qiu, Y.; Li, X.; Yu, S.; Yao, S.; Liu, S.; Jiang, Z. Slope Belts of Paleouplifts Control the Pore Structure of Organic Matter of Marine Shale: A Comparative Study of Lower Cambrian Rocks in the Sichuan Basin. *Geofluids* **2021**, *2021*, 5517655. [CrossRef]

Journal of Marine Science and Engineering

Article

Nanopore Structure and Multifractal Characteristics of Continental Shale Oil Reservoir: A Case Study from Ziliujing Shales in the Sichuan Basin

Youzhi Wang [1], Wei Li [2,3,*], Xiandong Wang [1], Zhiguo Wang [1], Weiqi Ma [1], Yanping Zhu [1], Mengdi Sun [2,3,*], Bo Liu [2,3], Lijuan Cheng [2,3] and Xiaofei Fu [2,3]

[1] Exploration and Development Research Institute, PetroChina Daqing Oilfield Company Limited, Daqing 163712, China; wangyouzhi@petrochina.com.cn (Y.W.); wangxd2004@petrochina.com.cn (X.W.); scv1977@petrochina.com.cn (Z.W.); maweiqi@petrochina.com.cn (W.M.); zhuyanping@petrochina.com.cn (Y.Z.)
[2] Key Laboratory of Continental Shale Hydrocarbon Accumulation and Efficient Development, Northeast Petroleum University, Ministry of Education, Daqing 163318, China; liubo@nepu.edu.cn (B.L.); chenglijuan211@163.com (L.C.); fuxiaofei2008@sohu.com (X.F.)
[3] Institute of Unconventional Oil & Gas, Northeast Petroleum University, Daqing 163318, China
* Correspondence: pingyaoliwei@nepu.edu.cn (W.L.); sunmd@nepu.edu.cn (M.S.)

Abstract: Thermal maturity of the shales from the Ziliujing Formation of the Jurassic age in the Sichuan Basin is in the hydrocarbon generation window, which makes it a candidate for shale oil and gas development. The meso- and macropore characteristics and heterogeneity of shales are important factors affecting the occurrence and development of oil and gas. However, the meso- and macropores of the Ziliujing shales have not been systematically studied. Thus, the mineral compositions and total organic carbon (TOC) of samples from this formation, as well as its pore structure, are analyzed by low-temperature N_2 adsorption technique. Moreover, the heterogeneity of the pores was determined by multifractal analysis. The results show that the Ziliujing shales can be classified into three types according to the distributions of mineral compositions of carbonate and mixed and argillaceous shales. Results revealed that the smallest meso- and macropore volume (PV), the smallest specific surface area (SSA), and the largest average pore diameter (APD) occur in the carbonate shales. However, the largest PV and SSA and the smallest APD are observed in the argillaceous shales. The porosity of carbonate shales is mainly concentrated between 5 nm and 30 nm. Compared with carbonate shales, the porosity with pore sizes less than 30 nm of mixed and argillaceous shales shows a rapid increase. Furthermore, inorganic minerals are the main factors affecting the pore distributions, while TOC shows a weak effect. Herein, clay minerals significantly increase the mesopore volume and the pore number with a size of less than 30 nm. The D_q-q curves reveal that the meso- and macropore distributions of Ziliujing shales show multifractal behavior, but the multifractal characteristics of pores of various shales are distinctly different. The information dimension D_1, the Hurst exponent H, and the width of the right side D_0–D_{10} are key indicators to distinguish the local variations within the pore structure of different types of shales. The carbonate shales have the largest multifractal spectra width and the smallest D_1 and H, while the opposite trend is found for the argillaceous shales. Clay minerals reduce the heterogeneity of the meso- and macropore distributions and increase the pore connectivity. Nevertheless, the carbonate minerals exhibit a reverse trend. Finally, it was found that TOC does not impact pore complexity as much. Collectively, this study supports our understanding of the occurrence of shale oil within various reservoir facies, thereby providing a guideline for future explorations in the Ziliujing Formation of the Jurassic age in the Sichuan Basin.

Keywords: Sichuan Basin; shale oil and gas; nanopores; low-temperature gas adsorption; multifractal analysis

Citation: Wang, Y.; Li, W.; Wang, X.; Wang, Z.; Ma, W.; Zhu, Y.; Sun, M.; Liu, B.; Cheng, L.; Fu, X. Nanopore Structure and Multifractal Characteristics of Continental Shale Oil Reservoir: A Case Study from Ziliujing Shales in the Sichuan Basin. *J. Mar. Sci. Eng.* **2023**, *11*, 1989. https://doi.org/10.3390/jmse11101989

Academic Editors: János Kovács and Michael Lazar

Received: 25 July 2023
Revised: 7 October 2023
Accepted: 10 October 2023
Published: 15 October 2023

1. Introduction

Recent advancements in technology have accelerated exploration and production from continental and lacustrine shale oil reservoirs in China [1–3]. The recoverable resource potential of shale oil in China is estimated at 34.98×10^8 t, which offsets the shortage of China's crude oil storage [4]. Large continental shale oil resources have been discovered in the Songliao, Ordos, Junggar, and Bohai Bay Basins [5]. As for the mentioned basins, significant achievements in occurrence, geological characteristics, sweet spot evaluation, and development technologies of shale oil reservoirs have been obtained [6–10]. Recently, several sets of organic-rich black shales with thermal maturity ranging from 1.3% to 1.5% in the Ziliujing Formation of the Jurassic age in the Sichuan Basin have been discovered [11]. The thermal maturity of the Ziliujing shale oil reservoir is in the oil and gas window, which shows great potential for shale oil and gas production. At present, the research regarding the Ziliujing shale reservoir is focused on the analysis of the main controlling factors of hydrocarbon enrichment [12,13]. Herein, shale oil reservoirs are considered to be fine-grained sediments with a wide range of pore sizes (nm to μm) and ultralow porosity [14]. The nanopores, including meso- and macropores, are important channels for shale oil occurrence and diffusion, which are important parameters to decide the future development of shale oil [15]. Therefore, systematic research on the meso- and macropore characteristics of the Ziliujing shales is needed.

Various methods have been reported to study pore characteristics of shale oil and gas reservoirs [16–21]. Jiang et al. [22] studied the pore characteristics of lacustrine shales with vitrinite reflectance (R_o) of 0.68–1.02% in the Ordos Basin through low-temperature nitrogen (N_2) adsorption (LNA) and high-pressure mercury intrusion porosimetry (MIP). They concluded that mesopores (2–50 nm) mainly controlled the distributions of pore volume and specific surface area of shales, and clay mineral contents and R_o were the main factors controlling pore development. Zhao et al. [23] employed small-angle neutron scattering, and fluid invasion methods were adopted to characterize the pore structure of typical American shale oil reservoirs. They found that large amounts of inaccessible porosity at pore size <10 nm occurred in typical American shales. Zhang et al. [24] analyzed the pore size distributions of shale oil reservoir samples from Dongying Sag, Bohai Bay Basin, by using LNA, scanning electron microscopy (SEM), MIP, and nuclear magnetic resonance (NMR) techniques. The results showed that a wide range of pore sizes with small-scale throats from nanometers to millimeters existed in shale oil reservoirs. Hou et al. [25] investigated the influences of clay minerals on the pore structure of lacustrine shales by SEM. They found that the type, occurrence, and composition of clay minerals are the key factors affecting pore characteristics. The results revealed that the pore-filling clay minerals, as well as layered ones, played a negative role in the pore development of lacustrine shales. Zheng et al. [26] characterized the multi-scale pore structure of tight shale oil reservoirs from the Triassic Yanchang Formation, Ordos Basin, by combining LNA, MIP, helium pycnometry, field emission scanning electron microscopy (FE-SEM), and nano-CT. They found that the pore sizes of the studied shales varied from 2 nm to >20 μm, but the dominant pores were in the range of 20–100 nm. Among these methods, the LNA technique is simple and convenient and has been widely used to characterize nanopore characteristics of shales [27,28]. Despite the popularity of this method, the evaluation of the nanopore structure of Ziliujing shales by the LNA method has not been carried out so far.

The pore heterogeneity or complexity is another critical parameter affecting the flow characteristics of fluids within pores of shales [29–31]. However, the traditional Euclidean geometry theory makes it difficult to characterize pore complexity [32]. Thus, fractal theory provides a scientific tool to quantitatively describe the complexity of the pore system by fractal dimension [33–35]. The pore size distributions of shales show fractal characteristics, but the pores with different size ranges exhibit multiple fractal characteristics [36]. The multifractals, described as an extension of single fractals, decompose self-similar measures into intertwined fractal subsets that can characterize the local complexity of measured variables [37]. Thus, in comparison with single fractal, multifractal can better explain the local and differential distributions of the pore structure of shales with an acceptable accuracy [38]. Currently, limited research has been carried out in applying the multifractal approach to analyze the pore heterogeneity of shale oil reservoirs [39,40]. Liu et al. [41] applied multifractal methods to analyze gas adsorption isotherms for meso- and macropore and micropore heterogeneity of the Bakken shale. Their results showed that clay and organic matter reduced micropore complexity but increased meso- and macropore complexity. Likewise, Liu et al. [42] investigated multifractal characteristics of the pore structure of Middle Bakken based on FE-SEM image. They found that the heterogeneity of entire pores in images weas larger than that of the sub-pore groups in the same image. Zhang et al. [43] applied multifractal analysis on the samples that were polished by the ion beam via scanning electron microscopy to quantify the heterogeneity of major pores of shale oil reservoirs from Dongying Sag, Bohai Bay Basin. They reported that the interparticle pores possessed major complexity of pore morphologies and the simplest pore structure, while the dissolution pores had the simplest pore morphologies but had the most complex pore distributions. Guan et al. [44] conducted a multifractal analysis of N_2 adsorption and Hg pore size distributions of lacustrine shales from Cangdong Sag, Bohai Bay Basin. The results revealed that the pore distributions obtained from LNA are more homogeneous than those obtained from MIP, which implied that the pores with larger diameters could increase the heterogeneity of pore distributions. In this regard, the thermal maturity (R_o) of shale oil reservoirs described above is mostly less than 1.05%. Nevertheless, the R_o of the shale oil reservoirs in some basins in China is measured over 1.05%. However, the investigation on the multifractal characteristics of nanopore structure of shale oil reservoirs with medium to high thermal maturity ($1.05\% < R_o < 2.0\%$) is scarce, which may limit a better understanding of shale oil occurrence with $R_o > 1.05\%$. In this case, the R_o values of the Ziliujing shale oil reservoirs mostly vary from 1.3% to 1.5%, and the research on the multifractal heterogeneity of nanopores from N_2 adsorption of this reservoir has not been reported. Hence, this study can help us to understand the occurrence of hydrocarbons and support exploration efforts within reservoirs with $R_o > 1.05\%$ in general.

In this paper, the lacustrine shale samples with different depths from the Ziliujing Formation in the Sichuan Basin were investigated. The aims are to (1) investigate the meso- and macropore characteristics, including pore volume (PV), specific surface area (SSA), and pore size distributions of shales by low-temperature N_2 adsorption; (2) analyze the effects of mineral compositions and total organic carbon (TOC) on meso- and macropore structures of different types of shales within this formation; (3) address the heterogeneity of meso- and macropore distributions by multifractal theory and discuss the governing factors on meso- and macropore heterogeneity. This research should be beneficial for more successful exploration and development of shale oil in the Ziliujing Formation of the Jurassic age in the Sichuan Basin.

2. Experiments and Methods

2.1. Sample Preparation

The shale samples were collected from drill cores of the Ziliujing Formation of the Jurassic age in the Sichuan Basin, with details listed in Table 1. The depth of the samples varies in the range of 1866.5–2017.0 m. Details about the geological settings of the study area and this formation can be found in our previous study [45]. To better address the meso- and macropore characteristics, a schematic diagram has been presented in Figure 1. First, the studied shale samples were grinded to measure TOC, and inorganic mineral compositions were determined via LECO test and X-ray diffraction (XRD) analysis, while meso- and macropore parameters were obtained by LNA. Then, the roles of TOC and mineral compositions on pore distributions of the studied shale samples were investigated. Finally, the heterogeneity of meso- and macropore distributions of different shale samples was studied by using multifractal theory. Moreover, major identified facies in the samples were chosen for more detailed analysis by quantitative evaluation of minerals through scanning electron microscopy (QEMSCAN).

Table 1. Depths and mineral compositions of shale samples.

Samples	Depth (m)	TOC (%)	Mineral Contents (%)							
			Q	FS	PC	CA	SR	PY	MC	CL
Dn9	1866.5	1.138	18.1	1.6	2.3	53.8	0.6	3.8	1.2	18.6
D97	2006.9	0.396	16.2	1.3	2.1	57.9	0.4	2.1	1.0	19.0
D62	1997.4	1.646	29.2	1.6	2.6	29.6	1.0	5.8	1.2	29.0
D123	2014.7	1.011	27.8	0.0	1.8	35.3	2.7	3.4	0.8	28.2
D130	2016.6	0.744	30.7	2.0	2.9	28.8	0.9	4.5	0.7	29.5
D132	2017.0	1.889	30.1	2.1	3.9	26.9	1.0	5.0	1.5	29.5
Dn17	1890.5	0.427	26.8	2.5	3.8	2.4	1.4	5.1	1.9	56.1
D17	1983.9	1.459	25.2	2.6	4.9	2.4	1.3	2.4	2.2	59.0
D26	1986.8	0.731	29.6	3.1	3.4	2.0	0.6	4.0	1.3	56.0
D71	1999.7	2.417	27.6	2.0	4.0	4.6	1.6	7.2	1.8	51.2
D80	2002.8	1.719	26.3	2.8	4.1	3.7	3.3	2.5	1.6	55.7
D91	2005.6	1.843	30.7	2.7	4.3	1.1	1.4	2.1	2.7	55.0

Note: Q—quartz; FS—feldspar; PC—plagioclase; CA—calcite; SR—siderite; PY—pyrite; MC—muscovite; CL—clay.

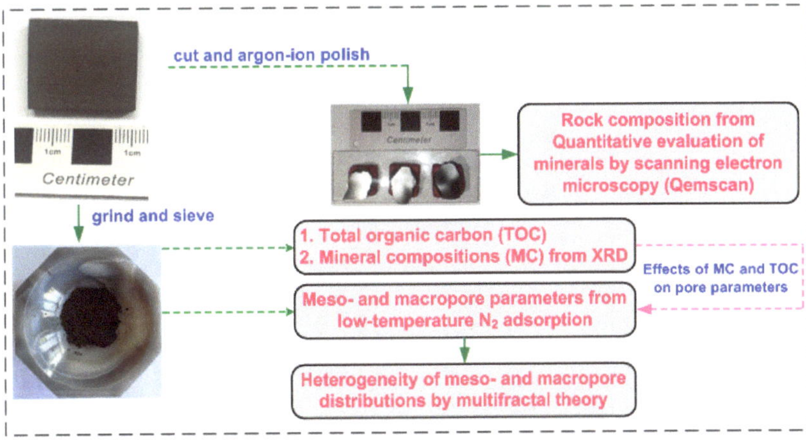

Figure 1. Schematic diagram for this study on the Ziliujing shale samples.

2.2. TOC Measurement, XRD Analysis, and QEMSCAN

The organic content and inorganic mineral compositions of the shale samples were characterized by TOC and XRD, respectively. The TOC of the shale samples was determined according to the GB/T 19145-2003 standard [46]. Based on SY/T 5163-2018 [47], the inorganic mineral compositions of the shale samples were determined by using a Japanese RIKEN SmartLab X-ray diffractometer. Before the experiment, the shale samples were grinded to a particle size >200 mesh using an agate mortar. The XRD experiments were performed at 40 kV and 200 mA, and the scanning range of minerals varied from 3° to 80° with a step size of 0.02°. A Quanta 650 equipment of FEI Company was chosen to conduct QEMSCAN analysis following the standard SY/T 6189-2018 [48]. Before the QEMSCAN, the shale samples were polished by an argon ion beam for 4 h to achieve a proper surface roughness of 1 μm in a scanning area of 1.5 × 1.5 mm^2.

2.3. Low-Temperature Nitrogen Adsorption Measurements

In this work, the pore size ranges of macropore and mesopore were determined according to the IUPAC classification [49], which has been broadly used in the pore characterization of coals and shales [27,30,36]. Thus, the pore sizes of macropore and mesopore were >50 nm and 2–50 nm, respectively. The meso- and micropore parameters were analyzed using a fully automatic specific surface analysis tester (ASAP 2460, Micromeritics,

Norcross, GA, USA). The N_2 at 77 K as a molecular probe was chosen to detect the pore characteristics of the shale samples. Before the experiments, the shale samples were placed in a vacuum drying oven and were dried at 110 °C for 12 h to remove the residual moisture and gas from the samples. The relative pressure (P/P_0) of the low-temperature N_2 adsorption/desorption isotherm was between 0.019 and 0.995. Based on the low-temperature N_2 adsorption/desorption curves, the specific surface area (SSA) and pore volume (PV) of the shale samples were obtained by the BET model and BJH model, respectively. The pore size distribution was obtained based on the DFT model, and the average pore diameter (APD) was obtained by the BJH model.

2.4. Multifractal Analysis

The multifractal singularity spectrum (α-$f(\alpha)$) and generalized dimensional spectrum (q-D_q) are equivalent to multifractal descriptions [50]. In this paper, relatively simple q-D_q spectra were used to characterize the pore heterogeneity of the shale samples [51]. As suggested by the work of Caniego et al. [52], the relative pressure P/P_0 is taken as the interval J, and the interval J of length L is divided into $N(\varepsilon) = 2^k$ boxes of scale ε based on the dyadic scaling down. In order to allow the smallest subinterval to contain measured pore volumes, the k values derived from low-temperature N_2 adsorption were from 0 to 3. The generalized dimensional spectra require the determination of four functions: mass probability $p_i(\varepsilon)$, partition function $\chi(q, \varepsilon)$, mass exponent $\tau(q)$, and generalized fractal dimension D_q.

(1) The probability mass distribution $p_i(\varepsilon)$ is the key to multifractal analysis to define a mass probability at different scales. These components will help to quantitatively characterize the local features of the pore volume distribution [53] and is defined in each box as:

$$p_i(\varepsilon) = \frac{N_i(\varepsilon)}{N_t} \tag{1}$$

where N_i is N_2 adsorption amount of the ith (i = 1, 2, 3...) box, and N_t is the total N_2 adsorption amount.

(2) The partition function $\chi(q, \varepsilon)$ can be calculated from $p_i(\varepsilon)$ function by using:

$$\chi(q, \varepsilon) = \sum_{i=1}^{N(\varepsilon)} p_i^q(\varepsilon) \tag{2}$$

where q is the order of statistical moments in the range of $[-\infty\ +\infty]$. In this paper, q varies from -10 to 10. When $q >> 1$, the information on the high probability of pore volume is amplified [50]. When $q << -1$, the information on the low probability of pore volume is amplified [34]. Therefore, the multifractal method divides the pore volume with different pore sizes into local high- or local low-porosity regions.

(3) The mass exponent $\tau(q)$ for fractals is a power law relationship between $\chi(q, \varepsilon)$ and ε [52]:

$$\chi(q, \varepsilon) \propto \varepsilon^{-\tau(q)} \tag{3}$$

where $\tau(q)$ can be obtained from the slope of the double logarithmic curve of $\chi(q, \varepsilon)$ versus ε.

(4) The generalized fractal dimension D_q. D_q is related to $\tau(q)$ as follows [53]:

$$\tau(q) = (q-1)D_q \tag{4}$$

In combination with Equations (3) and (4), the D_q can be given by:

$$D_q = \lim_{\varepsilon \to 0} \frac{1}{q-1} \frac{\log[\chi(q, \varepsilon)]}{\log(\varepsilon)} \tag{5}$$

where $q \neq 1$. For $q = 1$, D_1 can be obtained by L'Hôpital rule [51]:

$$D_1 = \lim_{\varepsilon \to 0} \frac{\sum_{i=1}^{N(\varepsilon)} \chi_i(1,\varepsilon) \log[\chi_i(1,\varepsilon)]}{\log(\varepsilon)} \quad (6)$$

The D_q values at $q = 0$, $q = 1$, and $q = 2$ are named capacity dimension D_0, information dimension D_1, and correlation dimension D_2, respectively. For monofractal, $D_0 = D_1 = D_2$ and the D_q spectrum is a horizontal line. For multifractal, the D_q spectrum is a monotonically decreasing function of q and $D_0 > D_1 > D_2$. According to Riedi et al. [54], D_2 can be replaced by Hurst exponent H as:

$$D_2 = 2H - 1 \quad (7)$$

where H indicates the autocorrelation of pore volume distribution over the set of pore sizes related to long-range spatial variation [36,51,55].

3. Results and Discussion

3.1. TOC Content and Mineral Compositions of Shale Samples

The TOC contents and mineral compositions of the shale samples are listed in Table 1. The TOC values vary from 0.396% to 2.417%, with an average value of 1.285%. This means both the organic-rich and organic-poor shale samples can be identified. Furthermore, previous studies [56–60] have generally categorized the shale types based on the contents of siliceous minerals (quartz and feldspar) and carbonate and clay minerals from XRD analysis. Thus, the studied shale samples can be classified into three types, as presented in the ternary plot (Figure 2), namely carbonate and mixed and argillaceous shales. This is consistent with the findings by He et al. [13], who have categorized the primary types of Ziliujing shales into the same three groups based on XRD results of 70 samples. Hence, the 12 samples chosen are considered to be representative of the studied interval in the reservoir. For carbonate shales, the calcite contents are larger than 50%, while both the contents of quartz and clay are less than 20% (Table 1). The contents of calcite, quartz, and clay in mixed shales are close, within the ranges of 26.9–35.3%, 27.8–30.7%, and 28.2–29.5%, respectively (Table 1). For argillaceous shales, the clay contents are higher than 50% (51.2–59.0%), and the quartz contents vary from 26.3% to 29.6%, while calcite contents are below 5% (Table 1). The mineral compositions of different samples show a significant impact on pore distributions and pore heterogeneity, which will be discussed in the following Section 3.3.

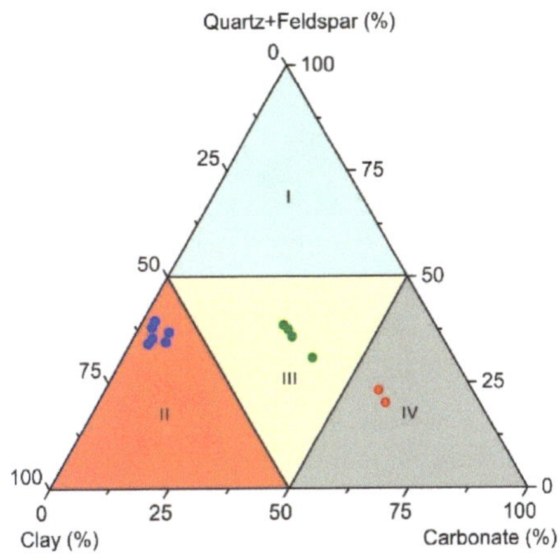

Figure 2. Mineral composition characteristics of the shale samples. (I: siliceous shales; II: argillaceous shales; III: mixed shales; IV: carbonate shales. The red dots represent carbonate shales, the green dots represent mixed shales, and the blue dots represent argillaceous shales).

The QEMSCAN images of the three typical types of shales can be seen in Figure 3. The carbonate shale (sample D97) is mainly composed of calcite (65.18%) with scarce organic matter (0.47%) (Figure 3a). This result is consistent with the XRD analysis. However, the content of quartz from QEMSCAN in sample D97 is only 7.9%, which is less than that obtained from XRD analysis. The heterogeneity of shale is strong at the nano and micro scale, which means, in comparison to QEMSCAN, XRD analysis of the power samples will be more accurate and free from such heterogeneity. In this regard, QEMSCAN can directly observe the distributions of mineral compositions, except for the quantitative contents of minerals. For carbonate shale (Figure 3a), the calcites exhibit a layered structure with illite embedded in calcite particles, and the pyrite aggregates are randomly distributed in the samples. In terms of the mixed shale (sample D123), the clay and quartz contents are 29.45% and 28.93%, respectively (Figure 3b), while the clay and quartz contents of the argillaceous shale (sample D71) with a lesser amount of calcite (3.80%) is 50.38% and 23.85%, respectively (Figure 3c). Both the mixed and argillaceous shales exhibit dense massive textures, which demonstrates an ultralow porosity and permeability in the shale samples should be expected.

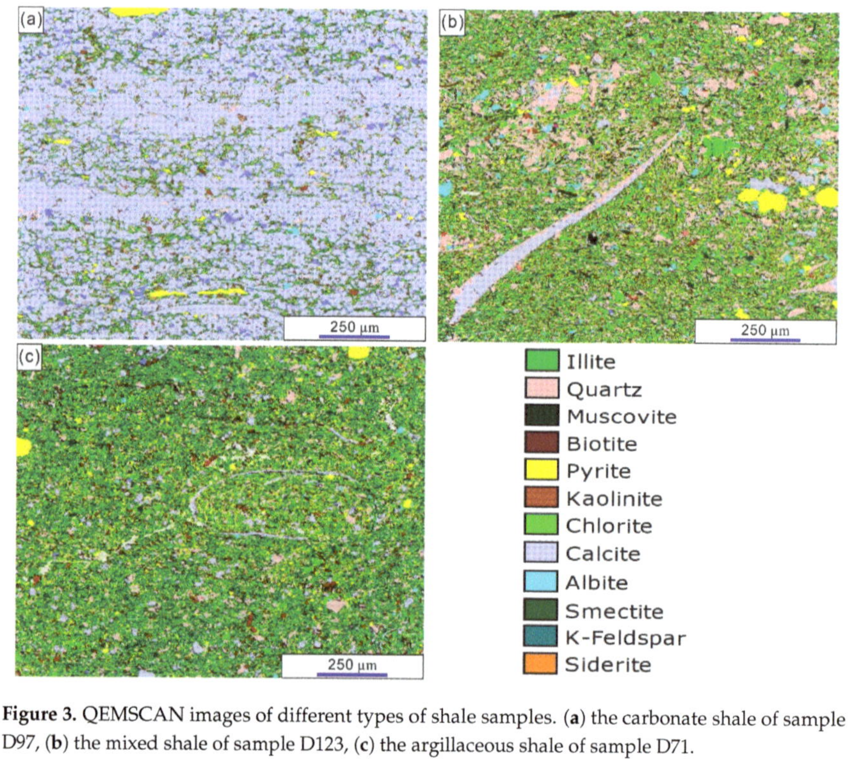

Figure 3. QEMSCAN images of different types of shale samples. (**a**) the carbonate shale of sample D97, (**b**) the mixed shale of sample D123, (**c**) the argillaceous shale of sample D71.

3.2. Meso- and Macropore Characteristics of the Shale Samples

Figure 4 shows the low-temperature N_2 adsorption/desorption isotherms of the shale samples. According to the gas adsorption classification [61], the isotherms of all the samples belong to type IV curves. The desorption isotherms of all the samples are located in the upper part of the adsorption isotherms, and a clear inflection point occurs at a relative pressure of 0.5, where the adsorption hysteresis occurs. The types of the adsorption hysteresis loop can be used to determine the pore shape of the adsorbent [62]. The adsorption hysteresis loops of all the samples are of H3 type (Figure 4), suggesting that the pore morphology is dominated by wide slit-type pores. However, there are obvious differences in the N_2 adsorption amounts of different types of shales. The N_2

adsorption values of both carbonate and mixed shales are less than 11 cm^3/g (Figure 4a), while those of argillaceous shales are higher than 15 cm^3/g (Figure 4b). The variations of N_2 adsorption amounts indicate that obvious differences exist in the pore structure of different shale samples. The pore structure parameters of the shale samples derived from the low-temperature N_2 adsorption/desorption isotherms are presented in Table 2.

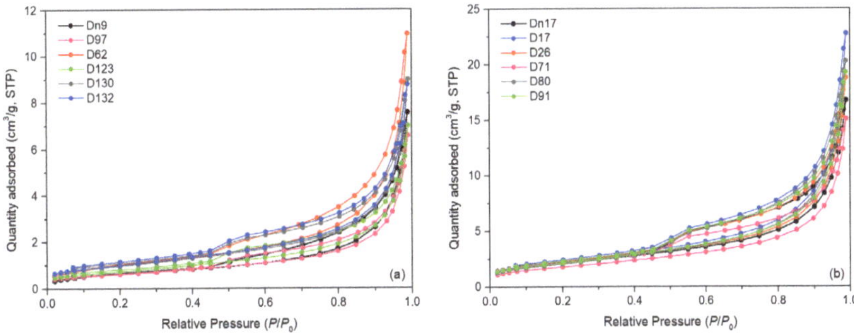

Figure 4. Low-temperature N_2 adsorption/desorption isotherms of the shale samples. (**a**) LNA curves of carbonate and mixed shale samples, (**b**) LNA curves of argillaceous shale samples.

Table 2. Pore parameters of the shale samples derived from LNA methods.

Samples	Shale Type	S_{BET} (m^2/g)	Pore Volume (cm^3/g)			APS (nm)
			V_T	V_{meso}	V_{macro}	
Dn9	carbonate	2.32	0.0125	0.0083	0.0041	14.01
D97		2.23	0.0109	0.0073	0.0036	14.99
D62	mixed	3.68	0.0181	0.0124	0.0057	12.07
D123		2.71	0.0115	0.0075	0.0041	11.43
D130		3.50	0.0146	0.0094	0.0052	11.97
D132		3.66	0.0143	0.0099	0.0044	11.03
Dn17	argillaceous	7.46	0.0277	0.0206	0.0071	10.09
D17		8.07	0.0376	0.0265	0.0111	11.98
D26		7.62	0.0310	0.0227	0.0083	10.92
D71		6.48	0.0246	0.0181	0.0065	10.15
D80		7.52	0.0338	0.0254	0.0085	11.27
D91		7.77	0.0319	0.0237	0.0083	10.85

Note: V_T—total pore volume; V_{meso}—mesopore volume (2–50 nm); V_{macro}—macropore volume (>50 nm); APS—average pore size.

As shown in Table 2, the mesopore volume of all the samples is higher than the macropore volume. This change is similar to the pore variations in samples from the Longmaxi Formation [22,35]. However, there are clear differences in the pore parameters of different types of shales. The carbonate shales exhibited the lowest SSA (2.23–2.32 m^2/g) and PV (0.0109–0.0125 cm^3/g) but the largest APD (14.50 nm on average). Compared with carbonate shale, the SSA and PV of mixed shales were found to increase by 1.13 m^2/g and 0.003 cm^3/g on average, respectively, and the meso- and macropore volumes increased by 25.46% and 25.08%, respectively, while the APD decreased to 11.62 nm on average. The SSA and PV of argillaceous shales showed a significant increase within the ranges of 6.48–8.07 m^2/g and 0.0246–0.0376 cm^3/g, respectively. In comparison with mixed shales, the mesopore volume of argillaceous shales increased by 2–4 times, and the macropore volume increased by 71.77% on average, while the APD was found to be the lowest (10.88 nm on average).

The pore size distributions (PSDs) can reveal subtle variations in the pore structure of different shale samples (Figure 5). From Figure 5, the peaks of PSDs of carbonate shales are concentrated at 9 nm, and the porosity is found between 2 nm and 30 nm. The PSDs of mixed shales become wider, while the peaks are left shifted to 3.4 nm. Except for the D123 sample, the number of pores with a diameter of less than 35 nm of mixed shales increased significantly, resulting in an increased mesopore volume and SSA. The pore number with a diameter less than 8 nm in D123 samples increased, but the opposite trend is recorded for pore diameter >8 nm. As a result, the meso- and macroporosity increased slightly. The PSDs of argillaceous shales are similar to those of mixed shales. However, the PV in the whole pore size range of clay shales increased, especially for the pores with size <30 nm. Thus, the mesopore volume and SSA sharply increased, but the APD decreased. The results indicate that the increase in mesopore volume of continental shale will prompt the formation of pores with small size and enlarged SSA, which is consistent with the previous studies [16,22,35].

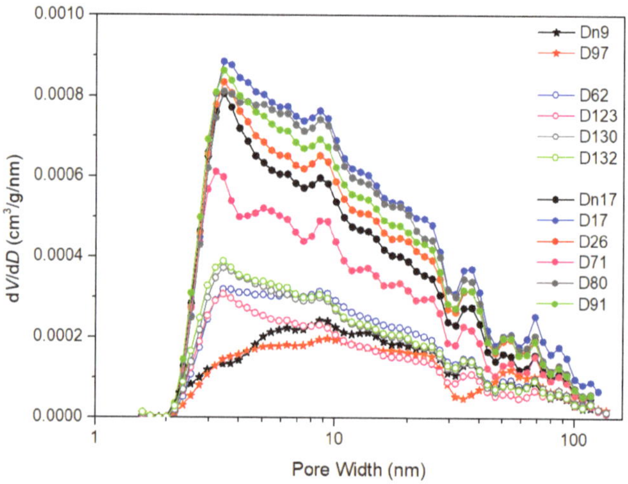

Figure 5. Pore size distributions of the shale samples derived from N_2 adsorption methods.

3.3. *Effect of Mineral Compositions on Meso- and Macropore Structures*

Many studies have shown that mineral compositions and TOC are the main factors that would affect shale pore structure parameters [16,22,63,64]. Figure 6 depicts the relationship between the mineral compositions with SSA, V_T, V_{meso}, and V_{macro}. A strong positive correlation can be seen between clay mineral contents and SSA, TPV, V_{meso}, and V_{macro}. All the complex correlation coefficients, R^2, are greater than 0.89 (Figure 6(a_1,a_2)). Nevertheless, a significant negative correlation ($R^2 > 0.70$) is observed between carbonate content and the above pore parameters (Figure 6(b_1,b_2)). Furthermore, we did not observe any clear relationship between the amounts of quartz and feldspar (QF) and pore parameters (Figure 6(c_1,c_2)). As described in Figure 6(a_2,b_2), the variations of R^2 demonstrate that the influence of clay and carbonate on V_{meso} is greater than that on V_{macro}. The mentioned results illustrate that clay and carbonate mineral content are the main factors affecting meso- and macropore distributions, but negatively. Previous works [22,27,35,43] have stated that shales with high clay mineral content generally own high meso- and macropore volumes. In this regard, Rexer et al. [65] explained that carbonate precipitation would hinder pore development, which means the higher the content of carbonate minerals, the lower the volume of mesopores and macropores should be. In this study, the amounts of QF among different shale samples were found to be similar, which leads to a very weak correlation between QF and pore parameters.

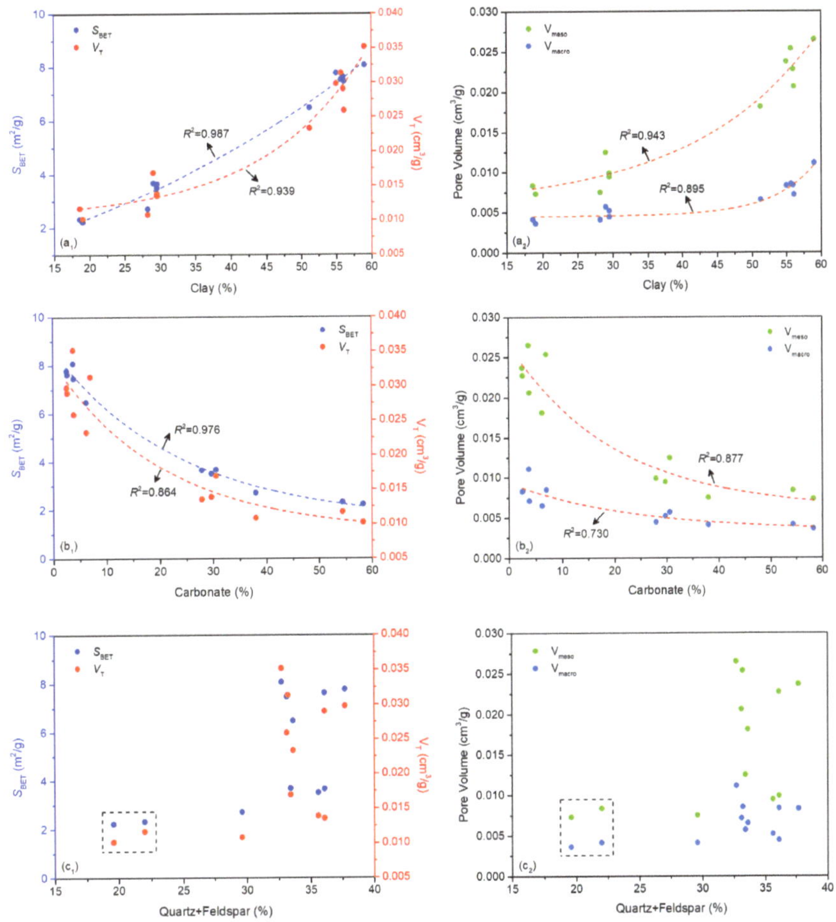

Figure 6. Relationships of mineral compositions and SSA, TPV, V_{meso}, and V_{macro}. (**a$_1$–c$_1$**) Relationships of clay, carbonate, and QF with SSA and TPV, respectively. (**a$_2$–c$_2$**) Relationships of clay, carbonate, and QF with meso- and macropore volumes, respectively. (Dotted boxes represent pore parameters of carbonate shales).

The relationships between TOC and SSA and pore volume are plotted in Figure 7. Ambrose et al. [66] found that TOC played an important role in the distributions of PV and SSA. Liu et al. [67] analyzed the relationships of TOC with V_{meso} and V_{macro} of continental shales from the Yanchang Formation in the Ordos Basin by low-temperature N_2 adsorption method. They found that TOC encouraged the development of V_{meso}, V_{macro}, and SSA. Similar trends were observed in marine shales by Wu et al. [68]. However, no correlations can be observed here between TOC and pore parameters. Jiang et al. [22] considered that the weak correlations between TOC and pore parameters might originate from six different factors, that is, organic matter preservation process to form TOC, organic matter biogenic origin, thermal maturity, intercrystalline pores between pyrites, secondary enlargement of quartz, and the occurrence of irreducible fluids. Therefore, the ambiguous relationships between TOC and pore parameters from N_2 adsorption might be due to multiple factors that need further study.

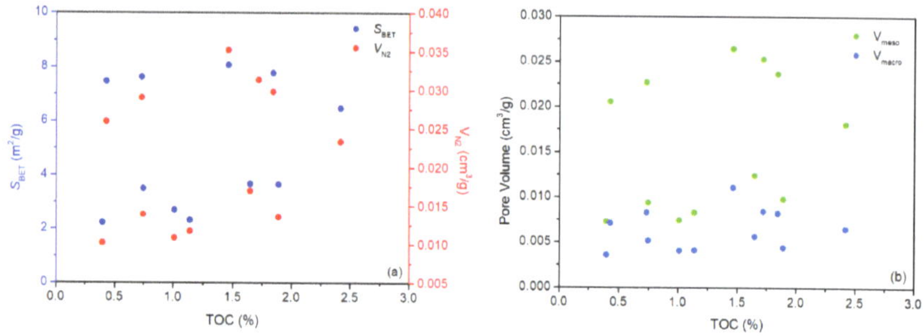

Figure 7. Relationships of TOC with SSA, V_T, V_{meso}, and V_{macro}. (**a**) Relationships of TOC with SSA and V_T, respectively. (**b**) Relationships of TOC with V_{meso} and V_{macro}, respectively.

3.4. Multifractal Characteristics of Meso- and Macropore Structures

According to Equations (1)–(6), the generalized dimension spectra (D_q vs. q) of the shale samples in the range of moment $q = -10$ to $q = 10$ are plotted in Figure 8. The relevant multifractal parameters are listed in Table 3. It can be seen that the D_q spectra of all the samples are monotonically decreasing as a function of q, and $D_0 > D_1 > D_2$. The results suggest that the pore size distributions of meso- and macropores of shales exhibit multifractal behavior. According to multifractal theory, it is known that the generalized spectra D_q and the corresponding characteristic parameters can describe the inner complexity or heterogeneity in the size-dependent distribution of pore volume [36,51].

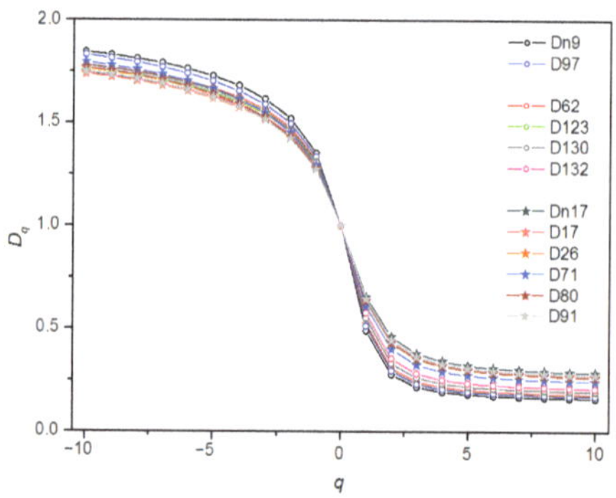

Figure 8. Generalized dimension D_q, versus q from $q = -10$ to $q = +10$ for PSDs of the shale samples.

Table 3. Multifractal parameters of pores derived from N_2 adsorption isotherms for shale samples.

Shales	Samples	D_0	D_1	H	D_{10}	D_{-10}	D_0–D_{10}	D_{-10}–D_0	D_{-10}–D_{10}
carbonate	Dn9	1.000	0.489	0.638	0.160	1.846	0.840	0.846	1.686
	D97	1.000	0.512	0.647	0.171	1.831	0.829	0.831	1.660
mixed	D62	1.000	0.526	0.654	0.179	1.781	0.821	0.781	1.602
	D123	1.000	0.552	0.665	0.193	1.768	0.807	0.768	1.575
	D130	1.000	0.551	0.665	0.193	1.781	0.807	0.781	1.588
	D132	1.000	0.575	0.677	0.208	1.743	0.792	0.743	1.535
argillaceous	Dn17	1.000	0.655	0.729	0.284	1.748	0.716	0.748	1.464
	D17	1.000	0.638	0.717	0.264	1.738	0.736	0.738	1.474
	D26	1.000	0.638	0.719	0.269	1.765	0.731	0.765	1.496
	D71	1.000	0.609	0.700	0.242	1.796	0.758	0.796	1.554
	D80	1.000	0.634	0.715	0.263	1.796	0.737	0.796	1.533
	D91	1.000	0.643	0.721	0.272	1.750	0.728	0.750	1.478

As seen in Figure 8 and Table 3, the multifractal characteristics of pores in different types of shales demonstrate significant discrepancies. The D_q spectra morphology and the width D_{-10}–D_{10} depict the variability of local porosity along the pore size intervals [53]. The wider the D_q spectra, the greater the local fluctuation in PSDs and the higher the complexity in pore structure. The carbonate shales exhibited the largest curvature of D_q spectra (Figure 8) and the largest values of D_{-10}–D_{10} (Table 3), indicating that carbonate minerals increase the multifractality of PSDs as well as the complexity of pore structure. The average values of D_{-10}–D_{10} for mixed and argillaceous shales are 1.575 and 1.499, respectively, denoting a higher degree of heterogeneity in PSDs of mixed shales.

Figure 8 shows that the left branches in D_{-10}–D_0 of the mixed shales and argillaceous shales are crossed and overlapped and cannot be distinguished well. However, the right branch D_0–D_{10} of the argillaceous shale is located above the mixed shales, and the D_q values of the right branch of carbonate shales are the smallest (Figure 8). As listed in Table 3, the D_0–D_{10} values of carbonate shales are greater than 0.82, and the average values for mixed shales and argillaceous shales are 0.81 and 0.72, respectively. Thus, D_0–D_{10} can be used as an effective parameter to distinguish the inner variations in PSDs of different types of shales. The left and right branches of the D_q spectrum represent different information on the variables. The right branch D_0–D_{10} ($q > 0$) and the left branch D_{-10}–D_0 ($q < 0$) correspond to the dominance of high and low concentrations of pore volume, respectively [36,51]. In combination with Figure 5, the pore volume distributions with size <50 nm (mesopores) lead to differential distributions among various shale samples. Thus, the right side D_0–D_{10} for $q > 0$ reflects the various distributions of mesopores and the left side D_{-10}–D_0 for $q < 0$ represents the inner complexity in the distribution of macropores.

D_1 and H characterize the concentration of PSDs and the autocorrelation of pore connectivity, respectively [36,51]. The smaller D_1 and H, the greater the fluctuation in the inner distributions of pores would be, the narrower the PSDs, and the lower pore connectivity along pore size intervals should be expected. On the contrary, uniformity in the PSDs in each pore size interval and stronger connectivity among different pore size intervals can be understood. The D_1 and H values of carbonate shale are the smallest, indicative of concentrated PSDs. This is compatible with the results in Section 3.3. Compared with carbonate shale, the D_1 and H values of mixed shales increased, varying from 0.52–0.58 and 0.65–0.68, respectively. This infers weakened pore clustering and enhanced pore connectivity. Combined with Figure 5, the pore volume with a size less than 10 nm in mixed shale obviously increased, thus widening the PSDs and enhancing the connectivity between the pores. As a result, the values of D_1 and H increase. The D_1 and H values of argillaceous shales continue to increase with the average values of 0.64 and 0.72, respectively. The results show that argillaceous shales have a relatively homogeneous distribution of pore volume over different pore diameter intervals. The PSDs of argillaceous shales in Figure 5 suggest that the rapid increase in mesopore volume may be attributed to the strong autocorrelation and pore connectivity among the pore size range, which is the main reason for the difference in D_1 and H of different types of shales.

3.5. Relationships of Pore Parameters, Mineral Compositions, TOC, and Multifractal Parameters from N_2 Adsorption

The correlations of pore parameters (V_T, V_{meso}, V_{macro}), mineral compositions, TOC, and multifractal parameters were conducted by a linear regressive analysis (Table 4). At the $p = 0.01$ level, V_T, V_{meso}, V_{macro}, and clay show strong linear positive correlations to D_1 and H and negative correlations to D_0–D_{10} ($R > 0.8$). On the contrary, carbonate content displays a strongly negative correlation to D_1 and H and a positive correlation to D_0–D_{10} ($R > 0.9$). QF also exhibits some positive or negative correlation to the above-mentioned multifractal parameters, although the degree of correlation decreases with correlation coefficients R ranging from 0.582–0.645 at the $p = 0.1$ level. V_T, V_{meso}, V_{macro}, and QF exhibit strong negative correlations to D_{-10}–D_{10} at the $p = 0.05$ level, while carbonate and clay are positively and negatively correlated to D_{-10}–D_{10} at the $p = 0.01$ level, respectively.

A lower correlation ($0.392 < R < 0.525$) can be found between V_T, V_{meso}, V_{macro}, and clay and D_{-10}–D_0. Carbonate shows a somewhat positive correlation to D_{-10}–D_0 at the $p = 0.1$ level, while a strong negative correlation exists between QF and D_{-10}–D_0. Moreover, a weak relationship between TOC and multifractal parameters is recorded ($0.394 < R < 0.442$).

Table 4. Relationships of V_T, V_{meso}, V_{macro}, mineral compositions, TOC, and multifractal parameters.

	D_1	H	D_0–D_{10}	D_{-10}–D_{10}	D_{-10}–D_0
V_T	0.875 **	0.893 **	−0.895 **	−0.782 *	−0.414
V_{meso}	0.886 **	0.906 **	−0.909 **	−0.782 *	−0.392
V_{macro}	0.807 **	0.820 **	−0.819 **	−0.754 *	−0.463
carbonate	−0.952 **	−0.946 **	0.941 **	0.912 **	0.646 †
clay	0.966 **	0.974 **	−0.974 **	−0.879 **	−0.525
QF	0.645 †	0.601 †	−0.582 †	−0.744 *	−0.764 *
TOC	0.442	0.415	−0.398	−0.416	−0.394

Note: **, significant at $p = 0.01$ confidence level; *, significant at $p = 0.05$ confidence level; †, significant at $p = 0.1$ confidence level.

The multifractal parameters (D_1, H, and D_0–D_{10}) for moment $q > 0$ denote the fractal characteristics of high concentrations of pore volume. As analyzed in Section 3.3, the variations in mesopore volume are the main factor affecting the pore distributions of different shale samples. Consequently, a strongly positive relationship between mesopore volume and multifractal parameters (D_1, H and D_0–D_{10}) for $q > 0$ is observed. The changes in pore distributions and regression results suggest that, in the shale samples studied, the increasing mesopore volume due to increased clay contents is accompanied by lower fluctuation, lower clustering, higher autocorrelation, and lower heterogeneity in the inner size-dependent distribution of pores as indicated by multifractal parameters for $q > 0$. However, the increased carbonate mineral contents play a reverse role.

Although a lower correlation between QF and the multifractal parameters for $q > 0$ exists when compared with clay and carbonate, the increased QF contents contribute to the widening of pore distribution, improve pore connectivity, and reduce the complexity of pore distribution. The multifractal parameters (D_{-10}–D_0) for $q < 0$ describe the complexity of small concentrations of pore volume, which characterizes the inner heterogeneous distribution of macropores. Although the correlation between macropore volume and D_{-10}–D_0 is low, the increasing macropore volume caused by clay mineral content still reduces the complexity in the inner macropore distributions as estimated by the negative correlation coefficient. However, carbonate enhances the complexity and fluctuation of the macropore distributions. In summary, D_1, H, and D_0–D_{10} can be considered to be the parameters that are important to distinguish the internal differences in meso- and macropore distributions along pore size intervals of the Ziliujing shales.

As mentioned above, limited parameters are not enough to demonstrate the multifractal heterogeneity of macropores. Previous research [69,70] showed that the pore size range of shale oil reservoirs varied from nanometers to millimeters, and the pores with a diameter larger than 100 nm (seepage-pores) are crucial to the transport and development of shale oil [71–73]. Thus, it is necessary to fully characterize the pore properties of seepage-pores by the combination of fluid-invasion methods and image analysis techniques, including MIP, low-field nuclear magnetic resonance, micro-CT, SEM, and others. Moreover, further comparison of the multifractal characteristics of pore structure from different methods of shale oil reservoirs is required.

3.6. Comparision of Multifractal Parameters of N_2 Adsorption among Different Shale Oil Reservoirs

Figure 9 depicts the comparison of multifractal parameters of N_2 adsorption among different shale oil reservoirs. The D_1 values of the Bakken shales in the US vary from 0.702 to 0.920, with an average of 0.797 [41]. The Ek2 shales in Cangdong Sag, Bohai Bay Basin, possess the largest D_1 values ranging from 0.927 to 0.982 [44]. The smallest D_1 values are recorded in the Ziliujing shales, and all the D_1 values are less than 0.7. On the contrary, the largest D_{-10}–D_{10} values can be observed in the Ziliujing shales, and the increasing order for the average value of D_{-10}–D_{10} is followed by Ek2 shales, Bakken shales, and Ziliujing shales. The results indicate that the meso- and macropore structures of the Ziliujing shales are much more heterogeneous than those of the Bakken shales and Ek2 shales.

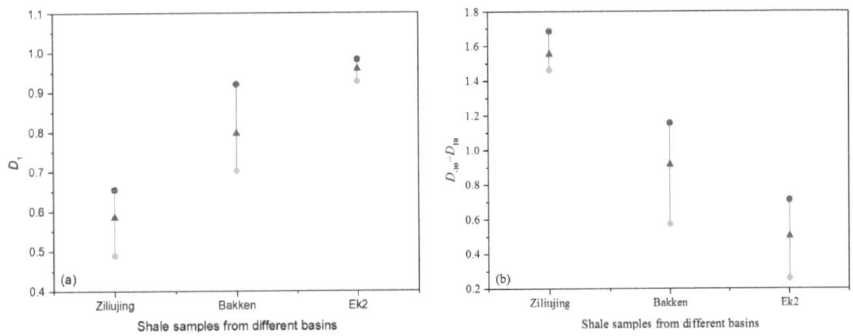

Figure 9. Comparison of multifractal parameters of nanopore heterogeneity of shale oil reservoirs from different basins (**a**) D_1 data of shale samples from different basins. (**b**) D_{-10}–D_{10} data of shale samples from different basins. (The red circle represents the maximum value, the green circle represents the minimum value, and the blue triangle represents the average value. The D_1 and D_{-10}–D_{10} data of the Bakken shale from Liu et al. [41], and the D_1 and D_{-10}–D_{10} data of the Ek2 shale from Guan et al. [44]).

The reasons for this discrepancy of nanopore heterogeneity of the shale oil reservoirs in different basins can be attributed to thermal maturity (R_o), mineral compositions, sedimentary environment, etc. [41,44]. Guan et al. [44] illustrated that the nanopore complexity subsequently increased when R_o is greater than 0.65%. Both the R_o of the Bakken shales from Liu et al. [41] and the Ek2 shales from Guan et al. [44] are less than 1.0%. The R_o of the Ziliujing shales in our study mostly varies from 1.3% to 1.5% [11]. Therefore, compared with the Bakken and Ek2 shales, the higher R_o may tend to increase the meso- and macropore heterogeneity of the Ziliujing shales. The impact on nanopore heterogeneity from inorganic mineral compositions of shale oil reservoirs from different basins is generally a complex inner relation. In this regard, Liu et al. [44] reported that quartz and clay played an opposite effect on the meso- and macropore heterogeneity of the Upper Bakken and Middle Bakken shales, which may be related to depositional sub-facies [70]. Thus, the effect of mineral compositions on nanopore heterogeneity of shale oil reservoirs could vary in different basins. This means that the impact of such components on pore heterogeneity, specifically composition, should be analyzed based on the geological setting and lithofacies in different basins.

4. Conclusions

In this study, the meso- and macropore characteristics of shales of the Ziliujing Formation of the Jurassic age in the Sichuan Basin were investigated by combining low-temperature N_2 adsorption techniques and multifractal method. Based on the results, the conclusions are as follows:

(1) The mineral compositions of the studied Ziliujing shale samples reveal that the shales can be divided into three types: carbonate shale, mixed shale, and argillaceous shale. The calcite contents of carbonate shale were found to be higher than 50%. The differences in calcite, quartz, and clay mineral contents were found to be small for mixed shale. For argillaceous shale, the clay mineral contents were found to be larger than 50%, and the calcite contents were less than 5%.

(2) There are clear differences in meso- and macropore characteristics among different types of shales. The carbonate shale exhibited the lowest specific surface area and pore volume but the largest average pore diameter. The argillaceous shale had the highest specific surface area and pore volume but the smallest average pore size. The pore parameters of the mixed shale were found to be in between. The mesopore volume is the main control of the differential distribution of pores in different types of shales. Clay minerals significantly increased the mesopore contents, while carbonate minerals had the opposite effect. TOC and the combined contents of quartz and feldspar had a weak effect on the pore distributions.

(3) The meso- and macropore distributions of the Ziliujing shales exhibited multifractal behaviors. The information dimension D_1, the Hurst exponent H, and the width of the right side D_0–D_{10} of the D_q spectrum were found effective parameters to distinguish the local variations within the pore distributions of different types of Ziliujing shale samples. As indicated by the change of the multifractal parameters, for the Ziliujing shales studied, clay minerals and quartz and feldspar contents caused a reduction in the pore clustering, enhanced pore connectivity, and weakened the complexity of pore distribution, while carbonate minerals played the opposite role. Therefore, the multifractal approach helps to improve the understanding of internal heterogeneity and differences in meso- and macropore distributions of different types of shales.

(4) A further step is needed to characterize the pores of the Ziliujing shales with sizes varying from nanometers to micrometers in combination with different methods, such as mercury injection and gas adsorption (CO_2 and N_2), an integration of small-angle neutron scattering techniques (SANS and USANS), and micro-CT. Additionally, the analysis of multifractal properties of seepage-pores heterogeneity is necessary, which has implications for permeability calculation, fluid transport, and future development plans.

Author Contributions: Conceptualization and methodology, W.L.; validation, Z.W., W.L. and W.M.; formal analysis, M.S. and L.C.; investigation, Y.Z.; resources, X.W. and X.F.; data curation, X.W.; writing—original draft preparation, Y.W.; writing—review and editing, Y.W. and W.L.; visualization, B.L. and M.S.; supervision, X.F.; project administration, Y.Z.; funding acquisition, Y.Z. All authors have read and agreed to the published version of the manuscript.

Funding: The research work is funded by Heilongjiang Province "Hundred Million" Engineering of Major Projects in Science and Technology (SC2020ZX05A0023), The Scientific Research and Technological Development Project of China National Petroleum Corporation (2022DJ1809).

Institutional Review Board Statement: Not applicable.

Informed Consent Statement: Not applicable.

Data Availability Statement: Not applicable.

Acknowledgments: This work is supported by the National Natural Science Foundation of China (No. 41902176), Outstanding Youth Fund of (No. YQ2021D005), University Nursing Program for Young Scholars with Creative Talents in Heilongjiang Province (No. UNPYSCT-2020145), Heilongjiang Postdoctoral Financial Assistance (No. LBH-Z19121), and Youth Science Fund of Northeast Petroleum University (No. 2018QNL-24).

Conflicts of Interest: The authors declare no conflict of interest.

References

1. Xu, Y.; Lun, Z.M.; Pan, Z.J.; Wang, H.T.; Zhou, X.; Zhao, C.P.; Zhang, D.F. Occurrence space and state of shale oil: A review. *J. Pet. Sci. Eng.* **2022**, *211*, 110183. [CrossRef]
2. Yu, Y.J.; Wang, H.Y.; Liu, D.X.; Zhao, Q.; Li, X.B.; Wu, J.; Xia, Z.Y. Development Status and Feasibility Evaluation Index System of Continental Shale Oil Demonstration Area in China. *Earth Sci.* **2023**, *48*, 191–205. (In Chinese with English abstract)
3. Zou, C.N.; Ma, F.; Pan, S.Q.; Zhang, X.S.; Wu, S.T.; Fu, G.Y.; Wang, H.J.; Yang, Z. Formation and distribution potential of global shale oil and the development of continental shale oil theory and technology in China. *Earth Sci. Front.* **2023**, *30*, 128–142. (In Chinese with English abstract)

4. Du, J.H.; Hu, S.Y.; Pang, Z.L.; Lin, S.H.; Hou, L.H.; Zhu, R.K. The types, potentials and prospects of continental shale oil in China. *China Pet. Explor.* **2019**, *24*, 560–568. (In Chinese with English abstract)
5. Jin, Z.Y.; Wang, G.P.; Liu, G.X.; Gao, B.; Liu, Q.Y.; Wang, H.L.; Liang, X.P.; Wang, R.Y. Research progress and key scientific issues of continental shale oil in China. *Acta Pet. Sin.* **2021**, *42*, 821–835. (In Chinese with English abstract)
6. Sun, L.D.; He, W.Y.; Feng, Z.H.; Zeng, H.S.; Jiang, H.; Pan, Z.J. Shale oil and gas generation process and pore fracture system evolution mechanisms of the continental Gulong Shale, Songliao Basin, China. *Energy Fuels* **2022**, *36*, 6893–6905. [CrossRef]
7. Guo, Q.L.; Pan, S.Q.; Yang, F.; Yao, Y.; Zheng, H. Characterizing shale oil occurrence in the Yanchang formation of the Ordos Basin, Assisted by Petrophysical and Geochemical Approaches. *Energy Fuels* **2022**, *36*, 370–381. [CrossRef]
8. Lv, J.H.; Jiang, F.J.; Hu, T.; Zhang, C.X.; Huang, R.D.; Hu, M.L.; Xue, J.; Huang, L.L.; Wu, Y.P. Control of complex lithofacies on the shale oil potential in ancient alkaline lacustrine basins: The Fengcheng Formation, Mahu Sag, Junggar basin. *Geoenergy Sci. Eng.* **2023**, *224*, 211501. [CrossRef]
9. Zhi, D.M.; Xiang, B.L.; Zhou, N.; Li, E.T.; Zhang, C.J.; Wang, Y.C.; Cao, J. Contrasting shale oil accumulation in the upper and lower sweet spots of the lacustrine Permian Lucaogou Formation, Junggar Basin, China. *Mar. Pet. Geol.* **2023**, *150*, 106178. [CrossRef]
10. Meng, M.M.; Peng, J.C.; Ge, H.K.; Ji, W.M.; Li, X.M.; Wang, Q.Y. Rock fabric of lacustrine shale and its influence on residual oil distribution in the Upper Cretaceous Qingshankou Formation, Songliao Basin. *Energy Fuels* **2023**, *37*, 7151–7160. [CrossRef]
11. Yang, Y.M.; Yang, J.J.; Yang, G.; Tao, S.Z.; Ni, C.; Zhang, B.; He, X.D.; Lin, J.P.; Huang, D.; Liu, M.; et al. New research progress of Jurassic tight oil in central Sichuan Basin. *Pet. Explor. Dev.* **2016**, *43*, 873–882. (In Chinese with English abstract) [CrossRef]
12. Cheng, S.J.; Zhang, H.X.; Lu, J.G.; Yang, Y.M.; Liu, C.W.; Li, W.; Zou, X.L.; Yang, J.J.; Tang, H.P.; Yao, Y.T.; et al. Controlling factors of Jurassic Da'anzhai Member tight oil accumulation and high production in central Sichuan Basin, SW China. *Pet. Explor. Dev.* **2015**, *42*, 186–193. (In Chinese with English abstract)
13. Wenyuan, H.; Xuefeng, B.A.; Qi'an, M.; Junhui, L.; Dazhi, Z.; Youzhi, W. Accumulation geological characteristics and major discoveries of lacustrine shale oil in Sichuan Basin. *Acta Pet. Sin.* **2022**, *43*, 885–898. (In Chinese with English abstract)
14. Li, Z.; Lei, Z.; Shen, W.; Martyushev, D.A.; Hu, X. A comprehensive review of the oil flow mechanism and numerical simulations in shale oil reservoirs. *Energies* **2023**, *16*, 3516. [CrossRef]
15. Liu, B.; Wang, Y.; Tian, S.S.; Guo, Y.L.; Wang, L.; Yasin, Q.; Yang, J.G. Impact of thermal maturity on the diagenesis and porosity of lacustrine oil-prone shales: Insights from natural shale samples with thermal maturation in the oil generation window. *Int. J. Coal Geol.* **2022**, *261*, 104079. [CrossRef]
16. Chalmers, G.R.; Bustin, R.M.; Power, I.M. Characterization of gas shale pore systems by porosimetry, pycnometry, surface area, and field emission scanning electron microscopy transmission electron microscopy image. *AAPG Bull.* **2012**, *96*, 1099–1119. [CrossRef]
17. Martyushev, D.A.; Ponomareva, I.N.; Osovetsky, B.M.; Kazymov, K.P.; Tomilina, E.M.; Lebedeva, A.S.; Chukhlov, A.S. Study of the structure and development of oil deposits in carbonate reservoirs using field data and X-ray microtomography. *Georesursy* **2022**, *24*, 114–124.
18. Sun, M.D.; Zhao, J.L.; Pan, J.Z.; Hu, Q.H.; Yu, B.S.; Tan, Y.L.; Sun, L.W.; Bai, L.F.; Wu, C.M.; Blach, T.P.; et al. Pore characterization of shales: A review of small angle scattering technique. *J. Nat. Gas Sci. Eng.* **2020**, *78*, 103294. [CrossRef]
19. Meng, M.; Ge, H.; Shen, Y.; Ji, W.; Li, Z. Insight into water occurrence and pore size distribution by nuclear magnetic resonance in marine shale reservoirs, southern China. *Energy Fuels* **2023**, *37*, 319–327. [CrossRef]
20. Martyushev, D.A.; Ponomareva, I.N.; Chukhlov, A.S.; Davoodi, S.; Osovetsky, B.M.; Kazymov, K.P.; Yang, Y. Study of void space structure and its influence on carbonate reservoir properties: X-ray microtomography, electron microscopy, and well testing. *Mar. Pet. Geol.* **2023**, *151*, 106192. [CrossRef]
21. Liu, Q.; Sun, M.D.; Sun, X.D.; Liu, B.; Ostadhassan, M.; Huang, W.X.; Chen, X.X.; Pan, Z.J. Pore network characterization of shale reservoirs through state-of-the-art X-ray computed tomography: A review. *Gas Sci. Eng.* **2023**, *113*, 204967. [CrossRef]
22. Jiang, F.J.; Chen, D.; Wang, Z.F.; Xu, Z.Y.; Chen, J.; Liu, L.; Huyan, Y.Y.; Liu, Y. Pore characteristic analysis of a lacustrine shale: A case study in the Ordos Basin, NW China. *Mar. Pet. Geol.* **2016**, *73*, 554–571. [CrossRef]
23. Zhao, J.H.; Jin, Z.J.; Hu, Q.H.; Jin, Z.K.; Barber, T.J.; Zhang, Y.X.; Bleuel, M. Integrating SANS and fluid-invasion methods to characterize pore structure of typical American shale oil reservoirs. *Sci. Rep.* **2017**, *7*, 15413. [CrossRef]
24. Zhang, P.F.; Lu, S.F.; Li, J.Q. Characterization of pore size distributions of shale oil reservoirs: A case study from Dongying sag, Bohai Bay basin, China. *Mar. Pet. Geol.* **2019**, *100*, 297–308. [CrossRef]
25. Hou, L.H.; Wu, S.T.; Jing, Z.H.; Jiang, X.H.; Yu, Z.C.; Hua, G.L.; Su, L.; Yu, C.; Liao, F.R.; Tian, H. Effects of types and content of clay minerals on reservoir effectiveness for lacustrine organic matter rich shale. *Fuel* **2022**, *327*, 125043. [CrossRef]
26. Zheng, H.; Yang, F.; Guo, Q.L.; Pan, S.Q.; Jiang, S.; Wang, H. Multi-scale pore structure, pore network and pore connectivity of tight shale oil reservoir from Triassic Yanchang Formation, Ordos Basin. *J. Pet. Sci. Eng.* **2022**, *212*, 110283. [CrossRef]
27. Clarkson, C.R.; Solano, N.; Bustin, R.M.; Bustin, A.M.M.; Chalmers, G.R.L.; He, L.; Melnichenko, Y.B.; Radliński, A.P.; Blach, T.P. Pore structure characterization of North American shale gas reservoirs using USANS/SANS, gas adsorption, and mercury intrusion. *Fuel* **2013**, *103*, 606–616. [CrossRef]
28. Mastalerz, M.; Hampton, L.; Drobniak, A.; Loope, H. Significance of analytical particle size in low-pressure N_2 and CO_2 adsorption of coal and shale. *Int. J. Coal Geol.* **2017**, *178*, 122–131. [CrossRef]

29. Yang, F.; Ning, Z.F.; Liu, H.Q. Fractal characteristics of shales from a shale gas reservoir in the Sichuan Basin, China. *Fuel* **2014**, *115*, 378–384. [CrossRef]
30. Liang, L.X.; Xiong, J.; Liu, X.J. An investigation of the fractal characteristics of the Upper Ordovician Wufeng Formation shale using nitrogen adsorption analysis. *Mar. Pet. Geol.* **2015**, *27*, 402–409. [CrossRef]
31. Sun, L.N.; Tuo, J.C.; Zhang, M.F.; Wu, C.J.; Chai, S.Q. Pore structures and fractal characteristics of nano-pores in shale of Lucaogou formation from Junggar Basin during water pressure-controlled artificial pyrolysis. *J. Anal. Appl. Pyrolysis* **2019**, *140*, 404–412. [CrossRef]
32. Martyushev, D.A.; Chalova, P.O.; Davoodi, S.; Ashraf, U. Evaluation of facies heterogeneity in reef carbonate reservoirs: A case study from the oil field, Perm Krai, Central-Eastern Russia. *Geoenergy Sci. Eng.* **2023**, *227*, 211814. [CrossRef]
33. Friesen, W.I.; Mikula, R.J. Mercury porosimetry of coals: Pore volume distribution and compressibility. *Fuel* **1988**, *67*, 1516–1520. [CrossRef]
34. Wang, M.; Xue, H.T.; Tian, S.S.; Wilkins, R.W.T.; Wang, Z.W. Fractal characteristics of upper Cretaceous lacustrine shale from the Songliao Basin, NE China. *Mar. Pet. Geol.* **2015**, *67*, 144–153. [CrossRef]
35. Yang, R.; He, S.; Yi, J.Z.; Hu, Q.H. Nano-scale pore structure and fractal dimension of organic-rich Wufeng-Longmaxi shale from Jiaoshiba area, Sichuan Basin Investigations using FE-SEM, gas adsorption and helium pycnometry. *Mar. Pet. Geol.* **2016**, *70*, 27–45. [CrossRef]
36. Guan, M.; Liu, X.P.; Jin, Z.J.; Lai, J.; Sun, B.; Zhang, P.P.; Chen, K.F. The evolution of pore structure heterogeneity during thermal maturation in lacustrine shale pyrolysis. *J. Anal. Appl. Pyrolysis* **2022**, *163*, 105501. [CrossRef]
37. Lopes, P.; Betrouni, F. Fractal and multifractal analysis: A review. *Med. Image Anal.* **2009**, *13*, 634–649. [CrossRef]
38. Liu, K.Q.; Ostadhassan, M. Quantification of the microstructures of Bakken shale reservoirs using multi-fractal and lacunarity analysis. *J. Nat. Gas Sci. Eng.* **2017**, *39*, 62–71. [CrossRef]
39. Zhang, P.F.; Yin, Y.J.; Lu, S.F.; Li, J.Q.; Chang, X.C.; Zhang, J.J.; Pang, Y.M.; Chen, G.; Liu, Y.Q.; Li, Z. Insights into pore structures and multifractal characteristics of shale oil reservoirs: A case study from Dongying Sag, Bohai Bay Basin, China. *Energy Fuels* **2022**, *36*, 8224–8237. [CrossRef]
40. Lu, C.G.; Xiao, X.M.; Xue, Z.Q.; Chen, Z.X.; Li, G.; Feng, Y. Fractal and multifractal characteristics of nanopores and their controlling factors in marine-continental transitional shales and their kerogens from Qinshui Basin, Northern China. *Nat. Resour. Res.* **2023**, *32*, 2313–2336. [CrossRef]
41. Liu, K.; Ostadhassan, M.; Zou, J.; Gentzis, T.; Rezaee, R.; Bubach, B.; Carvajal-Ortiz, H. Multifractal analysis of gas adsorption isotherms for pore structure characterization of the Bakken Shale. *Fuel* **2018**, *219*, 296–311. [CrossRef]
42. Liu, K.Q.; Ostadhassan, M.; Gentzis, T.; Fowler, H. Image analysis of the pore structures: An intensive study for Middle Bakken. *J. Nat. Gas Sci. Eng.* **2019**, *61*, 32–45. [CrossRef]
43. Zhang, P.F.; Lu, S.F.; Li, J.Q.; Chang, X.C.; Li, J.J.; Li, W.B.; Chen, G.; Wang, S.Y.; Feng, W.J. Broad ion beam-scanning electron microscopy pore microstructure and multifractal characterization of shale oil reservoir: A case sample from Dongying sag, Bohai Bay basin, China. *Energy Explor. Exploit.* **2020**, *38*, 613–628. [CrossRef]
44. Guan, M.; Liu, X.P.; Jin, Z.J.; Lai, J. The heterogeneity of pore structure in lacustrine shales: Insights from multifractal analysis using N_2 adsorption and mercury intrusion. *Mar. Pet. Geol.* **2020**, *114*, 104150. [CrossRef]
45. Bai, X.F.; Huang, S.P.; Wang, X.D.; Wang, Z.G.; Wang, Y.Z.; Ma, W.Q.; Zhu, Y.P.; Sun, M.D.; Liu, B.; Cheng, L.J.; et al. Microscopic Analysis of Natural Fracture Properties in Organic-Rich Continental Shale Oil Reservoirs: A Case Study from the Lower Jurassic in the Sichuan Basin, China. *J. Mar. Sci. Eng.* **2023**, *11*, 1036. [CrossRef]
46. *GB/T 19145–2003*; Determination of total organic carbon in sedimentary. General Administration of Quality Supervision, Inspection and Quarantine of the People's Republic of China: Beijing, China, 2003.
47. *SY/T 5163–2018*; Analysis Method for Clay Minerals and Ordinary Non-Clay Minerals in Sedimentary Rocks by the X-ray Diffraction. National Energy Administration: Beijing, China, 2018.
48. *SY/T 6189–2018*; Quantitative Analysis Method for Rock Minerals by X-ray Energy Spectrum. National Energy Administration: Beijing, China, 2018.
49. Anovitz, L.M.; Cole, D.R. Characterization and Analysis of Porosity and Pore Structure. *Rev. Miner. Geochem.* **2015**, *80*, 61–164. [CrossRef]
50. Vidal Vázquez, E.; Paz Ferreiro, J.; Miranda, J.G.V.; Paz González, A. Multifractal analysis of pore size distributions as affected by simulated rainfall. *Vadose Zone J.* **2008**, *7*, 500–511. [CrossRef]
51. Li, W.; Liu, H.F.; Song, X.X. Multifractal analysis of Hg pore size distributions of tectonically deformed coals. *Int. J. Coal Geol.* **2015**, *144–145*, 138–152. [CrossRef]
52. Caniego, F.J.; Martín, M.A.; San José, F. Singularity features of pore-size soil distribution: Singularity strength analysis and entropy spectrum. *Fractals* **2001**, *9*, 305–316. [CrossRef]
53. Paz Ferreiro, J.; Vidal Vázquez, E. Multifractal analysis of Hg pore size distributions in soils with contrasting structural stability. *Geoderma* **2010**, *160*, 64–73. [CrossRef]
54. Riedi, R.H.; Crouse, M.S.; Ribeiro, V.J.; Baraniuk, R.G. A multifractal wavelet model with application to network traffic. *IEEE Trans. Inf. Theory* **1999**, *45*, 992–1019. [CrossRef]
55. San José, M.F.; Martín, M.A.; Caniego, F.J.; Tuller, M.; Guber, A.; Pachepsky, Y.; García-Gutiérrez, C. Multifractal analysis of discretized X-ray CT images for the characterization of soil macropore structures. *Geoderma* **2010**, *156*, 32–42. [CrossRef]

56. Wang, G.C.; Carr, T.R. Methodology of organic-rich shale lithofacies identification and prediction: A case study from Marcellus Shale in the Appalachian basin. *Comput. Geosci.* **2012**, *49*, 151–163. [CrossRef]
57. Wang, G.C.; Carr, T.R. Organic-rich Marcellus shale lithofacies modeling and distribution pattern analysis in the Appalachian basin. *AAPG Bull.* **2013**, *97*, 2173–2205. [CrossRef]
58. Guo, X.W.; Qin, Z.J.; Yang, R.; Dong, T.; He, S.; Hao, F.; Yi, J.Z.; Shu, Z.G.; Bao, H.Y.; Liu, K.Y. Comparison of pore systems of clay-rich and silica-rich gas shales in the lower Silurian Longmaxi formation from the Jiaoshiba area in the eastern Sichuan Basin, China. *Mar. Pet. Geol.* **2019**, *101*, 265–280. [CrossRef]
59. Feng, Z.Q.; Zhou, S.W.; Wu, W.; Tian, J.Q.; Xie, C.; Cai, Y.W. Pore characteristics and methane adsorption capacity of different lithofacies of the Wufeng formation-Longmaxi formation shales, Southern Sichuan Basin. *Energy Fuels* **2020**, *34*, 8046–8062. [CrossRef]
60. Xia, P.; Li, H.N.; Fu, Y.; Qiao, W.L.; Guo, C.; Yang, Z.; Huang, J.Q.; Mou, Y.L. Effect of lithofacies on pore structure of the Cambrian organic-rich shale in northern Guizhou, China. *Geol. J.* **2021**, *56*, 1130–1142. [CrossRef]
61. Kruk, M.; Jaroniec, M. Gas Adsorption Characterization of Ordered Organic-Inorganic Nanocomposite Materials. *Chem. Mater.* **2001**, *13*, 3169–3183. [CrossRef]
62. Sing, K.S.W.; Everett, D.H.; Haul, R.A.W.; Moscou, L.; Pierotti, R.A.; Rouquerol, J.; Siemieniewska, T. Reporting physisorption data for gas/solid systems with special reference to the determination of surface area and porosity. *Pure Appl. Chem.* **1985**, *57*, 603–619. [CrossRef]
63. Loucks, R.G.; Reed, R.M.; Ruppel, S.C.; Hammes, U. Spectrum of pore types and networks in mudrocks and a descriptive classification for matrix-related mudrock pores. *AAPG Bull.* **2012**, *96*, 1071–1098. [CrossRef]
64. Bu, H.L.; Ju, Y.W.; Tan, J.Q.; Wang, G.C.; Li, X.S. Fractal characteristics of pores in non-marine shales from the Huainan coalfield, eastern China. *J. Nat. Gas Sci. Eng.* **2015**, *24*, 166–177. [CrossRef]
65. Rexer, T.F.; Mathia, E.J.; Aplin, A.C.; Mark Thomas, K. High-pressure methane adsorption and characterization of pores in Posidonia shales and isolated kerogens. *Energy Fuels* **2014**, *28*, 2886–2901. [CrossRef]
66. Ambrose, R.J. Shale gas-in-place calculations Part I: New pore-scale considerations. *SPE J.* **2012**, *17*, 219–229. [CrossRef]
67. Liu, X.J.; Xiong, J.; Liang, L.X. Investigation of pore structure and fractal characteristics of organic-rich Yanchang formation shale in central China by nitrogen adsorption/desorption analysis. *J. Nat. Gas Sci. Eng.* **2015**, *22*, 62–72. [CrossRef]
68. Wu, J.S.; Yu, B.S.; Zhang, J.C.; Li, Y.X. Pore characteristics and controlling factors in the organic-rich shale of the Lower Silurian Longmaxi Formation revealed by samples from a well in southeastern Chongqing. *Earth Sci. Front.* **2013**, *20*, 260–269.
69. Zhang, P.F.; Lu, S.F.; Li, J.Q.; Chen, C.; Xue, H.T.; Zhang, J. Petrophysical characterization of oil-bearing shales by low-field nuclear magnetic resonance (NMR). *Mar. Pet. Geol.* **2018**, *89*, 775–785. [CrossRef]
70. Liu, K.Q.; Ostadhassan, M.; Sun, L.W.; Zou, J.; Yuan, Y.J.; Gentzis, T.; Zhang, Y.X.; Carvajal-Ortiz, H.; Rezaee, R. A comprehensive pore structure study of the Bakken Shale with SANS, N_2 adsorption and mercury intrusion. *Fuel* **2019**, *245*, 274–285. [CrossRef]
71. Yang, R.; Jia, A.Q.; Hu, Q.H.; Guo, X.W.; Sun, M.D. Particle size effect on water vapor sorption measurement of organic shale: One example from Dongyuemiao Member of Lower Jurassic Ziliujing Formation in Jiannan Area of China. *Adv. Geo-Energy Res.* **2020**, *4*, 207–218. [CrossRef]
72. Zhao, J.L.; Sun, M.D.; Pan, Z.J.; Liu, B.; Ostadhassan, M.; Hu, Q.H. Effects of pore connectivity and water saturation on matrix permeability of deep gas shale. *Adv. Geo-Energy Res.* **2022**, *6*, 54–68. [CrossRef]
73. Khurshid, I.; Al-Shalabi, E.W.; Afgan, I.; Al-Attar, H. A numerical approach to investigate the impact of acid-asphaltene sludge formation on Worm-holing during carbonate acidizing. *J. Energy Resour. Technol.* **2022**, *144*, 063001. [CrossRef]

Journal of
Marine Science and Engineering

MDPI

Article

Sedimentary Environment Interpretation and Organic Matter Enrichment of the Lower Cambrian Shuijingtuo Shale in the Yichang Slope, South China: Insight from Sedimentary Geochemical Proxies with Major/Trace Elements

Sile Wei [1,*], Mingyi Hu [1], Sheng He [2], Wei Yang [3], Qing He [2], Quansheng Cai [1] and Ping Li [1,*]

[1] School of Earth Sciences, Yangtze University, Wuhan 430100, China; humingyi65@163.com (M.H.); cqsh0713@163.com (Q.C.)
[2] Key Laboratory of Tectonics and Petroleum Resources, Ministry of Education, China University of Geosciences, Wuhan 430074, China; shenghe@cug.edu.cn (S.H.); qinghe@cug.edu.cn (Q.H.)
[3] Guangzhou Marine Geology Survey, China Geological Survey, Ministry of Natural Resources, Guangzhou 510075, China; yangwei@cug.edu.cn
* Correspondence: weisile@outlook.com (S.W.); lip953945568@gmail.com (P.L.)

Abstract: The vertical geochemical variations in total organic carbon (TOC) content and major and trace elements of the Lower Cambrian Shuijingtuo Formation from the Yichang Slope in the Upper Yangtze were investigated to assess the environmental conditions (redox conditions, water mass restriction, terrigenous input, relative sea-level terrestrial, and paleoproductivity) and to determine the primary controlling factors for organic matter enrichment. The Shuijingtuo Shale is divided into three intervals. The lithofacies of interval I are mainly black siliceous shale with high TOC content, and interval II is mainly black siliceous shale with moderate TOC content. Interval III consists of black, clay-rich siliceous shale and dark-gray calcareous shale and is characterized by a low TOC content. The effects of diagenesis and hydrothermal activity on the elements were evaluated prior to analyzing the environmental condition. There are good positive correlations between TOC and U/Al or Mo/Al ratios, suggesting that major/trace elements still retain the geochemical signature of the sedimentary environment. Meanwhile, the geochemical proxies consisting of Al, Fe, Mn, and Ti indicate that the study area did not experience hydrothermal deposits. The redox proxies (U/Th, C_{org}:P_{tot}, and Mo_{EF}–U_{EF}) indicate that the interval I samples were formed under a strong reducing condition. The diagram of TOC vs. Mo indicates that the water mass was moderately restricted during the deposition of interval I shales. Proxies of terrigenous input and relative sea-level (Zr/Al and Zr/Rb) suggest that the organic-rich shales at the bottom were deposited under a relatively high sea-level and experienced minimal input of terrigenous debris from the source area. Additionally, the paleoproductivity indicated by Si_{bio} and (Ni + Cu + Zn)/Al was high for interval I samples. During the interval II period, the relative sea-level began to decline, the seawater still remained in reducing conditions, and there was no change in the productivity, but the input of terrigenous debris increased significantly. In the interval III depositional period, the relative sea-level continued to decrease, the seawater shifted to a dysoxic condition, and the paleoproductivity was also at a lower level. The evolution of the sedimentary environment indicates that the high TOC content in the interval I samples is mainly attributed to the strong reducing condition, the preservation condition and debris dilution together control the organic matter content within the interval II samples, and the low TOC content within the interval III samples is constrained by a combination of the poor preservation conditions and lower paleoproductivity.

Keywords: organic matter enrichment; Shuijingtuo Formation; environmental condition; marine shale

Citation: Wei, S.; Hu, M.; He, S.; Yang, W.; He, Q.; Cai, Q.; Li, P. Sedimentary Environment Interpretation and Organic Matter Enrichment of the Lower Cambrian Shuijingtuo Shale in the Yichang Slope, South China: Insight from Sedimentary Geochemical Proxies with Major/Trace Elements. *J. Mar. Sci. Eng.* **2023**, *11*, 2018. https://doi.org/10.3390/jmse11102018

Academic Editor: Antoni Calafat

Received: 31 August 2023
Revised: 8 October 2023
Accepted: 18 October 2023
Published: 20 October 2023

1. Introduction

Shale gas, as an important unconventional gas reservoirs with extensive resources, has gained increasing attention in countries such as North America and China [1–3]. Gas in shale reservoirs originates from the thermal cracking of petroleum retained during the thermal evolution of organic matter, which leads to the simultaneous formation of numerous organic pores that serve as storage space [4–6]. Therefore, the enrichment of organic matter is closely related to hydrocarbon potential and reservoir quality, which is of great significance in evaluating the exploration potential of shale gas.

Organic matter enrichment in marine sediments is a complex biogeochemical process influenced by various sedimentary environmental factors, such as primary productivity, redox condition, and detrital input [7–10]. Based on the relative contributions of these influencing factors, previous research proposed three models to explain organic matter enrichment, including the productivity model, the preservation model, and the dilution model [11,12]. The productivity model suggests that the organic matter content primarily depends on primary productivity or the productivity of marine organisms [7,8]. In marine environments, the higher productivity of marine organisms such as algae and microorganisms leads to a higher amount of organic matter formed upon their death. However, this model does not explain all cases, as it neglects the influence of preservation conditions and dilution effects on organic matter enrichment [11]. The preservation model assumes that the redox conditions of the seawater environment play a predominant role in the organic matter enrichment [13,14]. Under oxidizing conditions, organic matter is oxidatively decomposed, whereas under reducing conditions, organic matter is more likely to be preserved [14,15]. Thus, if the seawater environment is reducing, the organic matter content will be higher [16]. This model is more applicable when explaining situations where the seawater environment is reducing, particularly in deep ocean settings [17]. The dilution model suggests that the enrichment of organic matter is related to the balance between organic matter input and detrital dilution rate [7,18]. If the deposition rate of organic matter is lower than the dilution rate of terrigenous debris, organic-poor rocks are formed. This condition commonly occurs in marine environments with high inputs of debris, such as offshore and continental shelf areas [11]. These models emphasize the importance of different factors in organic matter enrichment [7]. However, the relative significance of these factors may vary in a certain sedimentary environment. Therefore, a comprehensive analysis of different environmental factors is necessary to gain a more in-depth understanding of the mechanisms of organic matter enrichment.

The Lower Cambrian Shuijingtuo Shale has become an important layer for shale gas exploration in China in recent years, following the success of the Wufeng–Longmaxi Shale [19–21]. It is mainly distributed in the Yangtze block and characterized by a wide sedimentary range, considerable thickness, and high organic matter content [3,20]. Previous studies on the Lower Cambrian Shale have focused on shale pore structure, shale gas adsorption capacity, and organic–inorganic diagenesis [6,20,22–25]. Among these investigations, the content of organic matter has been recognized as an important influencing factor. However, there has been limited research on the mechanisms of organic matter enrichment. This study takes the Lower Cambrian Shuijingtuo Formation in the Yichang Slope as an example to explore the main controlling factors of organic matter enrichment based on a comprehensive analysis of the sedimentary environment. It has significant implications for the selection of favorable exploration areas and the evaluation of high-quality shale intervals in the Lower Cambrian Shuijingtuo Formation.

2. Geological Settings

The Yichang Slope belt is situated in the upper region of the Yangtze block, which together with the Cathaysia block constitutes the South China Plate [17]. Global paleogeographic reconstructions suggested that during the late Ediacaran and early Cambrian transition period (539–514 Ma), the South China Plate was an isolated craton, located in the mid-latitudes of the northern hemisphere, north of the Gondwana continent and considered to be effectively connected with the open ocean (Figure 1a) [26]. From northeast to southwest, the water depth gradually increases within the Yangtze block, and three marine sedimentary sequences exist, including platform carbonates, shelf calcareous mudstone, slope, and basin shale deposits (Figure 1b). During the early Cambrian Stage 3 (521 Ma), a global transgression event occurred that resulted in the development of organic-rich black shales on the previously deposited carbonate platform (Figure 1c). Subsequently, as the global sea-level dropped 518 million years ago, the lithology of the upper section of the Shuijingtuo Formation transitioned to gray calcareous shale.

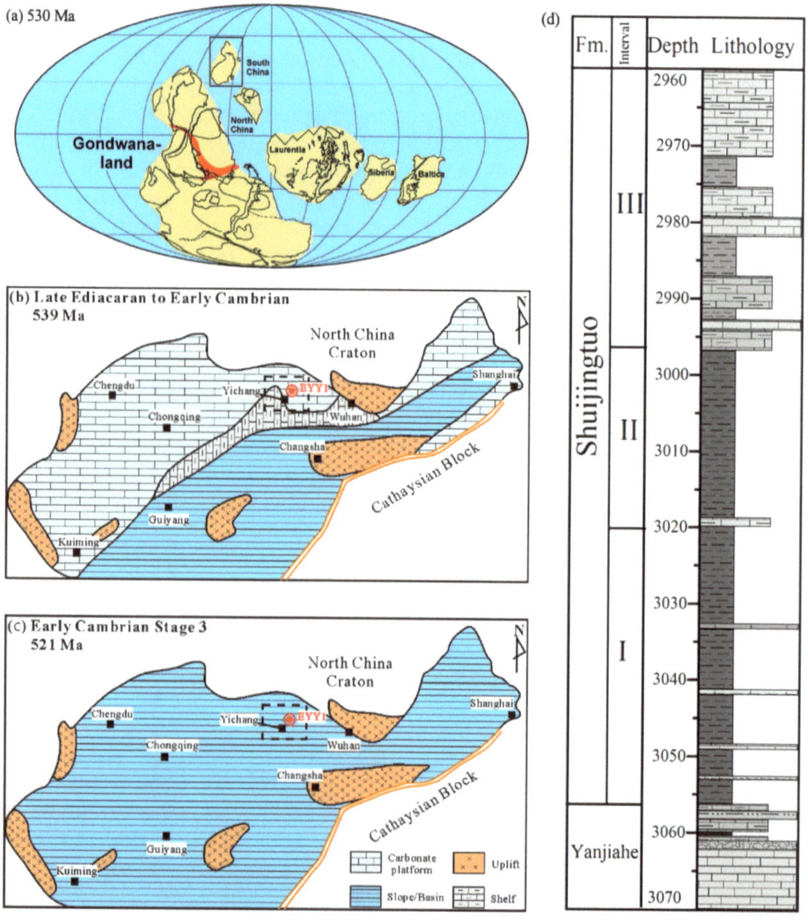

Figure 1. (**a**) Paleogeographic position of the South China Plate during the late Ediacaran and early Cambrian transition period (modified from Li et al. [26]); (**b**) paleogeographic maps of the Yangtze Block at 539 Ma (modified from Jiang et al. [27] and Wang et al. [28]); (**c**) paleogeographic maps of the Yangtze Block in early Cambrian Stage 3 (521 Ma) showing the study area and the location of the investigated well (modified from Jiang et al. [27] and Wang et al. [28]); (**d**) simplified lithologic profile of the Shuijingtuo Formation in the study area.

Based on high-resolution petrology, biostratigraphy, isotope dating, and inorganic geochemistry analysis of South China, the Lower Cambrian Shuijingtuo Formation in the study area can be divided into three intervals (Figure 1d). Interval I consists of black siliceous shale interbedded with thin layers of calcareous mudstone, containing trilobites such as *Tsunyidiscus*, radiolarians, sponge fossils, along with coexisting brachiopods, *Sunella bispinata*, and algae, *Vendotaenia* sp. Interval II comprises black shale that contains abundant trilobites and brachiopods, as well as brachiopods *Eohadrotreta, Palaeobolus, Spinobolus, Lingulellotreta,* and *Eoobolus*. Interval III is composed of dark-gray calcareous mudstone and carbonaceous limestone.

3. Samples and Methods

A total of 26 Shuijingtuo Formation samples were collected from the EYY1 Well, located in the Yichang Slope, the upper region of the Yangtze block.

3.1. Total Organic Carbon Content

Total organic carbon (TOC) content was conducted using a Rapid CS Cube analyzer. Shale powder samples with particle size less than 75 μm (200 mesh) were placed in silver paper and treated with hydrochloric acid (HCl) at a concentration of 7% by mass to remove the carbonates. The pretreated samples were then dried at 80 °C for 10 h and then subjected to complete combustion using a carrier gas consisting of 99.99% oxygen (O_2) at a temperature of 930 °C. The TOC content was calculated from the amount of carbon dioxide released during the oxidation.

3.2. Thin Section Observation

Thin sections of the Shuijingtuo Shale from the EYY1 Well were observed using transmitted polarized light microscopy (Leica DM4500P) to identify petrographic features including mineral compositions and distributions.

3.3. Major Element Analysis

The analysis of major elements was performed using an X-ray fluorescence spectrometer (XRF) (Zsx Primus II, Rigaku, Japan) at the Wuhan Sample Solution Analytical Technology Co., Ltd., Wuhan, China. The powdered samples were first dried in an oven at 105 °C for 12 h to eliminate free and adsorbed water and then heated to determine the loss on ignition. The X-ray tube utilized was a 4.0 kW end window Rh target, operating at a voltage of 50 kV and a current of 60 mA. All major elemental analysis lines were kα. The data were corrected using the theoretical α coefficient method, and the relative standard deviation was less than 2%. All the samples were processed according to the Chinese National Standard GB/T 14506.28-2010.

3.4. Trace Element Analysis

Trace element analysis was performed by utilizing the Agilent 7700e ICP-MS and at the Wuhan Sample Solution Analytical Technology Co., Ltd., Wuhan, China. The procedure involved drying 200-mesh sample powder at 105 °C for 12 h. Then, 50 mg of the powder was weighed and placed in a Teflon bomb. Slowly, 1 mL each of HNO_3 and HF were added to the bomb. The bomb was placed in a stainless steel pressure jacket and heated at 190 °C for more than 24 h. After cooling, the bomb was opened, and the remaining solution was evaporated on a hotplate at 140 °C. A total of 1 mL of HNO_3 was added and evaporated again. Subsequently, 1 mL each of HNO_3, MQ water, and an internal standard solution of 1 ppm were mixed, sealed, and heated at 190 °C for over 12 h. Finally, the solution was transferred to a polyethylene bottle and diluted to 100 g with 2% HNO_3 for ICP-MS analysis. All the samples were processed according to the Chinese National Standard GB/T 14506.30-2010.

Enrichment factor (EF) is used to reflect the degree of enrichment and depletion of each element and is calculated as follows: element $X_{EF} = (X/Al)_{sample}/(X/Al)_{PAAS}$ [29], where X and Al represent the weight percent concentrations of specified trace elements and aluminum, respectively. Shale samples are normalized using the Post-Archean Australian Shale (PAAS) as a standard reference [30]. $X_{EF} > 1.0$ and $X_{EF} < 1.0$ indicate enrichment and depletion of element X, respectively.

4. Results

4.1. Lithofacies and TOC Content

Based on the core description, thin section observation, and mineral composition, three major lithofacies were identified in the lower area of the Shuijingtuo Formation, including black siliceous shale; black, clay-rich siliceous shale; and dark-gray calcareous shale (Figure 2 and Table 1).

The black siliceous shale is mainly characterized by high silica content (Figure 2a1–a3, b1–b3). The mineral compositions and contents can be referred to in Wei et al. [23]. The black siliceous shale has the highest quartz content and the lowest carbonate mineral content, ranging from 55 to 75 vol.% (average of 62 vol.%) and 4 to 13 vol.% (average of 6 vol.%), respectively. The thin section shows that the black siliceous shale contains an abundance of radiolarians and sponge spicules (Figures 4 and 5 in Wei et al. [23]). The range of TOC content in the black siliceous shale varies widely, with high TOC content (average of 6.02 wt.%) mainly distributed in interval I (Table 1, Figure 2a1–a3) and moderate TOC content (average of 3.94 wt.%) distributed in interval II (Table 1, Figure 2b1–b3).

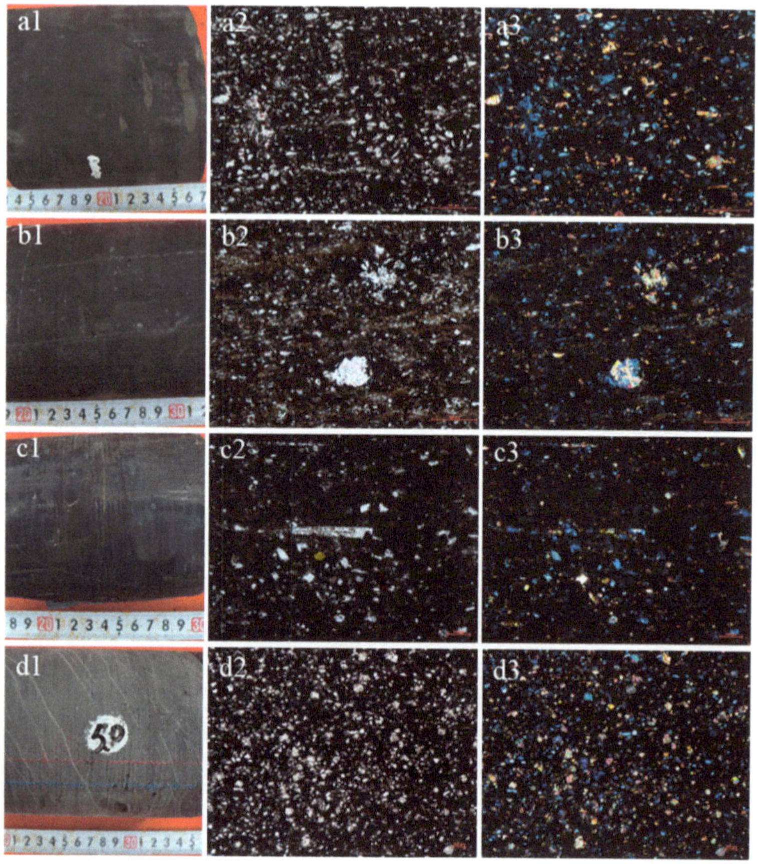

Figure 2. Core and thin section images of the Shuijingtuo Shale. (**a1–d1**) are the core images; (**a2–d2**) are the thin section images observed with the plane-polarized light; (**a3–d3**) are the thin section images observed with cross-polarized light. (**a1–a3**) are the images of the black siliceous shale with high TOC content, 3026.38 m, TOC = 6.20 wt.%; (**b1–b3**) are the images of the black siliceous shale with moderate TOC content, 3007.75 m, TOC = 3.02 wt.%; (**c1–c3**) are the images of the clay-rich siliceous mudstone, 2996.32 m, TOC = 2.27 wt.%; (**d1–d3**) are the images of the dark-gray calcareous mudstone, 2978.08 m, TOC = 1.58 wt.%.

Table 1. Major and trace elements of the Shuijingtuo Shale.

Depth (m)	Interval	TOC (wt.%)	Lithofacies	Al_2O_3 (%)	SiO_2 (%)	CaO (%)	TiO_2 (%)	Fe_2O_3 (%)	MnO (%)	P_2O_5 (%)	Th (ppm)	U (ppm)	Mo (ppm)	Cu (ppm)	Ni (ppm)	Zn (ppm)	Zr (ppm)	Rb (ppm)
2975.53	III	1.63	calcareous S. *	6.21	31.47	25.75	0.28	3.45	0.06	0.08	5.35	3.6	4.2	22.8	26.4	32.0	55.6	56.4
2978.08	III	1.58	calcareous S.	7.73	29.35	22.56	0.36	4.76	0.07	0.11	7.12	4.9	6.2	26.4	33.7	67.2	73.5	69.0
2983.63	III	2.47	clay-rich siliceous S.	12.81	59.62	4.87	0.61	3.32	0.02	0.09	11.4	8.0	16.7	44.0	45.7	48.3	122.0	113.0
2984.98	III	2.21	clay-rich siliceous S.	12.21	60.80	4.98	0.57	3.25	0.02	0.10	12.00	9.4	34.8	60.1	83.2	114	114.0	110.0
2989.32	III	1.80	calcareous S.	7.00	40.68	19.32	0.31	3.90	0.05	0.08	6.10	6.9	14.9	30.7	33.1	37.4	65.0	57.4
2994.23	III	3.27	clay-rich siliceous S.	12.47	55.15	5.61	0.53	4.64	0.04	0.11	10.20	10.3	11.6	58.5	46.4	80.8	107.0	108.0
2996.32	III	2.27	clay-rich siliceous S.	13.02	60.44	2.82	0.55	4.85	0.05	0.10	11.10	7.5	9.18	60.7	44.2	72.6	119.0	112.0
3003.50	II	2.72	siliceous S.	9.49	68.67	3.98	0.41	2.10	0.02	0.08	8.05	15.5	49.1	87.4	107.0	200.0	85.1	80.2
3004.88	II	3.08	siliceous S.	7.21	68.48	5.60	0.29	2.27	0.02	0.11	5.65	21.4	54.0	66.7	113.0	78.8	60.9	56.1
3007.34	II	2.93	siliceous S.	10.71	65.38	2.97	0.42	3.91	0.02	0.13	7.49	21.1	60.2	81.7	119.0	131.0	92.9	88.0
3007.75	II	3.02	siliceous S.	10.42	65.46	3.75	0.42	2.80	0.03	0.14	7.46	18.6	57.8	79.5	106.0	130.0	90.6	83.0
3008.78	II	3.47	siliceous S.	9.45	63.06	5.50	0.38	2.75	0.03	0.18	7.04	23.9	60.5	86.3	116.0	126.0	83.8	75.2
3011.50	II	4.66	siliceous S.	10.11	65.53	3.57	0.45	2.62	0.02	0.11	7.98	31.6	81.4	77.6	118.0	236.0	99.3	82.6
3020.54	I	4.33	siliceous S.	9.20	62.42	5.28	0.50	3.77	0.02	0.18	7.44	47.1	69.9	37.8	127.0	159.0	131.0	70.8
3023.83	I	5.22	siliceous S.	9.40	63.76	4.05	0.52	3.63	0.03	0.17	7.29	44.8	63.2	28.8	151.0	88.6	127.0	69.5
3024.33	I	5.24	siliceous S.	9.27	62.71	4.79	0.51	3.55	0.02	0.15	6.94	47.1	63.6	30.2	142.0	93.0	123.0	67.7
3026.38	I	6.20	siliceous S.	8.84	62.86	5.20	0.48	2.83	0.03	0.16	6.55	57.4	72.2	32.0	148.9	169.0	108.0	64.9
3030.18	I	5.78	siliceous S.	8.68	62.26	4.99	0.47	3.25	0.03	0.15	6.51	33.6	57.4	31.6	145.2	100.0	110.0	61.8
3034.92	I	5.76	siliceous S.	8.09	66.27	3.85	0.43	3.00	0.02	0.12	6.60	65.7	140	38.0	206.0	101.0	98.1	61.1
3037.53	I	4.76	siliceous S.	8.54	57.55	7.26	0.46	3.67	0.03	0.12	6.30	54.2	64.1	31.8	151.9	100.0	100.0	60.6
3041.33	I	5.88	siliceous S.	7.64	64.59	2.89	0.40	5.96	0.02	0.14	5.94	49.2	110	55.1	201.0	105.0	93.0	58.2
3043.48	I	5.08	siliceous S.	7.22	69.82	3.42	0.38	2.44	0.01	0.13	5.60	43.2	81.1	28.8	158.0	54.7	85.8	54.1
3044.76	I	6.06	siliceous S.	5.97	72.98	2.57	0.32	2.21	0.01	0.18	4.68	57.4	102	29.6	185.0	78.3	74.6	44.0
3053.28	I	7.73	siliceous S.	8.13	65.61	2.46	0.44	3.79	0.01	0.13	6.28	43.5	84.6	36.6	136.8	65.4	103.0	68.5
3053.88	I	8.25	siliceous S.	6.58	68.04	3.05	0.36	2.96	0.01	0.10	5.43	42.3	80.9	32.4	153.4	63.0	83.0	53.7
3054.20	I	7.55	siliceous S.	8.57	64.78	2.56	0.46	3.69	0.02	0.15	6.87	56.1	75.2	35.4	173.0	88.1	113.0	69.6

*: The abbreviation "S." stands for shale.

The black, clay-rich siliceous shale presents mainly in interval III of the examined Shuijingtuo Shale and is characterized by moderate to high quartz content (45~48 vol.%, average of 47 vol.%) and moderate clay content (30~34 vol.%, average of 32 vol.%) (Figure 2c1–c3). There is a significant reduction in the radiolarians and sponge spicules in the black, clay-rich siliceous shale compared to the black siliceous shale (Figures 4 and 5 in Wei et al. [23]). The TOC content of the black, clay-rich siliceous shale ranges from 2.21 wt.% to 3.27 wt.% (Table 1).

The dark-gray calcareous shale is also present in interval III. Unlike the other lithofacies, the dark-gray calcareous mudstone has the highest content of carbonate minerals (average of 36 vol.%) and the lowest TOC content (1.67 wt.%) (Figure 2d1–d3). The thin section shows that the siliceous organism is absent in the dark-gray calcareous shale (Figures 4 and 5 in Wei et al. [23]).

Overall, the lithofacies of interval I are mainly composed of black siliceous shale with high TOC content, and interval II is mainly composed of black siliceous shale with moderate TOC content. Interval III consists of black, clay-rich siliceous shale and dark-gray calcareous shale and is characterized by low TOC content.

4.2. Major and Trace Elements

The results of the major and trace elements are listed in Table 1 and the Supplementary Material. The composition of marine shales can be considered to be a mixture of three end-member oxides: SiO_2, Al_2O_3, and CaO, which are closely associated with minerals such as quartz, clay, and carbonates [31]. In general, the Shuijingtuo Shale show varying degrees of enrichment in SiO_2 compared to Al_2O_3 and CaO (Figure 3). The interval I and interval II shale samples have relatively consistent silica concentrations of approximately 64.90 wt.% and 66.10 wt.%, respectively, which are significantly higher than the average of 48.22 wt.% of the interval III samples (Table 1). Conversely, the average CaO content in interval III is 12.78 wt.% (up to 25.75 wt.%), which is much higher than the averages of 4.03 wt.% in interval I and 4.23 wt.% in interval II (Table 1).

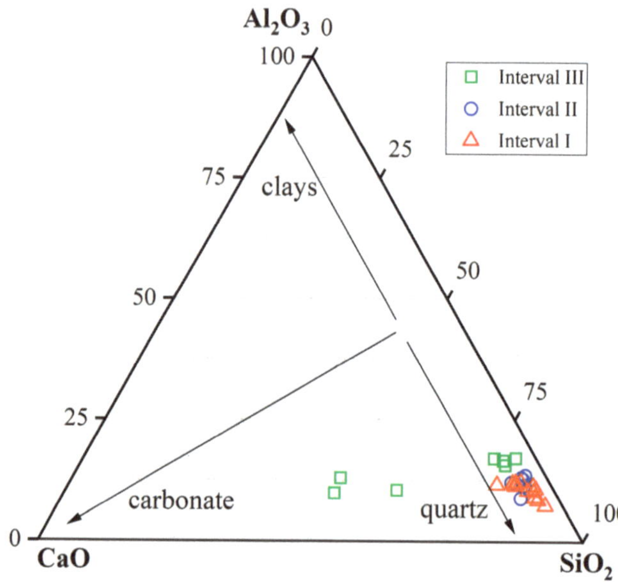

Figure 3. Ternary diagram showing the relative percentages of the major elements SiO_2, Al_2O_3, and CaO.

5. Discussion

5.1. Effect of Thermal Diagenesis and Hydrothermal Activity on Element

5.1.1. Thermal Diagenesis Influence

Organic matter deposited in sediments undergoes maturation due to high temperature and pressure underground, leading to the generation of hydrocarbons, which are then expelled through a rock's pore system. According to Abanda and Hannigan [32], thermal diagenesis and hydrocarbon expulsion can result in a significant loss of trace elements, as organic matter can host up to about 20% of total trace element content. However, Raiswell and Berner [33] argued that although thermal diagenesis does cause a loss of organic matter, and the concentration of trace elements in shale may be lower compared to the original depositional environment, the shale still retains its geochemical signature of deposition. The positive correlation between TOC and U/Al or Mo/Al ratios, proposed by Ross and Bustin [31], can be utilized to assess the impact of thermal diagenesis on the reactivation of elements.

The thermal maturity of the investigated Shuijingtuo Shale is approximately in the dry gas window (~1.93~2.20% Ro), which means that the Shuijingtuo Shale has generated and expelled a certain amount of liquid hydrocarbon [6]. The relationships between TOC content and U/Al as well as Mo/Al ratios are illustrated in Figure 4. It shows a strong positive correlation between both U/Al and Mo/Al ratios and TOC content, suggesting minimal remobilization of elements. Therefore, despite the high thermal maturity of the Shuijingtuo Shale, there is no obvious evidence of element partitioning or remobilization occurring within the examined shale, implying that the elements can be used to reconstruct the paleoenvironmental conditions.

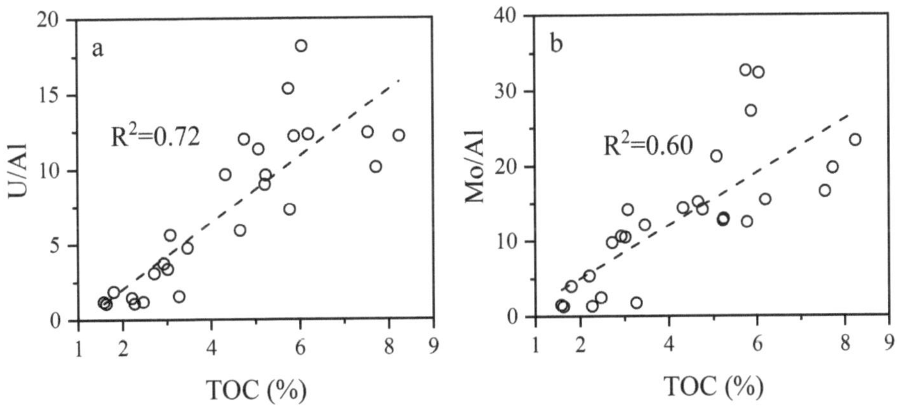

Figure 4. Relationships between TOC content and U/Al (**a**) as well as Mo/Al (**b**).

5.1.2. Hydrothermal Activity Influence

During the early Cambrian, the southern end of the Yangtze block experienced a period of intense rift tectonic activity, resulting in extensive hydrothermal deposition. The combination of major and trace elements, rare-earth elements (REEs), and sulfur isotopes, along with petrological and geological investigations, collectively provide compelling evidence of hydrothermal venting occurrences during this period [34]. Previous studies have revealed the presence of submarine hydrothermal deposits within the Lower Cambrian shale, spanning the southwestern Guizhou Province, the Xiuwu basin within Jiangxi Province, and the western Hunan Province [35,36]. The hydrothermal activity, enriched with volatile components, can alter the chemical conditions of seawater, which would lead to the enrichment and migration of trace elements [37,38]. Therefore, it is necessary to investigate whether the study area has been influenced by hydrothermal activity.

The Al/(Al + Fe + Mn) ratio, Al–Fe–Mn ternary diagram, and Fe/Ti versus Al/(Al + Fe + Mn) plot are widely used to discriminate hydrothermal from non-hydrothermal deposits [23,39–43]. Hydrothermal rocks are typically characterized by Fe- and Mn-rich and Al- and Ti-poor compositions compared to non-hydrothermal rocks [44]. The Al/(Al + Fe + Mn) ratios of the purely hydrothermal rocks are low (<0.01), whereas those of the purely biogenic rocks are greater than 0.60 [39]. The average Al/(Al + Fe + Mn) ratio is 0.66 within the Shuijingtuo Shale, indicating the absence of hydrothermal activity in the study area. The Al–Fe–Mn ternary diagram shows that all analyzed samples fall within the non-hydrothermal zone, characterized by the enrichment of Al relative to Fe and Mn (Figure 5). Furthermore, the Fe/Ti versus Al/(Al + Fe + Mn) plot shows that the data points of the Shuijingtuo Formation are far from the hydrothermal endmember and are situated between the biological and terrigenous endmembers (Figure 6). Therefore, based on the combination of various geochemical indicators, it can be concluded that the study area did not experience hydrothermal activity, which also agrees with the research of Zhang et al. [45], Luo et al. [46], and Wei et al. [47].

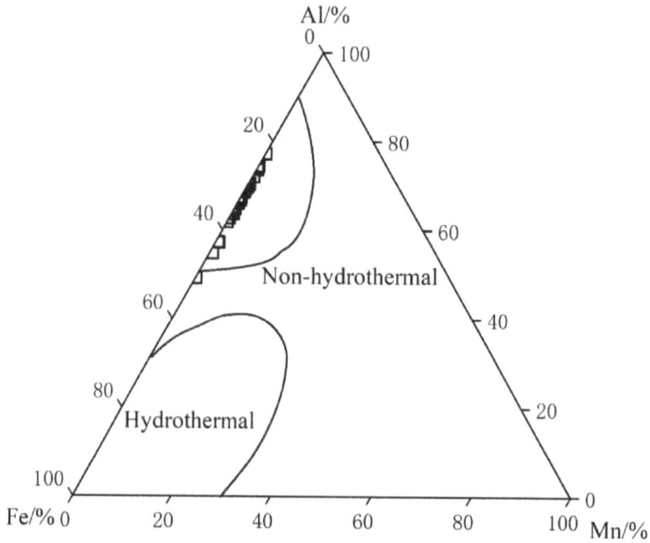

Figure 5. Al–Fe–Mn ternary diagram distinguishing between hydrothermal and non-hydrothermal deposits (modified from Boström and Peterson [39] and Yamamoto [40]).

5.2. Sedimentary Environments

5.2.1. Redox Condition

The ratio of uranium to thorium (U/Th) is a sensitive redox proxy that is most commonly used to interpret the redox conditions of ancient depositional environments [48–50]. Th is relatively immobile in low-temperature surface environments and is typically found in the detrital fraction associated with heavy minerals or clays in shale, where some U is also found [48]. U exists primarily in the hexavalent (6+) and tetravalent (4+) states. Most 6 + U is dissolved and lost during weathering processes, while 4 + U is insoluble and can be fixed under reducing conditions [48,51]. As the process of fixation continues, U is additionally supplied through diffusion in seawater, leading to an increase in the total U/Th ratio. Jones and Manning [48] suggested that a U/Th ratio of >1.25 corresponds to a suboxic and anoxic environment, ~0.75–1.25 corresponds to a dysoxic environment, and <0.75 corresponds to an oxic environment.

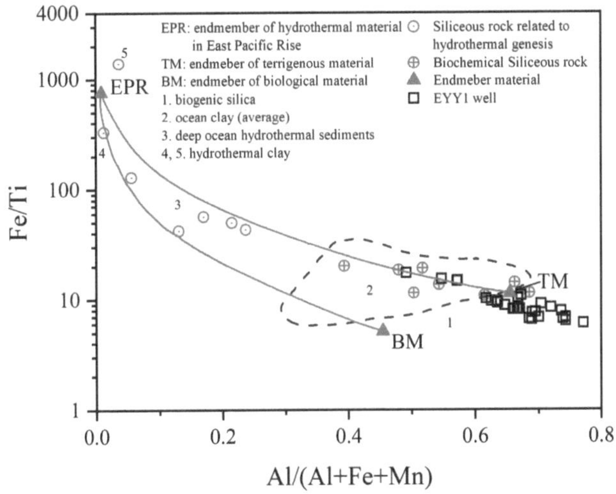

Figure 6. Fe/Ti versus Al/(Al + Fe + Mn) diagram distinguishing between hydrothermal and non-hydrothermal deposits.

The ratio of organic carbon (C_{org}) to organic phosphorus (P_{org}) (C_{org}:P_{org}) in marine sediments is another effective geochemical proxy to assess the redox conditions [14,52]. Most of the P is delivered to the sediment bound in organic matter. The preservation of organic matter and cycle of marine P are closely related to the oxygen level in the water column [52]. In an oxygen-depleted depositional environment, organic matter undergoes degradation, during which P_{org} is released and diffuses into seawater [53]. On the contrary, under oxic conditions, a large amount of P_{org} is captured and bound to sediment through trapping mechanisms such as adsorption, complexation reactions, and biological sequestration [54]. Hence, a higher C_{org}:P_{org} ratio represents a lower oxygen level. Generally, P_{org} is substituted with total P (P_{tot}), as P_{org} accounts for less than 20% of P_{tot} in marine sediments [52]. A C_{org}:P_{tot} ratio > 100 is considered to indicate an anoxic condition, ~50–100 is considered to indicate a dysoxic condition, and <50 is considered to indicate an oxic condition [52].

Figure 7 illustrates the profiles of U/Th and C_{org}:P_{tot} ratios. Both the U/Th and C_{org}:P_{tot} ratios within the interval I shale samples exhibit a considerably wide range, from 5.16 to 12.26 with an average of 7.91 and from 118 to 406 with an average of 216, respectively. These ratios suggest that the depositional environment during interval I was strongly anoxic (Figure 8). The U/Th and C_{org}:P_{tot} ratios within the interval II shale samples vary from 1.93 to 3.96 with an average of 3.07 and from 94 to 203 with an average of 139, respectively (Figure 7). Although the values of U/Th and C_{org}:P_{tot} ratios in the interval II shale samples are lower than those in interval I, they still exceed the critical values of the anoxic condition. Therefore, interval II was also deposited under an anoxic condition, albeit with a weaker reducing intensity. In contrast, the U/Th and C_{org}:P_{tot} ratios of the interval III shale samples demonstrate a relatively low value, ranging from 0.67 to 1.13 and from 73 to 155, respectively (Figure 8). These values suggest a dysoxic depositional environment, characterized by low to moderate levels of oxygen.

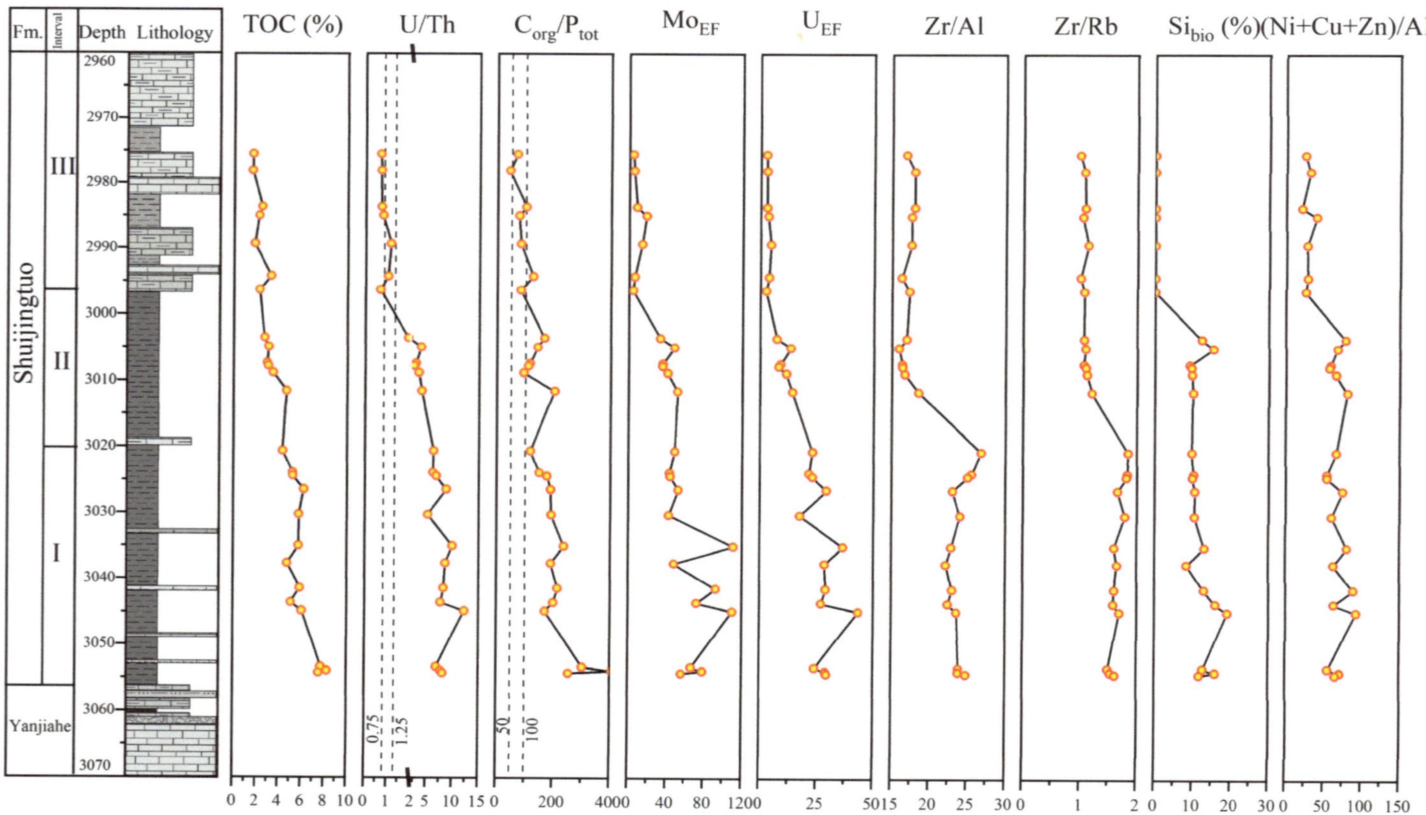

Figure 7. Stratigraphic profiles of redox conditions proxies (U/Th, C_{org}:P_{tot}), water mass restriction proxies (Mo_{EF}, U_{EF}), terrigenous input and relative sea-level proxies (Zr/Al, Zr/Rb), and paleoproductivity proxies (Si_{bio}, (Ni + Cu +Zn)/Al) of the EYY1 well.

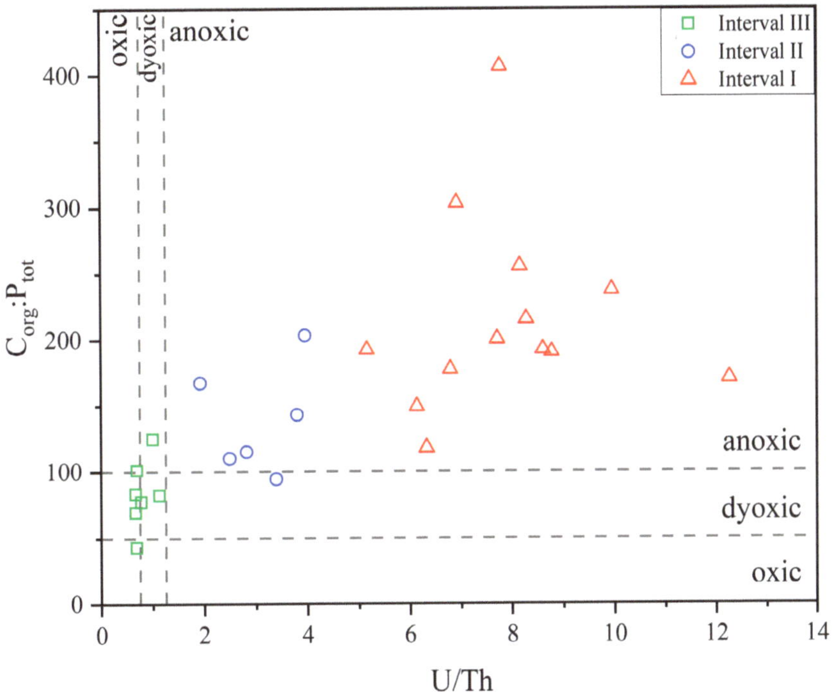

Figure 8. Cross plot of the U/Th and C_{org}:P_{tot} ratio reflecting the redox conditions of the Shuijingtuo Shale.

The variations in redox conditions in the EYY1 Well of this study exhibit comparability with another well in the study area, the EYY3 Well [17]. These two wells demonstrate similar changes in redox intensity vertically, with both showing an anoxic condition at the bottom [17]. However, the difference lies in the middle and upper parts, where dysoxic and oxic conditions are observed in the EYY3 Well, respectively, with a significantly lower reducing intensity compared to the EYY1 Well in this study [17]. This difference is probably attributed to the fact that the EYY3 Well is located in the platform depression zone of the Yichang Slope, where the water depth is shallower compared to the EYY1 Well, which is located in the continental shelf [23,55].

5.2.2. Water Mass Restriction

Patterns of molybdenum–uranium (Mo_{EF}–U_{EF}) covariation are extensively utilized to assess the water mass restriction and redox conditions, primarily because of their low concentrations in the upper continental crust, near-uniform concentrations in seawater worldwide, and predominantly authigenic uptake and enrichment from seawater [28,56–58]. The similarity in Mo–U relationships between modern marine systems and paleomarine systems allows the former to be used to explain paleoenvironmental parameters.

The examined Shuijingtuo Shale demonstrates a significant level of enrichment in Mo and U compared to the PAAS background sample (Figures 7 and 9). It shows that the Mo_{EF} is more enriched than the U_{EF}, which was also reported in the Wufeng–Longmaxi Shale in the Sichuan Basin [14,59,60]. A possible explanation for the low U_{EF} is that some U may have been lost during the hydrocarbon expulsion process, despite the fact that U is not typically the most abundant trace metal in kerogen [57]. Samples from interval III exhibit U_{EF} values that are close to 3, while the Mo_{EF} varies considerably, with a maximum value of up to 18 (average of 8.62, standard deviation 5.36). When the U_{EF} is higher than 3, the Mo_{EF} and the U_{EF} tend to increase equally on a logarithmic scale in interval I and interval II (Figure 9). With the exception of a few samples, the majority of the Shuijingtuo samples follow the unrestricted marine trend (Figure 9). The interval II samples conform almost completely to the trend, whereas the interval I samples scatter around the unrestricted marine trend (Figure 9). Therefore, there is some variation in the fidelity of ancient ocean systems in following the present-day unrestricted marine trend.

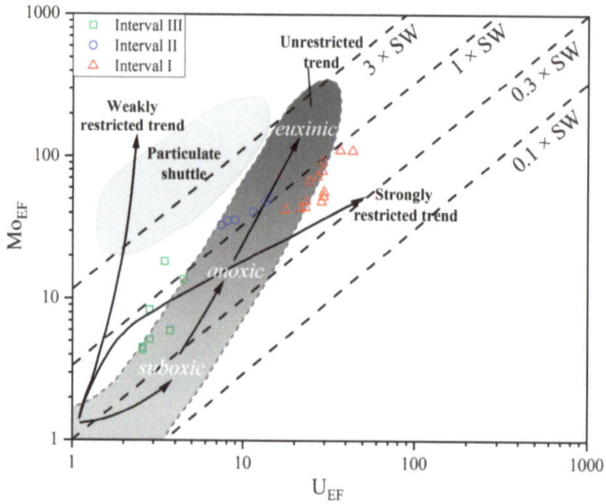

Figure 9. Mo_{EF}–U_{EF} covariation in the Shuijingtuo Shale (modified from Tribovillard et al. [57]).

The paleoredox interpretations based on the Mo_{EF}–U_{EF} covariation are in excellent agreement with those based on the geochemical proxies U/Th and C_{org}:P_{tot} ratios (Figure 9).

The diagram of TOC vs. Mo can be used to evaluate the degree of water mass restriction that cannot be accounted for by the Mo_{EF}–U_{EF} covariation [57,58,61]. Among the analyzed Shuijingtuo samples, interval I samples have the highest TOC and Mo content (Figure 10), as the organic matter can promote Mo adsorption by the sediment under highly anoxic conditions [57]. The Mo/TOC ratio of the interval I shales (ranging from 10 to 24, average of 14) falls between the Framvaren Fjord (Mo/TOC = 9) and the Cariaco Basin (Mo/TOC = 25) (Figure 10), suggesting that the water mass was moderately restricted during the deposition of interval I shales. Interval II samples are mainly located on the left of the Saanich Inlet trend line (Figure 10), indicating a low degree of water mass restriction. The TOC–Mo plot makes it difficult to assess the degree of mass restriction of interval III because the data points cross multiple lines of different degrees of restriction (Figure 10). However, based on the redox conditions and the lower Mo and TOC values, it is reasonable to infer that the water mass restriction of interval III is weak.

5.2.3. Terrigenous Input and Relative Sea-Level

Despite the relative stability of the marine sedimentary environment, continental shelves and slopes close to the source area experience an influx of terrigenous debris that can alter the depositional rate, rock composition, and abundance of organic matter [18,62]. Detrital-derived elements, such as aluminum (Al), zirconium (Zr), and rubidium (Rb), are commonly used as proxies for the sediment transport distance and transport energy [29,63]. Al is found in the aluminosilicate fraction of sediments, predominantly in clay minerals, which have the capacity to be transported over long distances. Similar to Al, Rb is a light element that exists in K-feldspar, mica, and clay minerals and is also capable of long-distance transport [63,64]. Conversely, Zr is mainly associated with heavy minerals such as zircon ($ZrSiO_4$), which has a limited transport distance [63]. Therefore, trends in Zr (expressed as Zr/Al and Zr/Rb ratios) can reflect the changes in terrigenous input and relative sea-level.

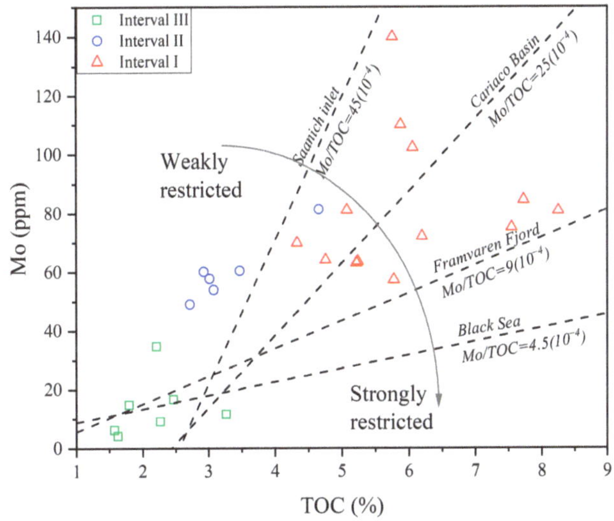

Figure 10. Cross plot of TOC content versus Mo showing the water mass restriction of the Shuijingtuo Shale (modified from Tribovillard et al. [57]).

The profiles of the Zr/Al and Zr/Rb ratios show similar variations (Figure 7). In the initial interval I period, the Zr/Al ratio ranges from 22.19 to 26.84, with an average of 23.93. After the deposition of interval I, the Zr/Al ratio within the interval II and interval III samples of the EYY1 Well sharply decrease and remain relatively stable, ranging from 15.97 to 18.54 (average of 17.09). The Zr/Al ratios and their variations at the bottom of the EYY1 Well are consistent with the EYY3 Well, both exhibiting relatively high values with a small fluctuation range [17]. Additionally, the lithology of both wells is predominantly composed of siliceous shale. However, in the middle and upper parts, the Zr/Al ratio in the EYY3 Well gradually decreases, which is distinct from the sharp decrease observed in the EYY1 Well. The comparison between the EYY1 and EYY3 Wells illustrates that the decline of relative sea-level has a greater impact on the continental shelf area. In the EYY1 Well, the profile of the Zr/Rb ratio exhibits a similar variation pattern to that of the Zr/Al. In interval I, it shows high values and a small fluctuation range (1.50~1.85, average of 1.67), while in intervals II and III, it rapidly decreases and remains relatively stable (0.99~1.20, average of 1.08).

The trend observed in the Zr/Al and Zr/Rb ratios suggests that the relative sea-level was at its highest and most stable during the interval I depositional period of the EYY1 Well. It experienced minimal input of terrigenous debris from the source area. During the interval II and interval III depositional periods, the study area underwent a sharp decline in relative sea-level, accompanied by a large influx of terrigenous sediments.

5.2.4. Paleoproductivity

Biogenic silica (Si_{bio}) was widely used as a geochemical proxy to assess paleoproductivity [65,66]. The common sources of silica in shales include the dissolution of siliceous skeletons (biogenic silica), transition of smectite to illite (clay-derived silica), pressure dissolution of detrital quartz, alteration of feldspar, alteration of volcanic ash, and precipitation of submarine hydrothermal venting [67–69]. In our previous study, we ruled out hydrothermal origin and suggested that no volcanic ash alteration occurred within the Shuijingtuo Shale in the study area [23]. The silica in the examined shale samples is mainly biological and of terrestrial depositional origin. The Si_{bio} was calculated as the absolute difference between Si measured in a sample and the Si versus Al regression line (detrital trend line). The detailed method can be found in Wei et al. [23].

Nickel (Ni), copper (Cu), and zinc (Zn) are trace metals that are essential micronutrients for the growth of marine organisms [29]. The sum of Ni, Cu, and Zn serves as a more dependable indicator of paleoproductivity. This is primarily due to its lower susceptibility to postdepositional alterations and diagenesis, as opposed to analyzing each element individually. To eliminate the effects of debris dilution, (Ni + Cu + Zn)/Al is suggested to be used for the qualitative assessment of paleoproductivity.

The Si_{bio} profile shown in Figure 7 has a distinct pattern of variation. The profile shows that Si_{bio} has a maximum at the bottom of interval I, then begins to decrease slightly during the middle and late stage of interval I, continues through the interval II period, and then declines rapidly to a minimum of nearly zero in interval III. Overall, Si_{bio} remains at a high level during the interval I and interval II periods and is almost absent during the interval III period. The variation in (Ni + Cu + Zn)/Al is similar to that in Si_{bio} (Figure 7). As a whole, (Ni + Cu + Zn)/Al ratios maintain high values (average value 68.32) within the interval I and interval II samples, which are significantly higher than those of the interval III samples (average value 28.23). For the same continental shelf sedimentation in the upper Yangtze block, the value of Ni + Cu + Zn in the study area, where there is no hydrothermal activity, is significantly lower than that in areas with hydrothermal activity, reaching a maximum value of 3470 ppm [70]. This is because hydrothermal activity can replenish a large number of trace elements into seawater. The lower values of Ni + Cu + Zn can also indicate, to some extent, the absence of hydrothermal activity in the study area. The trend of Si_{bio} and (Ni + Cu + Zn)/Al reflects changes in paleoproductivity, which experiences a rapid decline from high productivity levels in the interval I and interval II periods to lower levels in the interval III period. These changes in paleoproductivity are also supported by variations in the abundance of siliciclastic biogenesis observed in the thin sections reported by Wei et al. [23].

5.3. Sedimentary Environment of the Shuijingtuo Shale

The rapid rise in relative sea-level was one of the significant global geologic events that occurred during the late Ediacaran to the early Cambrian. Analysis of the TOC–Mo diagram indicates that during the interval I stage, the study area transformed into a moderate restriction sea due to the surrounding paleohighs in the upper Yangtze block (Figure 11). The restricted sea was characterized by strong anoxic conditions, indicated by the redox proxies (U/Th and C_{org}:P_{tot}) observed in the interval I samples (Figure 11). The significant abundance of siliceous organisms, especially the presence of sponge spicules fossils, and the high biogenic silica content in the interval I samples suggest that the connectivity between the upper Yangtze and the open ocean greatly increased the paleoproductivity. Simultaneously, the high relative sea-level reduced the input of terrestrial debris, as evidenced by the high Zr/Al and Zr/Rb ratios (Figure 11).

During the interval II stage, the relative sea-level in the study area began to decline, leading to a series of changes in the sedimentary environment. The most noticeable change is that the water mass restriction shifted from a moderately restricted sea to a weakly restricted sea, while the seawater remained under reducing conditions, albeit with lower reduction than in the interval I stage. Additionally, the input of detrital material into the study area increased significantly due to the longer migration distance of terrestrial debris caused by the fall in the relative sea-level.

With the continuous decline of the relative sea-level, the water depths during the interval III period were at their lowest, exhibiting characteristics of a shallow shelf environment. The wide and shallow sedimentary environment has developed a marine stratigraphy characterized by dark-gray calcareous mudstone interbedded with shales. The low values of the U/Th and C_{org}:P_{tot} indicate that during the interval III period, the seawater was under a dysoxic depositional condition. Furthermore, the input of terrigenous debris, indicated by the Zr/Al and Zr/Rb ratios, was kept at a stable, high level. Additionally, the nutrients (Ni + Cu +Zn)/Al necessary for microbial growth and development were reduced.

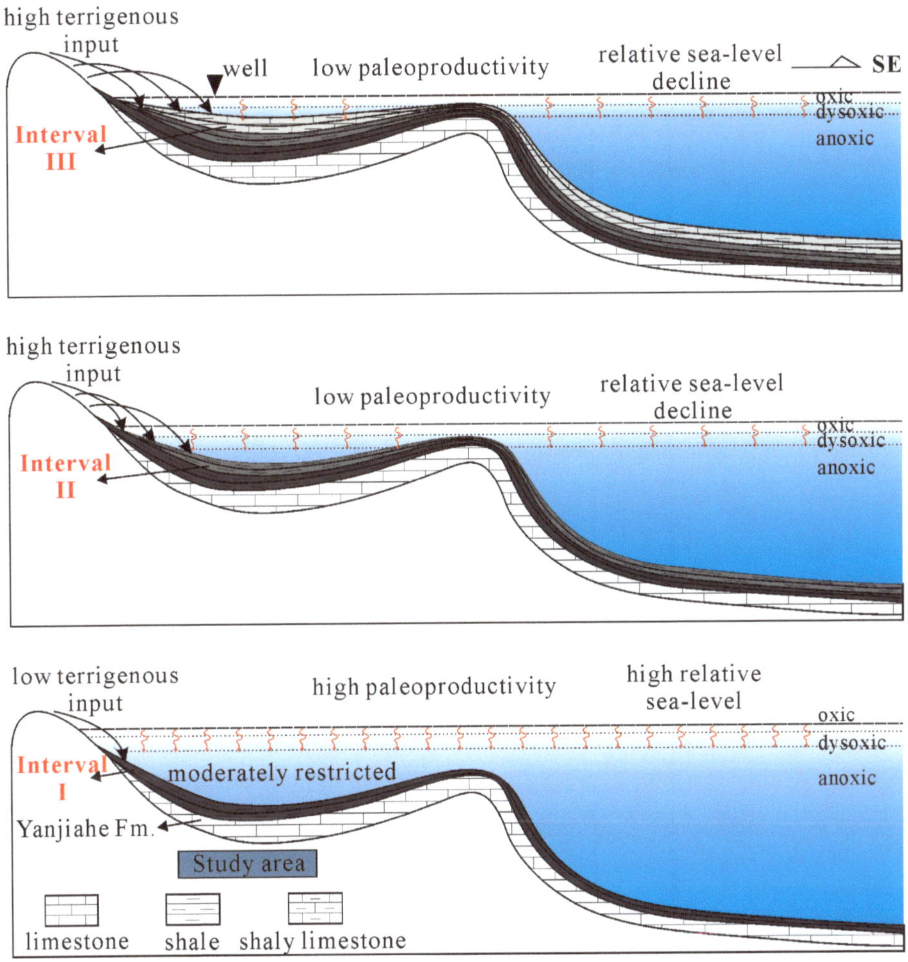

Figure 11. Schematic interpretation of the evolution of the sedimentary environment during the deposition of the Shuijingtuo Shale.

5.4. Main Factors Controlling Organic Matter Enrichment

The predominant factor controlling the high TOC content of the interval I shale samples is the strong reducing condition. The paleoproductivity of the interval I period was high and remained relatively stable, which does not match the decreasing trend of TOC content with decreasing burial depth (Figures 7 and 12e,f). This suggests that although paleoproductivity was high during the interval I period, it was not the primary controlling factor for TOC enrichment. Similarly, the input of terrigenous debris remained low during this depositional period and showed an increasing trend in the late interval I period, which does not agree with the variations in TOC content (Figures 7 and 12c,d). However, there is a strong synergistic trend between the redox proxies and TOC content (Figure 12a,b), indicating that the intense anoxic conditions caused by the deep seawater were conducive to preserving a substantial amount of organic matter. Therefore, through the comprehensive analysis, it can be concluded that the high TOC contents in the shale at the bottom of the

Shuijingtuo Formation during the interval I depositional period were primarily controlled by the strongly reducing depositional environment.

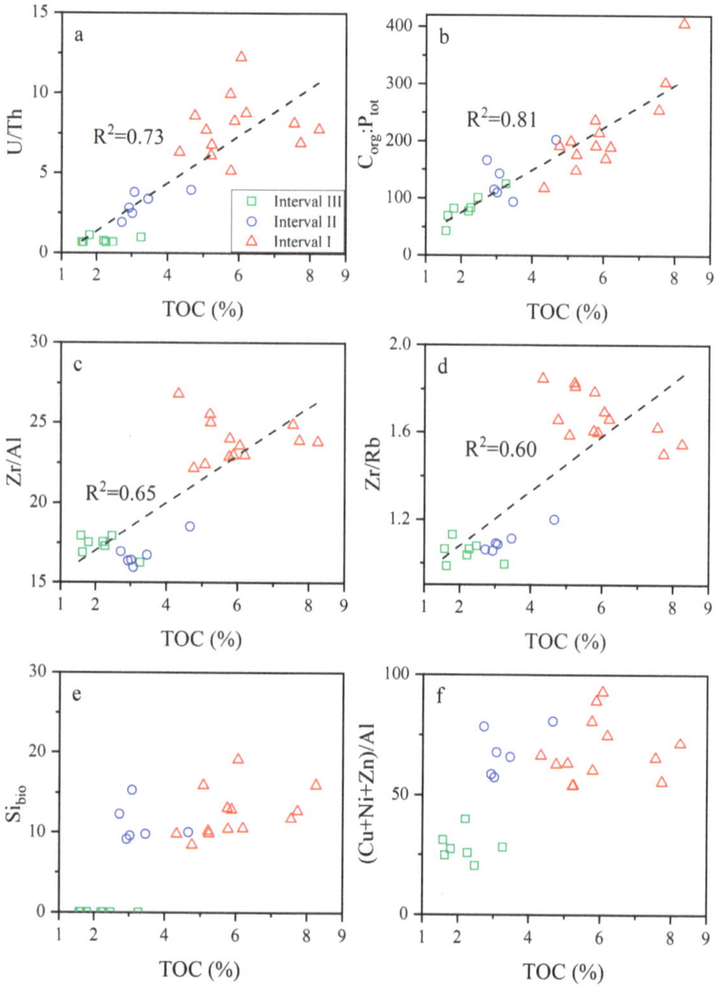

Figure 12. Relationships between the TOC content and the sedimentary geochemical proxies.

The variations in TOC content during the interval II period were controlled by favorable preservation conditions and the dilution effect of terrigenous debris. Overall, both the Si_{bio} and (Ni + Cu + Zn)/Al in the interval II samples are approximately equal to or little higher than those of the interval I samples (Figures 7 and 12e,f), indicating that there was no significant change in paleoproductivity. However, the TOC content within the interval II samples is lower than that in the interval I samples, indicating that despite the higher paleoproductivity levels, there was no corresponding increase in TOC content. After the deposition of interval I strata, oxygen concentration decreased due to the decline in the relative sea-level, but the environment still remained in a reducing state during the interval II period. The strong correlations between the TOC content and the redox indicators (Figure 12a,b) suggest that the reducing conditions remain the main controlling factor for the organic matter enrichment. The increased input of terrigenous debris caused by the fall in the relative sea-level diluted the original concentration of organic matter (Figure 12c,d), which is another factor leading to the decrease in TOC content during the interval II period. Therefore, the changes in TOC content within the interval II samples are primarily controlled by the reducing environment, followed by the input of terrigenous detritus.

The low TOC content within the interval III samples is constrained by a combination of the poor preservation conditions and the lower paleoproductivity. During the deposition of the interval III, the relative sea-level continued to fall, while the input of terrigenous debris remained relatively stable compared to interval II (Figure 7). It is clear that the dilution effect of terrigenous debris does not adequately account for the upward decrease in TOC content within the interval III samples (Figure 12c,d). The redox geochemical indicators suggest that the interval III samples were deposited under a dysoxic condition, where the preservation conditions for organic matter were relatively poor. This shows that TOC content has a good synergistic variation with the U/Th and C_{org}:P_{tot} within the interval III samples (Figure 12a,b), indicating that samples with relatively high oxygen concentrations, i.e., poor preservation conditions, have lower TOC content. Moreover, paleoproductivity was also at a lower level. There is a weak positive correlation between TOC content and (Cu + Ni + Zn)/Al values (Figure 12e,f). Based on the aforementioned analysis, we conclude that the low TOC content within the interval III samples is primarily associated with poor preservation conditions and lower paleoproductivity levels.

6. Conclusions

1. The Lower Cambrian Shuijingtuo Shale in the Yichang Slope is divided into three intervals. The lithofacies of interval I is mainly black siliceous shale with high TOC content, and interval II is mainly black siliceous shale with moderate TOC content. Interval III consists of black, clay-rich siliceous shale and dark-gray calcareous shale and is characterized by a low TOC content.
2. The major and trace elements in the examined gas-matured samples still retain the geochemical signature of the sedimentary environment, because the thermal diagenesis has little effect on element partitioning or remobilization. This is supported by the strong positive correlations observed between both U/Al and Mo/Al ratios and TOC content. The geochemical indicators used to distinguish between hydrothermal and non-hydrothermal activity reveal that the hydrothermal deposits, which were widely distributed in the southern Yangtze block, were absent in the Yichang Slope.
3. Based on the evolution of the sedimentary environment, the primary controlling factors for the organic matter enrichment in the Shuijingtuo Formation at different stages are determined. The high TOC content in the interval I samples is mainly attributed to the strong reducing condition, which was caused by the rapid rise of the relative sea-level in the late Ediacaran to early Cambrian. The subsequent decline in the relative sea-level after the interval II period led to a decrease in oxygen concentration. Additionally, the input of abundant terrigenous debris diluted the original organic matter concentration. Thus, preservation condition and debris dilution together control the organic matter content within the interval II samples. As the relative sea-level continued to fall, the preservation conditions and primary productivity reached lower levels, resulting in lower TOC content in the interval III samples.

Supplementary Materials: The following supporting information can be downloaded at: https://www.mdpi.com/article/10.3390/jmse11102018/s1, Table S1: results of all the major and trace elements.

Author Contributions: Conceptualization, S.W., M.H., S.H. and P.L.; methodology, S.W., M.H., S.H., W.Y. and P.L.; formal analysis, S.W., Q.H. and Q.C.; investigation, Q.H., W.Y. and Q.C.; resources, M.H. and S.H.; data curation, Q.H., W.Y. and P.L.; writing—original draft preparation, S.W. and P.L.; writing—review and editing, S.W., M.H., S.H., Q.H., W.Y., Q.C. and P.L.; funding acquisition, S.W., S.H. and Q.C. All authors have read and agreed to the published version of the manuscript.

Funding: This research was funded by the Natural Science Foundation of Hubei Province (No. 2022CFB652) and CNPC Innovation Found (No. 2021DQ02-0101).

Institutional Review Board Statement: Not applicable.

Informed Consent Statement: Not applicable.

Data Availability Statement: Data supporting the results can be found in the manuscript text, tables, and figures.

Conflicts of Interest: The authors declare no conflict of interest.

References

1. Curtis, J.B. Fractured shale-gas systems. *AAPG Bull.* **2002**, *86*, 1921–1938.
2. Hao, F.; Zou, H.; Lu, Y. Mechanisms of shale gas storage: Implications for shale gas exploration in China. *AAPG Bull.* **2013**, *97*, 1325–1346. [CrossRef]
3. Zou, C.; Yang, Z.; Dong, D.; Zhao, q.; Chen, Z.; Feng, Y.; Li, J.; Wang, X. Formation, distribution and prospect of unconventional hydrocarbons in source rock strata in China. *Earth Sci.* **2022**, *47*, 1517–1533. (In Chinese)
4. Ko, L.T.; Loucks, R.G.; Zhang, T.; Ruppel, S.C.; Shao, D. Pore and pore network evolution of Upper Cretaceous Boquillas (Eagle Ford-equivalent) mudrocks: Results from gold tube pyrolysis experiments. *AAPG Bull.* **2016**, *100*, 1693–1722. [CrossRef]
5. Loucks, R.G.; Reed, R.M.; Ruppel, S.C.; Jarvie, D.M. Morphology, genesis, and distribution of nanometer-scale pores in siliceous mudstones of the Mississippian Barnett Shale. *J. Sediment. Res.* **2009**, *79*, 848–861. [CrossRef]
6. Wei, S.; He, S.; Pan, Z.; Zhai, G.; Yang, R.; Dong, T.; Yang, W. Characteristics and evolution of pyrobitumen-hosted pores of the overmature Lower Cambrian Shuijingtuo Shale in the south of Huangling anticline, Yichang area, China: Evidence from FE-SEM petrography. *Mar. Petrol. Geol.* **2020**, *116*, 104303. [CrossRef]
7. Sageman, B.B.; Murphy, A.E.; Werne, J.P.; Ver Straeten, C.A.; Hollander, D.J.; Lyons, T.W. A tale of shales: The relative roles of production, decomposition, and dilution in the accumulation of organic-rich strata, Middle–Upper Devonian, Appalachian basin. *Chem. Geol.* **2003**, *195*, 229–273. [CrossRef]
8. Katz, B.J. Controlling Factors on Source Rock Development—A Review of Productivity, Preservation, and Sedimentation Rate. In *The Deposition of Organic-Carbon-Rich Sediments: Models, Mechanisms, and Consequences*; SEPM Society for Sedimentary Geology: Broken Arrow, OK, USA, 2005; Volume 82, pp. 7–16.
9. Zhao, J.; Jin, Z.; Jin, Z.; Geng, Y.; Wen, X.; Yan, C. Applying sedimentary geochemical proxies for paleoenvironment interpretation of organic-rich shale deposition in the Sichuan Basin, China. *Int. J. Coal Geol.* **2016**, *163*, 52–71. [CrossRef]
10. Algeo, T.J.; Maynard, J.B. Trace-element behavior and redox facies in core shales of Upper Pennsylvanian Kansas-type cyclothems. *Chem. Geol.* **2004**, *206*, 289–318. [CrossRef]
11. Harris, N.B. *The Deposition of Organic-Carbon-Rich Sediments: Models, Mechanisms, and Consequences—Introduction*; SEPM Society for Sedimentary Geology: Broken Arrow, OK, USA, 2005; Volume 82.
12. Bohacs, K.M.; Grabowski, G.J.; Carroll, A.R., Jr.; Mankiewicz, P.J.; Miskell-Gerhardt, K.J.; Schwalbach, J.R.; Wegner, M.B.; Simo, J.A. Production, Destruction, and Dilution—The Many Paths to Source-Rock Development. In *The Deposition of Organic-Carbon-Rich Sediments: Models, Mechanisms, and Consequences*; SEPM Society for Sedimentary Geology: Broken Arrow, OK, USA, 2005; Volume 82.
13. Rimmer, S.M.; Thompson, J.A.; Goodnight, S.A.; Robl, T.L. Multiple controls on the preservation of organic matter in Devonian–Mississippian marine black shales: Geochemical and petrographic evidence. *Palaeogeogr. Palaeoclimatol. Palaeoecol.* **2004**, *215*, 125–154. [CrossRef]
14. Wei, C.; Dong, T.; He, Z.; He, S.; He, Q.; Yang, R.; Guo, X.; Hou, Y. Major, trace-elemental and sedimentological characterization of the upper Ordovician Wufeng-lower Silurian Longmaxi formations, Sichuan Basin, south China: Insights into the effect of relative sea-level fluctuations on organic matter accumulation in shales. *Mar. Petrol. Geol.* **2021**, *126*, 104905. [CrossRef]
15. Shen, W.; Wang, Y.; Zheng, Z.; Shen, S.; Xie, H.; Qin, H.; Bai, M. Paleo-Environmental Variation and Its Control on the Organic Matter Accumulation in Black Shale of the Permian Gufeng Formation in the Lower Yangtze Area, South China. *Front. Earth Sci.* **2022**, *10*, 899947. [CrossRef]
16. Canfield, D.E. Factors influencing organic carbon preservation in marine sediments. *Chem. Geol.* **1994**, *114*, 315–329. [CrossRef] [PubMed]
17. Chen, L.; Zhang, B.; Chen, X.; Jiang, S.; Zhang, G.; Lin, W.; Chen, P.; Liu, Z. Depositional environment and organic matter accumulation of the Lower Cambrian Shuijingtuo Formation in the middle Yangtze area, China. *J. Petrol. Sci. Eng.* **2022**, *208*, 109339. [CrossRef]
18. Ibach, L.E.J. Relationship Between Sedimentation Rate and Total Organic Carbon Content in Ancient Marine Sediments1. *AAPG Bull.* **1982**, *66*, 170–188.
19. Liu, Z.; Xu, L.; Wen, Y.; Zhang, Y.; Luo, F.; Duan, K.; Chen, W.; Zhou, X.; Wen, J. Accumulation characteristics and comprehensive evaluation of shale gas in Cambrian Niutitang Formation, Hubei. *Earth Sci.* **2022**, *47*, 1586–1603. (In Chinese)
20. Chen, X.; Chen, L.; Jiang, S.; Liu, A.; Luo, S.; Li, H.; Li, P.; Chen, P. Evaluation of shale reservoir quality by geophysical logging for Shuijingtuo Formation of Lower Cambrian in Yichang area, central Yangtze. *J. Earth Sci.* **2021**, *32*, 766–777. [CrossRef]
21. Wang, Y.; Zhai, G.; Liu, G.; Shi, W.; Lu, Y.; Li, J.; Zhang, Y. Geological characteristics of shale gas in different strata of marine facies in South China. *J. Earth Sci.* **2021**, *32*, 725–741. [CrossRef]
22. Zhang, T.; Luo, H.; Meng, K. Discussion on main controlling factors of the differences of gas content for Cambrian shale in different regions of South China. *Earth Sci. Front.* **2023**, *30*, 1–13. (In Chinese)

23. Wei, S.; Hu, M.; He, S.; Shu, Y.; Dong, T.; He, Q.; Yang, W.; Cai, Q. Effects of Quartz Precipitation on the Abundance and Preservation of Organic Matter Pores in Cambrian Marine Shale in South China. *J. Mar. Sci. Eng.* **2023**, *11*, 1267. [CrossRef]
24. Yang, W.; He, S.; Zhai, G.; Tao, Z.; Yuan, X.; Wei, S. Maturity assessment of the Lower Cambrian and Sinian shales using multiple technical approaches. *J. Earth Sci.* **2021**, *32*, 1262–1277. [CrossRef]
25. Meng, M.; Ge, H.; Shen, Y.; Ji, W.; Li, Z. Insight into water occurrence and pore size distribution by nuclear magnetic resonance in marine shale reservoirs, southern China. *Energy Fuels* **2023**, *37*, 319–327. [CrossRef]
26. Li, Z.X.; Bogdanova, S.V.; Collins, A.S.; Davidson, A.; De Waele, B.; Ernst, R.E.; Fitzsimons, I.C.W.; Fuck, R.A.; Gladkochub, D.P.; Jacobs, J.; et al. Assembly, configuration, and break-up history of Rodinia: A synthesis. *Precambrian Res.* **2008**, *160*, 179–210. [CrossRef]
27. Jiang, G.; Wang, X.; Shi, X.; Xiao, S.; Zhang, S.; Dong, J. The origin of decoupled carbonate and organic carbon isotope signatures in the early Cambrian (ca. 542–520 Ma) Yangtze platform. *Earth Planet. Sci. Lett.* **2012**, *317–318*, 96–110. [CrossRef]
28. Wang, Z.; Tan, J.; Hilton, J.; Dick, J.; Wen, Z. Trace element enrichment mechanisms in black shales during the early cambrian (ca. 521–514 Ma), South China. *Mar. Petrol. Geol.* **2023**, *149*, 106083. [CrossRef]
29. Tribovillard, N.; Algeo, T.J.; Lyons, T.; Riboulleau, A. Trace metals as paleoredox and paleoproductivity proxies: An update. *Chem. Geol.* **2006**, *232*, 12–32. [CrossRef]
30. Taylor, S.R.; McLennan, S.M. The Continental Crust: Its Composition and Evolution. 1985. Available online: https://www.osti.gov/biblio/6582885 (accessed on 17 October 2023).
31. Ross, D.J.K.; Bustin, R.M. Investigating the use of sedimentary geochemical proxies for paleoenvironment interpretation of thermally mature organic-rich strata: Examples from the Devonian–Mississippian shales, Western Canadian Sedimentary Basin. *Chem. Geol.* **2009**, *260*, 1–19. [CrossRef]
32. Abanda, P.A.; Hannigan, R.E. Effect of diagenesis on trace element partitioning in shales. *Chem. Geol.* **2006**, *230*, 42–59. [CrossRef]
33. Raiswell, R.; Berner, R.A. Organic carbon losses during burial and thermal maturation of normal marine shales. *Geology.* **1987**, *15*, 853–856. [CrossRef]
34. Steiner, M.; Wallis, E.; Erdtmann, B.-D.; Zhao, Y.; Yang, R. Submarine-hydrothermal exhalative ore layers in black shales from South China and associated fossils—Insights into a Lower Cambrian facies and bio-evolution. *Palaeogeogr. Palaeoclimatol. Palaeoecol.* **2001**, *169*, 165–191. [CrossRef]
35. Zhang, K.; Li, X.; Wang, Y.; Liu, W.; Yu, Y.; Zhou, L.; Feng, W. Paleo-environments and organic matter enrichment in the shales of the Cambrian Niutitang and Wunitang Formations, south China: Constraints from depositional environments and geochemistry. *Mar. Petrol. Geol.* **2021**, *134*, 105329. [CrossRef]
36. Zhang, K.; Jiang, Z.; Yin, L.; Gao, Z.; Wang, P.; Song, Y.; Jia, C.; Liu, W.; Liu, T.; Xie, X.; et al. Controlling functions of hydrothermal activity to shale gas content-taking lower Cambrian in Xiuwu Basin as an example. *Mar. Petrol. Geol.* **2017**, *85*, 177–193. [CrossRef]
37. Li, Y.; Fan, T.; Zhang, J.; Zhang, J.; Wei, X.; Hu, X.; Zeng, W.; Fu, W. Geochemical changes in the Early Cambrian interval of the Yangtze Platform, South China: Implications for hydrothermal influences and paleocean redox conditions. *J. Asian Earth Sci.* **2015**, *109*, 100–123. [CrossRef]
38. Wang, Z.; Tan, J.; Boyle, R.; Hilton, J.; Ma, Z.; Wang, W.; Lyu, Q.; Kang, X.; Luo, W. Evaluating episodic hydrothermal activity in South China during the early Cambrian: Implications for biotic evolution. *Mar. Petrol. Geol.* **2020**, *117*, 104355. [CrossRef]
39. Boström, K.; Peterson, M.N.A. The origin of aluminum-poor ferromanganoan sediments in areas of high heat flow on the East Pacific Rise. *Mar. Geol.* **1969**, *7*, 427–447. [CrossRef]
40. Yamamoto, K. Geochemical characteristics and depositional environments of cherts and associated rocks in the Franciscan and Shimanto Terranes. *Sediment Geol.* **1987**, *52*, 65–108. [CrossRef]
41. Cai, Q.; Hu, M.; Kane, O.I.; Li, M.; Zhang, B.; Hu, Z.; Deng, Q.; Xing, N. Cyclic variations in paleoenvironment and organic matter accumulation of the Upper Ordovician–Lower Silurian black shale in the Middle Yangtze Region, South China: Implications for tectonic setting, paleoclimate, and sea-level change. *Mar. Petrol. Geol.* **2021**, *136*, 105477. [CrossRef]
42. Dong, T.; He, Q.; He, S.; Zhai, G.; Zhang, Y.; Wei, S.; Wei, C.; Hou, Y.; Guo, X. Quartz types, origins and organic matter-hosted pore systems in the lower cambrian Niutitang Formation, middle yangtze platform, China. *Mar. Petrol. Geol.* **2021**, *123*, 104739. [CrossRef]
43. Adachi, M.; Yamamoto, K.; Sugisaki, R. Hydrothermal chert and associated siliceous rocks from the northern Pacific their geological significance as indication od ocean ridge activity. *Sediment Geol.* **1986**, *47*, 125–148. [CrossRef]
44. Halbach, M.; Halbach, P.; Lüders, V. Sulfide-impregnated and pure silica precipitates of hydrothermal origin from the Central Indian Ocean. *Chem. Geol.* **2002**, *182*, 357–375. [CrossRef]
45. Zhang, Y.; Huang, D.; Zhang, L.; Wan, C.; Luo, H.; Shao, D.; Meng, K.; Yan, J.; Zhang, T. Biogenic silica of the Lower Cambrian Shuijingtuo Formation in Yichang, western Hubei Province—Features and influence on shale gas accumulation. *Earth Sci. Front.* **2023**, *30*, 83–100. (In Chinese)
46. Luo, H.; Shao, D.; Meng, K.; Zhang, Y.; Song, H.; Yan, J.; Zhang, T. Origin of excess barium in the Cambrian shale of Yichang area, western Hubei, and its implication for organic matter accumulation. *Earth Sci. Front.* **2023**, *30*, 66–82. (In Chinese)
47. Wei, K.; Chen, X.; Wang, C.; Liu, A.; Zeng, X.; Li, Z. Origin of siliceous rocks in west Hunan and Hubei Provinces during Late Ediacaran-Early Cambrian, and its geological significance of shale gas. *Bull. Geol. Sci. Technol.* **2020**, *39*, 20–30. (In Chinese)
48. Jones, B.; Manning, D.A.C. Comparison of geochemical indices used for the interpretation of palaeoredox conditions in ancient mudstones. *Chem. Geol.* **1994**, *111*, 111–129. [CrossRef]

49. Gambacorta, G.; Trincianti, E.; Torricelli, S. Anoxia controlled by relative sea-level changes: An example from the Mississippian Barnett Shale Formation. *Palaeogeogr. Palaeoclimatol. Palaeoecol.* **2016**, *459*, 306–320. [CrossRef]
50. Zhao, L.; Liu, S.; Li, G.; Zhang, M.; Liang, X.; Li, J.; Xu, J. Sedimentary Environment and Enrichment of Organic Matter During the Deposition of Qiongzhusi Formation in the Upslope Areas—A Case Study of W207 Well in the Weiyuan Area, Sichuan Basin, China. *Front. Earth Sci.* **2022**, *10*, 867616. [CrossRef]
51. Hoefs, J. *Stable Isotope Geochemistry*; Springer: Berlin/Heidelberg, Germany, 1997; Volume 201.
52. Algeo, T.J.; Ingall, E. Sedimentary Corg:P ratios, paleocean ventilation, and Phanerozoic atmospheric pO2. *Palaeogeogr. Palaeoclimatol. Palaeoecol.* **2007**, *256*, 130–155. [CrossRef]
53. Ingall, E.; Kolowith, L.; Lyons, T.; Hurtgen, M. Sediment carbon, nitrogen and phosphorus cycling in an anoxic fjord, Effingham Inlet, British Columbia. *Am. J. Sci.* **2005**, *305*, 240–258. [CrossRef]
54. Sannigrahi, P.; Ingall, E. Polyphosphates as a source of enhanced P fluxes in marine sediments overlain by anoxic waters: Evidence from 31P NMR. *Geochem. Trans.* **2005**, *6*, 52–59. [CrossRef]
55. Luo, S.; Chen, X.; Li, H.; Wang, C. Shale gas accumulation conditions and target optimization of the Lower Cambrian Shuijintuo Formation in Yichang area, Western Hubei. *Earth Sci.* **2019**, *44*, 3598–3615. (In Chinese)
56. Algeo, T.J.; Tribovillard, N. Environmental analysis of paleoceanographic systems based on molybdenum–uranium covariation. *Chem. Geol.* **2009**, *268*, 211–225. [CrossRef]
57. Tribovillard, N.; Algeo, T.J.; Baudin, F.; Riboulleau, A. Analysis of marine environmental conditions based onmolybdenum–uranium covariation—Applications to Mesozoic paleoceanography. *Chem. Geol.* **2012**, *324–325*, 46–58. [CrossRef]
58. Zhang, G.; Chen, D.; Ding, Y.; Huang, T. Controls on Organic Matter Accumulation from an Upper Slope Section on the Early Cambrian Yangtze Platform, South China. *Minerals* **2023**, *13*, 260. (In Chinese) [CrossRef]
59. Yang, X.; Yan, D.; Zhang, B.; Zhang, L.; Wei, X.; Li, T.; Zhang, J.; She, X. The impact of volcanic activity on the deposition of organic-rich shales: Evidence from carbon isotope and geochemical compositions. *Mar. Petrol. Geol.* **2021**, *128*, 105010. [CrossRef]
60. Shu, Y.; Lu, Y.; Hu, Q.; Wang, C.; Wang, Q. Geochemical, petrographic and reservoir characteristics of the transgressive systems tract of lower Silurian black shale in Jiaoshiba area, southwest China. *Mar. Petrol. Geol.* **2021**, *129*, 105014. [CrossRef]
61. Shen, W.; Zhu, X.; Li, J.; Yan, B. Mechanism of organic matter accumulation in black shale of the Datangpo Formation: Insights from paleo-environmental variation during the Cryogenian non-glaciation. *Precambrian Res.* **2022**, *383*, 106889. [CrossRef]
62. Rimmer, S.M. Geochemical paleoredox indicators in Devonian–Mississippian black shales, Central Appalachian Basin (USA). *Chem. Geol.* **2004**, *206*, 373–391. [CrossRef]
63. Thibault, N.; Ruhl, M.; Ullmann, C.V.; Korte, C.; Kemp, D.B.; Gröcke, D.R.; Hesselbo, S.P. The wider context of the Lower Jurassic Toarcian oceanic anoxic event in Yorkshire coastal outcrops, UK. *Proc. Geol. Assoc.* **2018**, *129*, 372–391. [CrossRef]
64. Yi, F.; Yi, H.; Xia, G.; Wu, C.; Li, G.; Cai, Z.; Li, N. Factors controlling the organic matter accumulation across the Pliensbachian-Toarcian transition in the Qiangtang Basin, Tibetan Plateau. *Mar. Petrol. Geol.* **2021**, *133*, 105304. [CrossRef]
65. Dong, T.; Harris, N.B.; Knapp, L.J.; McMillan, J.M.; Bish, D.L. The effect of thermal maturity on geomechanical properties in shale reservoirs: An example from the Upper Devonian Duvernay Formation, Western Canada Sedimentary Basin. *Mar. Petrol. Geol.* **2018**, *97*, 137–153. [CrossRef]
66. Cai, Q.; Hu, M.; Kane, O.I.; Yang, Z.; Wen, Y.; Luo, Q.; Li, M.; Hu, Z.; Deng, Q. Petrological and geochemical characteristics of the Ordovician–Silurian black shale in eastern Sichuan and western Hubei, South China: Differential sedimentary responses to tectonism and glaciation. *J. Palaeogeogr.* **2023**, *12*, 129–152. [CrossRef]
67. Worden, R.H.; Morad, S. Quartz cementation in oil field sandstones: A review of the key controversies. In *Quartz Cementation in Sandstones*; Worden, R.H., Morad, S., Eds.; Blackwell Science: Hoboken, NJ, USA, 2000; pp. 1–20.
68. Rutman, P.; Hoareau, G.; Kluska, J.-M.; Lejay, A.; Fialips, C.; Gelin, F.; Aubourg, C.; Bilbao, E.H. Diagenesis and alteration of subsurface volcanic ash beds of the Vaca Muerta Formation, Argentina. *Mar. Petrol. Geol.* **2021**, *132*, 105220. [CrossRef]
69. Dong, T.; Harris, N.B.; Ayranci, K.; Yang, S. The impact of rock composition on geomechanical properties of a shale formation: Middle and Upper Devonian Horn River Group shale, Northeast British Columbia, Canada. *AAPG Bull.* **2017**, *101*, 177–204. [CrossRef]
70. Zhao, J.; Jin, Z.; Lin, C.; Liu, G.; Liu, K.; Liu, Z.; Zhang, Y. Sedimentary environment of the Lower Cambrian Qiongzhusi Formation shale in the Upper Yangtze region. *Oil Gas Geol.* **2019**, *40*, 701–715. (In Chinese)

Journal of
Marine Science and Engineering

MDPI

Article

A New Approach for Production Prediction in Onshore and Offshore Tight Oil Reservoir

Kaixuan Qiu [1,2], Kaifeng Fan [3], Xiaolin Chen [4], Gang Lei [5,6,*], Shiming Wei [7,*], Rahul Navik [1] and Jia Li [4,*]

1 Jiangmen Laboratory of Carbon Science and Technology, Jiangmen 529100, China; qiukx940908@163.com (K.Q.); rahulnavik@hkustgz-jcl.ac.cn (R.N.)
2 The University of Hongkong, Hongkong 999077, China
3 College of Oceanography, Hohai University, Nanjing 210098, China; 18871388135@163.com
4 Society Hub, The Hong Kong University of Science and Technology (Guangzhou), Nansha, Guangzhou 511400, China; hljchan334@connect.hkust-gz.edu.cn
5 Faculty of Engineering, China University of Geosciences (Wuhan), Wuhan 430074, China
6 Shenzhen Research Institute, China University of Geosciences, Shenzhen 518063, China
7 College of Science, China University of Petroleum (Beijing), Beijing 102249, China
* Correspondence: leigang@cug.edu.cn (G.L.); we_shiming@163.com (S.W.); leejia@ust.hk (J.L.)

Abstract: Rapid technological advances have accelerated offshore and onshore tight oil extraction to meet growing energy demand. Reliable tools to carry out production prediction are essential for development of unconventional reservoirs. The existed tri-linear analytical solutions are verified to be versatile enough to capture fundamental flow mechanisms and make accurate production predictions. However, these solutions are obtained in Laplace space with the Laplace transform and numerical inversion, which may lead to uncertainty in the solution. In this paper, a general analytical solution is derived in real-time space through integral transform and average pressure substitution. Namely, the partial differential equations describing subsurface fluid flow are firstly triple-integrated and then the obtained volume average pressure are replaced with the rate-dependent expressions. Furthermore, the ordinary differential equations related to oil rate are solved analytically in real-time space. To validate our model, this derived solution is verified against two numerical models constructed with two typical physical configurations. The great match indicates the accuracy and applicability of the analytical solution. According to the developed workflow, two field cases including offshore and onshore tight oilfield data are selected for history matching and production prediction. This new approach not only makes the obtained solution more simplified, but also helps field engineers diagnose flow patterns more quickly to better optimize production schemes.

Keywords: tri-linear flow; offshore tight oilfield; analytical solution; history matching; production prediction

Citation: Qiu, K.; Fan, K.; Chen, X.; Lei, G.; Wei, S.; Navik, R.; Li, J. A New Approach for Production Prediction in Onshore and Offshore Tight Oil Reservoir. *J. Mar. Sci. Eng.* **2023**, *11*, 2079. https://doi.org/10.3390/jmse11112079

Academic Editor: Timothy S. Collett

Received: 22 September 2023
Revised: 20 October 2023
Accepted: 25 October 2023
Published: 30 October 2023

1. Introduction

The global demand for oil/gas is increasing sharply over the past few decades [1]. Unconventional resources have attracted domestic and international attention due to its abundance in onshore and offshore oilfields, particularly for tight oil with matrix permeability less than 0.1 mD. [2]. According to the statistical results from the Energy Information Administration (EIA), the contribution from offshore oilfields has reached nearly 30% of global production [3]. Well performance is considered as a critical factor because of its close relationship with economics of unconventional reservoirs. Therefore, advanced techniques for unconventional reservoirs are constantly updated to generate the accurate production prediction tools by covering the relative petrophysical properties of subsurface fluid storage and flow [4–7].

Unconventional reservoirs, unlike conventional reservoirs, must be hydraulically fractured prior to commercial production of hydrocarbons. The highly conductive fracture networks are created for subsurface fluid to flow from the tight matrix to the wellbore. The widely used techniques for flow pattern analysis and production prediction in these reservoirs include numerical simulation and analytical models [8–12]. Certainly, numerical simulation methods have good flexibility and can also deal with various complicated reservoir seepage problems. However, there will be a time-consuming and computationally inefficient process when they are adopted to compute the complex fracture geometry with a large number of grids [13,14]. Generally speaking, the analytical model offers relative simplicity and it can cover the fundamental flow mechanism with simple solutions. A large number of analytical models have been proposed to study the matrix–fracture communication and well performance in unconventional reservoirs. El-Banbi [15] observed the long-term half-slope line on a log–log plot of oil rate against production time and derived a series of analytical solutions to analyze the transient linear behavior in tight oil reservoirs. On the basis of El-Banbi's model, Bello [16] applied the transient-linear-flow model in fractured shale reservoirs and developed many asymptotic analysis equations to describe observable linear flow regimes. For most cases of unconventional oil/gas reservoir development, the communication between the tight matrix and complex fractures is thought of as the transient linear flow in the stimulated reservoir volumes (SRV). Brown et al. [17] presented a general tri-linear flow model describing the sequential flow among the hydraulic fracture region, the stimulated region with complex fracture networks and the low-permeability matrix region beyond hydraulic fractures. Moreover, Stalgorova and Matter [18] provided another tri-linear flow analytical model with a different configuration to consider complex branch fractures. The surrounding low-permeability matrix regions in their model are parallel to the hydraulic fractures. One of the key advantages of the trilinear flow model is that it is sufficient to study the key characteristics of flow convergence toward a fractured horizontal well within the stimulated reservoir volume. Because the completion process associated with multifracture horizontal wells can often significantly alter rock stresses and induce natural fractures, it is important that the complex interplay of flow among matrix, natural fractures and hydraulic-fractures can be captured in pressure-transient models. More importantly, the two model configurations are both typical and have been validated to accurately predict production in unconventional reservoirs.

In this paper, we present a new approach based on integral transform and average pressure substitution to directly derive an analytical solution in real-time space [19–21]. Firstly, the fractured formation is divided into three regions based on the two typical tri-linear conceptual models. Then, the sequential transient linear flow among three regions can be expressed by partial differential equations (PDEs). Furthermore, the PDEs are firstly transformed into ordinary differential equations (ODEs) and then solved analytically in real-time space through integral transform and average pressure substitution rather than Laplace transform and numerical inversion [22]. Finally, the analytical solution is verified with two classical equivalent numerical models and applied to two field cases from the onshore and offshore oilfield for production prediction.

2. Theory and Model Development

Commercial production from unconventional oil reservoirs (ultra-low permeability) depends on advanced horizontal well techniques with multistage hydraulic fracturing. Because of uncertain fracture branching, some stimulated regions are created around hydraulic fractures, which are often defined and modeled as regions of high permeability. Therefore, it is vital to choose appropriate conceptual models for simulating fluid flow. Two typical tri-linear conceptual models presented, respectively, by Brown et al. [17] and Stalgorova and Matter [18] are adopted in this paper. The whole analytical models are constructed on basis of these two models. Figure 1a,b are both the schematic of the reservoir with a multi-stage fractured horizontal well. There are three regions including a hydraulic fracture region that is individually connected to the well, a stimulated region (gray color) with higher permeability around each hydraulic fracture and an un-stimulated region (white color) with lower permeability connected to the stimulated region. The difference of the two model is the connection between the un-stimulated region and the stimulated region. It means that different reservoir configurations are also considered in the work. Figure 1c,d present the plan view, which is an idealized representation of the trilinear-flow model used to develop fluid flow equations. The blank arrows represent the flow direction. For the two conceptual models, the systems are assumed to be symmetrical, and no-flow boundaries exist in the symmetry plane between fractures as a result of the interference of flow. Therefore, one quarter of the reservoir (contained in a red dashed box) is adopted to construct the analytical model.

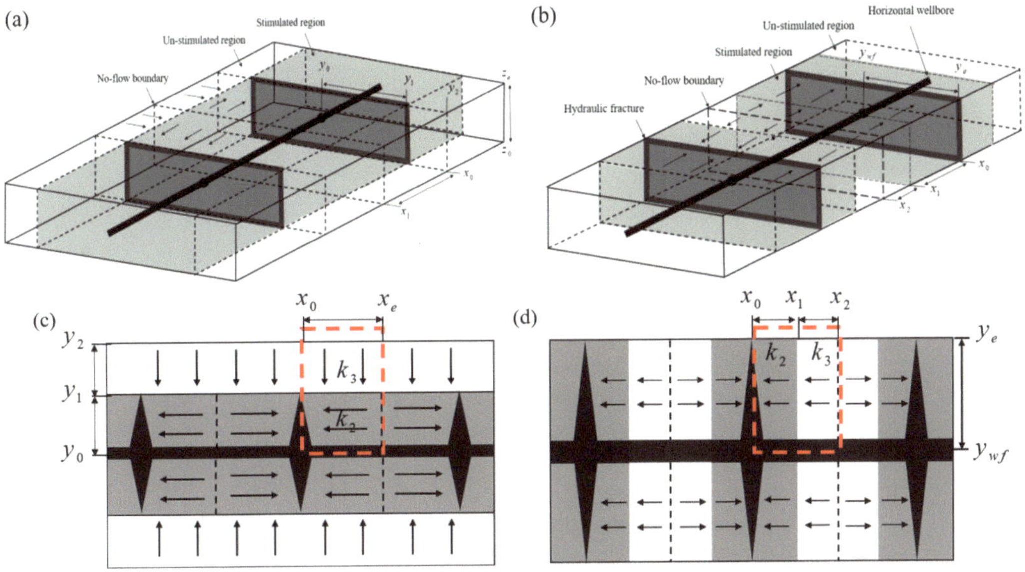

Figure 1. Schematic of a multi-stage fractured horizontal well based on two conceptual models. (**a**,**b**) 3D schematic. (**c**,**d**) Plan view.

Prior to develop the mathematical model, the model assumptions must be defined to simplify the derivation. The specific assumptions are as follows:

1. The reservoir is homogeneous, thick uniformly and isothermal.
2. Fluid flow is single phase and 1D linear in each region.
3. The whole production process is under constant bottom-hole pressure.
4. The entire flow system follows symmetry and continuity.
5. The effect of gravity and capillary forces are neglected.

Model 1: Based on Brown's conceptual model

Based on the above model assumptions, the mathematical models can be constructed through describing the transient linear flow process in different regions. Firstly, the governing equation for fluid flow in an un-stimulated region defined as Region 3 can be expressed as

$$\frac{\partial^2 p_{B3}}{\partial x^2} + \frac{\partial^2 p_{B3}}{\partial y^2} + \frac{\partial^2 p_{B3}}{\partial z^2} = \frac{(\phi \mu c_t)_{B3}}{k_{B3}} \frac{\partial p_{B3}}{t} \quad (1)$$

The pressure in Region 3 is considered to be the initial reservoir pressure before fluid flow occurs.

$$p_{B3}(x, y, z, 0) = p_i \quad (2)$$

The outer boundary of the Region 3 is assumed to be a no-flow boundary. Therefore, the appropriate mathematical expression for the outer boundary condition is given by

$$\left.\frac{\partial p_{B3}}{\partial x}\right|_{x=x_0} = 0 \tag{3}$$

$$\left.\frac{\partial p_{B3}}{\partial y}\right|_{y=y_2} = 0 \tag{4}$$

$$\left.\frac{\partial p_{B3}}{\partial z}\right|_{z=z_0,z_e} = 0 \tag{5}$$

Because of the symmetry, there is no fluid flow occurs at $x = x_e$.

$$\left.\frac{\partial p_{B3}}{\partial x}\right|_{x=x_e} = 0 \tag{6}$$

Based on the assumption of continuity, the inner boundary condition can be expressed as

$$\left.\frac{k_{B3}}{\mu}\frac{\partial p_{B3}}{\partial y}\right|_{y=y_1} = \left.\frac{k_{B2}}{\mu}\frac{\partial p_{B2}}{\partial y}\right|_{y=y_1} \tag{7}$$

Considering the fluid flow sequentially from the un-stimulated region to the stimulated region, the diffusivity equation to describe the flow process in stimulated region defined Region 2 can be written similarly as

$$\frac{\partial^2 p_{B2}}{\partial x^2} + \frac{\partial^2 p_{B2}}{\partial y^2} + \frac{\partial^2 p_{B2}}{\partial z^2} = \frac{(\phi\mu c_t)_{B2}}{k_{B2}}\frac{\partial p_{B2}}{t} \tag{8}$$

For Region 2, the initial condition is identical.

$$p_{B2}(x, y, z, 0) = p_i \tag{9}$$

The bottom of Region 2 is the horizontal wellbore, which is defined as an impermeable boundary. Therefore, the outer boundary condition is expressed as

$$\left.\frac{\partial p_{B2}}{\partial y}\right|_{y=y_0} = 0 \tag{10}$$

$$\left.\frac{\partial p_{B2}}{\partial z}\right|_{z=z_0,z_e} = 0 \tag{11}$$

Due to the symmetry, another outer boundary condition is similar to Equation (6):

$$\left.\frac{\partial p_{B2}}{\partial x}\right|_{x=x_e} = 0 \tag{12}$$

Because the stimulated region is connected to the hydraulic fracture region and the unstimulated region, the inner boundary condition of Region 2 can be written as

$$\left.\frac{k_{B2}}{\mu}\frac{\partial p_{B2}}{\partial y}\right|_{y=y_1} = \left.\frac{k_{B3}}{\mu}\frac{\partial p_{B3}}{\partial y}\right|_{y=y_1} \tag{13}$$

$$\left.\frac{k_{B2}}{\mu}\frac{\partial p_{B2}}{\partial x}\right|_{x=x_0} = \left.\frac{k_{B1}}{\mu}\frac{\partial p_{B1}}{\partial x}\right|_{x=x_0} \tag{14}$$

Similarly, the equations to describe the fluid flow in hydraulic fracture region defined as Region 1 can be expressed as

$$\frac{\partial^2 p_{B1}}{\partial x^2} + \frac{\partial^2 p_{B1}}{\partial y^2} + \frac{\partial^2 p_{B1}}{\partial z^2} = \frac{(\phi\mu c_t)_{B1}}{k_{B1}} \frac{\partial p_{B1}}{t} \tag{15}$$

The hydraulic fracture region is sole connection to well and the production is under constant bottom-hole pressure. Therefore, the initial conditions for Region 1 is given by

$$p_{B1}(x, y, z, 0) = p_i \tag{16}$$

$$p_{B1}(x, y = y_0, z, t) = p_{wf} \tag{17}$$

There is no fluid flow beyond the tip of hydraulic fracture and the location of $x = 0$. The boundary conditions can be written as

$$\left.\frac{\partial p_{B1}}{\partial x}\right|_{x=0} = 0 \tag{18}$$

$$\left.\frac{\partial p_{B1}}{\partial y}\right|_{y=y_1} = 0 \tag{19}$$

$$\left.\frac{\partial p_{B1}}{\partial z}\right|_{z=z_0, z_e} = 0 \tag{20}$$

The hydraulic fracture region is connected to the stimulated region. The flow is continuous at $x = x_0$.

$$\left.\frac{k_{B1}}{\mu}\frac{\partial p_{B1}}{\partial x}\right|_{x=x_0} = \left.\frac{k_{B2}}{\mu}\frac{\partial p_{B2}}{\partial x}\right|_{x=x_0} \tag{21}$$

Model 2: Based on Stalgorova and Matter's conceptual model

Comparing the two concept models, we can find that the flow direction in the two regions is vertical in Model 1 and parallel in Model 2. In other words, the mathematical model developed for Model 2 is nearly alike that of Model 1 except for the boundary conditions. Similarly, the governing equation for un-stimulated region defined as Region 3 is expressed as

$$\frac{\partial^2 p_{S3}}{\partial x^2} + \frac{\partial^2 p_{S3}}{\partial y^2} + \frac{\partial^2 p_{S3}}{\partial z^2} = \frac{(\phi\mu c_t)_{S3}}{k_{S3}} \frac{\partial p_{S3}}{t} \tag{22}$$

The pressure in Region 3 is the initial reservoir pressure when $t = 0$.

$$p_{S3}(x, y, z, 0) = p_i \tag{23}$$

As the boundary is defined as no-flow at the top and bottom of the reservoir, meanwhile, both ends of the y-direction can also be regarded as a no-flow boundary in Model 2.

$$\left.\frac{\partial p_{S3}}{\partial y}\right|_{y=y_{wf}, y_e} = 0 \tag{24}$$

$$\left.\frac{\partial p_{S3}}{\partial z}\right|_{z=z_0, z_e} = 0 \tag{25}$$

Because of the symmetry, there is no fluid flow that occurs at $x = x_2$.

$$\left.\frac{\partial p_{S3}}{\partial x}\right|_{x=x_2} = 0 \tag{26}$$

There is the contact surface between the unstimulated region and the stimulated region. Therefore, the last boundary condition can be given by

$$\frac{k_{S3}}{\mu}\frac{\partial p_{S3}}{\partial x}\bigg|_{x=x_1} = \frac{k_{S2}}{\mu}\frac{\partial p_{S2}}{\partial x}\bigg|_{x=x_1} \tag{27}$$

The flow in stimulated region defined as Region 2 can also be expressed by a series of equations. Firstly, the diffusivity equation for fluid flow in Region 2 can be expressed as

$$\frac{\partial^2 p_{S2}}{\partial x^2} + \frac{\partial^2 p_{S2}}{\partial y^2} + \frac{\partial^2 p_{S2}}{\partial z^2} = \frac{(\phi\mu c_t)_{S2}}{k_{S2}}\frac{\partial p_{S2}}{t} \tag{28}$$

Similarly, the initial condition is identical to Region 3.

$$p_{S2}(x,y,z,0) = p_i \tag{29}$$

The bottom of Region 2 is the horizontal wellbore, which is defined as impermeable boundary. Therefore, we can obtain the outer boundary conditions as follows

$$\frac{\partial p_{S2}}{\partial y}\bigg|_{y=y_{wf},y_e} = 0 \tag{30}$$

$$\frac{\partial p_{S2}}{\partial z}\bigg|_{z=z_0,z_e} = 0 \tag{31}$$

According to the continuity of flux and pressure in the interface, the inner boundary condition of Region 2 can be given by

$$\frac{k_{S2}}{\mu}\frac{\partial p_{S2}}{\partial x}\bigg|_{x=x_1} = \frac{k_{S3}}{\mu}\frac{\partial p_{S3}}{\partial x}\bigg|_{x=x_1} \tag{32}$$

$$\frac{k_{S2}}{\mu}\frac{\partial p_{S2}}{\partial x}\bigg|_{x=x_0} = \frac{k_{S1}}{\mu}\frac{\partial p_{S1}}{\partial x}\bigg|_{x=x_0} \tag{33}$$

Similarly, the equations to describe the fluid flow in hydraulic fracture region defined as Region 1 can be expressed as

$$\frac{\partial^2 p_{S1}}{\partial x^2} + \frac{\partial^2 p_{S1}}{\partial y^2} + \frac{\partial^2 p_{S1}}{\partial z^2} = \frac{(\phi\mu c_t)_{S1}}{k_{S1}}\frac{\partial p_{S1}}{t} \tag{34}$$

The initial condition is exactly same as Equation (29). Meanwhile, the pressure at $y = y_{wf}$ is the bottom-hole pressure and remains constant.

$$p_{S1}(x,y,z,0) = p_i \tag{35}$$

$$p_{S1}(x,y = y_{wf},z,t) = p_{wf} \tag{36}$$

There is also no fluid flow beyond the tip of hydraulic fracture and at top and bottom of the reservoir. The outer boundary conditions can be written as

$$\frac{\partial p_{S1}}{\partial x}\bigg|_{x=0} = 0 \tag{37}$$

$$\frac{\partial p_{S1}}{\partial y}\bigg|_{y=y_1} = 0 \tag{38}$$

$$\left.\frac{\partial p_{S1}}{\partial z}\right|_{z=z_0,z_e} = 0 \tag{39}$$

There is an interface between the stimulated region and the hydraulic fracture region. The flux is considered to be continuous, and then the boundary condition is expressed as

$$\left.\frac{k_{S1}}{\mu}\frac{\partial p_{S1}}{\partial x}\right|_{x=x_0} = \left.\frac{k_{S2}}{\mu}\frac{\partial p_{S2}}{\partial x}\right|_{x=x_0} \tag{40}$$

3. Model Derivation

The above equations to describe fluid flow process in these two models are all PDEs. Prior to deriving the analytical solution, the system of PDEs must be transformed into ODEs. In this section, the new approaches, i.e., the integral transform and average pressure substitution, are used to solve the above equations in real-time space. Firstly, Equation (1) in the Model 1 can be rewritten as

$$\int_{x_0}^{x_e}\int_{y_1}^{y_2}\int_{z_o}^{z_e}\frac{\partial}{\partial x}(\frac{\partial P_{B3}}{\partial x})dxdydz+\int_{x_0}^{x_e}\int_{y_1}^{y_2}\int_{z_o}^{z_e}\frac{\partial}{\partial y}(\frac{\partial P_{B3}}{\partial y})dxdydz+\int_{x_0}^{x_e}\int_{y_1}^{y_2}\int_{z_o}^{z_e}\frac{\partial}{\partial z}(\frac{\partial P_{B3}}{\partial z})dxdydz=\frac{(\phi\mu c_t)_{B3}}{k_{B3}}\frac{\partial}{\partial t}\int_{x_o}^{x_e}\int_{y_1}^{y_2}\int_{z_o}^{z_e}P_{B3}dxdydz \tag{41}$$

In order to obtain a simplified equation, the average pressure and effective pore volume are defined as

$$\overline{p_{B3}} = \frac{\iiint p_{B3}}{\iiint dxdydz} = \frac{\iiint p_{B3}}{V_{b,B3}} \tag{42}$$

$$V_{p,B3} = \phi V_{b,B3} \tag{43}$$

Based on the above definitions, Equation (41) can be re-written as

$$\int_{y_1}^{y_2}\int_{z_0}^{z_e}\left(\left.\frac{\partial p_{B3}}{\partial x}\right|_{x_e}-\left.\frac{\partial p_{B3}}{\partial x}\right|_{x_0}\right)dydz+\int_{x_0}^{x_e}\int_{z_0}^{z_e}\left(\left.\frac{\partial p_{B3}}{\partial y}\right|_{y_2}-\left.\frac{\partial p_{B3}}{\partial y}\right|_{y_1}\right)dxdz+\int_{x_0}^{x_e}\int_{y_1}^{y_2}\left(\left.\frac{\partial p_{B3}}{\partial z}\right|_{z_e}-\left.\frac{\partial p_{B3}}{\partial y}\right|_{z_0}\right)dxdy=\frac{(\phi\mu c_t)_{B3}V_{b,B3}}{k_{B3}}\frac{d\overline{p_{B3}}}{dt} \tag{44}$$

Substituting the initial conditions and boundary conditions, Equation (44) simplifies to

$$-\int_{x_0}^{x_e}\int_{z_0}^{z_e}\left(\left.\frac{\partial p_{B3}}{\partial y}\right|_{y_1}\right)dxdz = \frac{(\phi\mu c_t)_{B3}V_{b,B3}}{k_{B3}}\frac{d\overline{p_{B3}}}{dt} \tag{45}$$

According to Darcy's law

$$q_{B3} = \int_{x_0}^{x_e}\int_{z_0}^{z_e}\left(\left.\frac{k_{B3}}{\mu}\frac{\partial p_{B3}}{\partial y}\right|_{y_1}\right)dxdz \tag{46}$$

Therefore, Equation (45) can be rewritten as shown below

$$(c_t V_p)_{B3}\frac{d\overline{p_{B3}}}{dt} = -q_{B3} \tag{47}$$

Similarly, the simplified ODEs for Region 2 and Region 1 of Model 1 can be expressed as

$$(c_t V_p)_{B2}\frac{d\overline{p_{B2}}}{dt} = -q_{B2} + q_{B3} \tag{48}$$

$$(c_t V_p)_{B1}\frac{d\overline{p_{B1}}}{dt} = -q_{B1} + q_{B2} \tag{49}$$

For Model 2, Equation (22) can also be rewritten as

$$\int_{y_{wf}}^{y_e}\int_{z_0}^{z_e}\left(\frac{\partial p_{S3}}{\partial x}\bigg|_{x_2}-\frac{\partial p_{S3}}{\partial x}\bigg|_{x_1}\right)dydz+\int_{x_1}^{x_2}\int_{z_0}^{z_e}\left(\frac{\partial p_{S3}}{\partial y}\bigg|_{y_e}-\frac{\partial p_{S3}}{\partial y}\bigg|_{y_{wf}}\right)dxdz+\int_{x_1}^{x_2}\int_{y_{wf}}^{y_e}\left(\frac{\partial p_{S3}}{\partial z}\bigg|_{z_e}-\frac{\partial p_{S3}}{\partial y}\bigg|_{z_0}\right)dxdy=\frac{(\phi\mu c_t)_{S3}V_{b,S3}}{k_{S3}}\frac{d\overline{p_{S3}}}{dt} \quad (50)$$

Substituting the initial conditions and boundary conditions, Equation (50) can be simplified as

$$-\int_{y_{wf}}^{y_e}\int_{z_0}^{z_e}\left(\frac{\partial p_{S3}}{\partial x}\bigg|_{x_1}\right)dydz=\frac{(\phi\mu c_t)_{S3}V_{b,S3}}{k_{S3}}\frac{d\overline{p_{S3}}}{dt} \quad (51)$$

Note that

$$q_{S3}=\int_{y_{wf}}^{y_e}\int_{z_0}^{z_e}\left(\frac{k_{S3}}{\mu}\frac{\partial p_{S3}}{\partial x}\bigg|_{x_1}\right)dydz \quad (52)$$

Substituting Equation (52) into Equation (51) results in

$$(c_tV_p)_{S3}\frac{d\overline{p_{S3}}}{dt}=-q_{S3} \quad (53)$$

Similarly, the simplified ODEs for Region 2 and Region 1 of Model 2 can be expressed as

$$(c_tV_p)_{S2}\frac{d\overline{p_{S2}}}{dt}=-q_{S2}+q_{S3} \quad (54)$$

$$(c_tV_p)_{S1}\frac{d\overline{p_{S1}}}{dt}=-q_{S1}+q_{S2} \quad (55)$$

Comparing Equations (47)–(49) with Equations (53)–(55), we can find that the form of these equations are exactly the same, which also means that the obtained analytical solution through the new approach is identical for the two models. The next step is to replace the average pressure and the general form of average pressure representing the relationship between the pressure and the dimensionless flow rate, which can be expressed as [23]

$$\overline{p}=p_{wf}+\frac{4}{\pi^2}\left(p_i-p_{wf}\right)\sum_{n=1}^{\infty}\frac{q_{Dn}}{(2n-1)^2} \quad (56)$$

The linear flow process is from Region 3 to Region 2 and then Region 1, sequentially. Therefore, the average pressure in Region 1, Region 2 and Region 3 can be written, respectively, as

$$\overline{p_1}=p_{wf}+\frac{q_1}{J}\sum_{n=1}^{\infty}\frac{q_{Dn1}}{(2n-1)^2} \quad (57)$$

$$\overline{p_2}=\overline{p_1}+\frac{q_2}{T_{21}}\sum_{n=1}^{\infty}\frac{q_{Dn2}}{(2n-1)^2} \quad (58)$$

$$\overline{p_3}=\overline{p_2}+\frac{q_3}{T_{32}}\sum_{n=1}^{\infty}\frac{q_{Dn3}}{(2n-1)^2} \quad (59)$$

where q_{Dn1}, q_{Dn2} and q_{Dn3} are the dimensionless production rate from the n-th mode in Region 1, Region 2 and Region 3, respectively. q_1, q_2 and q_3 and $\overline{p_1}$, $\overline{p_2}$ and $\overline{p_3}$ are the initial production rate and average pressure in Region 1, Region 2 and Region 3, respectively. The productivity index is represented by J and the transmissibility between two regions are represented by T_{21} and T_{32}, which can be defined as $J=\frac{\pi^2}{4}\frac{q_1}{p_i-p_{wf}}$, $T_{21}=\frac{\pi^2}{4}\frac{q_2}{p_i-\overline{p_1}}$ and $T_{32}=\frac{\pi^2}{4}\frac{q_3}{p_i-\overline{p_2}}$.

Substituting the general equations of average pressure Equations (57)–(59) into Equations (47)–(49) or Equations (53)–(55), the ODEs in three regions can be rewritten as

$$\sum_{n=1}^{\infty}\frac{dq_{n1}}{dt}=\frac{1}{\tau_1}\sum_{n=1}^{\infty}(2n-1)^2q_{n2}(t)-\frac{1}{\tau_1}\sum_{n=1}^{\infty}(2n-1)^2q_{n1}(t) \tag{60}$$

$$\sum_{n=1}^{\infty}\frac{dq_{n2}}{dt}=\frac{1}{\tau_2}\sum_{n=1}^{\infty}(2n-1)^2q_{n3}(t)-\left(\frac{1}{\tau_2}+\frac{T_{21}}{J\tau_1}\right)\sum_{n=1}^{\infty}(2n-1)^2q_{n2}(t)+\frac{T_{21}}{J\tau_1}\sum_{n=1}^{\infty}(2n-1)^2q_{n1}(t) \tag{61}$$

$$\sum_{n=1}^{\infty}\frac{dq_{n3}}{dt}=-\left(\frac{1}{\tau_3}+\frac{T_{21}}{T_{32}\tau_2}\right)\sum_{n=1}^{\infty}(2n-1)^2q_{n3}(t)-\frac{T_{21}}{T_{32}\tau_2}\sum_{n=1}^{\infty}(2n-1)^2q_{n2}(t) \tag{62}$$

where τ_1, τ_2 and τ_3 indicate the linear flow time in Region 1, Region 2 and Region 3, respectively, which can be defined as $\tau_1=\frac{(c_tV_p)_1}{J}$, $\tau_2=\frac{(c_tV_p)_2}{T_{21}}$ and $\tau_3=\frac{(c_tV_p)_3}{T_{32}}$.

The initial oil rate depends on the flow rate from Region 1. Therefore, the initial conditions for three regions can be expressed as

$$q_1(t=0)=q_i \tag{63}$$

$$q_2(t=0)=0 \tag{64}$$

$$q_3(t=0)=0 \tag{65}$$

Solving Equations (60)–(62) by substituting Equations (63)–(65), the production rate for n-th mode can be obtained as

$$q_{n1}=(2n-1)^2\left(\beta_3r_3q_ie^{(2n-1)^2\lambda_1t}-\beta_2r_6q_ie^{(2n-1)^2\lambda_2t}+\beta_1r_9q_ie^{(2n-1)^2\lambda_3t}\right) \tag{66}$$

where λ_1, λ_2 and λ_3 are three eigenvalues and r_1~r_9 represent the nine elements in three eigenvectors. And β_1, β_2 and β_3 are the combination of parameters, which can be expressed as $\beta_1=\frac{r_1(r_1r_5-r_2r_4)}{(r_1r_9-r_3r_7)(r_1r_5-r_2r_4)-(r_1r_6-r_3r_4)(r_1r_8-r_2r_7)}$, $\beta_2=\frac{r_1r_8-r_2r_7}{r_1r_5-r_2r_4}\beta_1$, $\beta_3=\frac{\beta_2r_4r_7}{r_1}\beta_1$.

The production rate q_1 is the summation of all production rate terms. After many mathematical manipulations, the analytical solution can be obtained by simplifying Equation (66).

$$q_1=\left(\beta_3r_3q_ie^{\lambda_1t}-\beta_2r_6q_ie^{\lambda_2t}+\beta_1r_9q_ie^{\lambda_3t}\right)+\frac{\sqrt{\pi}\beta_3r_3q_i}{4\sqrt{|\lambda_1|t}}erfc\left(3\sqrt{|\lambda_1|t}\right)-\frac{\sqrt{\pi}\beta_2r_6q_i}{4\sqrt{|\lambda_2|t}}erfc\left(3\sqrt{|\lambda_2|t}\right)+\frac{\sqrt{\pi}\beta_1r_9q_i}{4\sqrt{|\lambda_3|t}}erfc\left(3\sqrt{|\lambda_3|t}\right) \tag{67}$$

Obviously, we can obtain that the oil rate depends on several variables from Equations (60)–(62), i.e., the linear flow time (τ_1, τ_2, τ_3) in Region 1, Region 2 and Region 3, productivity index J, two transmissibility T_{21} and T_{32} and initial production rate q_i. By fitting Equation (67) to the target data, the several variables can be obtained when the desired match is achieved. In addition, the analytical solution with output variables can be further used for production prediction.

4. Model Validation

The general analytical solution for two typical configurations has been derived. Therefore, the main objective of this section is to verify the accuracy of the analytical solution by constructing two numerical models based on the previous two physical configurations. Considering the assumption of symmetry, the two numerical models are both one quarter and are constructed with 27 grid cells in the x-direction, 50 grid cells in the y-direction and only 1 grid cell in the z-direction, which is illustrated in Figure 2. Due to the vast difference in the width dimensions of fractures and matrix, the additional refinement scheme around the hydraulic fractures and near the boundaries of discretized regions is employed, enhancing stability of the numerical solution while accurately capturing the transient responses within the fractures. The blue area represents the stimulated region (Region 2), while the gray area is the unstimulated region (Region 3). The hydraulic fracture region (Region 1) is relatively small and is represented by the grids in the first column. Furthermore, Regions 1 and 2 and Regions 2 and 3 are individually connected to ensure sequential flow among three regions. The input parameters for the two numerical models are summarized in Table 1.

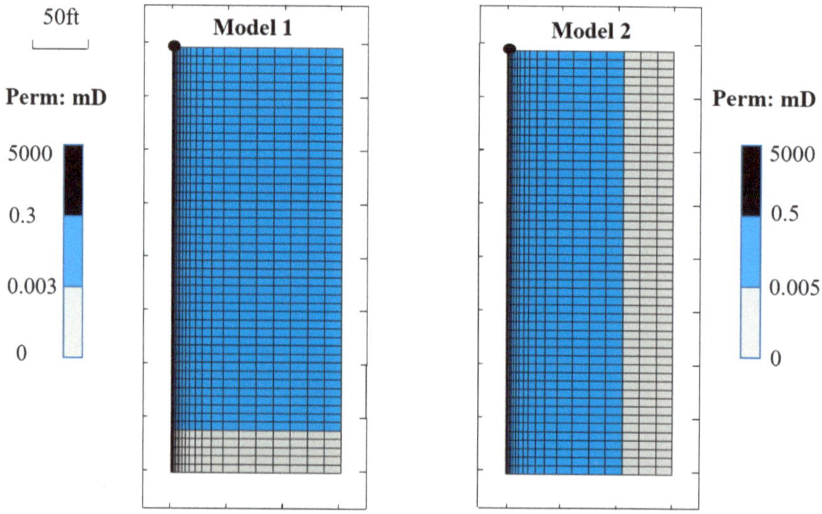

Figure 2. Schematic of the reservoir grids for two numerical models with typical configurations.

Table 1. Input parameters for the numerical models.

	Model 1	Model 2
Parameter	**Value**	
Model dimension (X × Y × Z) (ft)	150 × 400 × 10	150 × 400 × 10
Initial pressure (psi)	2500	2500
Bottom-hole pressure (psi)	500	500
Viscosity (cp)	0.0174	0.0174
Compressibility (10^{-5} psi)	9.75	9.75
Porosity	0.06	0.06
Permeability of hydraulic fracture	5000	5000
Permeability in Region 3 (mD)	0.003	0.005
Permeability in Region 2 (mD)	0.3	0.5

The comparison results between two numerical models and the analytical solution are presented in Figure 3. The black dots indicate the results of oil rate over time as calculated by the numerical models, while the red lines indicate the results of the analytical solution. Obviously, the fitting results are excellent. As shown in Figure 3a,b, four flow regimes are both identified. Because of the high-velocity flow in the hydraulic fracture region (Region 1), the two flow regimes happened in Region 1 are not presented. The first regime can be diagnosed as a transient linear flow in Region 2, which also corresponds to the −1/2 straight line presented on the log–log plot. Region 2 is the stimulated region with relatively high permeability, which is significant for short-term production of the reservoir, so the linear flow time in this region is 418 days in Model 1 and 244 days in Model 2, respectively. The exponential curve representing the second flow regime indicates that the pressure wave reaches the boundary of Region 2. As for the third flow regime, it indicates the transient linear flow in Region 3 with a low-permeability matrix, which contributes significantly to long-term production, so the linear flow time in this region is 1869 days in Model 1 and 1389 days in Model 2, respectively. Lastly, the fourth flow regime is outer boundary-dominated flow. The six output parameters from the fit to two numerical models are listed in Table 2.

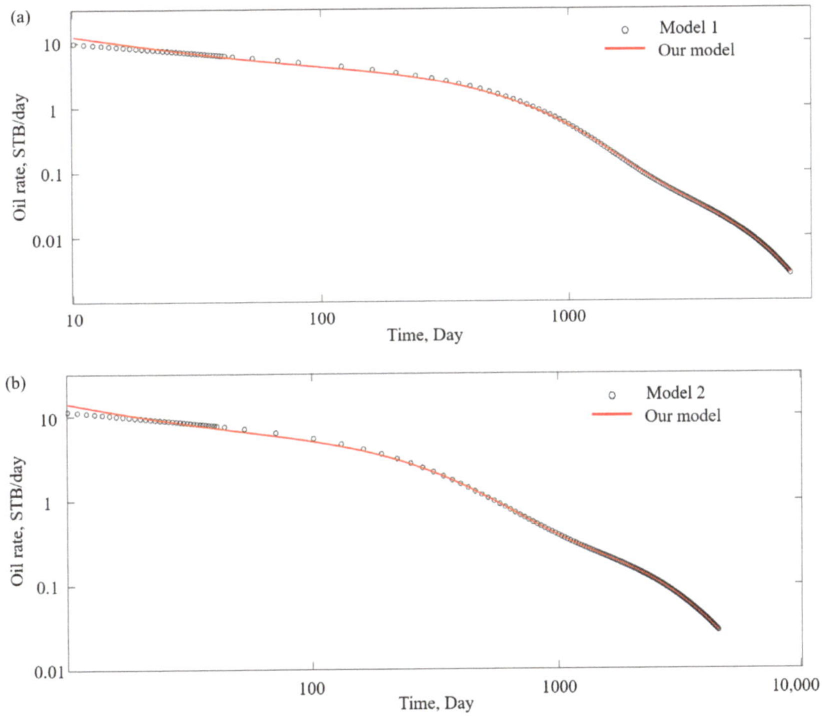

Figure 3. Comparative results with two numerical models.

Table 2. Output parameters obtained from the analytical solution for validation against two numerical models.

	Model 1	Model 2
Parameter	**Value**	
τ_1 (days)	1.2×10^{-4}	1.4×10^{-4}
τ_2 (days)	418	244
τ_3 (days)	1869	1389
T_{21}/J	0.012	0.01
T_{32}/T_{21}	0.022	0.065
q_i (STB/day)	10.96	12.02

5. Application to Field Cases

The accuracy of the new analytical solution has been verified against the numerical models based on two typical configurations. Before applying the analytical solution to field cases, a workflow for processing the raw field data must be developed because the type and number of flow regimes observed in a field data plot depends on the relative magnitude of the well and reservoir properties. Meanwhile, the dominance of linear flow regime observed in field data could result from (1) production from fractures whose lengths extend to reservoir boundaries, (2) transient drainage of low-permeability matrix blocks into adjoining fractures, and (3) linear shape of certain reservoirs [24]. Therefore, flow pattern analysis and history matching for production data from fractured reservoirs are necessary. One onshore field case and one offshore field case are extracted for production prediction according to the main workflow. The five main steps are listed as follows:

- Make a log–log plot of oil rate versus production time.
- Diagnose the flow regimes.
- Apply analytical solution to the production data.
- Output the six parameters after obtaining the desired matching.
- Predict the future production rate with the obtained parameters.

Onshore field case. Well O is one of tight oil wells located at Elm Coulee Field in eastern Montana, whose data are from the paper published by Kabir et al. [25] in 2011. Well O was selected primarily because of its relatively long production history and the availability of high quality pressure data. The well has been on production for 350 days at constant BHPs. Firstly, the original oil rate versus production time on the log–log plot can be obtained, which exhibits a $-1/2$ straight line and a nearly unit-slope line indicating linear flow and boundary-dominated flow in Figure 4a. Since the linear flow lasts for nearly 200 days, this linear flow can be diagnosed as the transient linear flow in stimulated region. Meanwhile, the boundary-dominated flow indicates that the pressure wave has reached the boundary of stimulated region. Following the work flow, we apply our analytical model to the field data. The result of production-rate history match and forecast are shown in Figure 4b on a log–log plot. The green marks (history match) and the red marks (field data) reach a high degree of fit except for the noisy data. Thus, we summarized the output parameters obtained from the history-match exercise in Table 3. After substituting the output parameters into the analytical solution, we can predict the future production rate until 3000 days, which are shown using black marks. As shown in Figure 4b, two more flow regions can be identified, which is the typical linear flow and boundary-dominated flow in the unstimulated region. The supply from the ultra-tight matrix can last for a long time, which is critical to the cumulative production of a tight oil well.

Table 3. Output parameters obtained from the analytical solution for application in two field cases.

	Well O	Volve Field
Parameter	Value	
τ_1	0.003 (days)	5×10^{-6} (Months)
τ_2	197(days)	14 (Months)
τ_3	2100 (days)	460 (Months)
T_{21}/J	0.203	0.0729
T_{32}/T_{21}	0.049	0.0045
q_i	1200 (STB/day)	1780 (MSTB/month)

Offshore field case. This offshore oil field data is obtained from Wang's published paper [26]. The production wells are drilled in the Volve oil field located in the central part of the North Sea. The target formation is tight sandstone of Middle Jurassic age at the depth of about 3000m below sea level. The field was developed in early 2008 and was in operation for roughly nine years. As illustrated in Figure 5a, the monthly oil rate versus production time is presented on the log–log plot with a $-1/2$ straight line. The linear flow time lasts for nearly 70 months, which means the flow regime can be diagnosed as transient flow in unstimulated region. After fitting our analytical solution to the field data, the results of history matching and production prediction are shown in Figure 5b. A great matching degree can be obtained, and six model output parameters are summarized in Table 3. Furthermore, we can predict the further monthly oil rate until 2500 months. We can observe that the monthly oil rate will be below 500 MSTB/month in the ninth year. That would explain why this oil field was decommissioned in 2016 combined with the high cost of offshore oil extraction.

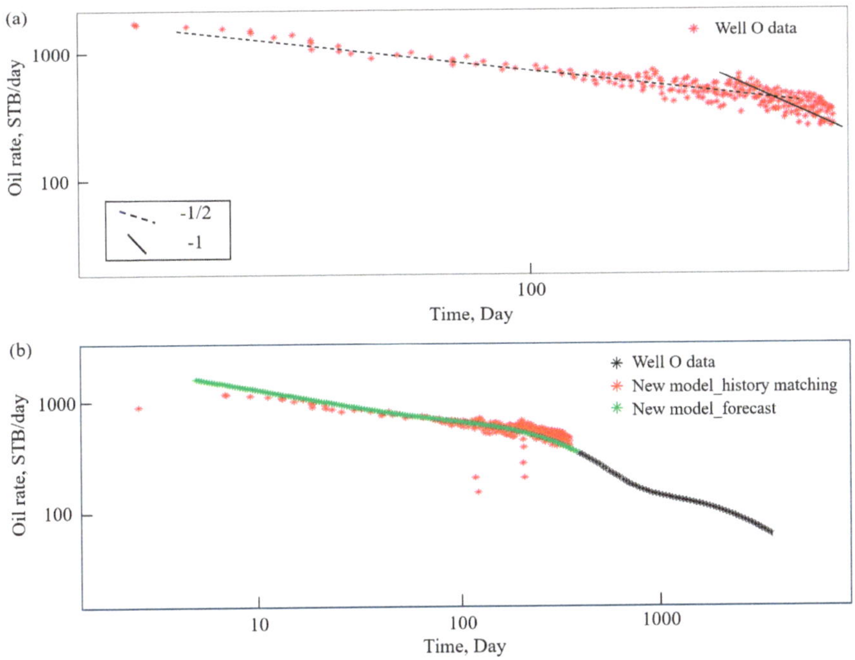

Figure 4. Production analysis in Well O. (**a**) Actual oil rate on the log–log plot; (**b**) the history matching and forecasting results on the log–log plot.

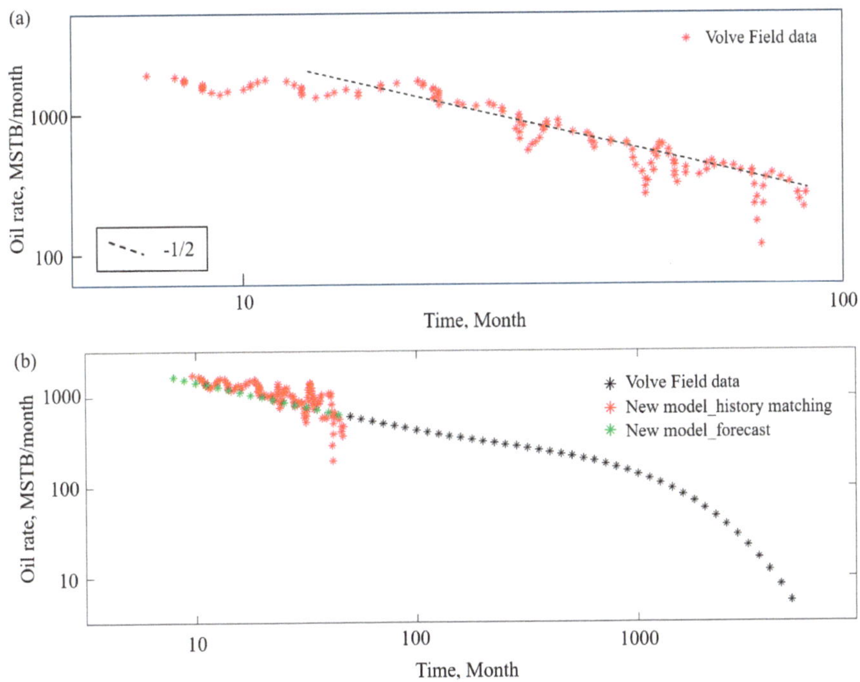

Figure 5. Production analysis in Volve oil field. (**a**) Actual oil rate on the log–log plot; (**b**) the history matching and forecasting results on the log–log plot.

6. Conclusions

In this study, we adopt a new approach to derive a general analytical solution in real-time space for two typical tri-linear physical configurations. Numerical models with identical physical configurations had been used to validate the accuracy of the analytical solution and then a production data analysis had been carried out to analyze short-term and long-term performance of two field cases based on the developed workflow. The specific conclusions can be drawn:

- Through bypassing the complex Laplace transform solution, the analytical solution is derived in real-time space and directly presents the oil-rate-versus-production-time relationship. Therefore, it is more convenient in field applications.
- The derived analytical solution is not only applicable to unconventional reservoirs without the region beyond the hydraulic fractures, but can also consider the contribution from the region beyond the hydraulic fractures because of the excellent agreement with two numerical models.
- The diagnosis of transient linear flow in offshore and onshore tight oilfields is critical for production data analysis. The same slope straight line may represent the different fluid-transfer mechanism, and the linear flow in unstimulated region cannot be ignored, which represents a significant contribution to long-term production.

Author Contributions: Methodology, K.Q.; Software, X.C., G.L. and S.W.; Investigation, K.F.; Resources, K.F. and J.L.; Data curation, K.F. and X.C.; Writing—original draft, K.Q.; Writing—review & editing, G.L., S.W. and R.N. All authors have read and agreed to the published version of the manuscript.

Funding: This study was supported by Basic Research Project from Jiangmen Science and Technology Bureau (Grant Nos. 2220002000356), China University of Petroleum (Beijing) (Grand No. 2462023BJRC007), National Natural Science Foundation of China (Grant Nos. 72140008) and The Guangdong Basic and Applied Basic Research Foundation (No. 2022A1515110376).

Institutional Review Board Statement: Not applicable.

Informed Consent Statement: Not applicable.

Data Availability Statement: All data have been provided in the paper.

Conflicts of Interest: We declare that we have no conflict of interest.

References

1. Zhao, Z.; Hou, L.; Luo, X.; Chi, Y.; Pang, Z.; Lin, S.; Zhang, L.; Liu, B. Heat-induced pore structure evolution in the Triassic chang 7 shale, Ordos basin, China: Experimental simulation of in situ conversion process. *J. Mar. Sci. Eng.* **2023**, *11*, 1363. [CrossRef]
2. Masoud, S.; Sohrab, Z.; Stephen, B. Molecular dynamics simulations in reservoir analysis of offshore petroleum reserves: A systematic review of theory and applications. *Earth-Sci. Rev.* **2019**, *192*, 194–213.
3. U.S. Energy Information Administration. Offshore Production Nearly 30% of Global Crude Oil Output in 2015. 2016. Available online: https://go.gale.com/ps/i.do?p=EAIM&u=anon~b06c66bc&id=GALE|A467716915&v=2.1&it=r&sid=sitemap&asid=c037960b (accessed on 1 October 2016).
4. Abassi, M.; Sharifi, M.; Kazemi, A. Development of new analytical model for series and parallel triple porosity models and providing transient shape factor between different regions. *J. Hydrol.* **2019**, *574*, 683–698. [CrossRef]
5. Lei, G.; Liu, T.; Liao, Q.; He, X. Estimating permeability of porous media from 2D digital images. *J. Mar. Sci. Eng.* **2023**, *11*, 1614. [CrossRef]
6. Fathy, D.; El-Balkiemy, A.F.; Makled, W.A.; Hosny, A.M. Organic geochemical signals of Paleozoic rocks in the southern Tethys, Siwa basin, Egypt: Implications for source rock characterization and petroleum system. *Phys. Chem. Earth* **2023**, *130*, 103393. [CrossRef]
7. Benetatos, C.; Catania, F.; Giglio, G.; Pirri, C.F.; Raeli, A.; Scaltrito, L.; Serazio, C.; Verga, F. Workflow for the validation of geomechanical simulations through seabed monitoring for offshore underground activities. *J. Mar. Sci. Eng.* **2023**, *11*, 1387. [CrossRef]
8. Clarkson, C.R. Production data analysis of unconventional gas wells: Review of theory and best practices. *Int. J. Coal Geol.* **2013**, *110*, 101–146. [CrossRef]
9. Dejam, M.; Hassanzadeh, H.; Chen, Z. Semi-analytical solution for pressure transient analysis of a hydraulically fractured vertical well in a bounded dual-porosity reservoir. *J. Hydrol.* **2018**, *565*, 289–301. [CrossRef]

10. Wei, S.; Xia, Y.; Jin, Y.; Chen, M.; Chen, K. Quantitative study in shale gas behaviors using a coupled triple-continuum and discrete fracture model. *J. Pet. Sci. Eng.* **2019**, *174*, 49–69. [CrossRef]
11. Rao, X.; Xin, L.; He, Y.; Fang, X.; Gong, R.; Wang, F.; Zhao, H.; Shi, J.; Xu, Y.; Dai, W. Numerical simulation of two-phase heat and mass transfer in fractured reservoirs based on projection-based embedded discrete fracture model (pEDFM). *J. Pet. Sci. Eng.* **2022**, *208*, 109323. [CrossRef]
12. Ahmadi, S.; Khormali, A. Optimization of the corrosion inhibition performance of 2-mercaptobenzothiazole for carbon steel in HCl media using response surface methodology. *Fuel* **2024**, *357*, 129783. [CrossRef]
13. Qiu, K.X.; Li, J.; Chen, D. Optimized long short-term memory (LSTM) network for performance prediction in unconventional reservoirs. *Energy Rep.* **2022**, *8*, 15436–15445. [CrossRef]
14. Khormali, A. Effect of water cut on the performance of an asphaltene inhibitor package: Experimental and modeling analysis. *Petrol. Sci. Technol.* **2022**, *40*, 2890–2906. [CrossRef]
15. El-Banbi, A. Analysis of Tight Gas Well Performance. Ph.D. Thesis, Texas A&M University, College Station, TX, USA, 1998.
16. Bello, R.O. Rate Transient Analysis in Shale Gas Reservoirs with Transient Linear Behavior. Ph.D. Thesis, Texas A&M University, College Station, TX, USA, 2009.
17. Brown, M.; Ozkan, E.; Raghavan, R.; Kazemi, H. Practical solutions for pressure-transient responses of fractured horizontal wells in unconventional shale reservoirs. *SPE Res. Eval. Eng.* **2011**, *14*, 663–676. [CrossRef]
18. Stalgorova, E.; Mattar, L. Practical analytical model to simulate production of horizontal wells with branch fractures. In Proceedings of the SPE Canadian Unconventional Resources Conference, Calgary, AB, Canada, 30 October–1 November 2012.
19. Ogunyomi, B.A.; Patzek, T.W.; Lake, L.W.; Kabir, C.S. History matching and rate forecasting in unconventional oil reservoirs with an approximate analytical solution to the double-porosity model. *SPE Res. Eval. Eng.* **2016**, *19*, 70–82. [CrossRef]
20. Qiu, K.X.; Li, H. A New Analytical Solution of the Triple-Porosity Model for History Matching and Performance Forecasting in Unconventional Oil Reservoirs. *SPE J.* **2018**, *23*, 2060–2079. [CrossRef]
21. Qiu, K.X. A practical analytical model for performance prediction in unconventional gas reservoir. *Front. Earth Sci.* **2023**, *11*, 1143541. [CrossRef]
22. Stehfest, H. Algorithm 368: Numerical Inversion of Laplace transforms [D5]. *Commun. ACM* **1970**, *13*, 47–49. [CrossRef]
23. Wattenbarger, R.A.; El-Banbi, A.H.; Villegas, M.E.; Maggard, J.B. Production Analysis of Linear Flow Into Fractured Tight Gas Wells. In Proceedings of the SPE Rocky Mountain Regional/Low-Permeability Reservoirs Symposium, Denver, CO, USA, 5–8 April 1998.
24. Ezulike, D.O.; Dehghanpour, H. A model for simultaneous matrix depletion into natural and hydraulic fracture networks. *J. Nat. Gas Sci. Eng.* **2014**, *16*, 57–69.
25. Kabir, C.S.; Rasdi, F.; lgboalisi, B. Analyzing production data from tight oil wells. *J. Can. Petrol. Technol.* **2011**, *50*, 48–58. [CrossRef]
26. Wang, B.; Sharma, J.; Chen, J.; Persaud, P. Ensemble machine learning assisted reservoir characterization using field production data—An offshore field case study. *Energies* **2021**, *14*, 1052. [CrossRef]

MDPI AG
Grosspeteranlage 5
4052 Basel
Switzerland
Tel.: +41 61 683 77 34

Journal of Marine Science and Engineering Editorial Office
E-mail: jmse@mdpi.com
www.mdpi.com/journal/jmse

www.ingramcontent.com/pod-product-compliance
Lightning Source LLC
Chambersburg PA
CBHW041428210926
51850CB00013B/3379
* 9 7 8 3 7 2 5 8 3 3 9 1 7 *